SAP® LeanIX

SAP PRESS

René de Daniel, Marianne Dahmen

SAP® LeanIX

The Practical Guide

Rheinwerk
Publishing

Editor Megan Fuerst
Acquisitions Editor Hareem Shafi
Copyeditor Melinda Rankin
Cover Design Abigail Damke
Photo Credit Shutterstock: 2450459697/© Zuyeu Uladzimir
Layout Design Vera Brauner
Production Hannah Lane, Kelly O'Callaghan
Typesetting SatzPro, Germany
Printed and bound in Canada, on paper from sustainable sources

ISBN 978-1-4932-2783-9
1st edition 2026

Rheinwerk Publishing, Inc.
2 Heritage Drive, Suite 305
Quincy, MA 02171
USA
info@rheinwerk-publishing.com
+1.781.228.5070

Represented in the E. U. by:
Rheinwerk Verlag GmbH
Rheinwerkallee 4
53227 Bonn
Germany
service@rheinwerk-verlag.de
+49 (0) 228 42150-0

Library of Congress Cataloging-in-Publication Control Number: 2026027632

Contents at a Glance

Contents

Foreword by André Christ

The success and future viability of organizations today depend more than ever on their ability to navigate complexity, innovate continuously, and execute transformation at scale. As technology becomes deeply embedded in every aspect of business, leaders are challenged to balance speed with stability, innovation with control, and cost efficiency with long-term value creation.

Yet, the reality many organizations face is fragmentation. Technology landscapes have grown organically over decades, resulting in thousands of applications, intricate dependencies, and increasing technical debt. What's visible on the surface is only a fraction of the true complexity. Without transparency and a shared understanding of this landscape, meaningful transformation remains out of reach.

This is where enterprise architecture plays a critical role.

Over the past years, we've witnessed a fundamental shift in the role of enterprise architects, from documentation specialists to strategic partners driving business transformation. Today, we're entering the next phase of this evolution, one fundamentally shaped by artificial intelligence (AI).

AI is not just another technology trend; it's redefining how enterprise architecture works. Instead of manually analyzing landscapes, architects can increasingly rely on AI to surface patterns, identify risks, and recommend actions. Questions that once required weeks of analysis, such as identifying applications nearing end of life, detecting capability gaps, or prioritizing transformation initiatives, can now be answered in seconds. Enterprise architecture is evolving from a discipline of documentation to a capability of decision intelligence.

But this shift doesn't diminish the role of the architect: It elevates it. With AI augmenting analysis, architects can focus on what truly matters: guiding decisions, aligning stakeholders, and shaping transformation outcomes. The combination of human judgment and machine intelligence will define the next generation of enterprise architecture practices.

This book captures that transformation in a practical and accessible way. It goes beyond introducing SAP LeanIX as a platform. It provides a structured, maturity-driven approach to building an enterprise architecture practice: from foundational transparency to advanced data-driven decision-making to AI-augmented enterprise architecture insights. Through concrete use cases such as application rationalization, ERP transformation, and risk management, it demonstrates how architecture becomes actionable and measurable. It also provides an introduction into the latest AI developments such as MCP server connections, AI agents, and the SAP AI Agent Hub to make your EA practice future ready.

A unique strength of this book is its use of a consistent, end-to-end case study. The EA7 company brings the concepts to life, illustrating how challenges, decisions, and outcomes unfold in a realistic enterprise setting. This narrative approach makes complex topics tangible and relatable.

Even more importantly, the book does not stop at theory. By providing access to an SAP LeanIX demo workspace, you can explore the use cases in action. This hands-on perspective is critical. Enterprise architecture is not learned through concepts alone; it's experienced through application. The ability to see, navigate, and interact with real examples significantly accelerates understanding and adoption.

Finally, this book reflects something that has always defined the enterprise architecture community: collaboration. Transformation is not achieved in isolation. It requires a shared language, aligned stakeholders, and a common understanding of the path forward. This book contributes to that shared understanding, combining methodology, practical guidance, and real-world application.

Whether you're at the beginning of your enterprise architecture journey or advancing towards an AI-augmented practice, this book will provide you with the structure, insights, and tools to move forward.

I hope it inspires you to turn complexity into clarity, and architecture into intelligent, actionable decisions.

André Christ
Co-Founder of SAP LeanIX

Preface

"If there's a book that you want to read, but it hasn't been written yet, then you must write it."
—Toni Morrison

During countless interactions with customers, we've heard that people need a detailed admin and user guide for SAP LeanIX. Who best could write such a book if not SAP LeanIX employees and enthusiasts? There is so much more to say and write about SAP LeanIX, but with this book in your hands, you're set up for a great start in the enterprise architecture discipline at your company.

Purpose and Scope

This practical guide is for SAP LeanIX enthusiasts, especially for administrators and senior enterprise architects. It navigates you from workspace provisioning and initial customizing through the first use cases, such as application rationalization, SAP transformation projects, and AI and EA governance, to advanced techniques like automations, calculations, and AI capabilities. It can accelerate your kick start and help spark curiosity for continuous learning and exploring in the SAP LeanIX customer community, using the various resources provided by the SAP LeanIX team and your own creativity to mold the solution to your needs and use cases while staying close to the core principles of the out-of-the-box meta model.

Word of Caution

SAP LeanIX evolves quickly due to its native software-as-a-service (SaaS) design, accelerated by AI advancements. Use this book as an initial step to understand the fundamental concepts, but always check in with the following sources for capability, feature, and design updates:

- SAP LeanIX website: *https://www.leanix.net/en/*
- SAP LeanIX user documentation: *https://help.sap.com/docs/leanix*
- SAP LeanIX updates: *https://updates.leanix.net/*
- SAP LeanIX product roadmap: *https://roadmap.leanix.net/tabs/10-on-roadmap*
- SAP LeanIX customer community: *https://community.leanix.net/*

Structure of the Book

The chapters are designed to mimic a typical user journey. Let's glance at what's covered in each:

- **Chapter 1: Understanding the New Enterprise Architecture Landscape**
  We start with an introduction into enterprise architecture as a discipline. This chapter explores the shifting role of enterprise architecture in today's fast-paced digital environment. You will learn how enterprise architecture has evolved from static frameworks to become a dynamic enabler of business transformation in the era of artificial intelligence (AI).
- **Chapter 2: Introduction to SAP LeanIX**
  This chapter is all about SAP LeanIX, including the history, current product structure, out-of-the-box meta model, and selected collaboration features. We'll explore the most commonly used fact sheets and showcase how to use surveys, to-dos, and more.
- **Chapter 3: Setting Up Your SAP LeanIX Workspace**
  This chapter discusses the technical setup, including workspace provisioning and administrator configuration.
- **Chapter 4: Modeling Core Architecture Objects**
  Next, we dive into data model configuration, catalog and discovery capabilities, and data import, creation, and maintenance. We'll provide step-by-step instructions to use each feature and to link, update, and enrich fact sheets.
- **Chapter 5: Reporting and Visualizations**
  This chapter covers visualization capabilities, including reports, diagrams, and dashboards with KPIs. We'll provide detailed instructions for accessing and creating these reports and dashboards.
- **Chapter 6: Creating Your First Architecture Roadmap**
  In this chapter, we discuss target architecture planning and road mapping. This includes a step-by-step approach to creating architecture roadmaps using SAP LeanIX time-based views and transformation features.
- **Chapter 7: Application Portfolio Management**
  Here, we explore our first use cases: architecture portfolio assessment, application modernization, and application rationalization. Important topics include data gathering, TIME classification, total cost of ownership (TCO), and business criticality.
- **Chapter 8: Enterprise-Wide Planning with SAP Enterprise Architecture Framework**
  In this chapter, we'll showcase SAP Enterprise Architecture Framework in action. The chapter details how to work with reference business and solution architectures, model capabilities, and design stakeholder-centric architecture views.
- **Chapter 9: Additional Use Cases and Extensions**
  We describe some additional options for SAP LeanIX in this chapter, including self-built software management and the use of SAP LeanIX in mergers and acquisitions (M&A), Digital Operations Resilience Act (DORA), and open finance.

- **Chapter 10: Advanced Techniques: Calculations and Automations**
  This chapter introduces selected advanced techniques: automations, calculations, and managed code.
- **Chapter 11: AI Capabilities in SAP LeanIX**
  In this chapter, we'll explore the latest AI and in-tool agent capabilities in SAP LeanIX, including embedded services, external agents, custom solutions, and more.
- **Chapter 12: Enterprise Architecture Governance**
  Governance is essential for sustainable enterprise architecture. This chapter covers how to implement governance processes using SAP LeanIX and assess governance maturity.
- **Chapter 13: AI Governance**
  This chapter covers one of the latest use cases for SAP LeanIX: AI governance. We'll discuss how to use SAP LeanIX to introduce AI agents into your enterprise architecture in order to achieve greater agility, efficiency, and competitive differentiation.

Downloadable Supplements

You can find bonus content on the webpage for the book at *www.sap-press.com/6206* under the **Product supplements** section. This includes the latest meta data model and a glossary of key terms.

Demo Workspace

With this guide, you will not only have a handbook and lookup guide in your possession but also viewing access to a dedicated and unique demo workspace that you can explore and browse at your own pace. The demo content in this workspace is centered around a fictional customer case (which we'll call EA7) that expands across the chapters and converts what you read into tangible outcomes in terms of, for example, fact sheets, reports, diagrams, and dashboards. The demo workspace will be curated and maintained by the authors as a long-term resource for you to experience the look and feel of SAP LeanIX and its capabilities based on SAP LeanIX modeling best practices.

To access the demo workspace, go to *https://app-7.leanix.net/LeanIXPracticalGuideDemoWorkspace*. Use the credentials in Table 1 to login.

User Name	Password
Username 1: demo.usera74401@leanixacademy.net	Password 1: slePMS0)9Z
Username 2: demo.usera74402@leanixacademy.net	Password 2: kTC(DHM$pt
Username 3: demo.usera74403@leanixacademy.net	Password 3: pOVucQf5-M

Table 1: Demo Workspace Credentials

How to Use This Guide

Use this guide however you need it. Newly assigned SAP LeanIX workspace owners and admins can use this guide as a resource to prepare for the workspace setup and rollout. Experienced admins can use this guide as a lookup and knowledge repository to find inspiration on how to do things differently. Aspirational and senior enterprise architects can learn about fundamental capabilities and features, then discuss with their SAP LeanIX admins what to test and introduce in their own SAP LeanIX workspaces. Students and young professionals can use this guide to level up their enterprise architecture and SAP LeanIX knowledge, helping them to become valuable members of enterprise architecture teams at future or current employers.

Prerequisites

Users with admin access to an SAP LeanIX workspace will benefit the most from this guide, but thanks to its many colorful screenshots and detailed, step-by-step guides, any reader, even one without access to a workspace or even an internet connection, can learn from this book. The book is designed to cater to a vast range of readers, including those completely new to the discipline and experienced experts.

Acknowledgments

First, we want to thank our families and friends for their support and understanding during countless writing sessions in the evenings, on weekends, on public holidays, and during vacation days. Furthermore, we want to thank our managers, Anna Gajda, Henrik Schürmann, and Carsten Schütz, for their support and guidance on this project. Without their generous backing and encouragement, we would not have been able to finish this project in such a short time. Thanks also to all our SAP LeanIX colleagues who have been working hard to develop and maintain SAP LeanIX as a Gartner Magic Quadrant leader for many years in a row. Special thanks to the product enablement team, the strategic projects and product success team, the product and engineering team, including user experience (UX), and the customer success team, as well as our colleagues who reviewed the chapter contents and Oktay Yavuz for his contribution. Last, but not least, thank you Megan Fuerst and Hareem Shafi for guiding us through a smooth publishing process with Rheinwerk Publishing.

Chapter 1
Understanding the New Enterprise Architecture Landscape

In this first chapter, we want to take you on the journey of enterprise architecture and its evolution over time. While exploring the wave of the AI-augmented era, we gain valuable insights into the shift currently happening in the market. With that gained, we set the scene and contextualize the overarching structure of this book.

You might be familiar with this scenario: You have three days to go to prepare the board-level portfolio review meeting. Your enterprise architecture tool is available and ready, but you have now already consumed the last two weeks to bring a consistent set of data in. You are following up with application owners daily about updating missing lifecycle information or reviewing functional and technical fit. There are still several applications for which the owners, as listed in the Excel file provided to you initially, have left the organization or moved on to other roles. You also found out that some of the applications were duplicates due to slightly different naming and others seem to be owned by another part of the organization now as a consequence of the recent reorganization.

You take a step back and think about the situation: This isn't a failure of the tool, nor is it a failure of the team. It's the structural condition of enterprise architecture management as it was designed. The human, in this case, the enterprise architect, is the integration layer between the data and the truth.

This is going to change, and that change is happening right now. We'll look at this evolution and explain the current enterprise architecture landscape in the following sections.

1.1 The Three Waves

The first idea for enterprise architecture sparked in the 1960s and 1970s, when organizations realized that they needed a structured and systemic approach to document and manage their information technology (IT) infrastructure. The term *enterprise architecture* was introduced by John Zachman in the 1980s. By the late 1990s, enterprise architecture landscapes had outrun any individual's ability to hold information in their head. There was a need to capture this information via a structured approach and to allow a common language between architects.

This was when The Open Group Architecture Framework (TOGAF), Zachman, ArchiMate, and others were created to address this issue. Enterprise architecture management tools emerged primarily to serve enterprise architects and their enterprise architecture deliverables, such as capability maps and architecture standards. When someone had a question, they'd ask the appropriate enterprise architect, who had the knowledge and could provide an answer. This mechanic worked well. Change was slow enough that businesses could update a repository on a quarterly basis and have it be close to reality. Governance gates provided the relevant oversight without becoming a bottleneck because deployment cycles were in quarters rather than sprints. The cost to keep the architecture repository up-to-date and to ensure curated information was proportional to the value the enterprise architect got out of it. We refer to this phase as *wave 1*.

The times changed: Digital transformation, cloud adoption, and continuous delivery significantly increased the speed of change. The pace of change broke the assumption that the enterprise architecture repository holds the truth. Enterprise architects couldn't catch up with this pace. The repository held the truth as of last quarter, and governance bodies relying on timely data started to drag down the process of change and were rather perceived as a bottleneck. Even the best tool at that time couldn't help. The fundamental issue was that the repository could not be updated in a timely manner with the changes happening, leading to outdated data, which had an impact on trust. As you might expect, regaining trust is much harder than earning it in the first instance.

The response to this first wave came from the platform side kicking off *wave 2*. Native software-as-a-service (SaaS) tools like LeanIX were coming to the market to address the previously outlined challenges. There were a few changes that shifted the game:

- An integration-first design addressed the issue of outdated data by integrating tools that kept up-to-date data in their specific remit. Good examples include integrations with IT service management solutions, cloud providers, and configuration management databases (CMDBs). Information relevant in the context of enterprise architecture was ingested in an automated way by integration of solutions into the broader ecosystem.
- Accessibility moved beyond enterprise architecture teams. Business stakeholders were able to access portfolio reports seamlessly. An easy-to-understand look and feel raised acceptance levels, and the architect as translator was not necessarily needed anymore.
- Reports became data driven and were rendered instantly based on repository data. Static diagrams were avoided.

Somehow the human bottleneck remained. The bottleneck shifted from data collection and governance to validating, curating, and interpreting information to be able to draw the right conclusions. The level of effort needed to keep information meaningful was high.

Nevertheless, a mindset shift from documentation to impact took place, which Forrester described as creating *outcome-driven architecture*. The question shifted from "Does the architecture exist and is it documented?" to "What decision does this architecture insight enable us to make?"

Another response to the governance framework of the first wave was the concept of *minimum viable architecture*, which is a pragmatic approach to enterprise architecture that concentrates on the most important aspects to make the next decision well, rather than on a complete model of everything.

Wave 2 did a lot of things right: It shifted the focus to outcomes, broader access to various stakeholder groups, and living data through an integration-first approach. However, the cognitive load problem remained. The fundamental bottleneck was still human: the effort to keep inventory information meaningful, to turn data into insights, making architecture knowledge accessible to stakeholders who didn't speak the notation language. Enterprise architecture tools reduced the burden but didn't remove it entirely. Enterprise architects promised strategic value, acting as trusted advisors to the C-suite, partnering for transformation decisions, but most of their time was still spent on the information layer beneath the strategy.

Wave 3 has just recently started, and our hypothesis is that AI doesn't make enterprise architecture faster, but rather changes what enterprise architecture is responsible for. In wave 1, the architect was the integration layer between reality and inventory. In wave 2, the tools improved, but the architect (although focused on outcomes and pragmatic approaches) remained in the integration layer to a large degree. In wave 3, we see a restructuring of the work an architect will do. The living inventory becomes real, featured by AI-assisted integrations in which a human in the loop will still provide judgement. This will lead to the following:

- Reactive searches for insights turned into proactive insights with recommendations to act upon
- Impact assessments taking not a matter of days but minutes to get a first-pass answer by querying the architecture knowledge graph
- Architecture decisions drafted in the context of an insight and handed over to an architect for judgment and approval
- Cognitive load handed over to enterprise architecture agents, with any required enterprise architecture agent skill that does not exist being designed, implemented, and made available to others

Figure 1.1 provides an overview of the three waves. Note that each wave builds on the previous wave while shifting the cognitive focus of enterprise architecture teams.

Further Resources

For more on the fundamentals and foundational concepts of enterprise architecture, browse the articles in the SAP LeanIX wiki at *https://www.leanix.net/en/wiki/ea#Fundamentals*.

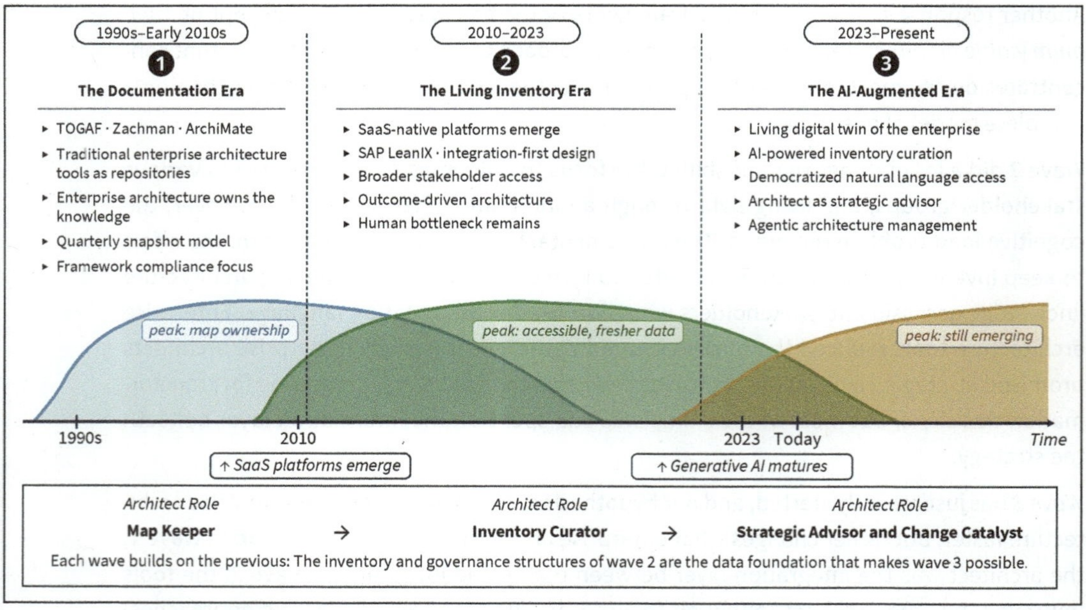

Figure 1.1: Evolution of Enterprise Architecture in Three Waves

1.2 The Shift

The historical barrier to architecture knowledge was not about access to enterprise architecture tools but about translation. Understanding the structure of the inventory and the required notation, learning the taxonomy, and knowing which report to best use and how to interpret the information were barriers and restricted architecture knowledge to the enterprise architecture team.

Imagine a chief information officer (CIO) who wanted to understand which applications handle customer data and lack a cloud migration path. He had two options: either ask an architect and wait, or navigate the enterprise architecture tool and try to interpret the results. Neither of these options were scalable, and as a result, architecture intelligence remained a specialized function.

Now imagine that a business stakeholder interacts with the enterprise architecture tool through natural language. That business stakeholder does not know what the functional fit "unreasonable" means exactly. The stakeholder types a question in his mother tongue and gets an answer without needing to understand the underlying taxonomy. Think about embedding this architecture intelligence into the working environment of people who will never open the SAP LeanIX UI, such as a finance lead reviewing the transformation program or a risk function assessing vendor dependency. They can ask an architecture question in the interface they use and get answers drawn from the inventory.

Therefore, architects are freed from inventory maintenance, routine analysis, and acting as the translation layer between data and stakeholders, leading to fundamental shifts in the core value proposition of enterprise architects. This shift should be seen as a unique opportunity for enterprise architects to add value to the organization. The following are some imminent shifts:

- **From describing the current state to shaping the target state**
  Instead of keeping the baseline accurate, architects will concentrate on questions such as "Where should this landscape be in three years, and what sequence of decisions will get us there?" The architect's focus will shift to these strategic questions.
- **From governing compliance to enabling speed**
  Governance in the old model was often a gate that slowed things down in exchange for oversight. In the AI-augmented model, governance becomes embedded in the inventory itself. The architect's focus shifts from enforcing the process to designing the guardrails.
- **From answering architecture questions to asking better business questions**
  Instead of arriving with architecture answers, the architect will ask the right business-related questions no one was thinking about before. Architects require judgment, context, and the ability to live with ambiguity.
- **From framework custodian to strategic advisor and catalyst for change**
  TOGAF, Zachman, and the like exist to produce better decisions, faster transformation, and lower risk. What the documentation era inadvertently achieved was making framework compliance its primary activity. In the era of AI, we can correct that inversion and architects can focus on the outcome.

Enterprise architects thus will need to develop and train new muscles around strategic thinking, judgement, and strategic advisory.

Your Monday Morning

Let's now imagine two versions of the same Monday morning:

- **Old Monday morning**
  You open SAP LeanIX and you see the following:
  - Fifteen fact sheets are flagged for review by application owners who haven't responded.
  - A portfolio report, due for Thursday, requires updates to its functional fit scores (eight of which are either missing or were last touched 14 months ago).
  - A message from a business unit lead asks why their transformation program doesn't appear in the roadmap view.

  Two hours later, no strategic work has started; the entire morning was spent on the information layer.

- **New Monday morning**
  You have an entirely different experience:
  - An AI-assisted architecture guide has already run overnight, and an insight surfaced three rationalization candidates with overlapping capability maps, drafted an architecture decision record, and suggested creating a survey to validate stakeholders as a next step.
  - Twelve application fact sheets were updated automatically via an enterprise architecture agent, which pulled current vendor lifecycle data and end-of-life announcements from the web.
  - The portfolio report is current because the integrations kept it current. The Thursday meeting is about what to decide, not what the data says.

Although you're working in the same team and the same workspace, you're facing fundamentally different work, which can be best described by the shifts taking place between the old enterprise architecture and the newer, AI-augmented enterprise architecture, as outlined in Table 1.1.

Old Enterprise Architecture	AI-Augmented Enterprise Architecture
Owns the architecture knowledge	Curates and interprets a living digital twin
Consulted by stakeholders	Embedded in strategic decisions
Measures outputs: diagrams, documents, gates	Measures outcomes: speed, cost, risk, capability
Framework-driven	Hypothesis-driven
Reactive governance	Proactive change catalyst
Specialist skill, accessed on request	Democratized intelligence, embedded in the organization

Table 1.1: Shift Between Old Enterprise Architecture and AI-Augmented Enterprise Architecture

This shift will not happen overnight and isn't automatic, but it can be achieved in a short period of time while focusing on human skills to be trained in safe environments, where failures are welcome and lead to the highest learning experiences.

Many organizations face an even more subtle issue: the risk of the in-between. Enterprise architects may be ready for and capable of using the new ways of working, but organizational mandates may be still based on the old contract. A new contract will be required to reposition the enterprise architecture team with its new value proposition.

With all this comes a reassessment of where the enterprise architecture discipline is located within the organizational hierarchy structure.

1.3 The Enterprise Architecture Mandate

Does it matter if enterprise architecture reports to the CIO or to the chief executive officer (CEO), chief financial officer (CFO), or chief technology officer (CTO)? As you can infer from the previous section, the enterprise architecture team serves as a critical advisor across C-suite offices, but regardless, they are located *somewhere* in the organization. The organizational chart varies from company to company, which is normal and expected.

In general, "The TOGAF Leader's Guide to Establishing and Evolving an EA Capability" differentiates among three common organizational chart positions for the enterprise architecture team:

- **Function-centric enterprise architecture**
  Enterprise architecture is part of every vertical function or part of the predominant function, with subject matter experts (SMEs) for the other peripheral functions.
- **IT-centric enterprise architecture**
  Enterprise architecture is part of the IT organization, which can be located under various CxO roles, such as IT under an independent CIO, IT under the CFO, and so on.
- **Strategy-centric enterprise architecture**
  Enterprise architecture is part of the CEO office, usually under strategy or operations.

Further Resources

You can view the TOGAF guide in its entirety at *http://s-prs.co/v620600*.

Depending on the placement of the enterprise architecture department, the initial use cases may vary. For instance, in a function-centric enterprise architecture, the focus is often (not always) on smooth operations per function and reduced cost of operations. In this setup, it is important that there is still a central team that keeps an eye on overarching objectives and synergies.

IT-centric enterprise architecture usually has a different focus. If IT is located under the CFO, cost optimization and rationalization is the predominant topic. If IT is located under the chief operating officer (COO) function, agility and efficiency in the IT landscape may play a pivotal role. If IT is reporting to the CTO, obsolescence and vulnerability risks and standards might be of higher interest.

Finally, if enterprise architecture is part of the strategy-centric organization, then business acceleration, transformation, and future-ready IT will be hot topics that the enterprise architecture team will direct most of its resources toward.

Ideally, a balanced enterprise architecture team will cover all these topics, but naturally, there will be a skew toward one or another, depending on the organizational anchoring of the enterprise architecture discipline.

1.4 Summary

In this chapter, we explored the three waves of enterprise architecture evolution and what they mean for the architect operating in the AI-augmented era. One key insight was that each wave is built on the foundations laid by the one before it:

- The living inventory of wave 2 required the meta model discipline and governance thinking of wave 1.
- The AI-augmented capabilities of wave 3 require a well-structured, trustworthy inventory to reason from.

That logic also shapes the structure of this book. We'll start with an in-depth exploration of SAP LeanIX and its core components in the next chapter.

Chapter 2
Introduction to SAP LeanIX

You have decided to level up your enterprise architecture with SAP LeanIX. Congratulations! Now, it's time to mold your workspace to your needs during the setup phases. This chapter will introduce the current meta model design and useful features and capabilities that will make this endeavor as smooth and collaborative as possible for you and your organization.

To begin, you'll learn about the origin of SAP LeanIX, understanding certain design decisions that were made in the past and that still influence the current solution, for instance, in the meta model and the development of features and capabilities.

We'll then explore which capabilities belong to which SAP LeanIX products and how the out-of-the-box meta model works with the various fact sheet types. We'll dive into selected fact sheet types to understand the modeling recommendations from the SAP LeanIX experts.

Finally, you'll get a taste of some useful collaboration features in SAP LeanIX and how they contribute to shared data quality and governance responsibility across the organization that will benefit everyone, from IT to business and service.

2.1 Fundamentals and Product Structure

Today, SAP LeanIX is one of the core solutions of SAP's business transformation management offerings. How did it develop into this strategic and important product for C-suite executives? Let's go back in history to understand where SAP LeanIX came from.

LeanIX was founded in 2012 by André Christ and Jörg Beyer in Bonn, Germany. Both had worked for years in a global logistics company and had built bridges between consulting, IT, and business users. A recurring pain point in their work was delayed strategic decision-making due to lack of information and documentation in the IT management discipline. Multiple consultants would spend months researching and interviewing IT and business colleagues to create an accurate and complete inventory of applications and IT infrastructure in use, only to find that by the time this information was to be used for decision making, it was already outdated. In addition, enterprise architecture back in that time was an ivory tower discipline with little connection to the business side of the company. Frameworks and tools were made for specialists and industry experts, and it was challenging to find a united language for cross-departmental exchange and decision making.

Christ and Beyer wanted to develop an easy-to-understand meta model that would facilitate IT and business communication, supported by intuitive and easy-to-use software that would encourage collaboration and exchange between the various disciplines in an organization. The goal was to develop an industry-agnostic solution that could be used anywhere in the world.

After several iterations, LeanIX, an abbreviation of *Lean Information Exchange*, was developed into three separate but integrated products:

- LeanIX Enterprise Architecture Management (EAM)
- Value Stream Management (VSM)
- LeanIX SaaS Management Platform (SMP)

In 2022, LeanIX's partnership with SAP intensified and LeanIX became a certified SAP-endorsed app, ultimately leading to the company's acquisition by SAP in November 2023, giving us SAP LeanIX. During the acquisition, SAP decided that SAP LeanIX would remain a software-as-a-service (SaaS) product and that the two other LeanIX software products, VSM and SMP, would be sunsetted. Core capabilities of both products have since been transferred and integrated into the SAP LeanIX enterprise architecture product. The design changed, leaving SAP LeanIX as a base product that also offered add-on products with additional capabilities:

- Base product: SAP LeanIX Application Portfolio Management
- Add-on product: SAP LeanIX Technology Risk and Compliance
- Add-on product: SAP LeanIX Architecture and Road Map Planning

We'll discuss each of these products in more detail in the following sections.

Note

In August 2026, André Christ, the remaining founder, decided to move on from SAP LeanIX and focus on new adventures. Dominik Rose, a longtime SAP LeanIX executive, took over the position of general manager for SAP LeanIX products.

Further Resources

You can find the most recent information about SAP LeanIX via the following resources:

- SAP LeanIX website: *https://www.leanix.net/en/enterprise-architecture*
- SAP LeanIX user documentation:
  - *http://s-prs.co/v620601*
  - *http://s-prs.co/v620602*
  - *http://s-prs.co/v620603*

2.1.1 SAP LeanIX Application Portfolio Management

The base product, *SAP LeanIX Application Portfolio Management*, is the foundation and starting point of the enterprise architecture work. This product includes the SAP LeanIX workspace provisioning and all its dimensions, for example:

- The artifact inventory configured with the out-of-the-box meta model
- The dashboard area with preconfigured example dashboards for onboarding and standard use cases
- The reporting area with template reports
- The diagramming area with access to SAP reference architecture diagrams
- Collaboration features such as to-dos, surveys, presentations, architecture decisions, and more

Figure 2.1 shows an overview of the **Inventory** tab in SAP LeanIX Application Portfolio Management. In the top ribbon, you can see tabs for other capability areas, such as **Dashboards**, **Reports**, **Diagrams**, and **More**; the latter is a dropdown menu, which you can expand to see additional capabilities. This is where you will find features such as **To-Dos**, **Surveys**, **Presentations**, **Reference Catalog**, and the **Extension Hub** (formerly known as the SAP LeanIX store).

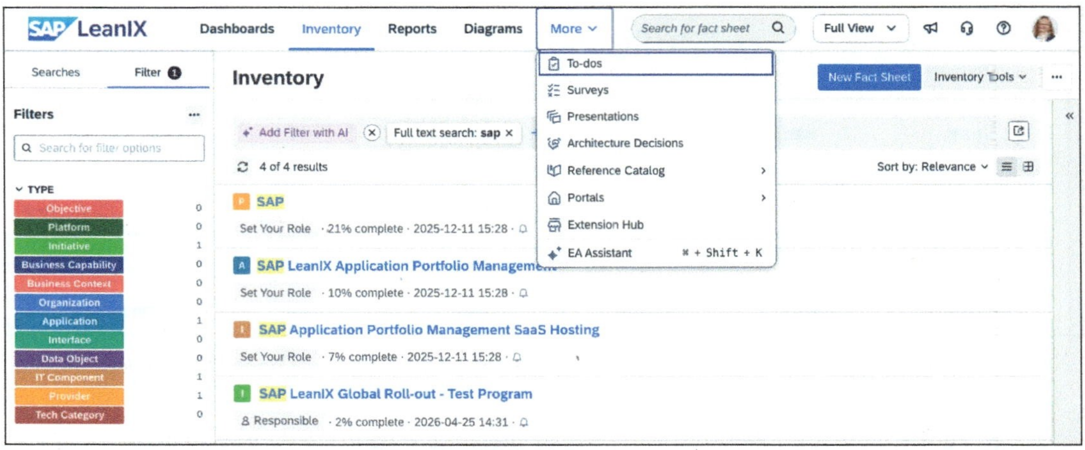

Figure 2.1: SAP LeanIX Application Portfolio Management: Top Ribbon Options

In addition, the following *discovery capabilities* are included:

- SaaS discovery
- SAP landscape discovery: public and private cloud, and on-premise
- SAP extension discovery: SAP Build, SAP BTP, Cloud Foundry environment, and SAP BTP, Kyma runtime
- AI agent and Model Context Protocol (MCP) server discovery

Figure 2.2 shows an overview of the available integrations for various discovery capabilities in SAP LeanIX Application Portfolio Management. To navigate to this overview, click the round **User** button in the upper-right corner, then click **Administration**. In the left-hand side menu, scroll to the **DISCOVERY AND INTEGRATIONS** area and click **Integrations**. In the upper-right corner, click the blue **Add Integration** button. A new overview will open that looks like Figure 2.2. On the left-hand side, you can filter the integrations based on classifications such as **EA Toolchain** or **Discovery**. Every integration has either a **More Information** or **Configure** button and a link to the SAP LeanIX user documentation.

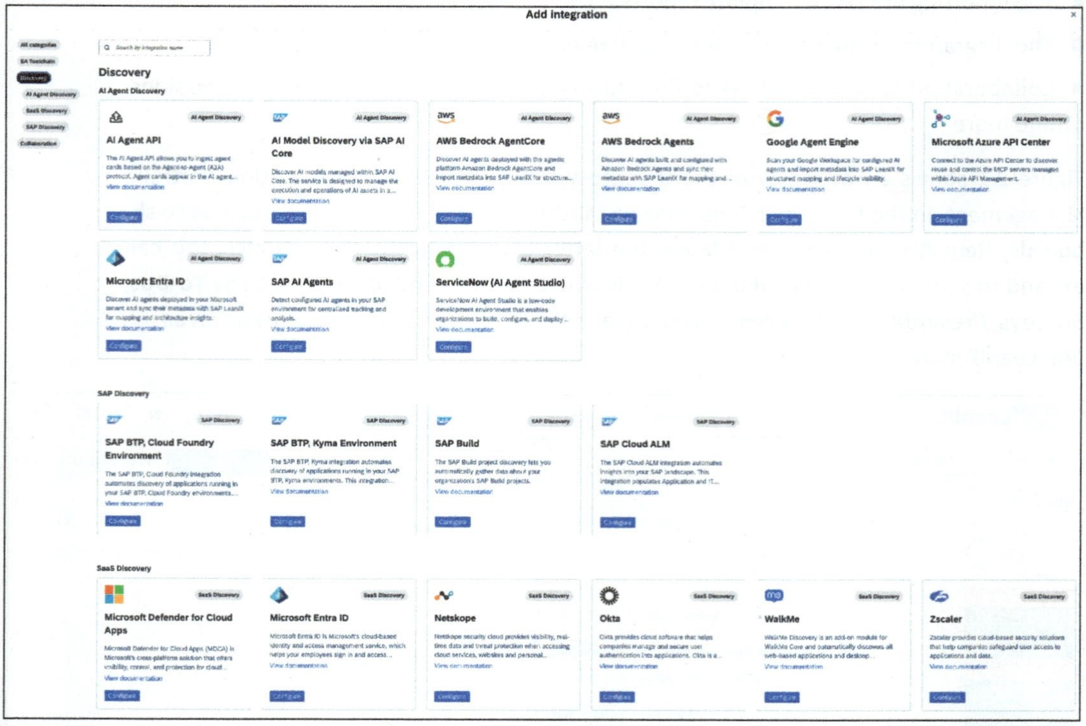

Figure 2.2: SAP LeanIX Application Portfolio Management: Discovery Capability Overview

Furthermore, the following *reference and integration capabilities* are available:

- SAP Reference Business Architecture business capability reference catalog
- Application and IT component catalog
- Tech category reference catalog based on the Technology Business Management (TBM) taxonomy
- SAP Signavio integration to retrieve business processes
- SAP SuccessFactors integration to receive the organization structure
- Apptio integration to enrich portfolio with cost data
- Collibra integration to sync data objects and SAP data products
- Custom integration capability (self-developed or via SAP Integration Suite)

In SAP LeanIX Application Portfolio Management, admins can configure four different reference catalogs in the **Administration** area. You can navigate there by clicking the round **User** button in the upper-right corner and then clicking **Administration**. As shown in Figure 2.3, in the left-hand side menu, scroll to the **DISCOVERY AND INTEGRATIONS** area and click **Reference Catalog**. There, under the related tabs, admins can configure the **Business Capability**, the **Application**, the **IT Component**, and the **Tech Category** catalogs.

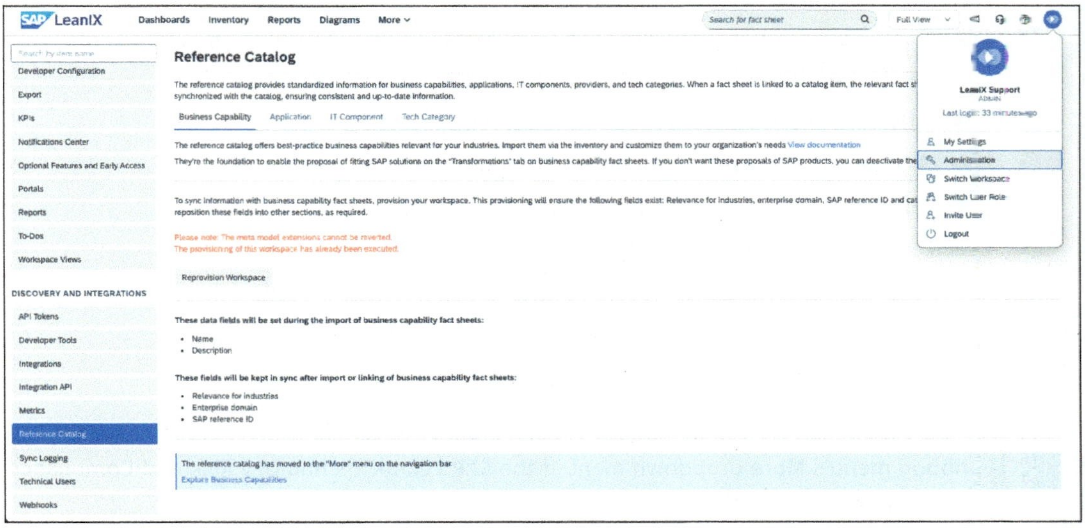

Figure 2.3: SAP LeanIX Application Portfolio Management: Reference Catalog Overview in Administration Area

Other relevant integrations can be seen in the **EA Toolchain** area, as shown in Figure 2.4.

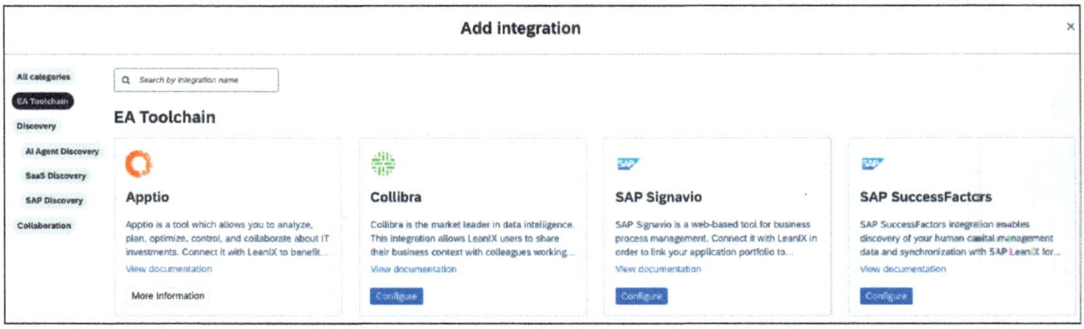

Figure 2.4: SAP LeanIX Application Portfolio Management Additional Integration Capabilities Overview

Collibra and **SAP Signavio** can be self-configured with the help of the user documentation. However, you need to request the **Apptio** integration via an SAP support ticket as customer support colleagues need to activate the integration in the backend and help with the setup.

You can always create an SAP support ticket if you have issues setting up an integration or want to discuss advanced configurations.

2.1.2 SAP LeanIX Technology Risk and Compliance

The *SAP LeanIX Technology Risk and Compliance* add-on product offers the following additional capabilities:

- ServiceNow and Jira Service Management integration for technology stack management/discovery
- GitHub enterprise integration for microservice discovery
- Microservice discovery via manifest file and script from any source
- Software bill of materials (SBOM) viewer
- Tech stack discovery from SBOMs
- Vendor lifecycle catalog for IT components to manage end-of-life risks
- Obsolescence risk and tech standards meta model extension
- Obsolescence risk management preconfigured dashboard

Figure 2.5 shows how to access the **Software Bill of Materials** viewer, available via the top ribbon menu's **More** dropdown menu if the SAP LeanIX Technology Risk and Compliance product has been purchased.

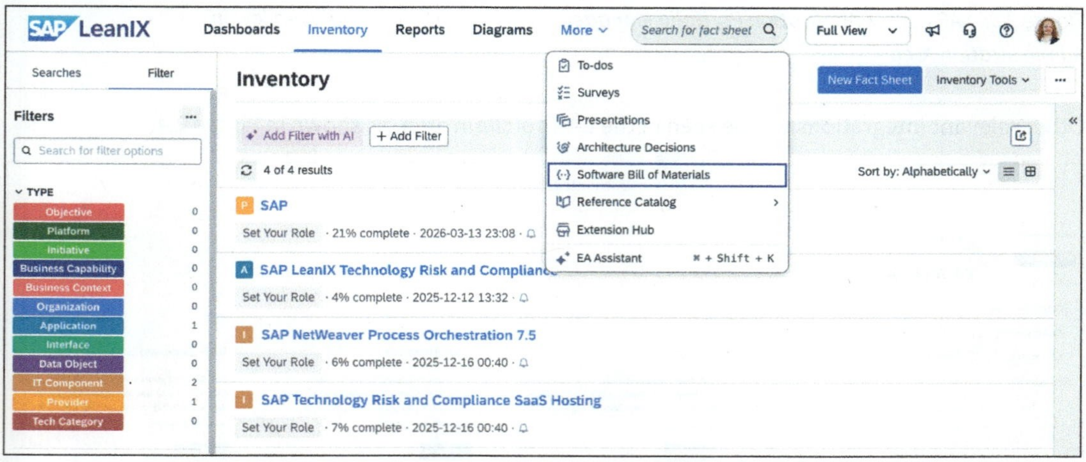

Figure 2.5: SAP LeanIX Technology Risk and Compliance: More Options

SAP LeanIX Technology Risk and Compliance also includes new capabilities via the **Inventory Tools** dropdown, as shown in Figure 2.6. The **Tech Stack Discovery** option via SBOM ingestion is an additional feature that reads out certain elements of an SBOM of a specific format and automatically creates an aggregated IT component to bring transparency into frameworks and libraries used in self-built software.

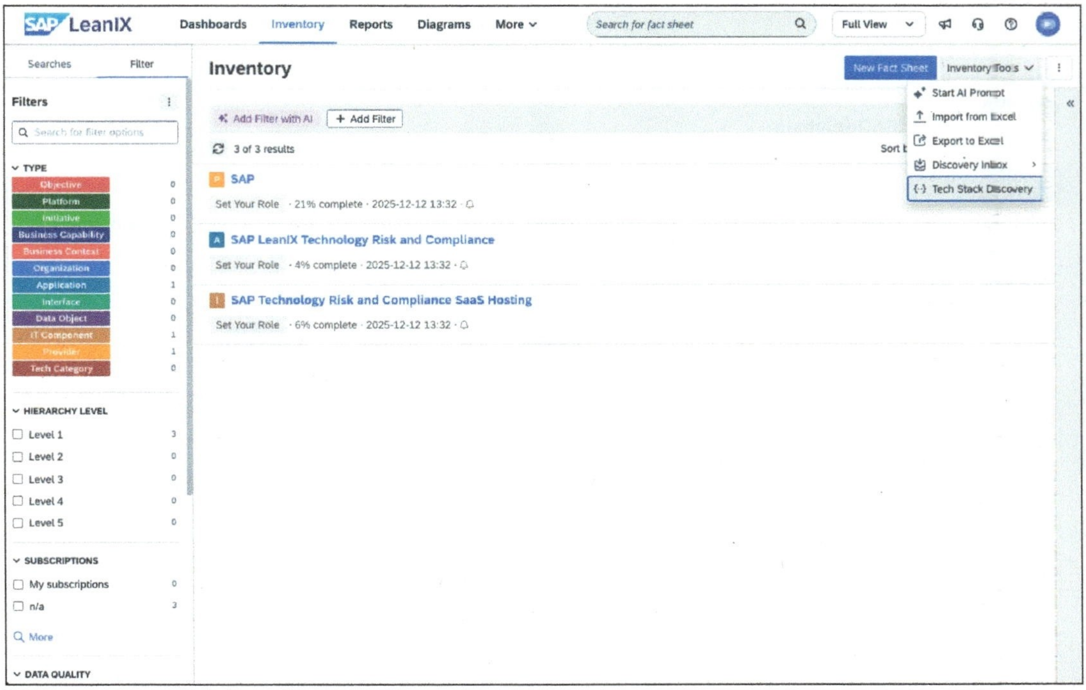

Figure 2.6: SAP LeanIX Technology Risk and Compliance: Tech Stack Discovery Capability Under Inventory Tools

Figure 2.7 displays the additional ServiceNow integration option, which becomes available when purchasing the SAP LeanIX Technology Risk and Compliance product. ServiceNow is widely used software for IT service and asset management. In ServiceNow, companies manage their IT assets from the logical down to the physical instance level. In SAP LeanIX, only the logical level, the product model level, is of interest to enterprise architects. This out-of-the-box integration is very versatile. It lets you sync artifacts from ServiceNow's configuration management database (CMDB) to SAP LeanIX and vice versa. Commonly exchanged artifacts encompass IT components, applications, business capabilities, and tech categories.

To activate the self-built software discovery option, workspace admins first must navigate to the **Administration** area. In the left-hand menu, select **Self-Built Software Discovery** under the **DISCOVERY AND INTEGRATIONS** section, as shown in Figure 2.8. First, the meta model extension must be provisioned, which will be triggered by clicking the blue **Provision** button. After that, the GitHub Enterprise integration can be configured, or a custom continuous integration/continuous delivery (CI/CD) pipeline can be connected via API and script. For more on this capability, see Chapter 9, Section 9.1.

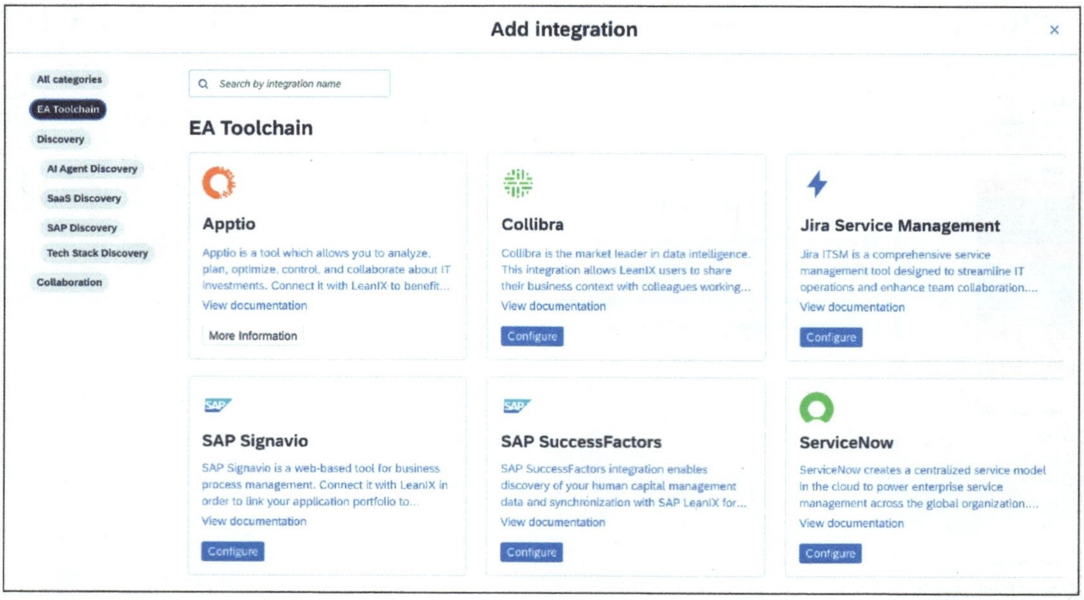

Figure 2.7: SAP LeanIX Technology Risk and Compliance: ServiceNow and Jira Integration Options in Administration Area

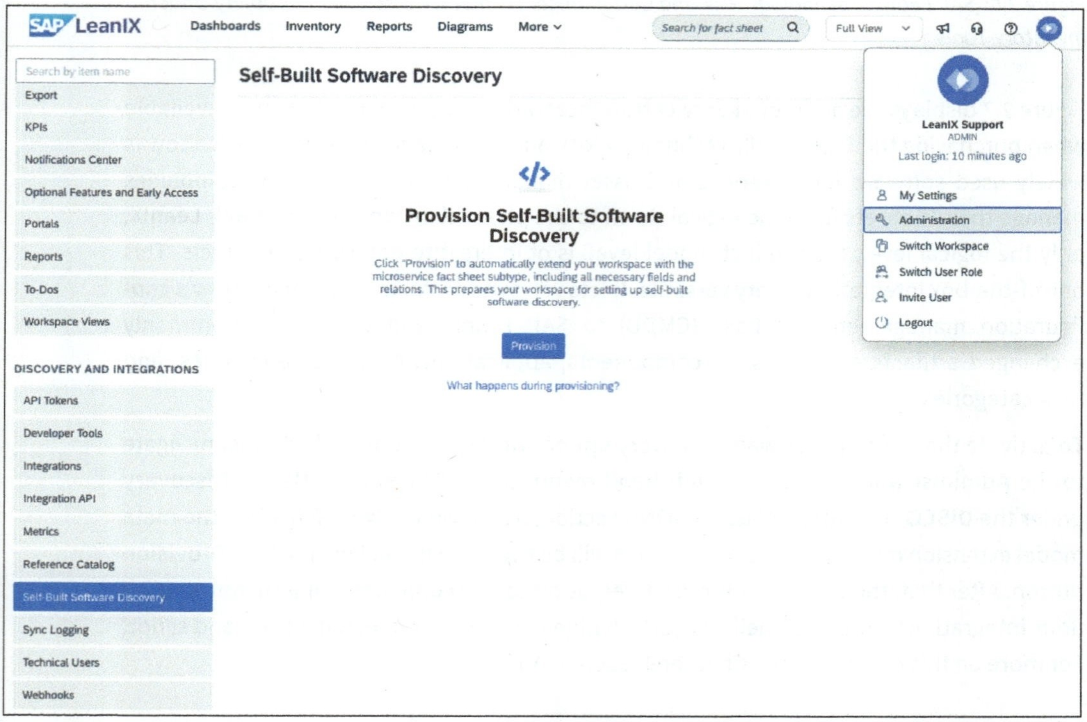

Figure 2.8: SAP LeanIX Technology Risk and Compliance: Self-Built Software Capability with GitHub Enterprise Integration and Microservice Meta Model Extension

Further Resources

For more information, see the following resources in the SAP LeanIX user documentation:

- Self-built software discovery: *http://s-prs.co/v620604*
- GitHub Enterprise integration: *http://s-prs.co/v620605*
- Manual ingestion via script and API call: *http://s-prs.co/v620606*

There is one additional feature in the IT component reference catalog that becomes available when activating the SAP LeanIX Technology Risk and Compliance product: vendor lifecycle. To access this, first ensure that the **Vendor Lifecycle** checkbox is selected under the **IT Component** tab of the **Reference Catalog** area under the **DISCOVERY AND INTEGRATIONS** section, as shown in Figure 2.9.

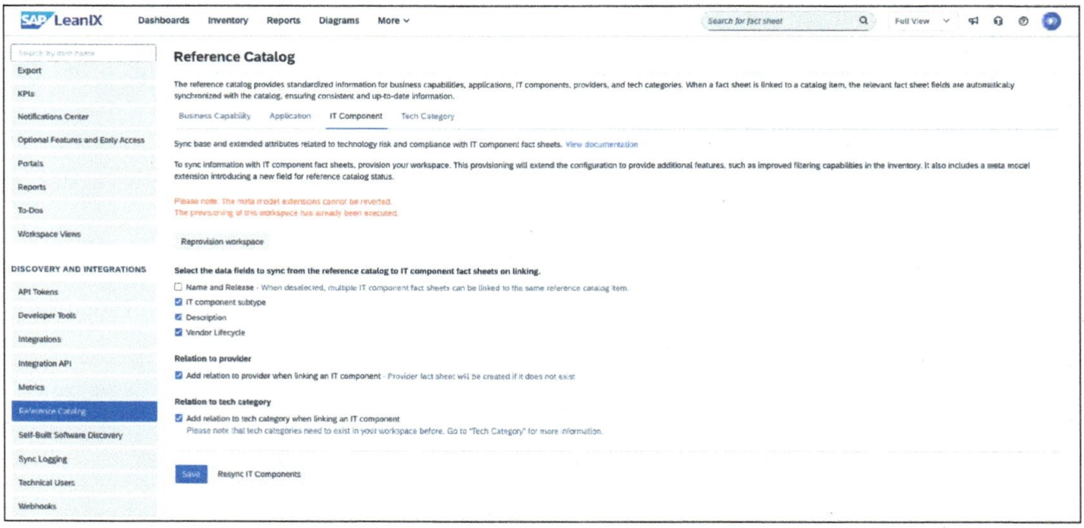

Figure 2.9: SAP LeanIX Technology Risk and Compliance: IT Component Catalog Settings

Figure 2.10 shows what the vendor lifecycle looks like in the IT component reference catalog. The regular IT component lifecycle, which is entered by the fact sheet owner, captures the time period during which the organization plans to use the IT component. The vendor lifecycle, however, shows additional information provided by the SAP LeanIX catalog research team. It indicates whether an IT component will still receive updates from the vendor, and is therefore **Active**, or has reached its **End of (Standard) Support**.

Admins in the **Administration** area can select which of the predefined dashboards will be visible for SAP LeanIX workspace users. Figure 2.11 shows the **Predefined Dashboards** options for the SAP LeanIX Application Portfolio Management base product plus the **Obsolescence risk management** option, which is available when SAP LeanIX Technology Risk and Compliance is added.

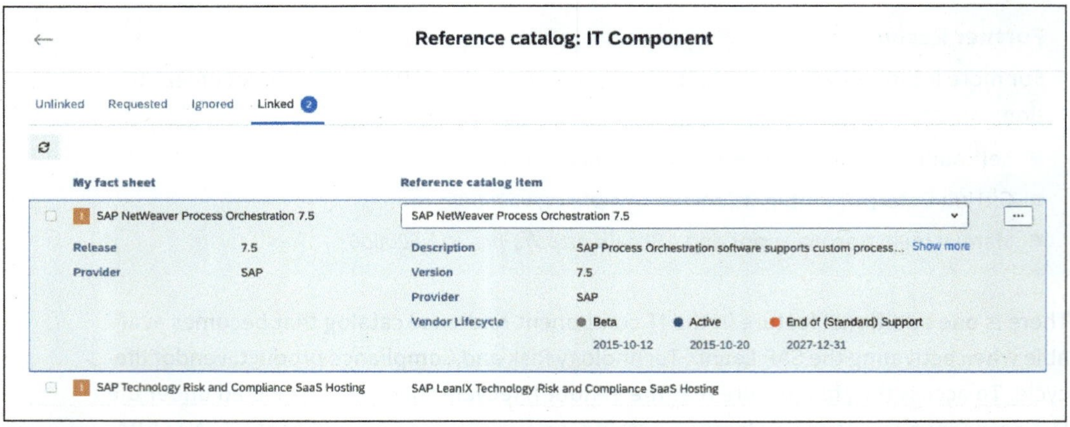

Figure 2.10: SAP LeanIX Technology Risk and Compliance: IT Component Catalog Vendor Lifecycle Example

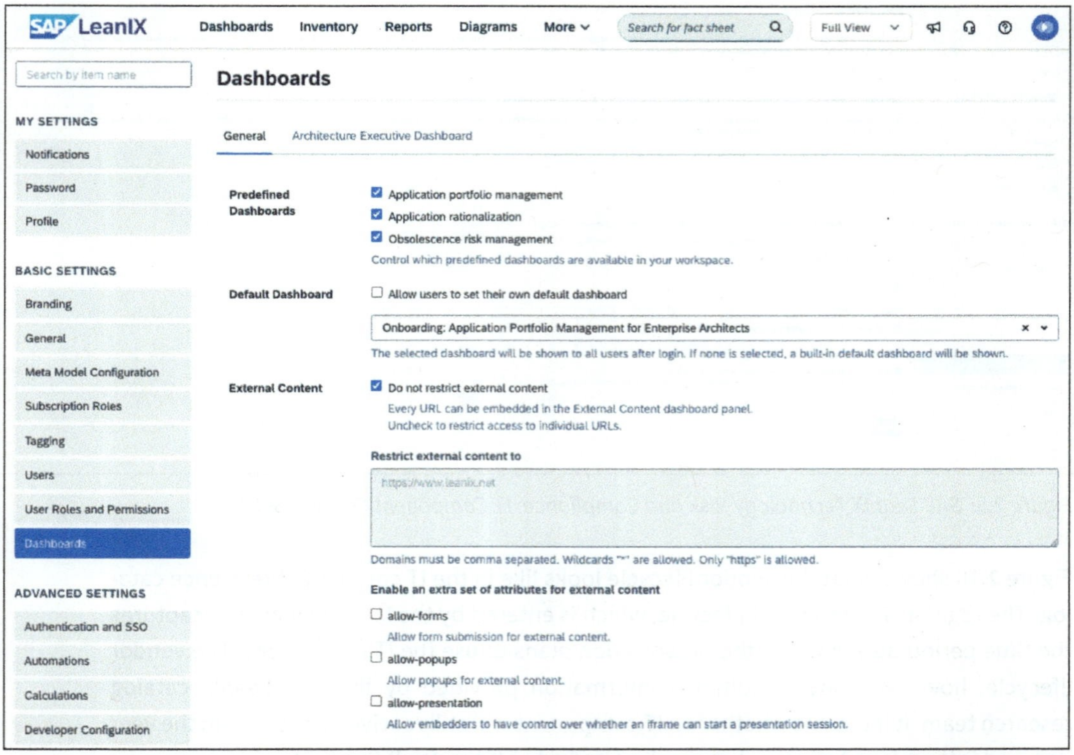

Figure 2.11: SAP LeanIX Technology Risk and Compliance: Dashboard Options in Administration Area

Users can see all dashboards that are available to them in the dashboard overview, accessed by navigating to the **Dashboards** tab in the top ribbon. Expand the left-hand side menu by clicking the left-pointing arrow in a circle on the separator line, then select **All Dashboards**. As shown in Figure 2.12, the **Obsolescence risk management** dashboard is available (first row of dashboards, fourth entry from the left).

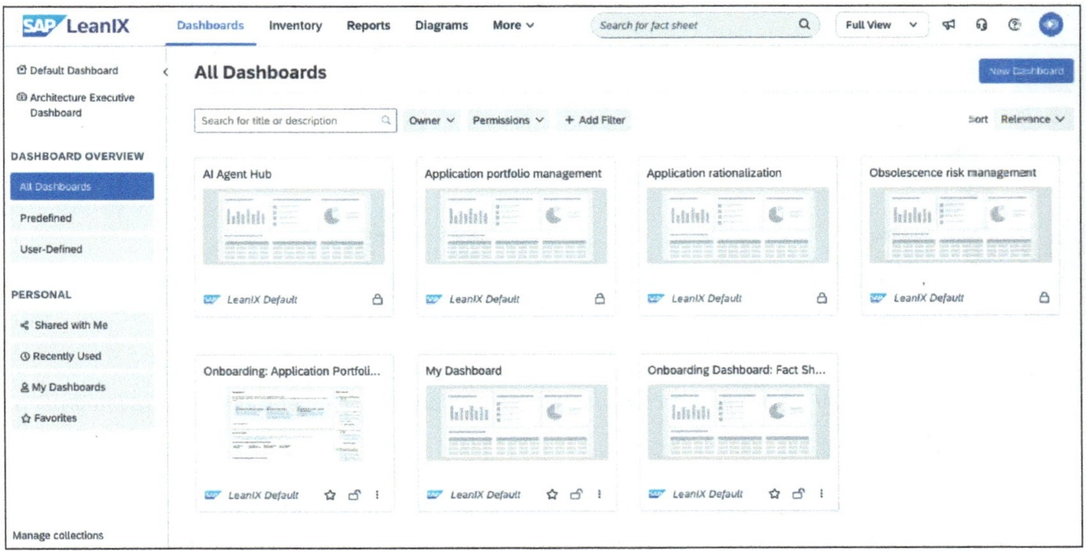

Figure 2.12: SAP LeanIX Technology Risk and Compliance: Obsolescence Risk Management Dashboard Shown in Dashboard Overview

2.1.3 SAP LeanIX Architecture and Road Map Planning

The final add-on product, *SAP LeanIX Architecture and Road Map Planning*, offers the following additional capabilities:

- SAP Cloud ALM for projects and Jira project management integration
- Preconfigured transformation templates
- Committed initiatives for time-slider transformation projections in reports
- Milestone extension of initiative fact sheets
- Impacts meta model extension
- Transformation explorer
- Target architecture diagrams

In Figure 2.13, the additional integrations that are available when activating the SAP LeanIX Architecture and Road Map Planning product are shown: **Jira for Architecture and Road Map Planning** and **SAP Cloud ALM for managing projects**. These can be found in the **Administration** area under the **DISCOVERY AND INTEGRATIONS** section in the **Integrations** menu item, after clicking the **Add Integration** button and filtering on the left-hand side for **Collaboration**.

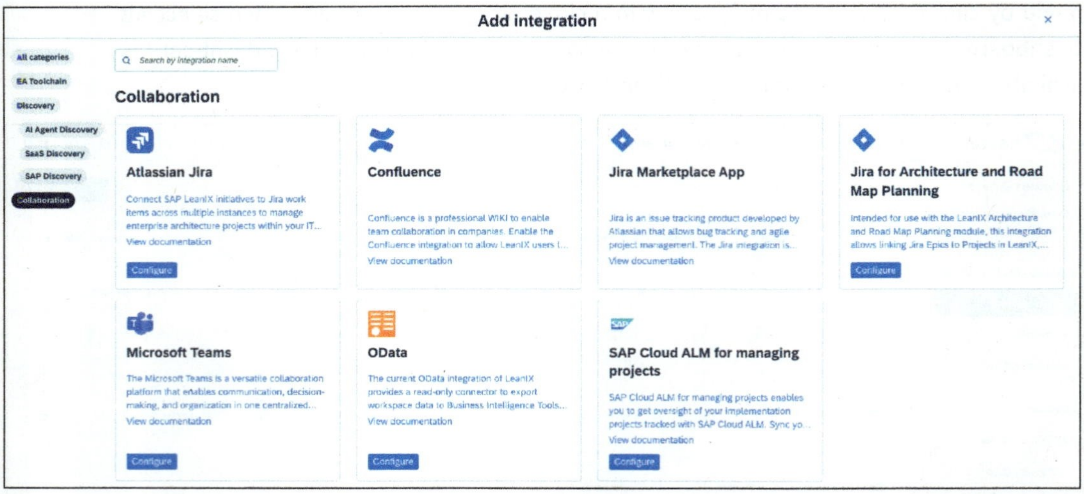

Figure 2.13: SAP LeanIX Architecture and Road Map Planning: SAP Cloud ALM and Jira Integration Capabilities

When the SAP LeanIX Architecture and Road Map Planning product is activated, the meta model extensions and features for transformations and milestones in the initiative fact sheets are not automatically provisioned. This might change in the future, but as of the time of writing (summer 2026), admins must navigate to the **Administration** area, navigate to the **ADVANCED SETTINGS** area in the left-hand side menu, and select **Optional Features and Early Access**, as shown in Figure 2.14. In the middle area, scroll to the **Use Case Extensions** area and click the **Activate** button next to **Architecture Road Map Planning: Transformations & Milestones**.

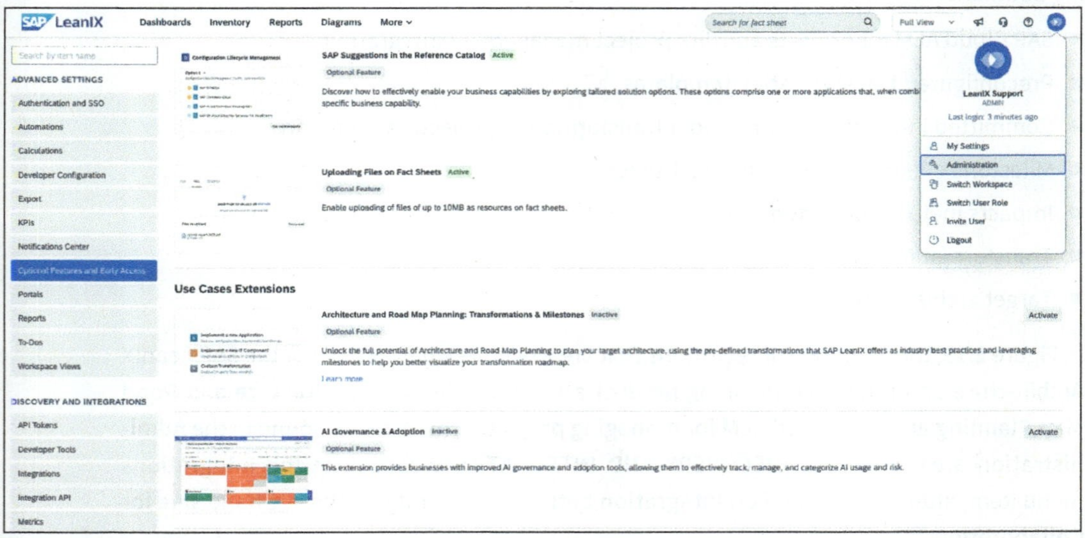

Figure 2.14: SAP LeanIX Architecture and Road Map Planning: Transformation and Milestone Meta Model Extension

Once transformations are activated in the workspace, admins will find a new menu item called **Transformations** in the **Administration** area under the **ADVANCED SETTINGS** section. As shown in Figure 2.15, admins can configure certain **Settings**, browse **Predefined Transformation Types**, create their own transformation templates in the **Configured Transformation Types** tab, or mass-update transformations in the **Excel Import & Export** tab.

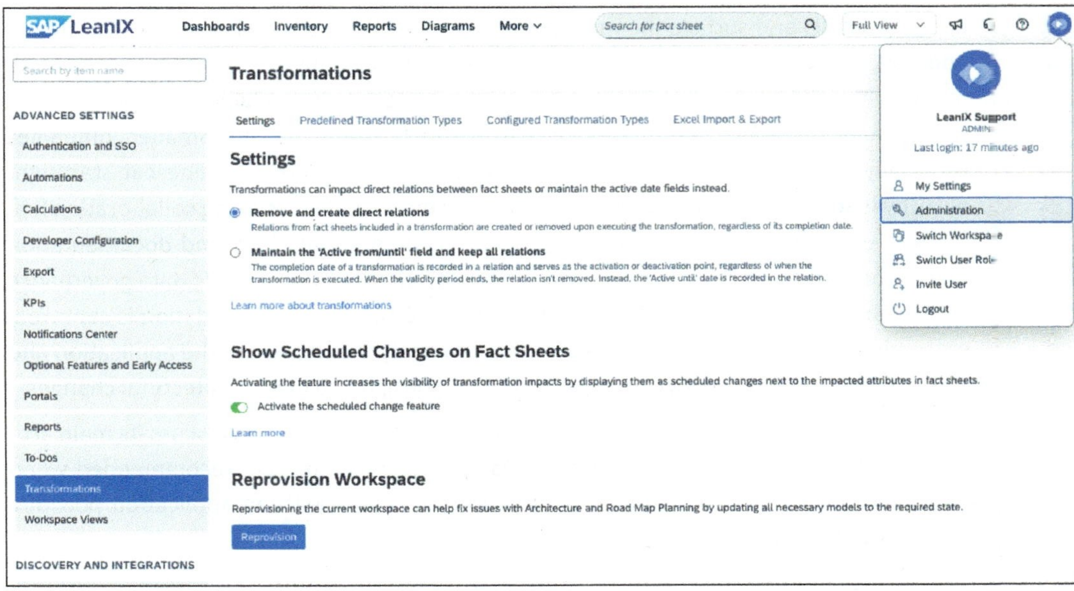

Figure 2.15: SAP LeanIX Architecture and Road Map Planning: Predefined Transformation Settings in Administration Area

Figure 2.16 displays the additional features that are available in the SAP LeanIX Architecture and Road Map Planning product in the top ribbon's **More** dropdown menu.

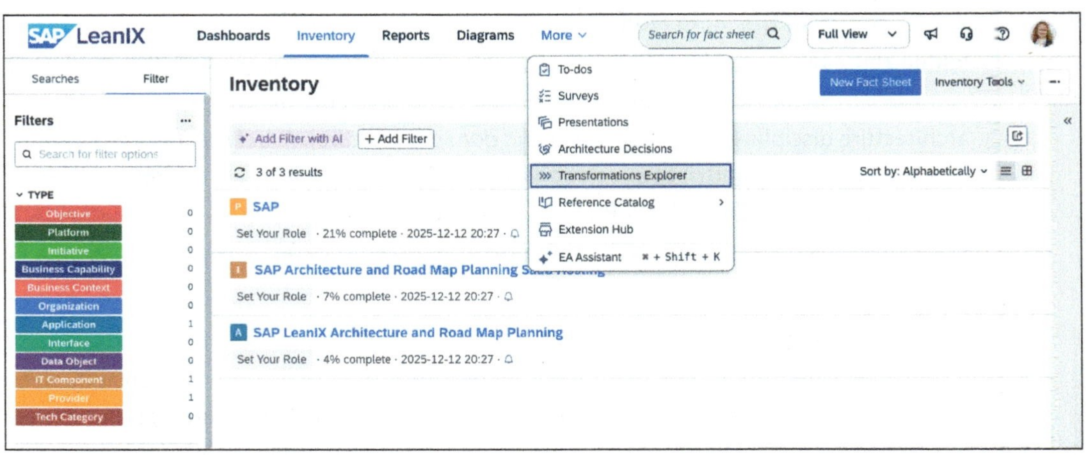

Figure 2.16: SAP LeanIX Architecture and Road Map Planning: Architecture Decisions and Transformations Explorer Capabilities Menu Options

2.2 Meta Model and Fact Sheets

It was the founders' wish to create simple, easy-to-understand, and easy-to-use software with an intuitive user interface. In contrast to other enterprise architecture concepts and frameworks, SAP LeanIX focuses on a simple meta model centered around the logical application. The logical artifacts are abstractions of the technical implementation to reduce the noise of details and surface the most important aspects for strategic decision-making. In addition, the software is designed for plug and play, meaning that a preconfigured meta model with example report visualizations and dashboards is available immediately after provisioning of the SAP LeanIX workspace to ensure fast time to value. Smaller companies especially, or those just starting with their enterprise architecture discipline, can start right away by loading their data into the out-of-the-box meta model and using collaboration features to enrich the data via crowdsourcing. After the first data load and documentation phase, the quality-tracking features of SAP LeanIX help to establish basic data governance responsibilities to ensure continuous data reviews. Once the data foundation is captured, enterprise architects and senior stakeholders can analyze and oversee the application and IT landscape to identify areas of improvement and design and plan architectural changes.

In the following sections, we'll look into the meta model in detail and discuss its main artifact, the fact sheet. We'll review selected fact sheet types that are recommended when beginning the enterprise architecture work in SAP LeanIX with an application portfolio assessment use case.

Further Resources

To learn in detail about every out-of-the-box and optional fact sheet type, check the SAP LeanIX user documentation under *http://s-prs.co/v620607*.

2.2.1 Basics of the Meta Model

Within the meta model, the conceptual building blocks of artifacts are called *fact sheet types* (e.g., the *application*) and are clustered by dimensions relevant for the enterprise architecture discipline (e.g., *application and data architecture*). A *fact sheet* is a grouping of relevant information (attributes, relations, meta data, tags, collaboration interaction, etc.) about a specific item (e.g., the SAP LeanIX application).

The SAP LeanIX meta model is structured along the following dimensions, as indicated in Figure 2.17:

- **Strategy and transformation**
  Used to capture strategic enterprise architecture objectives and to plan initiatives related to the modeled application and IT landscape, using aggregated platform bundling of applications or IT components for senior leadership alignment.

- **Business architecture**
  Used to bring context to the application and IT landscape by assigning organizational structures such as departments or regions/countries to artifacts via the organization fact sheet and assigning supported business capabilities to applications, as well as capturing various other business context types, such as processes, value streams, or products, and their relation to applications and IT components.
- **Application and data architecture**
  Includes the core and centerpiece of the SAP LeanIX product, the application, which is connected via logical interfaces and uses data objects or transfers data objects via interfaces to other applications.
- **Technical architecture**
  The IT layer underneath the application is centered around the IT component, which has a provider, also called a *vendor*, and is defined by a tech category.

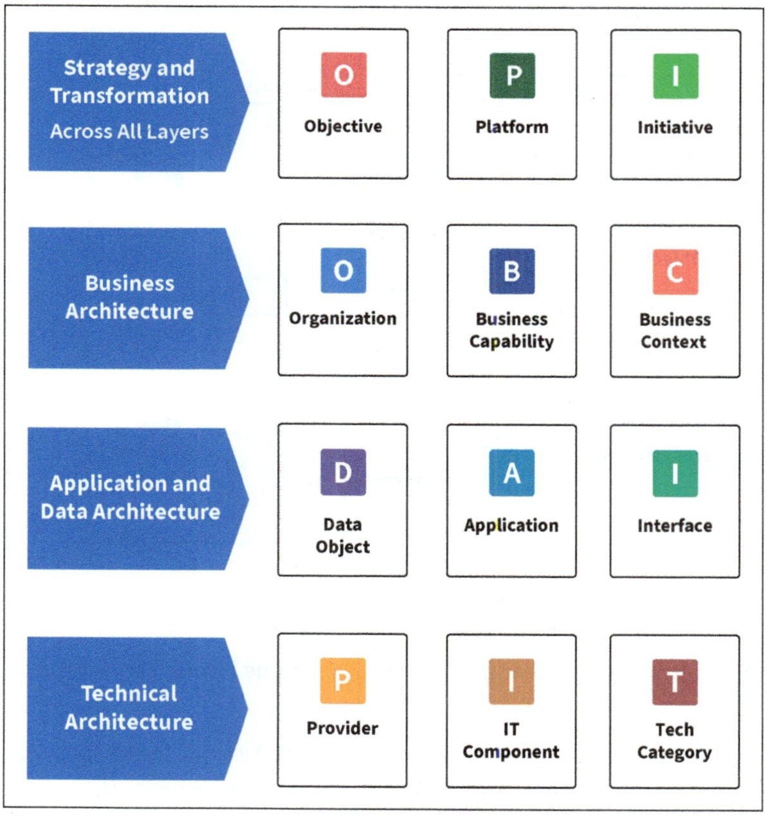

Figure 2.17: SAP LeanIX Current Meta Model: Version 4

> **Note**
> The colors of the fact sheet types correspond with the default color-coding of the artifacts of the respective types in diagrams and reports.

Beyond the out-of-the-box fact sheets, SAP LeanIX also provisions out-of-the-box relations between the preconfigured fact sheet types. Figure 2.18 illustrates the relational concept.

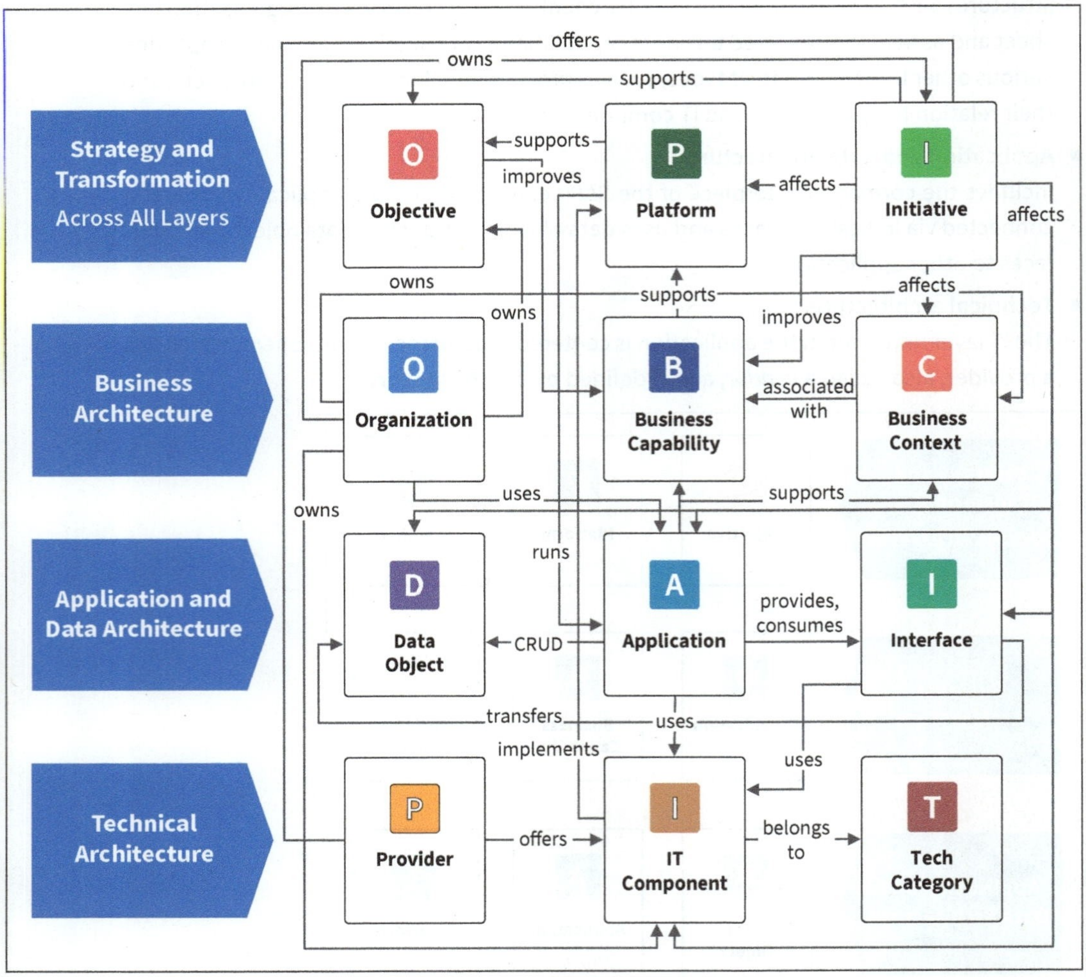

Figure 2.18: SAP LeanIX Out-of-the-Box Relational Concept of the Preconfigured Meta Model

Fact sheets are in general structured in the same way with the same layout (see, for example, Figure 2.19).

As a header, you will see the display name of the respective fact sheet. In the example shown in Figure 2.19, this is SAP LeanIX Application Portfolio Management. Under the name, you can see the fact sheet type with which this artifact was created (in this case, **Application**). Below the fact sheet type, various tabs capture different information about the artifact. The number of tabs vary depending on the SAP LeanIX products available.

The base product tabs include the following:

- **Fact Sheet**
  Captures data such as the attributes and relations, clustered in sections and subsections
- **Subscriptions**
  Captures the responsibility structure of the fact sheet, such as the responsible owner or interested observers
- **Comments**
  Used to publicly exchange notes and comments on the artifact
- **Recommendations**
  Intelligent feature that supports responsible persons and admins increasing data completeness via smart recommendations, for instance, AI-generated descriptions
- **To-Dos**
  Used to capture questions and assign action items between users of this specific fact sheet
- **Resources**
  Area to see linked diagrams, add links to external websites, and upload related files or logos
- **Metrics**
  Area to bring in external data points via API from third-party sources to visualize additional time series and charts on this fact sheet
- **Surveys**
  Overview of active or concluded surveys this fact sheets was a part of
- **Last update**
  Audit log of past changes to the fact sheet data

Figure 2.19: General Fact Sheet Structure

2.2.2 Application

The *application* is the core fact sheet type in SAP LeanIX. It represents a software application (sometimes also called a *software service* or *program*) that is used for business tasks by end users (human or agentic) to support a certain business capability. The application portfolio can be maintained, analyzed, and related to other entities such as underlying IT components, using or owning organizations, or supported business capabilities in SAP LeanIX.

The application is a logical artifact, meaning that it's an abstraction of the real technical implementation. Applications can be any of the following:

- Purchased, like commercial, off-the-shelf software (e.g., SAP LeanIX)
- Open-source applications (e.g., Jenkins)
- Custom-built software developed by your organization (e.g., a self-built contract compliance checker)
- Simple tools that don't seem to be applications (e.g., Excel spreadsheets that support specific business processes and business capabilities)

We recommend that you start with a high-level collection of logical applications if no respository already exists. Helpful features to discover applications include the following:

- **SaaS discovery**
  Offers discovery of SaaS applications by integrating with single-sign-on (SSO), secure access service edge (SASE), and cloud access security broker (CASB) solutions. See Chapter 4, Section 4.2.4 for more information.
- **SAP landscape discovery**
  Offers discovery of SAP products, systems, services, and custom-built extensions in use. See Chapter 4, Section 4.2.3 for more information.
- **Application reference catalog**
  Offers an in-tool application catalog curated by the SAP LeanIX research team for SAP applications, SAP AI agents, and business SaaS applications.
- **Excel import via formatted Excel tables**
  Offers manual bulk import of fact sheets, fact sheet data, and relations. See Chapter 4, Section 4.2.6 for more information.

After creating your collection of applications, you can proceed with application fact sheet modeling. It's important to define modeling guidelines to avoid confusion, duplicates, or unneeded noise in the inventory. Be clear about which logical level you need to model the different applications for strategic decision-making. Here are some best practices to keep in mind:

- In the beginning, it's best to stay at a high level, for example, only modeling SAP LeanIX and not add-on products such as SAP LeanIX Application Portfolio Management and SAP LeanIX Technology Risk and Compliance in the application fact sheet modeling design.

- Only model drilldowns into more granular levels such as subproducts, modules, regional instances, and so on if you need to do so from a strategic decision-making perspective or if you want to track data on a more granular level.
- For drilldowns, such as into specific modules or products, we recommend modeling them with a parent-child relation to the logical parent to make use of the hierarchy aggregation functionality in reports.
- When modeling parent-child relations, we recommend maintaining the business capability relation of applications at the lowest level and using reports to create aggregated views.

Caution

Keep in mind that every application counts toward the application quota for pricing. That could also be a reason to model applications using a flat hierarchy without parent-child relations and to forego the aggregation capability.

The application fact sheet has two recommended subtypes when activating additional capabilities:

- **AI agent**
  AI agents are prompted but autonomous systems using large language models (LLMs) to interact with their environments (e.g., applications) by performing complex tasks and to achieve a user-defined goal without explicit direction given. AI agents usually perform business tasks and pursue goals relevant to the business model of the organization and therefore create business value.

  When the AI agent feature is activated, then the AI agent subtype under the application fact sheet is created. You can activate the AI agent extension via the **Optional Features and Early Access** menu item in the **Administration** menu or via a support ticket. This might change in the future, so always check the SAP LeanIX user documentation for the latest information regarding how to activate certain features.

  Ideally, an AI agent is connected to the relevant business applications it originates from. It can have its own specific underlying IT components and connections to other AI agents or applications via logical interfaces.

  To avoid lots of manual work, we recommend using the AI agent discovery capability of SAP LeanIX to keep the agent repository as up to date as possible.
- **Microservice**
  Microservices are small, independently deployable, self-owned software components that provide functionality to self-built applications or feature extensions and can be consumed by multiple applications or services.

  The microservice subtype can be created and maintained manually, but it makes the most sense to use it with the SAP LeanIX Technology Risk and Compliance product. It can then be provisioned with activation of the GitHub Enterprise integration for microservice discovery.

In general, microservices are related to their corresponding logical or business application as they usually don't perform entire business tasks and can't function as standalone components.

For more details on this application subtype, refer to Chapter 9.

Caution

Keep in mind that in general, application subtypes also count toward the application quota. However, there might be exceptions for certain SAP LeanIX recommended subtypes. Check with your account executive about the most recent pricing rules before pursuing any meta model extension for the application fact sheet.

When modeling microservices in SAP LeanIX, ensure that modeling happens on a sensible level. SAP LeanIX is an enterprise architecture solution, not a CMDB, and users shouldn't be overwhelmed by too much detailed information that isn't relevant for decision-making.

2.2.3 Business Capability

Business capabilities mirror the capabilities an organization needs to fulfill its business requirements. They represent *what* an organization does but not *how* it is done. Business capabilities use business terms that can be understood by every employee and are again abstractions. They should be defined in a collaborative approach between IT and business. To define business capabilities, you can create a *business capability map* following these best practices:

- They should be designed in such a way that they are relatively stable over time, provide guidance for responsibility assignment, and support decision-making.
- It's easiest to start with the high-level core capabilities on level 1 and then break down to lower levels as needed.
- Always ensure that business capabilities are mutually exclusive to prevent confusion among SAP LeanIX users later.
- Also ensure that they truly represent capabilities and not end-to-end processes.

It's possible to import an existing business capability map using the **Excel Import** functionality. If your organization doesn't have a business capability map yet, SAP LeanIX can import the SAP reference architecture content, which contains curated business capability sets. The business capability reference catalog in SAP LeanIX is only accessible by admins. The broad template capability map can be filtered based on industries, relevant end-to-end processes, enterprise domains, and more. The SAP Reference Business Architecture capabilities go down to level 3 and, if desired, suggested SAP applications can be made visible. To navigate to the business capability reference catalog, click **More** in the top ribbon menu and expand the **Reference Catalog** menu item, then select **Explore Business Capabilities**, as shown in Figure 2.20. You'll reach the business capability explorer shown in Figure 2.21.

Read Chapter 4, Section 4.2.1 for more detailed information on how to use the business capability reference catalog.

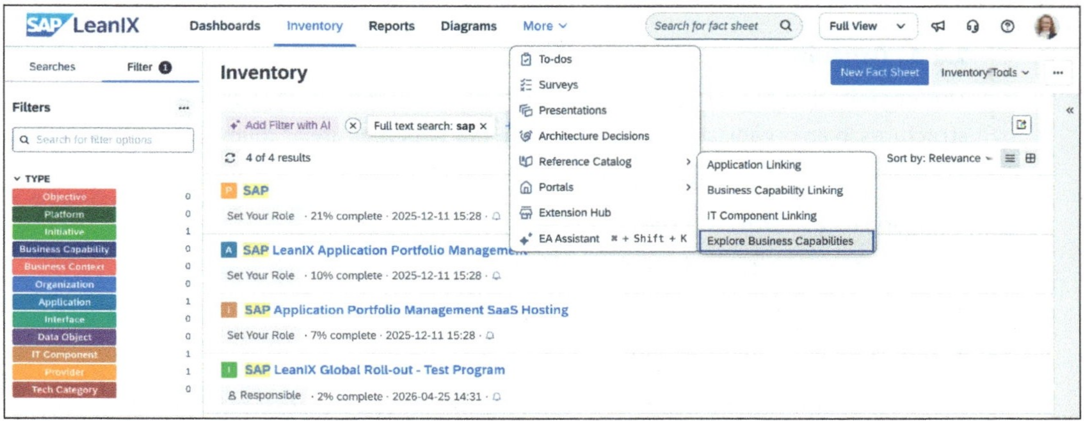

Figure 2.20: Navigating to the Business Capability Explorer

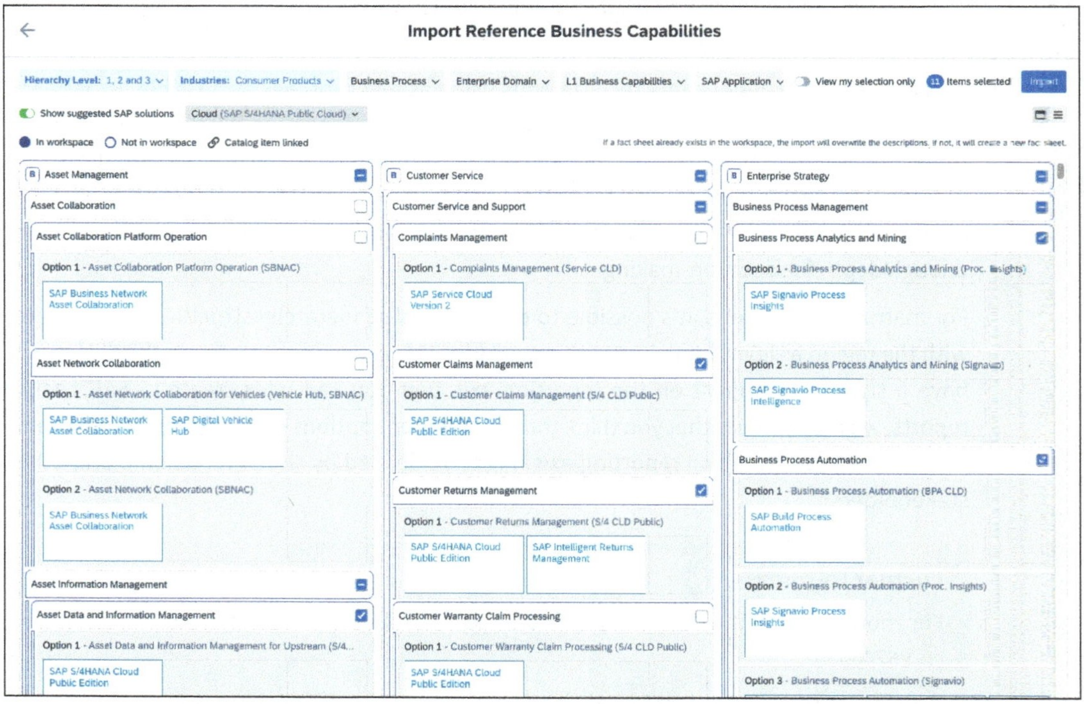

Figure 2.21: Overview of Business Capability Explorer in SAP LeanIX

Further Resources

For more information on business capabilities and examples of antipatterns, visit the SAP LeanIX user documentation at *http://s-prs.co/v620608*.

2.2.4 Organization

The organization fact sheets represent different dimensions of potential organizational structures in an organization. The out-of-the-box subtypes are as follows:

- Legal entity (e.g., SAP Germany)
- Region (e.g., North America)
- Business unit (e.g., product and engineering)
- Team (e.g., Team Overtake)
- Customer (e.g., buyer persona)

Just because these subtypes exist doesn't mean that every customer must use all of them in their SAP LeanIX workspace. Choose the ones that best fit your reporting and structuring purposes and are common and known dimensions among employees. It's also possible to rename or delete subtypes or create additional ones, if the predefined examples do not suit your organizational design.

As a best practice, SAP LeanIX recommends using as few subtypes as possible. Experience shows that if you aren't syncing the organizational dimensions from third-party source systems, updating the hierarchy manually might be very cumbersome after restructuring initiatives. The deeper you go into parent-child hierarchical structures, the more items you need to maintain manually. Therefore, stay as high level as possible while still providing enough depth for decision-making.

For matrix organizations, it's possible to create a nested hierarchy structure, for example, with the region as level 1 and business units as level 2 or vice versa. Because the design will have a significant impact on the reporting axis that you can later create in SAP LeanIX reports, we recommend that you think through different options carefully and create example reports to verify which reporting axis might be desired by senior leadership and other stakeholders.

Further Resources

For more information, check out the SAP LeanIX user documentation at *http://s-prs.co/v620609*.

If you're using SAP SuccessFactors, then you can automatically import your organizational structure using the out-of-the-box integration: *http://s-prs.co/v620610*.

2.2.5 IT Component

IT components include services, software, hardware, underlying SaaS hosting, platform as a service (PaaS), or infrastructure as a service (IaaS), as well as AI models that are needed to run or deliver an application.

Maintaining IT components can provide crucial insights into the following:

- Obsolescence risk and technical debt
- Operating costs
- Provider dependencies
- Location management
- Hosting structure

As a good starting point, you should only model what you need for a specific use case. For instance, to start, only model the IT components that have a relation to an active application in the SAP LeanIX workspace. Focus on the most important components; don't try to boil the ocean by bringing in all instances and deployments of legacy technology.

Note

Applications and IT components exist in a kind of symbiosis. That sometimes makes it difficult to decide whether something is an application or an IT component.

Operating systems, databases, virtual machines, hosting, technology services, and the like are typical examples of IT components. Take SAP LeanIX as an example. SAP LeanIX is the SaaS software modeled as an application. The corresponding IT component would be the SAP LeanIX SaaS hosting with the SAP provider. If your quoting tool is an Excel file with lots of formulas, then the quoting tool would be the application, and your specific version of Excel would be the corresponding IT software component related to the Microsoft provider. An application can also be connected to multiple IT components, such as the technical version and the on-premise hosting server, which could also come from a different provider.

IT components usually are named and referenced by their software release version, for instance, Spring Boot 3.4. Spring Boot is the IT component name entered in the corresponding **Name** field and represents the versionless IT component. The version number, 3.4, is optional information and can be entered into the **Release** field on the fact sheet. It specifies the version in the format *major_version.minor_version*.

Try to use as much automation as possible for IT component creation and modeling in SAP LeanIX. When using SAP LeanIX discovery capabilities such as SaaS discovery, SAP landscape discovery, SAP extension discovery, tech stack discovery from an SBOM, or creation from application reference catalogs, a good amount of IT components can be automatically detected in the discovery inbox and created plus linked to the IT component reference catalog for you. (Note that vendor lifecycle data is only available when you have purchased the SAP LeanIX Technology Risk and Compliance product.)

Figure 2.22 shows the settings of the application reference catalog in the **Administration** area, under **DISCOVERY AND INTEGRATIONS**, in the **Application** tab under **Reference Catalog**. The autocreation of IT components can be activated or deactivated when applications are created from the reference catalog by checking or unchecking the respective box.

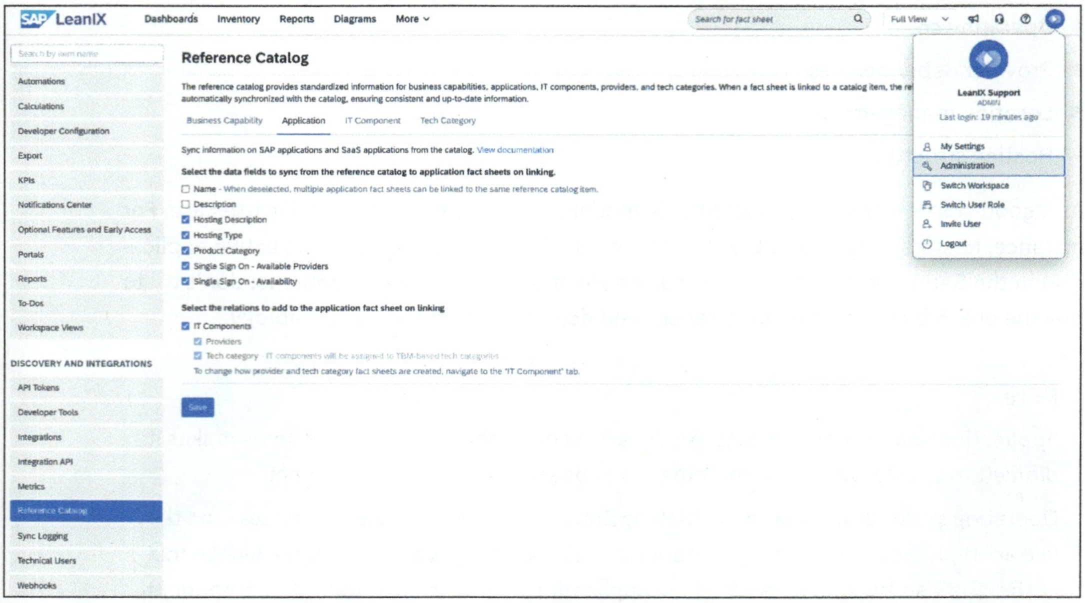

Figure 2.22: SAP LeanIX Application Reference Catalog: Settings for IT Component Autocreation

Nonetheless, as a human in the loop, you should review the **Discovery Inbox** under **Inventory Tools** (see Figure 4.11 in Chapter 4, Section 4.2.3) regularly as well as check for completeness and fill in gaps that cannot be automatically detected and created. We'll cover this in more detail in Chapter 4, Section 4.2.3. Integrations with CMDBs such as ServiceNow can also help with gathering IT components via the out-of-the-box integration.

Note

Remember that the ServiceNow integration is only available with the SAP LeanIX Technology Risk and Compliance product.

When modeling IT components manually, consider the following best practices:

- Start with versionless IT components if you have no existing repository, and detail major versions in a next step.
- Consider carefully for which use cases you need minor versions.

- Stay as high level and aggregated as possible for decision-making and insight generation to reduce possible manual maintenance and update work in the long term.
- Avoid modeling physical instances but stay on an aggregated logical level to not overcrowd the workspace.

The predefined subtypes of IT components include the following:

- SaaS (e.g., SAP LeanIX SaaS hosting)
- PaaS (e.g., SAP HANA Cloud, SAP Build)
- IaaS (e.g., Amazon Web Services [AWS])
- Software (e.g., SAP S/4HANA 2025; SAP S/4HANA would be the corresponding application)
- Hardware (e.g., IBM Power S1124)
- Service (e.g., maintenance/support service by third party; so-called managed IT services to implement or run an application)
- AI model (e.g., Claude, Gemini, GPT)

Further Resources

For more information, refer to the following SAP LeanIX user documentation pages:

- IT component modeling guidelines: *http://s-prs.co/v620611*
- IT component lifecycle catalog: *http://s-prs.co/v620612*
- How to create a fact sheet manually: *http://s-prs.co/v620613*

2.2.6 Provider

The provider fact sheet is kind of an extension of the IT component fact sheet, indicating from which provider/vendor a technology component was purchased. The creation of the provider fact sheet can also mostly be automated when you use the SAP LeanIX discovery capabilities and link IT components to the IT component reference catalog.

As shown in Figure 2.23, you can enable the autocreation of providers by going to the **Administration** area. Under the **DISCOVERY AND INTEGRATIONS** section, select the **Reference Catalog** menu item, go to the **IT Component** tab, and select the checkbox in the **Relation to provider** section.

Why is the provider fact sheet so important? First, you can use various SAP LeanIX template reports to visualize the dependency level of providers. Second, for the relation of IT component and applications, SAP LeanIX users can capture total annual cost, which can then be aggregated per provider in the **Provider Cost** report. You can find this by selecting the **Reports** option in the top ribbon menu bar, then selecting the **Cost reports** category under the **REPORT TYPES** section, as shown in Figure 2.24.

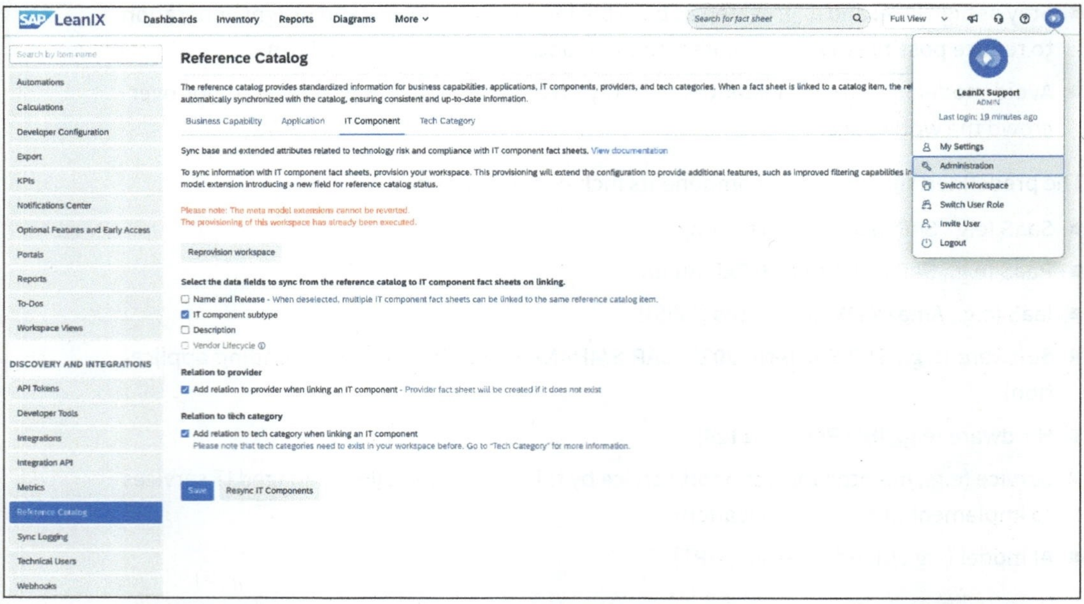

Figure 2.23: IT Component Reference Catalog Settings to Manage Provider Autocreation

Figure 2.24: Navigation to Predefined Provider Cost Report

2.2.7 Tech Category

Tech categories are used to classify and group IT components into standardized technology categories.

Companies can of course import their own tech categories, such as via Excel imports (see Chapter 4, Section 4.2.6 for more information). However, when using the recommended SAP LeanIX standard for tech categories, you can benefit from a few features that automate the linking between IT components and tech categories. The tech category model, similar to the business capability model, is quite stable over time and independent of industries. If you use your own tech category model, ensure the following aspects:

- *Breadth before depth:* It is better to have a wider map rather than a deeper map.
- *Don't overlap:* Tech category maps should be mutually exclusive and collectively exhaustive at all levels

If you want to explore the SAP LeanIX recommended standard, first navigate to the **Administration** area. Under the **DISCOVERY AND INTEGRATIONS** section, select menu item **Reference Catalog** and tab **Tech Category**. Here, admins can review and import selected tech categories of the TBM taxonomy (see Figure 2.25).

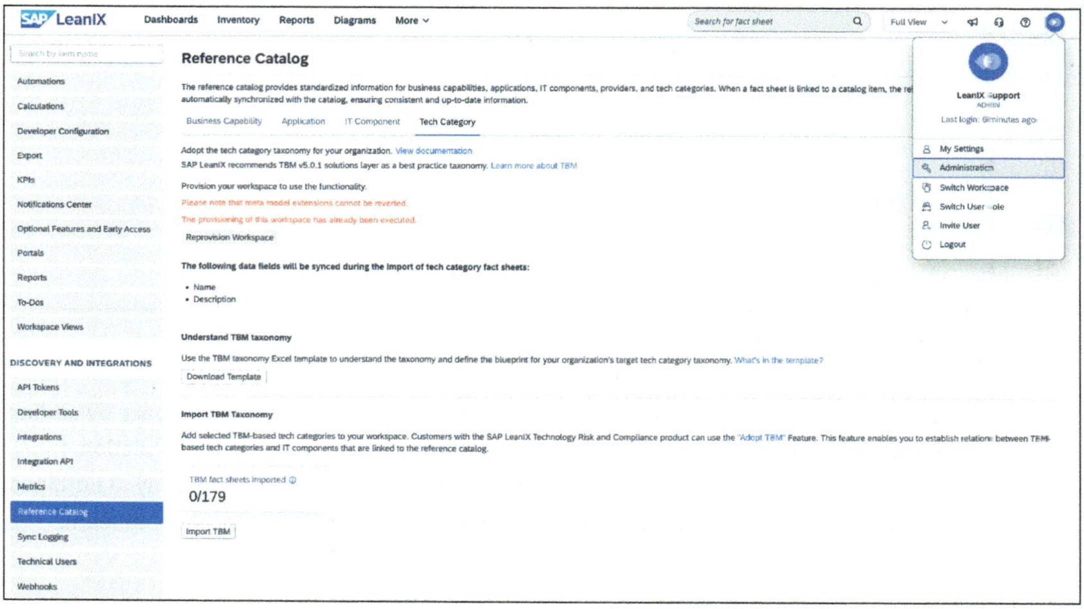

Figure 2.25: Tech Category Reference Catalog in Administration Area

> **Note**
>
> The TBM taxonomy is an open standard that is also used by the SAP LeanIX catalog research team.

Once you have reviewed the TBM catalog by downloading an overview (use the **Download Template** button under the **Understand TBM taxonomy** section) and then selecting the tech categories in scope and importing them to the SAP LeanIX workspace with the **Import TBM** button, you can configure the IT component reference catalog so that when an IT component is linked to the catalog, the relation to the tech category is automatically created.

To check the IT component reference catalog settings, switch to the neighboring **IT Component** tab in the **Reference Catalog** admin overview and ensure that the checkmark is set in the **Relation to tech category** section. Remember to click the **Save** button in the lower-left corner of the middle window area if you changed anything in the settings (see Figure 2.26).

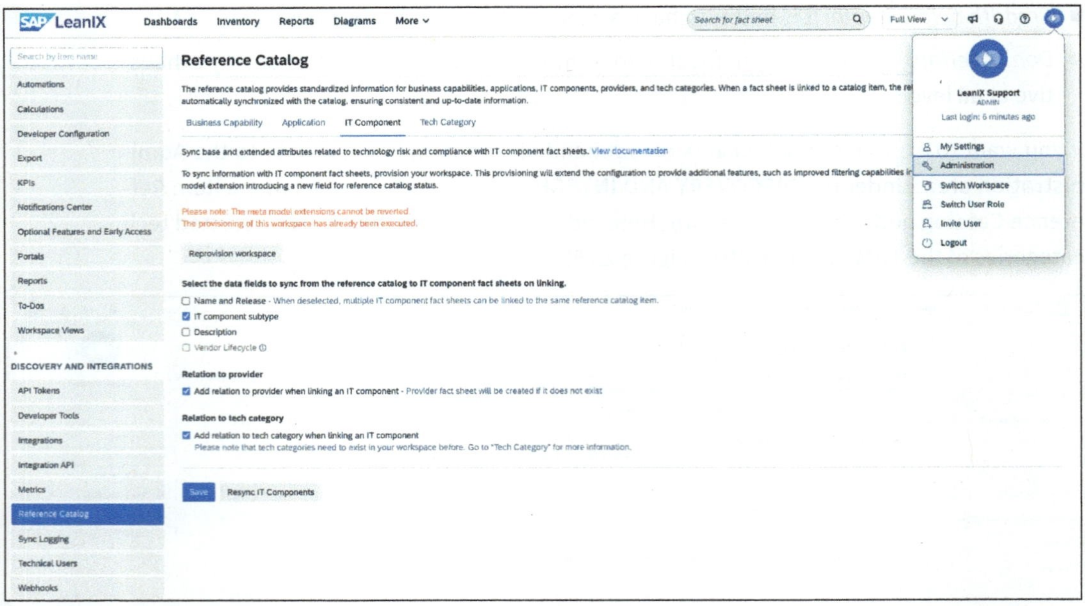

Figure 2.26: IT Component Catalog Tech Category Autolinking Settings

It's also possible to modify the TBM tech categories in the SAP LeanIX workspace by adding additional custom tech categories manually as fact sheets (see Chapter 4, Section 4.2.7). Just remember that if you add new categories, then they need to be manually maintained as the additions are not known nor included by the SAP LeanIX catalog research team.

2.2.8 Interface

The interface fact sheet is a delicate fact sheet type that's often implemented in a way different than originally intended, which might then impact the visualization opportunities in reports and diagrams in the SAP LeanIX workspace.

In SAP LeanIX, the interface is the *logical, business-oriented* interface, not the technically implemented interface. Therefore, it's a connection not between IT components but between logical applications. The logical interface relies on IT components and is implemented by IT components, which then provide insights into the middleware, software, services, and formats used.

The interface circle map in the **Reports** section only visualizes interfaces between application fact sheets. The same applies to the data flow diagram, which only functions with interfaces between applications.

SAP LeanIX suggests three different subtypes to capture different styles of interfaces:

- **Logical interface (business-oriented modeling)**
  Provides the connection and data flow between applications. Used to model communication patterns.
- **API (technical data flow dimension)**
  Provides an overview of available technical endpoints:
  - Connections between microservices (an application subtype for self-built software management)
  - If needed, system-to-system integration (the system is an optional fact sheet type; see Section 2.2.9)
- **MCP server**
  A collection of managed communication endpoints that expose specific information to AI capabilities or systems using standardized protocol interfaces. Typical modeling scenarios include the following:
  - Intelligent agent interactions
  - Protocol-based communication
  - Endpoints and access controls

In SAP LeanIX, interfaces show two dedicated types of relations to applications on the fact sheet:

- The provider application (can only be one)
- The consumer application(s) (can be multiple)

Interface Modeling False Friends

From interface provider to interface consumer is *not* the direction of the data flow within the interface.

The *interface provider* is the application that is the owner of the business logic, including change management, that triggers the data push or pull request. This can only be initiated by one application. The *consumer*, to which requests are sent, can be multiple applications consuming the data push or pull request.

Why is that important?

Whenever the scope or technical specification of an interface connection is updated, the interface owner must inform all impacted stakeholders down- or upstream. Changes can either occur on the logical interface level (e.g., the method for data push or data pull requests changed) or on a technical level (e.g., the API updated its authentication method). Both responsibilities can be captured in SAP LeanIX via the provider application and the responsible subscribers on a fact sheet for the interface subtypes logical interface and API.

Data objects can be related to applications, where create, read, update, delete (CRUD) rights are specified for the relation. Data objects can also be related to interfaces to indicate that an interface is transporting certain types of data.

Some interface-specific attributes are captured in the fact sheet data under the **Properties** section (see Figure 2.27):

- **Data Flow Direction** indicates whether data is **Incoming** into the interface provider application, **Outgoing** from the provider application, or the flow is **Bi-Directional**.
- **Interface Type** specifies whether the data exchanges happen in a **Synchronous**, **Asynchronous**, or **Batch** manner.
- **Frequency** shows how often the integration runs: hourly, daily, weekly, monthly, yearly, in real time, or on demand.

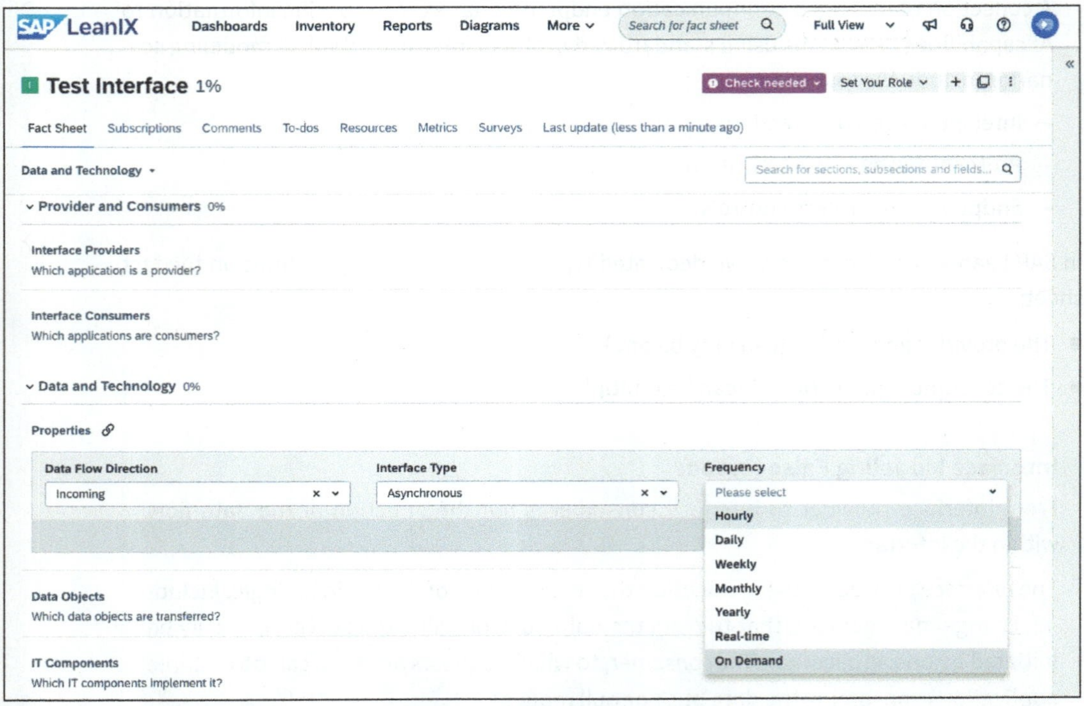

Figure 2.27: Interface Properties on Fact Sheet Details

Start at a high level with logical interfaces. For example, say that you have an interface established between SAP LeanIX and SAP Signavio, and SAP LeanIX is the provider application. It could happen that once you or the interface owner enters the detailed information into the fact sheet sections, you realize that different properties are needed for different data use cases. For example, in your integration, you sync processes from SAP Signavio to SAP LeanIX but application data from SAP LeanIX to SAP Signavio. With simplified modeling, you would mark the interface as **Bi-Directional** and add the data objects process and application without specifying the exact data flow direction per data object. However, it might be important during a transformation to know the exact data flow direction per data object. In this case, you would duplicate the interface and have one interface variant with the data flow direction incoming that has the process data object related to it and one interface variant with the data flow direction outgoing with the application data object related to it. Carefully analyze whether you need that level of detail in the enterprise architecture software for transformation planning and decision-making, and consider cleaning up and aggregating information after executed transformations.

SAP LeanIX can provide high-level insights into solution architecture, but it isn't a specialized solution architecture software. We do not recommend modeling the detailed technical data flow between systems and IT components in SAP LeanIX. The logical interface level between logical applications is the preferred modeling recommendation. If differentiation between properties is needed, create variants of the same logical interface for different use cases. For this, ensure clear naming convention guidance is in place so that any colleague can understand the scope of the logical interface modeled.

Further Resources

For more detailed examples, check out the SAP LeanIX user documentation at *http://s-prs.co/v620614*.

2.2.9 System

The system fact sheet is an optional meta model extension that is used in the SAP discovery feature or in ServiceNow integrations. It represents the deployment environment (production, quality, test, etc.) and is a technical instance of a logical application.

Systems can become handy in two SAP LeanIX capabilities:

- **SAP landscape discovery**
  Systems are an additional discovery capability in SAP landscape discovery. They need to be actively turned on and can then be autodetected and created via the discovery inbox (accessed via the **Inventory Tools** option). This is especially helpful for bigger SAP transformations.
- **ServiceNow**
  System fact sheets correspond to service instances in ServiceNow and can be synced to the SAP LeanIX workspace to bring transparency into risk management. (This option is only available with SAP LeanIX Technology Risk and Compliance.)

The system fact sheet captures information such as environment category (development, production, testing, etc.), deployment models (public cloud, private cloud, on-premise), and instance-level information.

> **Note**
> It's best to only activate the system fact sheet if you plan to have it autofilled by SAP landscape discovery or the ServiceNow integration. We don't recommend maintaining systems manually in the SAP LeanIX workspace.

To activate the system fact sheet, first navigate to the **Administration** area. In the left-hand side menu under the **ADVANCED SETTINGS** section, click **Optional Features and Early Access**. In the middle window, scroll down to the **Use Case Extensions** section and find the **System Fact Sheet** paragraph. Click the **Review and Activate** button, as shown in Figure 2.28.

You can activate SAP landscape discovery (see Chapter 4, Section 4.2.3) with or without system detection. To ensure that the system fact sheet discovery function is turned on, select the checkbox denoting it as in scope during the integration setup (this box is not checked by default; look ahead to Chapter 4, Figure 4.9).

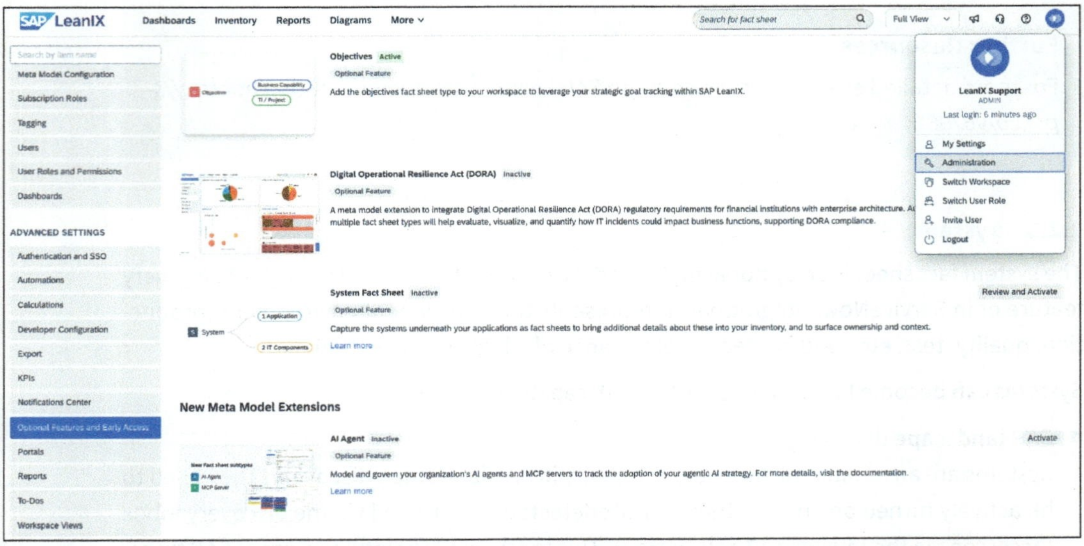

Figure 2.28: System Fact Sheet Activation in Administration Area Under Optional Features

2.3 Enterprise Architecture Features

SAP LeanIX is meant to be a collaborative, open, and transparent solution to collaborate across departments on enterprise architecture. Every employee engages daily with artifacts of the company's application and IT landscape, often directly by using programs or indirectly by consuming program outputs. It's in everyone's interest to have a reliable repository with high-quality data in order to understand the complex ecosystem and interdependency of solutions.

In this section, we'll look at a few SAP LeanIX features that will help enterprise architecture teams engage via the SAP LeanIX workspace with different stakeholders across the organization.

2.3.1 Subscriptions and Data Quality Governance

Having an enterprise architecture repository that is available and easy to understand is a huge benefit, but its data is only useful if the right data is maintained and up to date when needed for decision-making.

Therefore, in general, try to get as many artifacts as possible imported in an automated or semiautomated way via the discovery and integration features into source systems. To avoid duplicate data maintenance in different solutions, try to assign contact persons for each artifact directly from the source system if the necessary information is available. Some manual data maintenance might be inevitable in the SAP LeanIX workspace; keep that in mind when designing the authorization concept.

Fact sheet subscriptions are an essential feature that will help to keep data quality and accountability high. They assign responsibility and accountability to individual users on a fact sheet level for data maintenance. For guidance on how to configure subscription roles as an admin, check out Chapter 3, Section 3.3.3.

SAP LeanIX notification settings allow users to define when they want to be notified via email about changes on those fact sheets to which they have been assigned as responsible persons. This provides them with a clear call to action. Admins can predefine the available options in the **Administration** area. For more information, see Chapter 3, Section 3.3.3.

Ensure that all stakeholders who will be assigned as responsible persons to fact sheets know of their role in maintaining the overall data quality; prepare targeted training material for them. Try to keep the information as concise as possible and remember that the data maintenance in SAP LeanIX should be designed to be as easy as possible in order to encourage active engagement without taking too much time from other urgent business tasks. Therefore, only collect the data that you really need for decision-making. We do not advise collecting data just for the sake of data collection. On a regular basis, assess your data needs for your key use cases; if fields are no longer needed, back up the data and clean up the meta model or adjust the weights for the completion score (see Chapter 3, Section 3.3.3 on how to configure this feature).

The quality seal is also a useful feature to ensure regular data quality reviews. The quality seal can be activated by admins per fact sheet type in the meta model configuration. In day-to-day business, the persons responsible can set the quality seal to **Approved** when they have entered all required and mandatory information. A time period (monthly, per quarter, per half year, per year) can be set by the admins; after this period, the quality seal status will automatically change to **Check needed**, which triggers a notification to the person responsible on the fact sheet subscriptions to review the data, make needed updates, or confirm still valid data.

2.3.2 To-Dos and Comments on Fact Sheets

Another way to collaborate and exchange information on fact sheets in your SAP LeanIX workspace is via the **To-Dos** and **Comments** capabilities on individual fact sheets.

By default, every user can create, read, and update comments. This can be restricted by admins in the respective fact sheet type meta model configurations (see Chapter 3, Section 3.3.3 for more information).

To-dos and comments are always fact sheet–specific and have their own tabs on the fact sheet overview. To-dos can be used to delegate tasks via the **Add Action Item** or **Ask Question** buttons, as shown in Figure 2.29.

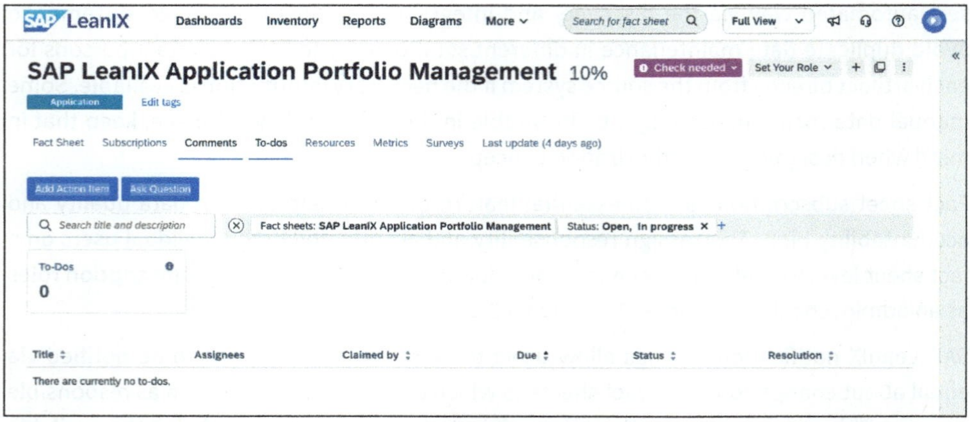

Figure 2.29: Comments and To-Dos Tabs in Fact Sheet Information

To-dos can be assigned to any workspace user. Either this user or a different user can claim the task, reassign the task, and ultimately choose to **Mark as Complete** via a click of a button (see Figure 2.30).

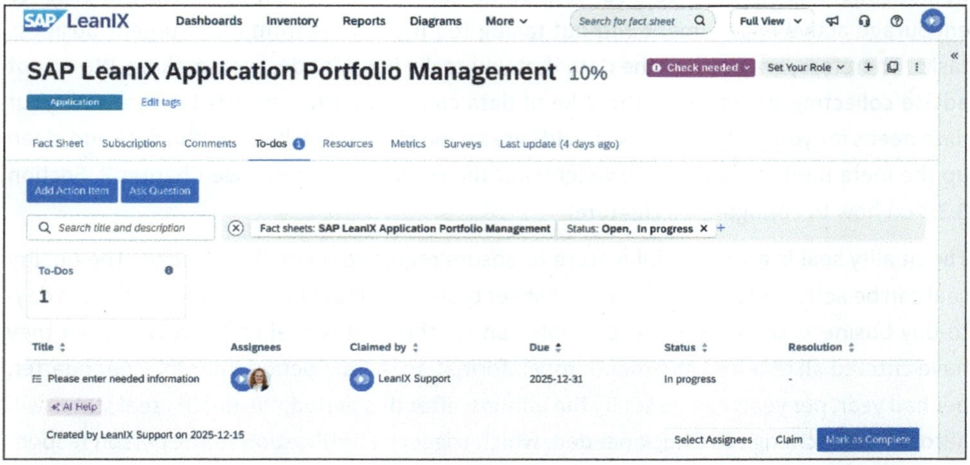

Figure 2.30: To-Do Management on Individual Fact Sheets

Admins have a full overview of all to-dos in the **Administration** area under the **ADVANCED SETTINGS** section and the **To-Dos** menu item (see Figure 2.31). There, admins can also perform to-do cleanup tasks on behalf of users if needed.

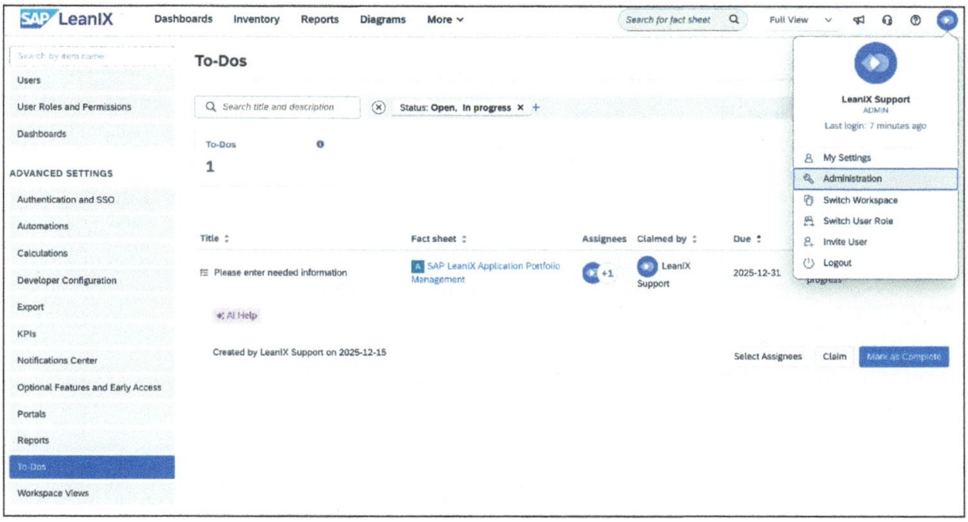

Figure 2.31: To-Do Management in Administration Area

> **Further Resources**
>
> For more information about and examples of working with to-dos, visit the SAP LeanIX user documentation at *https://help.sap.com/docs/leanix/ea/to-dos.*

2.3.3 Surveys

Another good way to verify information in fact sheets or crowd-source information is via the **Surveys** option, which can be found in the SAP LeanIX workspace top ribbon under **More**. Surveys are a good alternative to sending out individual emails or chasing colleagues in an organization via chat functionalities for information.

This feature requires subscription roles to be entered in fact sheets as they are used to determine the survey addressees.

> **Further Resources**
>
> For the latest information on surveys, check the SAP LeanIX user documentation at *https://help.sap.com/docs/leanix/ea/surveys.*

Admins can define in the **Administration** area which roles are allowed to design, send, and analyze surveys. By default, this feature is only available for admins and actively must be made available for other roles. For that, navigate to the **BASIC SETTINGS** section and select the **User Roles and Permissions** menu item (see Figure 2.32).

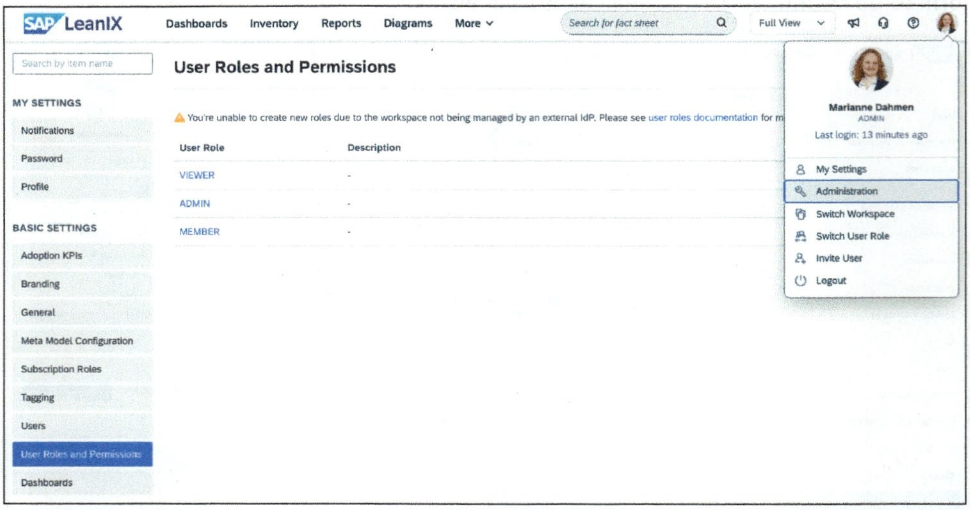

Figure 2.32: User Role Management in Administration Area

Once there, click the **MEMBER** user role link in the middle table, then click the **Permissions** tab in the upper ribbon, as shown in Figure 2.33.

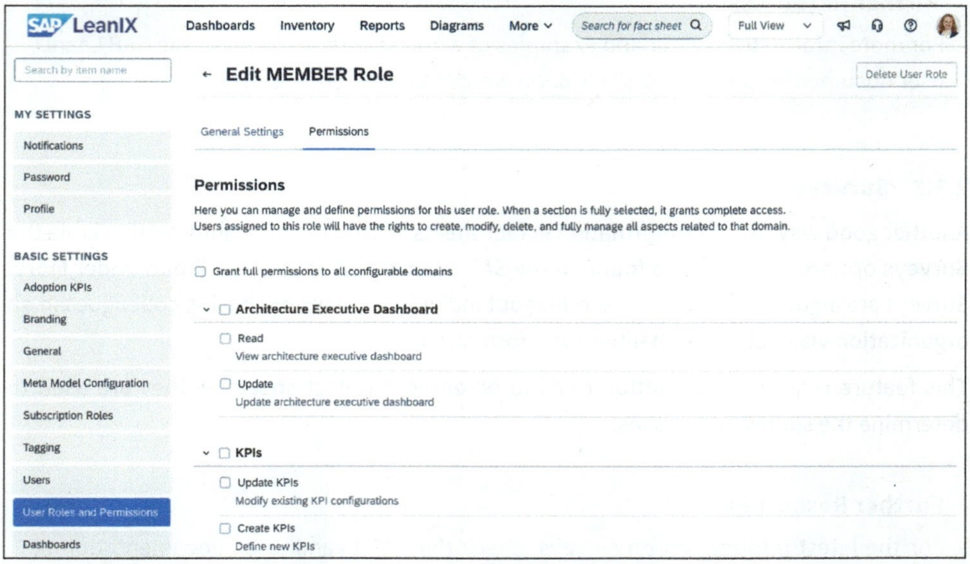

Figure 2.33: MEMBER User Role Permissions Update in Administration Area

Scroll down until you reach the **Surveys** section, then check the checkbox next to **Manage Surveys** if you want to grant member access to the survey management capability. Don't forget to save your changes at the bottom by clicking the **Save Changes** button.

To design a survey, click the **More** button in the top ribbon and select **Surveys** (see Figure 2.34).

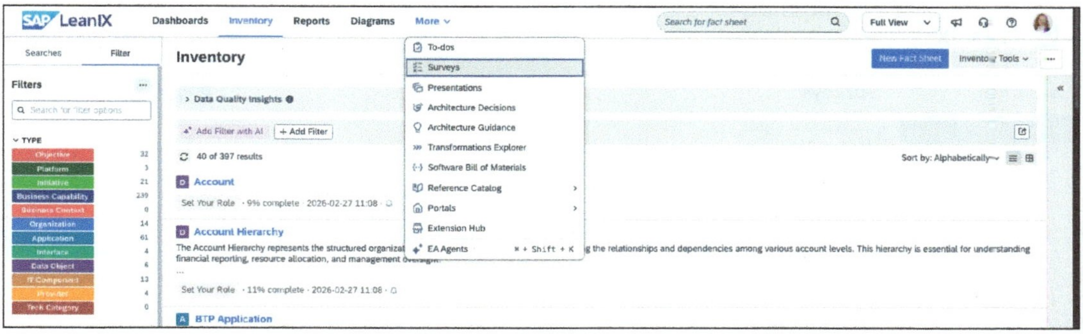

Figure 2.34: Navigating to Survey Management Area

Further Resources

At the time of writing, the SAP LeanIX product and engineering team was still in the process of designing and testing a new way of creating surveys assisted by artificial intelligence.

For the latest information on this feature, see the SAP LeanIX public roadmap and the SAP LeanIX online user documentation:

- Public roadmap (search for "survey"): *https://roadmap.leanix.net/tabs/10-on-roadmap*
- User documentation: *https://help.sap.com/docs/leanix/ea/surveys*

You will be directed to the **Surveys** management overview page. On the left-hand side, you can switch between **Assigned To Me**, for those assigned to you as a user, and **Manage Surveys**, where you can design and manage new surveys. In the center pane, you can filter for different surveys by **Status**; by clicking, for instance, the tile called **Total** to see all surveys; or by clicking **Draft** to only see prepared but not yet sent surveys. In the upper-right corner, you can create a new survey by clicking the blue **+ New Survey** button or click **Safety Net** to configure the related settings (see Figure 2.35).

Survey creation follows a multistep, guided process:

1. **Scope**
   Select the fact sheet type and fact sheets in scope for the survey via an inventory search.
2. **Design**
   Select the survey elements in the survey editor.
3. **Preview**
   Test a preview of your survey to get an idea of the look and feel as a survey participant.
4. **Share**
   Select the subscription roles to which the survey should be sent, check fact sheets with missing subscribers, and draft the invitation email text.

5. **Insights**
   Manage live surveys, send reminders, and close surveys.

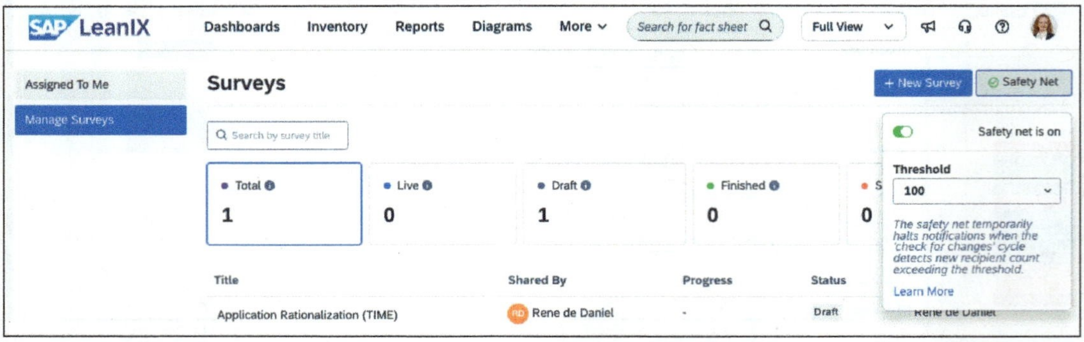

Figure 2.35: Manage Surveys Overview with Open Safety Net Settings

Figure 2.36 shows an example of the survey editor. You can either create a survey from scratch using the left-hand menu panels, which you can drag and drop onto the center area and then configure; or you can start with a template survey, by clicking the **Use Template** button in the upper-right corner.

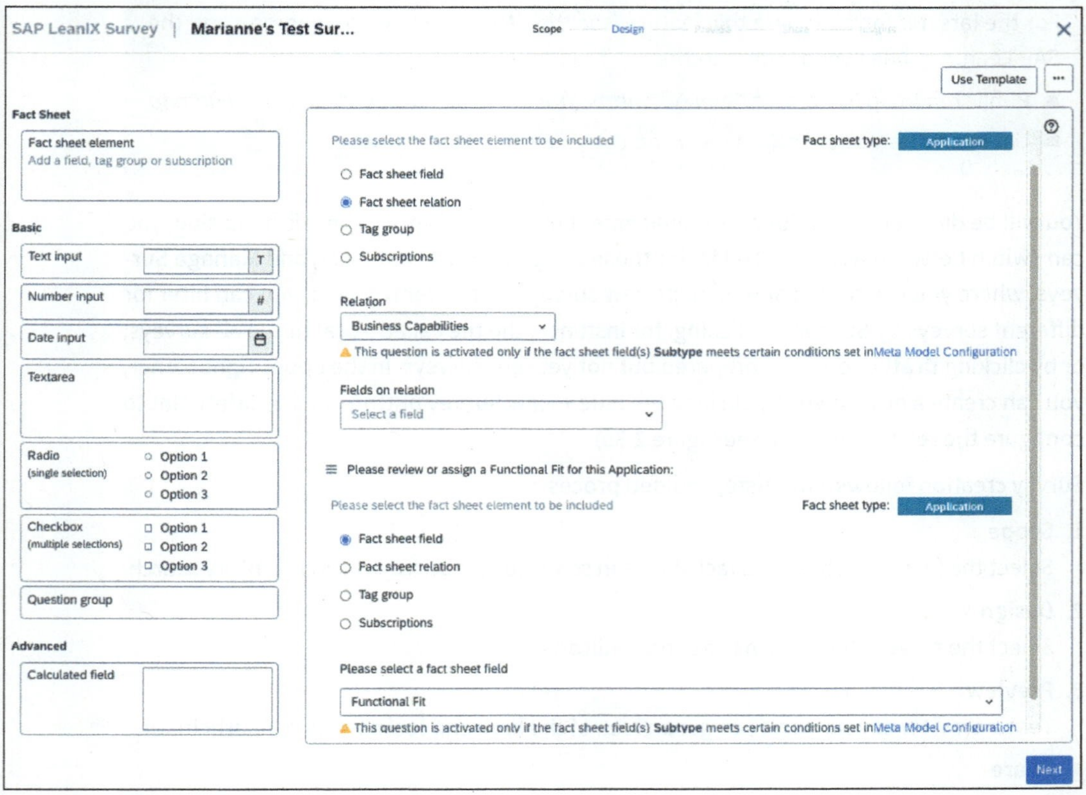

Figure 2.36: Survey Editor Example

Figure 2.37 shows the window for the **Share** step. You can only send surveys to maintained subscribers in the fact sheet. That is a hard prerequisite. You will get a warning about missing subscribers in the **Share Configuration** window in the top filter tiles. In this example, there are 10 in-scope filtered fact sheets with missing subscribers. Note that you don't have to finalize the survey in one sitting; when you click the **x** in the upper-right corner of the screen, your progress will be saved, and in the survey overview you will see this survey with the **Draft** status. Once done with the configuration, you can test the survey by using the **Test Survey** button in the lower-right corner. After a successful test and potential tweaks to the design, you can then click the blue **Go Live** button in the lower-right corner to make the survey live.

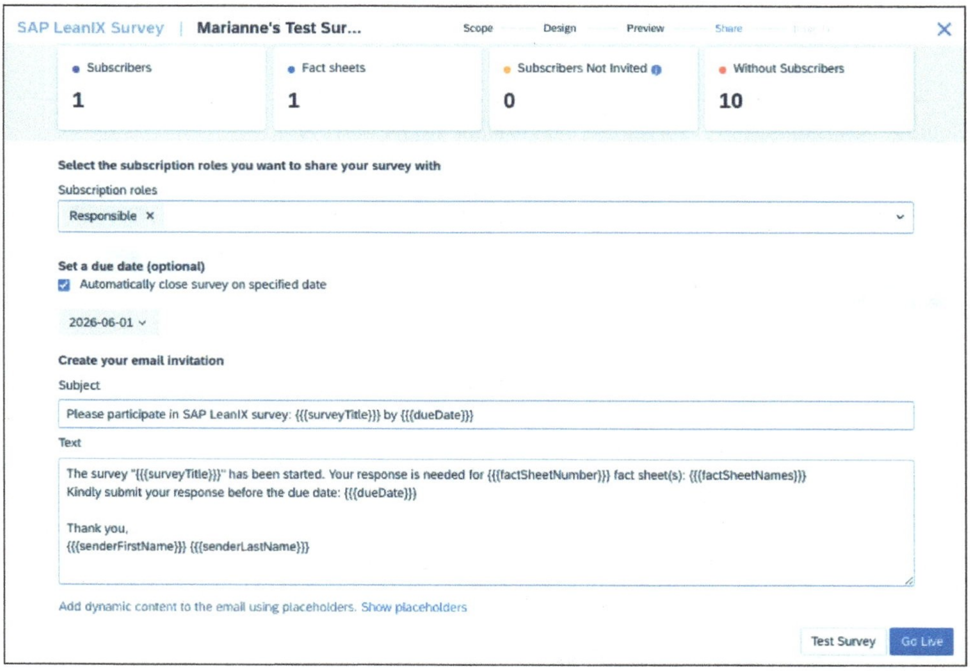

Figure 2.37: Configuring Survey Recipients and Invitation Email Text

Figure 2.38 shows features that allow you to manage running surveys. You can see the **Survey Progress**, select an option to **Send Reminder** emails, and download survey responses to Microsoft Excel (**Download .XLSX**) to manage the information in more detail.

Figure 2.39 shows an updated overview of the **Manage Surveys** tile, depicting two surveys that each have a different **Status**.

We'll revisit the topic of surveys in Chapter 7, Section 7.4, when we create an example as part of the use case.

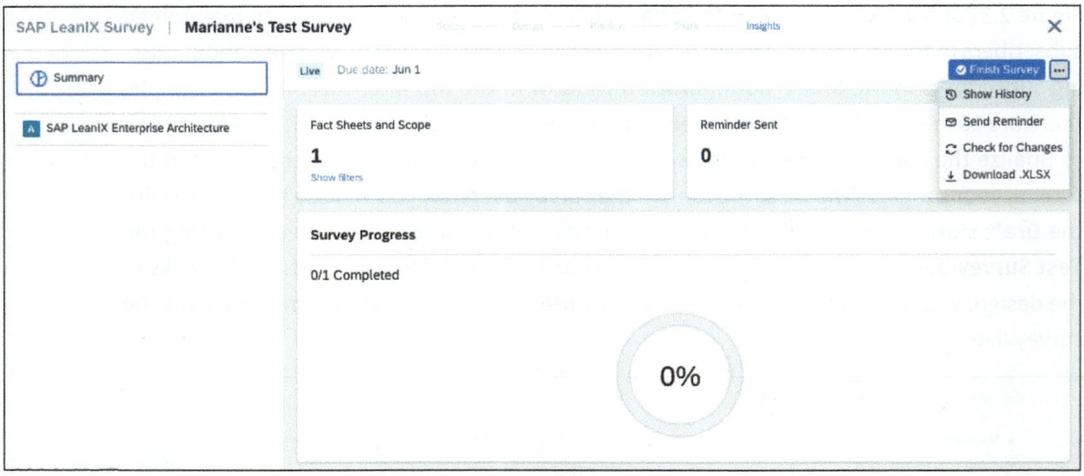

Figure 2.38: Insights Step of Survey Creation to Manage Live Surveys

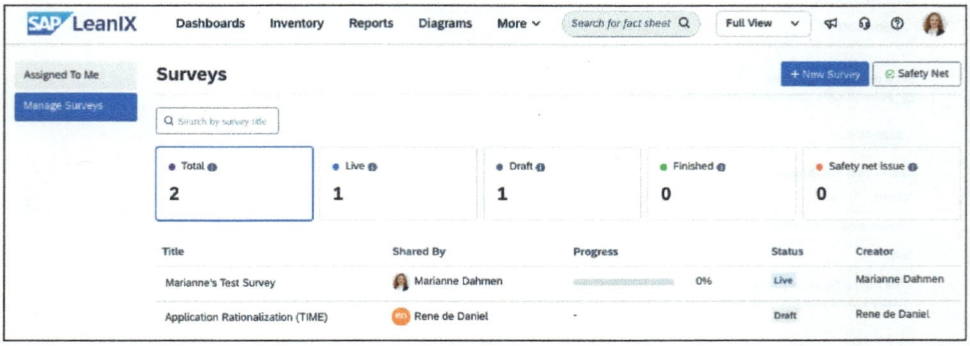

Figure 2.39: Surveys Overview with Surveys in Live and Draft Statuses

2.3.4 Workspace Views and Portals

An open and transparent SAP LeanIX workspace does risk overwhelming people with information that isn't relevant for them in their day-to-day work. SAP LeanIX therefore created features to reduce the mental load and noise for targeted stakeholder groups when working with SAP LeanIX on a casual basis.

Workspace views filter down the inventory to only selected fact sheet types that are relevant for a specific user group. They act like custom filters across dashboards, reports, diagrams, and fact sheets. This is a good method when you generally want to keep the users in the same SAP LeanIX workspace UI. Workspace views are not a method to restrict access. Users can always leave the view to navigate back to the full view. If you want to restrict access, custom roles and virtual workspaces are the features you should look into.

Workspace views can be set up and designed in the **Administration** area under the **ADVANCED SETTINGS** section, via menu item **Workspace Views** (see Figure 2.40).

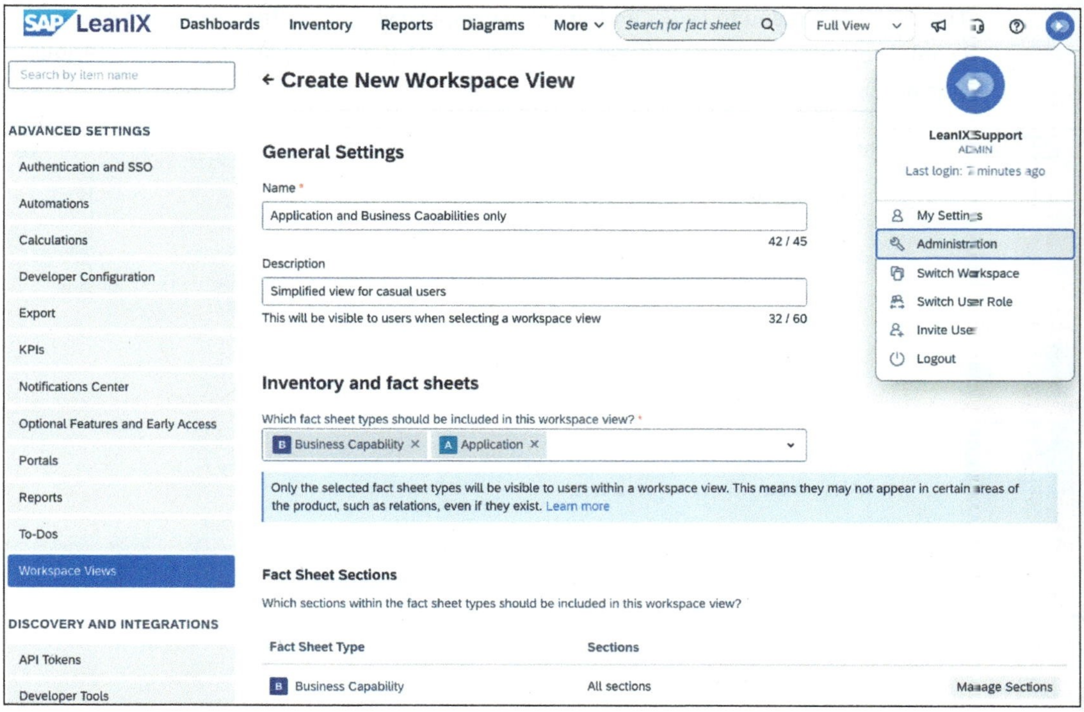

Figure 2.40: Workspace View Configuration in Administration Area

In this window, admins can configure the following items:

- **General Settings**
  - Name
  - Description
- **Inventory and fact sheets**
  - Fact sheet types in scope
  - Fact sheet sections per fact sheet type in scope
  - All fact sheets of a certain type or only a subset in scope
- **Reports, Diagrams, and Dashboards**
  - Reports in scope
  - Diagrams in scope
  - Dashboards in scope
- **Visibility**
  - Visible for everyone or for specific user roles
  - Set as default for specific user roles

You must give your workspace view a name, select at least one fact sheet type, and select at least one report and one dashboard to be able to save your new workspace view.

The dropdown menu to select and switch workspace views is in the top ribbon of the SAP LeanIX workspace. See the differences in Figure 2.41 (**Full View**) and Figure 2.42 (**Application only**).

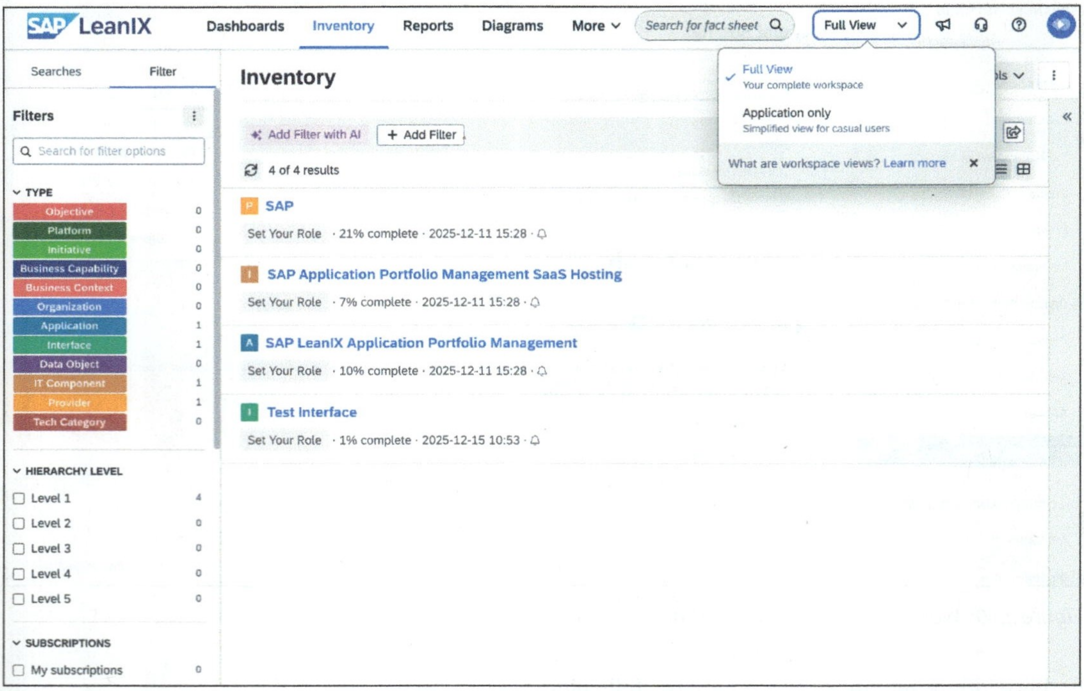

Figure 2.41: Full View Workspace View

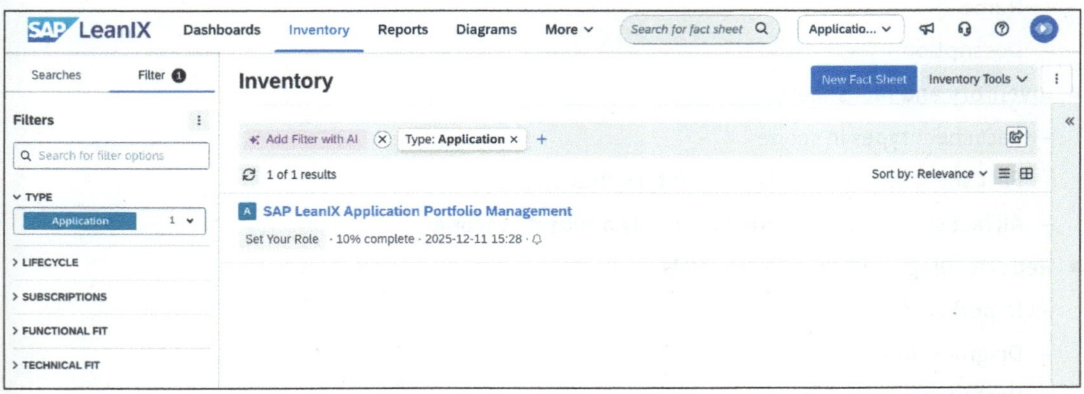

Figure 2.42: Configured Simplified Workspace View

Further Resources

For more information on workspace views, visit the SAP LeanIX user documentation at *http://s-prs.co/v620615*.

Portals, on the other hand, not only reduce the noise of information but also simplify the UI for authenticated users to focus only on a few relevant, preselected data points. Portals are customizable, sharable web views of selected fact sheet types and their data, reducing UI complexity to a minimum for casual SAP LeanIX users that don't need the full feature scope.

Portals can be designed by admins in the **Administration** area under the **ADVANCED SETTINGS** section, menu item **Portals** (see Figure 2.43).

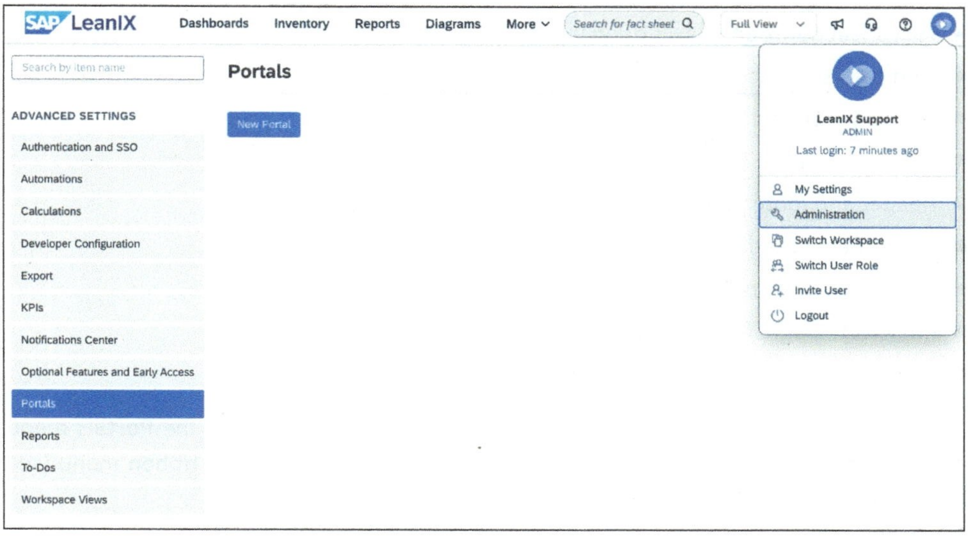

Figure 2.43: Portal Configuration in Administration Area

Portals are very versatile and flexible. Figure 2.44 shows an example of an application portal.

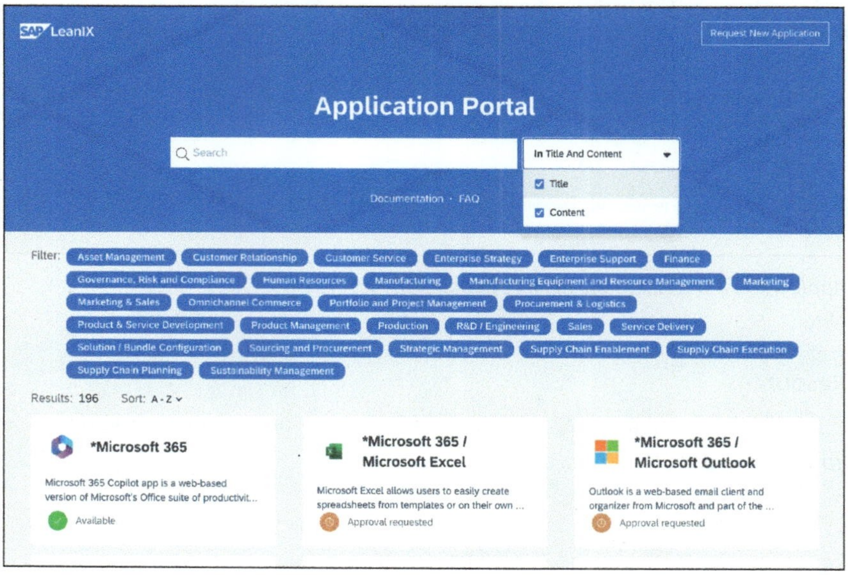

Figure 2.44: Example Application Portal

The following aspects can be configured in the **Administration** area, under the **Portals** configuration:

- General information
- User role accessibility
- Fact sheet type in scope (can only be one)
- Fact sheet filter to only include a subset of the inventory
- Fact sheet name and description input fields
- Icon colors
- Portal design and branding
- Action buttons
- Header and footer links (e.g., to documentation)
- Status field based on existing fact sheet field
- Search and filter options in the portal
- Preselected data for the detailed information tile view
- Wording configuration in the **Localization** tab

After portal creation, you might have to refresh your browser to make the **Portals** menu item appear under the **More** menu in the SAP LeanIX workspace's top ribbon menu (see Figure 2.45).

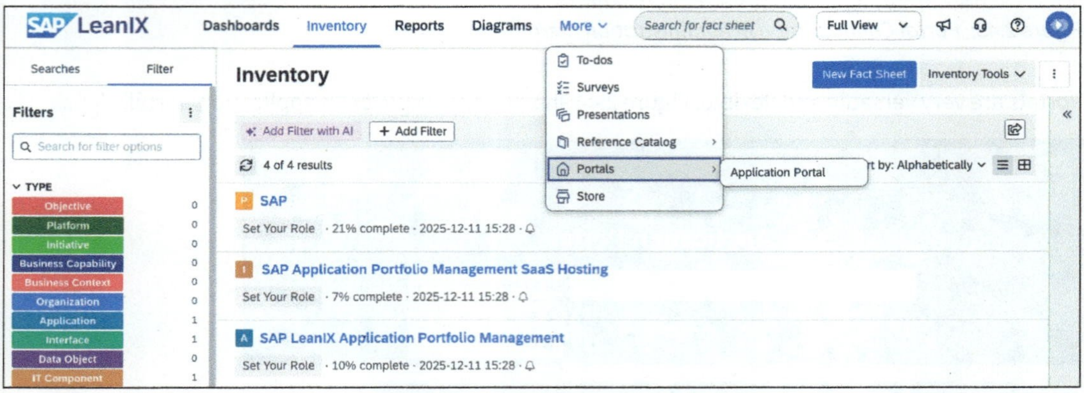

Figure 2.45: Application Portal Example in More Dropdown

Further Resources

For more information on portals, visit the SAP LeanIX user documentation at *https://help.sap.com/docs/leanix/ea/portals*.

2.4 Summary

In this chapter, you learned about the history of SAP LeanIX and the evolution of its products. You also saw an overview of the various features and capabilities available in the current SAP LeanIX solution range and where in the book you will find related deep dives. The SAP LeanIX out-of-the-box meta model is no longer a black box for you: You have dipped your toes into the basic concepts and modeling recommendations of the core fact sheet types needed for the initial application portfolio assessment use case. Remember that this practical guide is just a starting point to excite your interest in taking an insightful enterprise architecture journey in your organization using SAP LeanIX. Successful enterprise architects see this discipline as a team sport across the various departments and experts. That's why we touched on some of the collaboration features in SAP LeanIX to set you and your colleagues up for success.

Chapter 3 will be particularly relevant for administrators of the SAP LeanIX workspace as we will introduce the initial steps to provision and configure an SAP LeanIX workspace. Jump to Chapter 4 if you want to get an idea of how to load and import data into your workspace, but don't forget that you might have to adapt some naming conventions or wording (called *display names* in the translation model) in the out-of-the-box data model or configure some additional attributes for your organization's taste and use cases of interest (covered in Chapter 3).

Chapter 3
Setting Up Your SAP LeanIX Workspace

Congratulations, you now have access to your own SAP LeanIX workspace! In this chapter, we'll offer useful tips and tricks for setting up and configuring your SAP LeanIX instance. This is just the start of your exciting enterprise architecture journey.

This chapter introduces the basics of onboarding and then covers the administrative tasks that need to be performed after purchasing an SAP LeanIX workspace. We'll discuss workspace initialization, admin and core team invites, and user management, as well as different sections of the administration menu and what configurations you should handle early, and which can be dealt with at a later time with more advanced use cases.

3.1 Onboarding

SAP LeanIX offers a strong ecosystem of support functions, as shown in Figure 3.1. Let's walk through them:

- **Account executive**
  The account executive is the trusted advisor on anything contract- and commercial-related.
- **Solution advisor**
  You might meet the solution advisory team during solution demos or capability presentations, typically before signing the contract.
- **SAP LeanIX Community**
  The SAP LeanIX Community is a group of interested SAP LeanIX users and SAP LeanIX subject matter experts that can exchange knowledge and ideas on a curated, forum-style exchange platform.
- **SAP LeanIX Learning**
  The SAP LeanIX Learning platform (formerly SAP LeanIX Academy) includes community events as well.
- **Customer Success**
  The Customer Success organization offers vital postsale support, including the Customer Success onboarding manager for guided onboarding and the plentiful onboarding material provided to admin core teams.

- **Professional Services**
  For organizations with very specific and complex requirements for onboarding or other use cases, SAP LeanIX's Professional Services capability can help with hands-on work and enablement in the customer workspace. After onboarding, organizations can benefit from either dedicated or virtual Customer Success manager pooling.
- **Customer Success Management**
  Customer Success Management plays a crucial role in developing and guiding organizations during their SAP LeanIX journeys, helping them learn best practices, explore new functionalities, or explore ideas for new use case implementations or adoption initiatives.

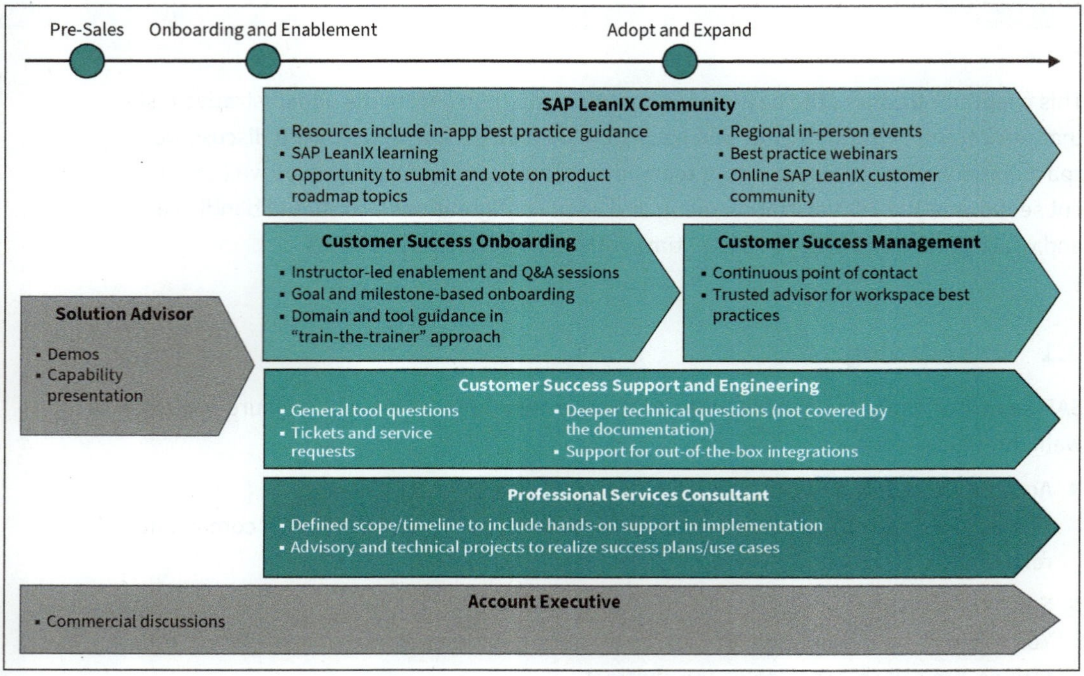

Figure 3.1: SAP LeanIX Customer Success Support Ecosystem

Further Resources

Did you know that SAP LeanIX offers various resources to help with your onboarding journey? To learn more, check out the following links:

- *http://s-prs.co/v620616*
- *http://s-prs.co/v620617*
- *http://s-prs.co/v620618*

3.2 Administrative Prerequisites

SAP LeanIX workspace provisioning can be triggered via SAP for Me if you have a valid SAP LeanIX license.

A successful SAP LeanIX workspace setup starts with the correct information in your SAP LeanIX contract. Make sure that the correct contact information is entered for a smooth and quick setup experience by following these steps:

1. Ensure that your organization has at least one designated SAP S-user administrator.
2. Verify that the technical administrator (IT contact) from the SAP LeanIX contract is the technical admin of the future SAP LeanIX workspace, rather than the CIO, CEO, or procurement contact. All provisioning-related information will be sent via email to this contact person.
3. Request S-users for all SAP LeanIX core admin team members.
4. Test your SAP for Me access and maintain all SAP LeanIX–relevant admin team contacts and roles in SAP for Me.
5. Create and assign a support user (S-user ID).
6. Learn how to create support tickets in the SAP for Me portal.

Further Resources

For detailed steps, check the SAP LeanIX user documentation on the SAP Help Portal at *http://s-prs.co/v620619*.

Contact SAP's Customer Interaction Center if you have questions regarding SAP for Me navigation, S-user IDs, and Universal IDs: *https://support.sap.com/en/contact-us.html*.

Depending on your chosen SAP LeanIX onboarding journey, you will receive information from one of the following:

- A dedicated Customer Success onboarding manager
- A Customer Success onboarding team
- A dedicated account executive and the Customer Success team for a self-learning guide

As soon as the contract for your SAP LeanIX workspace starts, an email with SAP LeanIX system access details, including the workspace URL, will be sent to the technical administrator named in the contract details. After clicking the URL link, the user must use the **Forgot Password?** functionality to assign a new password to the entered user name (email address) in order to log into the workspace.

3.3 Workspace Setup

Although SAP LeanIX offers a lot of preconfigured content, it will still be necessary to configure a few things with a dedicated admin team before rollout. For instance, the look and feel of the workspace can be adjusted to reflect the company's color schema and logo, or the naming of fact sheet types and fields can be adjusted in the display name translations to match the corporate jargon. In the following sections, we will briefly cover a few tasks that should be considered in the early admin configuration phase.

3.3.1 Admin Core Team Invitation

Once the technical administrator has received access to the SAP LeanIX workspace, it's best to invite other admin core team members, such as functional admins for the SAP LeanIX workspace or other core enterprise architecture team members who will support the setup and configuration of the SAP LeanIX workspace.

To access the **Invite User** form, as shown in Figure 3.2, click the round user button in the upper-right corner to open the user menu, and then click the **Invite User** menu item.

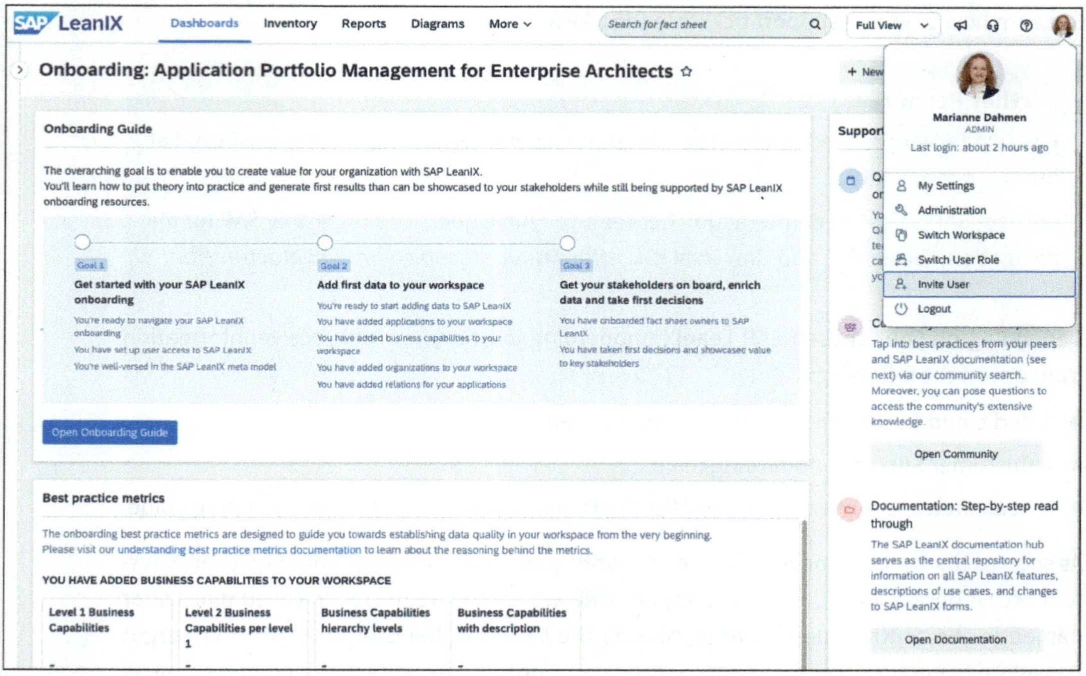

Figure 3.2: Invite New Users to Workspace

In the **Invite to Workspace** popup, as shown in Figure 3.3, the admin proceeds as follows:

1. Enter the new user's email address and select the **E-Mail Address** entry (multiple entries possible).

2. Define the role the new user(s) should have, choosing from the following options:
   - **Admin**: Administrators with full CRUD access on data and the administration menu
   - **Member**: Restricted CRUD access to data and features and no access to the administration menu
   - **Viewer**: Mostly read access to data, CRU access to collaboration features, and limited access to additional features (more access can be granted by admins as needed)
3. Customize the text in the email body.
4. Click the **Invite** button.

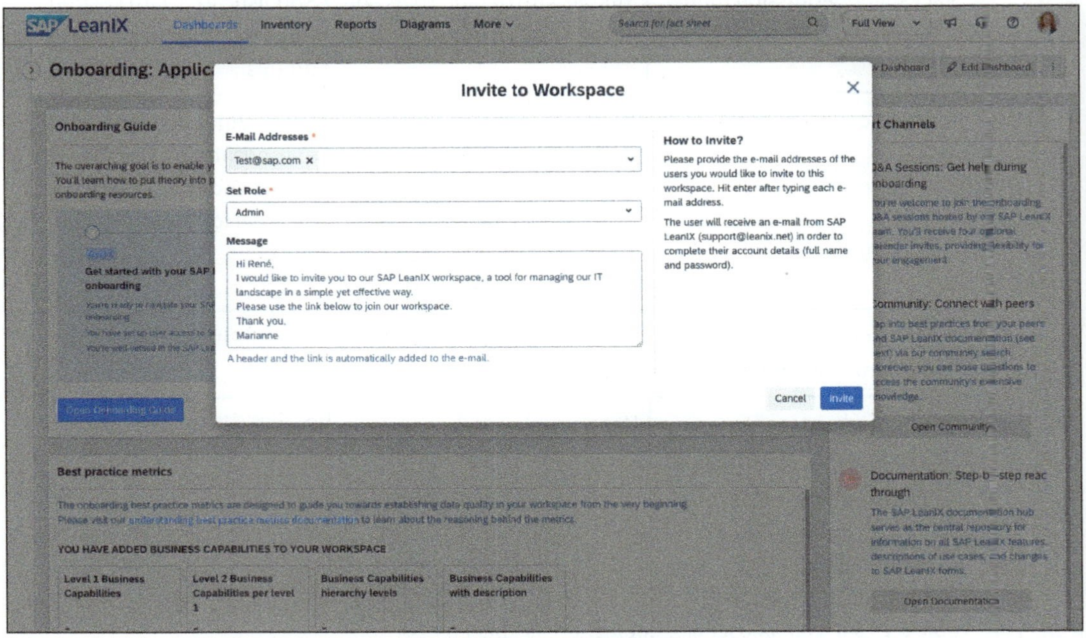

Figure 3.3: Personalize User Invitation and Fill Out User Invite Form

Only admins can see the **Administration** menu item and can define all special permissions, such as whether members can design and manage surveys, or role-based permissions to restrict the maintenance of certain information on fact sheets.

3.3.2 User Access Management and SSO

SAP LeanIX offers different options for user access and user management: either internally managed by the admin or externally managed by an identity provider (IdP). Table 3.1 presents a comparison of the two major concepts.

Comparison Point	Internal Admin Managed	External Identity Provider Managed
User management	Manually by admins via **Administration** menu in SAP LeanIX workspace	Via IdP-managed groups
Possible user roles	Admin, member, viewer	■ Admin, member, viewer ■ Custom roles
Fact sheet permission model	Simple with default roles only	Sophisticated with custom roles and IdP-assigned user groups
Authentication method	■ User name and password ■ SSO: Authentication only (with predefined default role that can be manually adjusted by admins in workspace per user)	SSO: Authentication and authorization (roles are maintained and transmitted via external IdP system)
System for cross-domain identity management (SCIM)	Only possible when using SSO	Possible if needed
Virtual workspaces (managing role-based access on the individual fact sheet level, e.g., for sensitive or confidential data)	Not possible	Possible if needed

Table 3.1: SAP LeanIX User Management Concepts Comparison

Consider the following questions when deciding if you want to manage roles internally (SAP LeanIX admin UI) or externally (via IdP):

- How big is the enterprise architecture team?
- Will SAP LeanIX be primarily an enterprise architecture team-managed solution, in which other stakeholders have viewing rights for transparency only, or does the enterprise architecture team plan to actively collaborate with other internal stakeholders from other departments using SAP LeanIX's collaboration functionality (subscribers, surveys, to-dos, etc.)?
- How big and diverse is the audience to be included in the workspace, and will they need separate permissions and CRUD access?
- Do you plan to capture and manage confidential information in fact sheets, such as costs or strategic initiatives that need to be hidden from certain viewers?

To configure the authentication and SSO setup, navigate to the **Administration** menu and choose **ADVANCED SETTINGS • Authentication and SSO**, as shown in Figure 3.4.

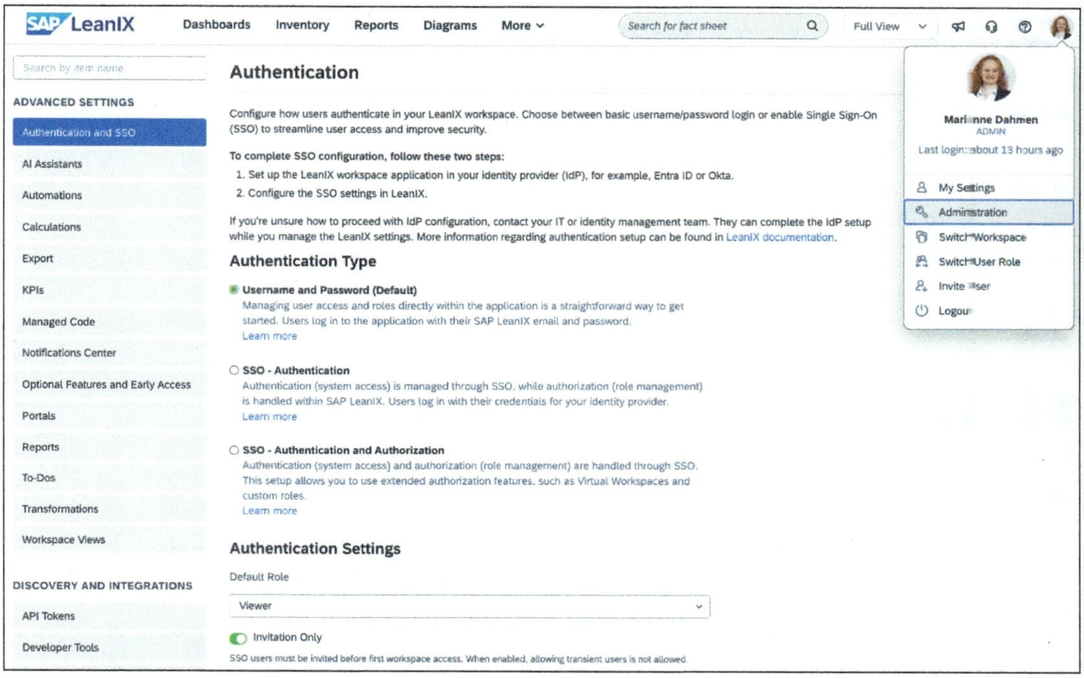

Figure 3.4: Configuring Authentication and SSO in Admin Menu

It's easiest to start with a small group of technical and functional workspace admins. We recommend enabling either the **Username and Password (Default)** or the **SSO – Authentication** setting and the **Invitation Only** setting. If your organization uses an external IdP, then setting up SSO will increase usability and user-friendliness significantly as users won't have to remember and type in their passwords every time they want to log in.

The **Default Role** field defines which user role a user is assigned when they log into the SAP LeanIX workspace for the first time. This setting is only relevant when you did not select **SSO – Authentication and Authorization** as then you must assign the roles via group assignment to your employees in your IdP system.

When deciding on your future strategy for user and access management, we recommend keeping it as simple as possible, based on your desired use cases and organizational setup.

You can switch to more advanced concepts at any time. If a more sophisticated user management concept is needed, you can create custom roles in the **User Roles and Permissions** admin menu (see Figure 3.5), but only if the **SSO – Authentication and Authorization** option with externally managed roles via IdP is activated. It's best to start with a simpler model during the workspace configuration phase and introduce a more complex one once the meta model is customized for the use case in scope before rolling out the workspace to your first users. Do proper testing with selected early adopters to ensure everything is working as expected.

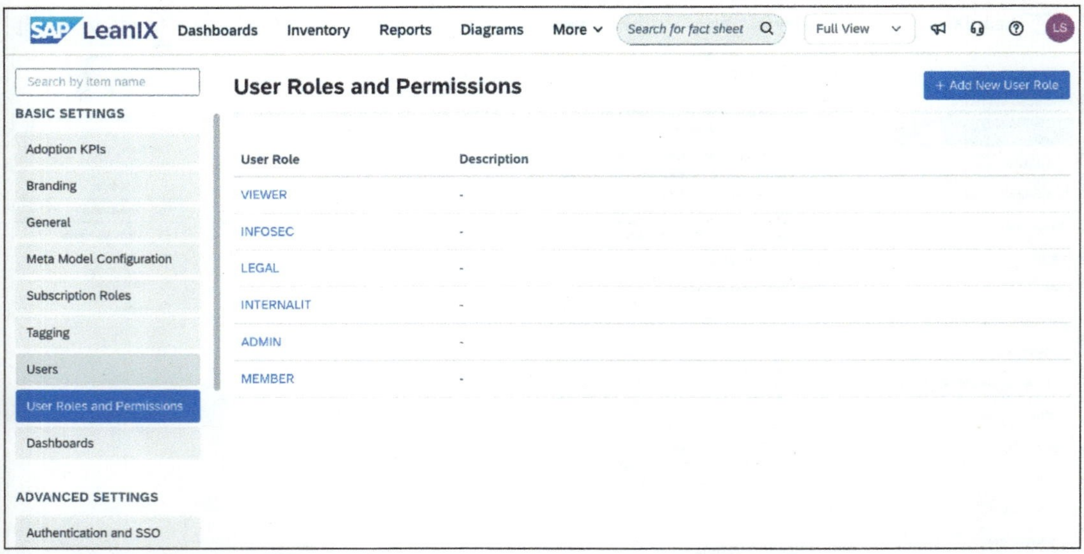

Figure 3.5: Activated New User Roles Feature in User Roles and Permissions

Further Resources

For more information on user and access management, check the SAP LeanIX user documentation on the SAP Help Portal at *http://s-prs.co/v620620*.

For more information on SSO, visit *http://s-prs.co/v620621*.

Admins can monitor users in the **Administration** area in the **Users** menu. The **All** tab (see Figure 3.6) is important if the **Username and Password (Default)** or **SSO – Authentication** options were selected. If a user should be assigned to a different user role than the default role, then it must be manually changed in the **Users** tab.

In the **Users** tab, admins can manage different views in the upper horizontal ribbon, showing **All**, **Active**, **Invited**, **Requested**, **Not Invited**, and **Archived** users.

Admins have a **Search Bar** under the ribbon menu where they can search by name or email address.

The little up and down arrows next to the table columns in the **Users** window indicate that you can sort the column in ascending or descending order based on the values of the column.

With the **Export** button in the upper-right corner of the **Users** window, admins can download an Excel workbook of all users with the respective table information.

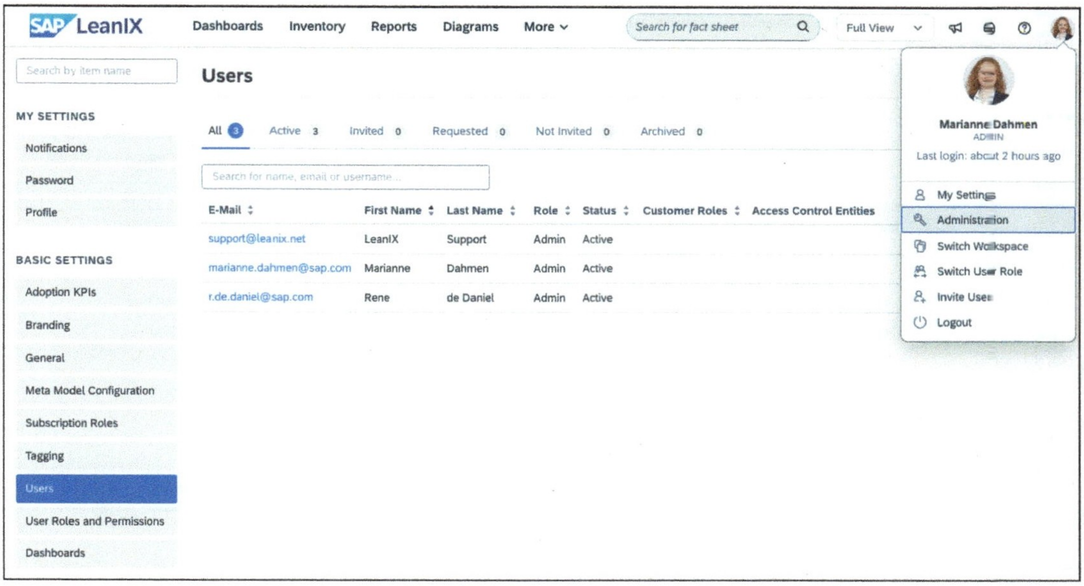

Figure 3.6: Users Tab in Administration Menu

3.3.3 Administration Menu

The **Administration** menu is structured into four sections, which we will explore in the sections ahead:

1. **MY SETTINGS** (available for all users: admin, member, viewer)
2. **BASIC SETTINGS** (admin only)
3. **ADVANCED SETTINGS** (admin only)
4. **DISCOVERY AND INTEGRATIONS** (admin only)

Further Resources

An SaaS solution can quickly change its look and feel with new updates and new features. Therefore, it's always good to check the SAP LeanIX user documentation on the SAP Help Portal to find the latest and newest information, for example, the latest about the **Administration** menu: *http://s-prs.co/v620622*.

My Settings

Under **MY SETTINGS**, users can manage their notification preferences in **Notifications**, passwords in **Password** (only relevant when SSO is not used), and personal details like preferred language and avatar image in **Profile** (see Figure 3.7). This section also can be seen by viewers and members when they select **My Settings** in the right-side menu under the round personal icon (compare to Figure 3.8).

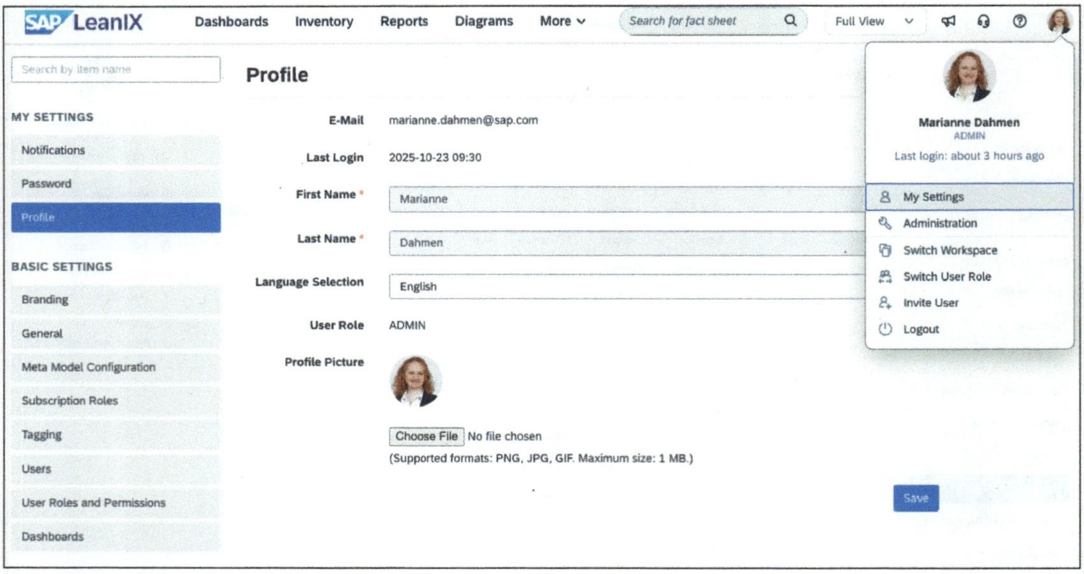

Figure 3.7: My Settings Shown from Admin Account

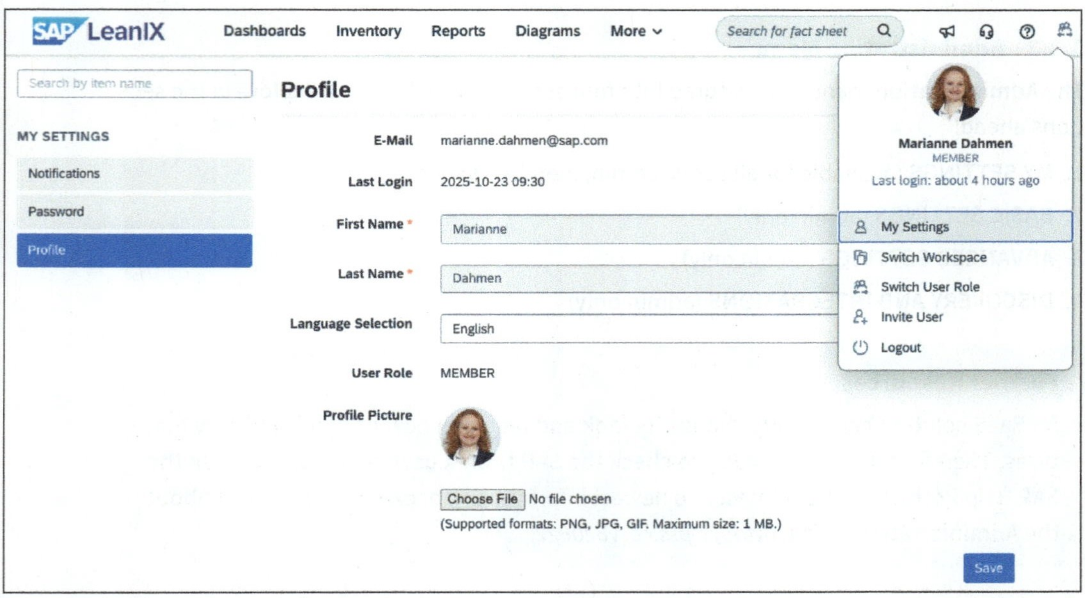

Figure 3.8: My Settings Shown from Member Account

The defaults and selectable options of the **Notifications** can be selected and limited by the admin in the **Notifications Center** under **ADVANCED SETTINGS**. We'll cover this later.

Basic Settings

The **BASIC SETTINGS** are important to go through in the beginning of the configuration phase as they cover topics like changing the logos and color schemes in the **Branding** tab, adjusting the out-of-the-box meta model in **Meta Model Configuration**, or defining default dashboards in the **Dashboards** tab.

Adoption KPIs

The default landing page of the **Administration** menu is the **Adoption KPIs** page (see Figure 3.9). Here, admins can see how users interact with the workspace and monitor, for instance, login frequency, page views by feature, fact sheet counts, report views, and much more. This is not only important during the rollout phase but also an ongoing monitoring tool to identify communication or training needs, if, for instance, a feature is not used at all.

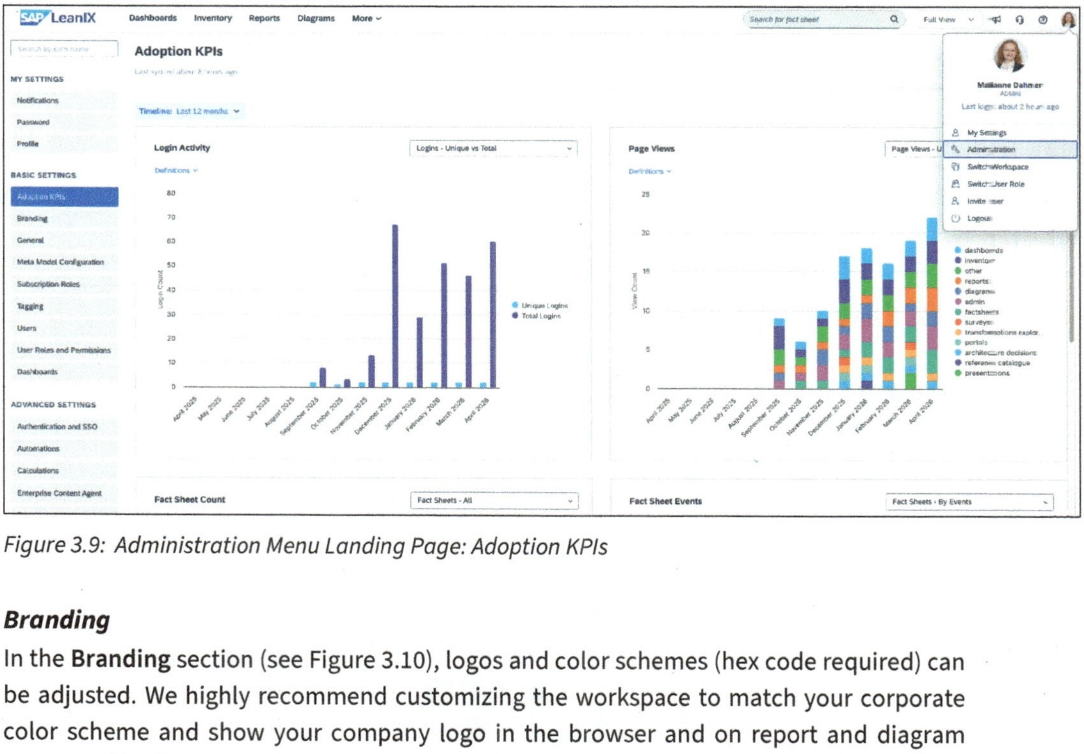

Figure 3.9: Administration Menu Landing Page: Adoption KPIs

Branding

In the **Branding** section (see Figure 3.10), logos and color schemes (hex code required) can be adjusted. We highly recommend customizing the workspace to match your corporate color scheme and show your company logo in the browser and on report and diagram exports. This allows you to comply with global brand marketing guidelines and fulfill corporate accessibility requirements related to backgrounds and font colors.

> **Caution**
>
> Be sure to pay attention to the maximum size and format requirements per logo upload.

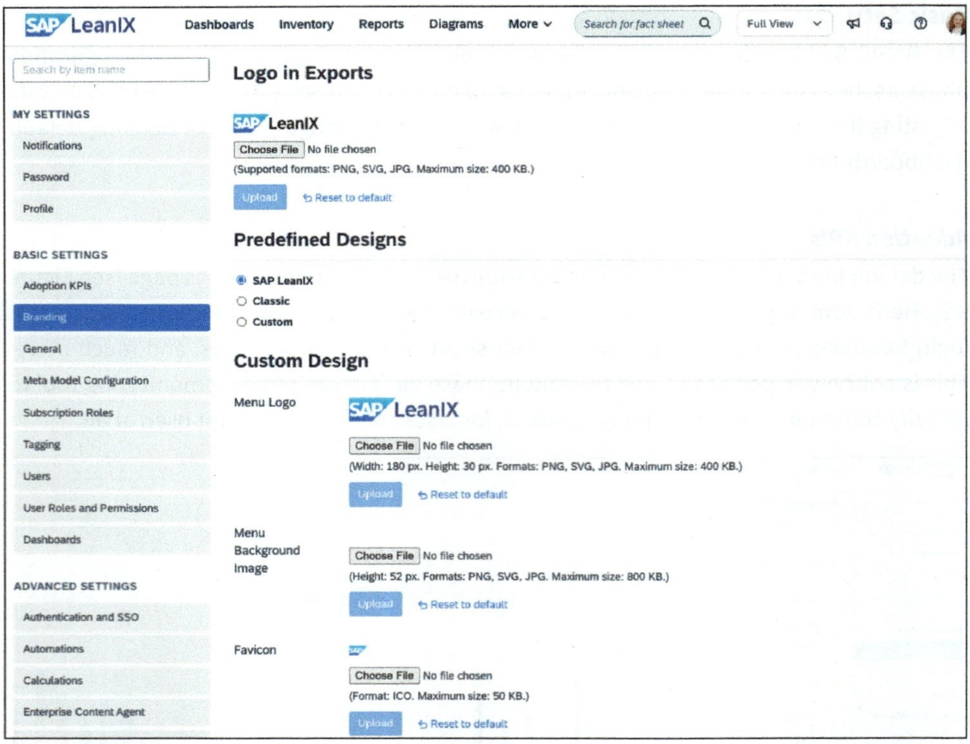

Figure 3.10: Branding Settings

The **Logo in Exports** area sets the logo that you will see on any report or diagram exports. In Figure 3.11, such a logo is shown in the upper-right corner.

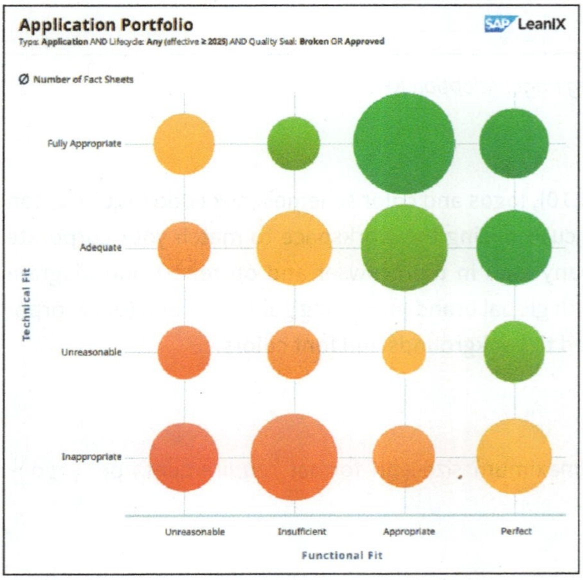

Figure 3.11: Example of Report PDF Export with Export Logo

The **Predefined Designs** options offer two SAP LeanIX options (blue logo with white background, and inverse with white logo and blue background), plus **Custom**. When selecting **Custom**, a new section at the bottom of the screen appears, where admins can define many color details for **Menu Bar**, **Links**, **Buttons**, and **Tour**, which is the help tour.

In the **General** menu (see Figure 3.12), many small but relevant settings can be defined for the entire workspace.

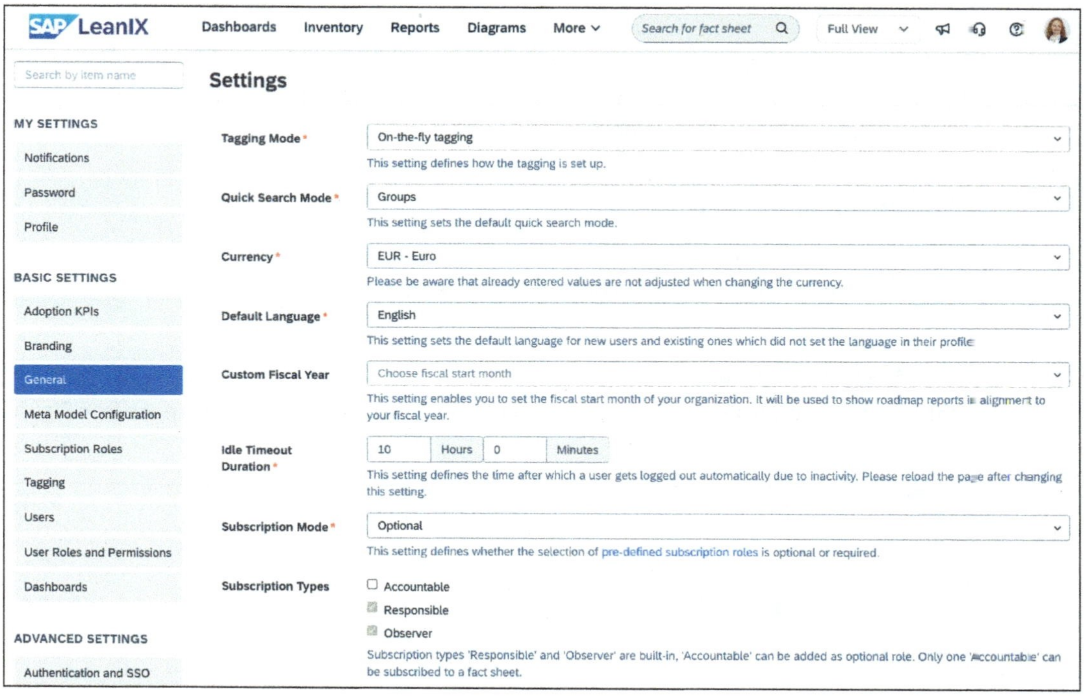

Figure 3.12: General Menu in Administration Area

The possible values are explained in the dropdown options. It's important to define the following:

- **Currency**
  This is the overall workspace currency (should match the consolidated corporate reporting currency).
- **Default Language**
  This is the default language (should align with your corporate language guidelines).
- **Idle Timeout Duration**
  This is the idle timeout duration (should match your security guidelines or SSO setup, if SSO is activated).
- **Getting Started Guide**
  This optional setting can point to an internal URL with a self-made user guide specific to selected use cases, if desired.

- **Support E-Mail** or **Support URL**
  Especially larger organizations often direct support requests first to an internal IT resource or to a Center of Excellence that has special permission to forward legitimate support requests to the SAP LeanIX Support Center.
- **Data Breach Contact**
  If the fields here are empty, the workspace will show an error message when opening the tab.

Do not select the **Danger Zone** checkbox. This is usually used by the SAP LeanIX support team and Customer Success engineers in alignment with the workspace admins when, in very rare cases, a data restoration needs to be performed.

> **Don't Forget to Save!**
>
> The **Save** button for the **General** tab is all the way at the end in the lower-right corner. Don't forget to click it once you have made all necessary changes.

Meta Model Configuration

Meta Model Configuration is a very important area in the SAP LeanIX workspace (see Figure 3.13).

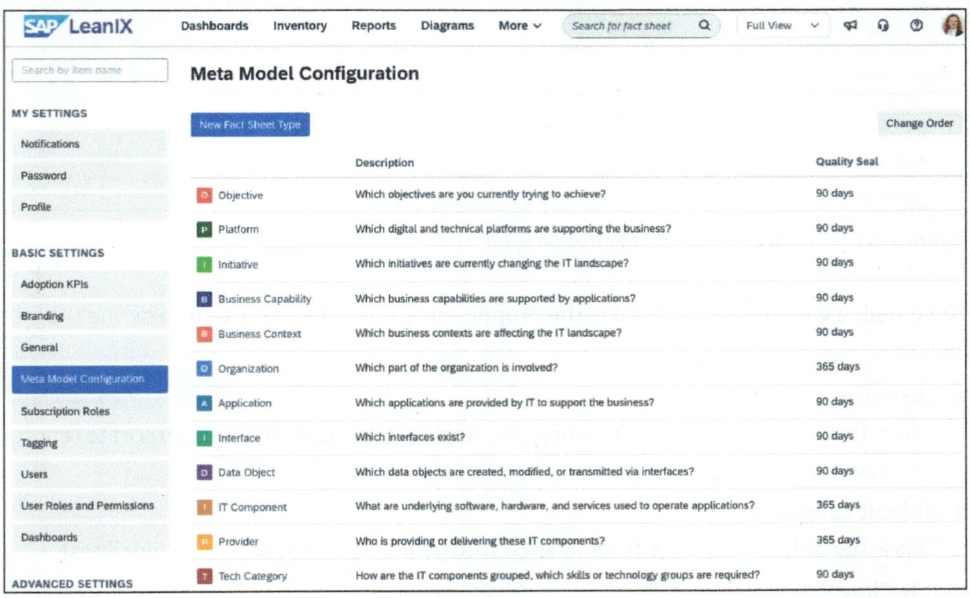

Figure 3.13: Meta Model Configuration in Administration Area

In general, the recommendation is to use the out-of-the-box meta model as closely as possible and to create new fact sheet types only if there is no other way to represent the data in the given structure. Under **Meta Model Configuration** in the **Administration** area, you can see the order of the meta model fact sheet types. There are two buttons at the top that can only be found here:

- **New Fact Sheet Type** (create a custom fact sheet type)
- **Change Order** (rearrange the order of the fact sheet types)

When hovering over the data table in the **Meta Model Configuration** view, you can see that it's possible to click the names of the out-of-the-box fact sheet types. This is how you can enter specific fact sheet configuration modes.

As you can see in Figure 3.14, there are different sections with various tabs.

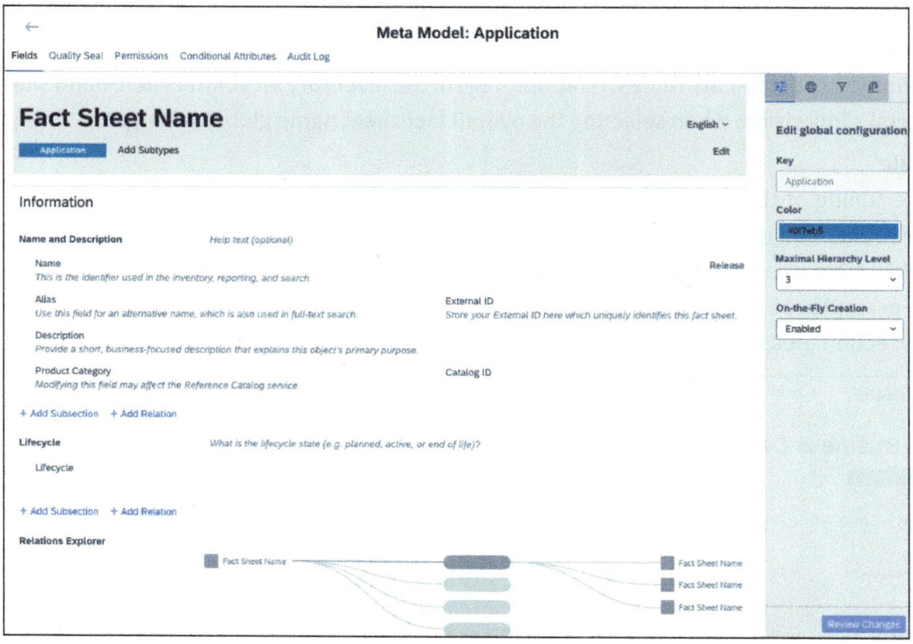

Figure 3.14: Example of Specific Application Fact Sheet Configuration

The top ribbon navigation includes the following options:

- **Fields**
  Captures the data and relational model of the selected fact sheet type
- **Quality Seal**
  Defines the settings and cadence for quality seal updates as well as mandatory data for data quality management
- **Permissions**
  Lets admins define specific access to and authorizations for the data model and collaboration features of this fact sheet type for defined user roles
- **Conditional Attributes**
  Holds logic for conditional data, meaning certain fields are only available if another field holds a defined value

- **Audit Log**
  Audit log containing the meta model configuration changes per fact sheet type

The center area is contained by a grey box when hovered over. This is how you can select sections or subsections to modify.

Finally, the right-hand menu for customization shows four icons at the top:

- **Sliders icon**
  Represents the edit area.
- **World globe**
  Maintenance of display names/labels, help text, descriptions, and translations of fields.
- **Funnel**
  Definition of default attributes, relations, tags in the inventory facet filters (left-hand side filters). Only visible when selecting the overall fact sheet name global settings.
- **Scale**
  Fine-tuning of the completion score weights (included and excluded values as well as their relational weights) when utilizing the completion score percentage on fact sheets. Only visible when selecting the overall fact sheet name global settings. See Figure 3.15 for examples of completion score percentage values on different fact sheet section and subsection areas.

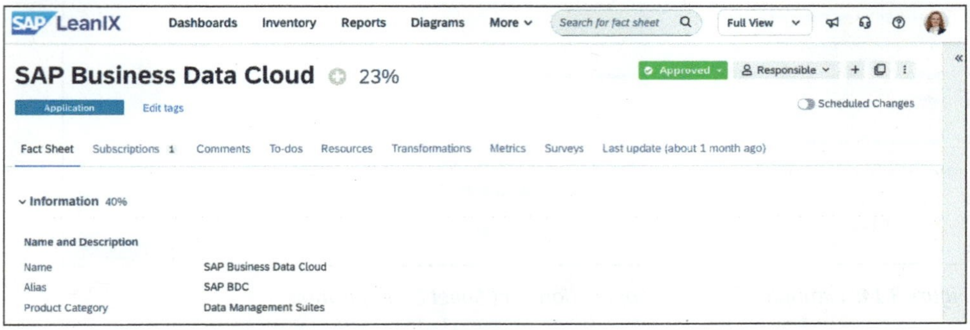

Figure 3.15: Example of Overall and Section-Related Completion Score

Completion Score Weights

Try not to get too complicated with the completion score weights. To begin, it's sufficient to only use the values 0 and 1:

- A value of *0* indicates that the section, subsection, attribute, or relation is not included in the calculation of the completion score, meaning that whether the end user filled out this section or field in the fact sheet has no impact on the percentage that you can see at the top of the fact sheet page.
- A value of *1* indicates that the section, subsection, attribute, or relation is included in the calculation of the completion score.

The completion score can nudge users to fill in the information in the fact sheet that is relevant for the selected use case. You can include this metric in data crowdsourcing events to encourage fact sheet owners to bring their completion score close to 100%.

The completion score can be enabled or disabled in the **General** tab of the **Basic Settings** in the **Administration** area.

The right-hand menu is dynamic and changes when selecting other sections, subsections, attributes, or relations in the center area. Other symbols you might see include the following:

- **Two arrows pointing in opposite directions (up and down)**
  Used to move fields to a different section or subsection or a different position within the section or subsection. Can also move attributes and relations into the **Unused fields and relations** section at the bottom of the page.
- **Gear**
  Indicates if the following is true for an attribute or relation:
  - **Included in views** (relevant for color-coding of reports and diagrams)
  - **Show in filters** (relevant for the left-hand side filter options, e.g., in the inventory)
  - **Display as read-only**
  - **Included in full text search** (only for text fields)
  - **Included in quick search** (only for text fields)

Caution

Some of the out-of-the-box fields and relations are crucial elements for the functionality of certain features in SAP LeanIX. Always read the **Notes** and **Warnings** that can be found in the right-hand menu before modifying or deleting something, as shown in Figure 3.16.

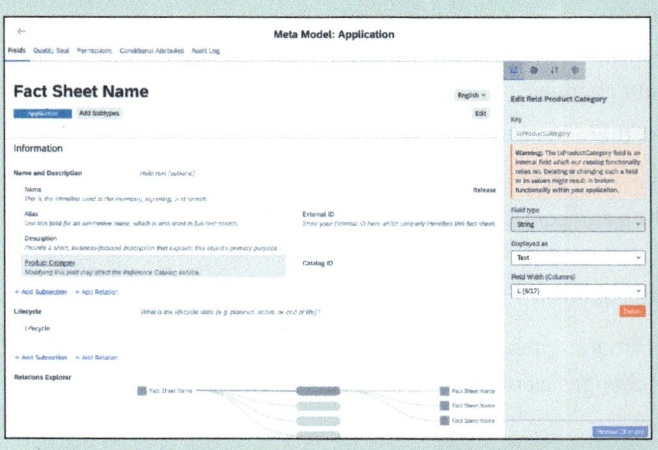

Figure 3.16: Meta Model Configuration Warning for Field with Technical Name lxProductCategory

For instance, the **lxProductCategory** field is part of the SAP LeanIX reference catalog functionality. Deleting this field might result in broken catalog functionalities.

To create new sections, subsections, fields, or relations, hover over an area in the **Fields** tab until a grey frame appears around that section, then click the section with the grey frame. When done, additional blue texts will appear at the bottom and below the grey frame (see Figure 3.17):

- **+ Add Section** (visible when selecting a top-level section)
- **+ Add Subsection** (visible under each subsection)
- **+ Add Relation** (visible under each subsection)
- **+ Add Field** (visible when selecting a subsection)

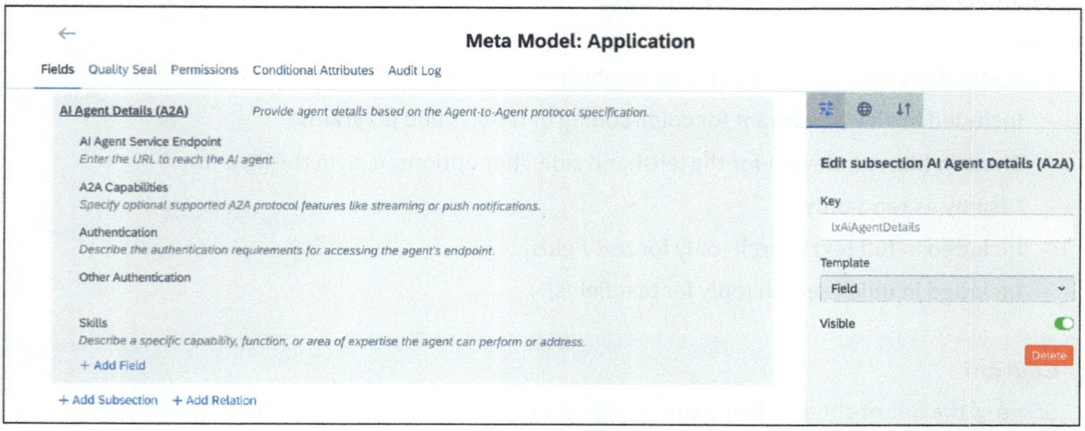

Figure 3.17: Data Model Modifications in Fact Sheet Configuration

Always Save Your Changes

In **Meta Model Configuration**, there are several steps, including a review, required to finalize and implement your changes:

1. In the right-hand menu, there are various buttons in the lower-right corner.
2. To create a new object, click the **Create** button.
3. When you're done editing and configuring the object, click the **Review Changes** button.
4. After the review, click the **Save** button.

You will see a green checkmark appear with a note in the upper-right corner of the screen, indicating that all changes were performed successfully.

Further Resources

For more details and examples of meta model configuration, read the SAP LeanIX user documentation with subsections on the SAP Help Portal at *http://s-prs.co/v620623*.

Subscription Roles

In this section, admins can define custom subscription roles for fact sheets. Out-of-the-box SAP LeanIX offers the following roles:

- **Accountable**
  Optional role that can be activated in the **General** tab of the administration menu; there can only be one subscriber per fact sheet
- **Responsible**
  Has more editing rights for the fact sheet, as well as data quality responsibilities due to owning the quality seal, and can receive special notifications
- **Observer**
  Has primarily viewing rights on fact sheets and can receive custom notifications

You can create custom subscription roles by clicking the blue **New Subscription Role** button in the **Subscription Roles** menu in the **Administration** area (see Figure 3.18).

Custom subscription roles are always based on out-of-the-box subscription roles. Many customers use the custom subscription roles to bring in their official role names, for instance, application owner as a responsible role, or technical owner as a responsible role.

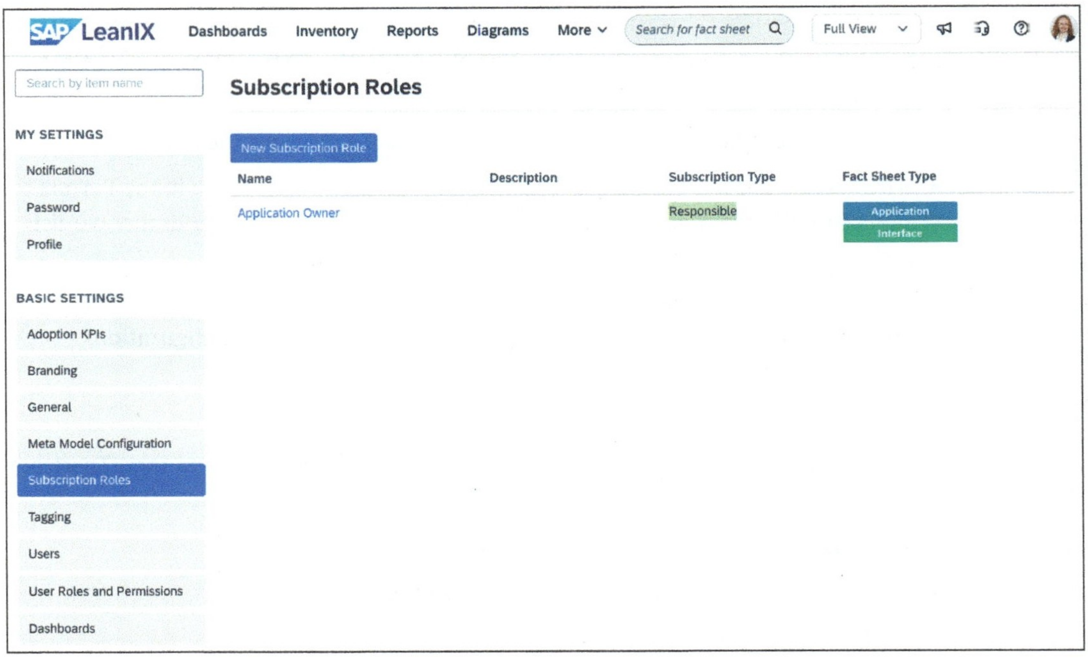

Figure 3.18: Custom Subscription Role Example: Application Owner

When adding a subscriber to a fact sheet, the user can then select values for **Type**, which corresponds to the out-of-the-box subscription roles, and **Role (optional)**, which corresponds to the custom subscription roles.

Maintain Subscribers for Fact Sheets to Benefit from Collaboration Features

Subscription roles are key filtering options to define target audiences for surveys or automations in SAP LeanIX. The general advice is to assign a responsible person to every fact sheet even though this is not always practical. Having assigned subscribers with their roles in the fact sheet information will help to do the following:

- Identify the right contact person when something needs to be discussed or clarified
- Increase data quality since the responsible persons own the quality seal and overall data quality responsibility on the fact sheet
- Filter and slice and dice the inventory, reports, and dashboards
- Select target users and audiences for collaboration features such as surveys, automations, and to dos
- Develop advanced permission concepts with subscription-based permissions for fact sheet editing or viewing

Caution

Subscription roles define the roles and responsibilities of a user for a specific fact sheet (accountable, responsible person, observer, or custom role based on an out-of-the-box subscription role). In contrast, user and permission roles assign the authorization of a user in the workspace (admin, member, viewer, or custom role based on an out-of-the-box user role).

Both concepts can be combined in advanced permission setups in the following cases:

- When defining the overall authorization of the user/permission role in the **Permissions** tab in **Meta Model Configuration** for a fact sheet type (top dropdown list under **Manage Permissions**)
- When defining advanced subscription-based permissions (bottom section for **Advanced Permissions** on **Manage Permissions** in **Meta Model Configuration**)

Tagging

Next is the **Tagging** section, where admins can create new tag groups and tags. There are several benefits of tags:

- They are very prominent and visible in the inventory.
- They can be used for search and filtering capabilities.

- They can be used for advanced permissions (e.g., members can see a specific field of a fact sheet with a certain tag).
- They can be used as axes in reports.
- They can be used as views for colorful heatmaps in reports (only in single mode).
- Tags can be used initially before creating new fields to see if end users are utilizing the information.
- They can be restricted to admins only or open so that any user can create tags in the **Other tags** tag group (setting in the **General** tab under **Basic Settings** in the **Administration** area).

However, there are also some shortcomings of tags:

- They can become crowded and messy, especially if categories are not mutually exclusive and collectively exhaustive.
- They require regular review and cleanup, such as when they're used to select target applications for a strategic initiative and the initiative is then completed or stopped.
- They can put too much emphasis on details due to their prominent visibility in the inventory.
- Their level of visibility can't be restricted. That is, tags are always visible; you can't define that a certain tag group is only visible to a certain permission role.

To create a tag group, first click the blue **New Tag Group** button in the **Tagging** menu (see Figure 3.19).

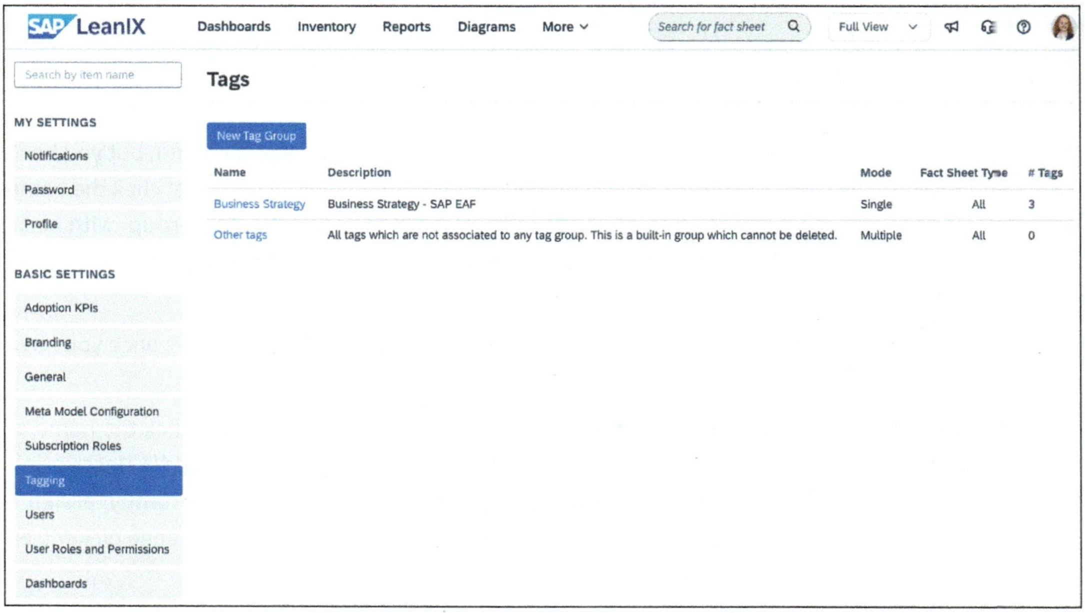

Figure 3.19: Tagging Menu in Administration Area

You will be redirected to a form to enter basic information such as the **Name** of the tag group, provide a **Description** (optional), define the **Fact Sheet Type**(s) in scope for this tag, and select the **Mode** (see Figure 3.20). **Single** mode means only one of the tags of this tag group can be assigned to a fact sheet. **Multiple** mode means many tags of the tag group can be assigned to the same fact sheet. When you're done, click the blue **Add** button at the end of the form and the tag group will be created.

Figure 3.20: Creating New Tag Group

You will see your custom tag group now in the overview of the **Tagging** menu, but you have only created the tag group at this point, not the tags. To create the tags, first click the name of your new tag group to open the edit mode. You will see two tabs: **Tag Group**, with basic information, and **Tags**. Select the **Tags** tab and click the blue **+ Add Tag** button in the upper-left corner of the center pane (see Figure 3.21).

The button transforms into a **Tag Name** entry field. Click the blue **OK** button once you have entered the name (see Figure 3.22).

You will now see the new tag in the **Tags** tab, but it's still unconfigured. Click the name of your newly created tag to open the edit form of the tag. Here, you can add a **Description** (optional), assign the tag to a new **Tag Group** if needed, and, most importantly, assign a **Color** to it by either entering a hex code or selecting the color from a color range picker (see Figure 3.23).

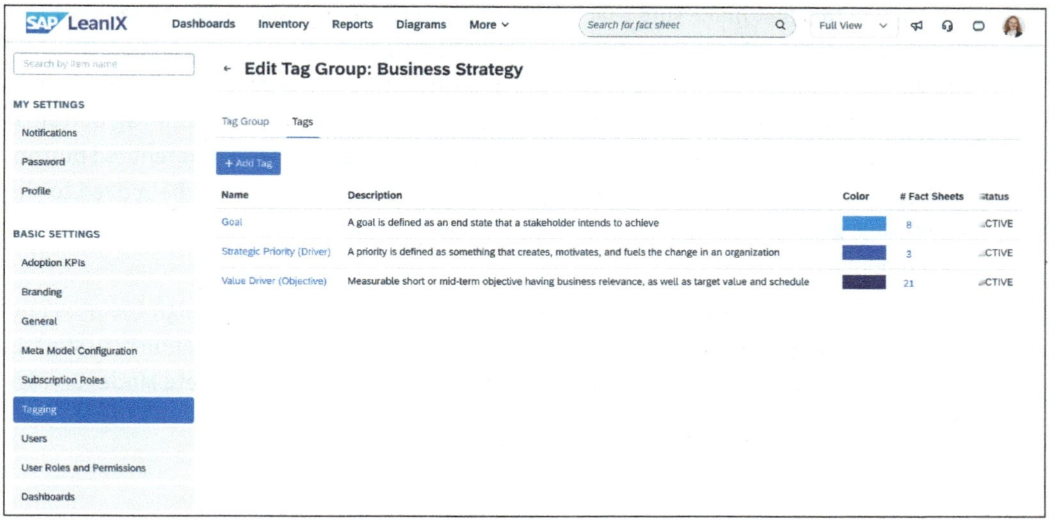

Figure 3.21: Tag Group Edit Mode to Modify Tags

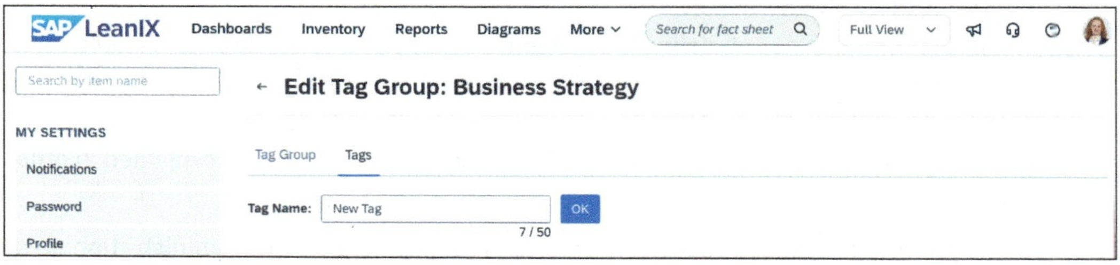

Figure 3.22: Creating New Tags

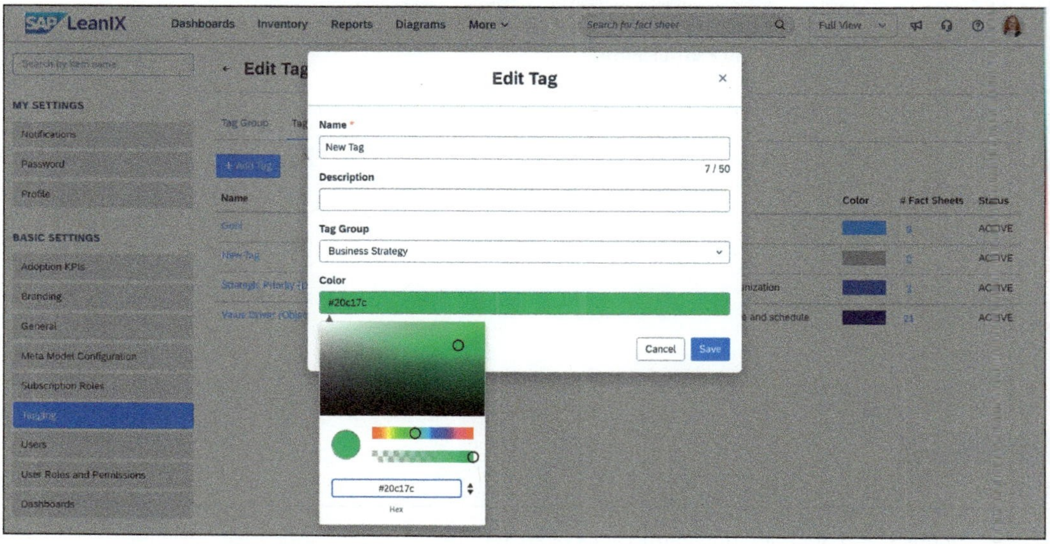

Figure 3.23: Editing Tag Settings

To delete a tag, first make sure that it isn't assigned to any fact sheets anymore. Use the **# Fact Sheets** column in the **Tags** tab to see how many fact sheets a tag is assigned. By clicking the blue number, you will be redirected to an inventory search filtered for this tag. Once you have removed the tag from the fact sheets, you can click the red **Delete Tag** button in the tag edit form. If the tag is still assigned to fact sheets, you will see a different red button, saying **Remove from tag group**. When you click this button, the tag will be moved to the **Other tags** group, which is the default out-of-the-box group for tags without a tag group assignment.

Tags can be a very useful tool in SAP LeanIX, but you always have to weigh the pros and cons of using tags over fields. Sometimes this can be an evolution; information may start as a tag and at a later point in time become a field on the fact sheet under **Meta Model Configuration**.

User

We covered the **Users** tab earlier, in Section 3.3.2.

User Roles and Permissions

In this section, admins can define the **General Settings** of a user role (e.g., default workspace views or labels, descriptions, and translations) and the **Permissions** (e.g., whether members are allowed to read and update diagram fact sheet shape templates). A lot of the functionalities and features that you see in the **Permissions** tab will be explained in other chapters in this book.

When you navigate to the **User Roles and Permissions** menu in the **Administration** area, you will see an overview of the available roles (see Figure 3.24).

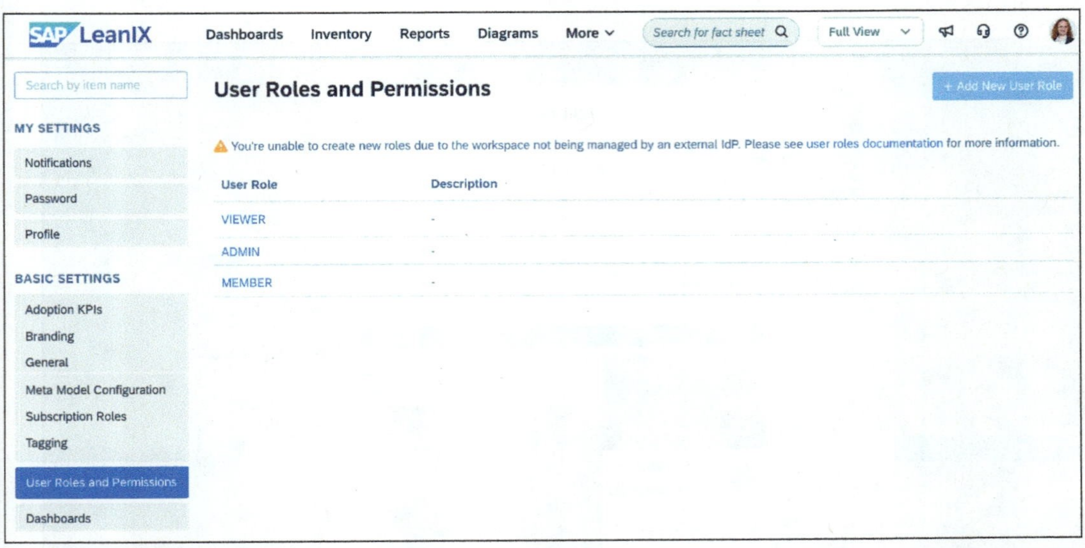

Figure 3.24: User Roles and Permissions Overview Page

By clicking the name of a **User Role,** you will enter the editing mode for this role and will see the **General Settings** and **Permissions** tabs (see Figure 3.25).

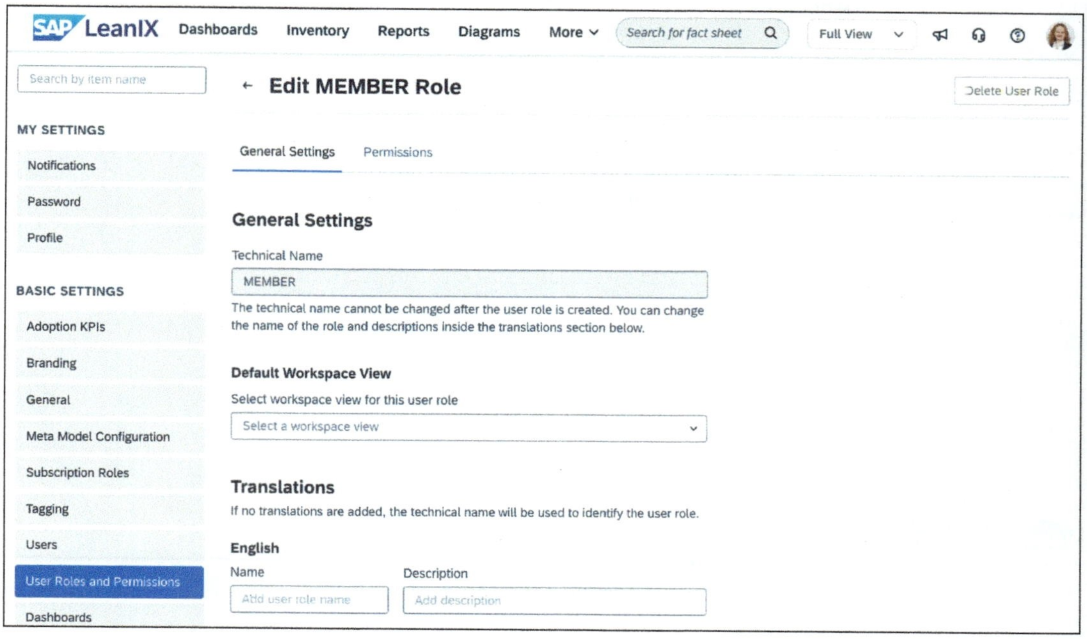

Figure 3.25: MEMBER Role in Editing Mode

> **Note**
>
> Custom user roles are only possible with externally IdP-managed role settings. Revisit Section 3.3.2 for more information.

Dashboards

Under the **Dashboards** tab, admins can activate or hide out-of-the-box curated dashboards for SAP LeanIX standard use cases such as **Application portfolio management** or **Application rationalization** (see Figure 3.26). In addition, admins can define **Default Dashboards** and whether users can set their own **default** dashboards. Furthermore, admins can restrict the content that can be embedded on dashboards and decide whether forms, popups, or presentations in iFrames are allowed to be displayed.

Settings for the **Architecture Executive Dashboard** are in a second tab. Here admins can define global filters and views.

As always, don't forget to **Save** your changes when you are done.

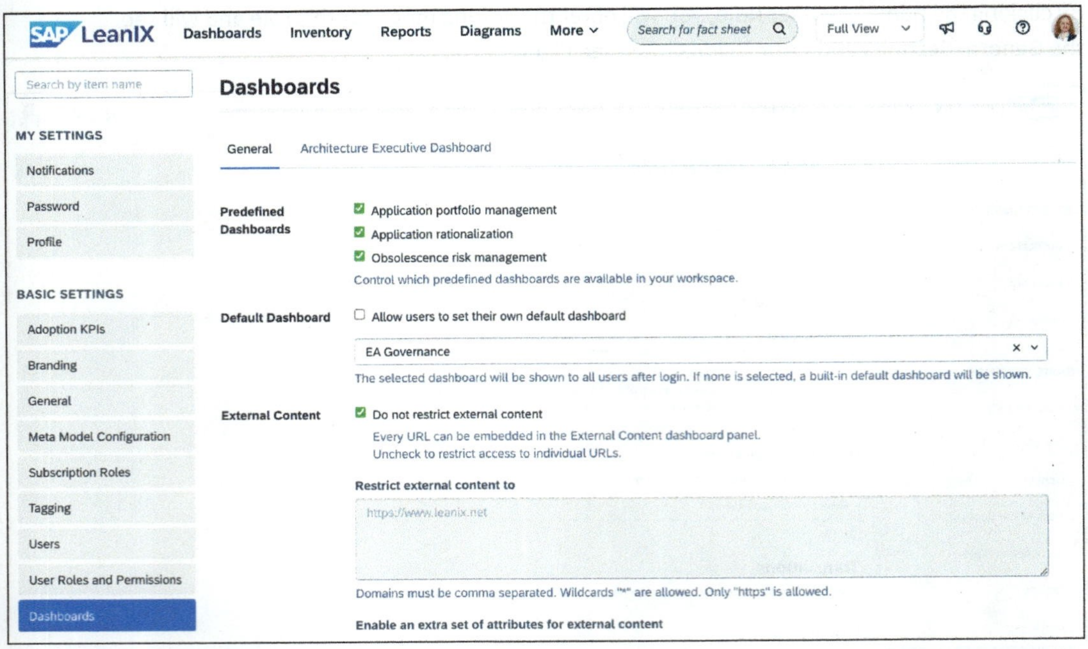

Figure 3.26: Dashboards Menu in Administration Area

Advanced Settings

For the initial basic setup of a new SAP LeanIX workspace, not all sections in the **ADVANCED SETTINGS** admin area need to be configured. However, a few sections are useful and will be explained here.

> **Further Resources**
>
> For more details and examples, read the SAP LeanIX user documentation about advanced settings on the SAP Help Portal at *http://s-prs.co/v620624*.

For a deep dive into **Automations** and **Calculations**, see Chapter 10.

Export

In the **Export** section, admins can create full extracts of the SAP LeanIX workspace data. Under the **Create New Export** tab, you can decide to perform the following:

- A full export of all active fact sheets and related objects
- An export of the change log within a specific timeframe
- An export of the archived workspace data (data is archived for 90 days before complete deletion)

We recommend archiving outdated data that is no longer used or needed for decision-making. Some organizations nonetheless require certain documentation and archiving timelines above 90 days. To comply with such rules, you can set up an independent process to regularly retrieve the archived data from SAP LeanIX and place it into a separate backup folder for auditing and archiving purposes.

Under the **User Exports** and **Admin Exports** tabs, admins can monitor who downloaded which data from the workspace over time.

Notifications Center

The **Notifications Center** settings are very important for collaboration purposes. Here, admins can activate or deactivate notifications for all users in the workspace and define default values, as well as determine which selections will be available for users in their individual **Notifications** area under **My Settings**. SAP LeanIX offers enabling the following options under the **Channels** tab:

- **Email** notifications
- **Microsoft Teams** notifications

Note

Enabling Microsoft Teams notifications requires setting up the Microsoft Teams integration in the **Integrations** section of **DISCOVERY AND INTEGRATIONS**.

Notifications are important once subscribers are assigned to fact sheets. Fact sheet subscriptions assign dedicated users to fact sheets as responsible persons, observers, or accountable persons (optionally; the accountable subscription type can be activated in the **General** section under **BASIC SETTINGS**). Subscribers to a fact sheet can be assigned by others (e.g., admins), subscribe themselves, or be created via an integration with another solution.

To avoid sending out mass emails by accident, admins should turn off the notifications while the workspace is still in the setup phase. Admins should turn off the button in the upper-right corner of the channel tile so that the toggle changes from green to grey. See Figure 3.27 for an example in which the email channel is turned on and the Microsoft Teams channel is turned off.

Remember to turn notifications back on once the workspace is about to be rolled out and made accessible to users.

To configure the default values and the available options for users, the admin must click the **Edit Settings** button in the lower-left corner of the **Channels** tile (visible in Figure 3.27). A new window with new tabs will open, as shown in Figure 3.28.

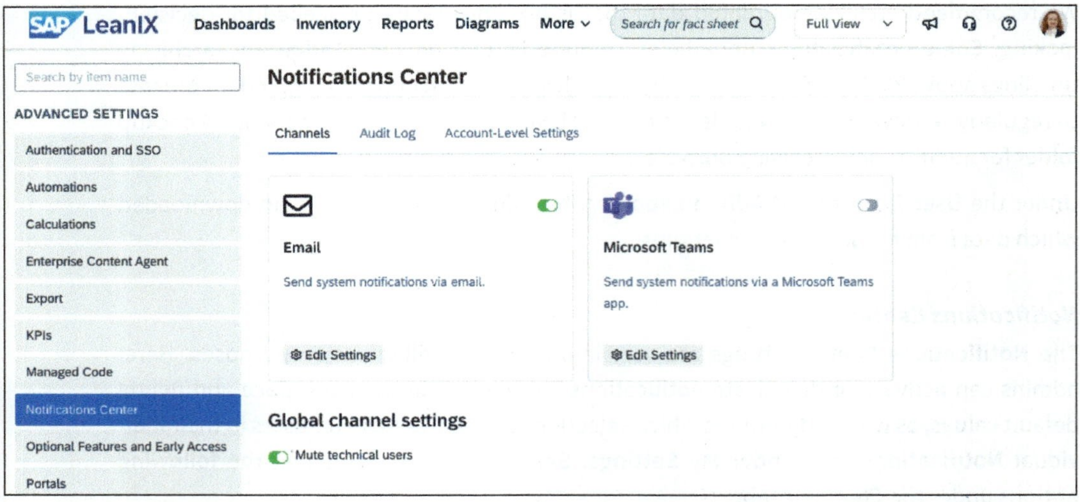

Figure 3.27: Activating and Deactivating Notification Channels

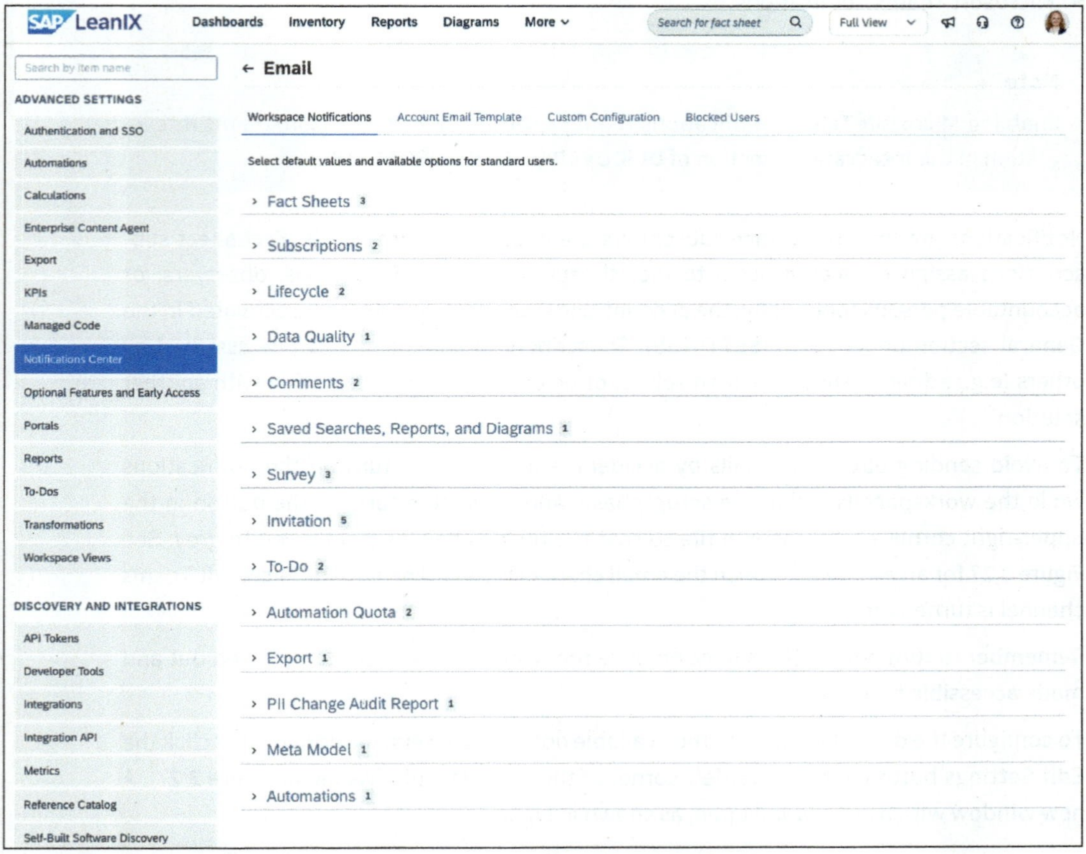

Figure 3.28: Email Notification Settings Page with Collapsed Sections

The first tab, **Workspace Notifications**, holds all available areas in which admins can preconfigure defaults and optional values. The number in the grey square next to the collapsed section title indicates how many settings the admin can configure. You might find more areas in a section that are greyed out, indicating that those values are SAP LeanIX standards and cannot be changed.

The configuration design is very similar and consists mostly of two parts: one area with the possible **Available Options** that can be checked on or off by clicking the corresponding box; and one area underneath with a dropdown menu to define the **Default** value for users, as made visible in the middle-bottom area of Figure 3.29.

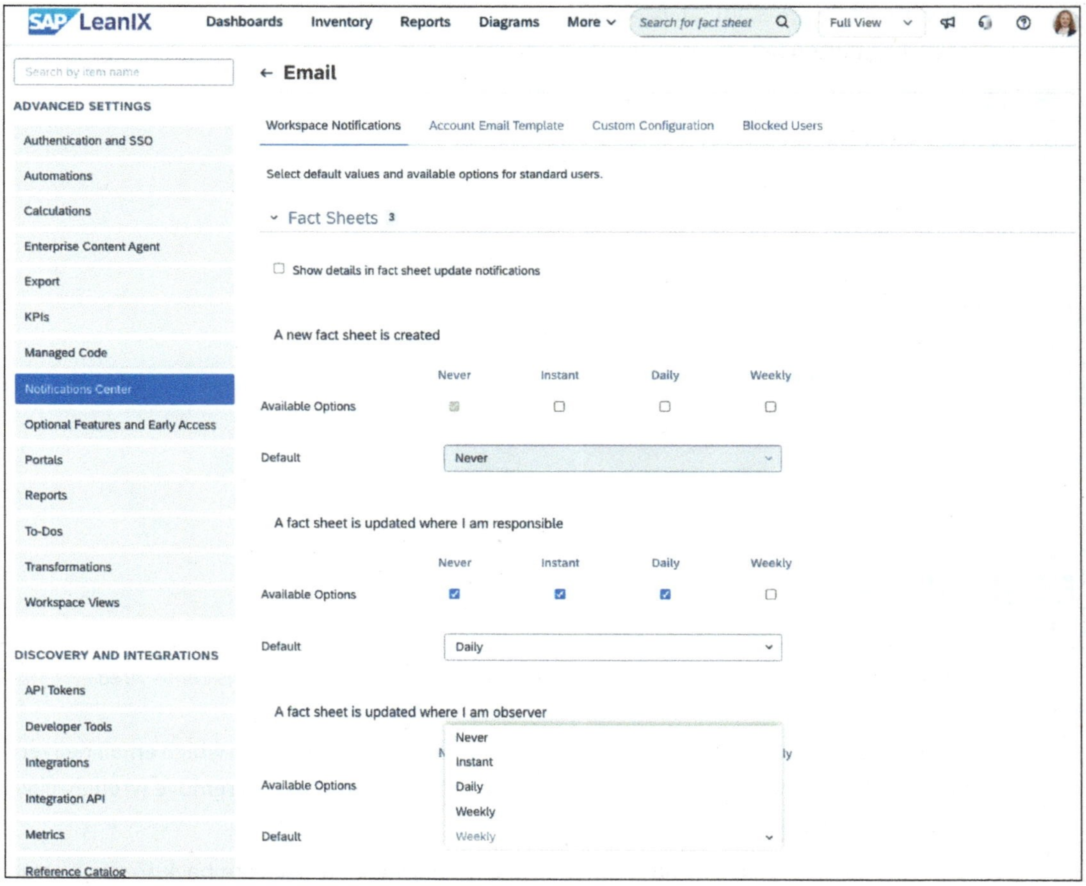

Figure 3.29: Email Notification Settings Configuration

The **Account Email Template** tab in the **Email** settings provides admins with the ability to customize the email design and layout of notifications sent. If you have multiple workspaces, be aware that changes in **Account Email Template** will be applicable to all SAP LeanIX workspaces in your account. To enable custom HTML designs, check the checkbox under **HTML Template** that reads **Enable custom email header and footer**. Compose your

HTML header and footer and copy them into the designated boxes. Always click the **Send test** button with your email address entered to verify the correctness of your HTML design before saving the settings.

Further Resources

For more information, check out the following resource: *http://s-prs.co/v620625.*

In the **Custom Configuration** tab, admins can configure **Exchange Online** or a custom Simple Mail Transfer Protocol (**SMTP**) server as an alternative to the default SAP LeanIX mail server (see Figure 3.30). Always perform a test SMTP configuration by clicking the corresponding button before activating and saving the settings.

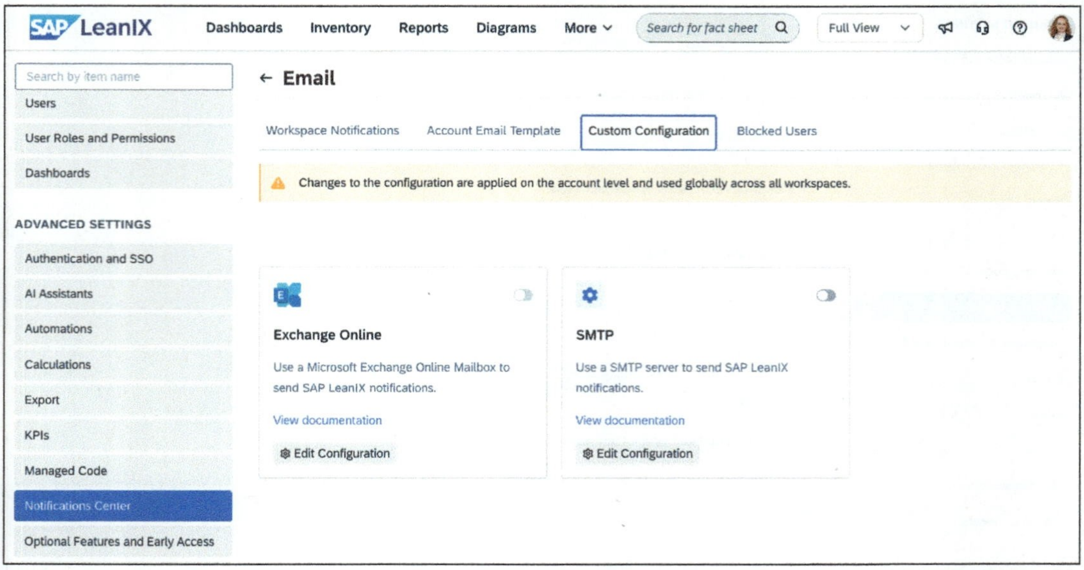

Figure 3.30: Custom Configurations in Email Settings Within the Notifications Center Area

The **Blocked Users** tab gives admins an overview of email addresses to which email delivery failed with a permanent error. Only the SAP LeanIX support team can remove wrongly classified email addresses from the blocked user list.

Because this was the last tab in the email settings, you can navigate back to the overall **Notifications Center** settings by clicking the **Back** button (left-pointing arrow) to the left of the **Email** title.

The **Audit Log** tab of the **Notifications Center** settings shows the event log for notifications from the past 14 days. This includes, for instance, the delivery status of email notifications (e.g., sending, pending, sent, failed, deferred, or unknown).

Tip

Users benefit largely from customized user manuals with the look and feel of your organization's branded workspace.

One top tip is to include a collaboration guide in which certain SAP LeanIX collaboration features are explained. Here, you can also include a note that users should customize their individual notification settings. Some users might find it useful to get immediate notifications if, for example, data was changed in a fact sheet to which they are subscribed as a person responsible. For others, a daily or weekly summary might be sufficient. As an admin, you can define the defaults and the options that you want to offer your users.

3.4 Summary

In this chapter, we focused on the administrative tasks that need to be performed when activating an SAP LeanIX workspace. We examined the workspace provisioning via SAP for Me and core team invites as well as different authentication and authorization setups. Different sections of the administration menu that are only accessible to workspace admins were discussed, and tips and tricks regarding useful configurations were provided with additional background information and plenty of further resources for additional learning.

In Chapter 4, we'll focus on the initial data intake. This is a central task for the core team in collaboration with the business as it will also involve data model adjustments needed to tailor the workspace to the use case in scope and to fit the organization's look and feel and lingo.

Chapter 4

Modeling Core Architecture Objects

All hands on deck! You'll now start to import your data into your SAP LeanIX workspace. This chapter helps you jump-start creating transparency and insights in SAP LeanIX Application Portfolio Management, a foundational use case in any enterprise architecture journey.

Now that you understand the principles and meta model of SAP LeanIX and have an idea of the initial administrative work to configure the workspace, let's dive deeper into repository creation in SAP LeanIX. For this, we will follow along with Chris Green, the principal enterprise architect of the fictional company Essentials for All 7 (EA7). You, as a valued team member, will support Chris in activating various discovery capabilities, using Excel export/import, and updating data in various ways.

Establishing Our Use Case

EA7 is a global consumer goods company headquartered in Geneva, known for its strong brand portfolio across the food, personal care, and household categories. With approximately 70,000 employees and operations in over 40 countries, EA7 competes in a market characterized by rapid consumer shifts, fierce competition, and increasing expectations for personalized, frictionless experiences.

Over the past decade, EA7 expanded through new product categories, regional growth, and several acquisitions. While this strengthened the company's commercial footprint, it also introduced operational complexity, inconsistent ways of working, and fragmented customer engagement models. These challenges now limit EA7's ability to innovate, scale, and execute its long-term strategy.

To remain competitive, EA7 must modernize its business architecture, harmonize global operations, and strengthen digital capabilities across all functions, especially direct-to-consumer and product-as-a-service, which are becoming differentiating factors in the consumer goods industry.

Enterprise architecture, supported by SAP LeanIX, plays a foundational role in enabling this transformation. Chris Green is now tasked to fill the SAP LeanIX enterprise architecture workspace with meaningful data. He will use the SAP LeanIX out-of-the-box meta model and will activate additional features as needed. You, as a valued team member, will fill the inventory with its first artifacts, link them, and create initial visual templates.

4.1 Fact Sheet Attributes

As admins, you familiarized yourself with the **Meta Model Configuration** options in the **Administration** menu under **BASIC SETTINGS** (see Chapter 3, Section 3.3.3).

Every out-of-the-box fact sheet type contains generic attributes, which are available for any fact sheet type, as well as specific sets of attributes and relations that fit each respective fact sheet type and its purpose in the meta model. Examples of specific attributes are shown in Figure 4.1.

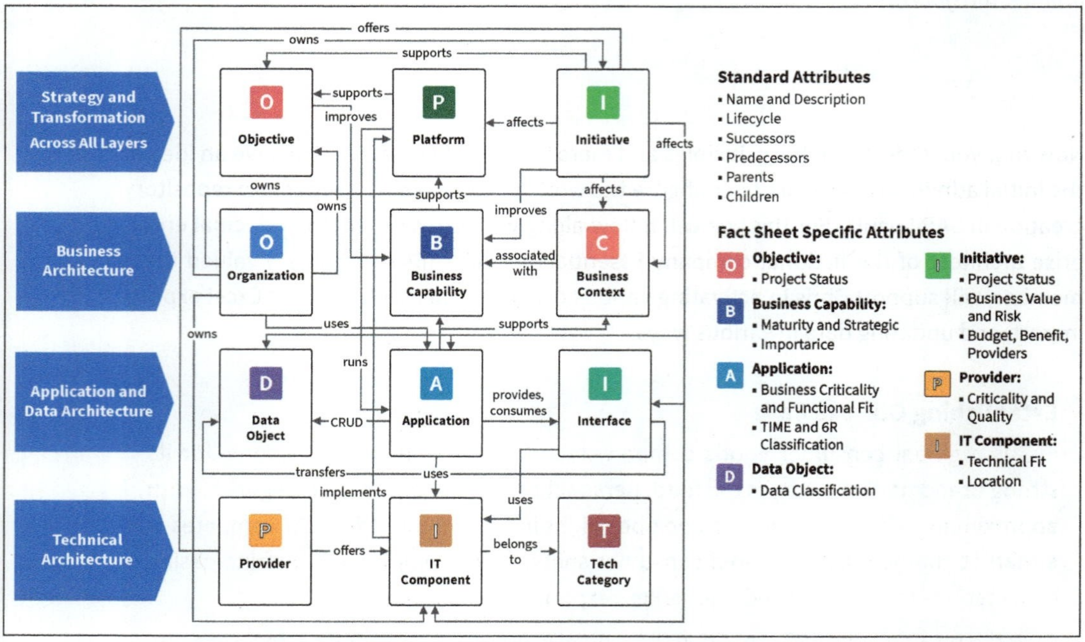

Figure 4.1: SAP LeanIX Meta Model Overview with Examples of Fact Sheet Attributes

Although attributes such as **Name**, **Description**, and **Lifecycle** are universal to all fact sheet types, others are only available for certain fact sheet types. For instance, the current and target **Maturity** is tracked on the **Business Capability** level. **Business Criticality** and **Functional Fit** are tracked on the **Application** level, whereas **Technical Fit** is captured on the **IT Component** level. Some out-of-the-box fields in **Initiatives** support the initiative planning and execution in SAP LeanIX, like **Project Status**, **Business Value**, and **Project Risk**, as well as **Budget Costs** (**OpEx (Budget)**/**CapEx (Budget)**) and **Benefit** (**Net Present Value (NPV)**).

For the initial intake of the inventory via discovery functionalities, you will use the out-of-the-box fields. Later in the project, you will most likely create additional custom fields such as *EA7 ID*, for this use case, a unique identifier for applications and IT components to support the cross-departmental architecture work. This ensures that the architect, IT, and business colleagues reference the same fact sheets when developing new strategies and discussing changes to the IT architecture landscape.

4.2 Discovering, Importing, and Creating Fact Sheets

Enterprise architect teams want to show value quickly to the senior leadership team when introducing a new enterprise architecture management SaaS solution. Until now, EA7 has mostly captured its application repository in Microsoft Excel spreadsheets, with ad hoc data-gathering activities that utilized lots of capacity when an update was needed. Therefore, EA7 decides to leverage the SAP LeanIX reference catalog and discovery capabilities to speed up and automate the logical application detection in the long term.

In the following sections, we'll walk through the steps to discover and import fact sheets, using examples of the different discovery capabilities. First, however, we'll discuss an important additional feature: the business capability reference catalog in SAP LeanIX.

4.2.1 Reference Business Capability Map

In the case of EA7, a good place to start is to use the SAP-recommended business capability reference catalog for consumer products filtered to high-tech results. Navigate to the business capability reference catalog to explore and import the relevant business capabilities. (Refer to Chapter 2, Section 2.2.3 to learn how to navigate to the business capability reference catalog.)

To filter the business reference capability catalog to the current consumer products industry and the target high-tech industry, click the **Industries** dropdown in the top ribbon and check the boxes for **Consumer Products** and **High Tech** (see Figure 4.2).

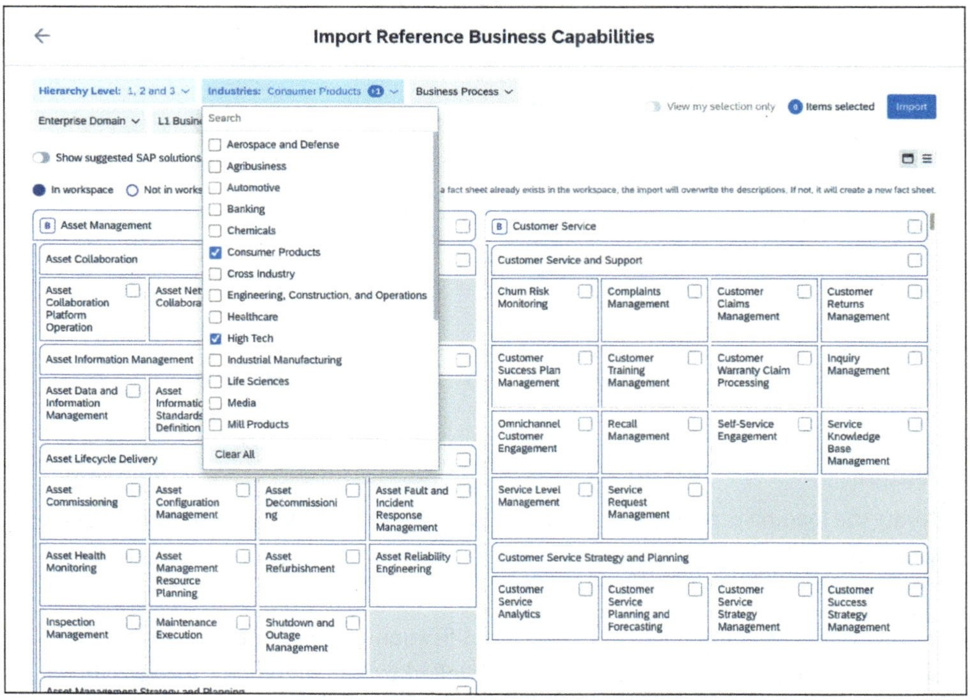

Figure 4.2: Example of Filtered Business Capability Reference Catalog

The enterprise architecture team must align with the business units on a common business capability map. It is important that the functional business units agree on the terminology and the selection as the application owners (very often business users) will later map or support the mapping of the application to the corresponding business capability that is supported by this application in their use case.

Once alignment has concluded, select the agreed-upon business capabilities in the business capability reference catalog UI and click the blue **Import** button in the upper-right corner to load the business capabilities into the SAP LeanIX workspace (see Figure 4.3). In the next popup window, click the blue **Confirm Import** button to acknowledge that this action might take a few minutes.

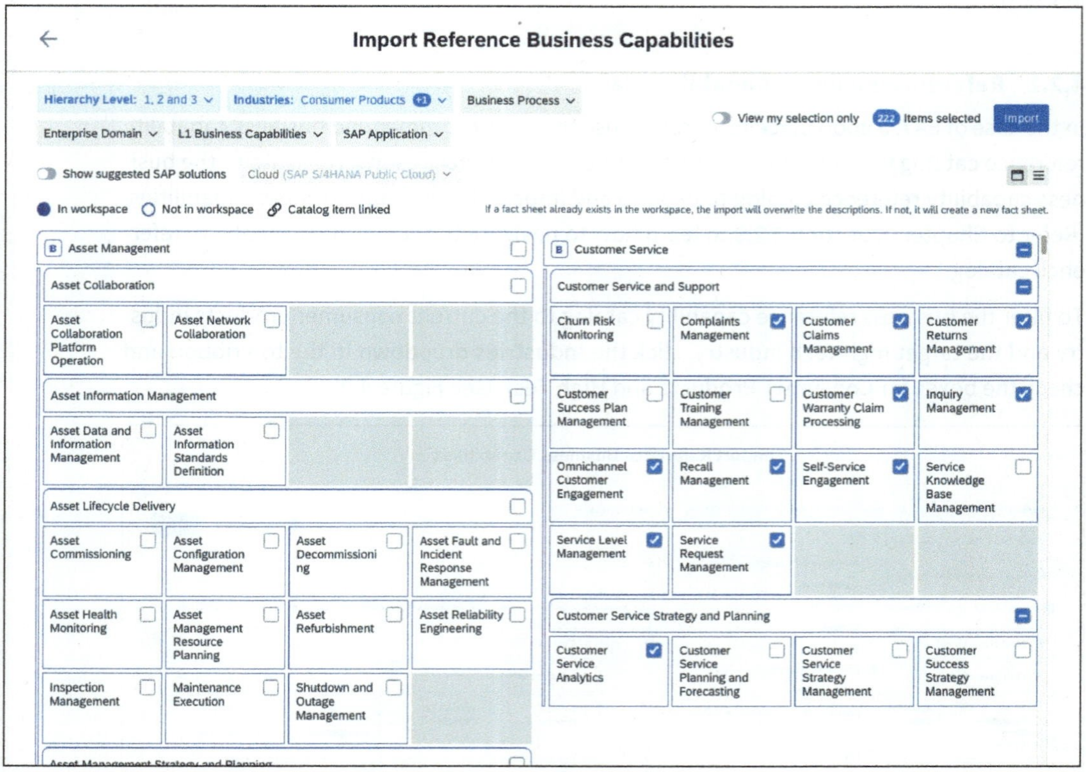

Figure 4.3: Selecting Business Capabilities for Import into Reference Catalog

> **Caution**
>
> During the loading process, you should not close the browser window or navigate elsewhere within the SAP LeanIX workspace in this browser tab.

Once the import is completed, you'll see a notification of the successful import, and the business capability tiles that are created and linked to the reference catalog will change from a white interior color to blue (see Figure 4.4).

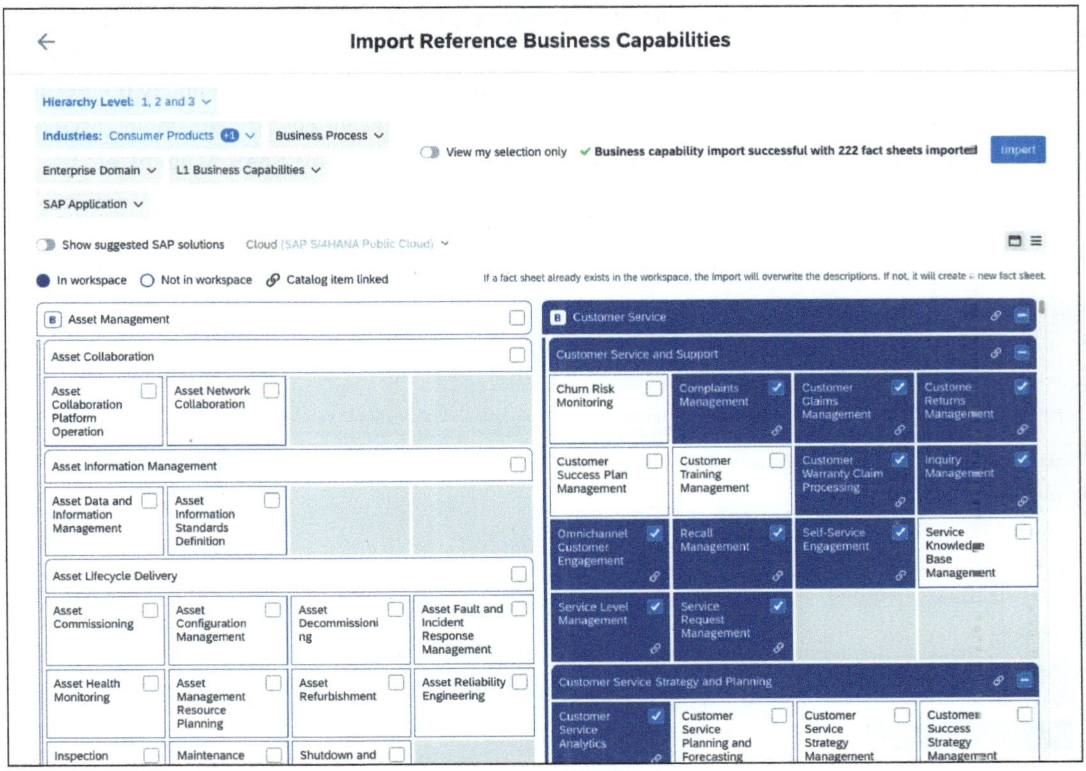

Figure 4.4: Successfully Imported Business Capabilities: Blue-Filled in Catalog

The business leaders of EA7 demanded a few extra business capabilities that were not part of the business capability reference catalog. Those additional ones were created manually using the Excel import functionality, which we'll discuss in Section 4.2.6.

Further Resources

To learn more about the business reference capability catalog, visit the online SAP LeanIX user documentation on the SAP Help Portal at *http://s-prs.co/v620626*.

4.2.2 Discovery Inbox

The SAP LeanIX discovery capabilities do not immediately create artifacts in your inventory; they're pushed to the discovery inbox for review first. The SAP LeanIX admin, or the role specified that can access the discovery inbox, can choose which artifacts should be created and which should be ignored.

That is particularly important because discovered applications in the discovery inbox that are not yet linked to the inventory do not count against the application quota for pricing. Artifacts in the discovery inbox, however, cannot be visualized in reports or diagrams or used in other collaborative features.

By default, only admins can access the discovery inbox. If you want to set up a custom role concept, you can grant access to the discovery inbox to other roles via the **Administration** menu, under **BASIC SETTINGS**, then **User Roles and Permissions**, and finally the **Permissions** tab (see Figure 4.5). Here, you can select the relevant checkboxes under **Discovery Inbox** to grant access for each capability that we'll discuss (**SAP**, **SaaS**, **AI Agent**, etc.).

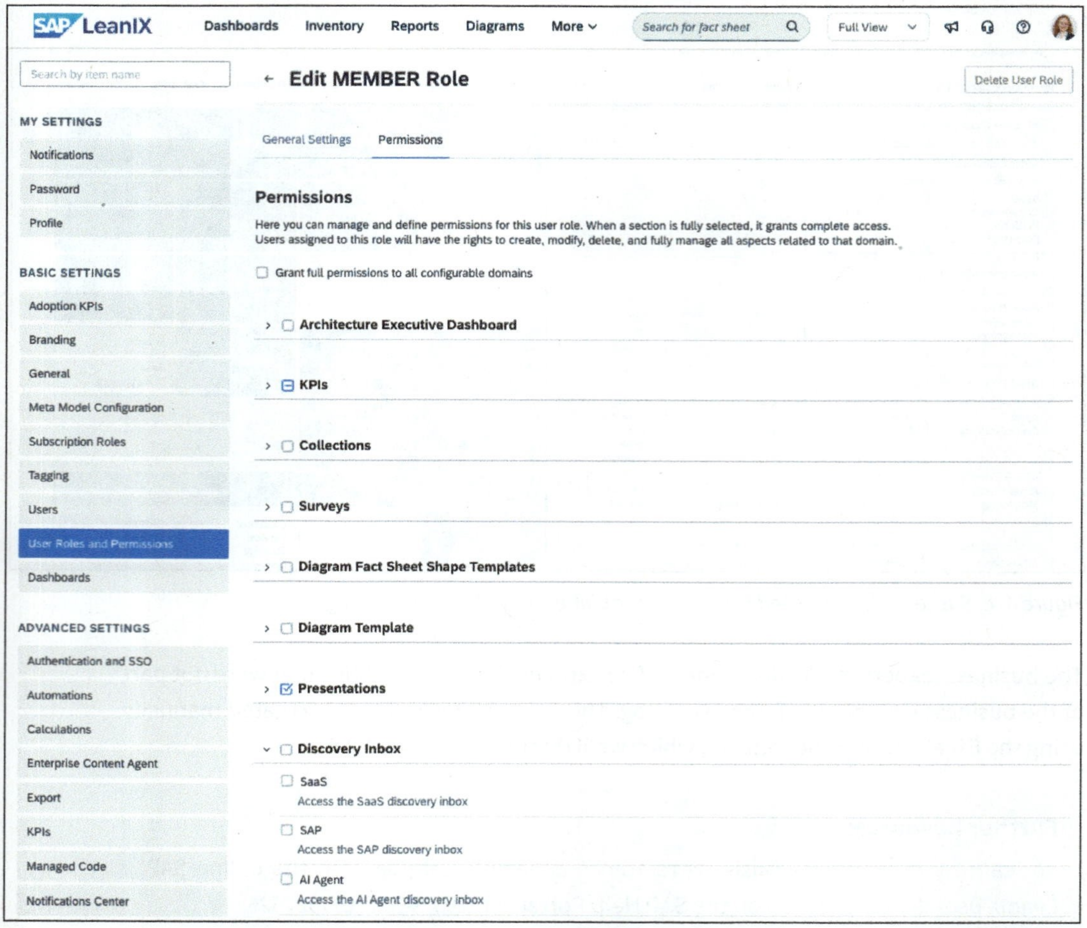

Figure 4.5: Admin Settings to Grant Read and Edit Access to Other Roles for Discovery Inbox

4.2.3 SAP Discovery via SAP Cloud ALM

The first SAP LeanIX discovery capability that you will set up is the one to autodetect SAP applications in use by EA7, which is called *SAP discovery via SAP Cloud ALM*.

The high-level process for SAP discovery's capabilities is depicted in Figure 4.6.

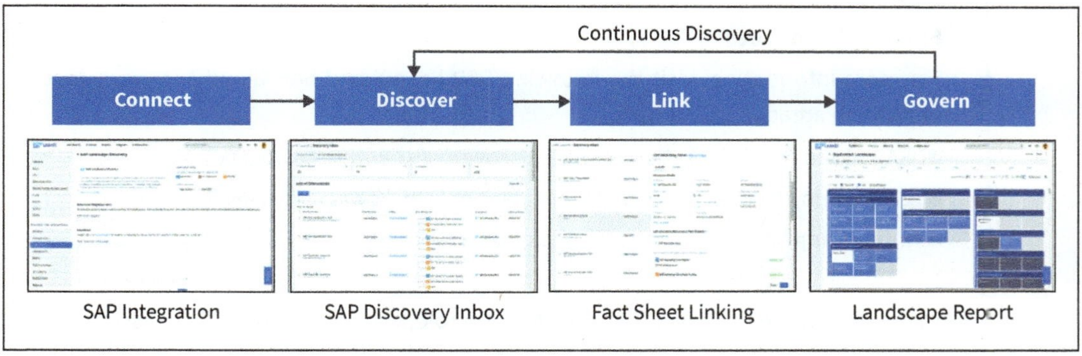

Figure 4.6: High-Level Steps of Discovery Features

First, the discovery source systems need to be connected. Next, the discovery inbox needs to be checked. Relevant and correct entries should be created or linked from within the discovery inbox, which will make the fact sheets appear in the inventory and count toward the application quota. Finally, the information created and linked to the inventory is used for reporting and planning purposes to analyze the as-is architecture and find redundancies, white spots, risks, or unused potential. We'll discuss the first three steps in more detail in the following sections.

Connect

Discovery can be performed on different levels (e.g., applications with underlying IT components, systems, self-built extensions, custom-built SAP Build projects). Here, we'll explore SAP discovery via SAP Cloud ALM, which will surface SAP applications and underlying IT components. SAP Cloud ALM, the successor to SAP Solution Manager, is SAP's cloud application lifecycle management solution. SAP Cloud ALM consists of multiple functional areas:

- SAP Cloud ALM for implementation, for IT transformation project planning and implementation
- SAP Cloud ALM for operations, for business operations monitoring
- SAP Cloud ALM for service, for end-to-end service delivery process management

For SAP discovery, SAP LeanIX integrates with SAP Cloud ALM as a gateway to the SAP backend systems. You don't need to be an active user of SAP Cloud ALM for the SAP discovery to work. SAP Cloud ALM just needs to be active with basic configuration performed.

To set up the integration, you will need the following roles:

- SAP LeanIX admin
- SAP Business Technology Platform (SAP BTP) cockpit admin

Further Resources

For more information, visit the following SAP LeanIX and SAP Cloud ALM user documentation areas:

- SAP Cloud ALM integration for SAP discovery: *http://s-prs.co/v620627*
- Configuring the SAP Cloud ALM integration: *http://s-prs.co/v620628*
- Integrating SAP LeanIX: *http://s-prs.co/v620629*

With the help of the EA7 SAP BTP cockpit admin role, you can configure SAP discovery.

For setup in the SAP BTP cockpit, the SAP BTP cockpit admin has to navigate to the subaccount that manages the SAP Cloud ALM API. Under the **Instances and Subscriptions** menu, the admin must create a new service instance for the SAP LeanIX connection by clicking the blue **Create** button in the upper-right corner.

In the popup form, the SAP BTP cockpit admin must enter the following information under the **Basic Info** tab:

- **Service**: Select **SAP Cloud ALM API**
- **Plan**: Select **Standard**
- **Runtime Environment**: Select **Other**
- **Instance Name**: Enter a meaningful instance name, such as "SAP LeanIX Landscape Discovery"

In the **Parameters** tab, the SAP BTP cockpit admin must enter the JSON code shown in Listing 4.1, adjusting the `<your-instance-name>` portion to reflect the previously defined **Instance Name** from the **Basic Info** tab.

```
{
    "xs-security": {
        "xsappname": "<your-instance-name>",
        "authorities": [
            "$XSMASTERAPPNAME.calm-api.landscape.read",
            "$XSMASTERAPPNAME.calm-api.subscriptions.read"
        ],
        "oauth2-configuration": {
            "credential-types": [
                "binding-secret"
            ]
        }
    }
}
```

Listing 4.1: Instance Parameter Template with Permissions Needed for SAP Cloud ALM API Service Instance

After the review, the service instance can be created. In the next step, the SAP BTP cockpit admin creates a new service binding and shares the service credentials JSON with the SAP LeanIX admin.

In the SAP LeanIX workspace, you as the SAP LeanIX admin navigate to the **Administration** area and the **Integrations** menu item. After clicking the blue **Add Integration** button, you can filter for **SAP Discovery** in the left-hand filter menu, then click the blue **Configure** button in the **SAP Cloud ALM** tile (see Figure 4.7).

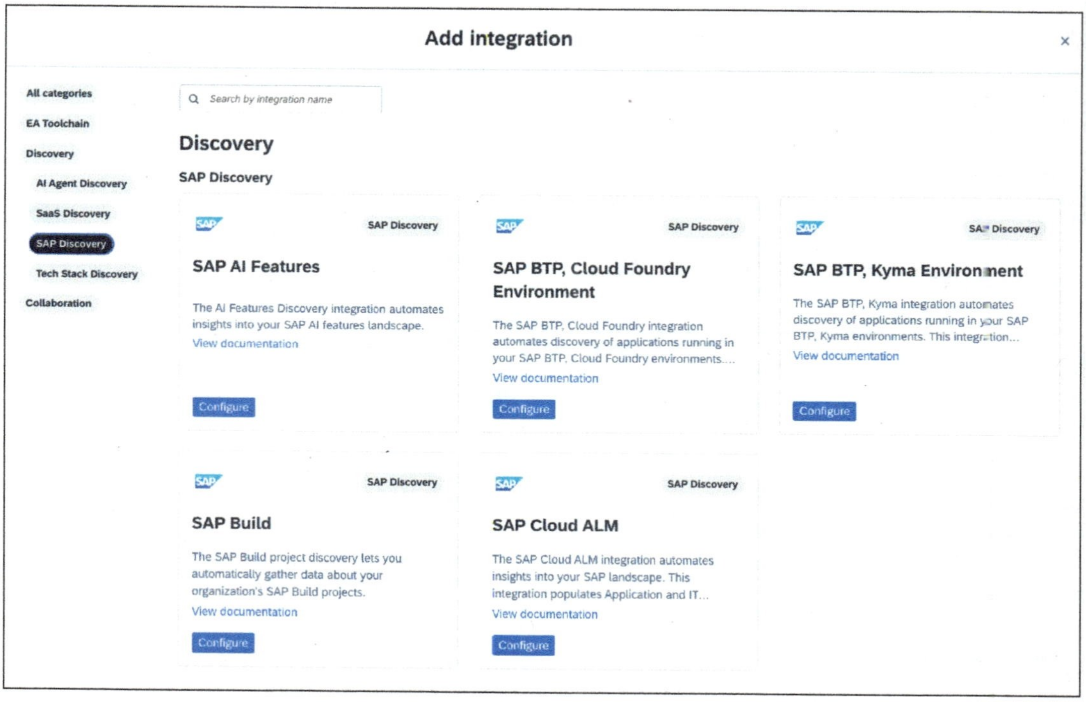

Figure 4.7: Activating Discovery Integrations in Administration Menu

Enter a meaningful name for the integration in the popup form under the **1 Start** step (see Figure 4.8). We've entered "EA7 SAP landscape discovery" in this example.

Click the **Next** button to arrive at the **Select Capabilities** step, as shown in Figure 4.9. You can start with the recommended basics in SAP discovery and extend to, for example, detecting system fact sheets (an optional additional fact sheet type; see our discussion in Chapter 2, Section 2.2.9) by selecting **SAP System Information** at a later point in time. For this demo, we won't activate the optional features for systems and integration flows; therefore, keep the preselected features as set under **PRIMARY CAPABILITIES**.

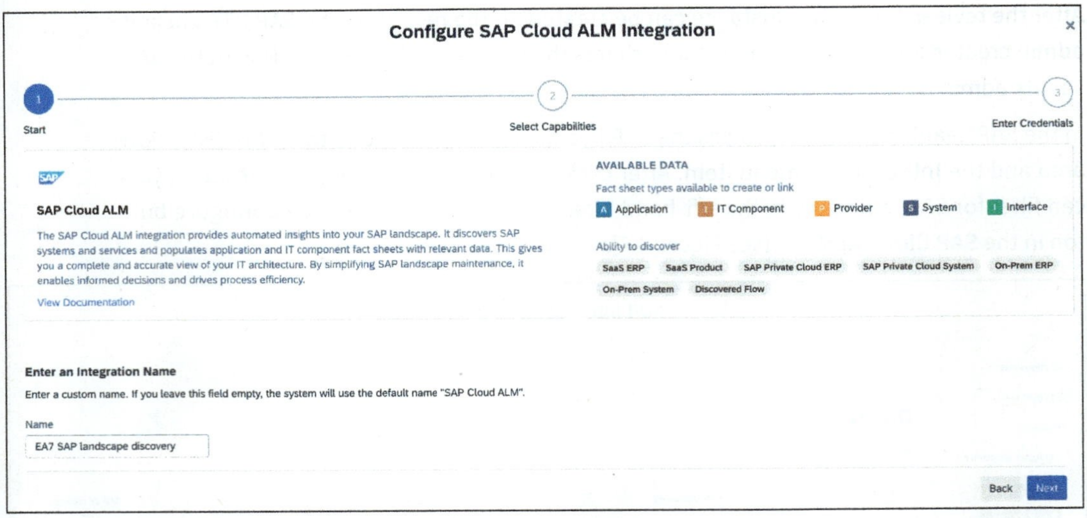

Figure 4.8: Screen 1: Start of Configuration Guided Form

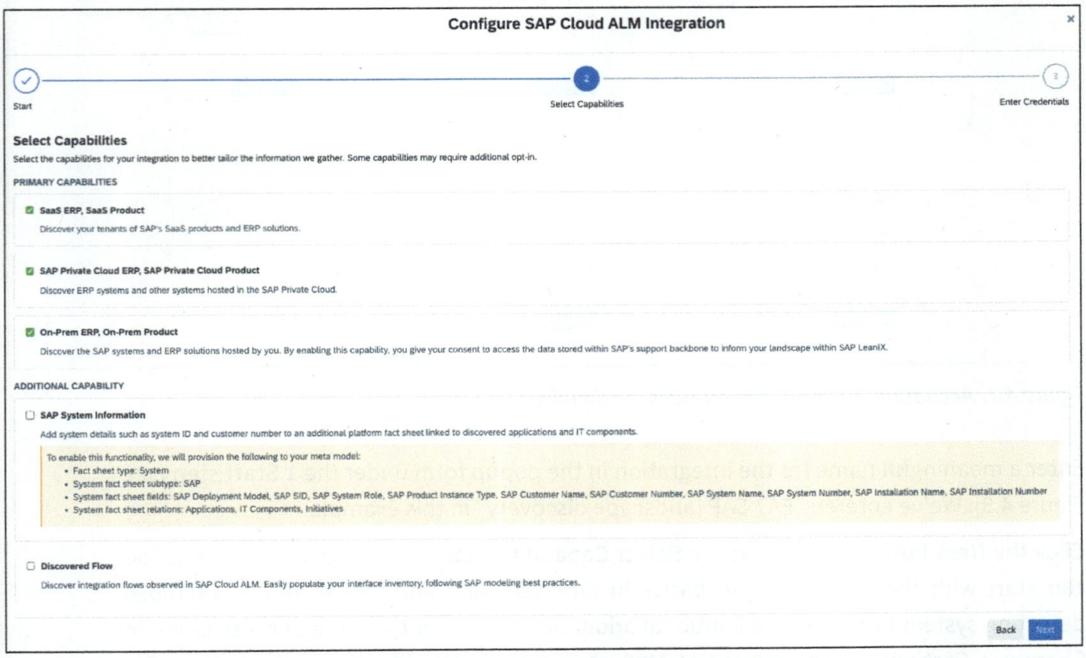

Figure 4.9: Screen 2: Select Capabilities of Configuration Guided Form

Click **Next** to arrive at the last tab of the form, **Enter Credentials**, where you can enter the binding credentials in JSON format that were received from the SAP BTP cockpit admin.

After clicking the blue **Save** button, the integration is established.

Discover

You can monitor the progress of the connection by going to the **Integrations** menu item and selecting the **SAP Cloud ALM** tile. It's easiest to refresh the browser window from time to time or to revisit the page after some time. Once all traffic lights are green, the integration is established and a first scan for the discovery inbox has started (see Figure 4.10).

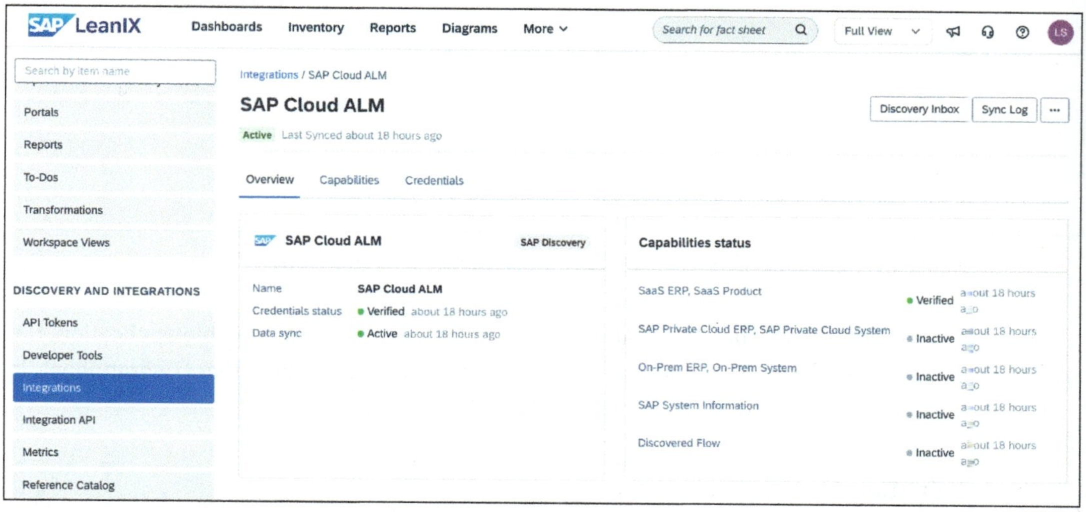

Figure 4.10: Integration Monitoring in Administration Menu via Traffic Light Signaling

To navigate to the discovery inbox, first click the **Inventory** menu item in the top ribbon. Then in the upper-right corner, click **Inventory Tools**, and then hover over **Discovery Inbox**. Click **SAP Discovery**, as shown in Figure 4.11.

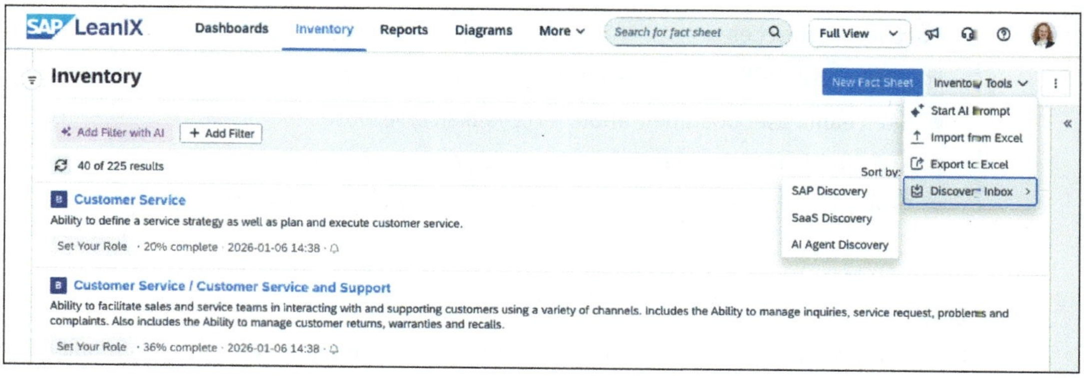

Figure 4.11: Navigating to SAP Discovery Inbox

SAP discovery will find multiple items in different environments (production, test, development, and demo). Those environments are best captured in the system fact sheet. It's fine to focus on the logical application and only on production (PROD) systems in the beginning. Filter the discovery inbox accordingly by selecting **PROD** in the **System role** dropdown (see Figure 4.12).

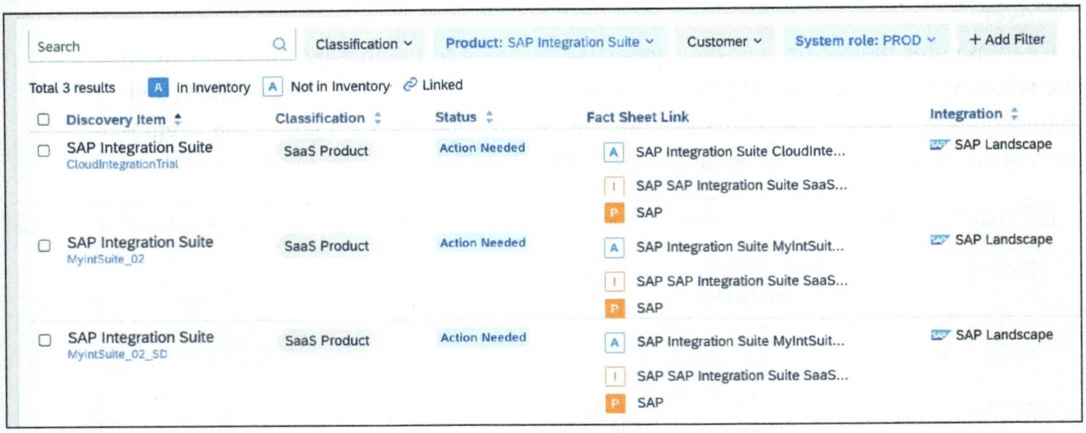

Figure 4.12: Example of Detected PROD Artifacts in SAP Discovery Inbox

As you can see in Figure 4.12, SAP discovery detects not only the possible application, but also the associated IT component and the provider associated with that IT component. In this example, you can see the SaaS hosting of SAP Integration Suite.

Caution

Always check the SAP discovery results for completeness. You can, for instance, compare the results with the system landscape overview in SAP Cloud ALM and have an experienced employee review the completeness of the data. If data is missing, create a ticket with the SAP support team to investigate the issue.

Further Resources

For an up-to-date list of in-scope and out-of-scope SAP solutions, visit the following links:

- SAP LeanIX user documentation: *http://s-prs.co/v620630*
- SAP Cloud ALM user documentation: *http://s-prs.co/v620631*
- Reviewing discovered SAP Cloud ALM items: *http://s-prs.co/v620632*

Link

By clicking a table line item, a detail pane appears on the right; here, you can modify the linking of the fact sheets. To do that, hover over the fact sheet tile in the **Discovered Fact Sheets** section and click the **Edit** (pen) icon that appears when hovering over an item under **Discovered Fact Sheets** to enter edit mode (see Figure 4.13).

After you click the **Edit** icon, a text entry field appears, as shown in Figure 4.14, in which existing fact sheets can be selected or new fact sheets can be created on the fly, including the linking to the reference catalog if a matching entry exists.

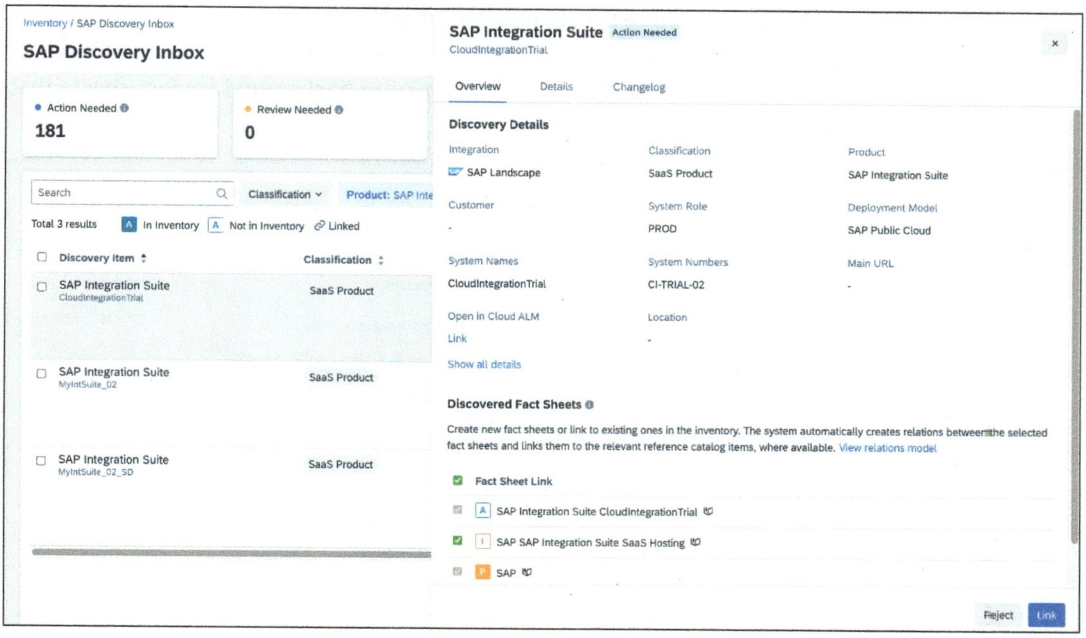

Figure 4.13: Modifying Application Linking Using Edit Functionality

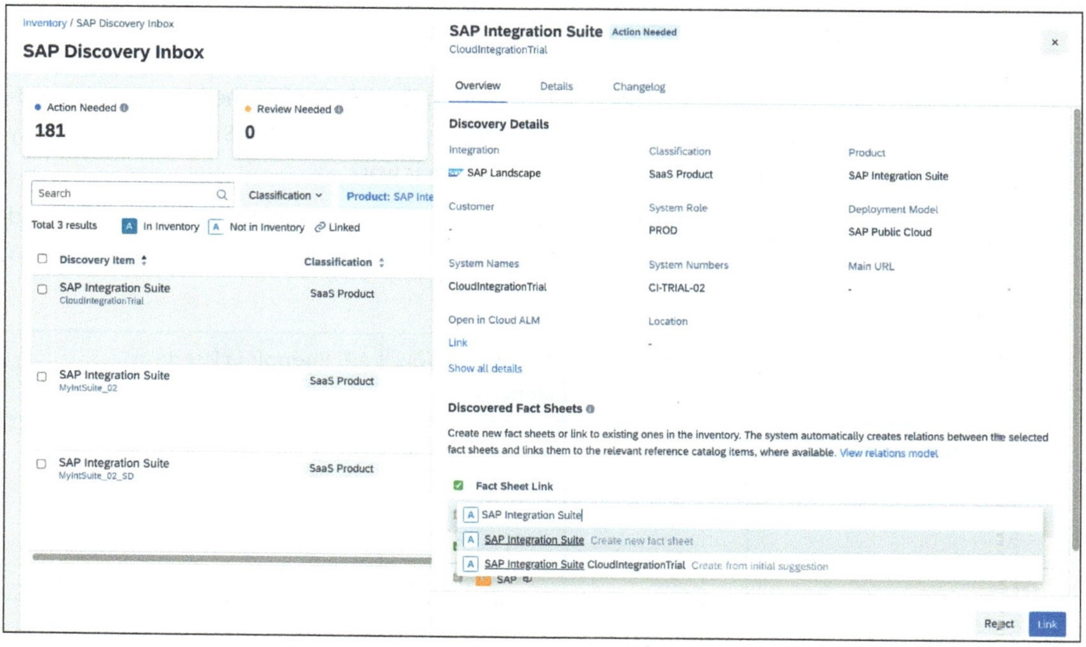

Figure 4.14: Creating Generic SAP Integration Suite Application Without System Indication in Its Name

Select the **SAP Integration Suite Create new fact sheet** option, then click the **Link** button. The status will change to **Linked**, and the indicator square in front of the name will change from a white background to the correct fact sheet color background for fact sheets that exist in the inventory workspace (see Figure 4.15).

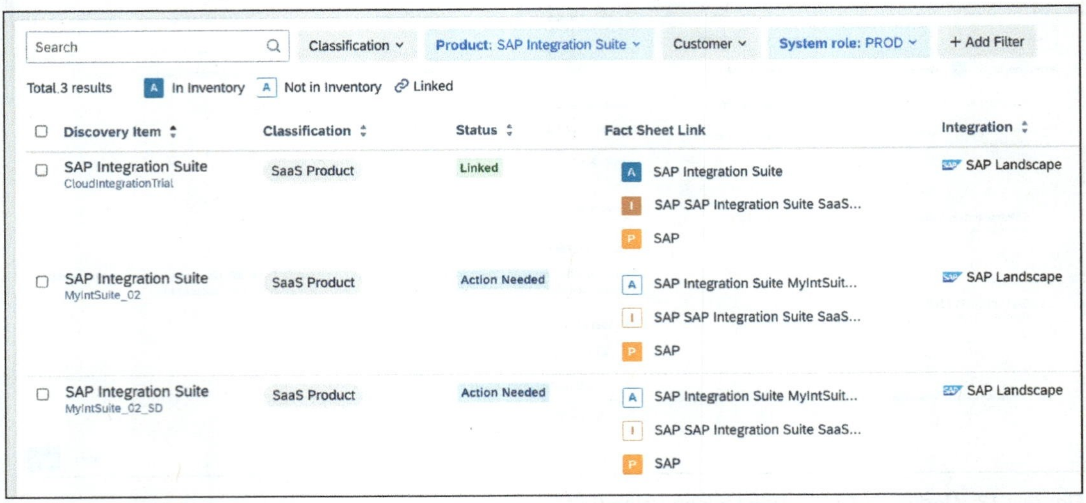

Figure 4.15: Linked Fact Sheet with Filled Icon in SAP Discovery Inbox

4.2.4 SaaS Discovery

You've now detected all active SAP solutions and created and linked the fact sheets accordingly. Next, you'll turn on SAP LeanIX SaaS discovery to find all SaaS applications of any vendor that are used by EA7's employees during their work.

SAP LeanIX offers several single sign-on (SSO), cloud access security broker (CASB), and secure access service edge (SASE) connections to discover SaaS solutions, as shown in Table 4.1.

Name	Integration Category	Authentication Method	API Endpoints Used
Microsoft Defender for Cloud Apps	CASB	REST API/OAuth	For API authentication: */<tenantID>/oauth2/v2.0/token* For SaaS discovery: */api/v1/discovery/discovered_apps*

Table 4.1: Overview of SaaS Discovery Sources

Name	Integration Category	Authentication Method	API Endpoints Used
Microsoft Entra ID	SSO	REST API/OAuth	For API authentication: */<tenantID>/oauth2/v2.0/token* For SaaS discovery: */servicePrincipals*
Netskope	CASB	REST API/OAuth	For API authentication: N/A For SaaS discovery: */api/v2/events/data/application*
Okta	SSO	API/OAuth	For API authentication: */oauth2/v1/token* For SaaS discovery: */api/v1/apps*
Zscaler	CASB	REST API/API token auth	For API authentication: */api/v1/authenticatedSession* For SaaS discovery: */api/v1/cloudApplications/lite* */api/v1/shadowIT/applications/export*
WalkMe	Add-on product	REST API/OAuth	For API authentication: */accounts/connect/token* For SaaS discovery: */public/v1/discovery/getCustomerAppsStats*

Table 4.1: Overview of SaaS Discovery Sources (Cont.)

To configure the integration, navigate to the **Administration** menu and go to the **DISCOVERY AND INTEGRATIONS** section, then the **Integrations** menu. Click the blue **Add Integration** button and filter for **SaaS Discovery** (see Figure 4.16).

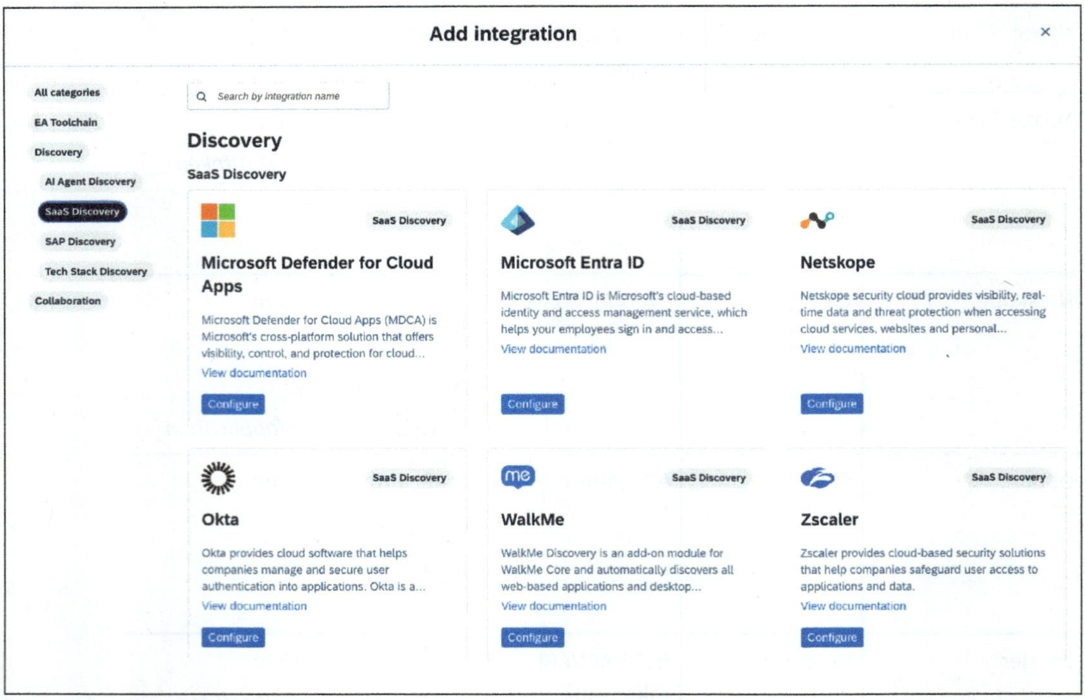

Figure 4.16: SaaS Discovery Source System Integration Configuration Overview in Administration Area

Select the tile for your preferred source system, then click the blue **Configure** button. A three-step wizard will guide you through the setup (see Figure 4.17).

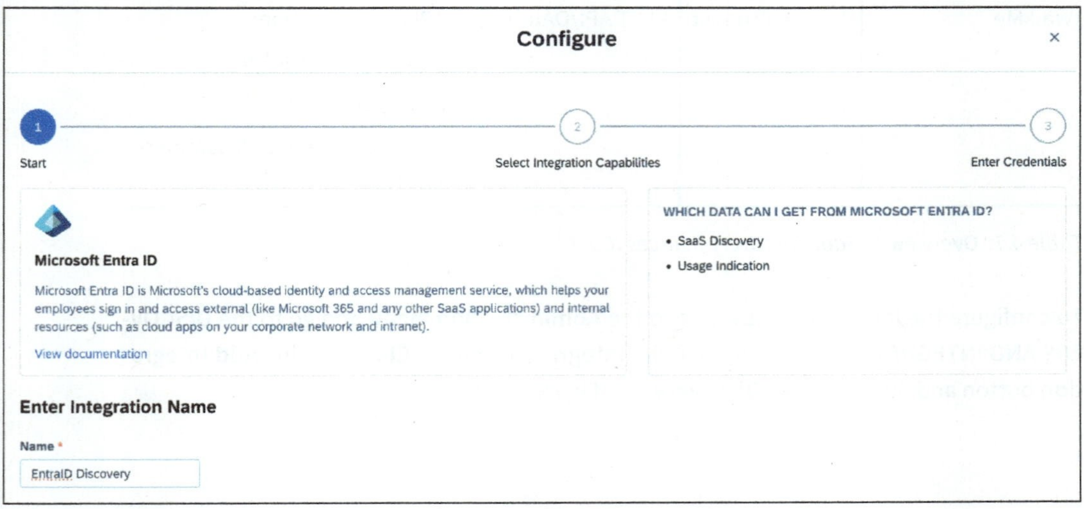

Figure 4.17: Three-Step Configuration Setup Wizard for SaaS Discovery

The integration connects via API to the source systems. Therefore, the source system admin must always create and enter some kind of credentials/token. In this example, the Microsoft Entra admin from EA7 successfully created the client secret for Microsoft Entra ID via the in-tool user guidance and shared it securely, including the directory and application ID, with the SAP LeanIX admin.

Further Resources

The steps needed to create the credentials/tokens vary slightly per system. Check the online SAP LeanIX user documentation at *http://s-prs.co/v620633* with the respective source system admin before scheduling meetings to set up the integration.

Input source systems might change over time (new ones added, previous ones deprecated). Therefore, always check which sources are available via the SAP LeanIX user documentation. For the latest information on SaaS discovery, visit the SAP LeanIX user documentation at *http://s-prs.co/v620634*.

By going to **Inventory • SaaS Discovery Inbox**, you can review the results, for both business SaaS and nonbusiness SaaS systems. Nonbusiness SaaS systems that are used by users but are unrelated to the core work can be rejected by the SAP LeanIX admins in the discovery inbox (e.g., Spotify), and business SaaS systems can be created and linked based on the recommendations from the SAP LeanIX application reference catalog. To do that, select the checkmark to the left of a line item and then click on the blue **Link** or the grey **Reject** button in the top-right corner, as shown in Figure 4.18.

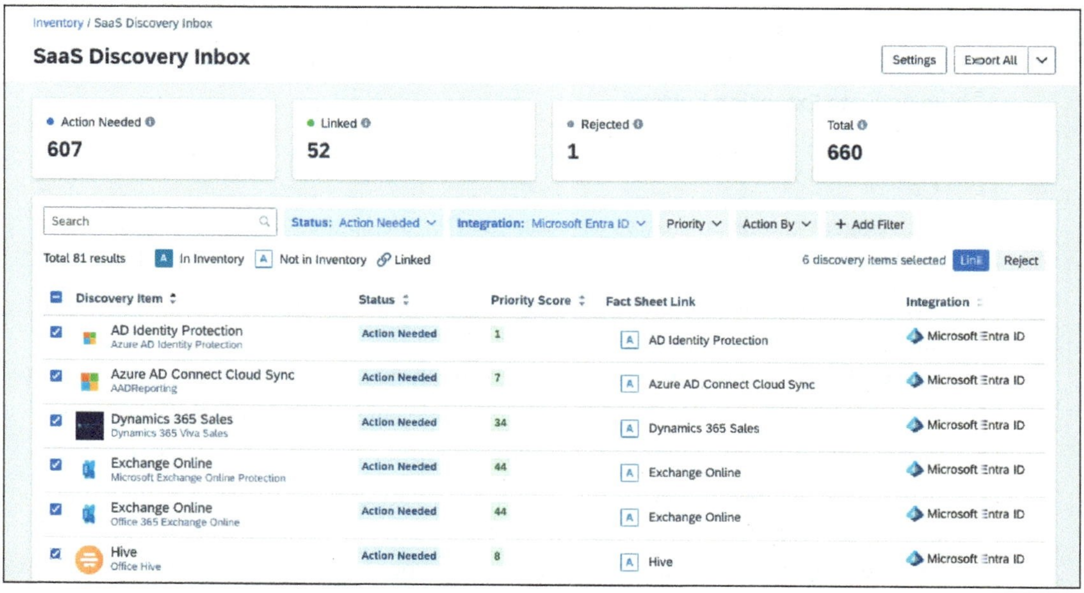

Figure 4.18: SaaS Discovery Inbox Example

Tip

It's likely that you'll find some shadow IT subscriptions from business users for business software that the IT department was not aware of. Document and share those findings with the purchasing department so that they can be taken into consideration in future renegotiations or rebundlings of contracts and licenses for the same vendor.

Linking to the same application will not create duplicates in the SAP LeanIX workspace. In the fact sheet, you can see that a logical application was connected to multiple SaaS discovery line items.

4.2.5 AI Agent Discovery

Finally, management often also wants to see all AI agents used. For that, activate the AI agent discovery, for instance, via Microsoft Entra ID with the Microsoft Entra admin.

Note

In the future, more sources for AI agents will be published. Therefore, always check the options in the SAP LeanIX user documentation on SAP Help Portal, or check the available tiles in the **Administration** area, under **Integrations**.

Navigate again to the **Administration** area and **Integrations**, and filter the overview to **AI Agent Discovery**. In the **Microsoft Entra ID** tile, click the blue **Configure** button. A window appears informing you that the meta model extension for AI agents needs to be activated (see Figure 4.19). Check the checkmark to agree to the meta model extension, then click the blue **Activate Extension** button.

Tip

The AI agent extension creates new subtypes for the application fact sheet: **Business Application** and **AI Agent**.

If you activate the extension after having already created application fact sheets in the inventory, those fact sheets will not automatically be mapped to the **Business Application** subtype. This needs to be corrected manually in a mass update (Excel export and import, or in the inventory table view in **Edit** mode) after the extension activation.

To avoid this extra work, it's best to activate all extensions that you plan to use at the beginning of your workspace setup before you create any fact sheets. Anything subsequently created by discovery capabilities or the reference catalog will then automatically be assigned to the correct application subtype. To learn how to configure the AI agent extension, see Chapter 13, Section 13.2.1.

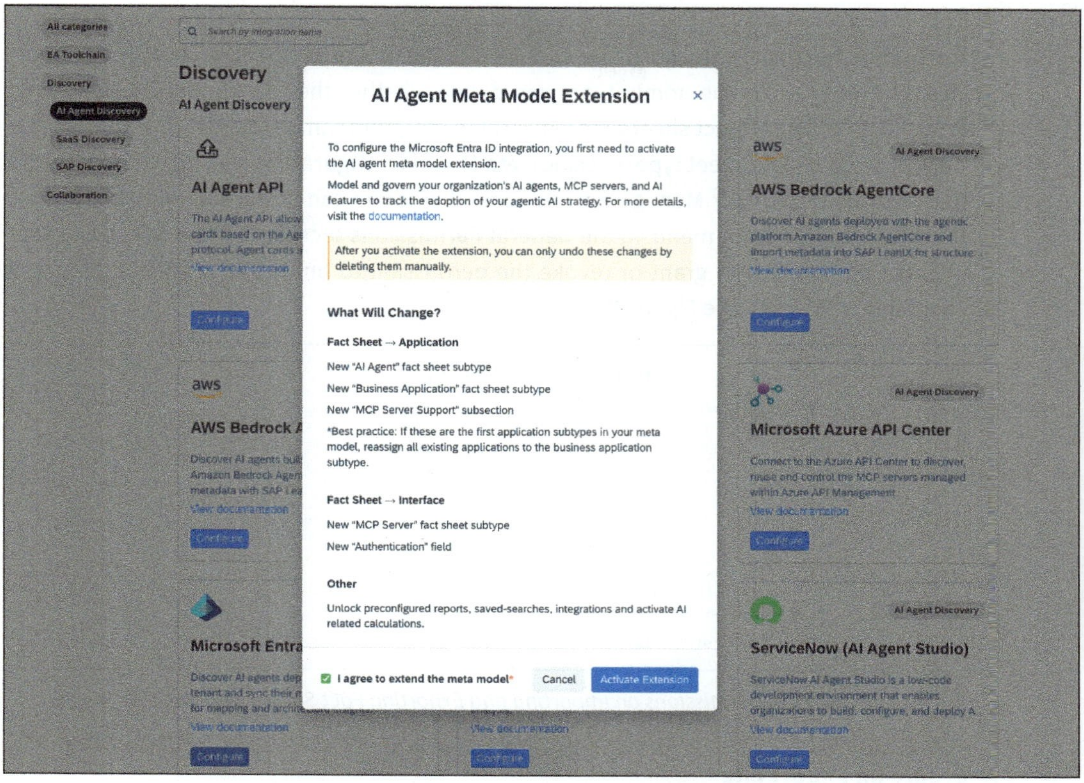

Figure 4.19: Administration Menu: AI Agent Meta Model Extension Activation

After activating the AI agent discovery, navigate back to **Inventory** by clicking the **Inventory** top ribbon menu item. From here, open the AI agent discovery inbox by clicking the **Inventory Tools** button, then hover over the **Discovery Inbox** menu item.

Like SAP discovery and SaaS discovery, SAP LeanIX admins can create fact sheets from the discovery inbox and link to SAP LeanIX reference catalog items. To learn more about ways to manage AI agents in SAP LeanIX, see Chapter 13.

4.2.6 Microsoft Excel Import

After activating SAP discovery, SaaS discovery, and AI agent discovery, you have already detected a sizeable amount of the EA7 application landscape without contacting anyone in the business organization. Now it's time to compare the loaded results with the previous Excel lists and create the remaining non-SAP on-premise or private cloud solutions, as well as self-built business applications, manually via Excel import.

In the following sections, we'll explain how to set up Microsoft Excel export files as templates for import in SAP LeanIX. We'll also discuss the three different modification scenarios that can be achieved via an Excel import: creating new fact sheets, updating them, and archiving them. We'll conclude by showcasing how to import the data into your workspace.

Set Up Permissions

Workspace admins are always allowed to download Excel exports and upload Excel import files. For other roles, the admin team can define whether the role should be allowed to export data or import fact sheet data via Excel. The setting can be found in the **Administration** menu in the fact sheet type–specific **Meta Model Configuration** area in the top ribbon's **Permissions** tab. Under **Manage Permissions**, the admin can select the **Permission Role** value from a dropdown menu. In the **General Permissions** section, the admin can check or uncheck checkboxes to grant or revoke the permission to **Import Fact Sheets** or **Export Fact Sheets to Excel** (see Figure 4.20).

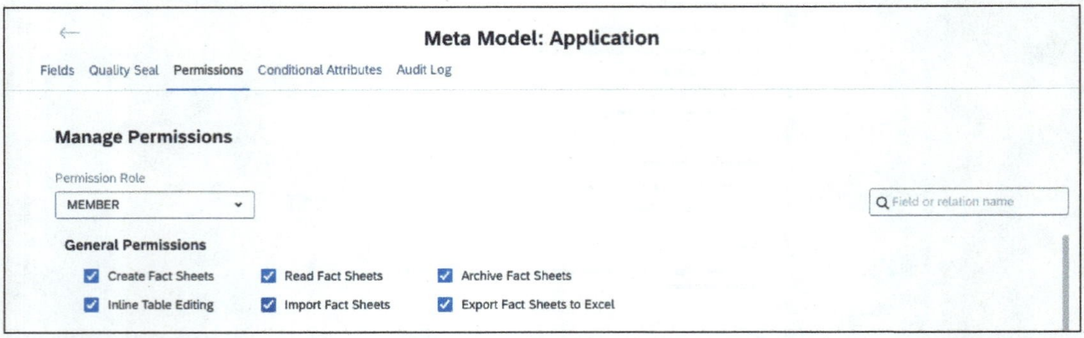

Figure 4.20: General Permissions on Importing and Exporting Fact Sheets via Excel

Create an Export File

To create an Excel import template, you must create an application export file first. To do that, navigate to the **Inventory** tab, filter for **Application**, and switch to the table view by clicking the grey **As table** icon in the upper-right corner, as shown in Figure 4.21.

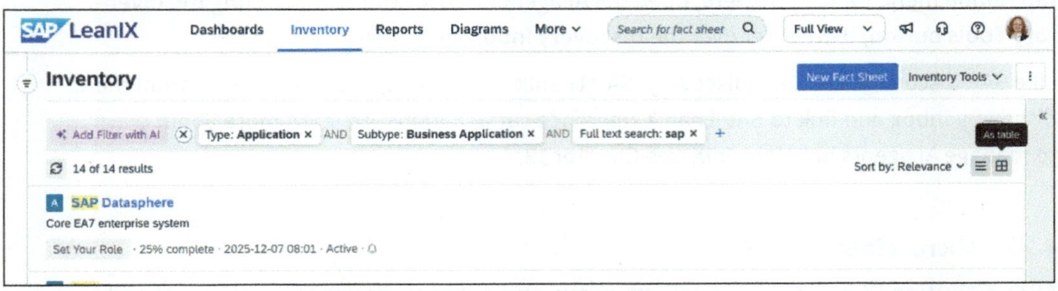

Figure 4.21: Switching Inventory to Table View

> **Excel Export and Import for Cross Fact Sheet Types**
>
> It's possible to export *mixed data*; that is, data from different fact sheet types. Just remember that you will only see the attributes, relations, and tags in the filters that are available for all fact sheet types.

If you want to add fact sheet type–specific columns, then you have to filter for that specific type and create the export for that subset. Mixing fact sheet types isn't possible in the import functionality, however. To import data via Excel, you must create import files for each individual fact sheet type.

Further Resources

The SAP LeanIX product and engineering team is working on a more intuitive UI with more support features to make Excel imports easier. Therefore, check the SAP LeanIX user documentation to learn more about the Excel import functionality and how it has evolved since the publication of this book: *http://s-prs.co/v620635.*

By default, you'll see the **Display Name** and **Description** fields in the overview. Because you created subtypes during the AI agent meta model extension activation in Section 4.2.5, you should make the **Subtype** field visible in the table.

To open the field menu, click the eye icon (**Select columns**) in the upper-right corner, as shown in Figure 4.22.

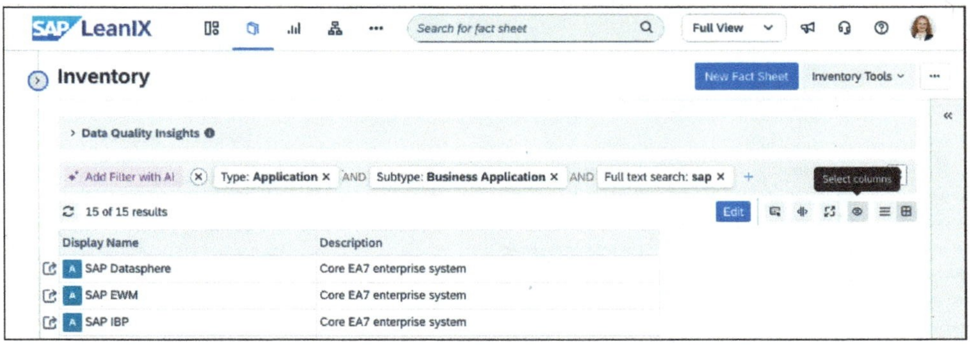

Figure 4.22: Open Select Columns Pane in Inventory Table View

The **Columns for Applications** side bar on the right appears, as shown in Figure 4.23. There you can see the **Active Columns** list, showing columns in white cells; these are visible in the table view. You can also see the **Inactive Columns** list, showing column names in grey cells. At the top of the **Inactive Columns** list, a search bar helps you find columns that will become visible in the table view. After you've searched for a column, to add it, click the **+** icon to the right of its name. To remove columns, click the **x** next to the name of the column you want to remove in the **Active Columns** list.

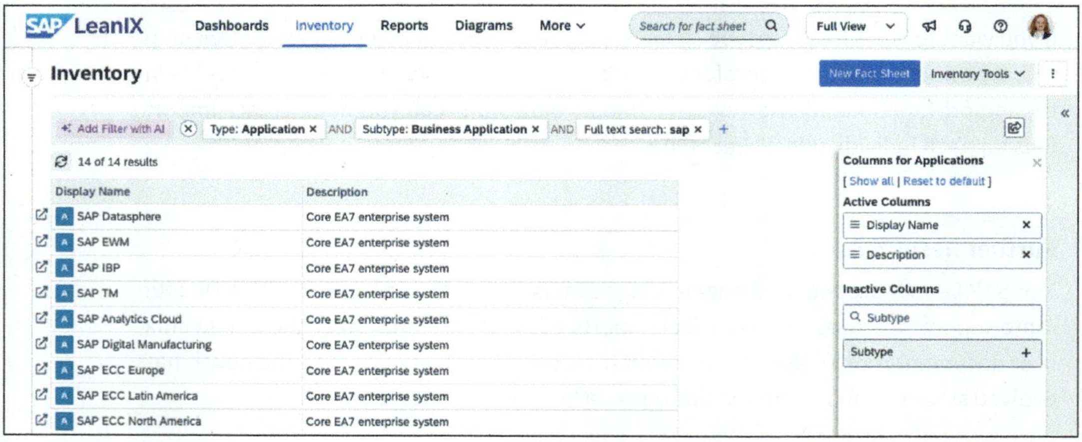

Figure 4.23: Changing Active and Inactive Columns in Column Selector Pane

You can then also change the layout of the table view in SAP LeanIX by dragging and dropping active columns to a different position in the top-down order in the side bar, which translates to a left-to-right order in the table (see Figure 4.24).

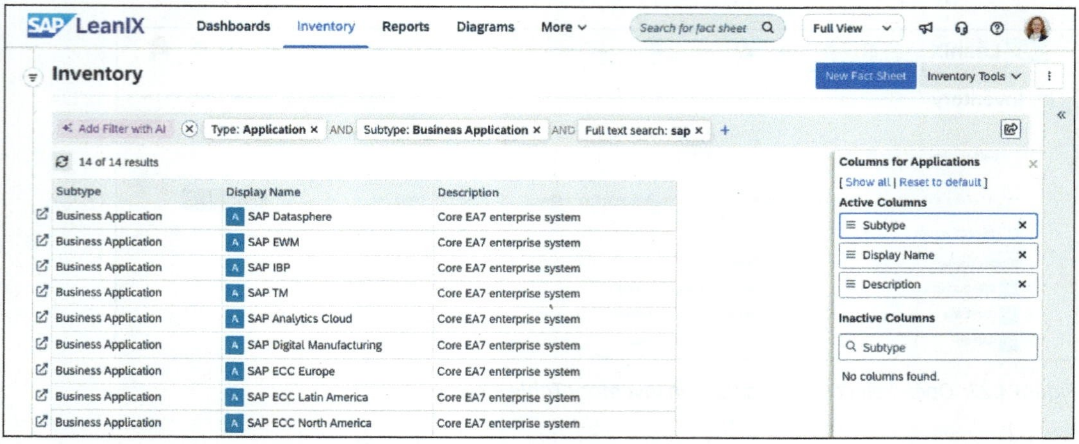

Figure 4.24: Changing Column Order in Inventory Table View

Once you're done with the column selection, click the **x** in the upper-right corner of the **Columns for Applications** side bar, then click the **Inventory Tools** button and select **Export to Excel** (see Figure 4.25).

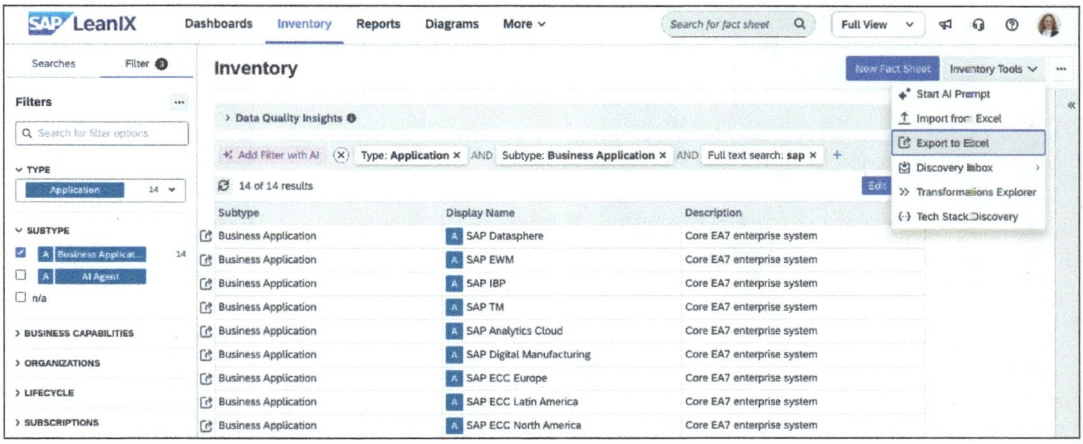

Figure 4.25: Starting Export of Data via Excel

A new popup appears in which you must confirm the export by clicking the blue **Export** button (see Figure 4.26).

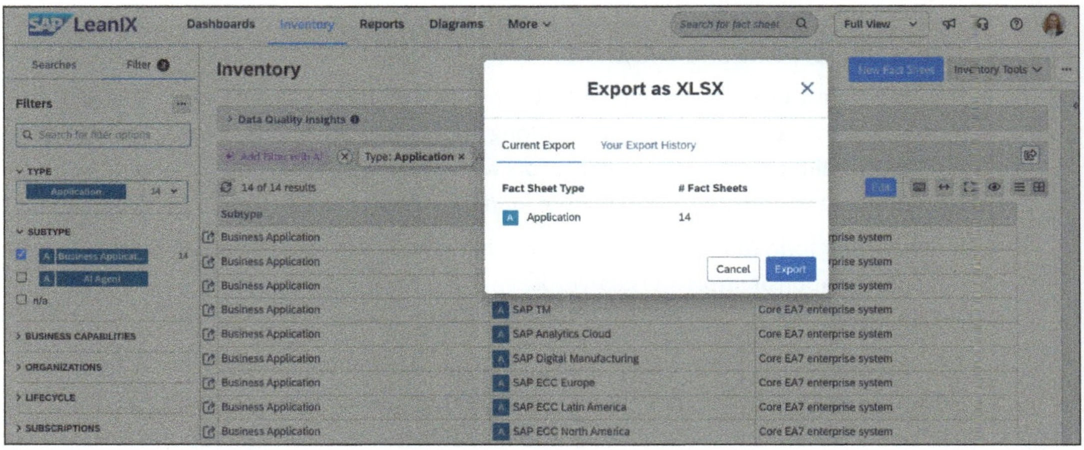

Figure 4.26: Confirming Transfer into Excel Tables

This will then trigger the formatting in the backend to change from the SAP LeanIX table format to the Microsoft Excel spreadsheet format. In the next step, the blue button changes into a grey **Download** button, which can be used to finally download the formatted Microsoft Excel file (see Figure 4.27).

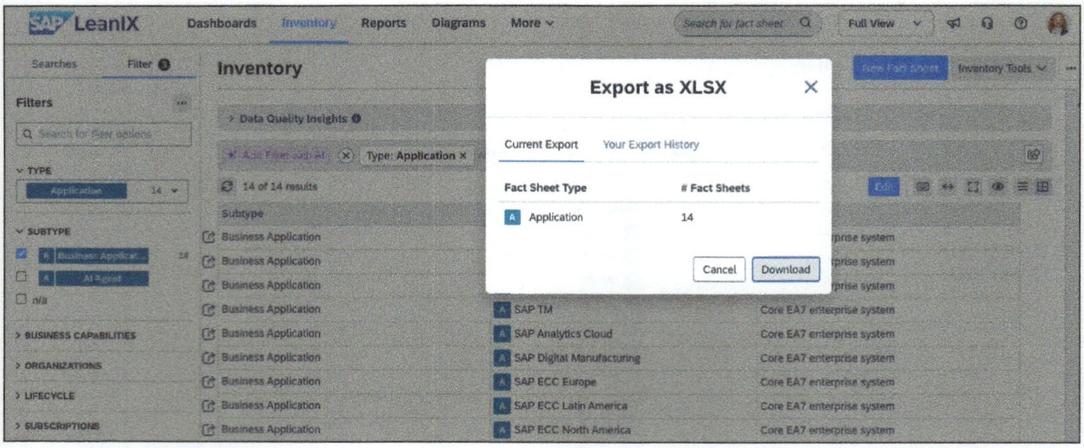

Figure 4.27: Downloading Excel Export

The Excel file can then be found in the defined download path, usually the default **Downloads** folder. The file consists of two worksheets: the **Export** sheet and the **ReadMe** sheet.

Excel Import Line and Cell Limits

As of the time of writing (summer 2026), the Excel import has a limit for data intake of 10,000 rows, plus character limits for string lengths in cells. If you experience problems with these limitations that result in errors with your data imports, you may want to use the SAP LeanIX Integration API feature for data intake.

For more information on the SAP LeanIX Integration API, check the SAP LeanIX user documentation at *http://s-prs.co/v620636*.

The export sheet has two table header rows. Row 1 contains the technical field name (known as the key in the meta model configuration), while row 2 contains the displayed field name (known as the translation/label in the meta model configuration).

The values of the fields can be found in the fact sheet configuration, as shown in Figure 4.28 (see the discussion of meta model configuration in Chapter 3, Section 3.3.3). Under the configurations slider icon on the right-hand side menu, you can find the **Key** field.

In the **Translations** tab with the globe icon, you can find the translation/label (entered in the white box) and the corresponding keys (headers in black font above the white boxes; see Figure 4.29).

For any field values in the table on the export sheet that are of type single or multiselect, the technical keys are used, not the labels.

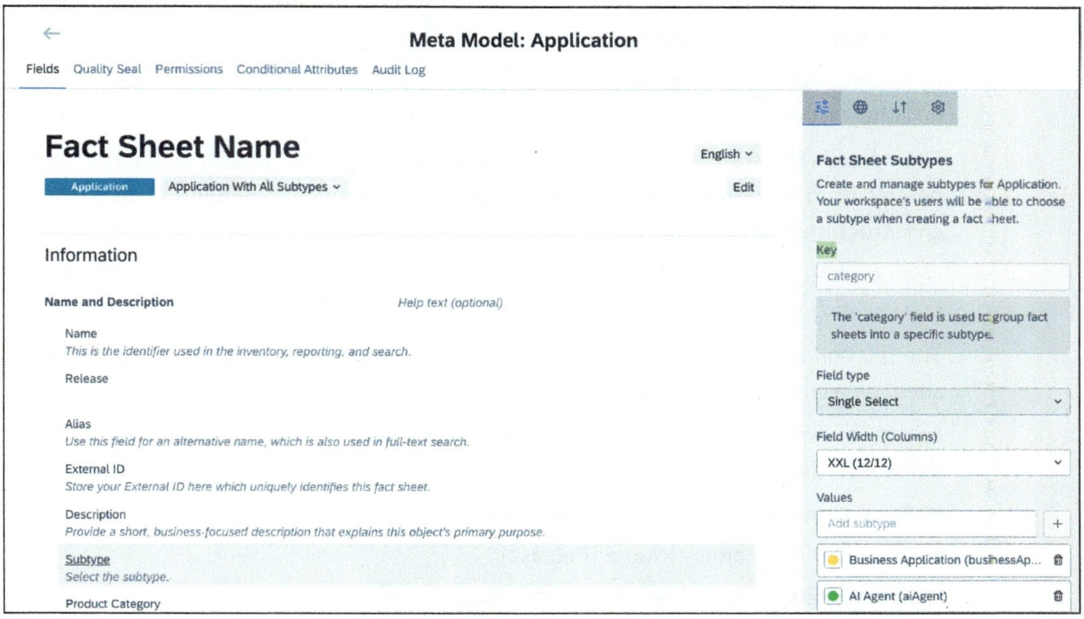

Figure 4.28: Finding Technical Keys in Fact Sheet Meta Model Configuration

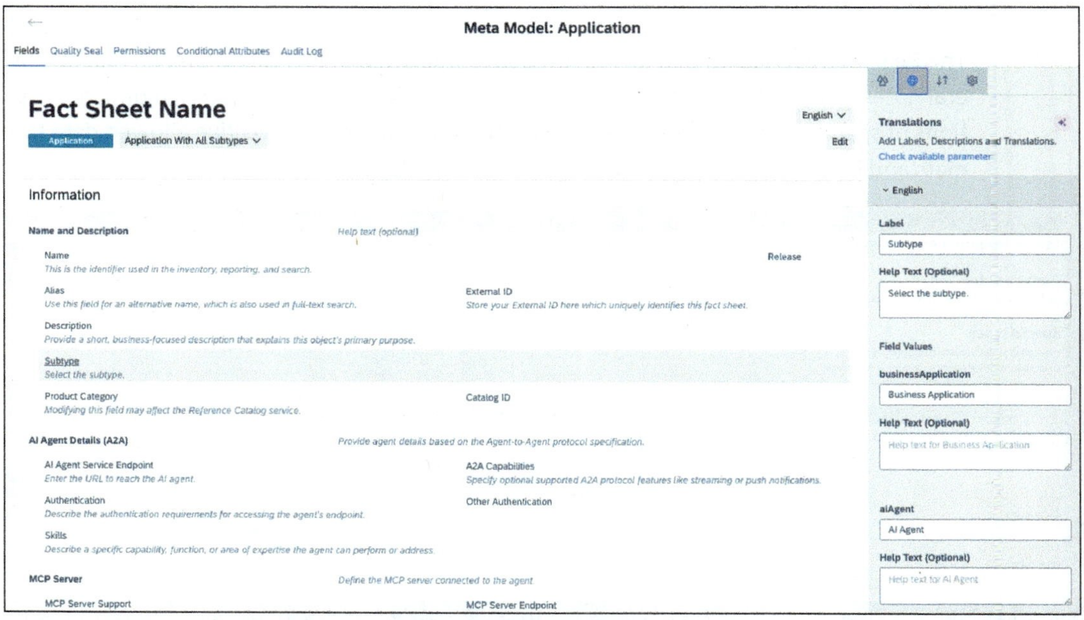

Figure 4.29: Finding Translations/Labels in Fact Sheet Meta Model Configuration

Name Versus Display Name

By default, the Microsoft Excel export file will include the name and the display name of fact sheets, even if you don't select their columns before you create the export file.

The *name* is the standalone name of the fact sheet, while the *display name* is a field derived from the parent-child relation and a concatenation of their names. The hierarchy structure is as follows:

<Name of the parent fact sheet> / <Name of subparent fact sheet> / <...> / <Full name of fact sheet>

The *full name* is also a derived field; it always includes the name, but for some fact sheet types it adds either prefixes or suffixes to it.

Special out-of-the-box fact sheet types for which the full name is not identical to the name include the following:

- Application: Full Name = [Name][Release]
- Interface: Full Name = [Name][Release]
- IT component: Full Name = [Provider][Name][Release]

For more information, check the SAP LeanIX user documentation at *http://s-prs.co/v620637.*

The ReadMe sheet, as shown in Figure 4.30, contains valuable information, including general notes, such as forbidden characters in names (< [less than] and *;* [semicolon]); descriptions of special types; and a list of the data columns and their respective allowed data formats and inputs (technical keys instead of displayed labels).

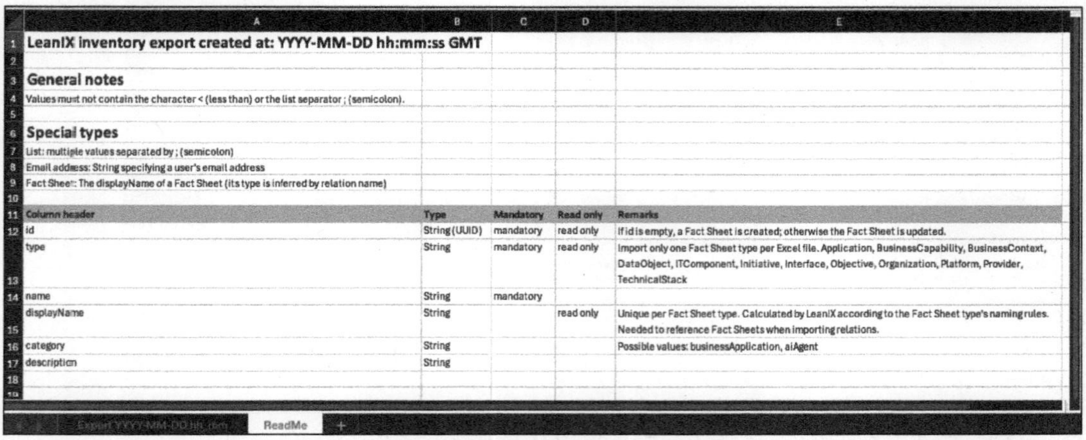

	A	B	C	D	E
1	**LeanIX inventory export created at: YYYY-MM-DD hh:mm:ss GMT**				
2					
3	**General notes**				
4	Values must not contain the character < (less than) or the list separator ; (semicolon).				
5					
6	**Special types**				
7	List: multiple values separated by ; (semicolon)				
8	Email address: String specifying a user's email address				
9	Fact Sheet: The displayName of a Fact Sheet (its type is inferred by relation name)				
10					
11	Column header	Type	Mandatory	Read only	Remarks
12	id	String (UUID)	mandatory	read only	If id is empty, a Fact Sheet is created; otherwise the Fact Sheet is updated.
13	type	String	mandatory	read only	Import only one Fact Sheet type per Excel file. Application, BusinessCapability, BusinessContext, DataObject, ITComponent, Initiative, Interface, Objective, Organization, Platform, Provider, TechnicalStack
14	name	String	mandatory		
15	displayName	String		read only	Unique per Fact Sheet type. Calculated by LeanIX according to the Fact Sheet type's naming rules. Needed to reference Fact Sheets when importing relations.
16	category	String			Possible values: businessApplication, aiAgent
17	description	String			
18					

Figure 4.30: ReadMe Worksheet in Excel Export

Create New Fact Sheets

To create new fact sheets via the Excel import functionality, add rows to the export sheet in Excel with an empty ID column. As noted in the ReadMe sheet, you should provide at least the mandatory information. In this simplified example, that means the type (fact sheet type: application) and the name.

If the fact sheet is not part of a parent-child hierarchy, the display name equals the name column. Because it isn't mandatory information, you can delete the display name column if desired.

Don't forget to save your changes in the Excel spreadsheet!

Tip

The **Display Name** is a derived field in SAP LeanIX. It provides context to show at which level in a parent-child hierarchy the fact sheet is located.

For the following scenarios, the user should know and use the **Display Name** of the relevant fact sheet:

- **Adding the parent fact sheet**
  Enter the **Display Name** of the parent fact sheet into the column of the parent-child relation (**relToParent**).
- **Adding or updating relations**
  Enter the **Display Name** of the fact sheet to be related in the respective column of the relation. If you want to enter a list of fact sheets, create a semicolon-separated list without any leading or tailing spaces in the display names of the fact sheets that you want to be related.

The **Display Name** column can be used as a lookup table while adding relations to fact sheets.

Remember, if you leave the **Display Name** column in place during the import process, then the **Display Name** column needs to be adjusted if the parent-child hierarchy changes. Row order matters. Therefore, for the import, ensure that the correct sorting is applied. The parent needs to be created first before you can maintain a parent-child relation to a child fact sheet. Therefore, if you use the **Display Name** column, sort it in **Ascending** order to have the parent before the child.

If you want to mass-archive fact sheets, the logic flows the other way around. You should archive the children before you can archive the parent; for this use case, that means sorting the **Display Name** column in **Descending** order.

Update Fact Sheet Data

To update fact sheet data, you can simply overwrite the value of the cell with the new value. Remember to use the technical keys for any single or multiselect fields or tags. To update relations, you must use the **Display Name** of the corresponding fact sheet. If you want to add a list of related fact sheets, the **Display Names** must be concatenated with ; (a semicolon), without leading or tailing spaces.

Don't forget to save your changes in the Excel spreadsheet!

Archive Fact Sheets

To (mass) archive fact sheets, add a column with "action" as the technical key in the first row and "Action" as the translation in the second row. Populate the action column cells with "archive" as the value for each fact sheet to be archived.

Don't forget to save your changes in the Excel spreadsheet!

The archived fact sheets will remain for 90 days in the so-called trash bin and can be recovered during that time if needed.

To navigate to the trash bin, go to **Inventory**. There, click the button with three dots (...) next to the **Filters** title in the left-hand menu. Finally, click the **Manage Filters** menu item, as shown in Figure 4.31.

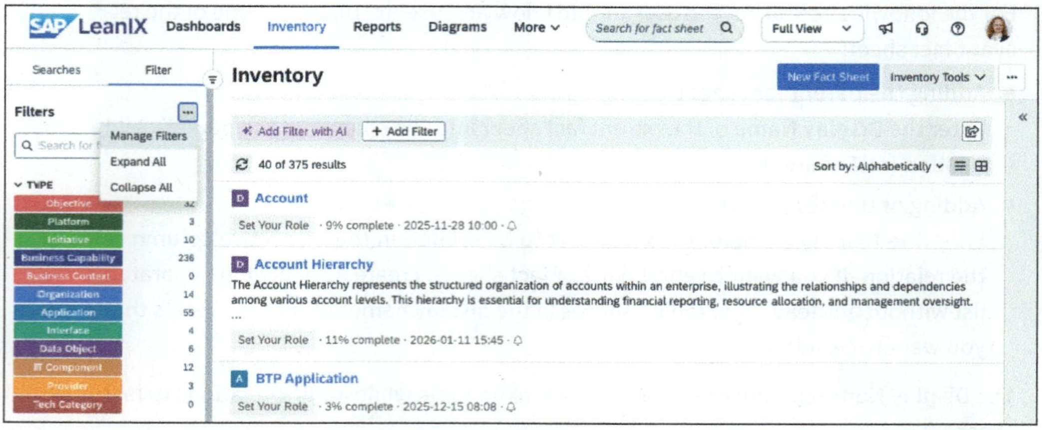

Figure 4.31: Switching to Manage Filters View to Unhide the Trash Bin

Enter "Trash Bin" in the **Search** bar. Under the title **Hidden**, hover over the cell called **Trash Bin** and click the + icon to make it visible in the filter pane. Then click the blue **Back** button in the upper-left corner of the left-hand menu (see Figure 4.32).

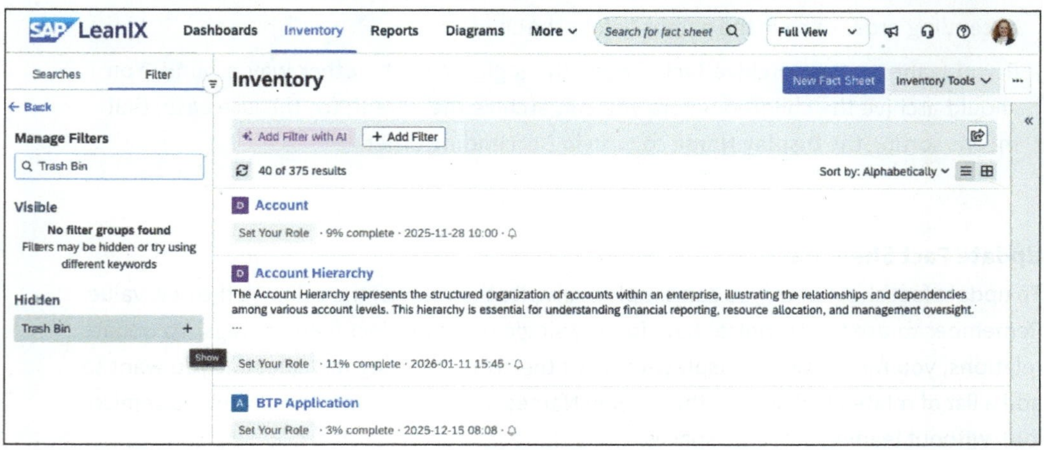

Figure 4.32: Unhiding Trash Bin in Filters

In the filter pane, scroll to the new section labeled **Trash Bin** and check the checkbox next to the value **Archived**. The inventory now shows all archived fact sheets that are not yet permanently deleted from the workspace (see Figure 4.33).

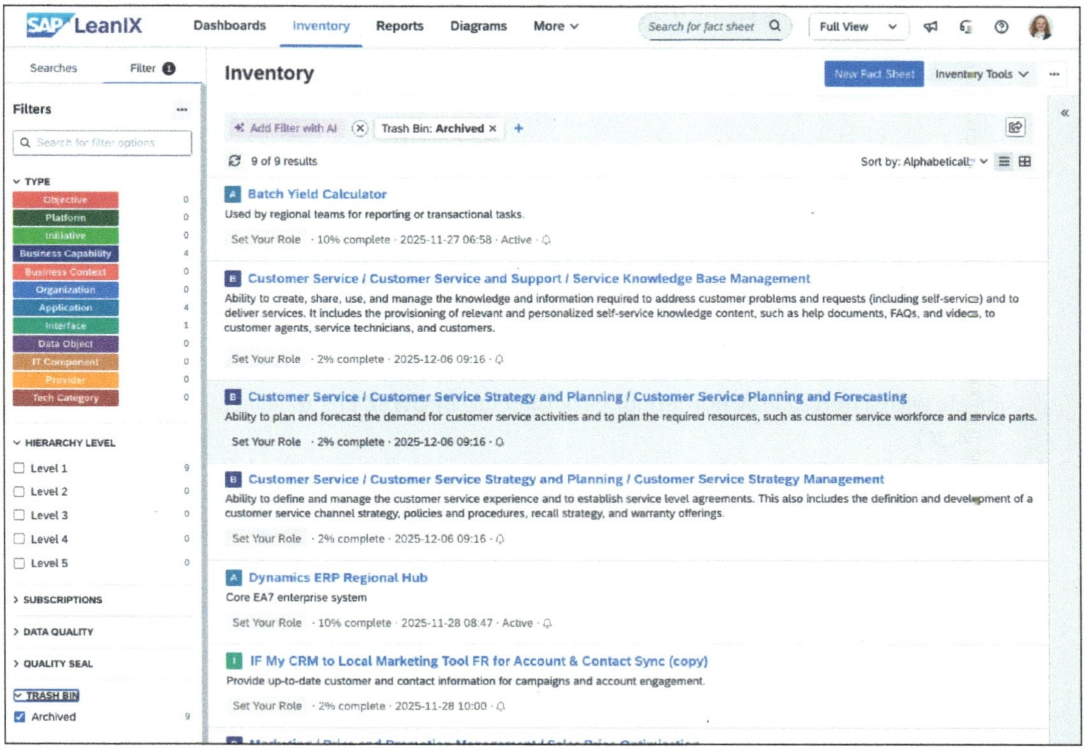

Figure 4.33: Filter for Archived Fact Sheets

To recover fact sheets that were archived but shouldn't have been, you have to open the fact sheet in the filtered inventory, click the **...** button in the upper-right corner, and select the **Recover** menu item (see Figure 4.34). This action will recover not only the fact sheet with the information maintained at its time of archival, but also the relations, tags, and so on.

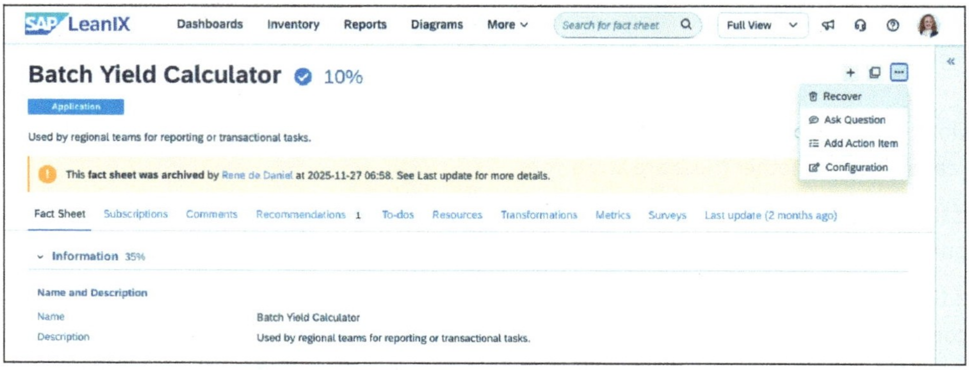

Figure 4.34: Recovering Mistakenly Archived Fact Sheets

Import Excel Data into Your Workspace

Now that the Excel sheet is prepared as an import sheet, you're ready for the upload.

Navigate to the **Inventory** tab and expand the **Inventory Tools** button in the upper-right corner, as shown in Figure 4.35. Then click **Import from Excel**.

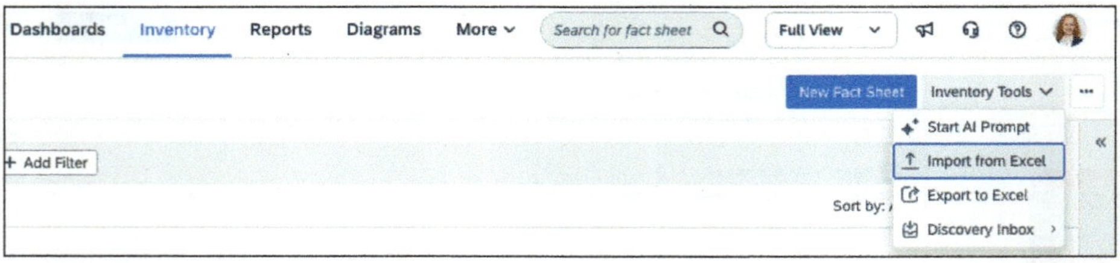

Figure 4.35: Navigating to Excel Import Form

A popup window appears. You can either drag and drop the file into the drop area (rectangle with a dotted grey border), or click the blue **BROWSE** link to select your file (see Figure 4.36).

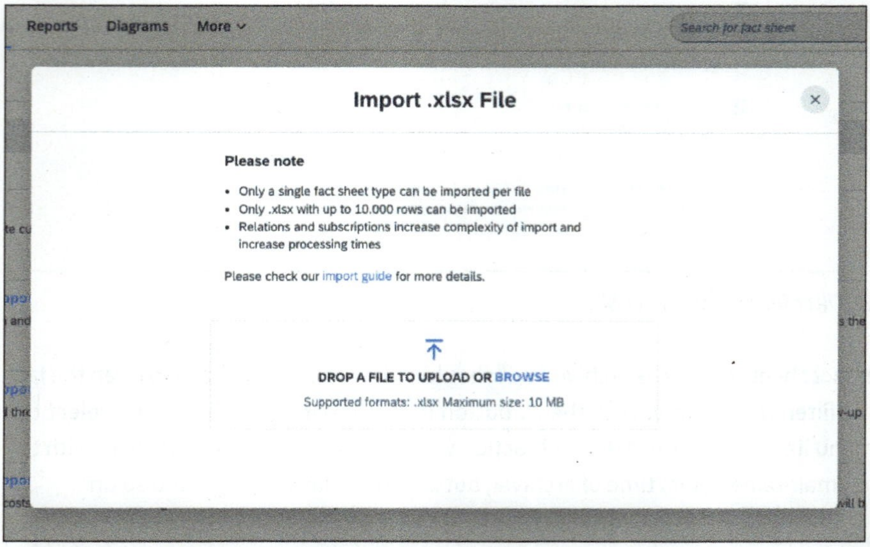

Figure 4.36: Excel File Upload Form

Once selected and uploaded, the import feature will run a test to determine if the data is readable and whether there are any errors.

If the result of the assessment is that there are zero errors, then you will see the blue **Import** button and can finalize the import (see Figure 4.37).

If there is an error in the import file, as shown in the example in Figure 4.38, then the error count will be bigger than zero and the blue **Import** button cannot be selected. Instead, you must click the **Download Summary Report** button to download an Excel analysis file, which will inform you about the error details.

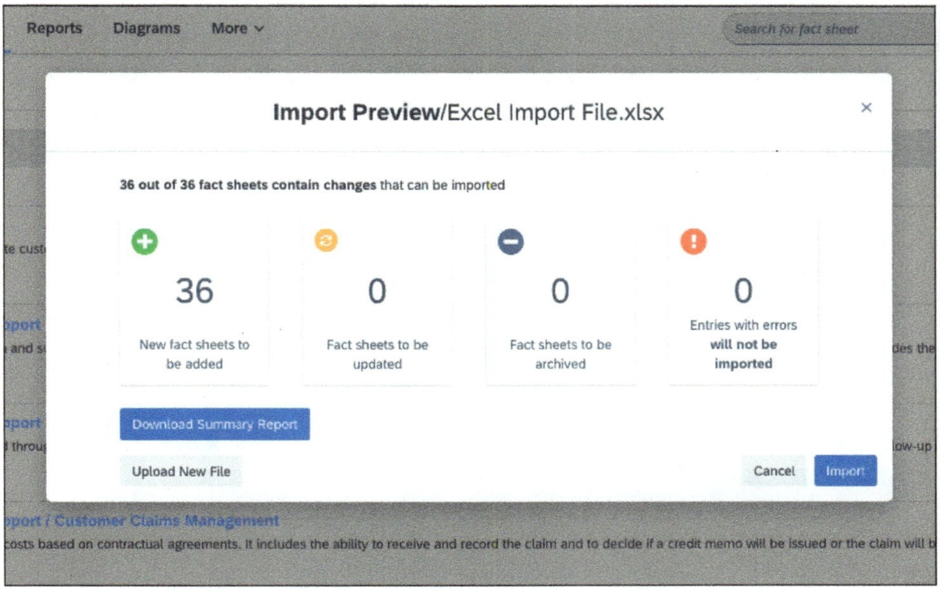

Figure 4.37: Error-Free Import File Analysis

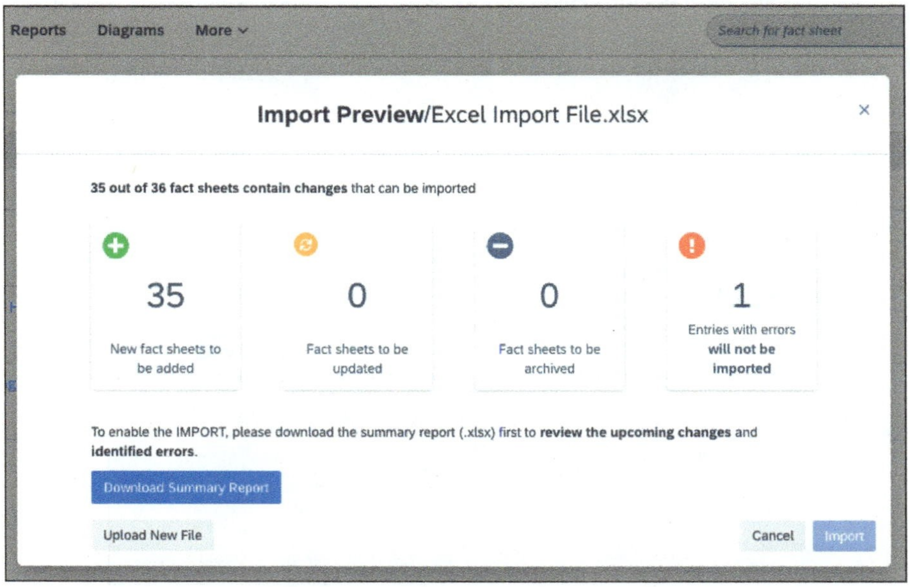

Figure 4.38: Import File Analysis with Errors

Once errors have been analyzed, you can decide whether you want to cancel the import by clicking the grey **Cancel** button, correct the import Excel file and restart the import process, or import only those fact sheets that contain no errors and correct the errors later individually. To do the latter, click the blue **Import** button to import the error-free fact sheets (see Figure 4.39).

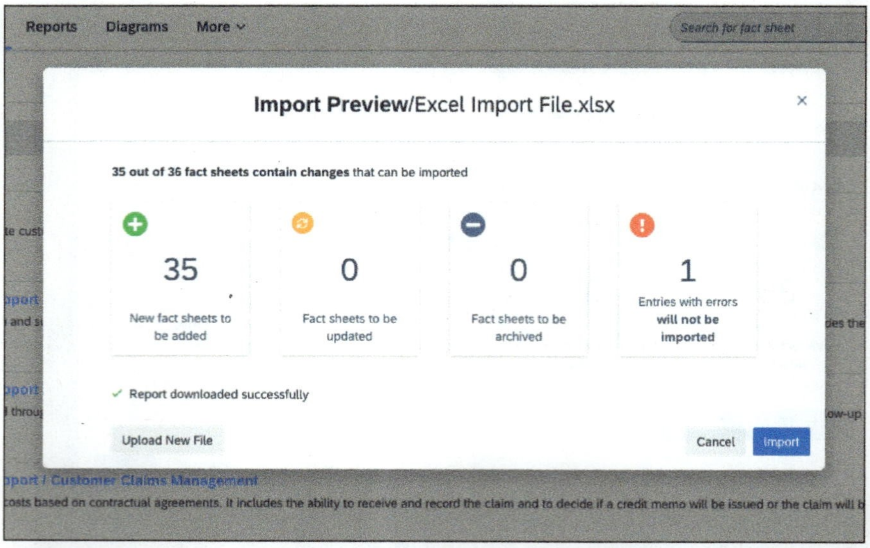

Figure 4.39: Importing Error-Free Fact Sheets and Omitting Entries with Errors

After processing completes, you'll see a notification indicating how many fact sheets were changed successfully. Click the blue **OK** button to close the popup window (see Figure 4.40).

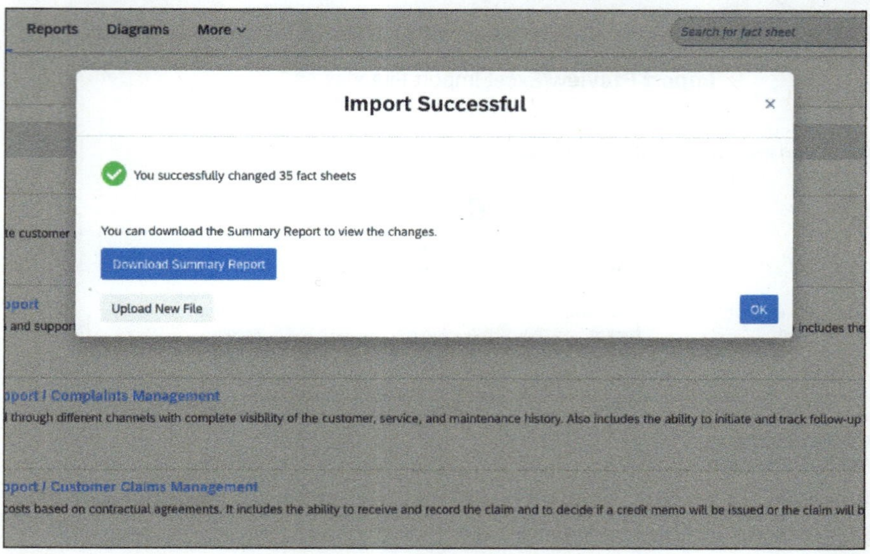

Figure 4.40: Notification of Successful Excel Import

Tip

There are a few things that you can double-check before uploading the import file to prevent errors. Examples that result in errors include the following (this is not an exhaustive list):

- Duplicates in display names
- Forbidden characters in names
- Duplicates in fields that are formatted as external IDs
- Using translations/labels instead of technical key values
- Using names instead of display names to create relations
- Adding leading or tailing spaces or spaces after the list item-separating semicolon (;)
- Mixed fact sheet types in one Excel import file
- Missing mandatory field column (ID, type, name; ID can be blank if new fact sheets are to be created)
- Import worksheet with data not the first worksheet in the import Excel file
- Read-only field columns included and changed that cannot be updated manually
- Date formats not entered in the format yyyy-mm-dd
- Entering a decimal when the field format is set to integer
- Using the comma instead of the period to write decimals (e.g., 1.7 is correct; 1,7 is wrong)

Further Resources

You can find the latest information on this feature on the SAP LeanIX user documentation online: *http://s-prs.co/v620638*.

4.2.7 Manually Creating a Fact Sheet

It's also possible to create fact sheets manually in the UI. Navigate to **Inventory** by clicking the respective top ribbon menu item. Then click the blue **New Fact Sheet** button in the upper-right corner (see Figure 4.41).

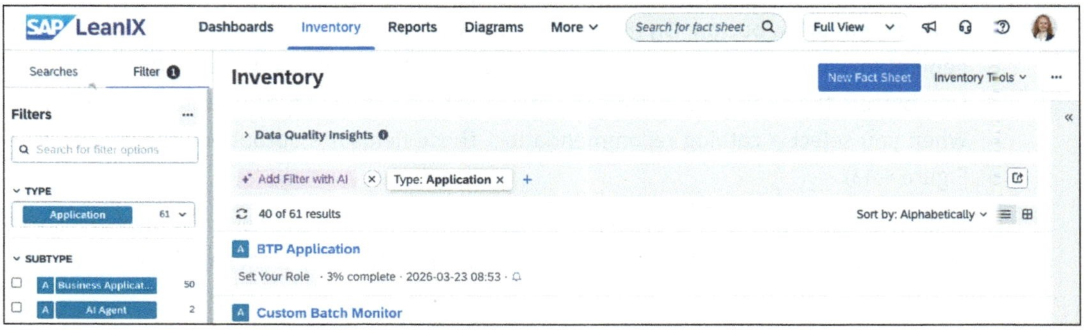

Figure 4.41: Creating Fact Sheet Manually from Inventory

An interactive form will appear. Enter the fact sheet that you want to create in the **Enter Fact Sheet Name** field. The form then searches the curated in-tool application and IT component catalog and suggests options under **Start from a recommendation**, listing similar entries from the catalog. If the artifact tile under the recommendation section has no fill color, this means that the fact sheet doesn't exist in your inventory yet. If the tile is filled with a fact sheet type color, then the fact sheet already exists in your inventory. Further down in the form, you can also see **Existing fact sheets in your inventory** with similar names (see Figure 4.42).

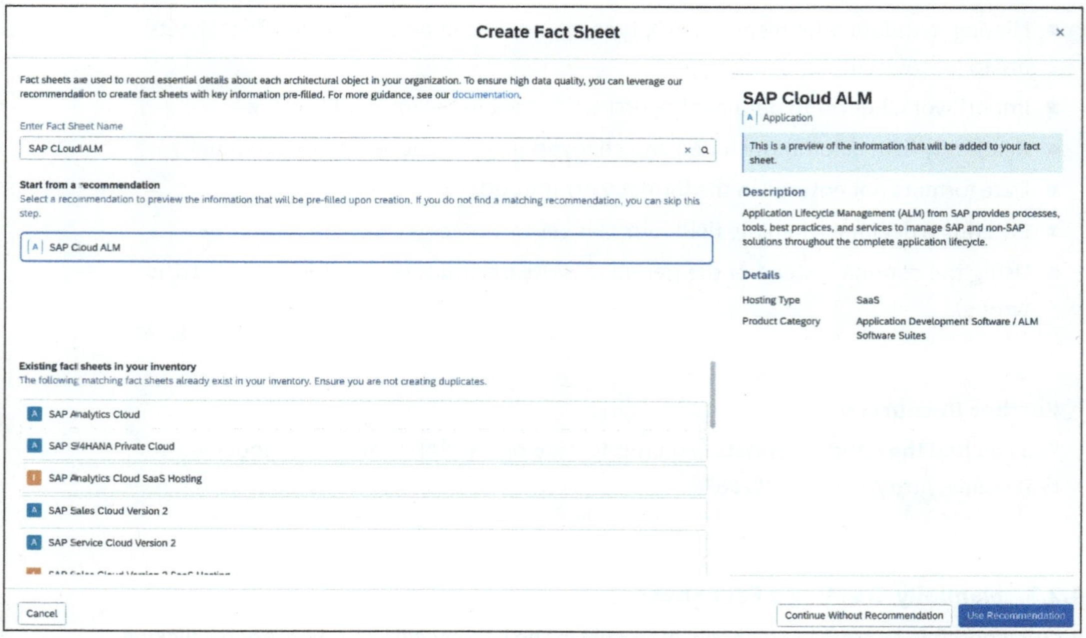

Figure 4.42: First Page of Create Fact Sheet Manually Form

In this example, click the recommended SAP Cloud ALM entry and then click the blue **Use Recommendation** button. On the next screen, you can then enter fact sheet type–specific general information. If you have previously chosen the **Continue Without Recommendation** button, then you would select the fact sheet type and subtype in the second window. When you select a catalog recommendation, those fields are autopopulated for you (see Figure 4.43).

Finally, either click the blue **Create and Open** button to open the newly created fact sheet or click the white **Create and Start Another** button to manually create another fact sheet.

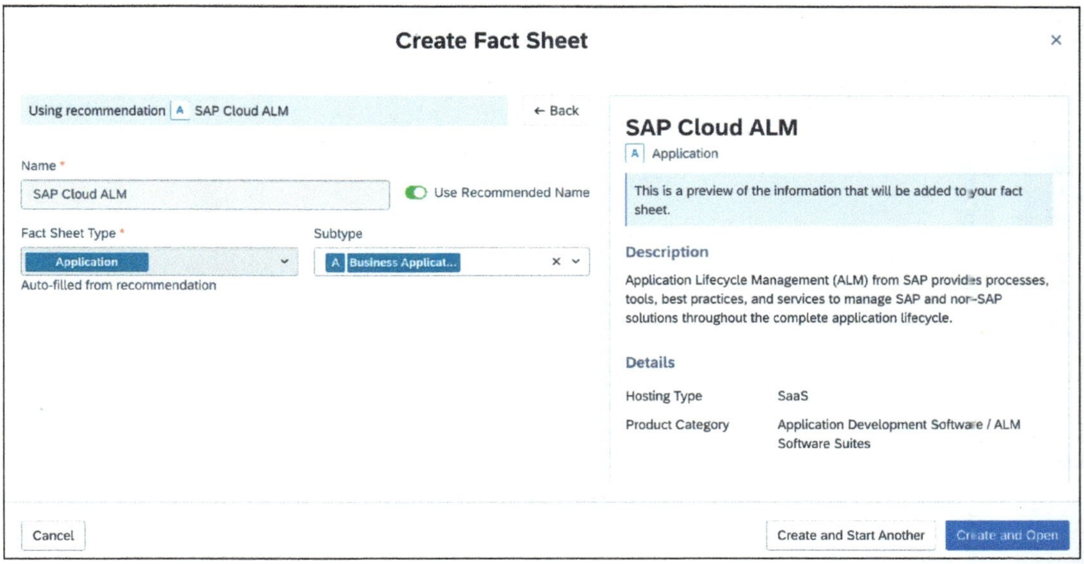

Figure 4.43: Entering General Information in Fact Sheet Creation Form

4.3 Linking, Updating, and Enriching Fact Sheets

Now that you've started to load data into your inventory, it's time to also think about tracking your data quality progress for the initial application portfolio assessment use case. A good in-tool help option to assess your workspace health and artifact completeness is the out-of-the-box **Onboarding: Application Portfolio Management for Enterprise Architects** dashboard. You can follow along with the in-tool guidance and selected KPIs to improve data completeness and quality in your inventory.

To navigate to the dashboard, click **Dashboards** in the top ribbon, then click the little arrow in the left-hand menu separator line. In the dashboard menu that appears, select the **All Dashboards** button under the **DASHBOARD OVERVIEW** menu header. In the main center menu, various tiles with out-of-the-box dashboards appear, including the **Onboarding: Application Portfolio Management for Enterprise Architects** dashboard (see Figure 4.44). To open it, click its tile.

The dashboard offers some best practice metrics and additional resources for the initial start with SAP LeanIX and the application portfolio assessment use case (see Figure 4.45).

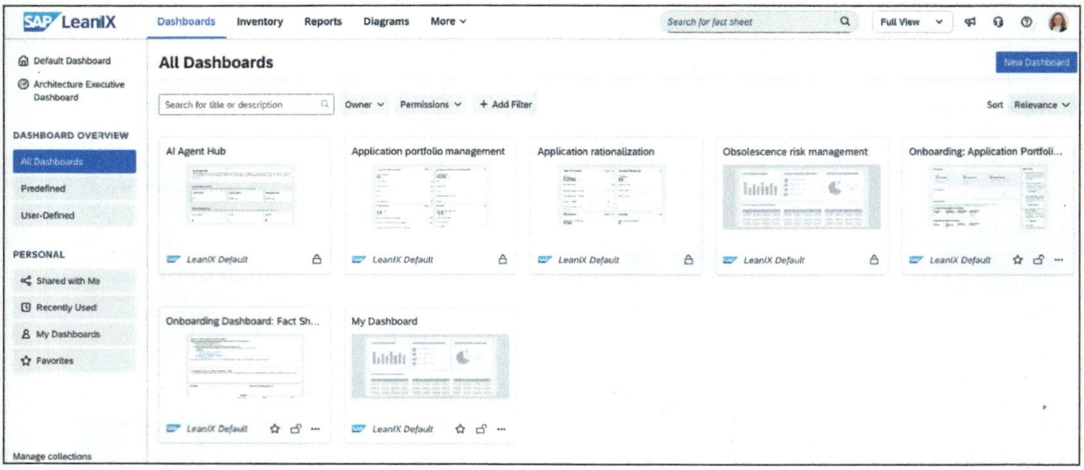

Figure 4.44: All Dashboards Overview in Workspace

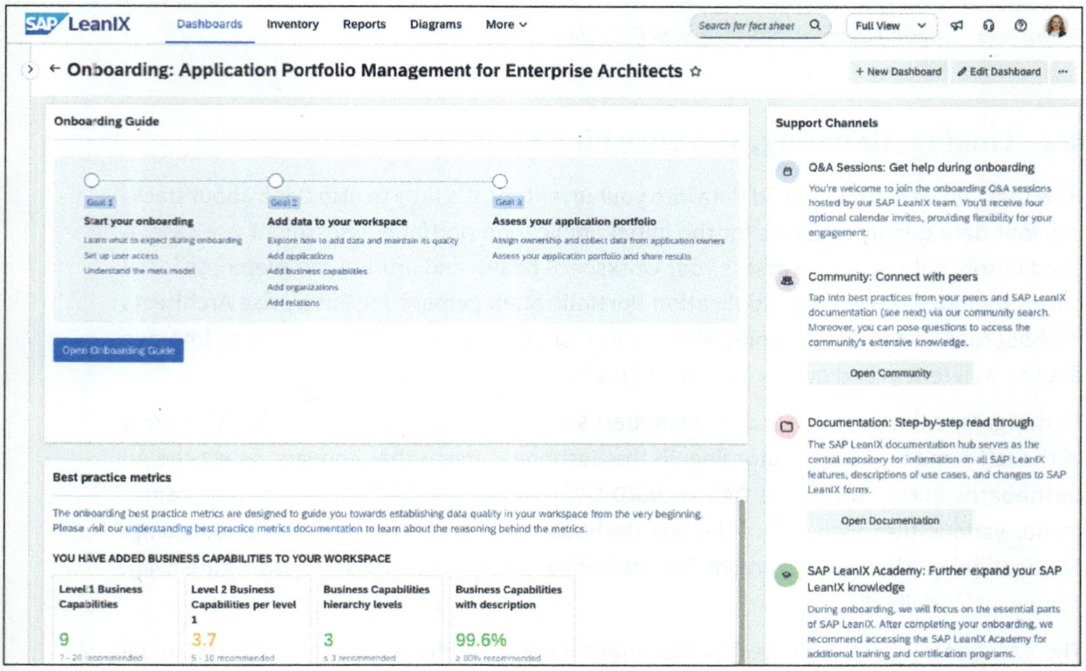

Figure 4.45: Onboarding Dashboard with Additional Resources and Out-of-the-Box Fact Sheet Quality KPIs

As a first internal goal, the EA7 enterprise architecture team decides to focus on business capabilities and applications. We have already loaded the newly defined business capability map (see Section 4.2.1), a solid baseline of applications, and some IT components that came automatically from the discovery capabilities and the linking to the reference catalogs. One KPI from the onboarding dashboard is called **Applications* linked to business capabilities**. EA7 wants to bring this KPI to 100% in this quarter. They also want to increase the data quality in the application fact sheet by entering values for the following metrics:

- **Business criticality**
  Rough assessment for the appropriate service level and disaster recovery requirements
- **Functional fit**
  Rough assessment of how well the application meets the business requirements today and in the near future (next two years)
- **Application lifecycle**
  Date when the application was rolled out (active date) and for how long it's planned to be used in the organization (phase-out or end-of-life date)

The next task for the EA7 enterprise architecture team is clear: connecting the applications to the business capabilities they support. How can this be done the most efficiently and accurately? Let's explore this task in the following sections.

4.3.1 Linking Applications to Business Capabilities

The applications need to be mapped to the correct business capabilities. Ideally, this step is done with or by the application owners as they know how and for what the application is used.

You can either start from scratch and let the app owners pick the right business capabilities from the inventory, or the enterprise architecture team can enter a suggestion based on SAP LeanIX workspace recommendations and let the application owner confirm the selection, such as via an SAP LeanIX survey.

When business capabilities are a known concept in your organization and app owners are trained in enterprise architecture basic modeling principles, then the first approach can bring good results. If enterprise architecture is new or a niche discipline in your organization, then the second approach might bring more complete and accurate data as app owners will be guided to review instead of being overwhelmed with a blank box to be filled at the start.

Let's take a closer look at workspace recommendations. In the **Search for fact sheet** field in the top ribbon, enter **SAP Cloud ALM** and click the suggested application line item in the search results (see Figure 4.46).

The **Recommendations** tab, as shown in Figure 4.47, offers help to increase the data quality of fact sheets. The recommendations are targeted to specific fact sheet types. When you open the recommendations on an application fact sheet, you will see suitable business capabilities recommended based on semantic analysis of the **Application Description** field data.

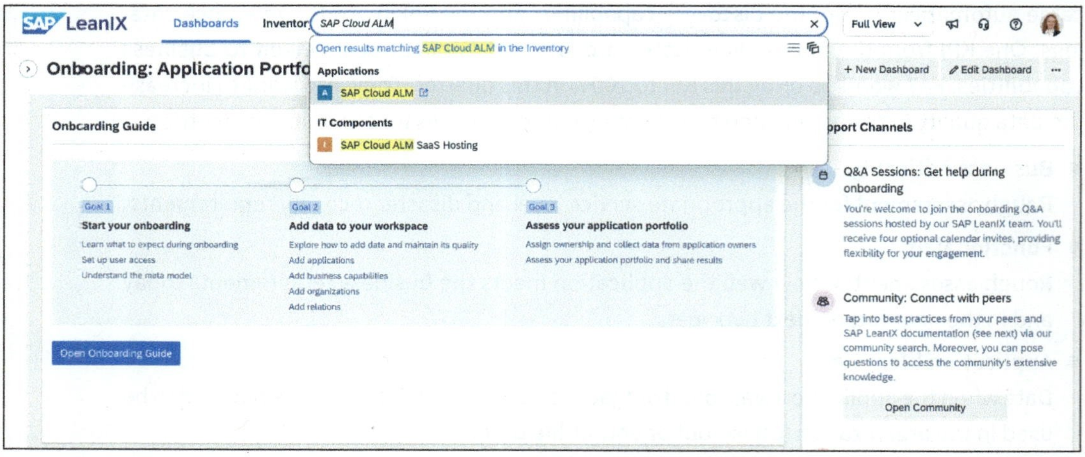

Figure 4.46: Using Quick Search to Navigate to SAP Cloud ALM Application Fact Sheet

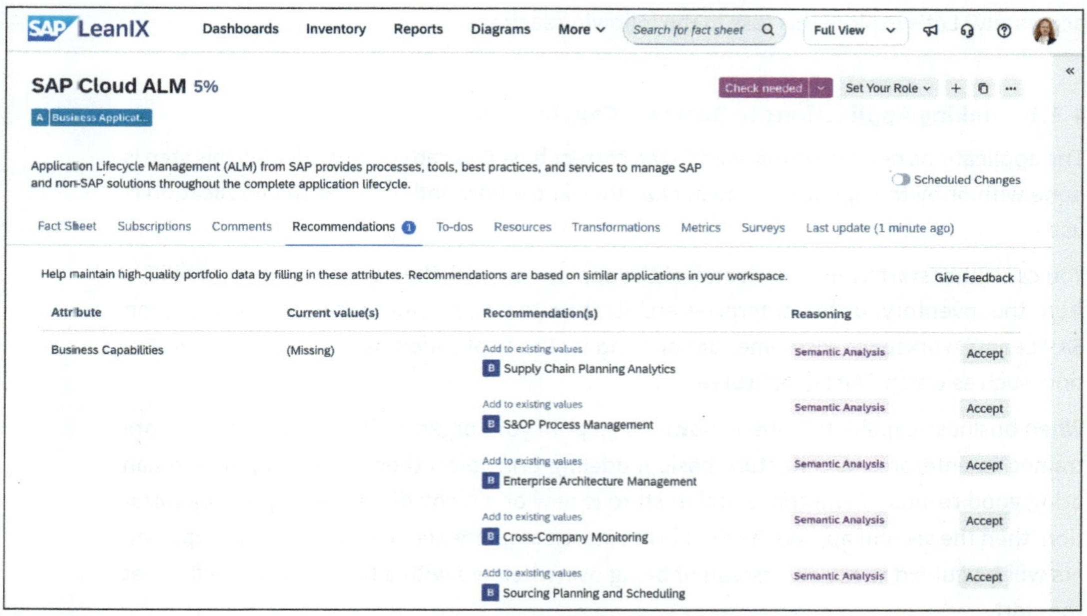

Figure 4.47: Recommendations Tab in Fact Sheet Information

If the **Description** field is empty, then the **Recommendations** feature will ask for a description to be entered first (see Figure 4.48).

> **Further Resources**
>
> For latest information on the recommendations feature, visit the SAP LeanIX user documentation at *http://s-prs.co/v620639*.

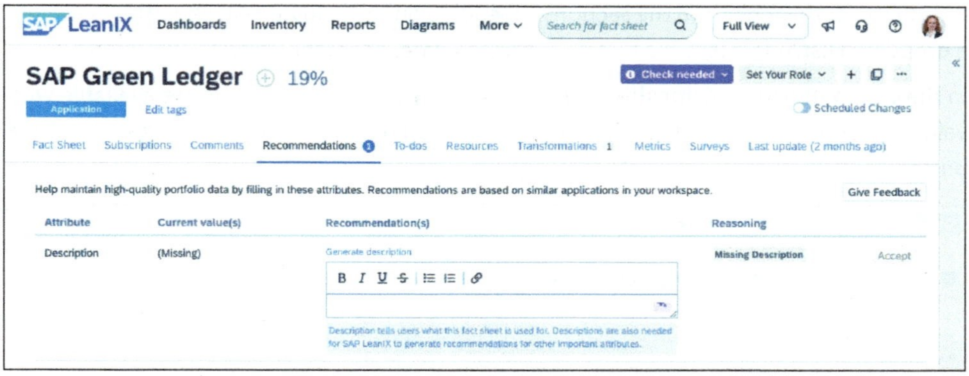

Figure 4.48: Entering Missing Description Info in Recommendations Tab

4.3.2 Edit Information in the Fact Sheet

Every role with editing rights can change data in fact sheets. This right can be restricted with either custom roles with special authorization or via advanced subscription-based authorizations; for example, you can make it so that only subscribed responsible persons can edit certain fields on a fact sheet. If nonresponsible persons make changes to the fact sheets, the quality seal will break and, depending on the notification settings, the responsible person will be notified about changes in the fact sheet.

Let's say you have editing rights to a fact sheet and you are tasked by the responsible person to update the **Business Criticality** and the **Functional Fit** attributes of an application. One way to maintain the information would be to access the fact sheet, scroll to the section on the **Fact Sheet** tab where you want to add data, hover over the area until the **Edit** button appears, and then click it to enter the information (see Figure 4.49).

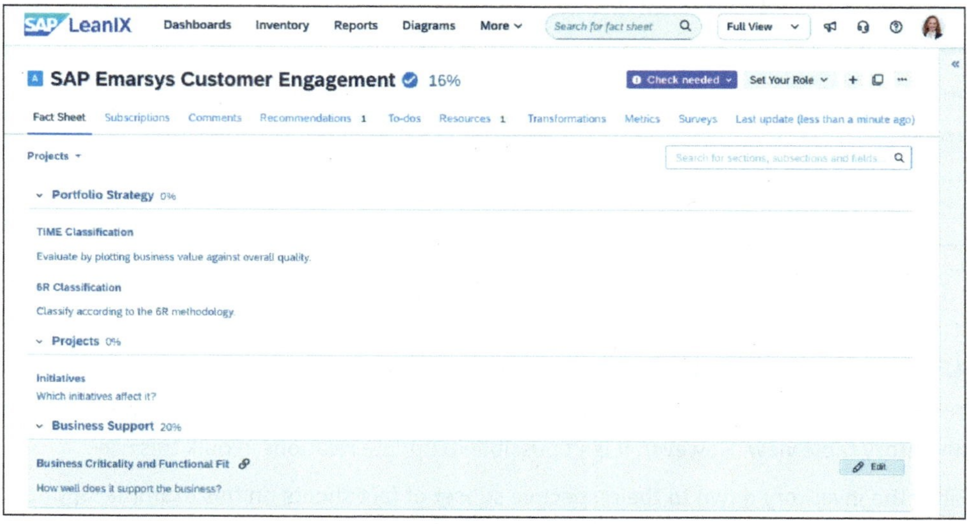

Figure 4.49: Editing Section in Fact Sheet Tab

An editable form appears in the section. Depending on the field format, you can select from a dropdown list, enter text in a text field, select a date, or enter numerical information.

In the case of **Business Criticality** and **Functional Fit**, dropdown menus are available with value explanations (see Figure 4.50). After the data input, you simply click the blue **Save** button.

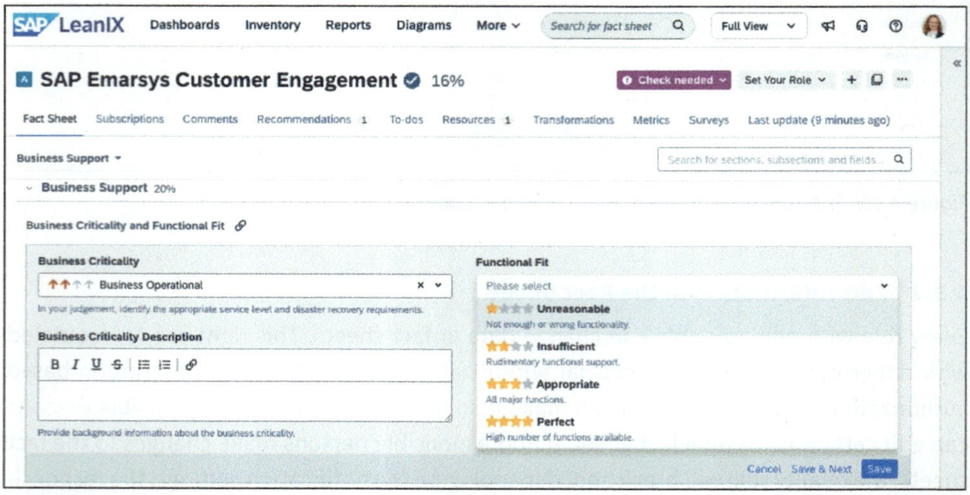

Figure 4.50: Entering Data Manually in Section Input Form

The information is saved in the fact sheet and visible in the previously empty section (see Figure 4.51).

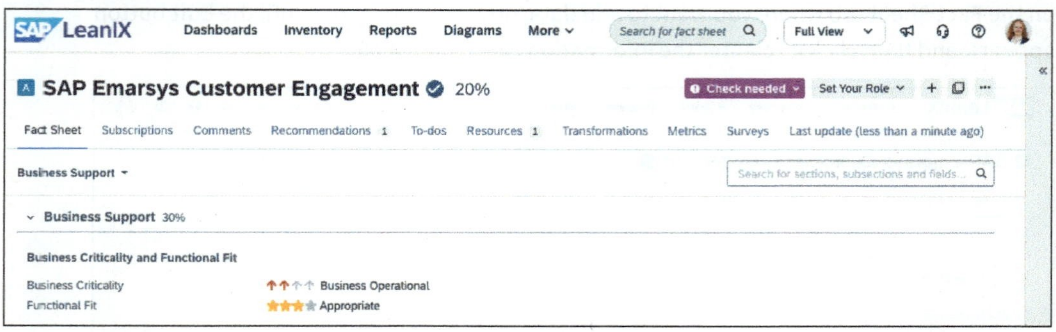

Figure 4.51: Entered Data Visible in Fact Sheet Tab Section

4.3.3 Mass Data Updates in Inventory Table View

To mass-update attributes, subscribers, or tag data, you can use the editing mode in the **Inventory** table view. However, it isn't possible to update relations in bulk this way.

Filter the inventory down to the respective subset of fact sheets (in this example, applications of subtype business application) and click **As table** in the upper-right corner of the center panel (see Figure 4.52) to switch to a table view.

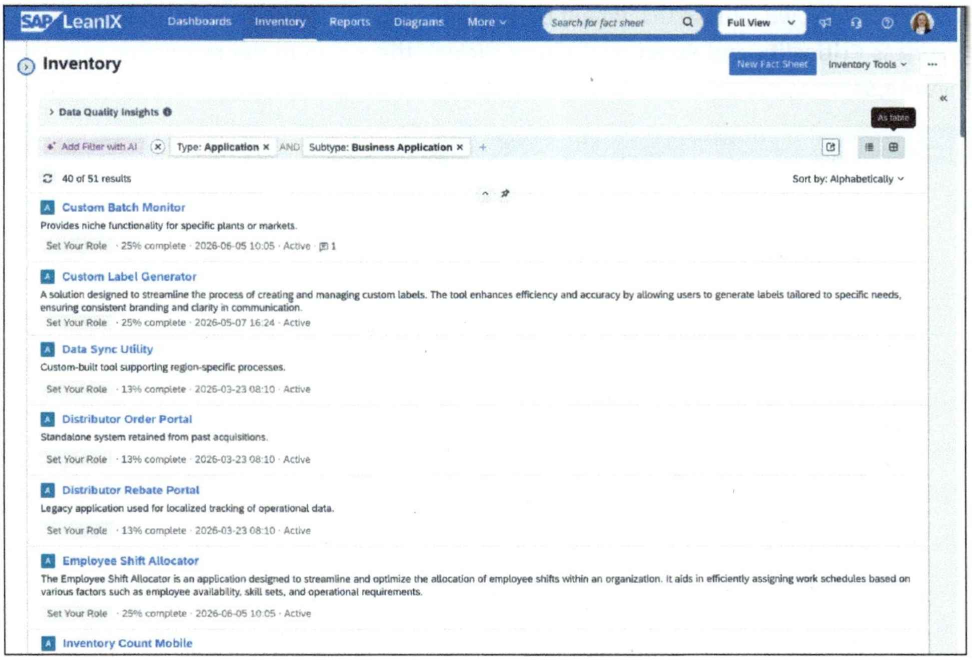

Figure 4.52: Switching to Table View in Inventory

Next, click the **Manage columns** button (see Figure 4.53).

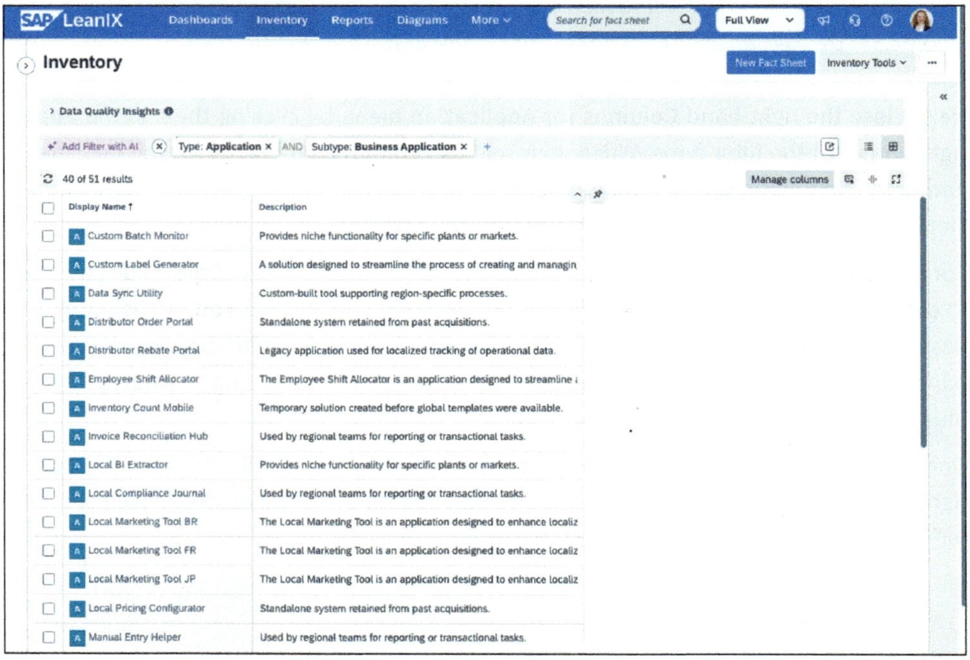

Figure 4.53: Opening Columns Pane in Inventory Table View

In this example, we remove the default **Description** field by clicking the **x** in its cell, and add **Business Criticality** and **Functional Fit** by clicking the + sign in the respective cells (see Figure 4.54).

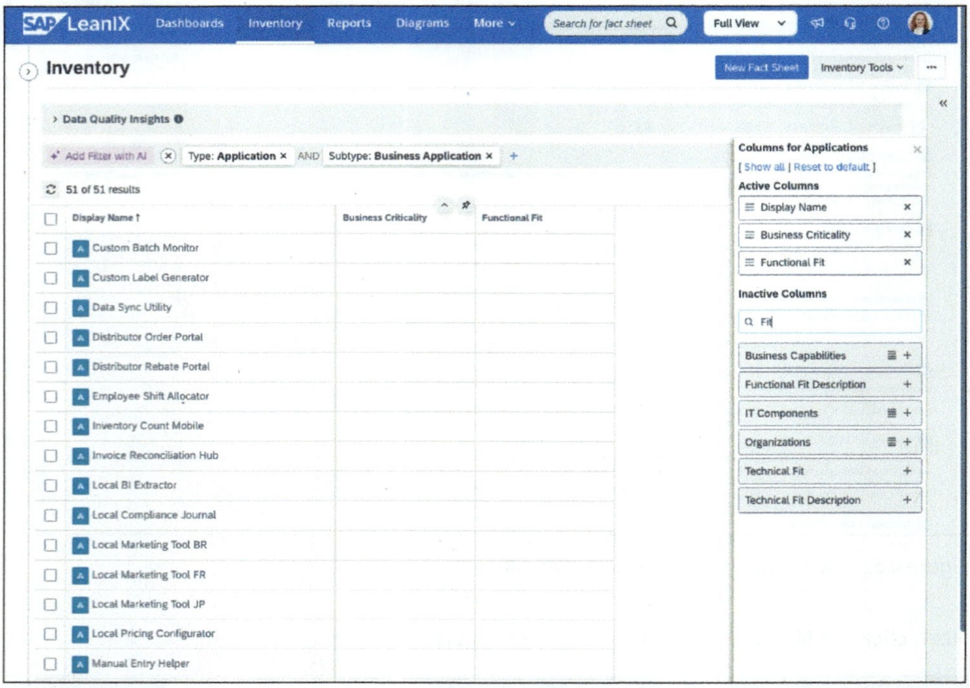

Figure 4.54: Deselecting and Selecting Columns Visible in Table View

Next, close the right-hand **Columns for Application** menu by clicking the **x** in the upper-right corner of the filter pane. When your role has editing permissions for the fact sheets and fields that you selected, you'll be able to modify the data in the table immediately. Fields that you cannot edit have a grey interior cell color.

For single-select fields, click in an empty cell and start typing the key of the value you want to enter. A list will appear in which the correct value can be selected. You can use copy and paste in the table and the dropdown functionality. In the top-right corner of the center pane, you can see how many fact sheets will be changed. After entering all values, click the blue **Save Changes** button (see Figure 4.55).

Once the updates are saved in the workspace, a notification box will appear in the upper-right corner, with a green checkmark informing you that the changes were saved successfully (see Figure 4.56).

SAP LeanIX User Documentation

For the latest updates on the bulk updating feature in the table view, check the SAP LeanIX user documentation at *http://s-prs.co/v620640*.

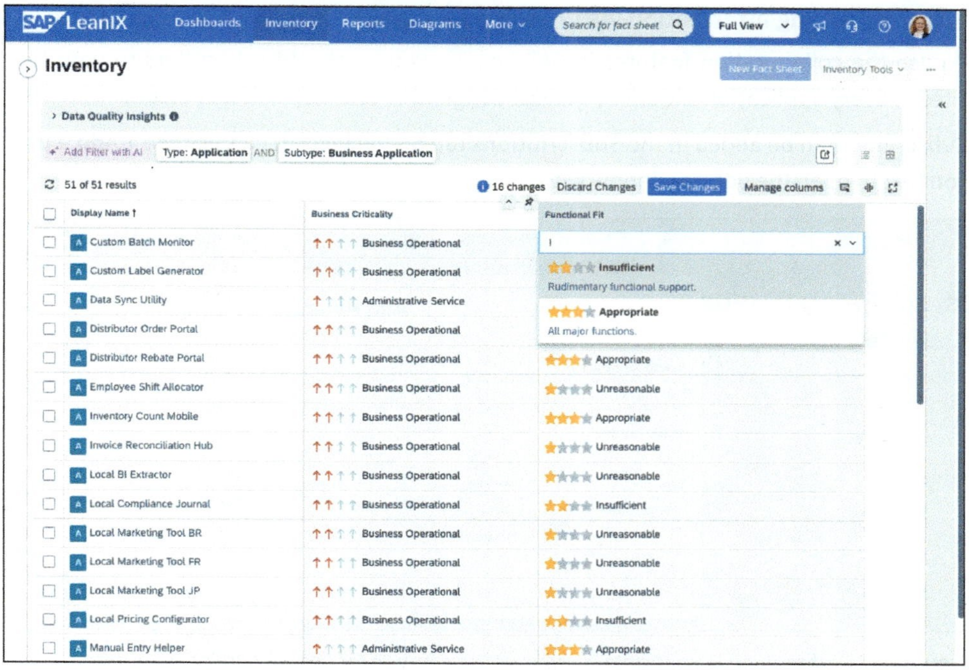

Figure 4.55: Entering Values in Edit Mode of Inventory Table View

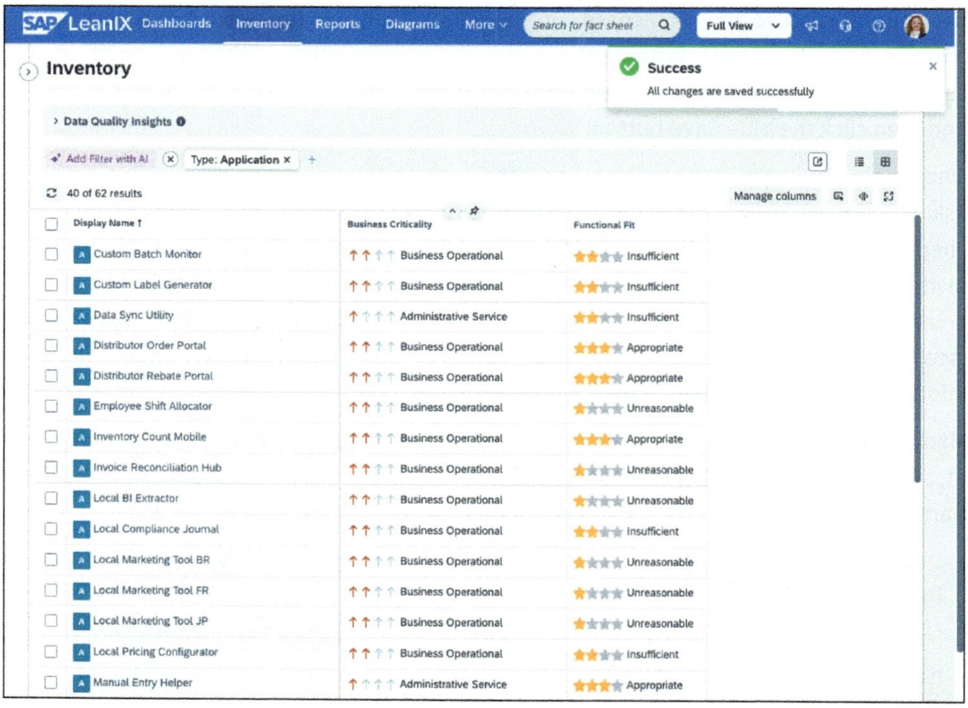

Figure 4.56: Notification of Successful Updates via Table View Edit Mode

4.3.4 Enriching Information via Crowdsourcing

To use the collaboration features in SAP LeanIX, you must add subscribers to your fact sheets. Every application should be at least assigned to one responsible person.

Subscribers can be added in the **Subscriptions** tab on the fact sheet. Click the **+ Add a person** text to open the popup window with the input form, as shown in Figure 4.57.

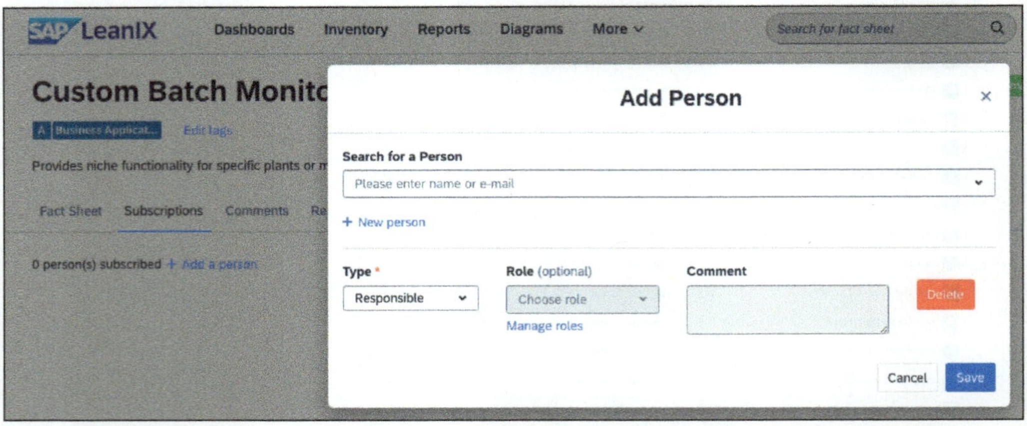

Figure 4.57: Subscribing Responsible Persons as Preparation for Crowdsourcing Information

You can then either **Search for a Person** using a name or email address, which will show you the users known to the workspace, or you can click **+ New person** to enter the email address, first name, and last name of an unknown person to the workspace. Don't forget to select the correct subscription **Type** from the dropdown and, if maintained, also the **Role**, and then click the blue **Save** button.

Once maintained, you can, for instance, send surveys to application responsible persons, asking them for lifecycle date information or to review the business capability mapping, or checking the business criticality and functional fit assignment. Remember that you must invite unknown users before you can send them notifications. Also, don't forget to turn on email notifications as admins in the notification center before you try to send out a survey (see Chapter 3, Section 3.3.3). Otherwise, the survey won't be sent out via email. For more information on surveys, see Chapter 2, Section 2.3.3.

Many customers set up regular yearly surveys to review the business criticality, functional fit, and technical fit or other assessments. Those values can then also trigger, for instance, a Gartner TIME Assessment via activated **Calculation Templates** (see Chapter 10).

Further Resources

For more information on calculation templates, check out the SAP LeanIX user documentation at *http://s-prs.co/v620641*.

4.3.5 Searching the Inventory

Now that your inventory is filled and you've enriched some additional fields and attributes, you need to learn how to filter the inventory. The filter menu will be the same for the inventory, reports, and diagrams, so you can reuse this knowledge in the following chapters.

Navigate to the **Inventory** top ribbon menu. Make sure that the left-hand filter menu is expanded. If you can't see it, hover over the separator line on the left until you see a little arrow in the upper-left corner, then click the arrow to expand the **Filters** menu (see Figure 4.58).

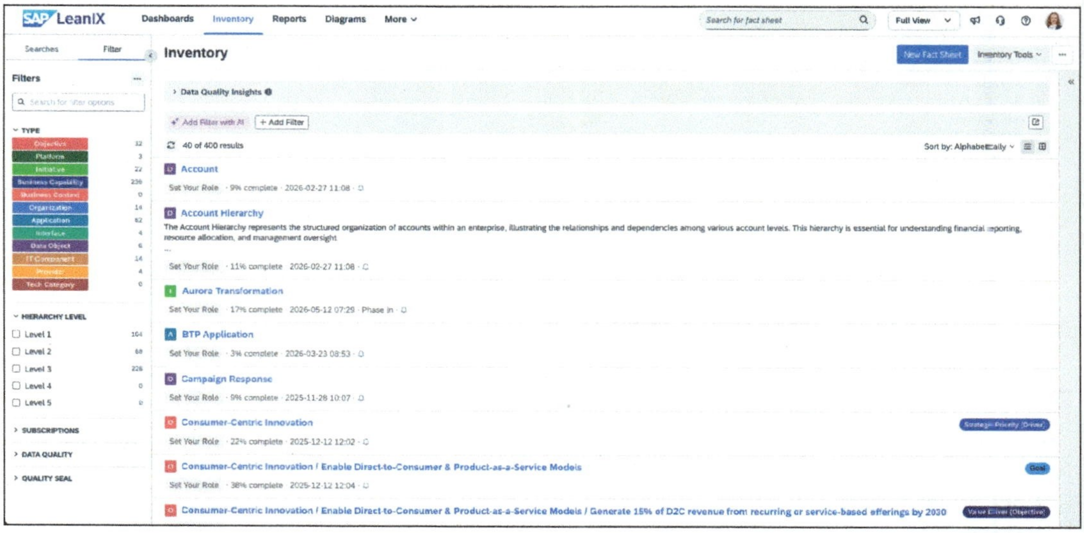

Figure 4.58: Filters Menu in Inventory

In the center pane, you'll now see all fact sheets in the inventory of any fact sheet type. At the top, you can see a filter line with a purple box that says **Add Filter with AI** and a white button labeled **+Add Filter**. For more information on the AI filtering assistant, see Chapter 11, Section 11.1. When you click **+Add Filter**, a **Smart Search** box will open for text searches. Administrators can modify which custom fields of field type string are included in these searches. Let's walk through a quick example.

Say that you want to create a custom text field of field type **String** in the application meta model configuration. (See Chapter 3, Section 3.3.3 to review how to navigate to the **Meta Model Configuration** area.) After entering the **Key**, **Field type**, and **Displayed as** values, click the blue **Create** button to create the text field (see Figure 4.59).

Click the tab with the globe icon to open the **Translations** panel. Expand **English** and define a **Label**, which will be the displayed name of the field (see Figure 4.60).

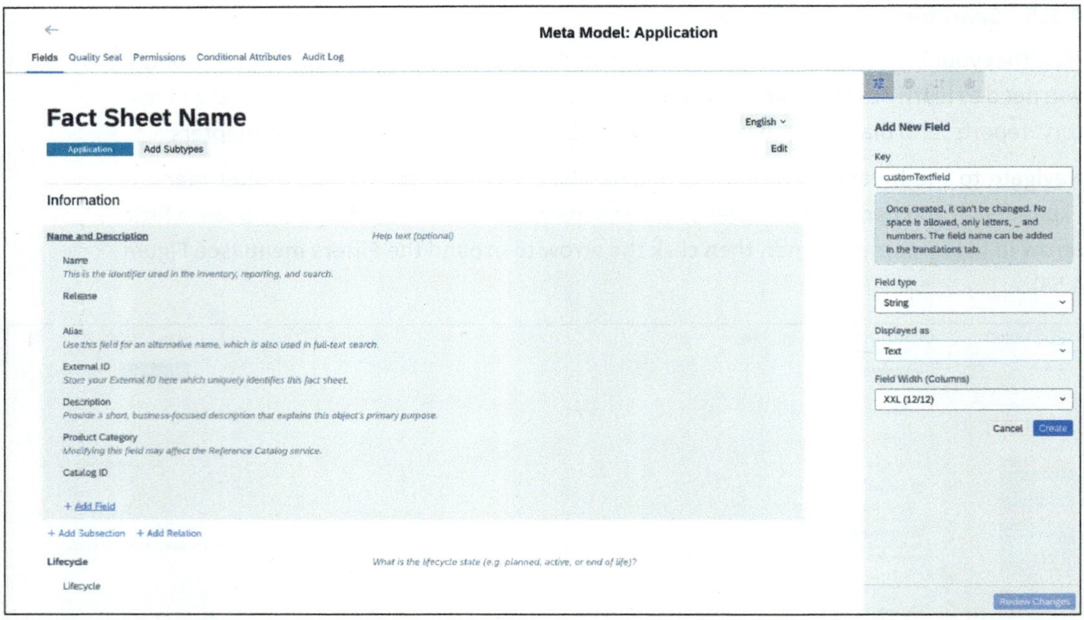

Figure 4.59: Creating Custom Text Field

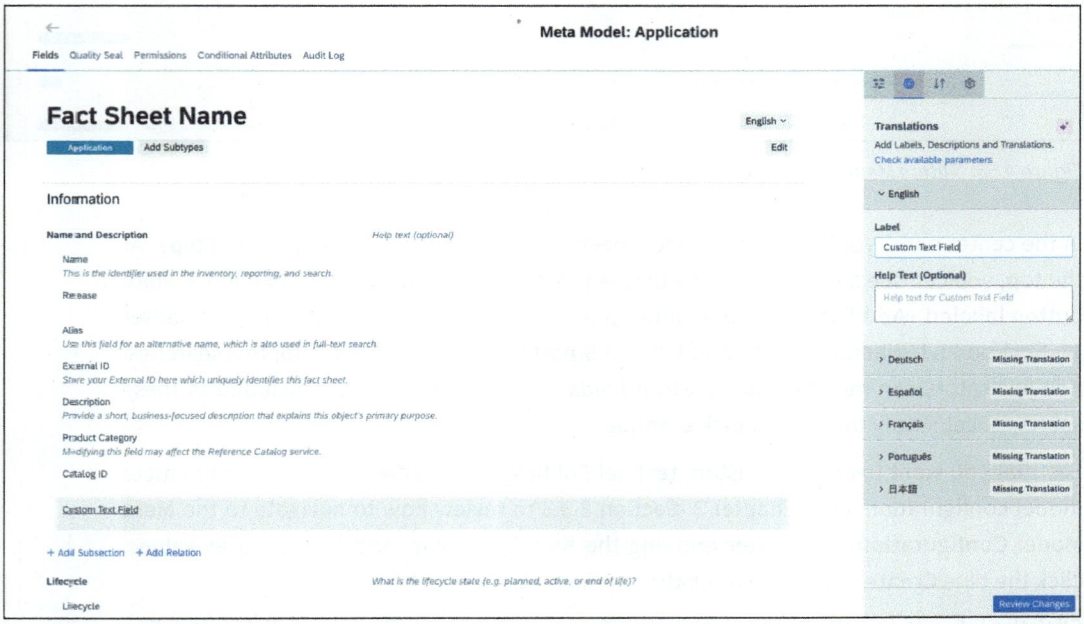

Figure 4.60: Updating Translation Model

Now, navigate to the tab with the gear button. For text fields, you have two important decisions to make. If you switch the **Include in full text search** toggle to green, then this field will be considered in the add filter smart search that we discussed before in the inventory overview. If you switch the **Include in quick search** toggle to green, then the text field will be considered as searchable in the top ribbon **Search for fact sheet** search field. You must change those settings before finalizing the field creation by clicking the blue **Review Changes** button in the lower-right corner (see Figure 4.61). It isn't possible to change the toggles at a later point in time.

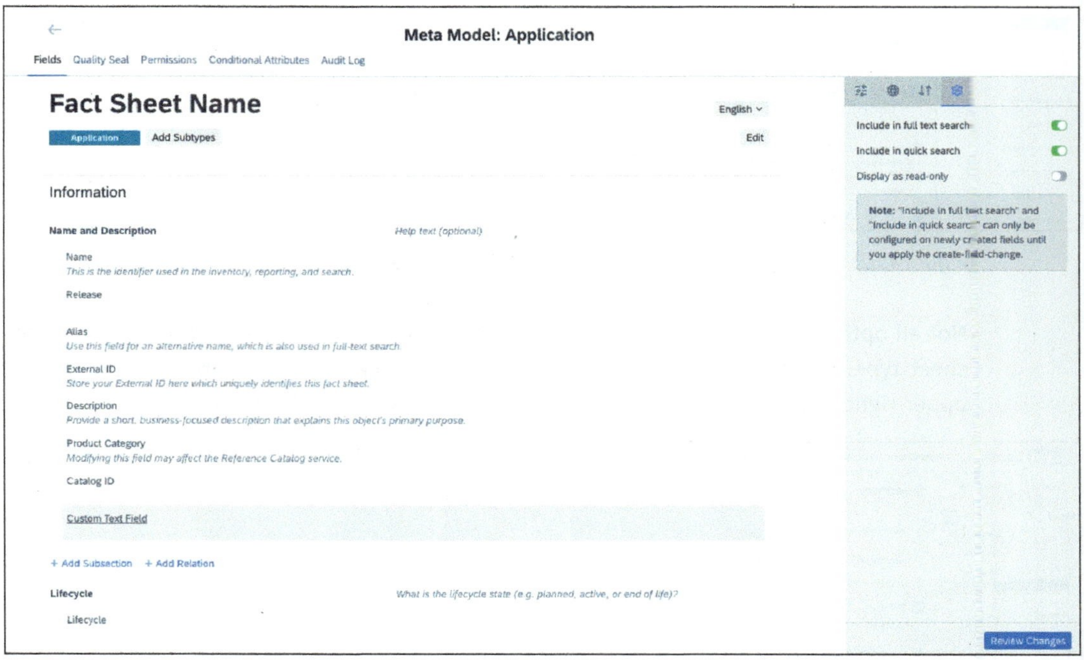

Figure 4.61: Text Search Settings to Be Determined Before Finalizing Text Field Creation

Going back to the inventory search, the smart search is a text-based search. The left-hand side **Filters** menu offers plenty of additional options for searches. The first option that you have is to filter for the fact sheet **TYPE**. Simply click one of the meta model fact sheet types shown under the **Filters** section to add this filter (in this example, we've selected **Application**). You will see that the filter changes in the left-hand menu and is added to the grey filter pane above the center panel (see Figure 4.62).

Underneath the **TYPE** section, you can see many more sections for filtering.

In general, the **Filters** menu can filter for the following results:

- Single-select and multiselect fields
- Relations and fields on relations
- Tags
- Subscriptions

- Lifecycle status and date ranges
- Status of, for example, the quality seal, catalog linking, or active integration

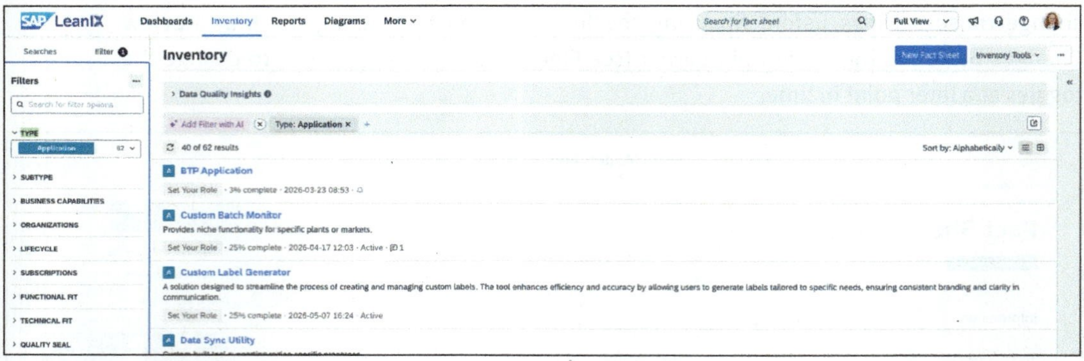

Figure 4.62: Filtering Inventory for Specific Fact Sheet Type

Not all options are visible by default. You can customize the visible filter criteria per fact sheet type. To modify the available filter sections, click **...** to the right of **Filters** in the upper-right corner of the left pane. Select **Manage Filters** (see Figure 4.63).

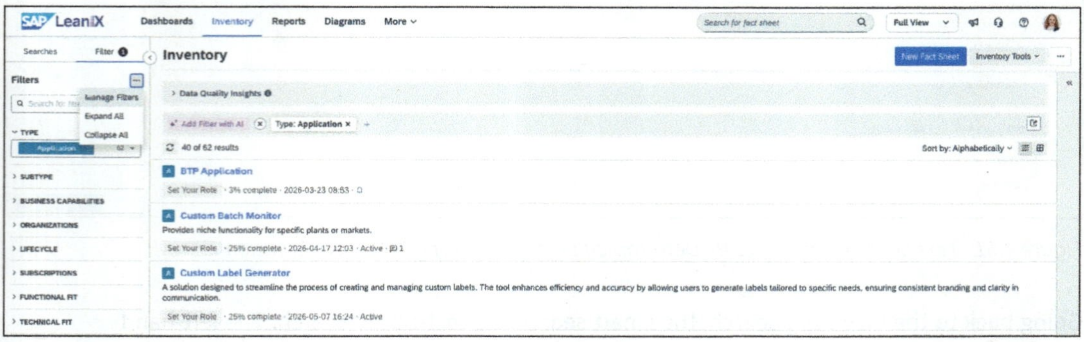

Figure 4.63: Manage Filters Settings in Left-Hand Filters Area

In the **Manage Filters** settings, you will see the **Visible** filter criteria at the top. You can rearrange the order via drag and drop. Underneath, you can see the **Hidden** filter criteria. To make these criteria visible in the **Filters** menu, hover over a white tile and click the **+** icon that appears at the right side of the tile. To remove a filter criterion from the **Visible** filters, click the **x** on the right side of the respective filter criteria tile. Once you have your filter menu configured, click the blue **Back** text with the left-pointing arrow in the upper-left corner of the **Manage Filters** pane (see Figure 4.64).

Most of the expanded filter sections will show a little gear icon (**Settings**) to access advanced settings when hovering over them. This gear button lets you change the filter operator from **OR** (default) to **AND** or **IS NOT** (see Figure 4.65).

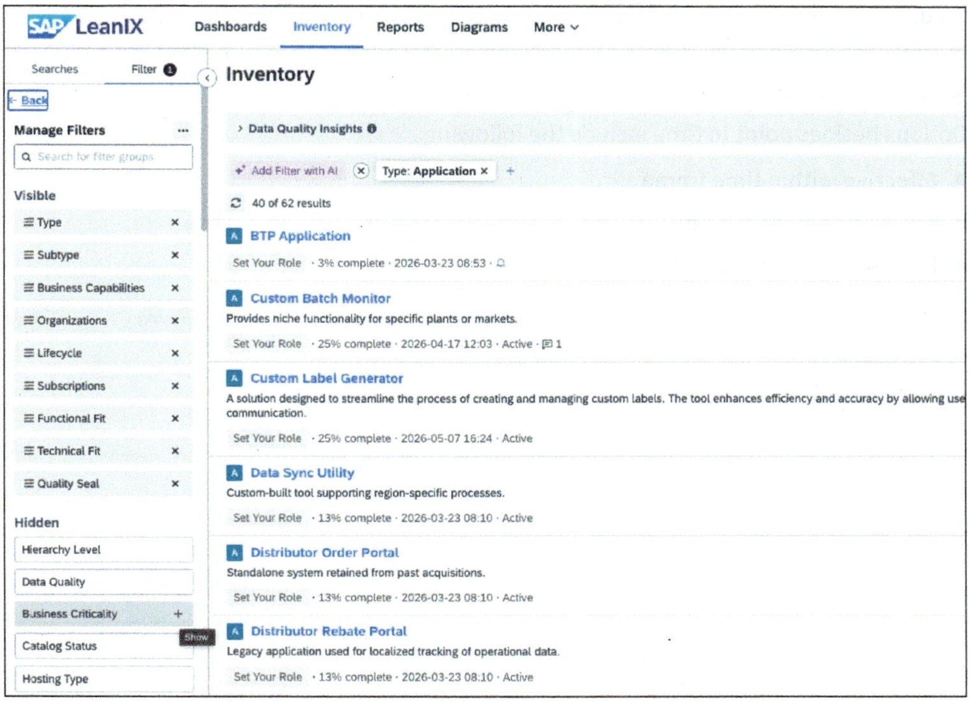

Figure 4.64: Configuring Filter Criteria in Manage Filters

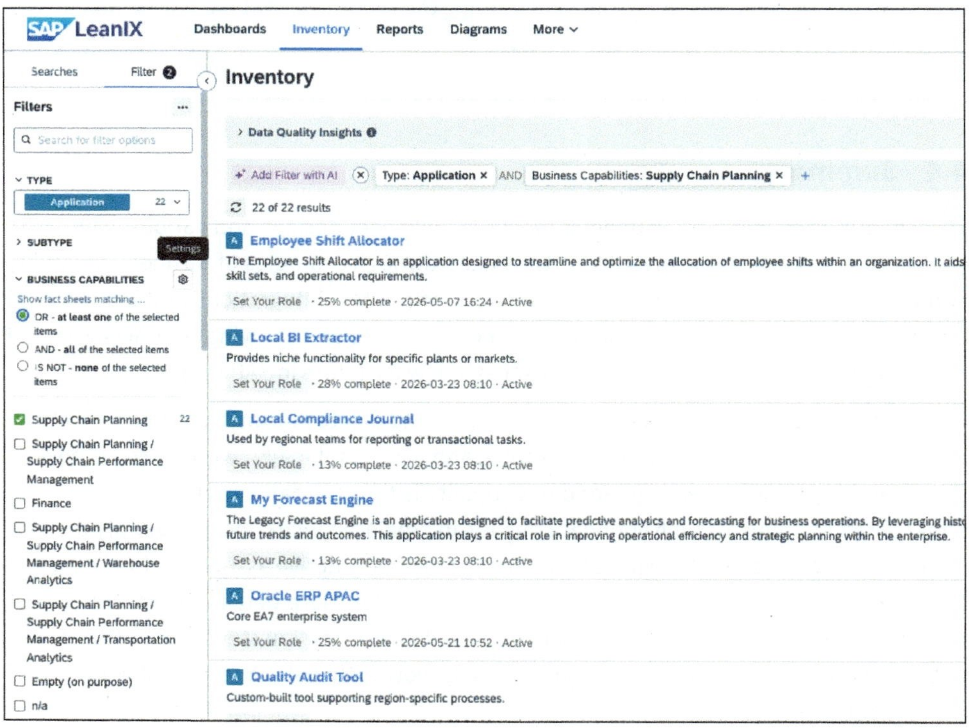

Figure 4.65: Changing Filter Operator from OR to AND or IS NOT

On date fields such as **Lifecycle**, you can also define whether you want to search for a **point in time**, for instance, all **Active** applications as of **Today** (see Figure 4.66), or a specific range.

Options besides **point in time** include the following:

- **Effective within time frame**
- **Starts within time frame**
- **Ends within time frame**

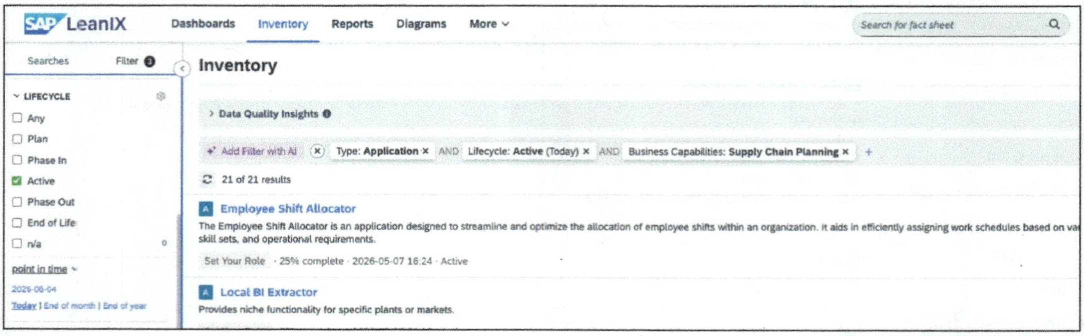

Figure 4.66: Filtering Options for Date Fields

> **Further Resources**
>
> For more information on searching and filtering the inventory, visit the SAP LeanIX user documentation at *http://s-prs.co/v620642*.

4.4 Summary

This chapter provided a basic understanding of fact sheet designs and various ways to fill your data repository in the inventory. We walked through examples of business capability and application fact sheets, but many more out-of-the-box integrations and features will help you fill in other fact sheet types as well. Keep learning and applying your newly gained knowledge to adjacent areas and features that you will explore in SAP LeanIX.

We also discussed different ways to input, update, and delete information in fact sheets, in inventory table views, and via Excel exports and imports. Finally, we looked at how to search and filter your inventory. Good data quality is the foundation of user trust in a system. Therefore, think through your data quality and maintenance concept. Identify the ultimate sources of truth and integrate them, or automatically derive as much data as possible (e.g., via automations or calculations; see Chapter 10 for more details) to ensure correct and up-to-date data without unnecessary manual data duplication steps. Check Chapter 11 to learn how artificial intelligence (AI) can help you with data quality, validation, and enrichment tasks.

In the next chapter, we'll focus on how to make sense of the data and generate insights via visualizations. This chapter prepared you to create a nicely filled inventory with relations and attributes, which is a mandatory prerequisite to use SAP LeanIX's full visualization capability. Don't rush this step during your solution setup: The better your data concept, the easier continuous data maintenance will be, and the higher user trust in the data will become.

Chapter 5

Reporting and Visualizations

You've imported and enriched your data: Now it's time to make sense of it. Visualizations play a crucial role in analyzing data. Reports can show dependencies, aggregate to a higher level, or drill down into lower levels. Diagrams can visualize complex ecosystems and data flows. SAP LeanIX offers out-of-the-box reporting and visualization formats that we'll explore in this chapter.

One important mantra in enterprise architecture is that you can only manage what you know. Visualizations are critical in enterprise architecture for multiple reasons:

- **Simplifying complexity**
  Reduce the noise in the data to focus on the relevant data points for strategic decision-making.
- **Identifying risks and opportunities**
  Visualizations are powerful ways to find white spots, red flags, or redundancies in one view.
- **Supporting governance and compliance**
  Audits and the like are often enriched with reports or diagrams.
- **Enabling cross-stakeholder communication**
  Visualizations need to adapt to the target audience and thus also adapt the level of aggregation.
- **Driving strategic decision-making**
  Data is gold. Enterprise architects handle critical IT landscape data that, in the right format, can make them important strategic advisors to the C-suite and department heads in order to support strategic initiatives, such as cost reduction or making the company AI-ready.

A core strength of SAP LeanIX is its ability to help you visualize contextualized information on different aggregated levels.

In the following sections, we'll look at the most common visualization methods in SAP LeanIX: reports and diagrams. Both belong to the bread-and-butter business of enterprise architects from a graphical output perspective. We'll then also touch on additional capabilities such as custom reports and fact sheet key performance indicators (KPIs) on dashboards.

5.1 Reports

Contextualizing and visualizing data and dependencies is an important method in the toolbox of enterprise architects. Reports are a great way to make dependencies and context easy to see visually.

We'll explore the different types of reports in SAP LeanIX, how to modify them, and how to create custom reports in the following sections.

5.1.1 Report Types

SAP LeanIX offers a huge variety of out-of-the-box report formats to visualize data. The most commonly used ones are as follows:

- **Landscape report**
  A center fact sheet type clustered by one dimension (either related fact sheet, tag, or value list attribute, e.g., single-select field). It's possible to filter the data, create colorful heatmaps, and aggregate and expand along the hierarchy chain of the axis fact sheet. Figure 5.1 shows an example of a business capability map in the landscape report format.

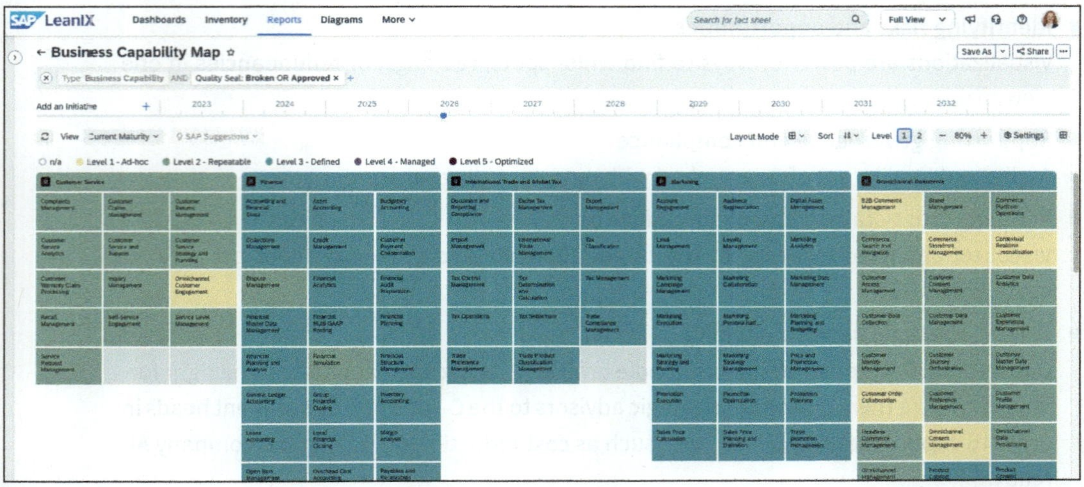

Figure 5.1: Landscape Report Example: Business Capability Map Color-Coded by Current Maturity Level

- **Matrix report**
  A center fact sheet type clustered by two dimensions. It's possible to filter the data, create colorful heatmaps, aggregate and expand along the hierarchy chain of the axis fact sheet, and even drill down into other related fact sheets. Figure 5.2 shows an application matrix example with organization and business capability relations as axes in the rolled-up layout format and heat-mapped for the functional fit values.

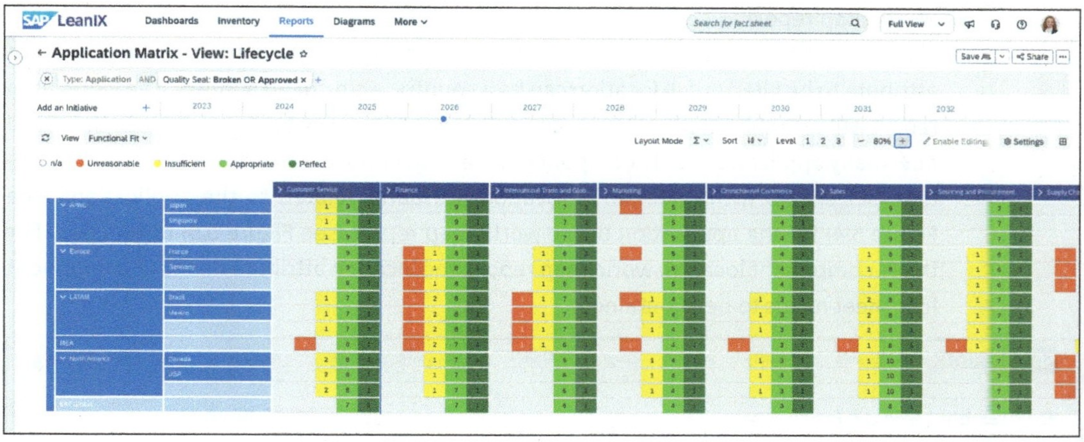

Figure 5.2: Matrix Report Example: Application Matrix

- **Portfolio report**
  A fact sheet count or fact sheet field average for integer fields, visualized as a bubble chart clustered by two dimensions (fields or tags). The size of the bubble represents either the count of fact sheets that belong in this dimension intersection, or the average or sum of an integer field (e.g., cost field). Figure 5.3 shows an example application portfolio report with axes representing the technical and functional fit and the bubble size corresponding with the fact sheet count.

Figure 5.3: Application Portfolio Report Example

- **World map report**
  Visualizes counts of fact sheets by heatmaps. This requires the out-of-the-box **Location** attribute to be filled out. A location can be a country, a city, or an address. The world map report will pick this information up and will use color-coding to show where in the world how many applications or IT components are used or hosted. You have to have organization fact sheets with maintained location attributes related to the applications (see Figure 5.4) for the application usage world map report (see Figure 5.5) to function. For the IT component location world map report, the location attribute on the IT component fact sheet needs to be maintained.

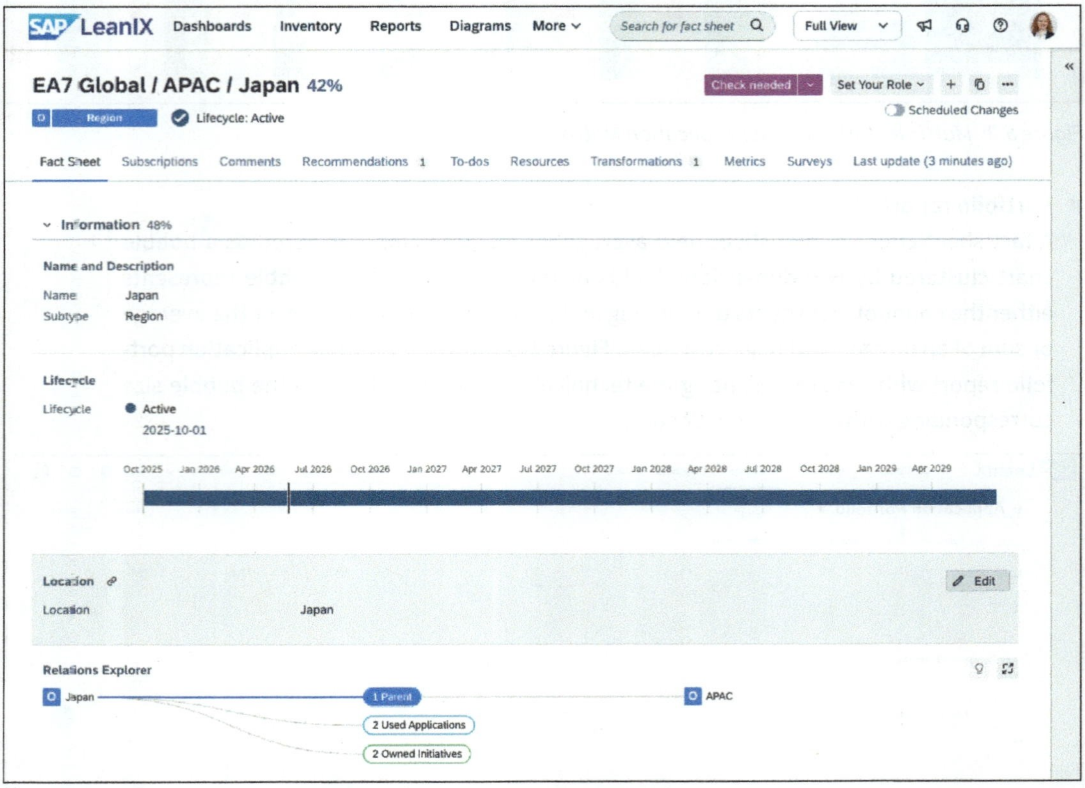

Figure 5.4: Location Attribute on Organization Fact Sheet

- **Roadmap report**
  A timeline report of a fact sheet type that allows clustering and drilldowns into related fact sheet types. This is useful when planning transformations or visualizing lifecycle stage changes over time. For this report to work, lifecycle dates of initiatives, applications, IT components, or other fact sheet types that should be analyzed need to be maintained in the fact sheet information. This report type will be explained in detail in Chapter 6.

- **Interface circle map report**
  Visualizes logical interface relations between the *provider* (technical owner) and *consumer* (triggered via a data push or pull request) applications. The circle map report only

shows interface relations between applications in which both sides, provider and consumer, are maintained (see Figure 5.6).

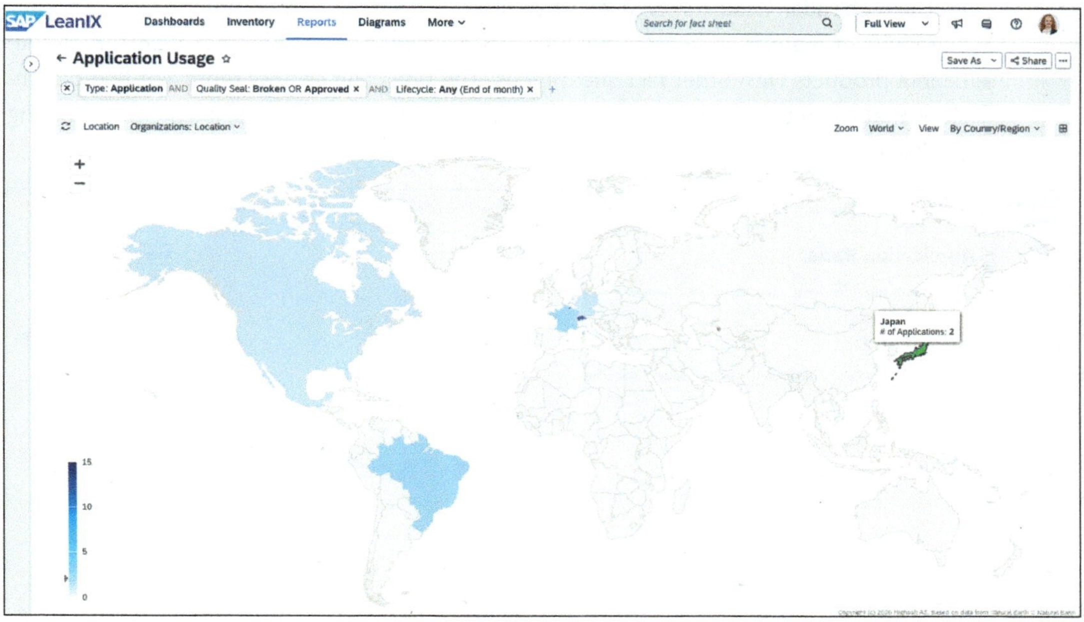

Figure 5.5: Application Usage World Map Report Example

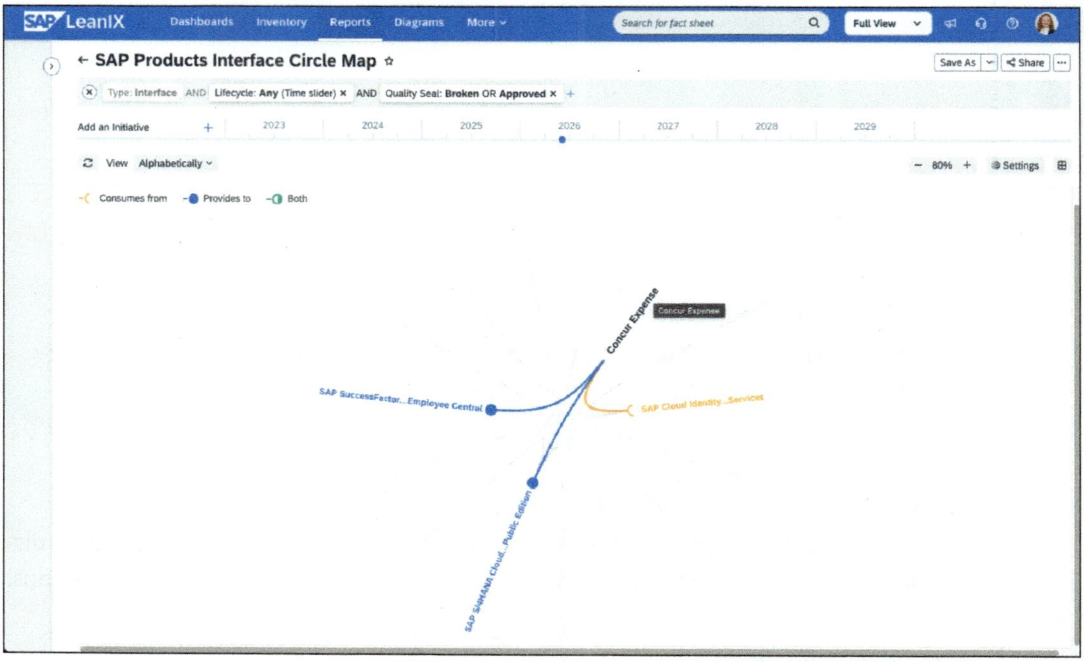

Figure 5.6: Interface Circle Map Example

Initially, the *radar report* was developed for use cases in the SAP LeanIX Technology Risk and Compliance product. It's comprised of rings and sectors and was initially introduced for the tech radar visualization used when analyzing aggregated IT components from software bills of materials (SBOMs). It's now also possible to use the radar report in all SAP LeanIX products with other fact sheet types. The rings and sectors represent two dimensions that can represent single-select fields or tags. With color-coding, it's possible to bring a third dimension into the visualization (see Figure 5.7).

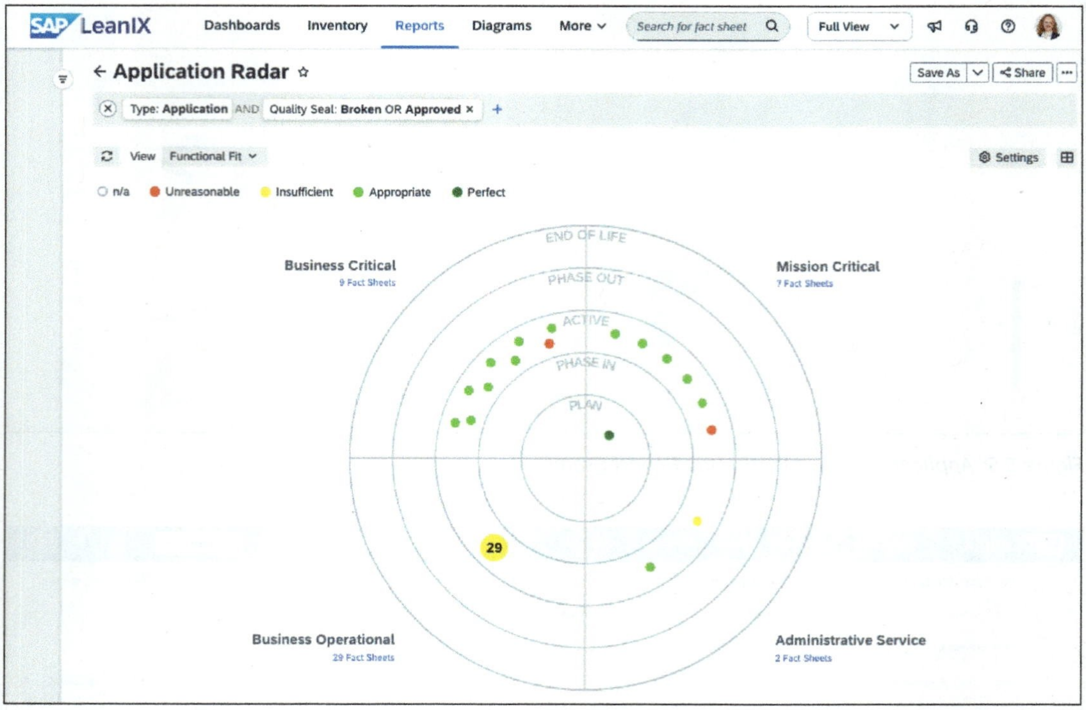

Figure 5.7: SAP LeanIX Radar Report Example

> **Further Resources**
> For the latest updates on the SAP LeanIX report feature, check the user documentation at *https://help.sap.com/docs/leanix/ea/reports*.

5.1.2 Creating and Modifying Reports

Let's now explore the **Reports** overview in the workspace, as shown in Figure 5.8. We'll discuss how to manage collections and create and customize reports in the following sections.

Managing Collections

On the left-hand side of Figure 5.8, you can see the collection structure (similar to a folder structure). Out of the box, you'll see the bigger collection groups: **REPORT TYPES**, with collections per report type, and **PERSONAL**, to filter for reports relevant to each respective user. As an admin, when you click **Manage Collections** in the lower-left corner, you can create new custom collection groups and collections, and the order of the collection groups can be modified via drag and drop (see Figure 5.9).

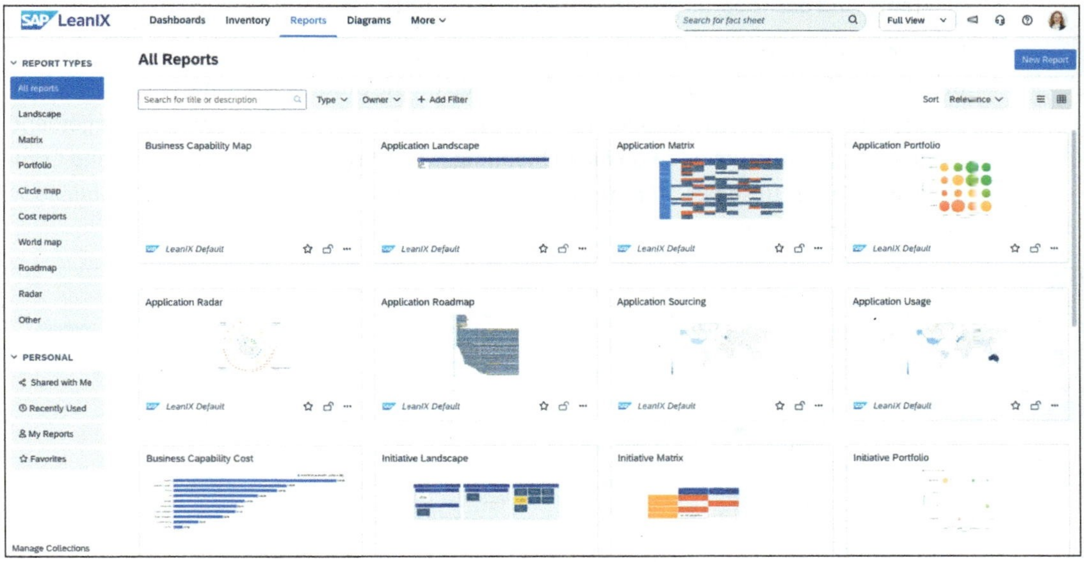

Figure 5.8: All Reports Overview in SAP LeanIX Workspace

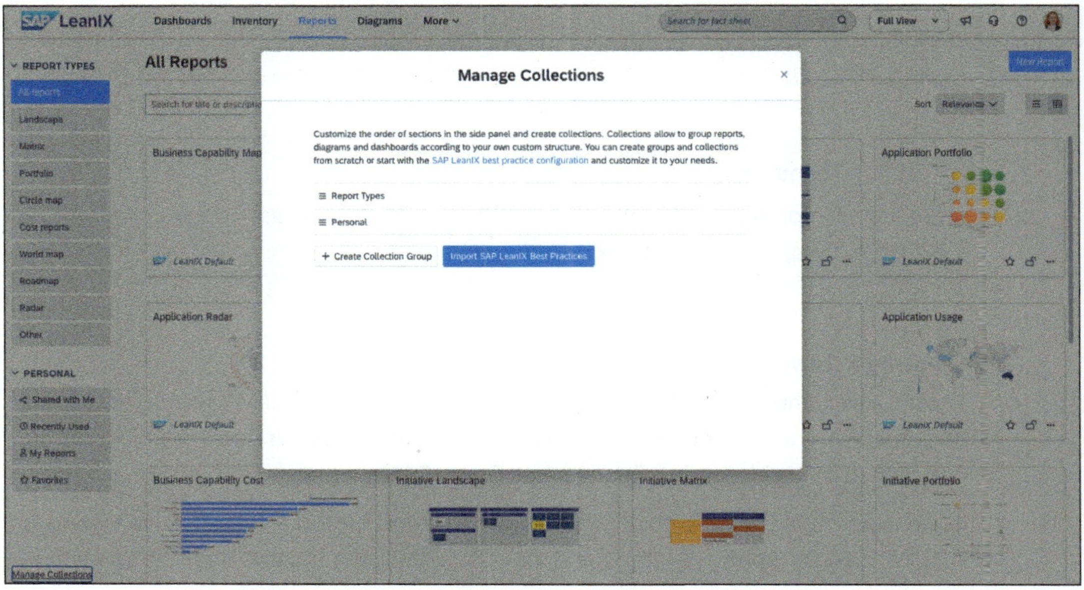

Figure 5.9: Manage Collections Form: Shared Between Reports and Diagrams

Admins can define which user roles can adjust collections or map content to them. By default, only admins can manage collections. In the **Administration** menu under **BASIC SETTINGS • User Roles and Permissions**, in the **Permissions** tab, admins can grant *administrative rights* (to create or delete collections) or *content management rights* (to map reports, diagrams, and dashboards to a collection) to other roles, such as members or viewers.

It's also possible to import predefined collections by clicking the blue **Import SAP LeanIX Best Practices** button, shown in Figure 5.9. Based on the product add-ons purchased, you'll then get a prepared report collection for SAP LeanIX Application Portfolio Management, SAP LeanIX Architecture and Road Map Planning, or SAP LeanIX Technology Risk and Compliance. Figure 5.10 shows an example of imported collections for SAP LeanIX Application Portfolio Management.

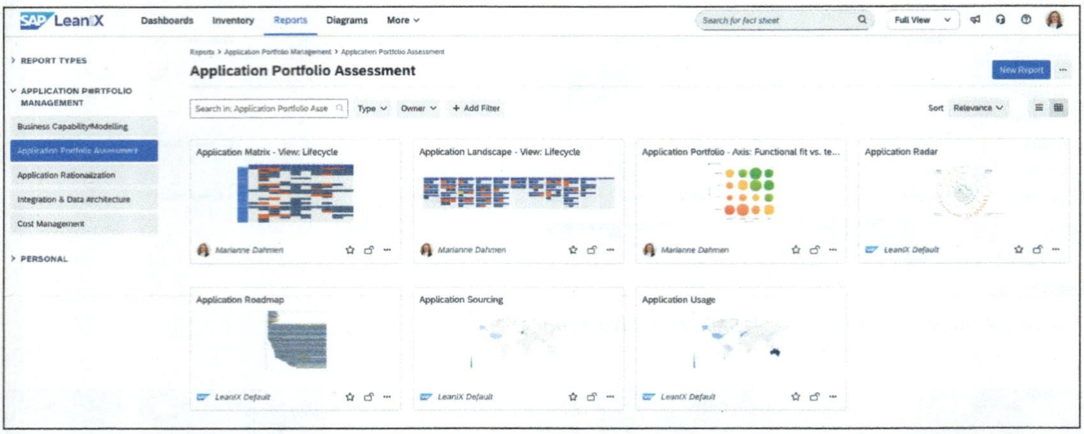

Figure 5.10: Application Portfolio Assessment: Best Practices Collection of Out-of-the-Box Reports

> **Further Resources**
>
> For more information on collections, visit the SAP LeanIX user documentation at *https://help.sap.com/docs/leanix/ea/collections*.

Creating a New Report

The top row in the center pane helps you navigate the reports. You can search for a specific title or words in a description, select certain report types or owners, sort in a different way, or switch from a grid view to a list view (see Figure 5.11).

In the upper-right corner, you will see a blue **New Report** button. This button isn't visible for all roles; it appears only for admins or roles with certain permissions. There are two types of reports:

- **Report templates**
  Admins define which fact sheet type of interest should be made available in which report style (e.g., interface matrix).

- **Report views**
  These are variants of a report template with different axes, color-coding, or aggregation levels.

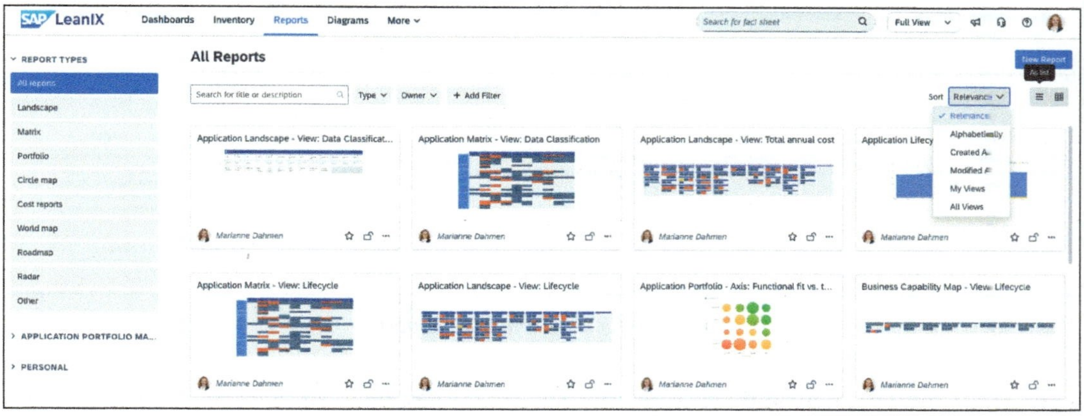

Figure 5.11: Filtering, Sorting, and Layout of Dashboard Overviews

The **New Report** button is for admins to create new report *templates*, not views. To check whether a report template exists, the admins should navigate to the respective **REPORT TYPES** collection (for instance, **Matrix** reports) and check whether a certain fact sheet type report exists, such as the interface matrix. Figure 5.12 shows that such a template doesn't exist yet. In this scenario, as an admin, you can click the blue **New Report** button to start creating the missing report template.

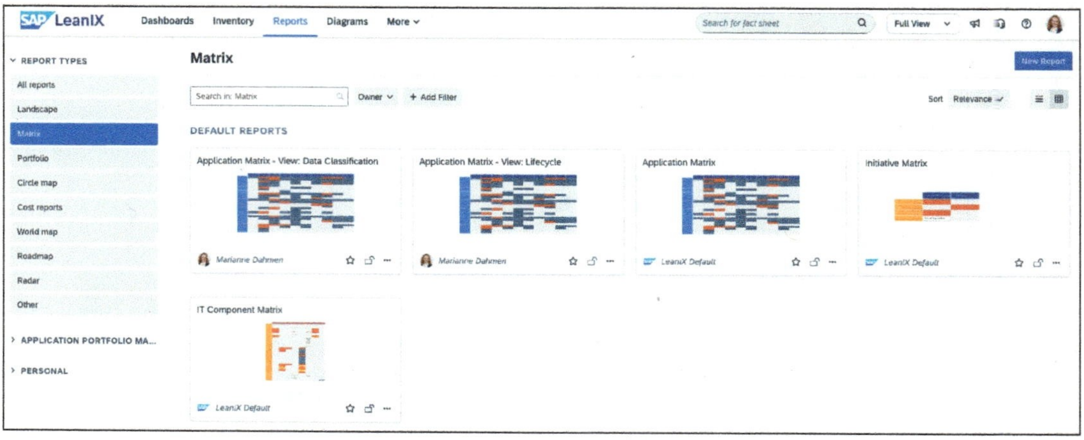

Figure 5.12: Checking for Existing Report Templates in Matrix Report Type

After clicking the button, you enter a full-screen guide, as shown in Figure 5.13. Here, you choose a report type; for this example, select **Matrix**.

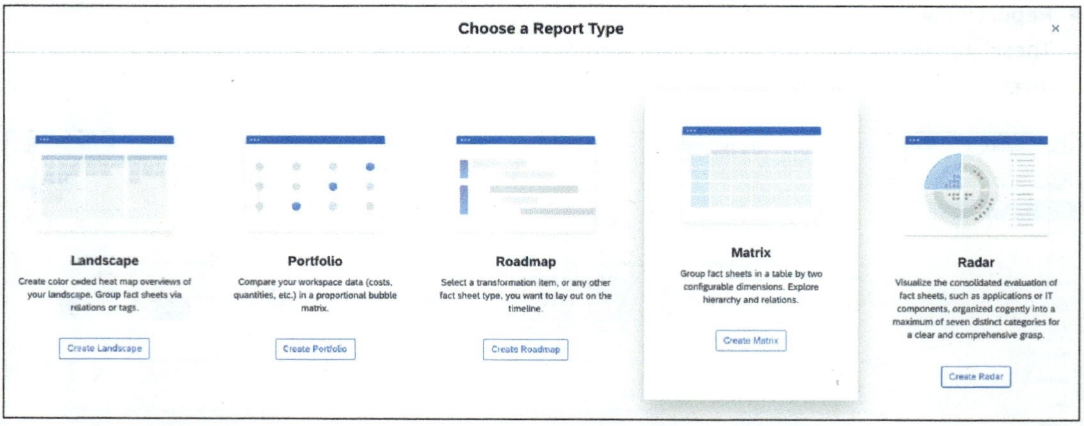

Figure 5.13: Choosing Report Type for New Template Report

In the next window, define a **Report Name** for the new template and choose an option from the **Select Base Fact Sheet Type** dropdown (see Figure 5.14). A description text is optional but recommended to guide users on how and when to use this template. This text will be visible in the grid view on the **Reports** overview page when hovering over a report tile. After filling out all information, click the blue **Create Report** button.

Figure 5.14: Entering New Report Details

You might see a warning indicating **No Data**. Don't worry, nothing is broken; you just have to fine-tune the report settings.

To access the report settings, click the grey **Settings** button with the gear icon in the upper-right corner (see Figure 5.15).

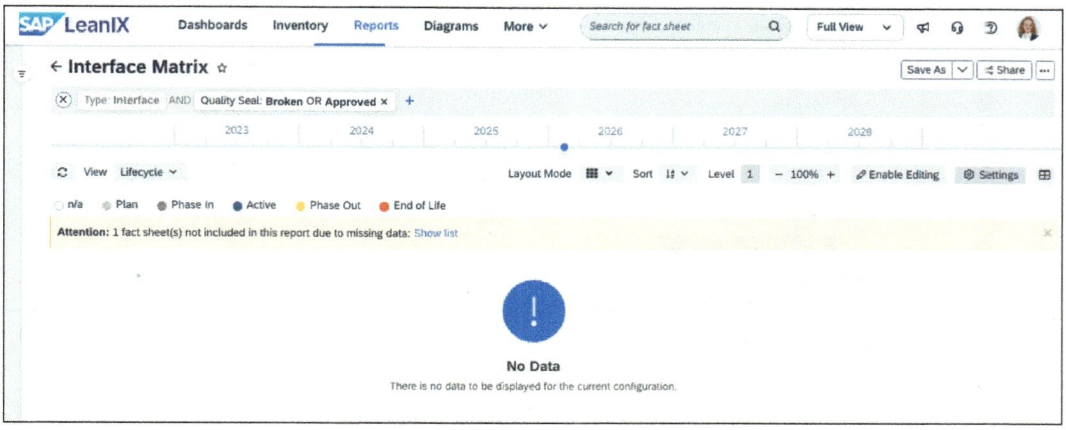

Figure 5.15: New Report Showing No Data

A popup window will appear in which the dimensions of the axes, labels, and other report-specific settings can be adjusted. By default, the interface matrix report will be set to have consumer applications on the x-axis and data objects on the y-axis. You haven't maintained any relations between data objects and interfaces yet, however; therefore, the report shows no data. Switch the axes to relations or attributes that you have maintained; in this example, select **Provider Applications** under **X-Axis** and **Consumer Applications** under **Y-Axis** (see Figure 5.16). When you're finished, click the blue **Apply** button in the lower-right corner.

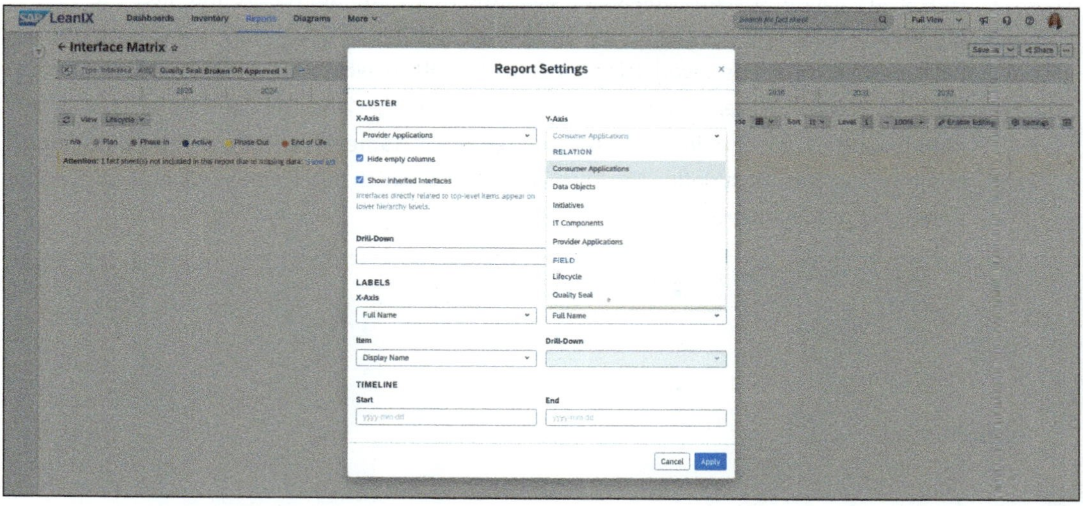

Figure 5.16: Adjusting Report Settings

You will now see the new report template as shown in Figure 5.17. Don't forget to save your new settings for the report template by clicking the arrow next to the **Save As** button in the upper-right corner and selecting **Save**. If you want to save a report variant and not override the template settings, use the **Save As** button instead (we'll explain this in detail later).

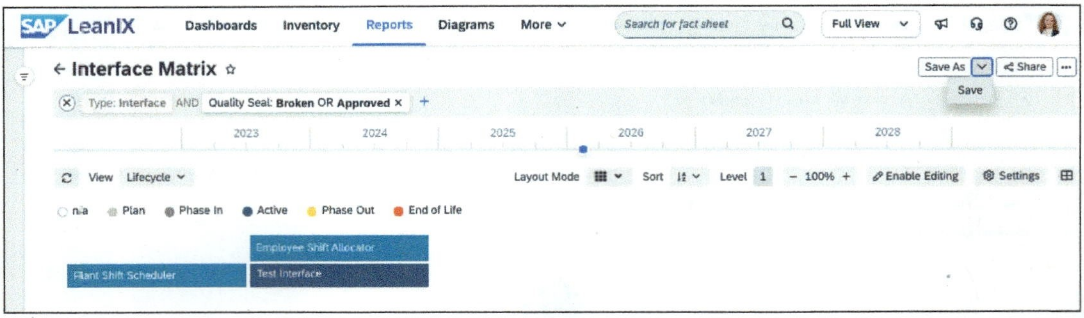

Figure 5.17: Saving New Report Settings

A notification window will ask you to confirm the overwriting of the predefined report template. Click the blue **Confirm** button (see Figure 5.18).

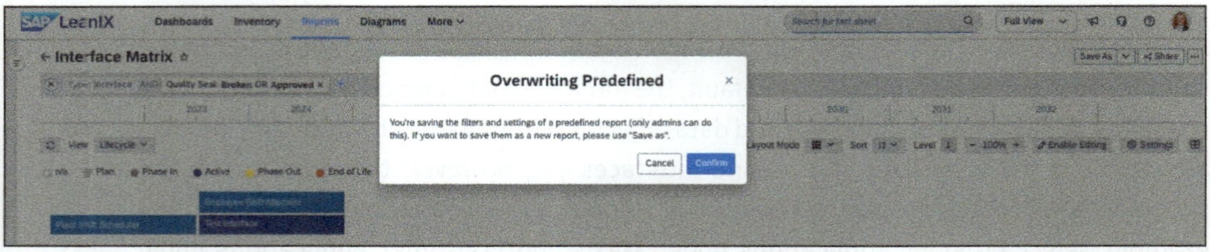

Figure 5.18: Confirming Saving Report Settings for New Template Reports

Now every user in the workspace can open the new template report and filter the data or modify the axes, color-coded views, and report settings however they want. Their individual report views will be saved using the **Save As** button, which will create a personal report variant tailored to their needs and questions. When clicking the **Save As** button, a new popup window appears in which the user enters a **Name** value for the report variant. In addition, the user must define the **Permission Type**, which determines who else will be able to see and modify the report (see Figure 5.19).

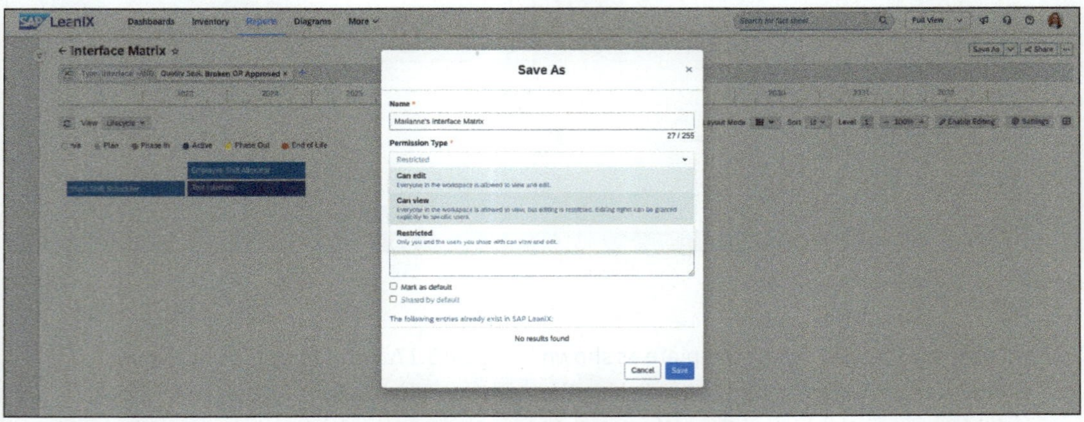

Figure 5.19: Saving Report View/Variant

As a best practice, educate users to only make report variants visible that are needed by a larger group of users, such as entire teams or departments. Otherwise, keep individual report variants **Restricted** to avoid overcrowding the **Reports** overview page. Restricted reports that were created by the current user or shared by a colleague can be found in the **PERSONAL** collection group in the left-hand pane on the **Reports** overview page. You can share the report with a colleague in the **Report Details** settings. To access the settings, click the **...** button in the report tile and select **Report Details** (see Figure 5.20).

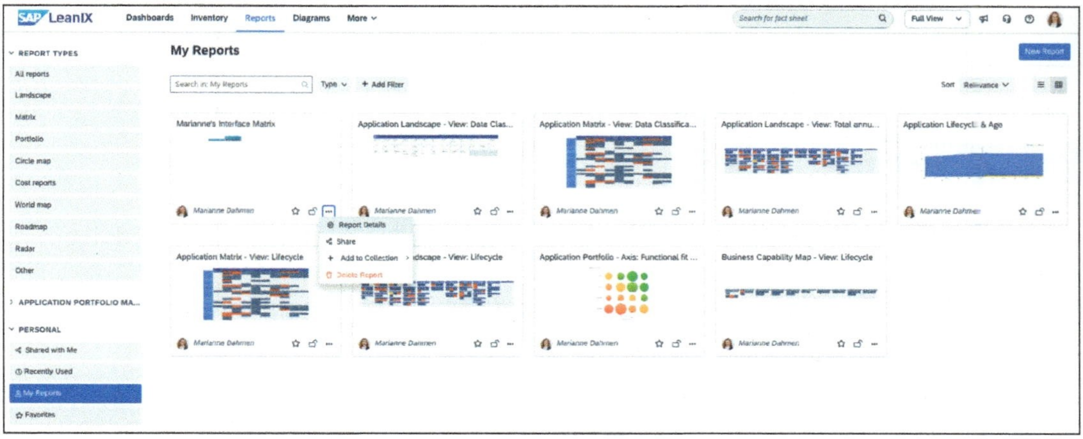

Figure 5.20: My Reports and How to Access Report Details

The **Report Details** menu item opens a new window with three tabs (see Figure 5.21):

- **Settings**
  Here you can change the **Name**, the **Description**, the **Owner**, and some object-related settings, like **Mark as default** or **Add to favorites**, by checking the checkboxes.
- **Permissions**
  Here you can change the **Permission Type** of the report (**Restricted, Can view, Can edit**) and invite users to collaborate on the diagram with individual access rights; for example, colleague A can edit, but colleague B can only view.
- **Shared With**
  This area shows an overview of the individual users you have shared the report with. The report appears at the top of their personalized list.

Similar to the **Report Details** window, you can also access **Diagram Details** and **Dashboard Details** windows with the same structure and same functionality. Only the object-related checkboxes in the **Settings** tab will vary slightly.

In the report details, navigate to the **Permissions** tab and enter the name or email address of your colleague in the **Add a Person** field. Select the correct entry and specify, via the dropdown next to the name, whether the user should have edit or only view rights. Click the blue **Add** button. Before leaving the window, click the blue **Save** button in the lower-right corner (see Figure 5.22).

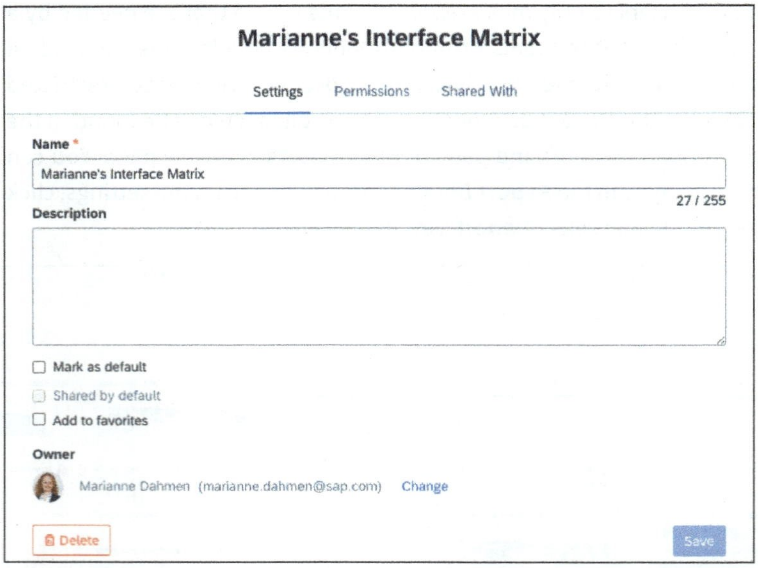

Figure 5.21: Report Details Window

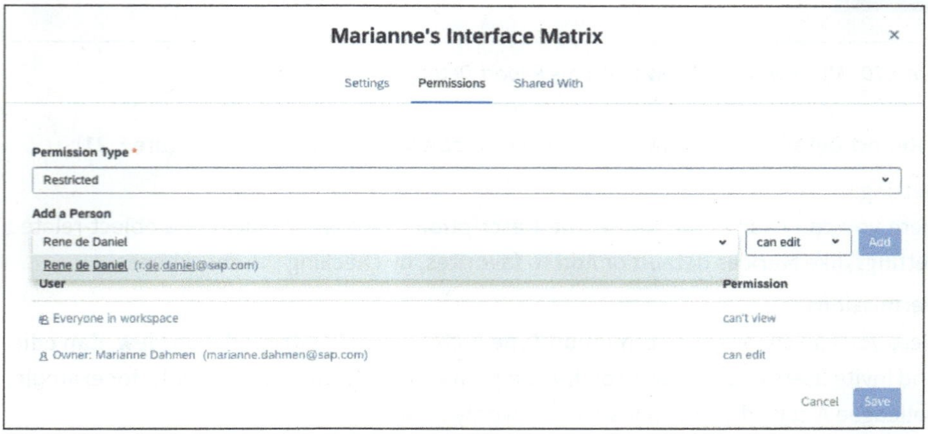

Figure 5.22: Sharing Restricted Report with Other Users

Designing Your Report

The header panes of reports look very similar for most report types. As shown in Figure 5.23, on the left-hand side is the **Filters** area, where you can filter the data in the report. You can filter on all dimensions, meaning not only the base fact sheet in the center but also the axis fact sheets. To switch between the options, click the arrow to the right of the fact sheet type field in the upper-left corner under the **Filters** title.

Similar to the inventory, in the top line you can see the applied filters, and you can switch the operators (**AND**, **OR**, **IS NOT**) by clicking them.

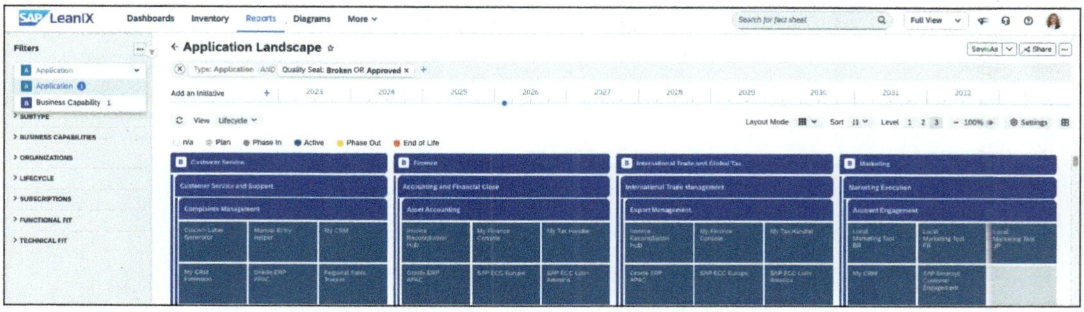

Figure 5.23: Report Canvas with Filter Options per Axis and Fact Sheet Types Involved

Underneath the filter line, you can see the timeline. This feature adds value when the lifecycle dates in the fact sheets are maintained. You can move the blue dot on the timeline into the past or into the future to simulate lifecycle changes in the base fact sheets in the report. If you hold the `Shift` button while clicking, you can also select a range to see the changes of lifecycles over time.

Below the timeline you can find the **View** dropdown, which allows for color-coding of reports based on fields or tags. To select single-select values in the **View** for color-coding, you have to ensure that the values have assigned colors in the meta model configuration in the field edit tab with the sliders icon (see Figure 5.24) and that the setting for **Include in views** is toggled on in the **Settings** tab (see Figure 5.25).

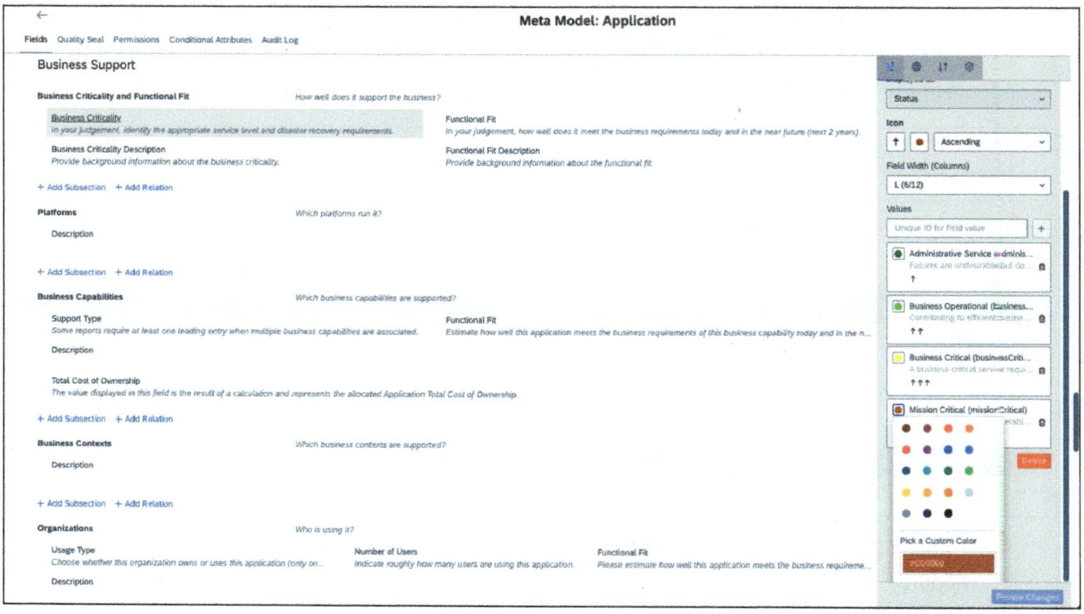

Figure 5.24: Fact Sheet Meta Model Configuration: Color Selection for List Values

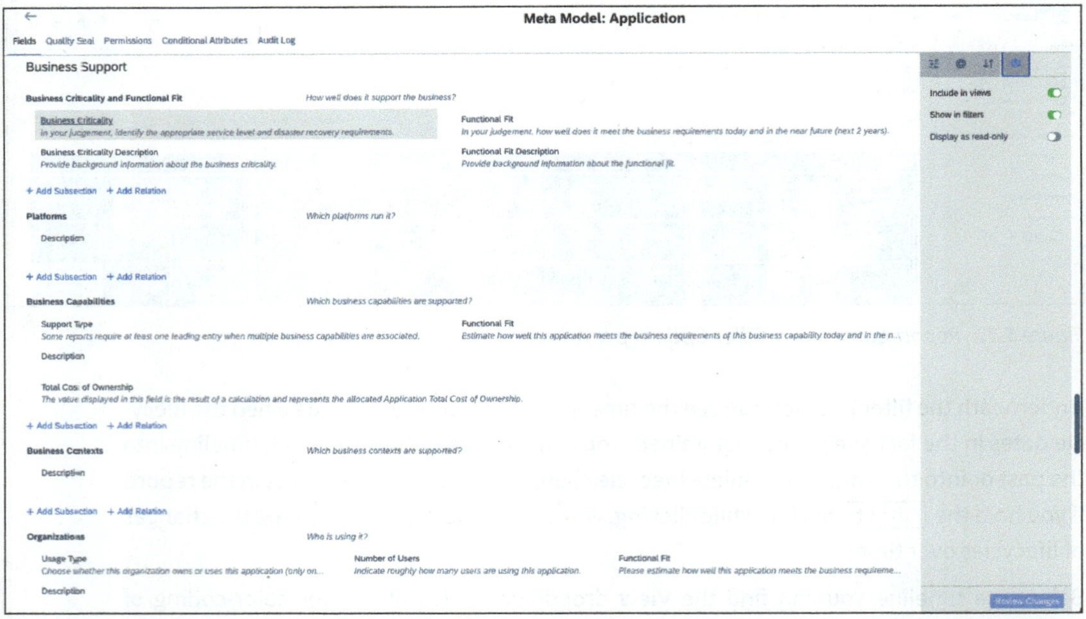

Figure 5.25: Fact Sheet Meta Model Configuration: Toggle on Include in Views for List Values

On the upper-right side, you can find various options to change the layout and aggregation level of the report. The default **Layout Mode** selection is **Detailed**, but **Stacked** or **Rolled-up** can be useful as well, depending on the question to be answered with the reports. Figure 5.26 shows an example of a rolled-up report, aggregated to level 1 business capabilities and zoomed in to 120%.

Figure 5.26: Example of Changed Layout (Rolled-Up) and Aggregation (Level 1) in Landscape Report

The **Settings** menu is slightly different per report type. Therefore, we recommend familiarizing yourself with the different report types and their settings to help users create the most meaningful reports for their needs.

Above the timeline in the upper-right corner is the **Save As** button, used to save a variant. You can also use the **Save** button in the dropdown to overwrite the current variant version. The **Share** button allows you to quickly share a report with one or more colleagues. The **...** button will open the extended menu, where you can navigate to the **Report Details** settings; to the **Discard All Changes** button, which will bring you to the last saved variant of this report view; or to the **Export** function, with options to download a **PDF File**, a **PNG Image**, or an **HTML Embed Code** (see Figure 5.27).

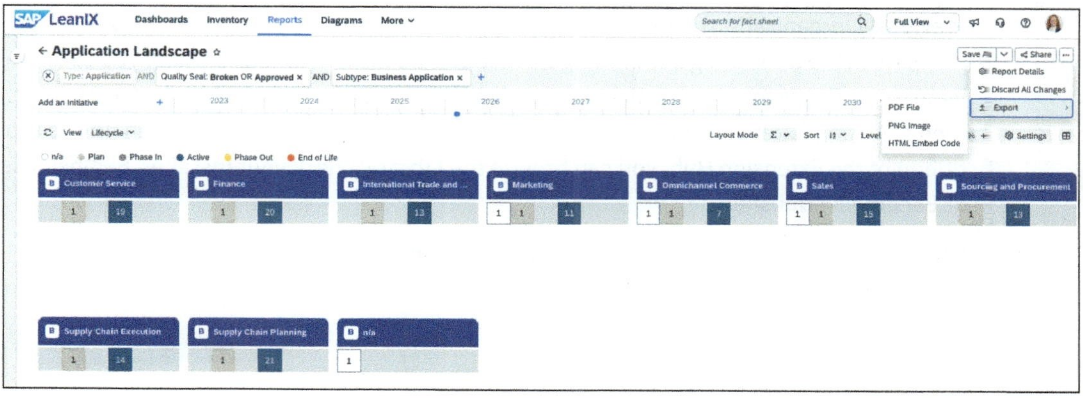

Figure 5.27: Top-Right Reports Menu and Exporting Report Options

5.1.3 Custom Reports

If the out-of-the-box reports aren't sufficient for your reporting purposes, you can create custom reports. Ideally, you should have a basic understanding of the following before starting the custom report journey:

- SAP LeanIX authentication via API tokens
- SAP LeanIX reports
- React, TypeScript
- GraphQL APIs
- REST APIs

You can also use an artificial intelligence (AI) support tool to navigate you through the process.

SAP LeanIX product and engineering developed a custom report toolset. It consists of the *SAP LeanIX reporting library*, published as a public library called *@leanix/leanix-reporting* on

GitHub (see *https://github.com/leanix/leanix-reporting*), and *SAP LeanIX custom report tools*, an npm package (see *http://s-prs.co/v620643*) with React/TypeScript templates to speed up your custom report development, testing, and uploading to the workspace.

SAP LeanIX reports contain HTML, JavaScript, and CSS files and are loaded into iFrames inside SAP LeanIX. The reporting library takes care of the communication with SAP LeanIX via the Reporting Library API. Various features are supported by the SAP LeanIX reporting library, for instance, filtering of fact sheet information, custom dropdowns, color-coded views, or formatting numeric values as currencies.

On GitHub, you can also find *AI_AGENT_GUIDE.md*, which can be reused to instruct an AI tool to support your custom report creation.

Previously, organizations and partners could share their custom reports via the SAP LeanIX store. Now, the publishing and sharing will happen via the new Extension Hub. You can navigate to the Extension Hub by clicking **More** in the top ribbon of your SAP LeanIX workspace, then selecting **Extension Hub**. You will be redirected to a new overview screen. In the top ribbon of the Extension Hub, you can browse the **Library**, check your **Installed Extensions**, or **Become a Publisher**.

Figure 5.28 shows an example of a custom report published in the Extension Hub.

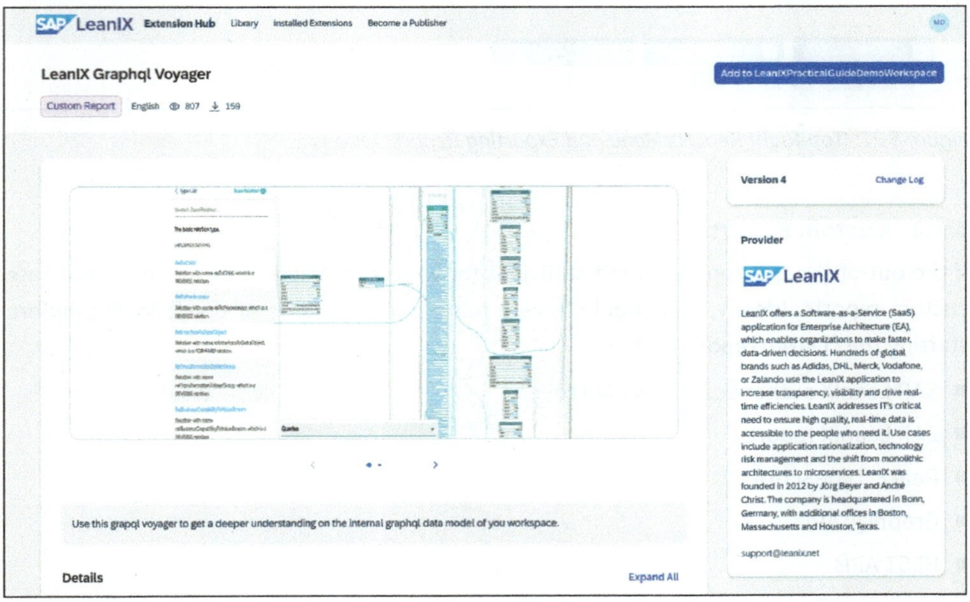

Figure 5.28: Example of Custom Report in SAP LeanIX Extension Hub

Further Resources

For the latest information on custom reports, visit the SAP LeanIX user documentation at *http://s-prs.co/v620644*.

For a detailed step-by-step guide, visit the following pages in the SAP LeanIX user docs:

- *http://s-prs.co/v620645*
- *http://s-prs.co/v620646*

5.2 Diagrams

Diagrams are core artifacts in enterprise architecture work and a very popular SAP LeanIX feature. They are often used to visualize complex concepts, including multiple fact sheet types in one visual. Shapes can be linked to inventory entries to increase architectural transparency across functions and usage.

We'll walk through the process to create and edit diagrams in the following sections.

Further Resources

For the latest information on the diagrams feature in SAP LeanIX, visit the SAP LeanIX user documentation at *https://help.sap.com/docs/leanix/ea/diagrams*.

5.2.1 Draw.io as the Technical Foundation

The diagram editor is based on draw.io capabilities. The available features are curated to ensure SAP LeanIX compatibility. To check which version of draw.io is currently used in SAP LeanIX, click the ? icon in the upper-right corner of the top ribbon in the SAP LeanIX workspace, then select the **About SAP LeanIX** menu item (see Figure 5.29).

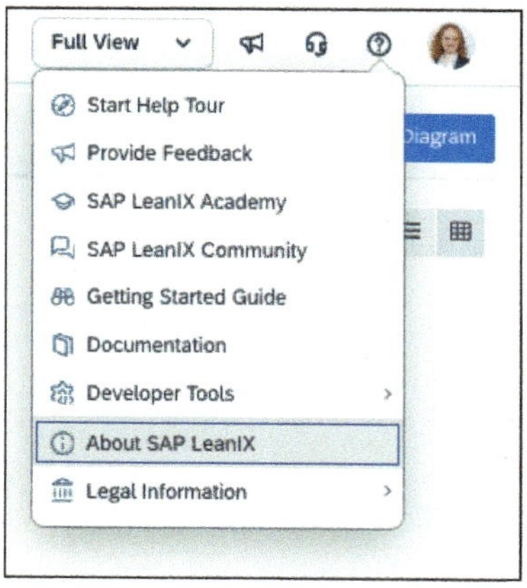

Figure 5.29: Accessing About SAP LeanIX from Top Menu

In the popup window, switch to the **Licenses** tab and scroll down to the draw.io version information (see Figure 5.30).

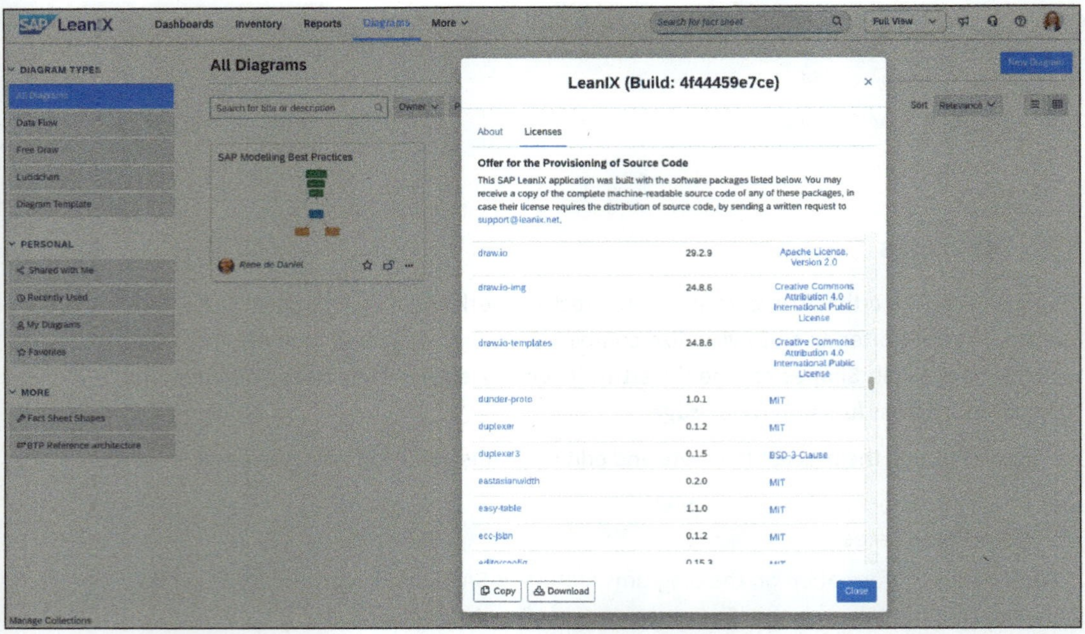

Figure 5.30: Browsing Licenses and Software Packages Used in SAP LeanIX

5.2.2 Navigating Diagrams

The **Diagrams** top ribbon tab, as shown in Figure 5.31, is structured similarly to the **Reports** top ribbon tab. On the left-hand side, you can find the collection groups and collection folders. In the center, you can see the grid with available diagram tiles. In the middle ribbon, you have various filter options to search for specific diagrams. Finally, on the right side, you can change the sorting logic or switch between grid or list view.

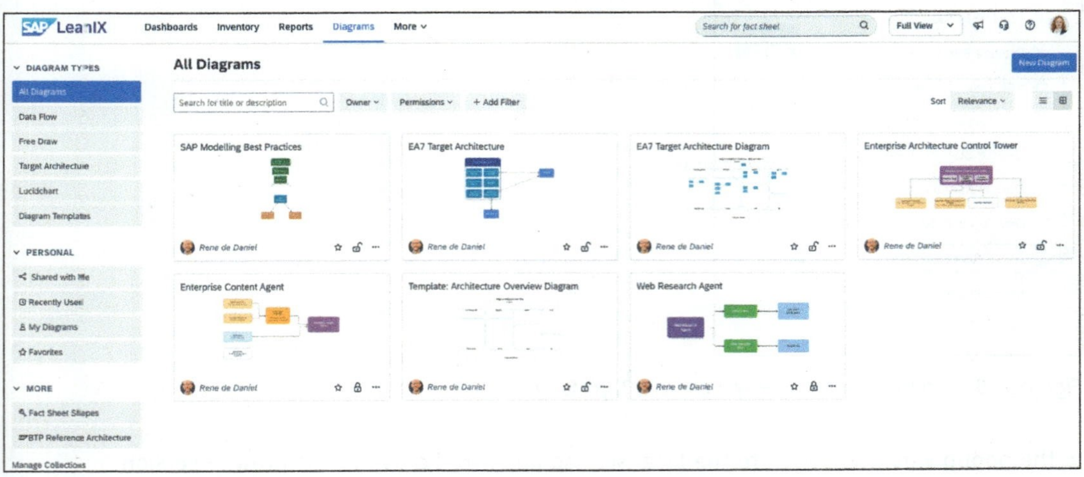

Figure 5.31: Top Ribbon Diagram Tab: Overview

5.2.3 Diagram Types

The blue **New Diagram** button in the upper-right corner opens a new window in which you can select the diagram style (see Figure 5.32).

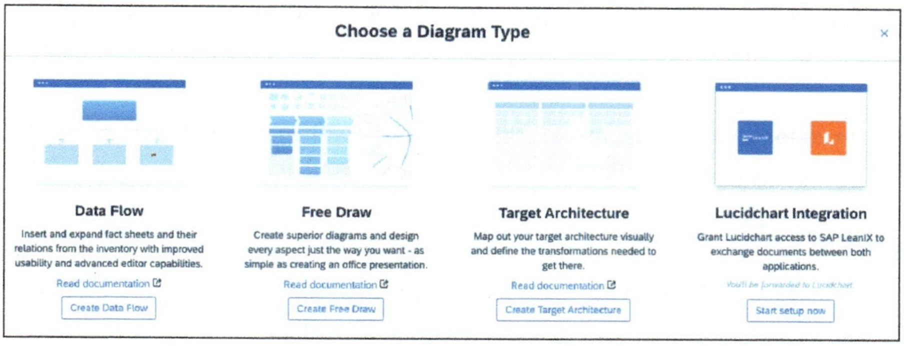

Figure 5.32: New Diagram Input Form to Choose Diagram Type

There are four options:

- **Data Flow**
  The data flow diagram is designed specifically to showcase applications that are connected via interfaces. The features in this diagram type are designed to quickly and semi-automatically create diagrams of connected application ecosystems. There are a few prerequisites:
  - *Mandatory:* Modeling of applications with provided and consumed logical interface relations
  - *Optional:* Interface lifecycle dates (connection line color) and properties (data flow direction attribute to create connecting arrow direction)
- **Free Draw**
  As the name indicates, with the free draw diagram, the user has complete flexibility in terms of how to design diagrams and what kind. Users can use shapes from different shape libraries and add symbols, text, and various connectors like layers and links to other diagrams.
- **Target Architecture**
  This type is only available with the SAP LeanIX Architecture and Road Map Planning product. This diagram type's editor has advanced functionalities to model changes and target state architecture visually in diagrams before changing inventory data. It supports the planning and visualization of target states and differential views that compare current and target states.
- **Lucidchart Integration**
  This is actually not a diagram style but an integration to import and export diagrams between SAP LeanIX and *Lucidchart*, a cloud-based diagramming platform offering features for visual and real-time collaboration between users beyond standard diagramming capabilities.

5.2.4 Diagram Editor

Let's explore the diagram editor with a free draw diagram example.

When you click the **New Diagram** button (see Figure 5.31) and select the **Free Draw** diagram type (see Figure 5.32), you are then guided to a form to enter a diagram **Name** and select the **Permission Type**. This form is similar to the form for saving a report variant (see Section 5.1.2).

The diagram editor, shown in Figure 5.33, has a top ribbon with menus and toolbars. On the left-hand side are various shape libraries that can be collapsed or expanded.

> **Note**
>
> The user interface of the diagram editor looks exactly the same for the data flow, free draw, and target architecture diagram types; there are just different features available in each variant. For instance, in the data flow diagram, you can only insert applications from the inventory and no other fact sheet type as it only shows applications as shapes.

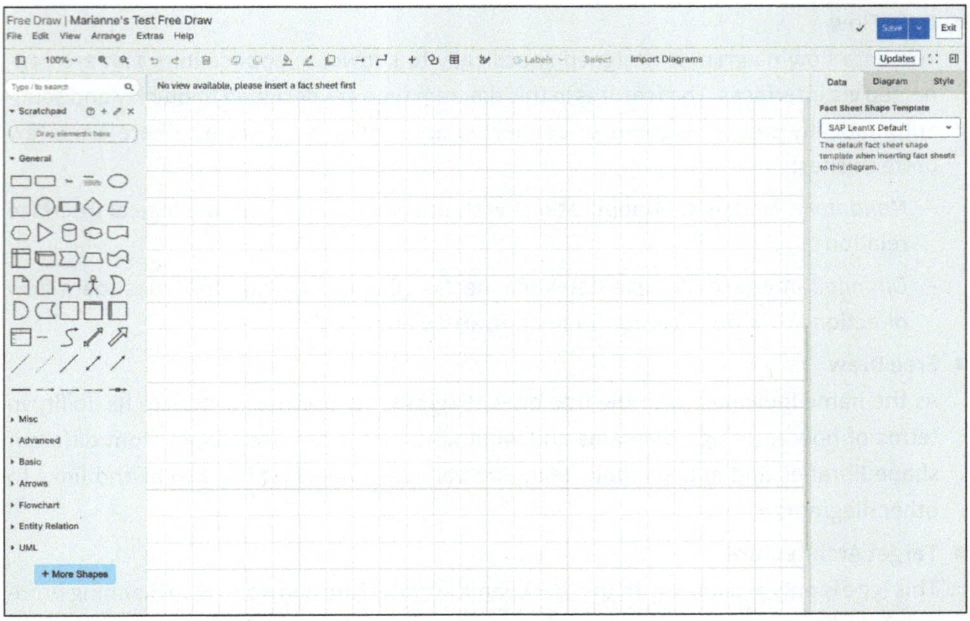

Figure 5.33: SAP LeanIX Diagram Editor Canvas

If you click the **+ More Shapes** button in the lower-left corner, then a new popup window opens in which the selected shape libraries can be customized. Simply check or uncheck the libraries that you want to use and click the **Apply** button in the lower-right corner (see Figure 5.34).

The center canvas is used to design the diagram. You can either drag and drop shapes onto the canvas or right-click to access actions like inserting artifacts from the SAP LeanIX inventory or selecting related fact sheets by filters. On the right-hand side is the format panel, where styles can be changed and fact sheets can be linked or unlinked.

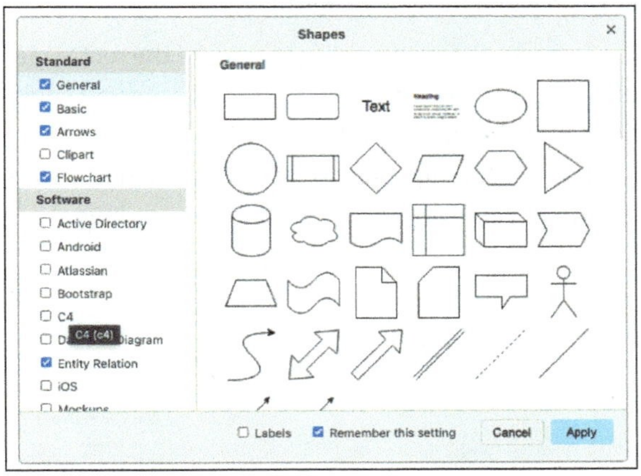

Figure 5.34: Customizing Shape Libraries in the Diagram Editor

When you right-click and select **Insert from Inventory**, a popup window appears with an almost exact copy of the inventory (see Figure 5.35).

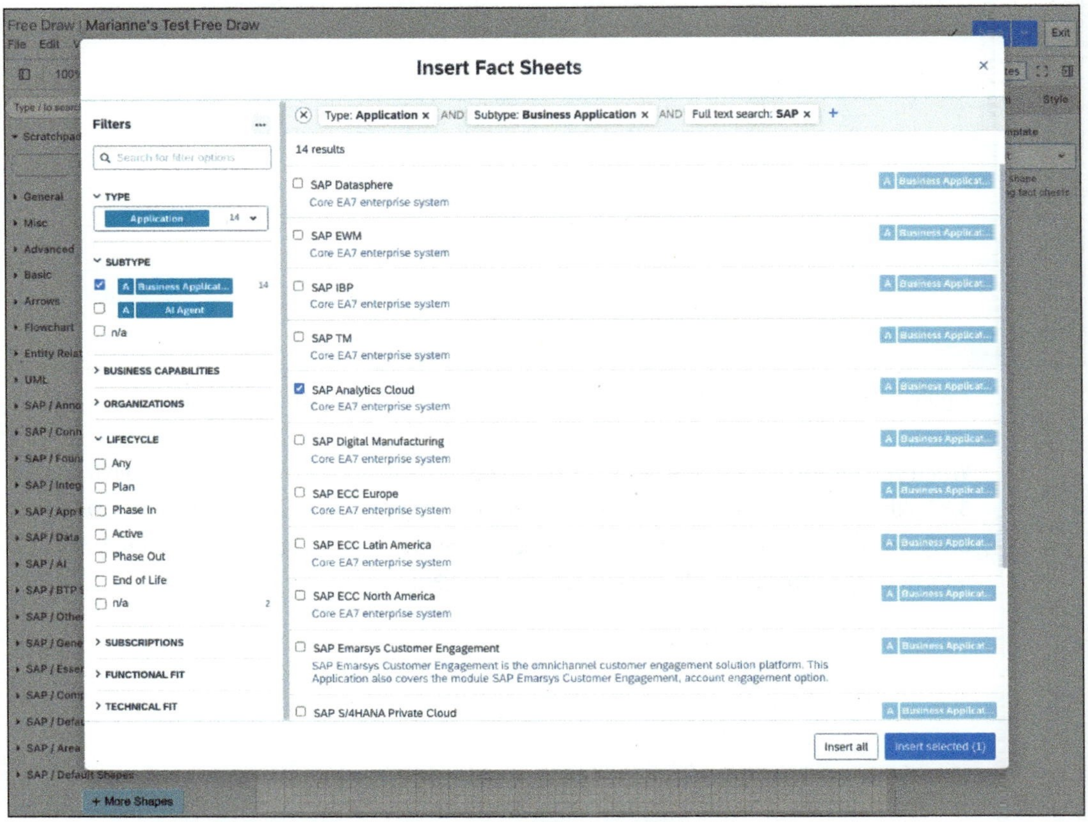

Figure 5.35: Selecting Fact Sheets in Inventory Search to Add in Diagram Editor

You can filter it down as you would for inventory filtering and check the artifacts that should be added and linked to the diagram. After filtering, you can either click **Insert all** to add all filtered artifacts or click **Insert selected** to only add the checked artifacts.

When you hover over a shape, you can see arrows directing to all four sides of the rectangle. When you hover over one of these arrows, you can see which related fact sheets you could bring in as dependencies. You can either connect and show all related fact sheets of a certain type by clicking it or you can filter it down further by clicking the **Filter** icon next to the desired related fact sheet type (see Figure 5.36). After you click the **Filter** icon, an **Inventory** view appears to filter the available values.

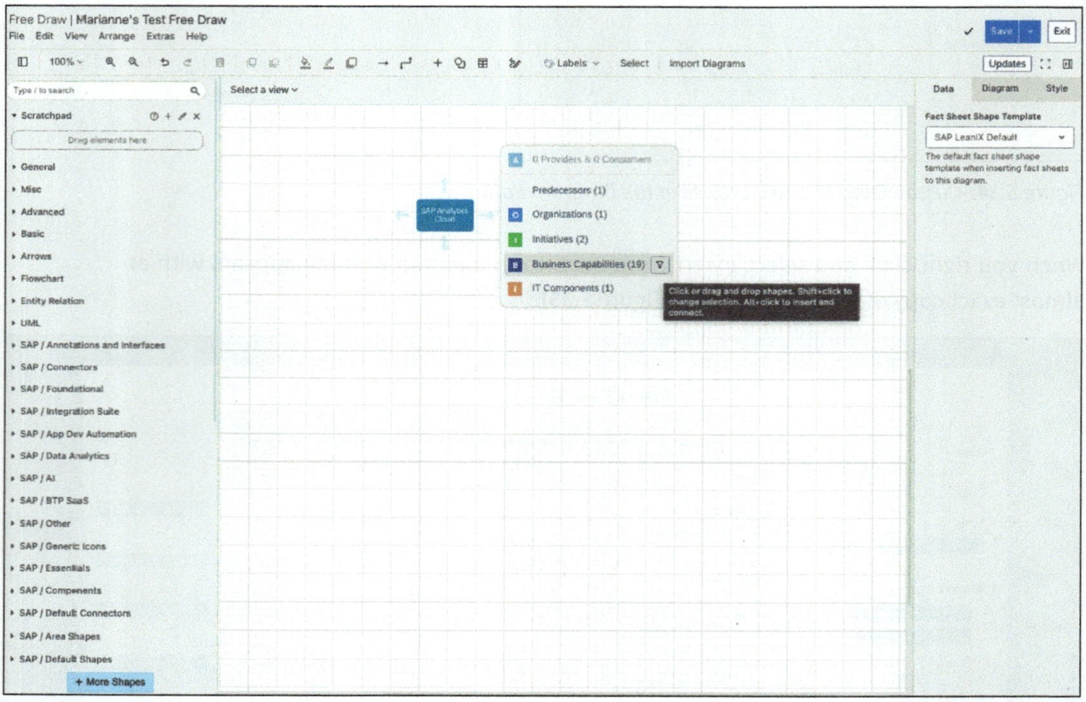

Figure 5.36: Insert Filtered Dependencies via Arrow Hovering in Canvas

When you right-click a shape, an action box appears with lots of options, as shown in Figure 5.37. What was previously done using the arrows is synonymous to using the **Show Dependency** or **Show Dependency by Filter** options. For the related business capability level 3, let's **Roll Up** to the parent fact sheet (see Figure 5.38).

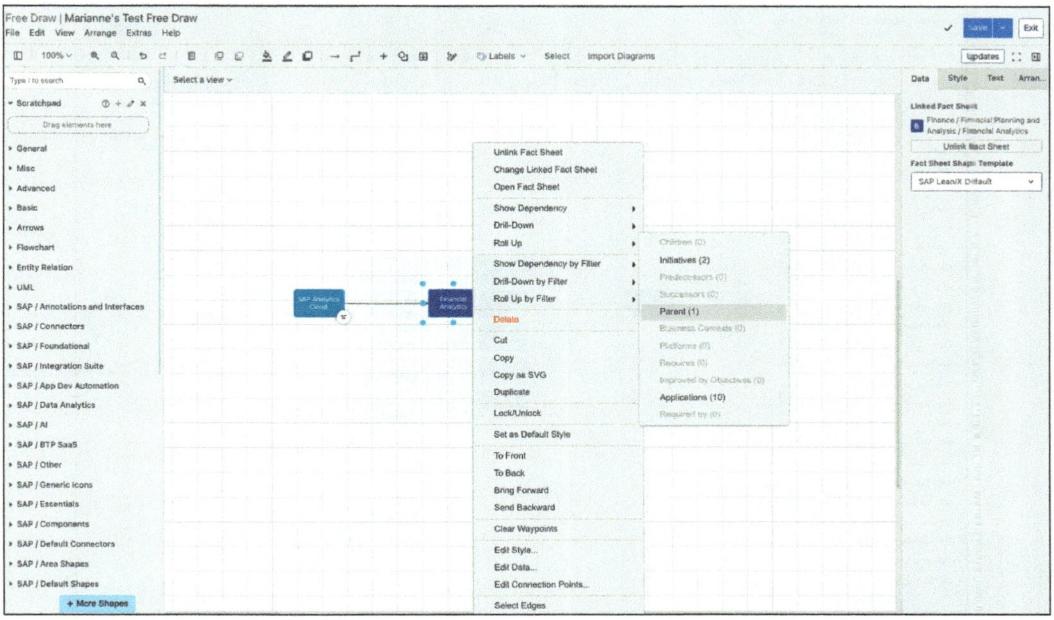

Figure 5.37: Diagram Editor Right-Click Action Box

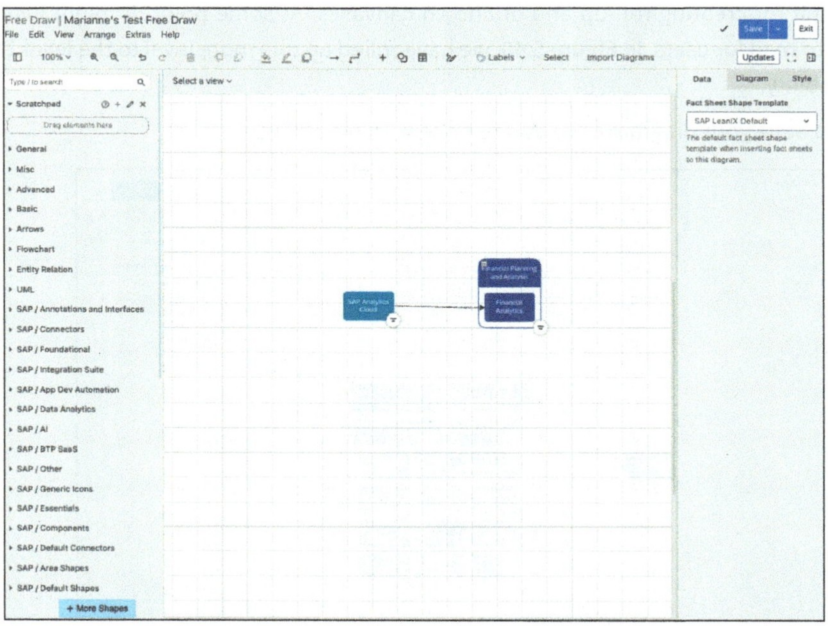

Figure 5.38: Level 3 Business Capability Wrapped in Level 2 Business Capability Container

You can now select the parent business capability shape and create a drilldown into the children fact sheets, to see what other business capabilities belong to the same cluster and are not related to the SAP Analytics Cloud fact sheet (see Figure 5.39).

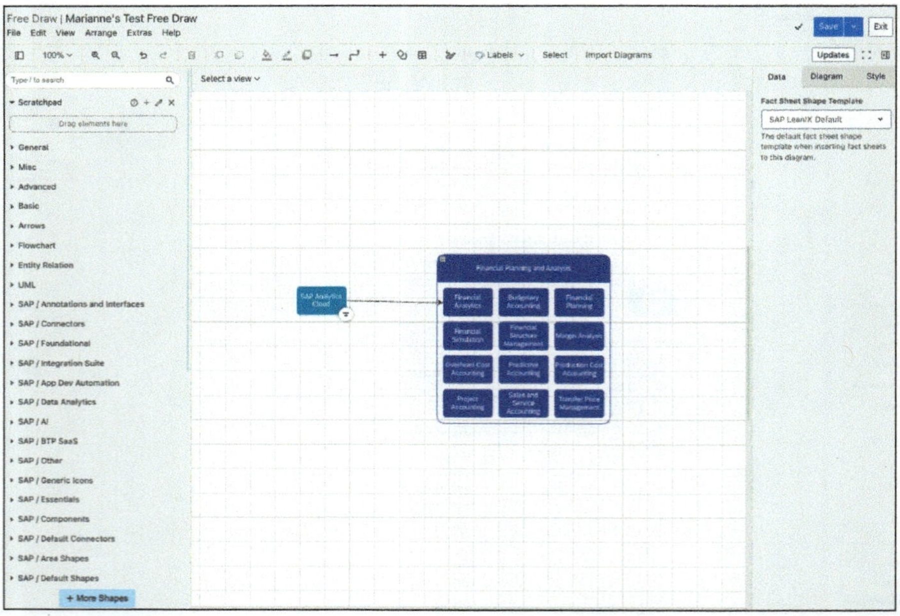

Figure 5.39: Drilldown in Business Capability 2 Container into Child Fact Sheets

There is no limit for creating roll-up and drilldown canvases. At some point, it might just become hard to read for users. In Figure 5.40, we have rolled up one more level to the level 1 business capability.

Don't forget to **Save** your diagram once you're done with modeling.

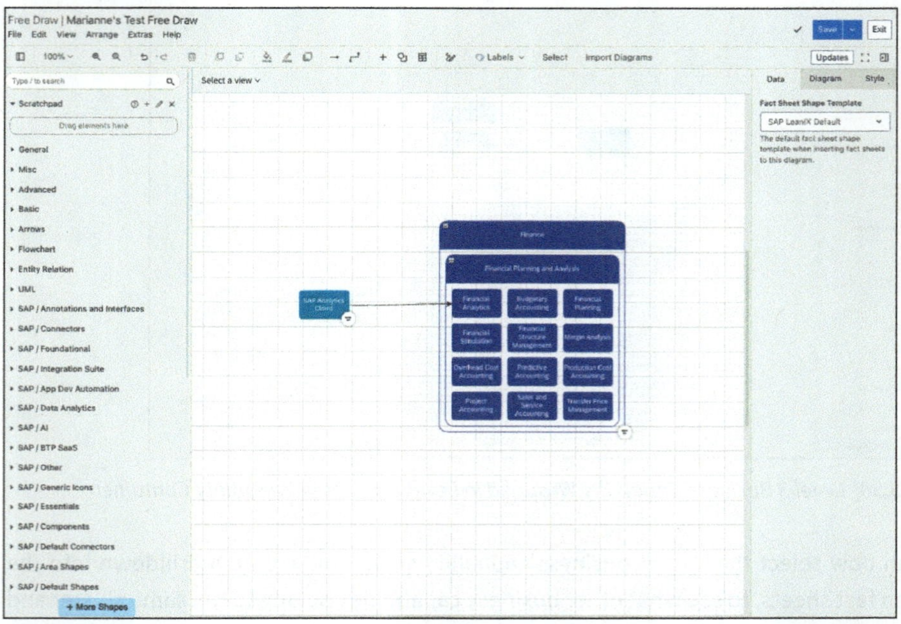

Figure 5.40: Rolled-Up Business Capability Container to Level 1 Business Capability Parent

5.2.5 Color-Coding Diagrams

When you open a diagram, it's possible to color-code the fact sheet types based on list values, similar to color-coding reports.

When in the view mode of a diagram, you can see the **Select a view** dropdown in the upper-left corner. When hovering over the available fact sheet types, you can see the various options for color-coding (see Figure 5.41).

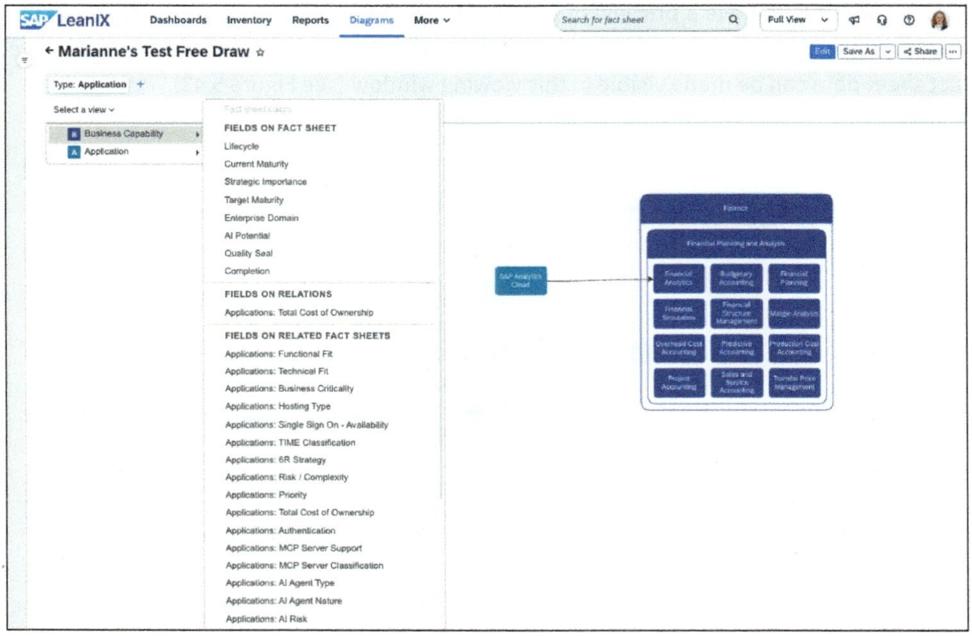

Figure 5.41: Select View Menu for Color-Coding Shapes by List Values or Tags

Let's test the **Current Maturity** business capability values (see Figure 5.42).

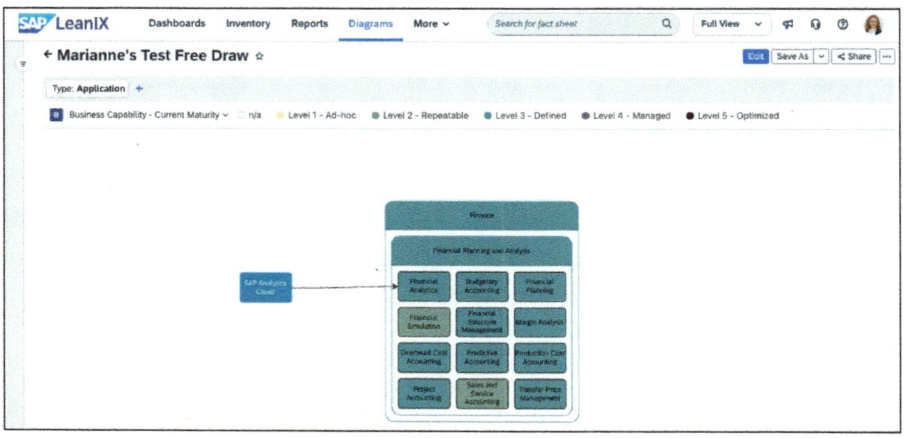

Figure 5.42: Color-Coded Diagram: Current Maturity Business Capability

This color-coding can happen on the fly without editing or modifying the diagram, and the view will not be saved automatically. It's possible to save a diagram variant with this specific color-coded view applied to it. To do so, simply click the **Save As** button in the upper-right corner of the canvas.

5.2.6 Fact Sheet Preview

It's also possible to have a preview of fact sheet data appear when clicking a shape in the diagram viewing mode. With the **Configure Content** button in the upper-right corner, more fact sheet data can be made visible in this viewing window (see Figure 5.43).

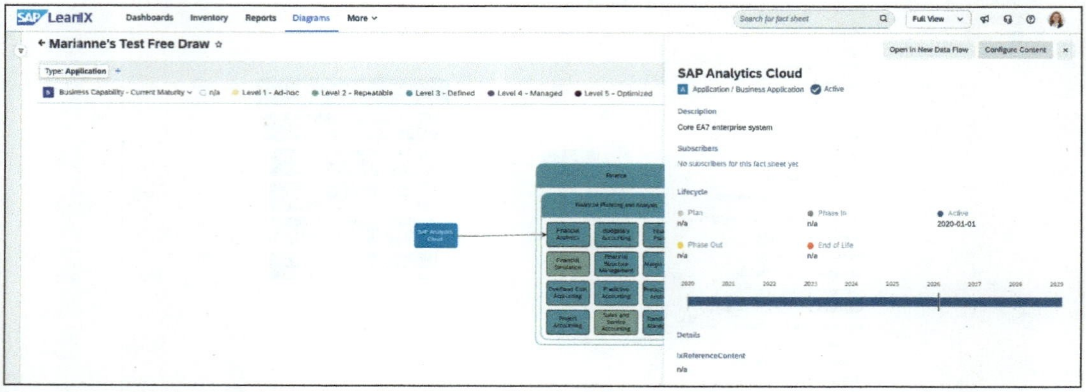

Figure 5.43: Fact Sheet Preview Window When Clicking on Linked Diagram Shape

It's also good to know that users can see all diagrams that the fact sheet is linked to in the **Resources** tab in the fact sheet information (see Figure 5.44).

Figure 5.44: Reference of Diagram in Fact Sheet Resources Tab

Further Resources

To learn more about advanced diagram features, visit the SAP LeanIX user documentation:

- Adding layers: *http://s-prs.co/v620647*
- Filtering diagrams: *http://s-prs.co/v620648*
- Inventory updates: *http://s-prs.co/v620649*

5.3 Dashboards

Dashboards are a great stakeholder- and use case-targeted communication channel as they are also the default landing pages when entering the SAP LeanIX workspace via the URL path. To revisit the administration settings for dashboards, including how to define a default dashboard and how to activate the SAP LeanIX out-of-the-box template dashboards, visit Chapter 3, Section 3.3.3.

We'll dive into creating and managing dashboards, as well as visualizing KPIs, in the following sections.

Further Resources

For the latest information on SAP LeanIX dashboards, visit the online user documentation at *https://help.sap.com/docs/leanix/ea/dashboards*.

5.3.1 Creating a New Dashboard

After clicking the **Dashboards** top ribbon menu item, you'll see the dashboard window with the default dashboard in the central grey frame (see Figure 5.45). On the left, you can find the standard menu, similar to that for reports and diagrams, which offers quick links to the **Default Dashboard**, the **Architecture Executive Dashboard**, and the **DASHBOARD OVERVIEW** and **PERSONAL** sections. At the bottom of the left-hand menu, you'll see the **Manage collections** hyperlink, which you would click to modify or change the order of the collections. Under the top ribbon menu, another subheader section contains the name of the dashboard currently displayed, a star icon next to the name that you can click to mark it as a favorite, and three grey buttons:

- **+ New Dashboard**
  To create a new dashboard
- **Edit Dashboard**
  To modify the current dashboard on display
- **... (Dashboard Details)**
  To change the dashboard settings and access rights

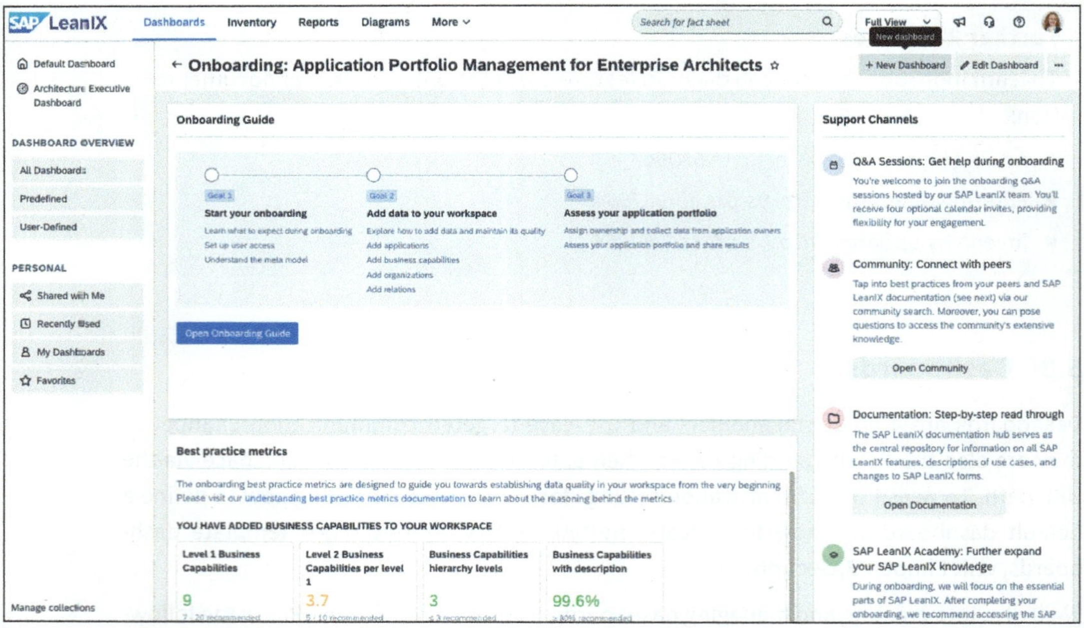

Figure 5.45: Creating New Dashboard from Dashboard Top Ribbon Menu

When you're in a collection, you won't see the grey buttons to manage the specific dashboard settings; there will only be a blue button to create a **New Dashboard** (see Figure 5.46).

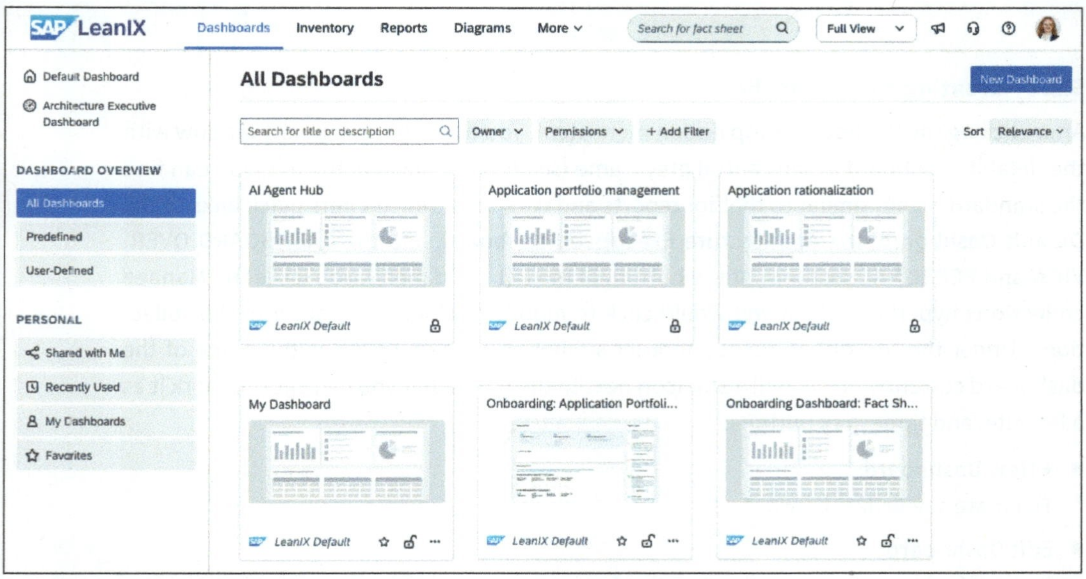

Figure 5.46: Creating New Dashboard from Collection Menu

After clicking the button to create a new dashboard, a new form appears in which you can define the **Name** of the dashboard, select the **Create an empty dashboard** or **Copy existing dashboard** option, and optionally enter text in the **Description** field (see Figure 5.47).

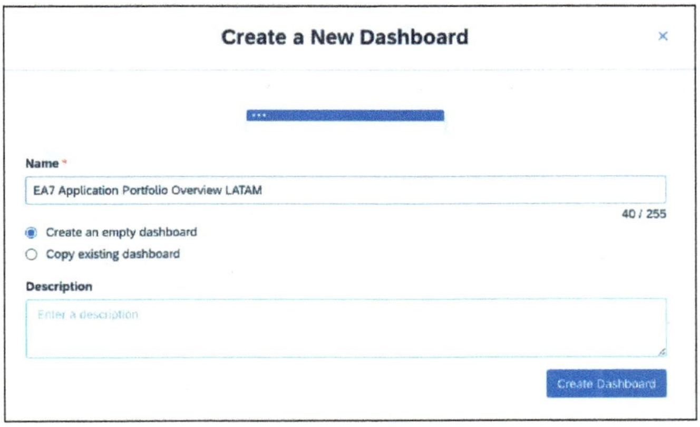

Figure 5.47: Create New Dashboard Form

You'll arrive at the dashboard editor, shown in Figure 5.48, which consists of a header pane with the dashboard name and the **Apply global filter** function. The center area is a grid into which dashboard panels from the right-hand menu can be placed via drag and drop.

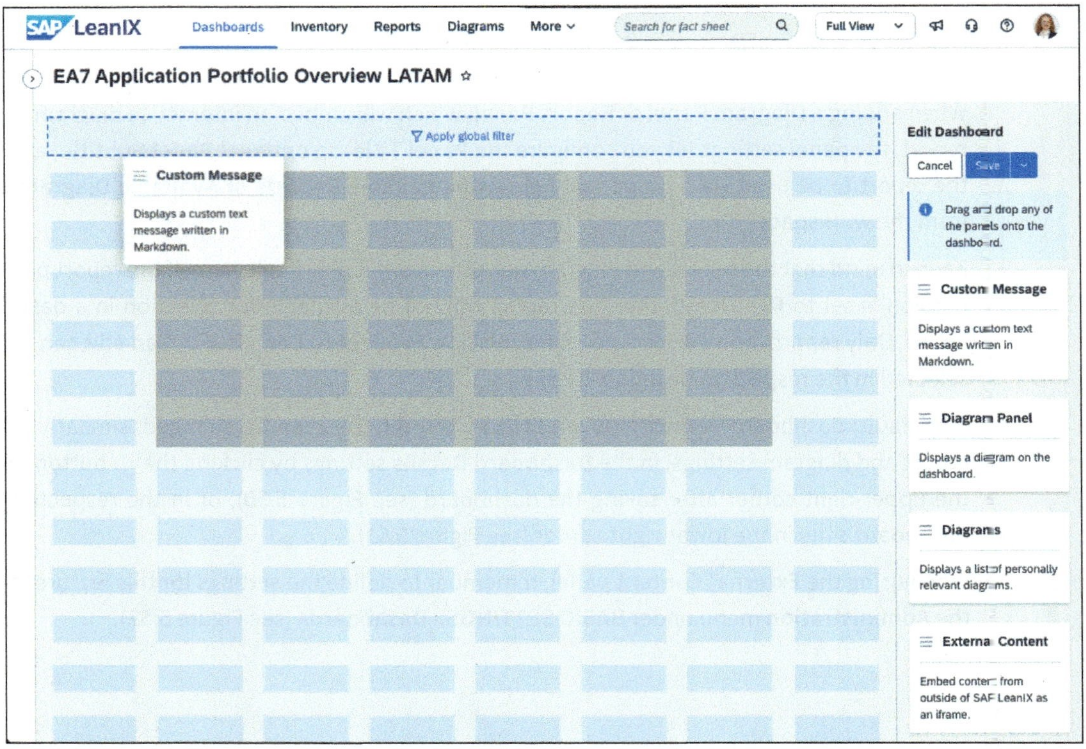

Figure 5.48: Dashboard Editor

It's also possible to resize the elements by hovering over the edges until the cursor changes into a double-sided arrow. Click and hold while moving the edge to create the desired width or length (see Figure 5.49).

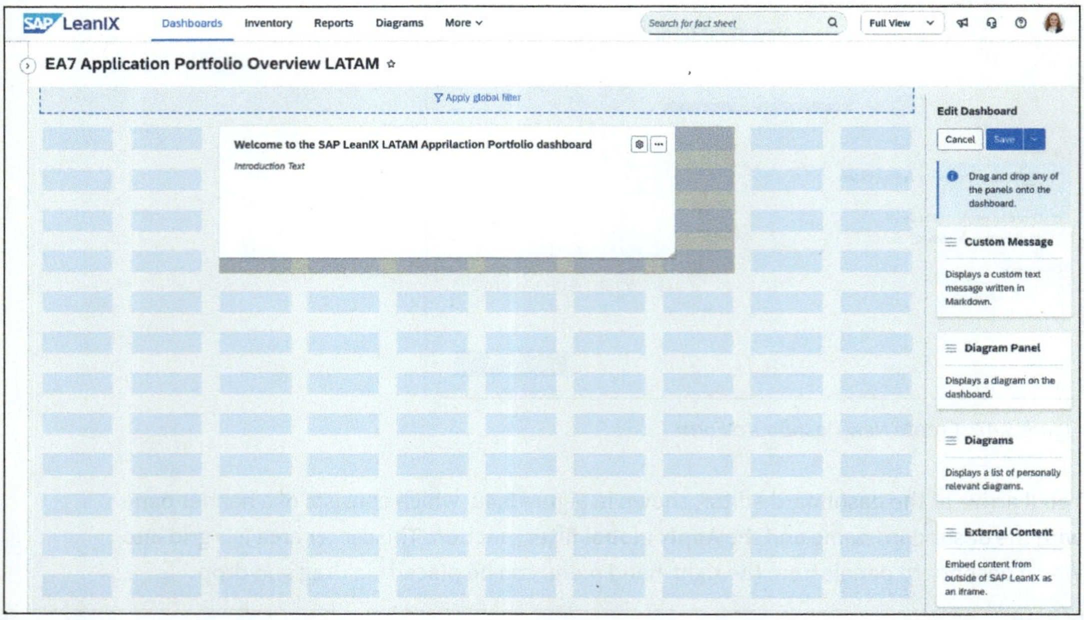

Figure 5.49: Resizing Elements in Dashboard Editor

When adding a **Diagram Panel** or **Report Panel**, which will embed an interactive diagram or report, the panel settings let you configure the **Panel Title**, an optional **Panel Subtitle**, and the report to be used via a dropdown field with **Available Reports** or **Available Diagrams** from the workspace.

Keep in mind that the report and diagram permissions will play a role here. If the report permission is set to **Restricted**, then the report will not be available for selection in a dashboard. Only reports and diagrams with **Permission Type** set to **Can view** or **Can edit** can be selected in the respective dashboard panel.

By default, dashboard permissions are set to **Can edit**. This can be changed, similarly to report and diagram settings, in the **Dashboard Details** settings by clicking the **...** button in the upper-right corner after saving the dashboard (see Figure 5.50), or in the respective dashboard's tile in the lower-right corner (see Figure 5.51).

When using the **External Content** panel, remember to define the settings for this feature in the **Administration** menu under **BASIC SETTINGS • Dashboards** (see Figure 5.52).

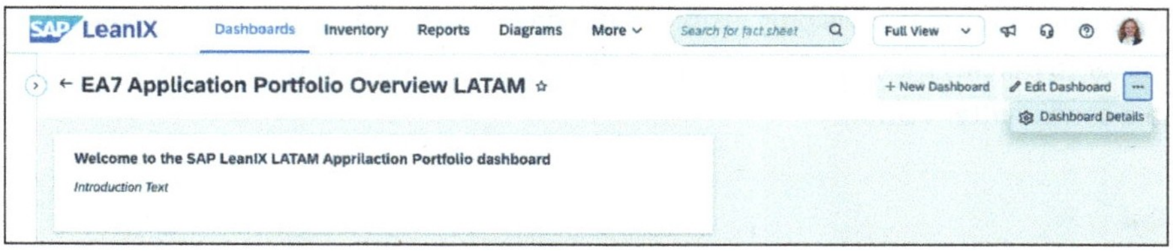

Figure 5.50: Dashboard Detail Settings Access from Dashboard View

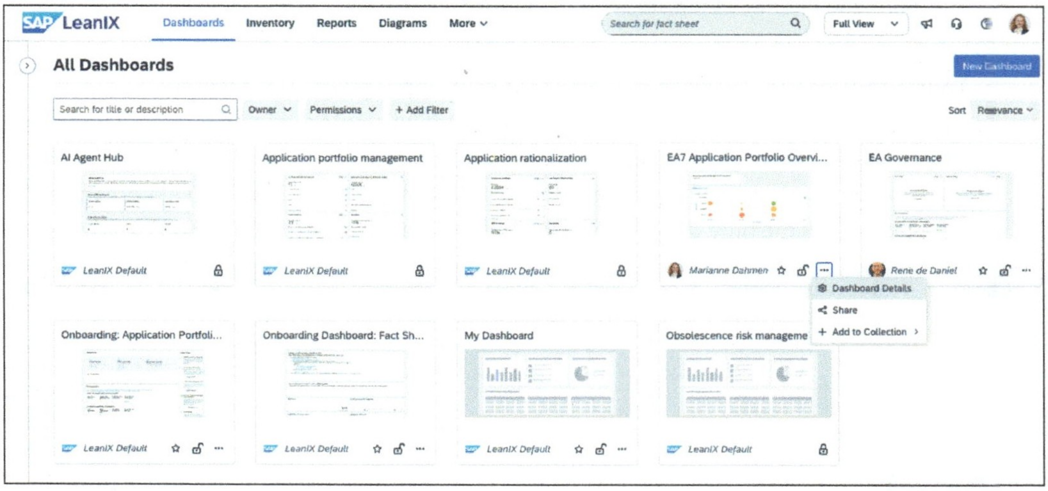

Figure 5.51: Dashboard Detail Settings Access from Dashboard Collection Overview

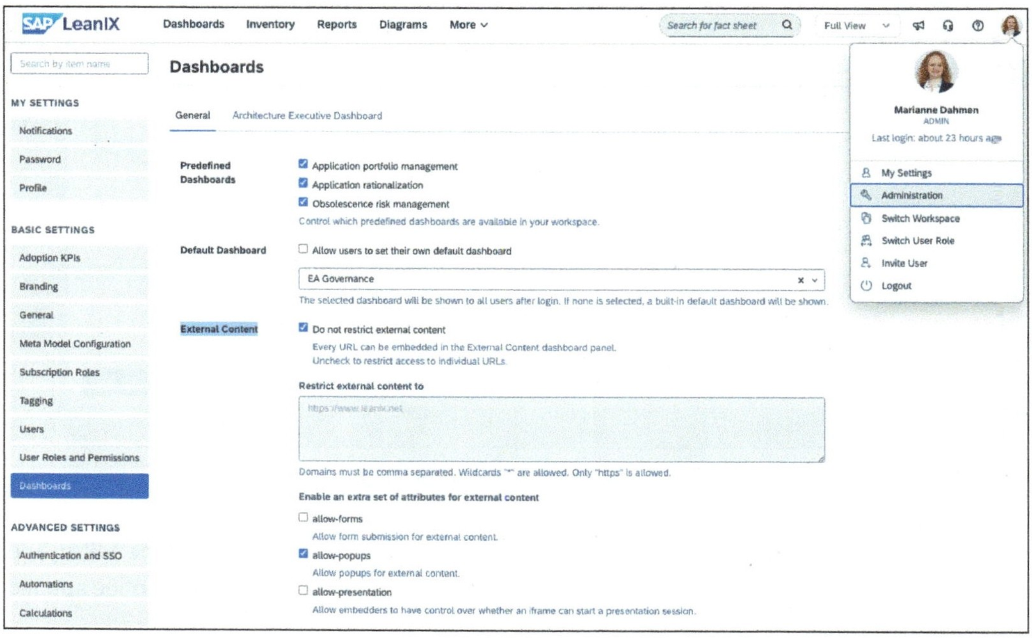

Figure 5.52: External Content Settings in Administration Menu

The **Apply global filter** feature is only visible in the dashboard edit mode. It helps the dashboard creator to quickly modify all reports, fact sheet charts, and saved searches on a dashboard via a central filter option.

Caution

The filters aren't visible on the dashboard in view mode, so make sure you add additional information about the filters in the dashboard name or the panel titles, or explain the filter in a **Custom Message** panel.

To use this feature, click the blue box that says **Apply global filter** under the dashboard name. An inventory window will appear in which the inventory can be filtered down using the known inventory filtering features. In this example, as shown in Figure 5.53, we want to filter for all **Applications** that have a connection to the **EA7 Global / LATAM** organization fact sheet. Click the blue **Use Fact Sheet Filter** button once you're done filtering.

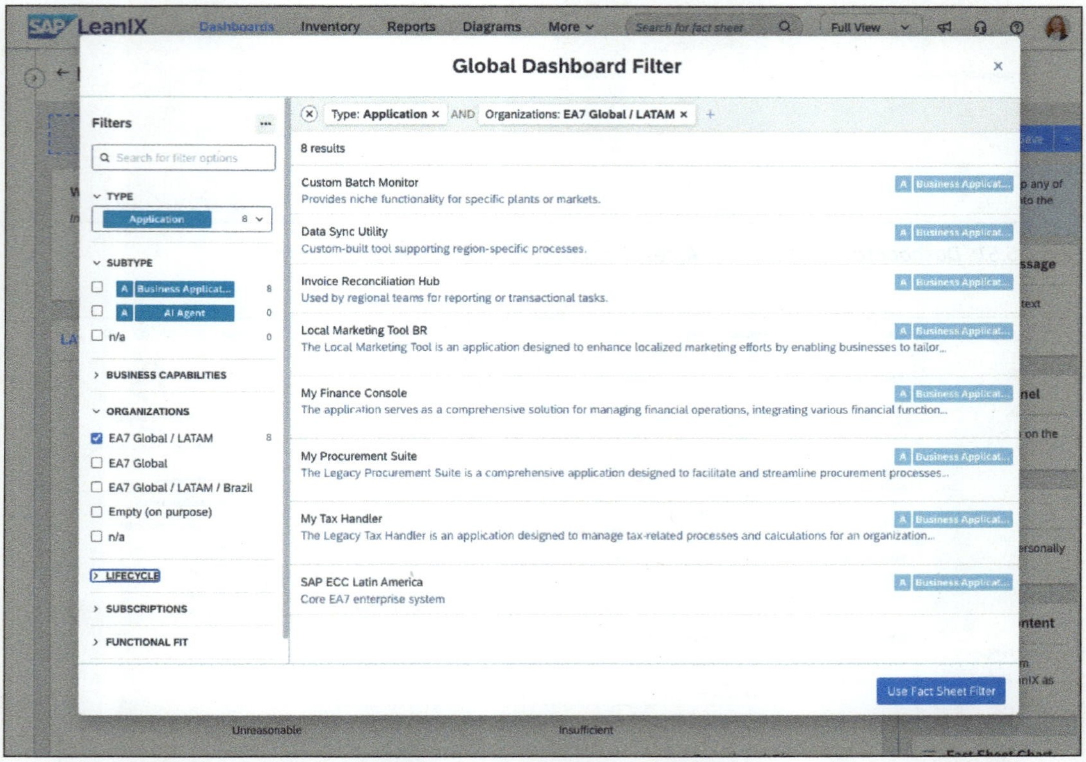

Figure 5.53: Defining Global Filters on Dashboards in Edit Mode

Back in the dashboard editor, you'll see the applied global filter at the top of the editor, and you'll also see warning messages on the panels to which the global filter cannot be applied (see Figure 5.54).

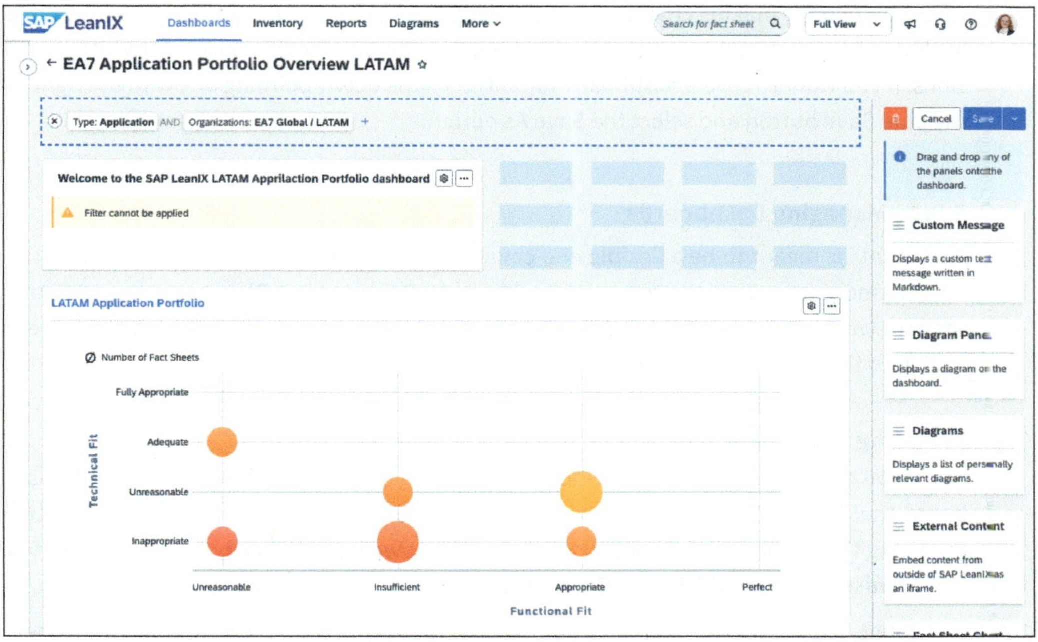

Figure 5.54: Dashboard Editor with Active Global Filter Feature

The global filter feature can help you to efficiently build nested dashboards targeting different legal entities, regions, or departments while keeping the same look and feel. To do that, define a global dashboard focused on a use case or fact sheet of interest. Once done, create copies of that dashboard and apply a global filter to focus on region-specific data. Adjust the name, panel titles, and custom messages as needed. You can link the dashboards via hyperlinks in a **Custom Message** panel.

To modify the panel settings or **Duplicate** or **Delete** a panel, use the **Settings** (gear) button or the **...** button in the upper-right corner of a panel tile (see Figure 5.55).

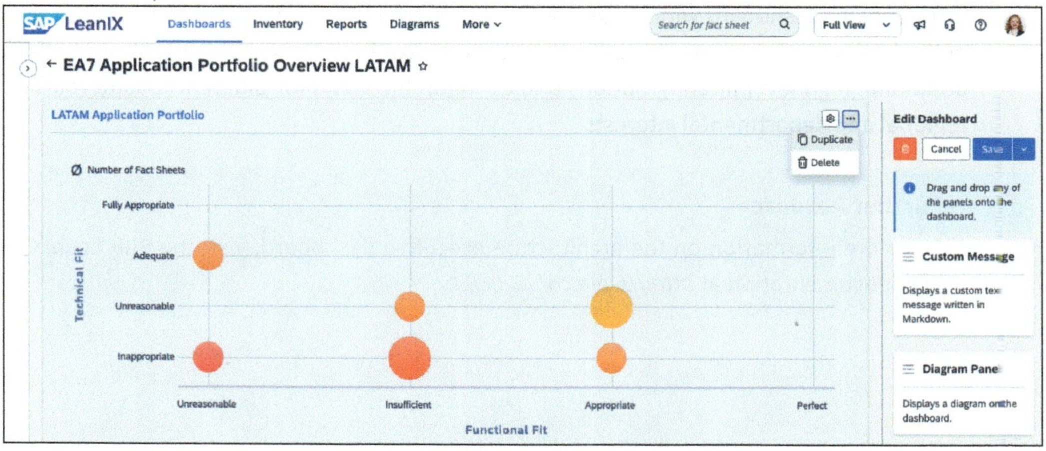

Figure 5.55: Adjusting Dashboard Panel Settings or Duplicating/Deleting Panels

Once you're done making changes, don't forget to save your dashboard using the blue **Save** button in the upper-right corner of the dashboard editor, and revisit the dashboard permissions. If you want to save a dashboard variant, click the down-pointing arrow to the right of the blue **Save** button and select the **Save As** option.

5.3.2 Managing Dashboards

SAP LeanIX is meant to be a flexible and collaborative solution. For ease of navigation, the UI is harmonized across the top-level **Dashboards**, **Reports**, and **Diagrams** menus. In this section, we'll briefly recap the familiar features for managing dashboards, which are very similar to those for report and diagram management.

Browsing Dashboards

When you click the top ribbon's **Dashboards** button, you'll be directed to the default dashboard that was defined by either the workspace admin or possibly the user, the latter assuming that the setting allowing users to set their own default dashboards was activated in **Administration • BASIC SETTINGS • Dashboards**.

By default, the left-hand menu will be collapsed. To expand it, click the arrow in the upper-left corner located on the vertical divider line.

You'll see a familiar menu with sections and subsections, called *collections*. In the lower-left corner, the admin will see the **Manage collections** link, which we discussed in Section 5.1.2 (refer to Figure 5.9).

Collections are shared across all areas, so you can also reuse a collection that you defined for the reports in the **Dashboards** menu. The **Default Dashboard** and **Architecture Executive Dashboard** links, in the upper-left corner of the collection menu, are special. The default dashboard can be defined by admins or users, depending on the settings in the **Administration** menu. For more information, revisit Chapter 3, Section 3.3.3, and read about the **Dashboard** menu under **BASIC SETTINGS**. The architecture executive dashboard is a template dashboard with predefined and clustered KPI panels that can be rearranged and adjusted with custom KPIs based on the organization's needs. This dashboard offers additional high-level filtering options and configurable views for different C-suite roles or regional and departmental interests.

Further Resources

For more information on the architecture executive dashboard, visit the SAP LeanIX user documentation at *http://s-prs.co/v620650*.

To see all dashboards that are visible to you, click **All Dashboards** in the **DASHBOARD OVERVIEW** section (see Figure 5.56).

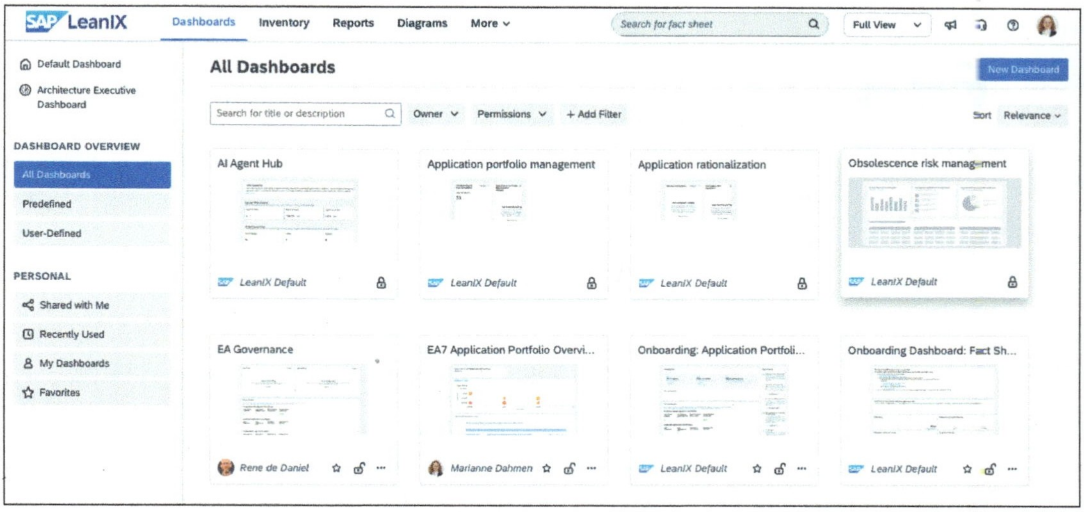

Figure 5.56: Navigating to Dashboard Overview

As you can see in Figure 5.56, the various available dashboards are represented as tiles in an overview page, which you can also filter and sort based on your needs via the options under the **All Dashboards** title in the center pane.

The SAP LeanIX out-of-the-box reports and the selected predefined use case dashboards from the **Administration** menu's **Dashboards** configuration can be found in the **Predefined** collection under the **DASHBOARD OVERVIEW** section. These dashboards cannot be modified by admins or users.

Admins or roles with assigned *manage content* permissions for collections can map a user-defined dashboard to a collection so that it is easier to find. To do that, click the **...** button in the lower-right corner of a dashboard tile, hover over **+ Add to Collection**, and select the desired entry (see Figure 5.57).

You can also mark your dashboards as favorites and use the collections under the **PERSONAL** section to find the dashboards that you created, that were shared with you, or that you recently used.

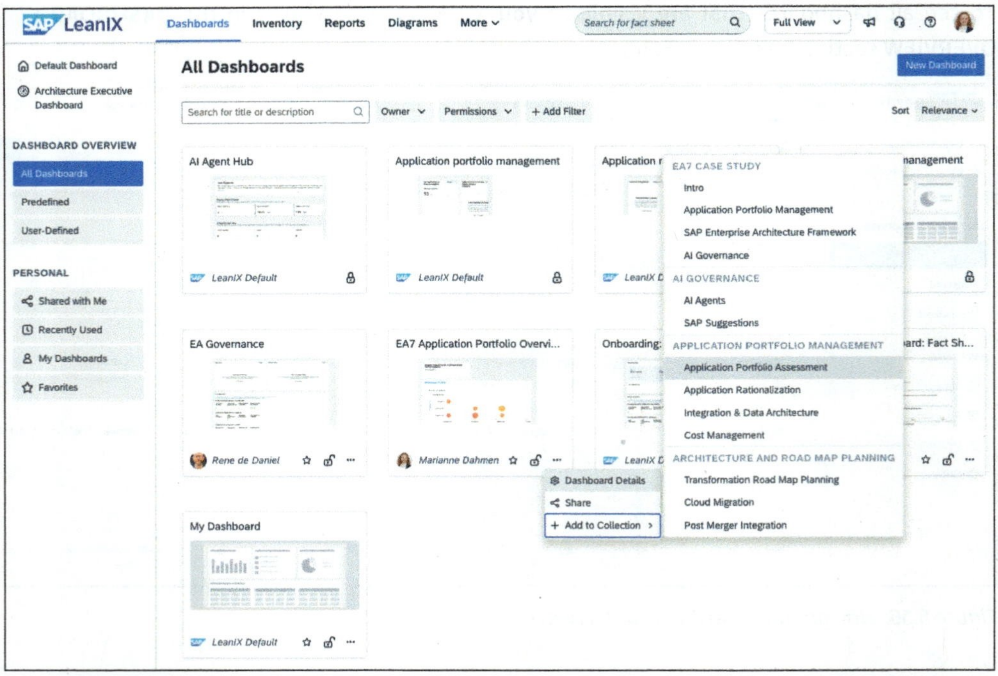

Figure 5.57: Adding Dashboard to Collection

Access Management

Access is managed in the dashboard details settings in the same way as for reports and diagrams. The lock icon on the dashboard overview tiles (which can also be found on the tiles in report and diagram overviews) will give you a first indication of access rights. Hover over the lock to see your access rights. When the lock is closed, you can only see but not edit the dashboard. When the lock is open, you have view and edit rights (see Figure 5.58).

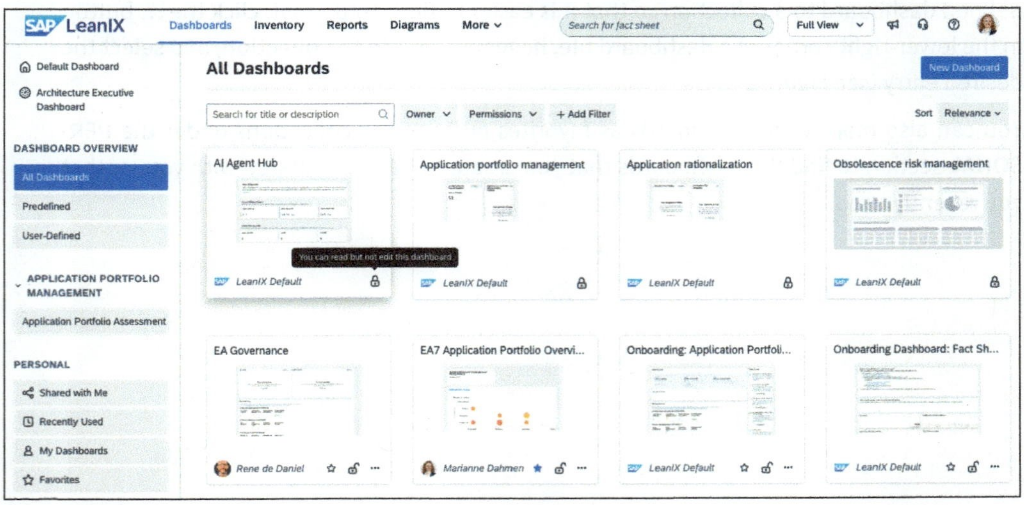

Figure 5.58: Interpreting Lock Icon for Quick Check of Access Rights

To access the **Dashboard Details** menu, click the **...** button in the lower-right corner of a dashboard tile and select **Dashboard Details**. You can also access this page from within a dashboard, in which case the **...** button will be in the upper-right corner of the dashboard.

A new window will open (see Figure 5.59), in which you can change settings on three tabs:

- **Settings**
  To modify the dashboard name, description, and owner, and to **Delete** a dashboard
- **Permissions**
  To define the **Permission Type**, and add new collaborators or viewers
- **Shared With**
  To share the dashboard with individuals so that it appears in the **Shared with Me** collection folder

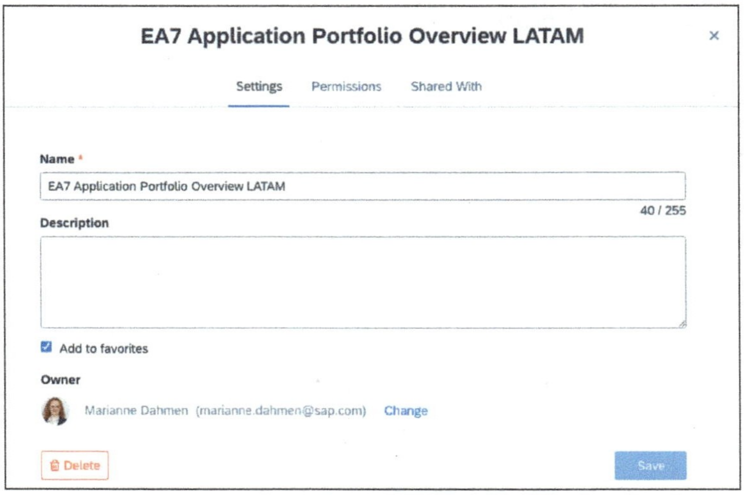

Figure 5.59: Dashboard Details Window

On the **Settings** and **Permissions** tabs, always remember to **Save** your changes by clicking the blue button in the lower-right corner of the window.

5.3.3 KPIs on Dashboards

One of the dashboard panels is called **KPI Panel**. With it, you can visualize calculated KPIs based on fact sheet data and relations (see Figure 5.60).

KPIs can be managed by admins in the **Administration** menu under **ADVANCED SETTINGS • KPIs.**

As you can see in Figure 5.61, SAP LeanIX comes with a set of predefined KPIs, some of which can be found in the out-of-the-box dashboards; however, it also offers the option to create new custom KPIs. The KPI overview in the center pane can be filtered by name, **KPI Type**, **Use Case / Groups** (expanded dropdown in Figure 5.61), and **Display on**, with which you can filter for only dashboard or only executive dashboard KPIs.

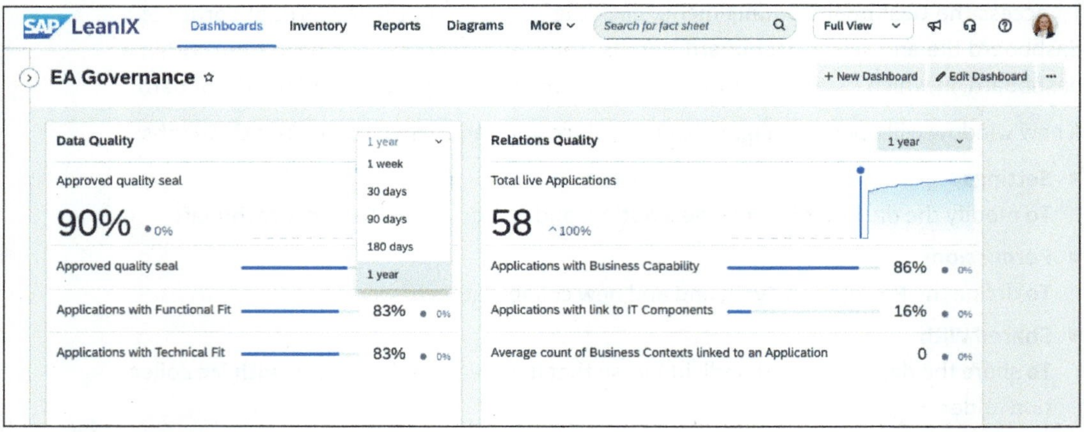

Figure 5.60: Example of KPI Panels on Dashboards

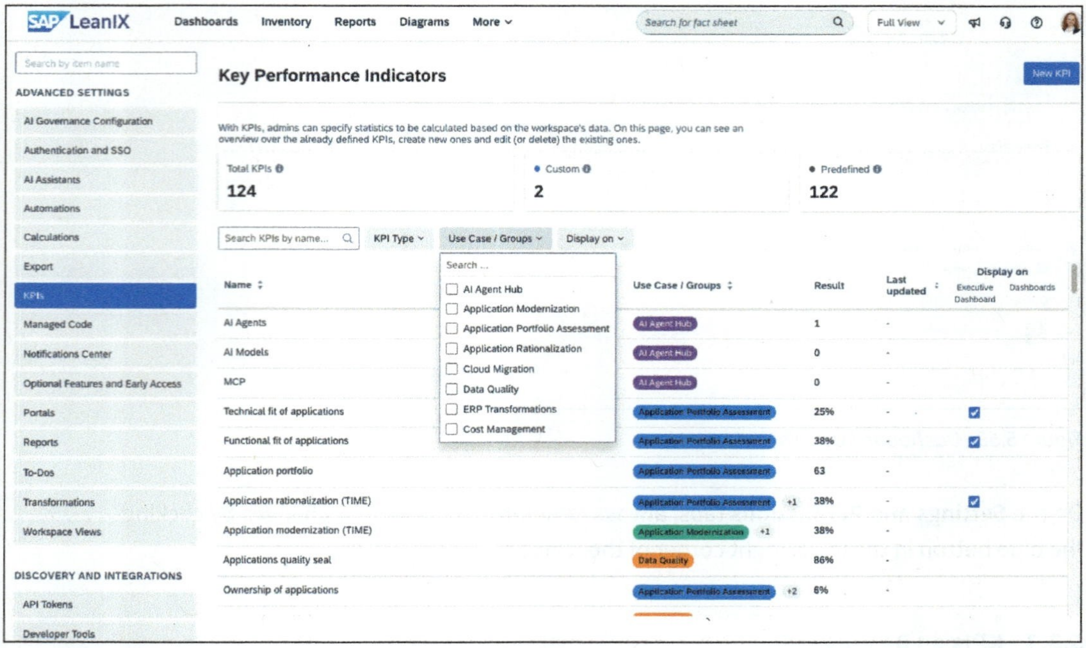

Figure 5.61: KPI Overview in Administration Menu

To create a new KPI, click the blue **New KPI** button in the upper-right corner. A form will open in SAP LeanIX that will guide you through the three-step configuration process.

In **1 General Information**, you must enter the **Name** of the KPI and select a **Use Case** from a dropdown menu (mandatory information). Optionally, you can also enter a short **Description** of the indicator (recommended). Click the blue **Next** button in the lower-right corner to navigate to the next screen.

In **2 Calculation Formula**, the KPI is defined. You can calculate either an **Absolute** number or a **Percentage**.

When selecting the **Absolute** option, you can choose among three options in the **Select Operator** dropdown:

- COUNT
  Count of fact sheets selected in the **Set Filters** area
- SUM
  Sum of a numerical field on the fact sheet type selected in the **Set Filters** area
- AVERAGE
  Average of a numerical field on the fact sheet type selected in the **Set Filters** area

See Figure 5.62 for an example of an absolute KPI.

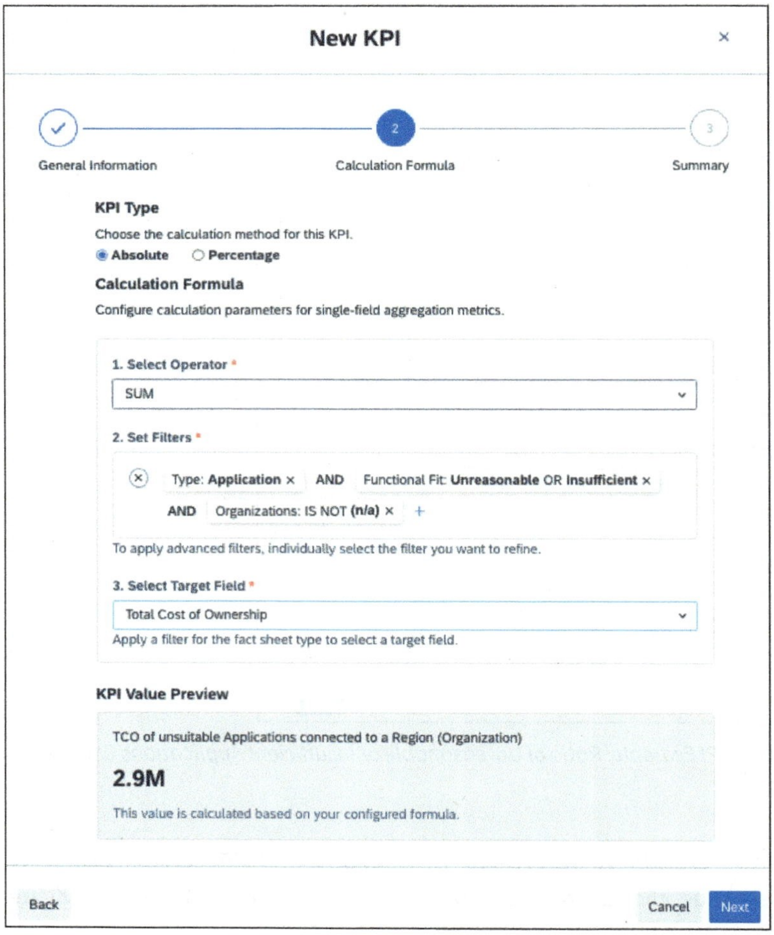

Figure 5.62: Absolute KPI Example: TCO of Unreasonable or Insufficient Applications Connected to at Least One Region

When selecting the **Percentage** KPI, you must define the **Numerator** and **Denominator** of the proportional metric formula. They can either hold as **Select Operator** a **COUNT** of fact sheets based on a certain filter or a **SUM** of a numeric field on a specific filtered fact sheet type.

See Figure 5.63 for an example of a percentage KPI.

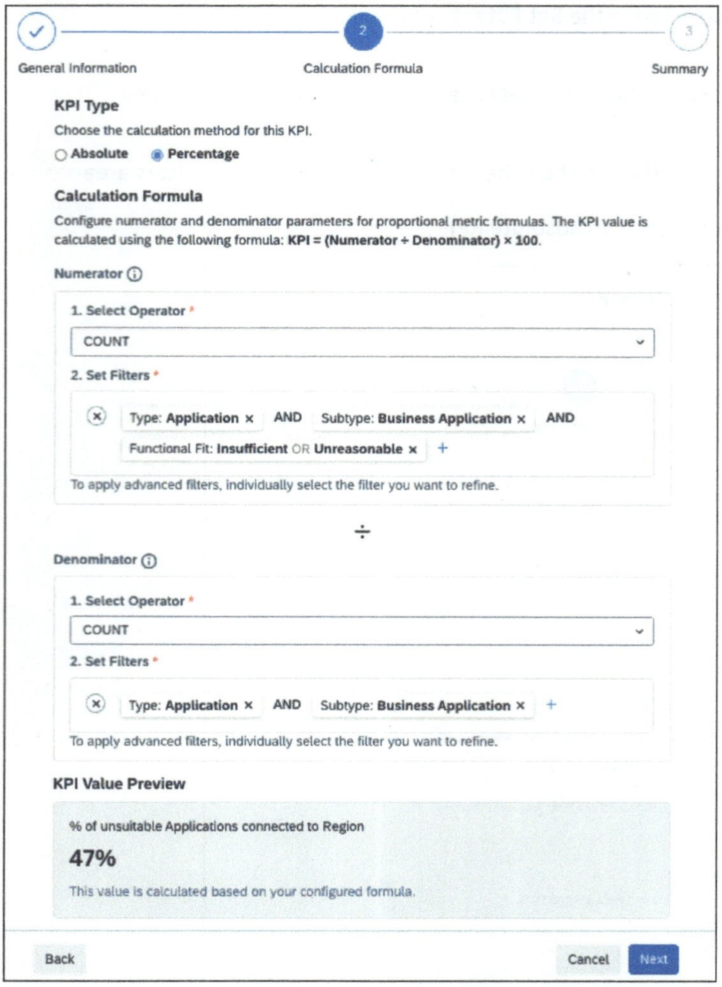

Figure 5.63: Percentage KPI Example: Ratio of Unreasonable or Insufficient Applications Connected to at Least One Region

Once you're done and the **KPI Value Preview** at the bottom of the screen shows the expected and desired value, click the blue **Next** button in the lower-right corner to navigate to the next screen.

In **3 Summary**, you can review the information entered. Use the grey **Back** button in the lower-left corner if you want to navigate back to steps 1 or 2. If everything looks correct, click the blue **Save** button in the lower-right corner to finish the KPI configuration.

Back in the KPI overview in the administration menu, you can manage the custom KPIs. Predefined KPIs cannot be modified or deleted, but custom ones can. When hovering over a KPI line item in the table, you will see two potential checkboxes to activate the **Display on** settings for the **Executive Dashboard** or regular **Dashboards**. There are also three buttons on the right side. The first one, the pen, lets you edit the KPI. The second one, the two sheets, lets you duplicate the KPI. The third one, the trash bin, lets you delete the KPI (see Figure 5.64).

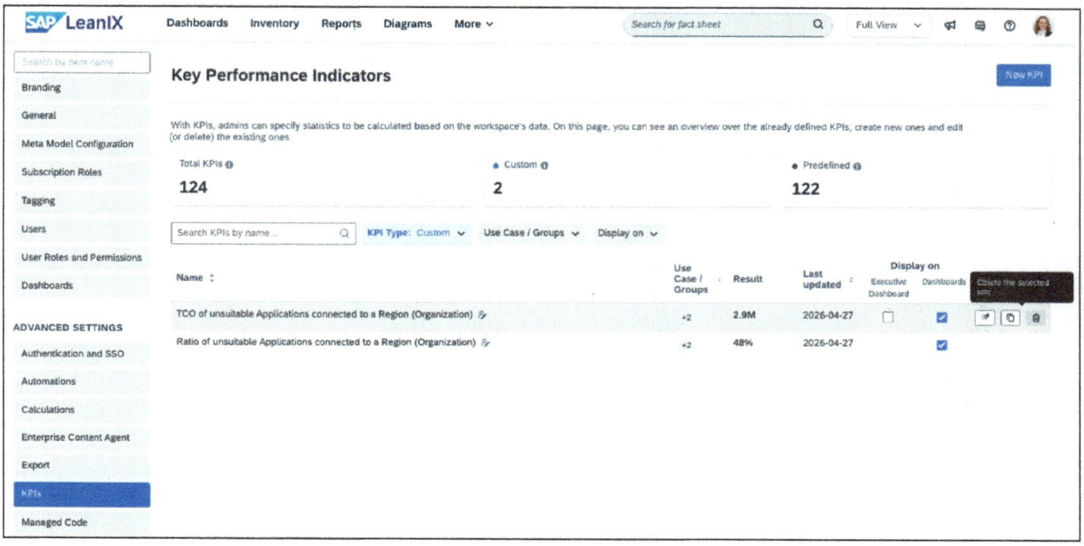

Figure 5.64: Custom KPI Overview in Administration Menu and Modification Options

> **Caution**
>
> Keep in mind that when you change the KPI calculation logic, it will not recalculate historic values. Therefore, it's best to create a new KPI if the calculation logic significantly changes in order to avoid confusion caused by historic data reflecting the old logic and new values following the new calculation logic.

Be sure to consider your access and permission concept. If a user role isn't allowed to see a certain fact sheet type, then the users in this user role also cannot select KPIs that include those fact sheet types in the KPI panel when creating dashboards. If, however, the KPI is already embedded in a KPI panel on a dashboard and this dashboard is public, then users with that user role will see the KPI, even though they are not allowed to see the fact sheets in the inventory.

> **Further Resources**
>
> For the latest updates on KPIs, visit the SAP LeanIX user documentation at *https://help.sap.com/docs/leanix/ea/kpis*.

5.4 Summary

In this chapter, we explored the basics of the SAP LeanIX visualization capabilities for reports, diagrams, and dashboards. You learned how to create, save, share, delete, navigate, and manage the different objects and learned how they are linked to the inventory and to real-time data that can be used as axes or color-coded views.

In the next chapter, we'll apply this knowledge in a use case for architecture and roadmap planning, in which we'll dive deeper into important features used in target architecture planning.

Chapter 6

Creating Your First Architecture Roadmap

Our examination of SAP LeanIX pivots now from fundamentals to more advanced concepts. It's time to immerse ourselves in one of the core disciplines of enterprise architecture: target architecture planning and road mapping. Are you ready for the future?

In this chapter, you'll learn how to use the initiative fact sheet type to support enterprise architecture projects and transformation initiatives. We'll discuss modeling best practices and specifics of the initiative fact sheet in combination with other fact sheet types, such as the provider fact sheet. We'll also explore what can be done using only the base SAP LeanIX product, SAP LeanIX Architecture Portfolio Management, and what additional capabilities are available with the addition of SAP LeanIX Architecture and Road Map Planning.

It's important to remember that SAP LeanIX targets not project management experts and professionals, but enterprise architects. Therefore, the features for target planning and initiatives are designed for the day-to-day use of IT and enterprise architecture, with a focus on capturing changes to the IT landscape and architecture, not on project management office operational work.

Transformation Example

EA7 has a big enterprise resource planning (ERP) transformation ahead, which it wants to plan from an enterprise architecture perspective with SAP LeanIX. (More details of this transformation will be explained in Chapter 8.) Before starting with the big initiative, Chris Green and his team want to explore the capabilities of the initiative fact sheet in SAP LeanIX as well as special features of the SAP LeanIX Architecture and Road Map Planning product. They therefore start with a test project called *SAP LeanIX Global Rollout.*

6.1 Initiative Fact Sheets

The initiative fact sheet is connected to all other fact sheet types in the SAP LeanIX out-of-the-box meta model (see Figure 6.1). An *initiative* usually supports a strategic *objective*, is owned by an *organization* (e.g., legal entity, department, or team), is executed with the help of an external or internal *provider* (e.g., consulting firm), improves certain *business capabilities*, and affects *applications*, *interfaces*, *data objects*, *IT components*, *tech categories*, *business context*, or *platforms*.

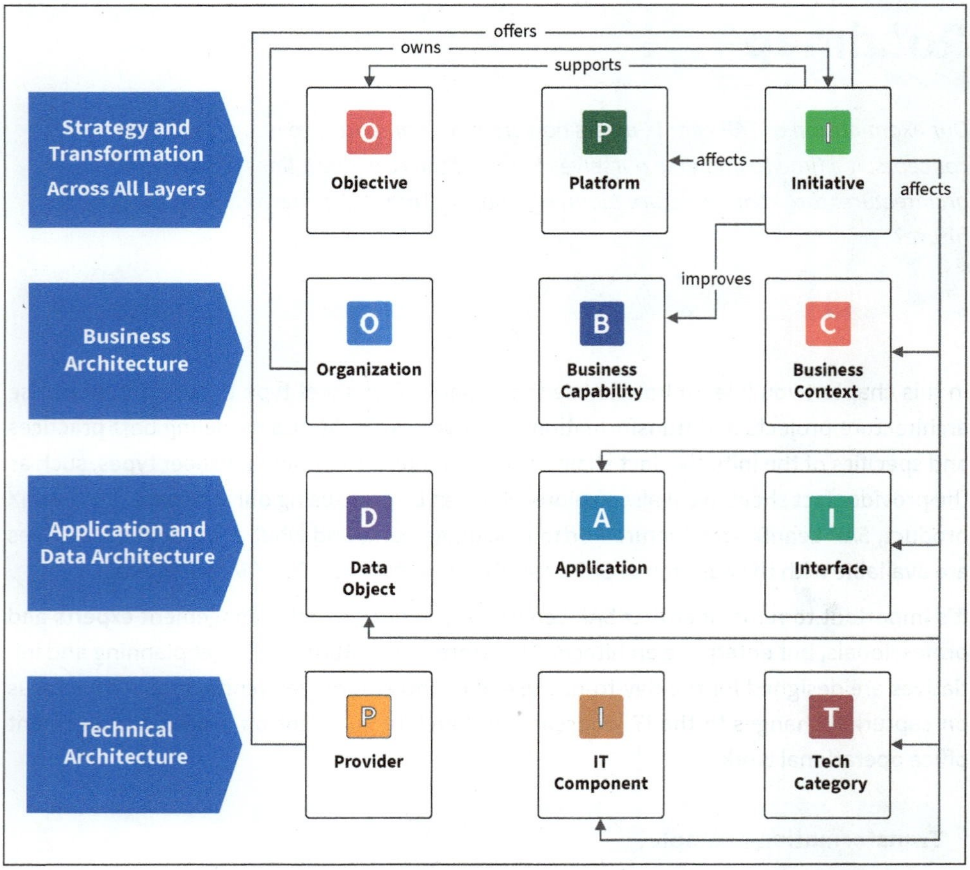

Figure 6.1: Initiative Fact Sheet and Its Relations

Initiatives have the following out-of-the-box subtypes:

- **Idea**
  Early ideation of potential changes
- **Program**
  Cluster or group of projects, or epic (defined ahead), ideally managed with a proper project management office setup and additional project management tooling
- **Project**
  Either standalone project or child of a program (see Figure 6.2)
- **Epic**
  Change unit in a product-oriented organization; usually used in combination with the Jira integration

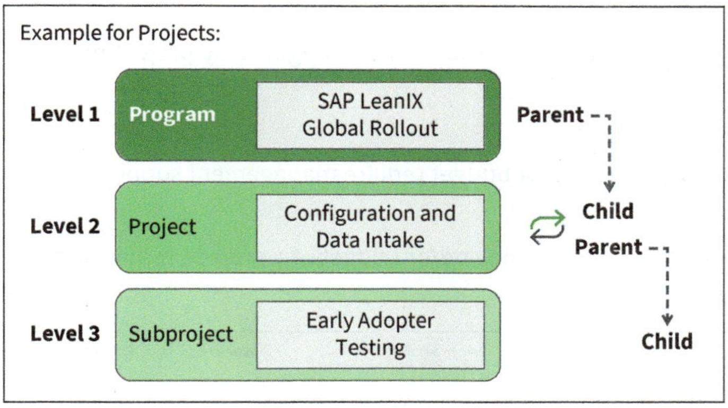

Figure 6.2: Initiative Hierarchy Example

Tip

Remember that you can always modify the translations/display names of subtypes in the fact sheet meta model configuration. You can also create new subtypes, if needed, or delete subtypes that you don't want to use. Revisit Chapter 3, Section 3.3.3 for more information.

The initiative fact sheet has a few special out-of-the-box fields. For demonstration purposes, an initiative fact sheet called **SAP LeanIX Global Roll-out – Test Program** was created. To learn how to create a fact sheet manually, read Chapter 4, Section 4.2.7. Open the fact sheet either via inventory filtering or by using the quick search in the top ribbon's **Search for fact sheet** field. Scroll down in the **Fact Sheet** tab until you arrive at the **Initiative Environment** section, where you can find fields to capture the **Business Value** or the **Project Risk** (see Figure 6.3).

Possible values for selection in the **Business Value** dropdown include the following:

- **Marginal Benefit**
  Small quality or financial improvements
- **Little Benefit**
  Some quality and financial improvements
- **Large Benefit**
  Remarkable quality and/or financial improvements
- **Significant Benefit**
  Significant quality and financial improvements

Possible values for the **Project Risk** dropdown include the following:

- **Low Risk**
  No risks or a minor risk that can be easily mitigated

- **Moderate Risk**
  Some effects on quality, timeline, or budget that can be mitigated by project management
- **High Risk**
  Remarkable effects on quality, timeline, or budget require management support
- **Severe Risk**
  Intolerable effects on quality and/or timeline and/or budget of a project, requiring dedicated management attention and support

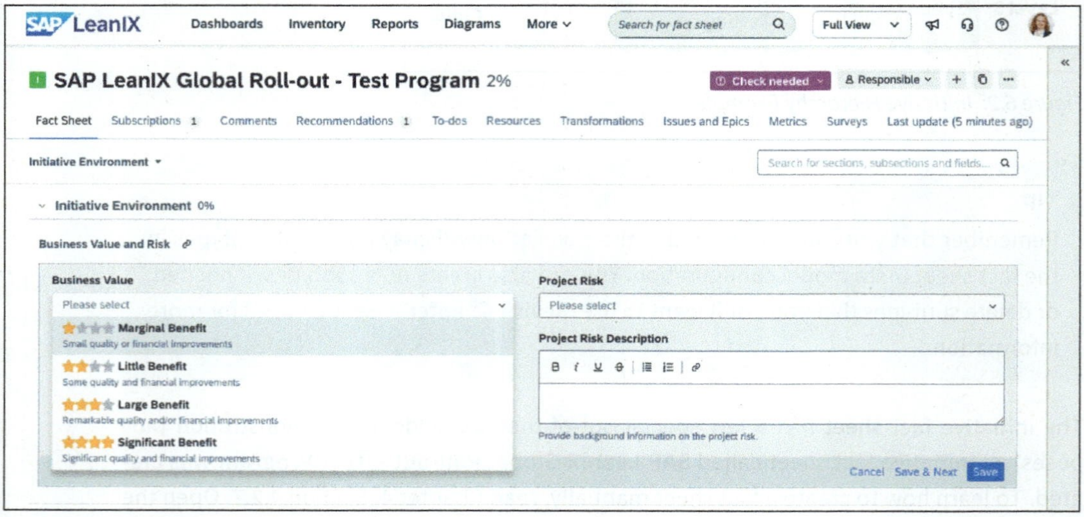

Figure 6.3: Business Value and Project Risk Fields

The next section with initiative-specific fields is the **Initiative Setup** section and its **Budget Costs** subsection, which includes fields for **OpEx (Budget)** and **CapEx (Budget)**; that is, operational expenditure and capital expenditure, respectively. Both values are numerical currency fields. The currency can be changed in the **Administration** menu under **BASIC SETTINGS • General**.

> **Tip**
>
> Remember that you can only show one currency, the group currency, in the SAP LeanIX workspace. If you have a multicurrency program, always enter the values translated into group currency into the initiative fields.

Another subsection in this area is **Benefit**, in which the **Net Present Value (NPV)** and the **Payback Period** can be entered.

Finally, you can see the **Provider** area, with fields for the relation in the **Initiative Setup** section. The provider can be an internal or external party running or supporting the project.

For the relation, it's possible to capture **Order No** (order number), **CapEx (Ordered)**, and **OpEx (Ordered)** values.

If you also have the SAP LeanIX Architecture and Road Map Planning product, you will find two more functionalities on the initiative fact sheet: the **Milestones** feature, where you can define project milestones, which will be visible in roadmap reports as visual markers; and the **Transformations** tab in the center section's title ribbon (see Figure 6.4).

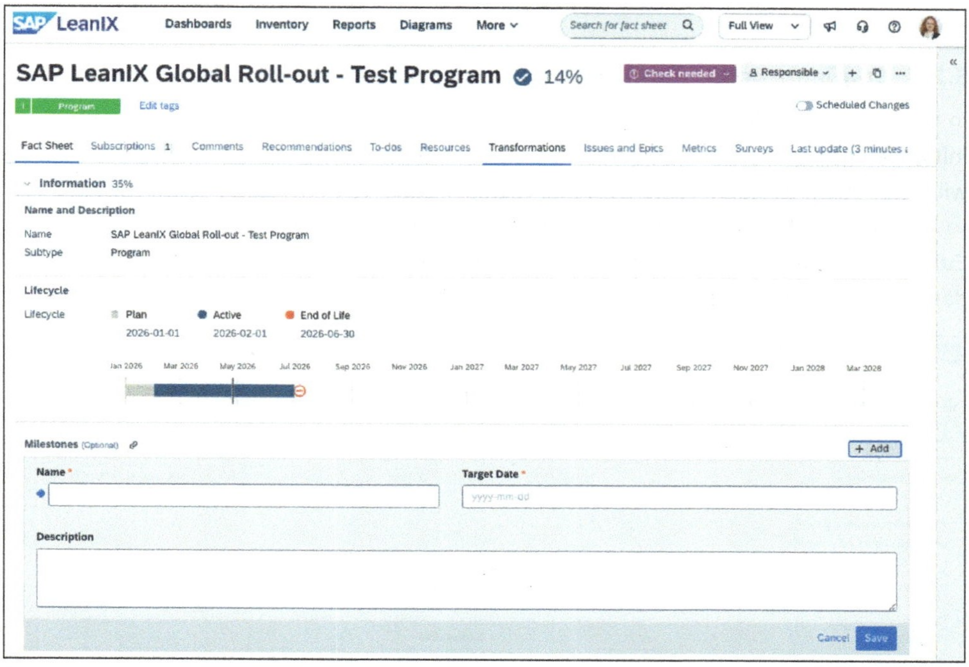

Figure 6.4: Milestones and Transformations Tab: Only Available in SAP LeanIX Architecture and Road Map Planning

A new feature still in development as of the time of the writing is a status showing whether an initiative is still in an early planning stage or whether it's already committed and highly likely to be implemented.

Further Resources

For the latest status of features and updates, check the SAP LeanIX user documentation:

- Meta model fact sheet modeling guidelines for initiatives: *http://s-prs.co/v620651*
- Features and capabilities of the SAP LeanIX Architecture and Road Map Planning product: *http://s-prs.co/v620652*

6.2 Basic Future Architecture Planning

It's possible to do very basic architecture planning in SAP LeanIX without using the SAP LeanIX Architecture and Road Map Planning product. For simple target architecture pictures, just using the **Active From** and **Active Until** fields for relations can provide some insights in reports and diagrams. We'll walk through the steps for this in the following sections, then use the insights to create a roadmap report.

6.2.1 Configure Active From Field

In this example, we'll look at the relation between the SAP LeanIX Global Rollout Program initiative and the region to which the SAP LeanIX application should be rolled out. France will be the first region in the rollout. Therefore, in the Initiative **Fact Sheet** tab, scroll down to the **Initiative Environment** section. Under **Affected Organizations**, add the **EA7 Global / Europe / France** organization fact sheet and maintain the **Active From** field by entering the date on which the rollout project should start (see Figure 6.5).

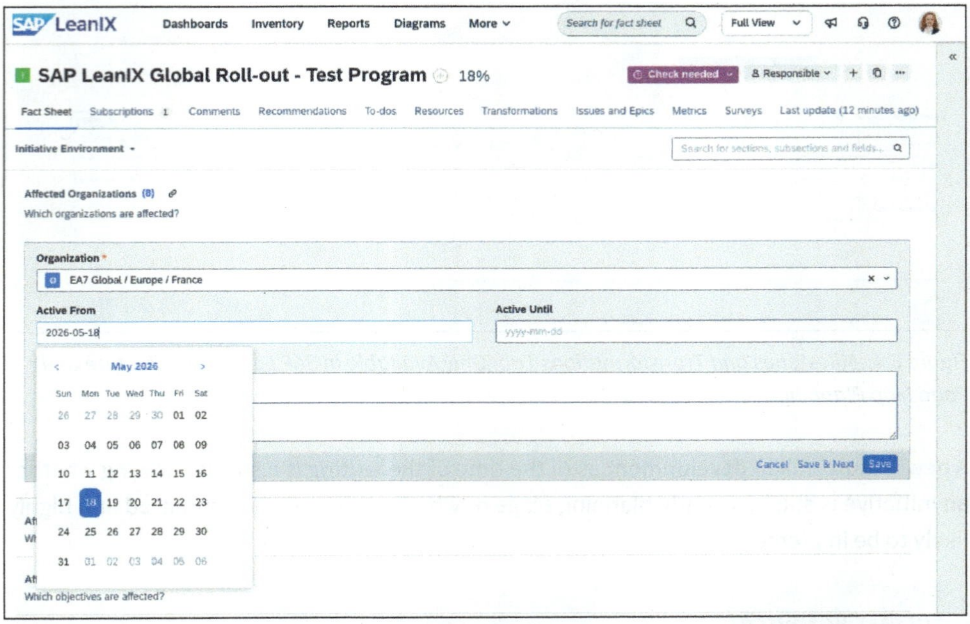

Figure 6.5: Maintaining Active From Date Field for Relation Between Initiative and Organization

To maintain all other **Active From** fields to **Affected Organizations**, we won't manually edit every line item in the fact sheet relation section; instead, switch to the inventory table view to perform mass edits. To do that, navigate to the **Inventory** tab in the top main ribbon. On the left-hand-side inventory filter, select **Initiative** in the **TYPE** filter. Next, filter for **Program** under **SUBTYPE** by clicking the box next to this element. To limit the results to those related to SAP LeanIX, filter either for the name via a text filter, or for an already related fact sheet. In this case, we've previously connected the initiative to SAP LeanIX enterprise architecture and can filter for **Applications** with that name that are related to the initiative/program. Finally, click the **Table** icon in the upper-right corner of the center pane to switch to the table view (see Figure 6.6).

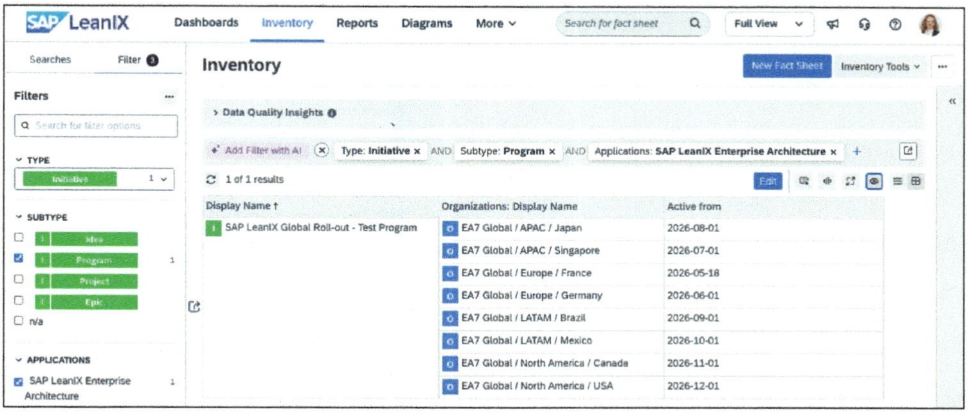

Figure 6.6: Inventory Table View Filtered for SAP LeanIX Global Roll-out Program

To see exactly the view shown in Figure 6.6, you must also select the correct columns. To do that, click the eye icon in the upper-right corner to customize the table columns. Remove the **Description** column and add the **Organizations** column. Click the menu icon next to the **x** in the columns' tiles to make the fields in the relation visible. Click the crossed-out eye icon next to **Display Name** and **Active from** in the **Organizations** tile to make those fields visible in the table view (see Figure 6.7).

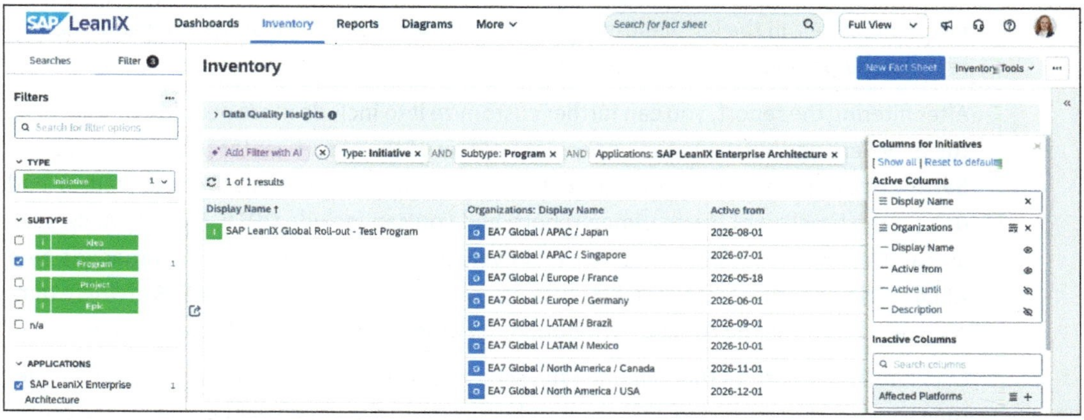

Figure 6.7: Configuring Table View to See Fields in Relations

Finally, click the grey **x** in the corner of the right-hand-side **Columns for Initiatives** menu to close the column customization, then click the blue **Edit** button to change into the edit mode of the inventory table view. Click an **Active from** cell of an organization, then select the correct date in the calendar widget that opens. You can also enter a date using the format YYYY-MM-DD (see Figure 6.8).

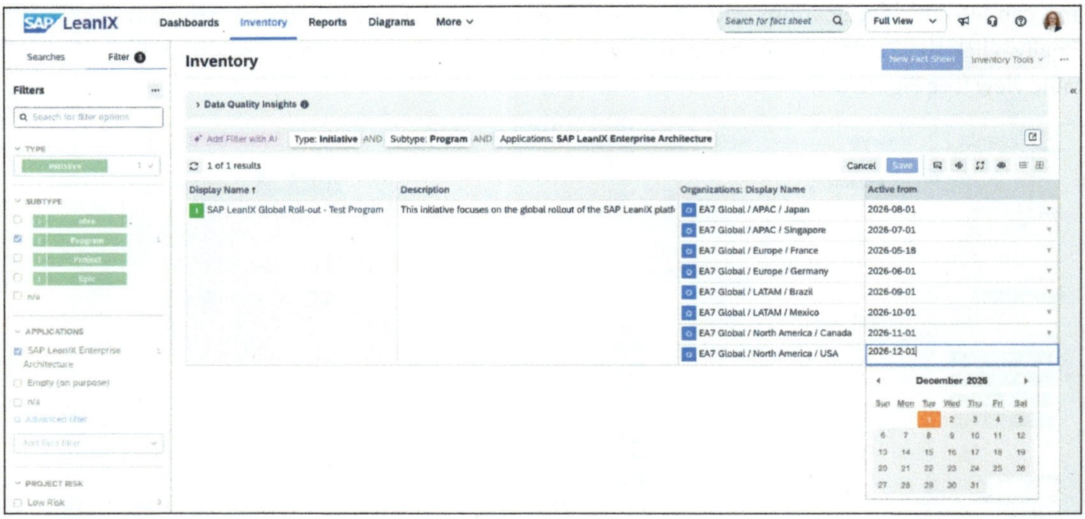

Figure 6.8: Selecting Active From Date in Table Edit Mode

6.2.2 Create Roadmap Report

With that information captured in the initiative fact sheet, you can create your first roadmap report.

Navigate to the **Reports** area by clicking the respective header in the top ribbon, then select **Roadmap** from the left-hand collection overview in the **REPORT TYPES** section. Click the **Initiative Roadmap** template tile in the center pane. Next, filter down the report to the relevant initiatives. In the left-hand-side filter menu, select **SUBTYPE** under **Program**, then filter related **Applications** for **SAP LeanIX Enterprise Architecture**.

After filtering the report, you can further customize it to include more details. For that, click the grey **Settings** button in the upper-right corner of the report center pane. In the **Report Settings** popup, as shown in Figure 6.9, you can configure, for instance, the start or end date of the timeline of the roadmap report, and you can include **OPTIONAL DETAILS**, such as **Tag Group**, **Subscription Role**, or **Field** values. You can also **CLUSTER** initiatives by a **RELATION** to another fact sheet, **FIELDS** values of single-select fields, or **TAGS**. In this example, we want to use the **DRILLDOWN** options, setting **Drilldown To** to **Organizations**.

Click the **Apply** button; this will open the screen shown in Figure 6.10, where you can fine-tune the report appearance. Use the + and - buttons in the upper-right corner of the center pane to zoom in or out in the timeline. Scroll left or right in the timeline to position the bar chart nicely in the middle. If desired, adjust the bar color-coding by changing the **View** option. Don't forget to save your report variant by clicking the **Save As** button in the upper-right corner next to the report title (see Figure 6.10).

Another visualization option is to use a different report format, such as the landscape report. First, open the default **Initiative Landscape** report; in Figure 6.11, this is the first tile in the left-hand **Landscape** submenu under **REPORT TYPES**.

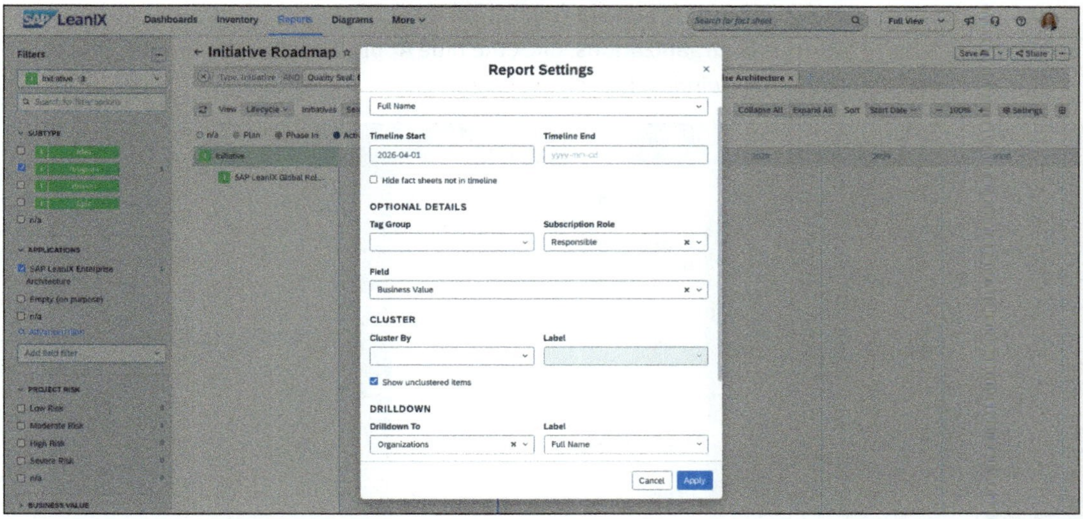

Figure 6.9: Configuring Roadmap Report

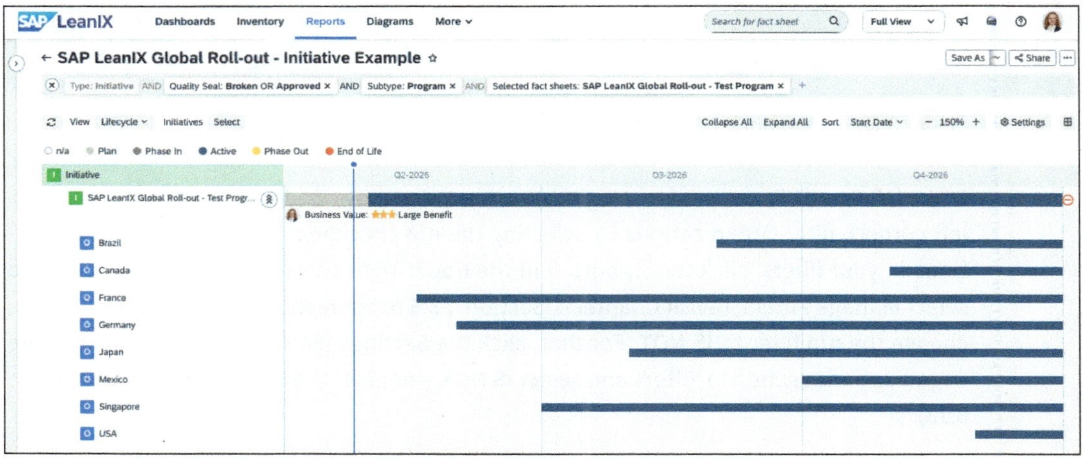

Figure 6.10: Zoomed in Roadmap Report Saved as Initiative Roadmap Variant

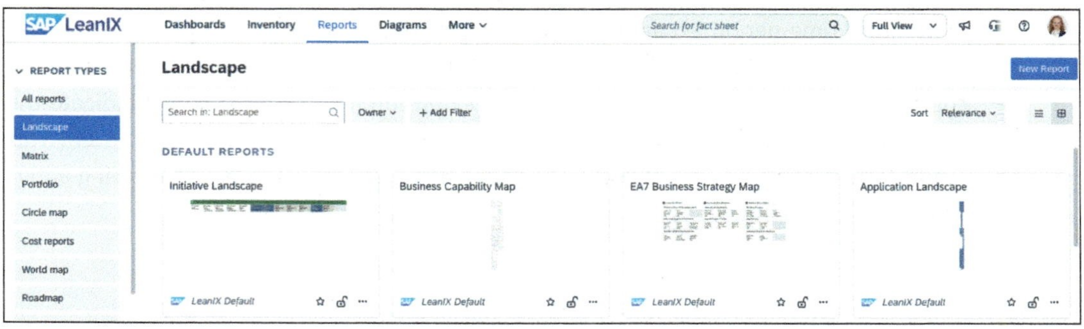

Figure 6.11: Opening Initiative Landscape Default Report

Click the **Settings** button in the upper-right corner to change the **Cluster By** relation from **Affected Platforms** to **Organizations**. Click the blue **Apply** button in the lower-right corner of the **Report Settings** window (see Figure 6.12).

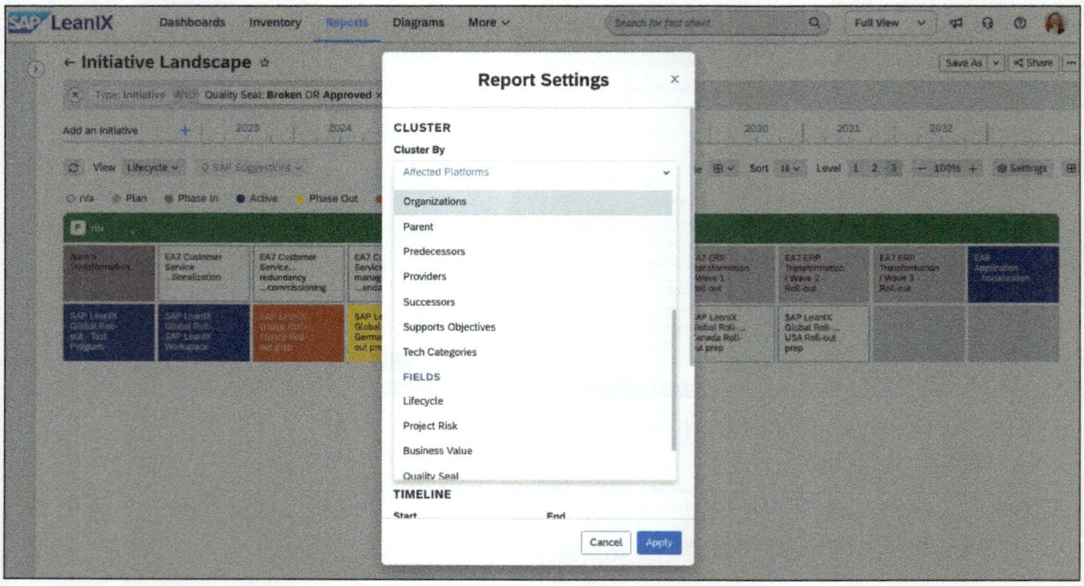

Figure 6.12: Changing Report Settings in Initiative Default Report

In the left-hand filter menu (you might have to expand it by clicking the arrow in the upper-left corner), filter **Organizations** by selecting the **n/a** checkbox. If you can't see **Organizations** in your filters, click the **...** button in the upper-right corner of the **Filters** menu and select **Manage Filters**; revisit Chapter 4, Section 4.3.5 to learn about the **Filters** menu. Now, change the **n/a** filter to **IS NOT**. For that, click the **Settings** (gear-shaped) icon next to the **Organizations** section in **Filters** and select **IS NOT – none of the selected items** (see Figure 6.13).

Finally, set a time slider range from today to the end of Q3 2026. To do that, hold the `Shift` key on your keyboard while selecting a point in time in the future on the timeline with your mouse. Every report tile with an extension box underneath it indicates a change during that time period, so an initiative will start, end, or change the attribute that was selected in the report view (in this example, the lifecycle). The **Report** tile will show the color-coded view at the target date, and the extension box underneath will indicate the color-coded view at the starting point of the selected time range (see Figure 6.14). This report format can help on a high level to gauge how busy regions are with ongoing enterprise architecture initiatives.

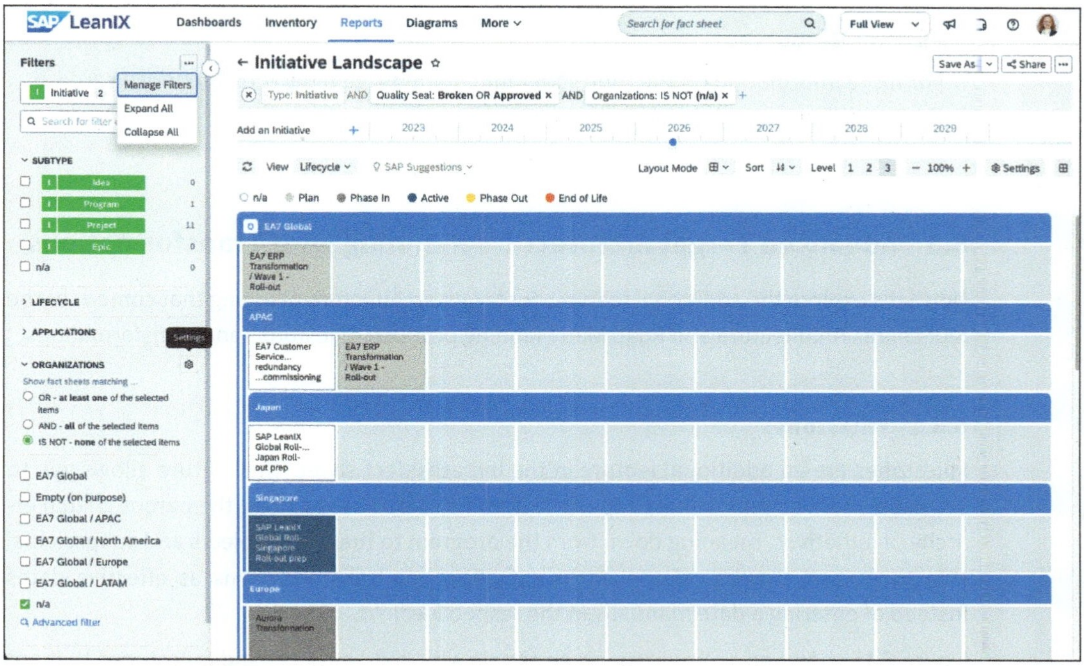

Figure 6.13: Filtering Report for Only Organizations with Initiative Elements

Figure 6.14: Initiative Landscape Clustered by Organizations with Time Slider Active

> **Further Resources**
> For more information on the time slider functionality, called the *differential view*, visit the SAP LeanIX user documentation at *http://s-prs.co/v620653*.

6.3 Advanced Target Architecture Planning with Transformations

Now let's explore the additional features for target architecture planning that come with the SAP LeanIX Architecture and Road Map Planning product: milestones and transformations.

6.3.1 Milestones

Milestones are an additional feature in the initiative fact sheet. This feature allows you to set visual markers on roadmap reports that can also be passed along the parent-child hierarchy of initiatives, meaning down from the program to the child projects and subprojects. Milestones can also be referenced in date fields or transformations as effective dates instead of entering a date manually in the respective field.

Figure 6.15 shows an example roadmap report with child projects and milestones that are also used as referenced lifecycle dates (dotted lines). If you have editing rights in the report and the correct access rights to the initiatives, then you can also move the milestones in the report via drag-and-drop and define new target dates that way. The changes will be saved into the initiative fact sheet upon clicking the blue checkmark to confirm after the milestone move.

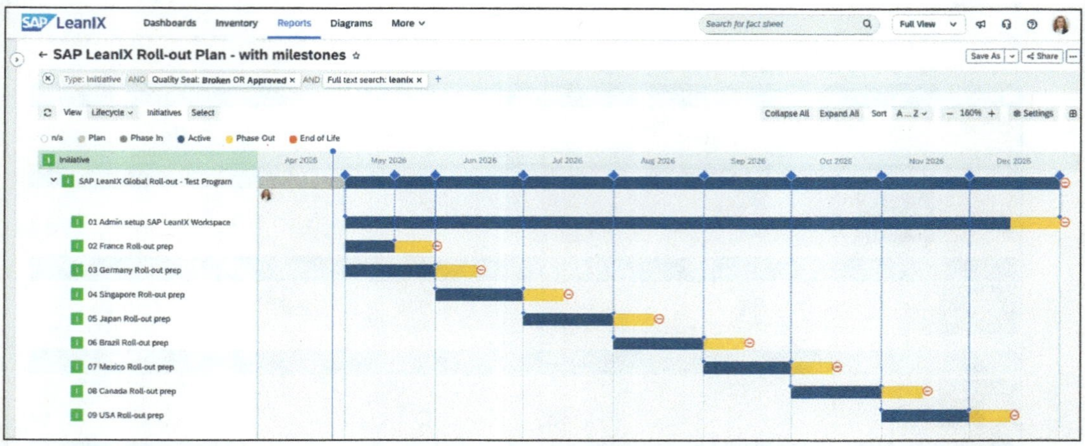

Figure 6.15: Roadmap Report Example with Child Projects and Referenced Milestones

Creating a milestone manually is done in the initiative fact sheet. In the **Fact Sheet** tab, under the **Information** section, you'll find the **Milestones** subsection, as shown in Figure 6.16. When you hover over this section, a grey **+ Add** button appears. A form will appear in which you must enter a milestone **Name** and select a **Target Date**; in this example, we've planned a **Program Kickoff** for May 1, 2026. The **Description** is optional.

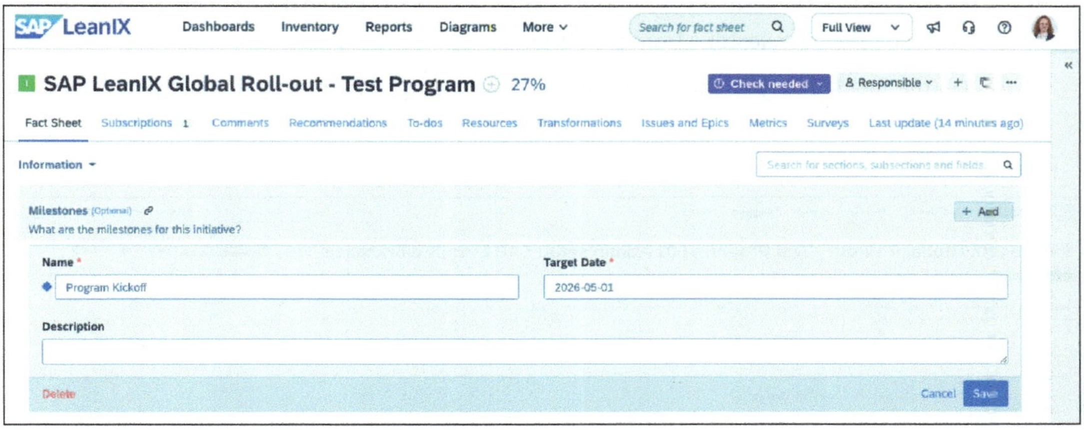

Figure 6.16: Creating Milestone Manually in Initiative Fact Sheet

To reference a milestone in a date field on an initiative child fact sheet, enter the edit mode on that specific field by hovering over the section with the respective date field and click the **Edit** button that will appear on the right, for instance, for the lifecycle initiative in the example in Figure 6.17.

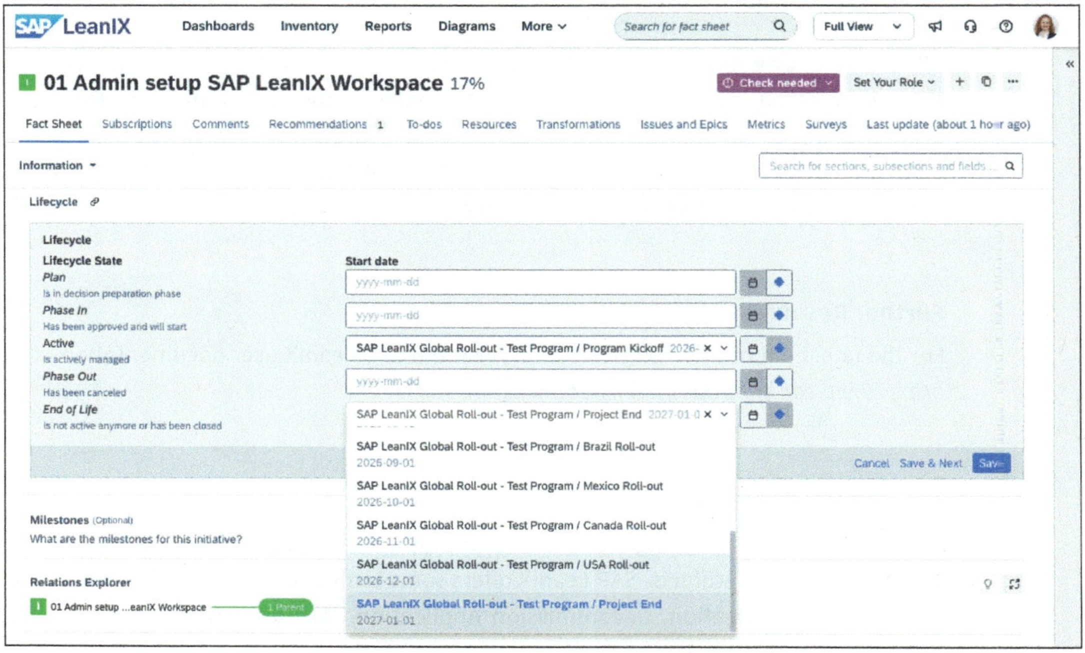

Figure 6.17: Selecting Milestone as Reference Date for Child Initiative Fact Sheet

Next to the date field, you can see two buttons, one depicting a calendar and another one depicting a blue diamond. The **Calendar** button is activated by default and allows you to enter a date in the format yyyy-mm-dd. When you click the blue diamond (**Use a Milestone**), you can select a milestone from the parent hierarchy via a dropdown.

On the fact sheet, you can determine whether a date was entered manually or whether a milestone was referenced. Simply check for a blue diamond next to the date (see Figure 6.18).

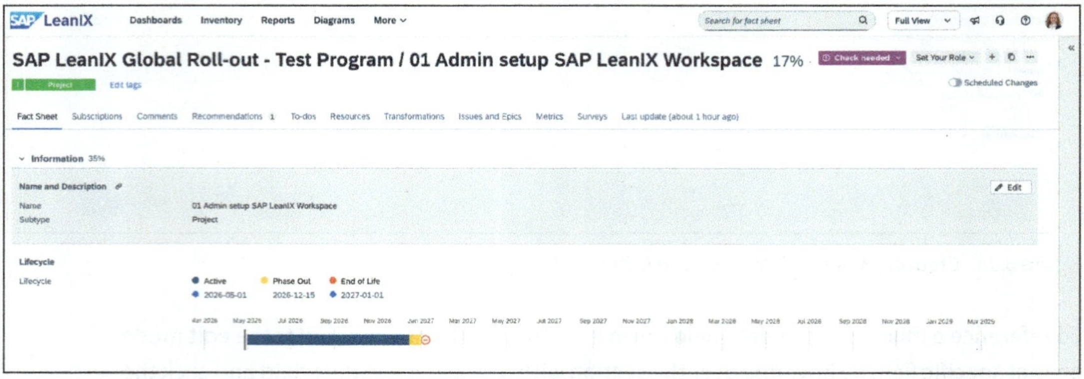

Figure 6.18: Mixed Dates in Initiative Lifecycle: Entered Manually and Referencing Milestones

Caution

When referencing milestones as lifecycle dates or target dates for transformations, those dates change whenever the milestone changes in the parent initiative. On the one hand, this can simplify maintenance work within larger projects, but on the other hand, proper communication is needed when different stakeholders are collaborating on programs and initiatives to avoid catching anyone by surprise.

Further Resources

For the latest information on milestones, visit the SAP LeanIX user documentation at *https://help.sap.com/docs/leanix/ea/milestones*.

6.3.2 Transformations

Transformations capture changes to the IT landscape in the form of planned changes to the SAP LeanIX inventory artifacts. SAP LeanIX offers some templates out of the box, including **Introduce New Application**, **Decommission Application**, **Introduce New Interface**, and **Change Data Transfer**.

Admins can configure transformations in the **Administration** menu under **ADVANCED SETTINGS • Transformations** (see Figure 6.19).

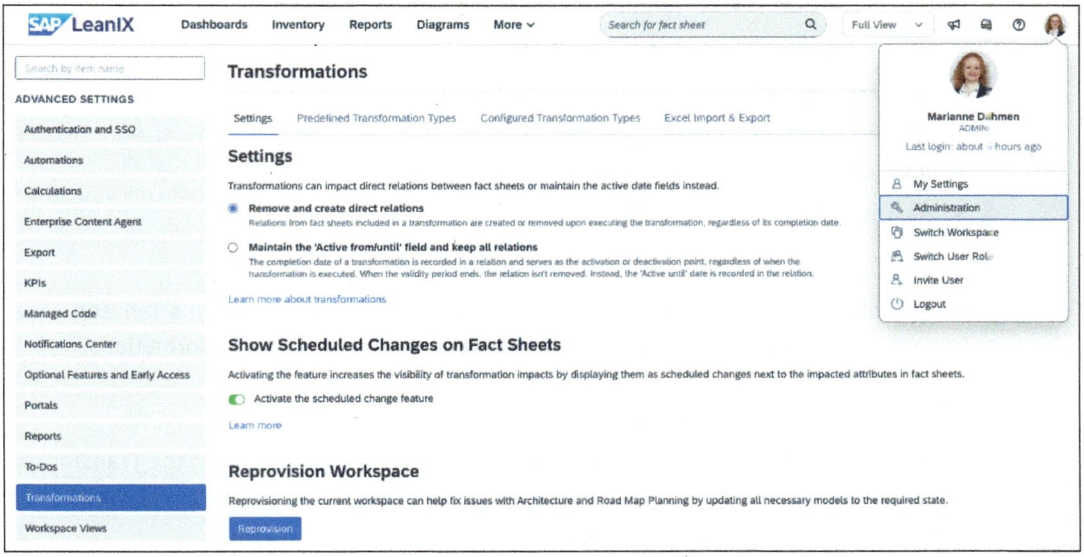

Figure 6.19: Transformation Configuration in Administration Menu

In the center pane are four tabs:

- **Settings**
  Here, you can do the following:
  - Define how relations are maintained, with the following options:
    - **Remove and create direct relations**, which will delete the old relations once a transformation is marked as executed
    - **Maintain the 'Active from/until' field and keep all relations**, which will keep old and new relations and maintain the **Active from** and **Active to** dates (see Section 6.2)
  - Define if scheduled changes should be visible in the fact sheet next to the field/attribute by toggling the **Activate the scheduled change feature** option on or off.
  - Reprovision the workspace of a long-standing customer to make it compatible with the latest features by clicking the **Reprovision** button.
- **Predefined Transformation Types**
  In this tab, you can turn the predefined transformation templates on or off.
- **Configured Transformation Types**
  Here, you can create custom transformation templates.
- **Excel Export & Import**
  In this tab, you can extract, create, and maintain transformations in bulk, which we'll discuss in more detail later in this section.

There are different ways to create transformations in SAP LeanIX, but one thing they all have in common is that you must always map the transformation to an existing initiative fact sheet. You can create transformations in the following ways:

- In the **Transformations** tab of an initiative
- Based on suggested SAP solutions from the business capability level 2 and level 3 fact sheets (see Chapter 8, Section 8.4.7)
- Via Excel import in the **Transformations** admin menu
- Via a custom script/integration
- Via the target architecture diagram (see Section 6.4)

We'll explain how to create transformations using the **Transformations** tab and Excel import in the following sections. We'll also explore options to view transformations.

Create Transformation Manually

To create a transformation manually in an initiative fact sheet, navigate to the **Transformations** tab and click the blue **Create Transformation** button in the middle of the center pane (see Figure 6.20).

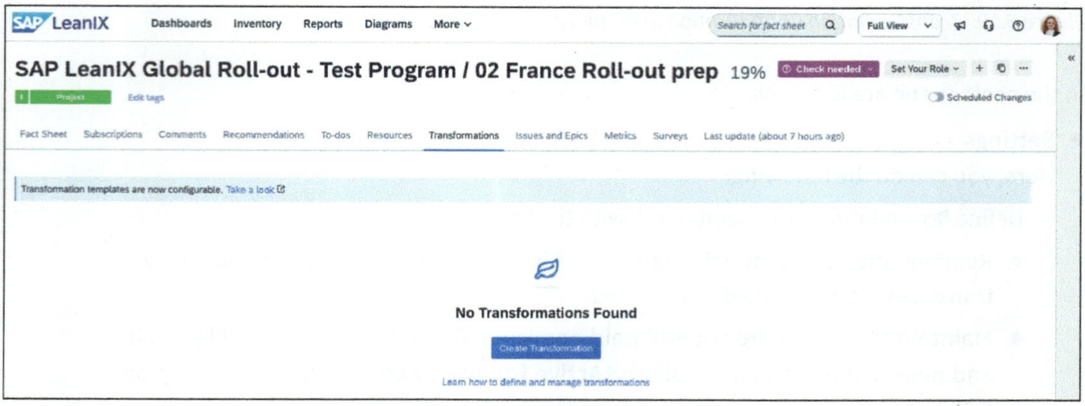

Figure 6.20: Manually Creating Transformation from Initiative Fact Sheet

In this example, we'll use the **Rollout Application** template (see Figure 6.21).

After selecting the relevant template, you'll see a form in which you can enter the transformation **Name**, enter the **Completion date** (in this example, we selected a milestone), enter a **Description**, select the **Application Name** from a dropdown menu, select the **Roll Out To** organization via a dropdown menu, and define the **Supporting** business capability that the application will be used for in the specified organizations (see Figure 6.22). Once all this information has been correctly entered, click the blue **Save** button in the lower-right corner of the form.

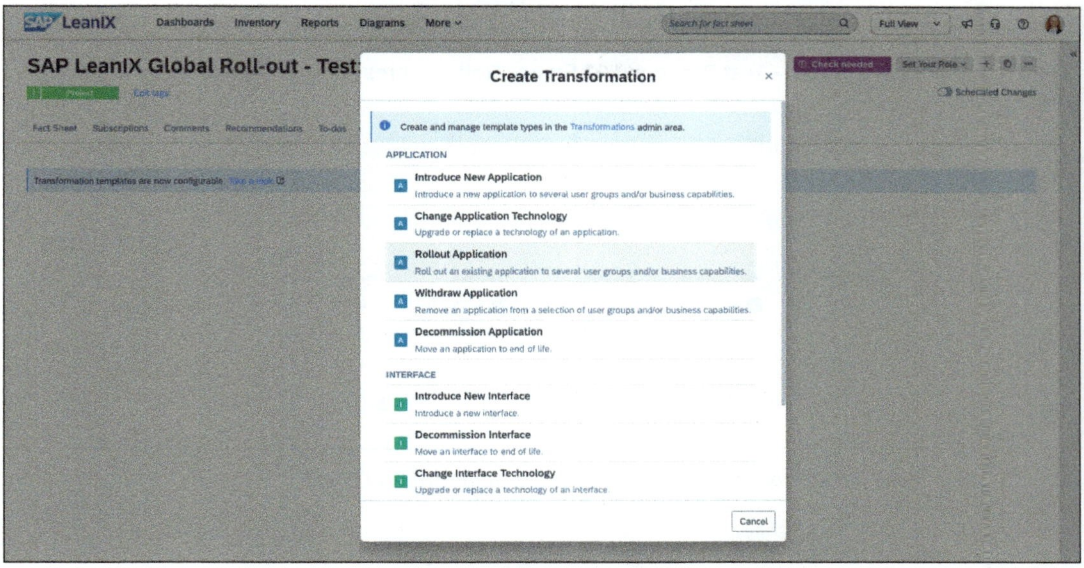

Figure 6.21: Selecting Rollout Application Template Transformation

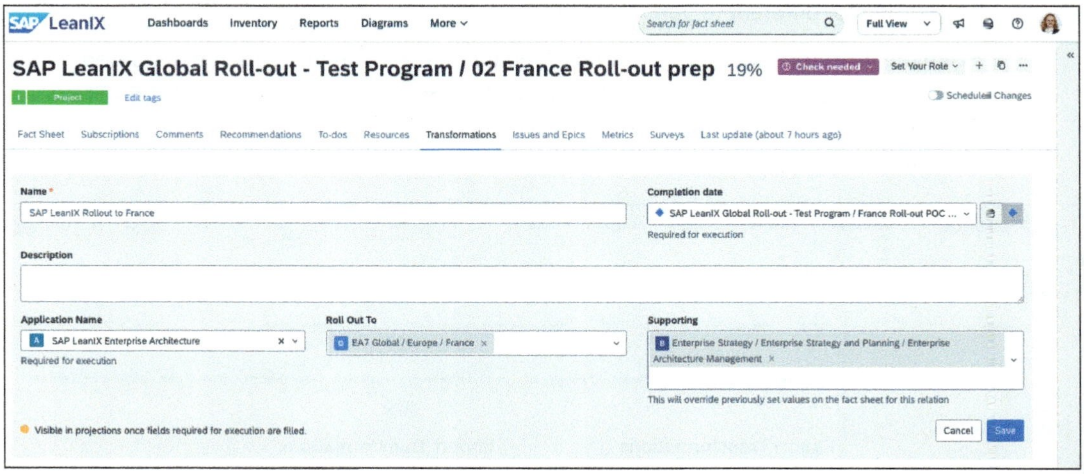

Figure 6.22: Template Transformation Input Form

After saving, you'll see the planned transformation on the **Transformations** tab of the initiative in an editable table view (see Figure 6.23). You'll also see the transformation on the other referenced fact sheets in the **Transformations** tab, in this example, on the SAP LeanIX application fact sheet, on the France Organization fact sheet, and on the Enterprise Architecture Management Business Capability fact sheet.

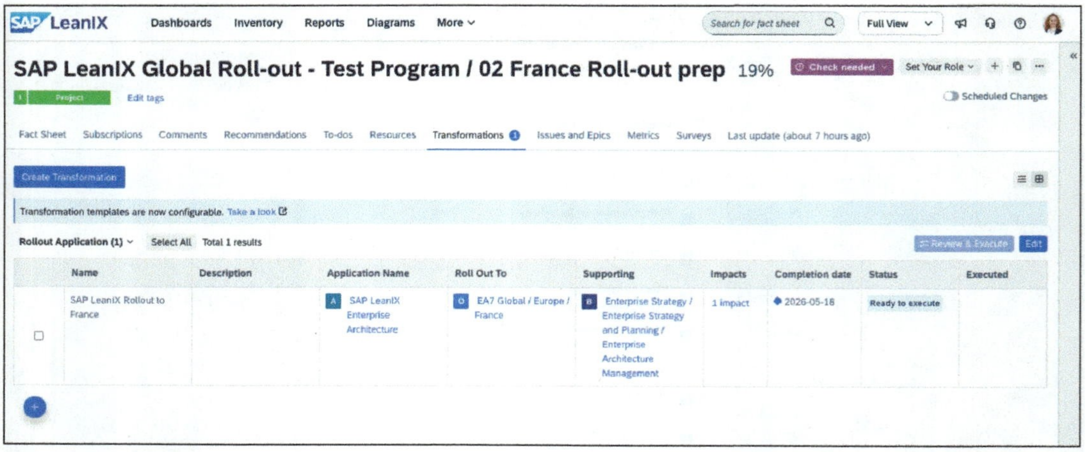

Figure 6.23: Transformation Overview Table on Initiative Fact Sheet

Create Transformation Using Excel Export/Import

For the creation of the other rollout transformations, we'll use the Excel export/import functionality as the impact, meaning the changes to the SAP LeanIX inventory, is exactly the same, just for different regions at different times.

Like the inventory Excel export/import, you'll first create an import template based on the export structure. Navigate to the **Administration** menu • **ADVANCED SETTINGS** • **Transformations** • **Excel Import & Export** tab. Click the blue **Export** button in the **Export Transformations** tile (see Figure 6.24).

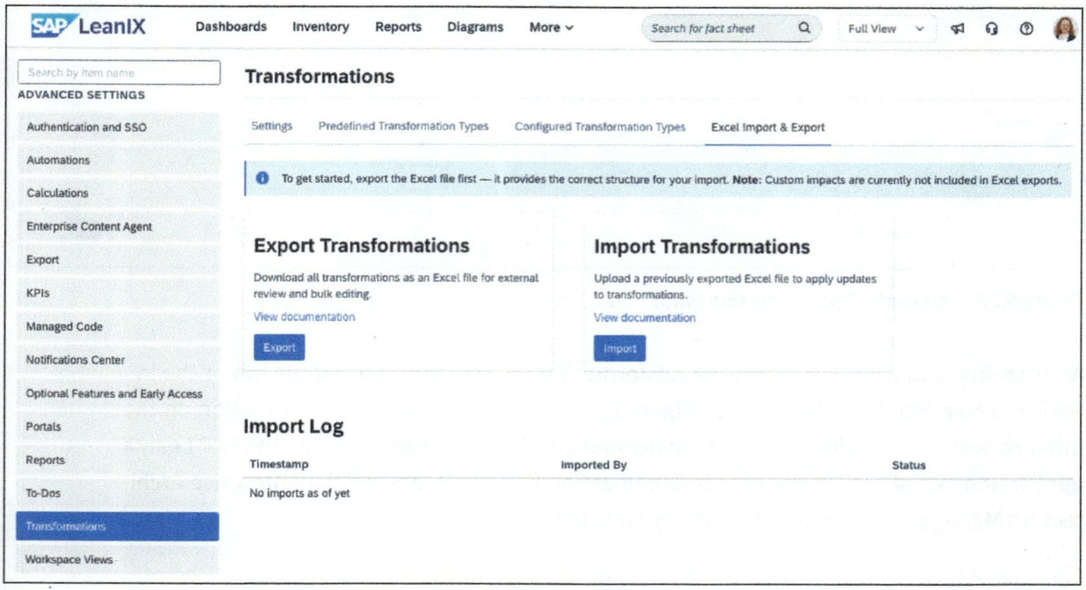

Figure 6.24: Export Transformations from Administration Menu

The downloaded Excel file has a **Readme** worksheet and a worksheet per defined transformation template (see Figure 6.25).

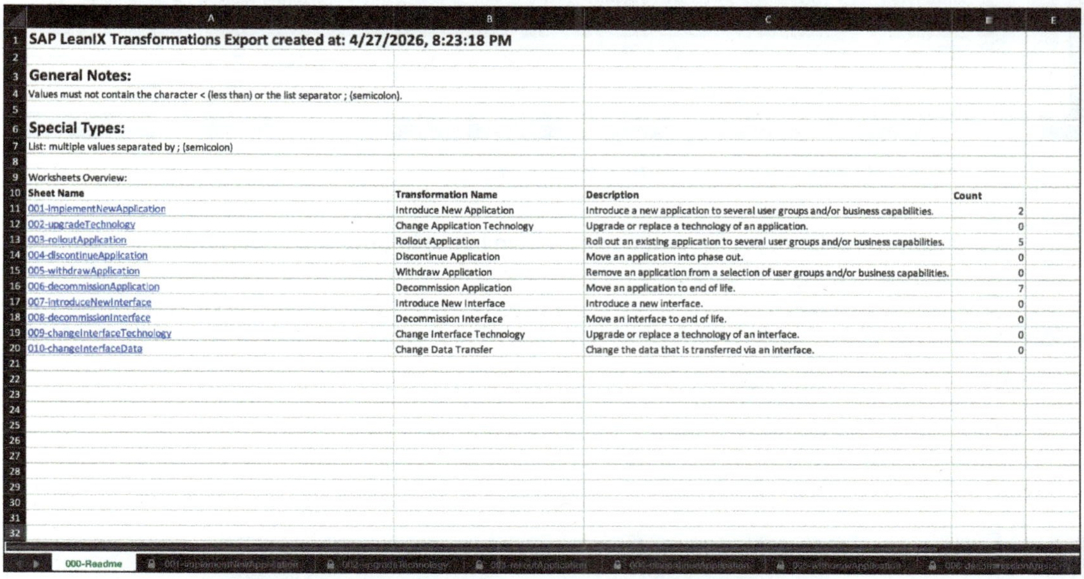

Figure 6.25: Excel Export File Structure

For this example, you'll need the structure of the worksheet **003-rolloutApplication**. Open a new Excel workbook to create your import template and copy the structure and data of the **003-rolloutApplication** worksheet. Make sure that you also name the new worksheet **003-rolloutApplication**.

Here are a few tips and tricks for the transformation import:

1. Only keep what you need! Delete rows that aren't relevant to you so that you don't accidentally modify them.
2. Don't delete any columns; otherwise, the import won't work.
3. Remember to always use the display names when referencing fact sheet names.
4. Don't use unsupported characters for transformation names (e.g., semicolon [;], less than [<]).
5. Leave the **ID** field blank to create new transformations.
6. To delete a transformation, enter **Delete** in the **Action** column (column A).

See Figure 6.26 for an example import structure.

Once you're done with the data entry, save the file and navigate back to **Excel Import & Export** under **Transformations** in the **Administration** menu. This time, click the blue **Import** button in the **Import Transformations** tile. A popup window will open; drag and drop your import file into it.

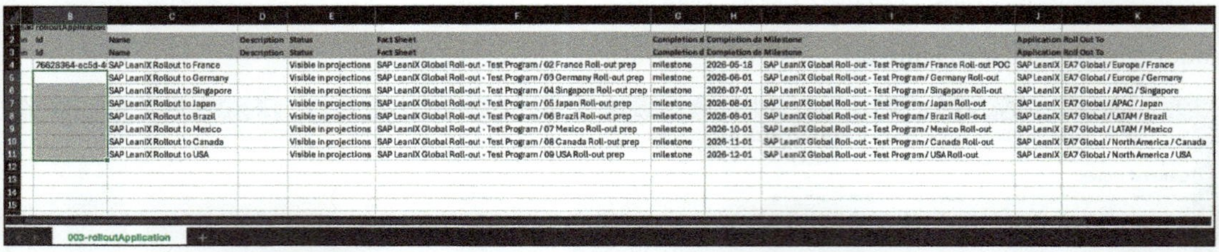

	Id	Name	Description	Status	Fact Sheet	Completion d	Completion da	Milestone	Application	Roll Out To
	Id	Name	Description	Status	Fact Sheet	Completion d	Completion da	Milestone	Application	Roll Out To
	76628364-ec5d-4	SAP LeanIX Rollout to France		Visible in projections	SAP LeanIX Global Roll-out - Test Program / 02 France Roll-out prep	milestone	2026-05-18	SAP LeanIX Global Roll-out - Test Program / France Roll-out POC	SAP LeanIX	EA7 Global / Europe / France
		SAP LeanIX Rollout to Germany		Visible in projections	SAP LeanIX Global Roll-out - Test Program / 03 Germany Roll-out prep	milestone	2026-06-01	SAP LeanIX Global Roll-out - Test Program / Germany Roll-out	SAP LeanIX	EA7 Global / Europe / Germany
		SAP LeanIX Rollout to Singapore		Visible in projections	SAP LeanIX Global Roll-out - Test Program / 04 Singapore Roll-out prep	milestone	2026-07-01	SAP LeanIX Global Roll-out - Test Program / Singapore Roll-out	SAP LeanIX	EA7 Global / APAC / Singapore
		SAP LeanIX Rollout to Japan		Visible in projections	SAP LeanIX Global Roll-out - Test Program / 05 Japan Roll-out prep	milestone	2026-08-01	SAP LeanIX Global Roll-out - Test Program / Japan Roll-out	SAP LeanIX	EA7 Global / APAC / Japan
		SAP LeanIX Rollout to Brazil		Visible in projections	SAP LeanIX Global Roll-out - Test Program / 06 Brazil Roll-out prep	milestone	2026-09-01	SAP LeanIX Global Roll-out - Test Program / Brazil Roll-out	SAP LeanIX	EA7 Global / LATAM / Brazil
		SAP LeanIX Rollout to Mexico		Visible in projections	SAP LeanIX Global Roll-out - Test Program / 07 Mexico Roll-out prep	milestone	2026-10-01	SAP LeanIX Global Roll-out - Test Program / Mexico Roll-out	SAP LeanIX	EA7 Global / LATAM / Mexico
		SAP LeanIX Rollout to Canada		Visible in projections	SAP LeanIX Global Roll-out - Test Program / 08 Canada Roll-out prep	milestone	2026-11-01	SAP LeanIX Global Roll-out - Test Program / Canada Roll-out	SAP LeanIX	EA7 Global / North America / Canada
		SAP LeanIX Rollout to USA		Visible in projections	SAP LeanIX Global Roll-out - Test Program / 09 USA Roll-out prep	milestone	2026-12-01	SAP LeanIX Global Roll-out - Test Program / USA Roll-out	SAP LeanIX	EA7 Global / North America / USA

Figure 6.26: Example Transformation Import Structure

Once you're done, you'll see a new line item in the **Import Log**, indicating that the import has been scheduled (see Figure 6.27).

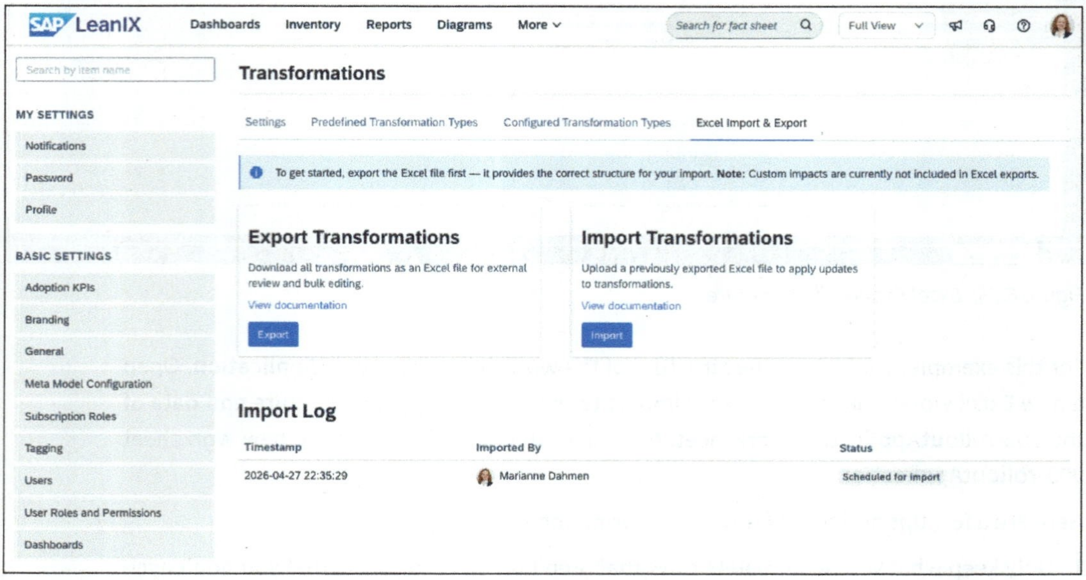

Figure 6.27: Importing Transformations in Admin Menu

> **Note**
>
> Depending on the complexity of the import, it might take several minutes for the status to change. You can continue working in the SAP LeanIX workspace and periodically check back to see if the status has been updated.

If everything went well, the **Import Log Status** will change to green text that says **Imported**. If there were errors during the import, the **Status** column will show red text that says **Imported with X errors**, where X is the number of errors. In this case, hover over the line item in the log table and click the **Summary Report** button that appears next to the entry with errors (see Figure 6.28).

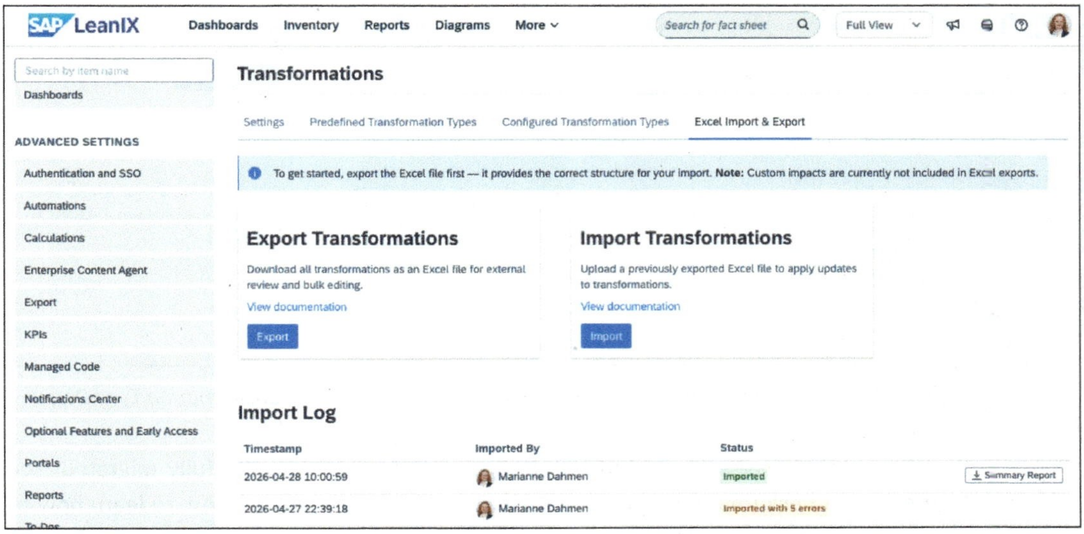

Figure 6.28: Transformation Import with Errors

In the summary report, you'll find the **Readme** worksheet and another worksheet called **000-Errors**. In this latter worksheet, you'll find information about the errors.

Further Resources

For more information on transformations export and import, visit the SAP LeanIX user documentation online:

- Export transformations: *http://s-prs.co/v620654*
- Import transformations: *http://s-prs.co/v620655*

Create a Custom Transformation Template

When you're planning a bigger transformation, for which you want to model certain changes recurringly or changes to custom fields that you created and that aren't covered by the out-of-the-box predefined transformations, you can create your own custom templates.

In this example, we'll create a custom transformation to speed up the planning and modeling of the SAP Process Integration and SAP Process Orchestration migration to SAP Integration Suite. All our interface scenarios that were previously managed by SAP Process Orchestration on-premise have to be moved to SAP Integration Suite in the public cloud. The migration will happen in waves.

For the demo, we created a test initiative fact sheet and a test interface fact sheet (see Figure 6.29), with SAP Process Orchestration 7.5 as the related IT component.

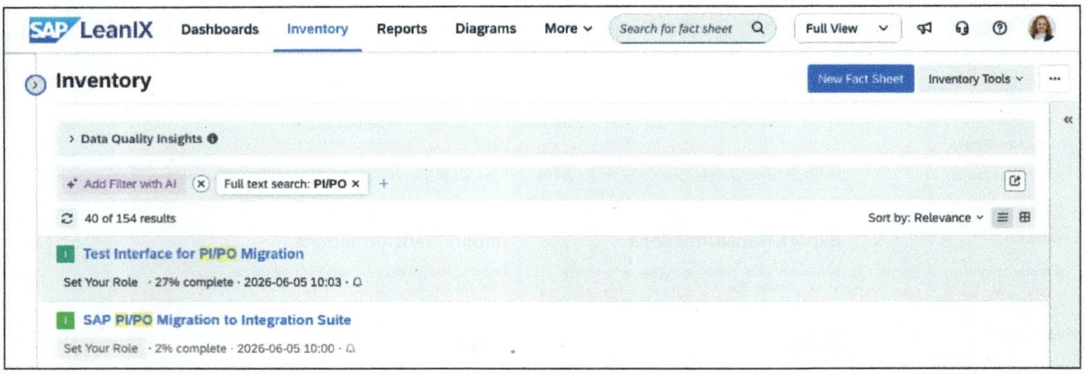

Figure 6.29: Prepared Initiative and Interface Fact Sheet for Custom Transformation Type Demo

We also created a new tag to indicate that an interface was successfully migrated (see Figure 6.30). Check Chapter 3, Section 3.3.3, under the "Tagging" section, to learn how to create new tag groups and tags.

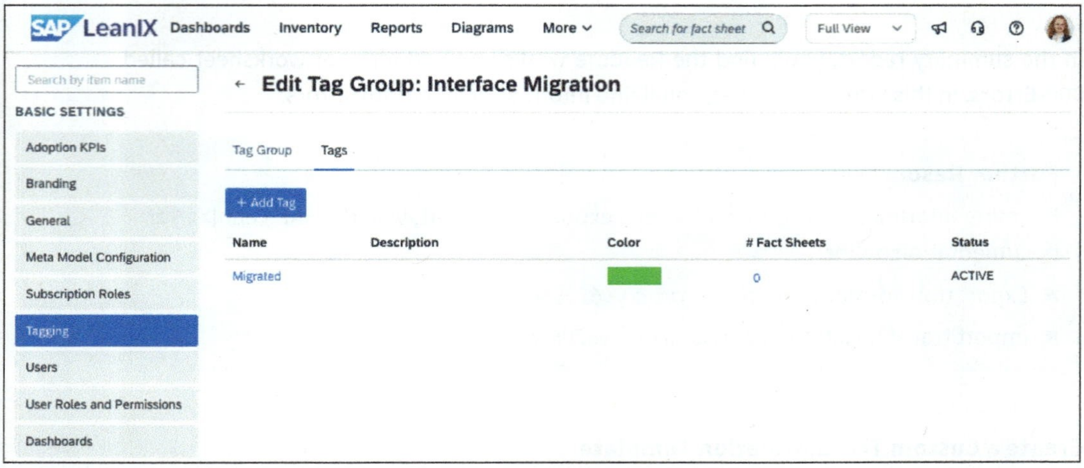

Figure 6.30: New Tag for Interface Fact Sheets for Custom Transformation Demo

To configure custom transformation templates, go into the **Administration** menu, navigate to the **Transformations** submenu under **ADVANCED SETTINGS**, open the **Configured Transformation Types** tab, and click the blue **Create Template** button (see Figure 6.31).

A new form will open in which you have to enter values for **Name of Transformation Type**; **Template Category**, referring to the fact sheet type this transformation is primarily about; and optionally, **Description** (see Figure 6.32). Click the blue **Save** button to finalize this step.

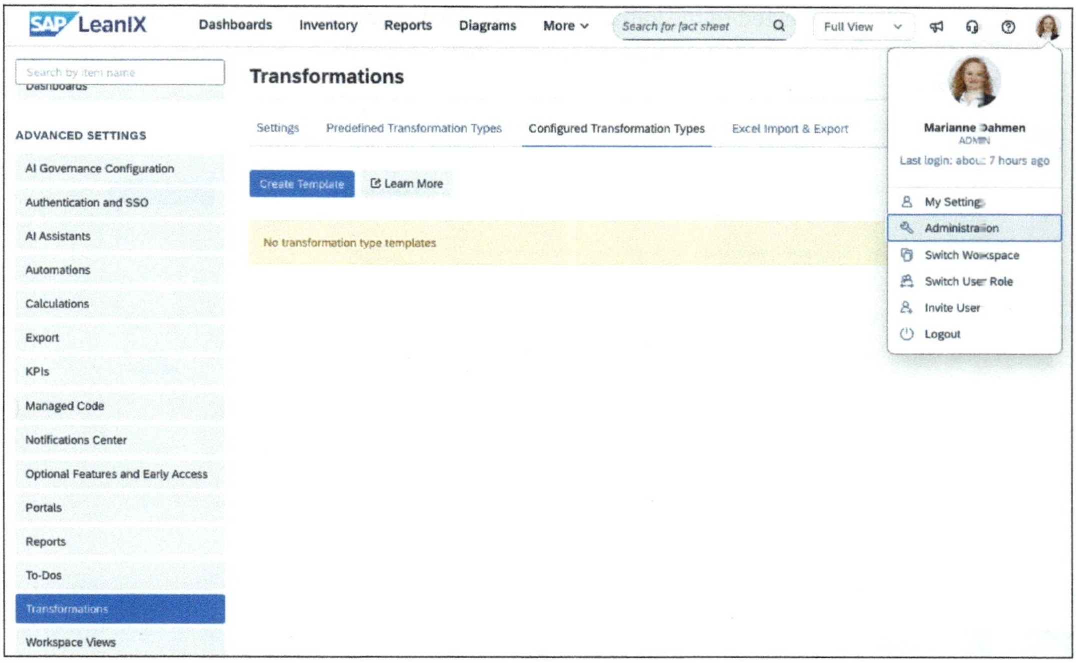

Figure 6.31: Create Custom Transformation Template in Administration Menu

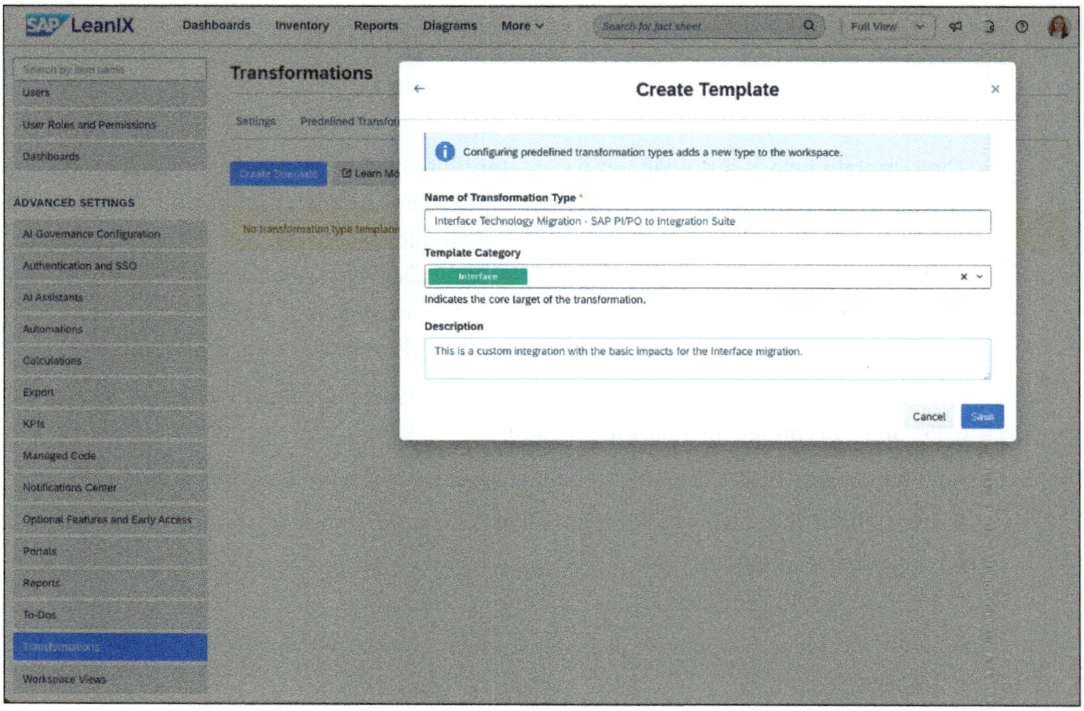

Figure 6.32: Custom Transformation Template General Information Form

You will be redirected to the initial template configuration screen. Click the blue **Add Template Field** button to start designing your custom transformation template (see Figure 6.33).

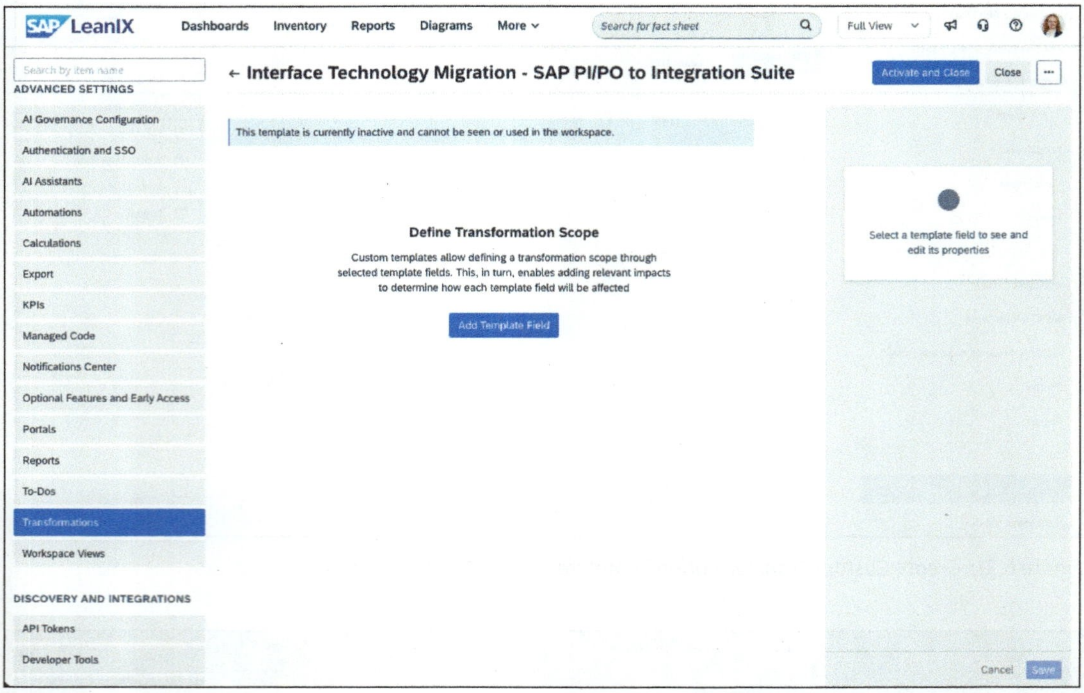

Figure 6.33: Initial Custom Transformation Configuration Screen

The template field is a predefined parameter that will later be used in impact modeling. First, define that **Interface** for **Fact Sheet Type** is one parameter for impact. Because you want to apply this transformation to multiple interface fact sheets per migration wave, for **Fact Sheet Selection Mode**, select **Multiple**. Also select **No** under **Allow Creation of New Fact Sheets** as you only want to use interfaces that exist in your inventory repository (see Figure 6.34). Click the blue **Save** button once you're done with the configuration.

The custom transformation configuration screen has now changed its design (see Figure 6.35), and you can see the following sections:

- **TEMPLATE CATEGORY**, for the fact sheet type that this transformation is about
- **DESCRIPTION**, showing the description of the custom transformation
- **TEMPLATE OVERVIEW**, showing defined template fields, with the option to create more by clicking **Add Template Field**
- **IMPACTS**, for an overview of designed impacts and option to add more by clicking **Add Impact**

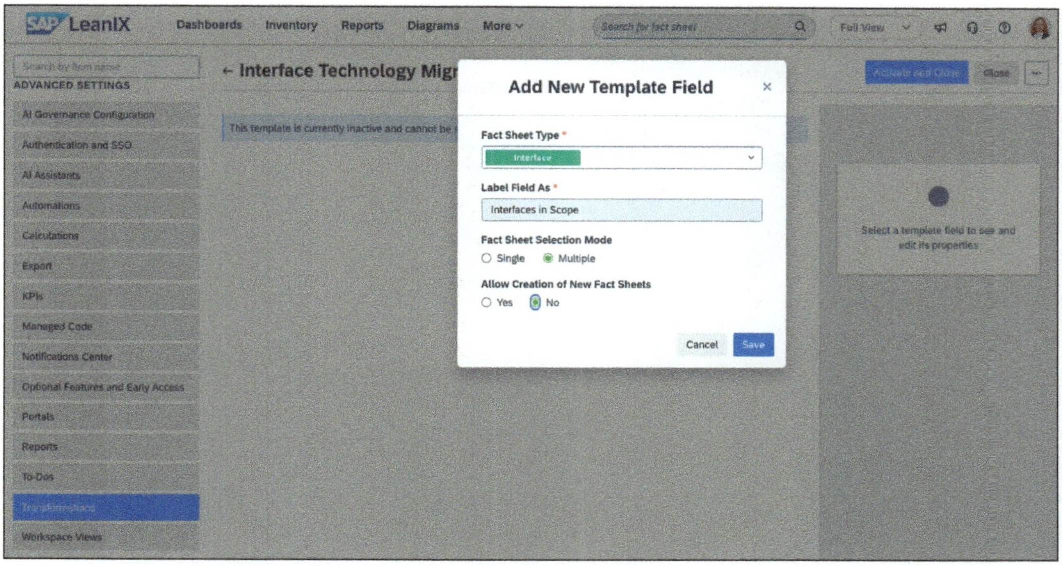

Figure 6.34: Creating First Template Field

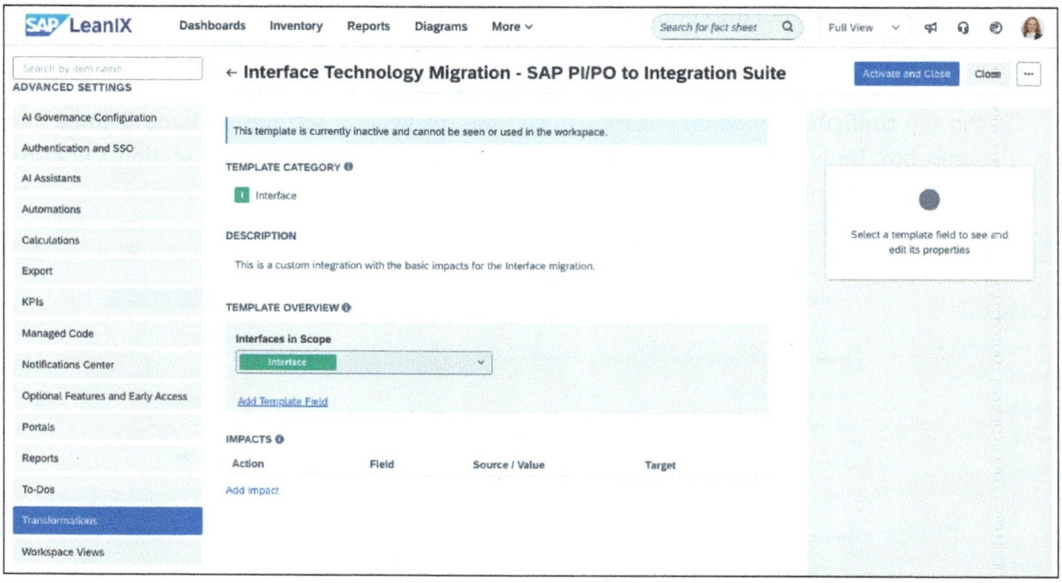

Figure 6.35: Custom Transformation Configuration Screen with Template Field

Let's now create the template field for the SAP Process Orchestration 7.5 IT component. For **Fact Sheet Type**, select **IT Component** from the dropdown. Set **Label Field As** to "SAP PO 7.5" so you can correctly identify it when creating custom impacts. Now set **Single** for **Fact Sheet Selection Mode** as there is only one IT component fact sheet you want to select here. Again, select that you don't want to create new fact sheets during the transformation. Don't forget to click **Save** when you're done (see Figure 6.36).

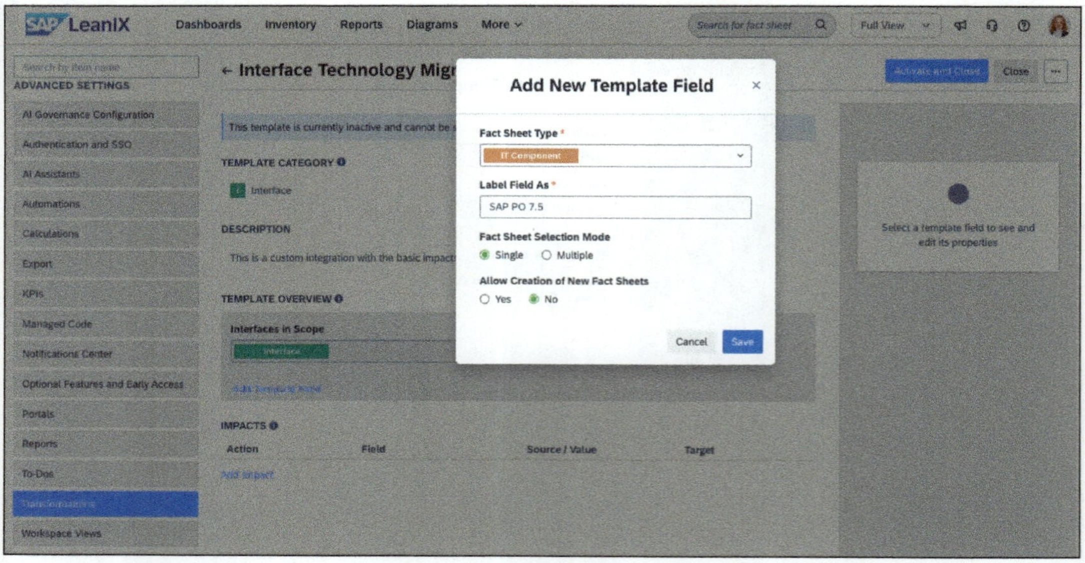

Figure 6.36: Creating Second Template Field

We're not done with the configuration of the IT component field yet. So far, you've only created the shell of the template field. Now you have to specify that it should only include the SAP Process Orchestration 7.5 IT component fact sheet. For that, use the right-hand menu in the configuration editor, where you can see the general settings that you defined in a grey box. Under **Allow Creation of New Fact Sheets**, you can see a **Filters** section. Click **Add Filter** (see Figure 6.37).

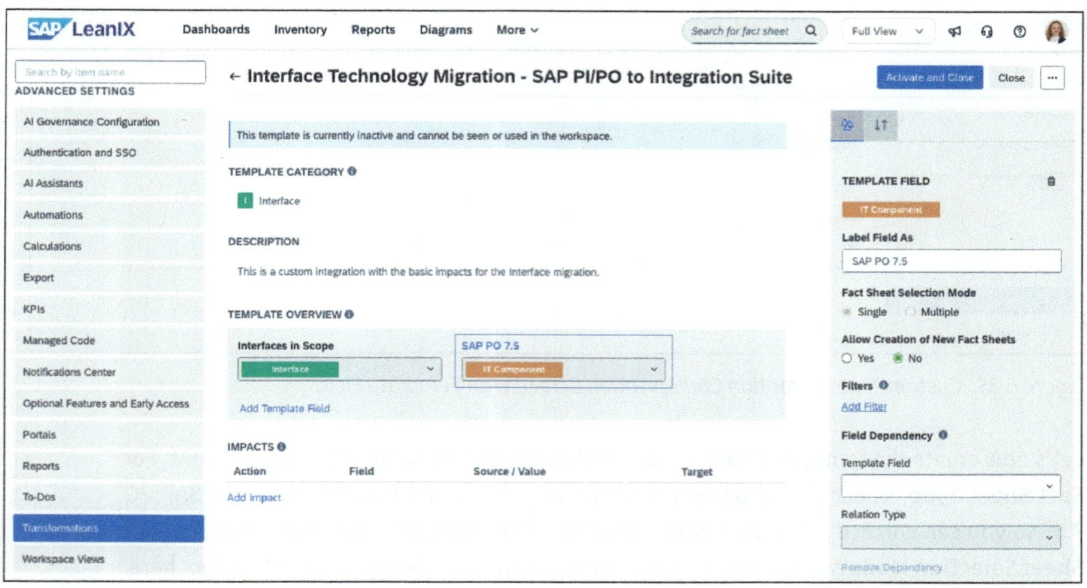

Figure 6.37: Adding Filter to Template Field

A new popup will open with a familiar inventory search option called **Filters** (see Chapter 4, Section 4.3.5). Filter the inventory down to the desired fact sheet scope using the left-hand filters, then click the blue **Use Fact Sheet Filter** button in the lower-right corner once you're done (see Figure 6.38).

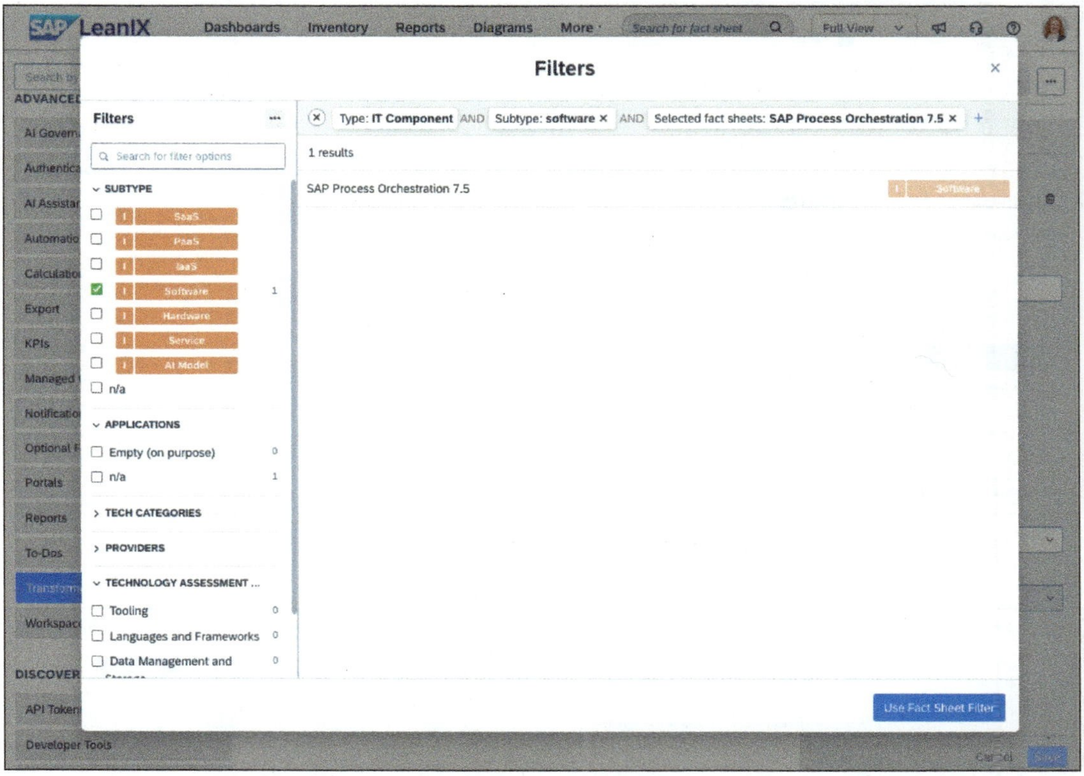

Figure 6.38: Inventory Filtering to Define Scope of Template Field

You'll now see that there are **Filters** applied to the **Template Field**. Save your changes by clicking the blue **Save** button in the lower-right corner of the right-hand menu (see Figure 6.39).

Repeat these steps for the template field for the SAP Integration Suite IT component.

Next, we'll add an impact. To do that, click the blue **Add Impact** link at the bottom of the configuration overview. In the popup, select **Remove relations** under the **RELATIONS** section (see Figure 6.40).

As the first **Template Field** value, select the **IT Component** via the dropdown; its relation to the interface fact sheets will be removed. In this case, this is **SAP PO 7.5**, which is the label assigned during the template field creation. Now define the **Relation Type** via the dropdown. Here, this is the relation to the **Interfaces** of the IT component that should be modified.

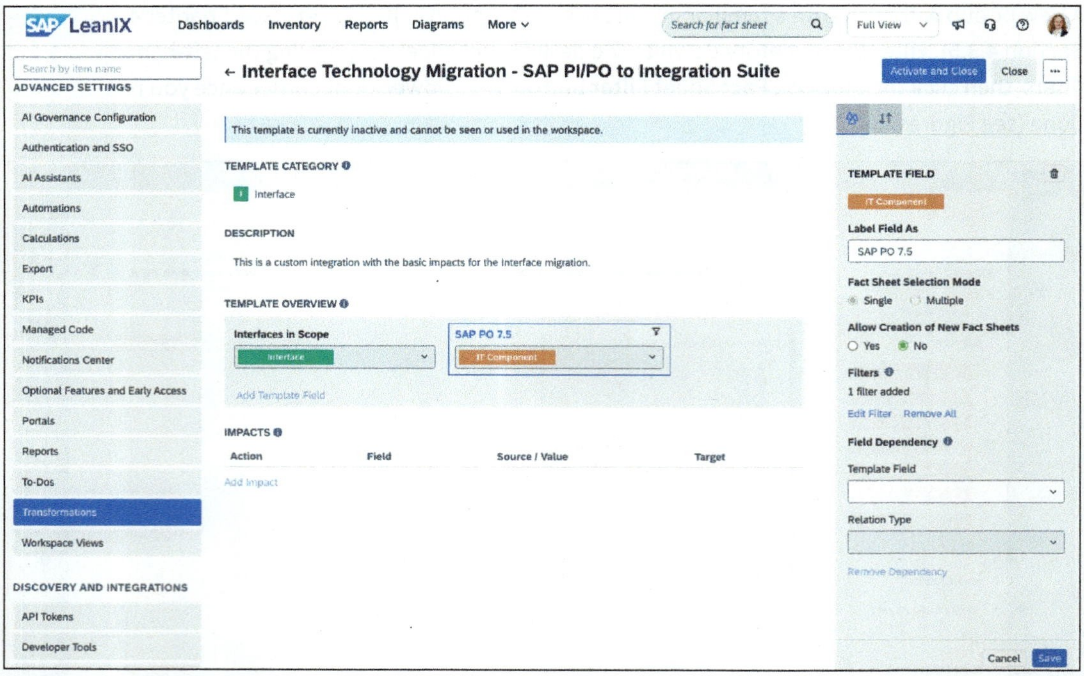

Figure 6.39: Saving Changes to Template Field Filtering

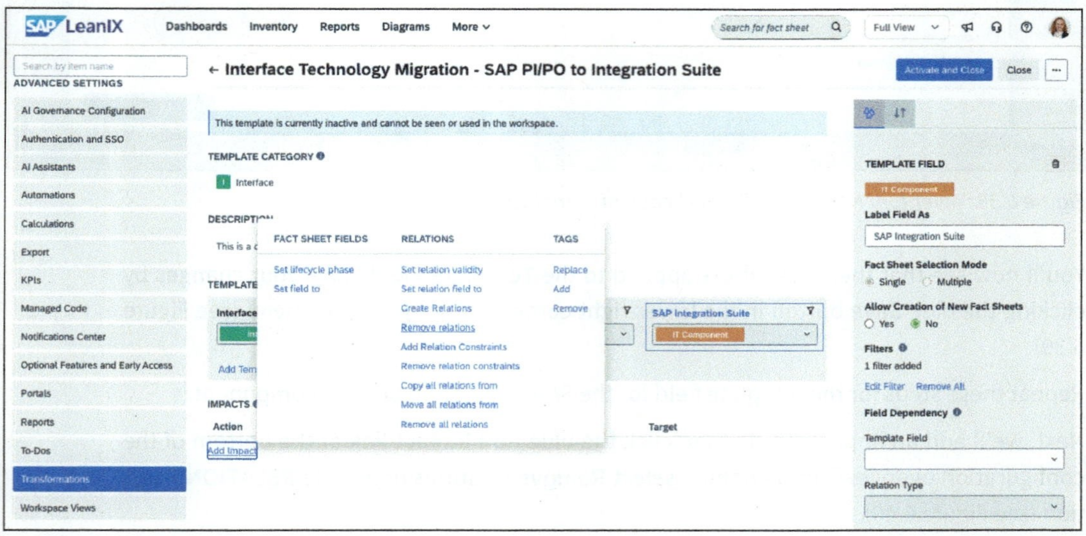

Figure 6.40: Adding Remove Relations Impact

Finally, select **Interfaces in Scope** as the second **Template Field** needed for this action, to specify from which interface fact sheets the relation should be removed (see Figure 6.41). Click the blue **Save** button once you're done with the entry.

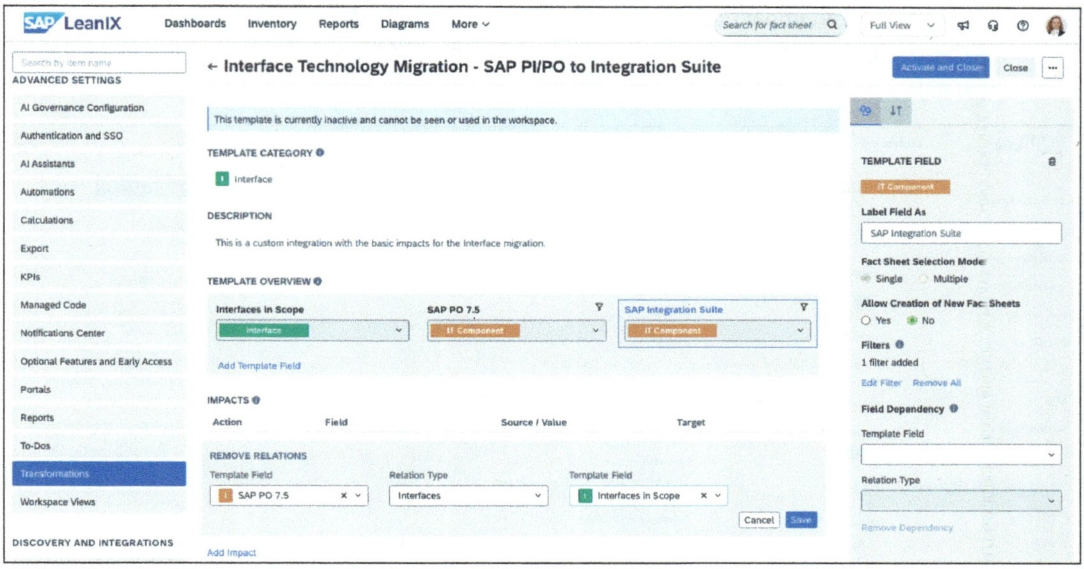

Figure 6.41: Configuring Remove Relation Impact

In a similar manner, create the **Create Relations** impact to model the switch to SAP Integration Suite.

Select **SAP Integration Suite** as the first **Template Field**, select **Interfaces** as **Relation Type**, and **Interfaces in Scope** as the second **Template Field** (see Figure 6.42). Click **Save**.

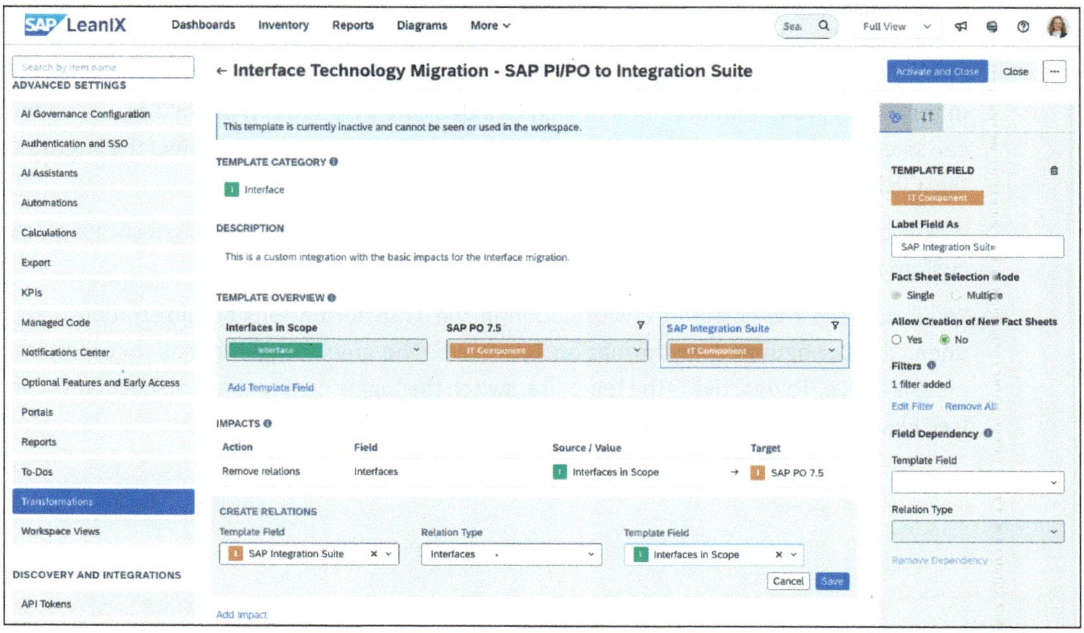

Figure 6.42: Adding Second Impact: Create Relations

Finally, add an impact to add a tag. Again, click the blue **Add Impact** link at the bottom of the configuration page and select **Add** under the **TAGS** section of the popup (see Figure 6.43).

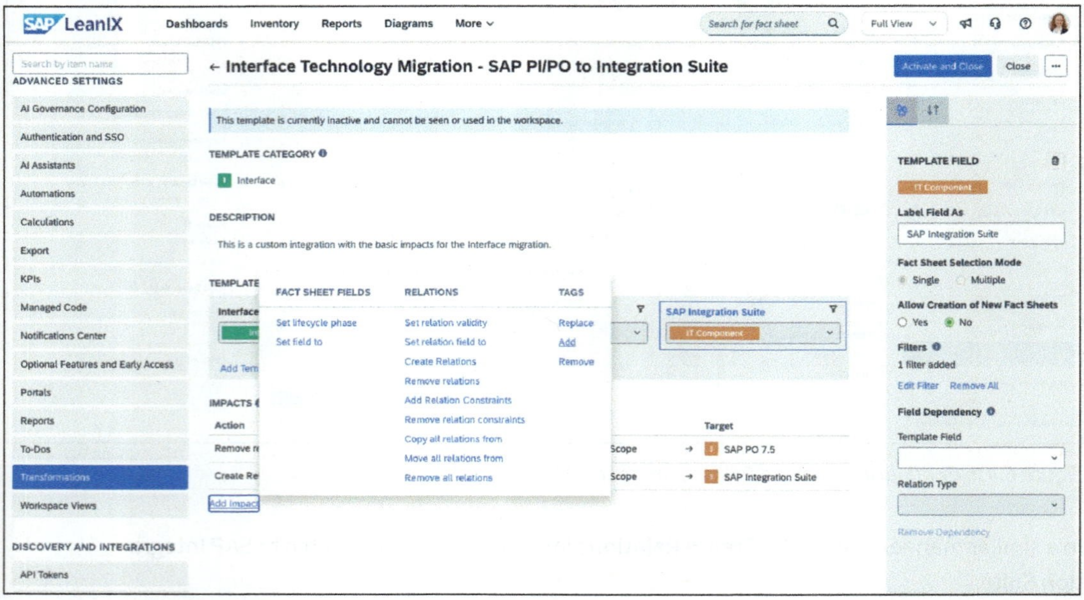

Figure 6.43: Adding Impact to Add Tag

The first **Template Field** addition specifies the fact sheets where the tag should be applied, in this case, **Interfaces in Scope**. Next, under **Tags type**, you can define whether the tag can be found in the **Pre-defined** tag groups (created by admins) or the **User defined** ones (tags in the *others* group that are not assigned to a dedicated tag group). Under **Tags**, you then can select one or more tags from the dropdown menu; for this example, select the **Migrated** tag. Click **Save** (see Figure 6.44).

When you're done, click the blue **Activate and Close** button in the upper-right corner of the custom transformation configuration screen.

You'll now see the custom transformation in the **Transformations** administration menu under the **Configured Transformation Types** tab. The green toggle means that the template is active. To deactivate the template, switch the toggle next to the fact sheet type icon (see Figure 6.45).

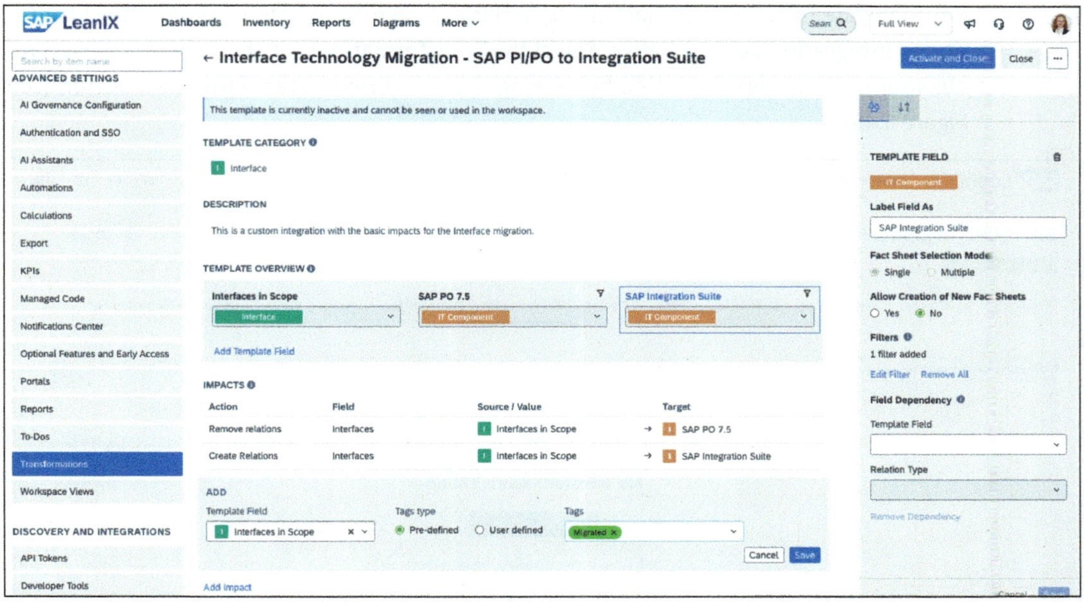

Figure 6.44: Entering Data for Add Tag Impact

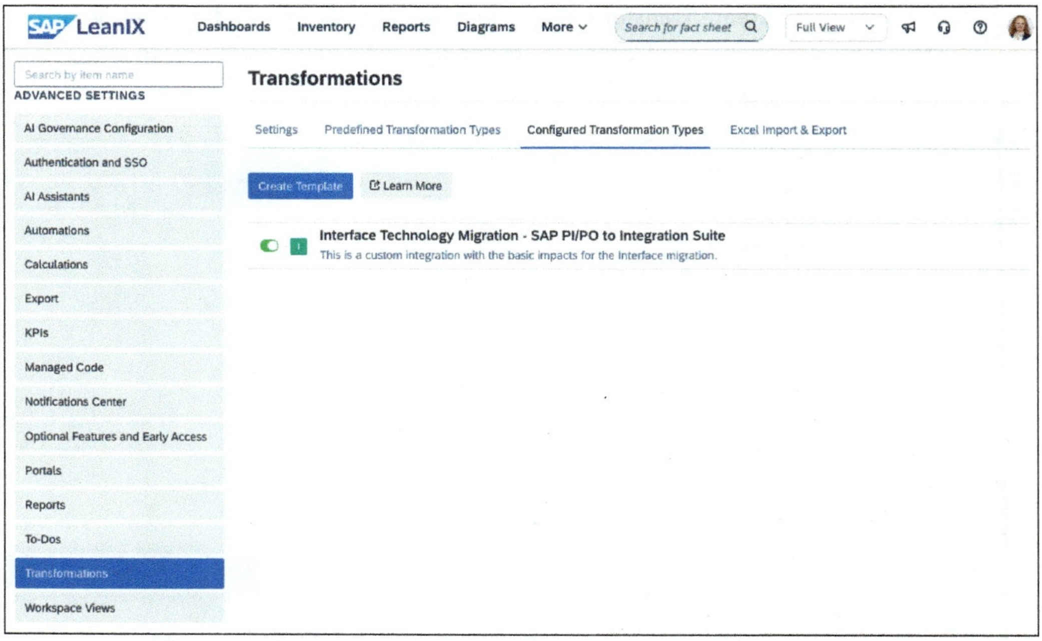

Figure 6.45: Configured and Active Custom Transformation Template

Now navigate to an initiative fact sheet; in this example, we created the **SAP PI/PO Migration to Integration Suite** initiative. Select the **Transformations** tab. From here, you can create your custom transformation by clicking the blue **Create Transformation** button (see Figure 6.46).

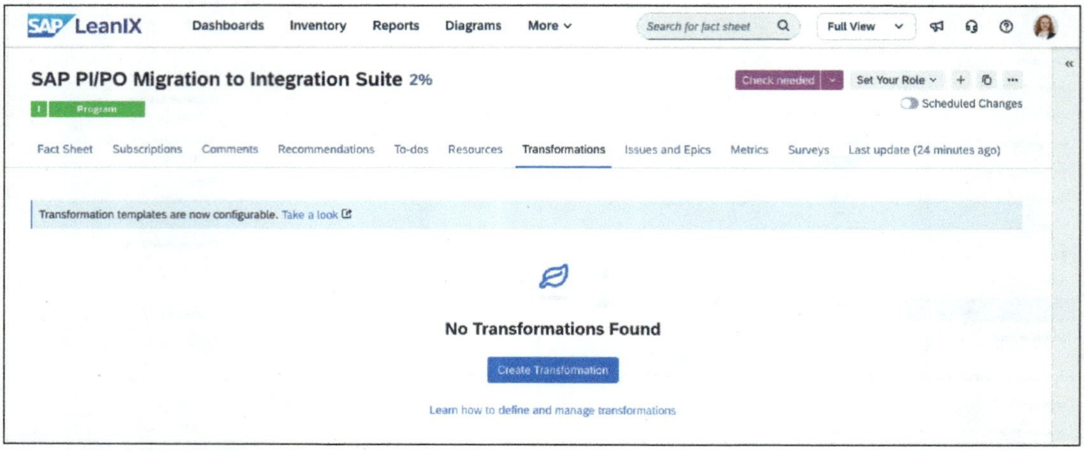

Figure 6.46: Adding Transformation to Initiative Fact Sheet

Scroll down in the **Create Transformation** popup window until you find your newly created custom transformation, called **Interface Technology Migration – SAP PI/PO to Integration Suite**. Click that entry (see Figure 6.47).

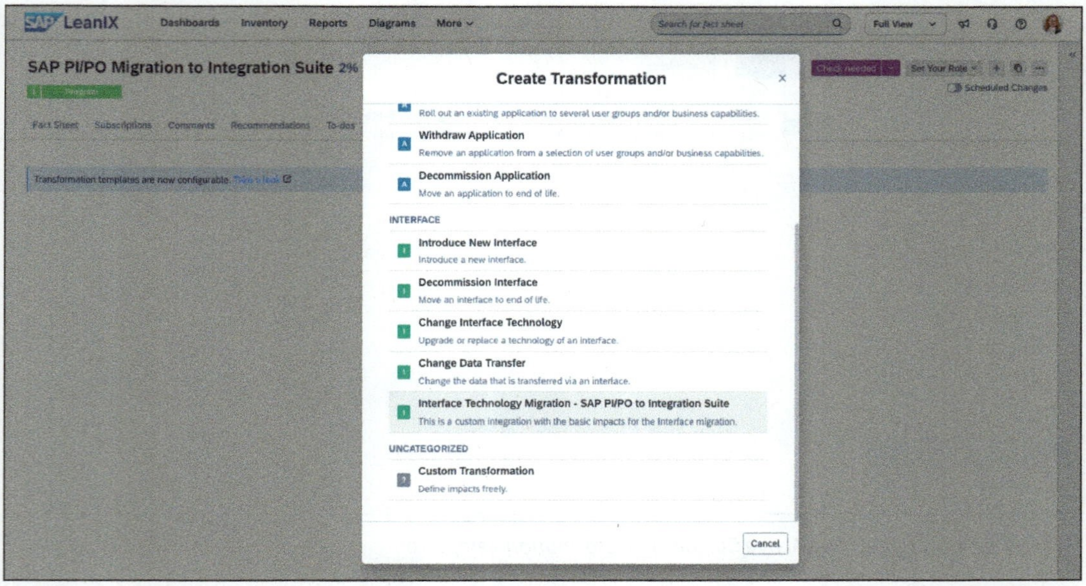

Figure 6.47: Selecting New Custom Transformation Template

A new form opens in which you can modify the **Name** of the transformation as needed, enter a **Completion date** and a **Description**, and then select the fact sheets in scope in the preconfigured dropdown menus. For **Interfaces in Scope**, we selected the **Test Interface for PI/PO Migration**, but you can also select more interfaces as needed. For **SAP PO 7.5**, we selected the **SAP Process Orchestration 7.5** IT component option; because we selected single mode, we can only choose one entry. For **SAP Integration Suite**, we selected the available **SAP Integration Suite** IT component fact sheet. Click **Save** to save the transformation (see Figure 6.48).

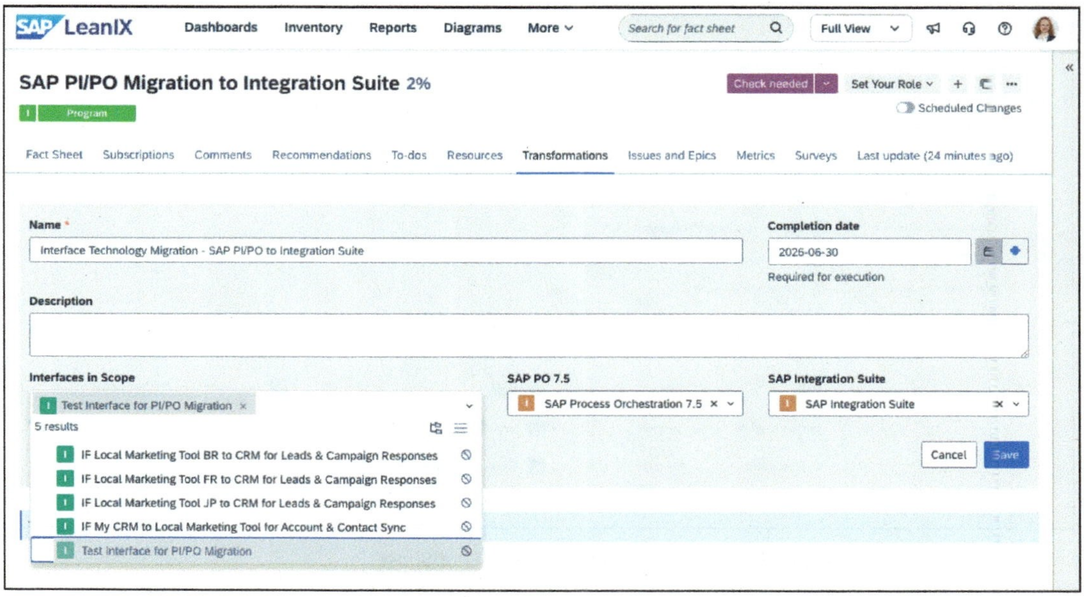

Figure 6.48: Entering Details of Custom Transformation in Initiative Fact Sheet

You will now see the transformation on the initiative fact sheet (see Figure 6.49).

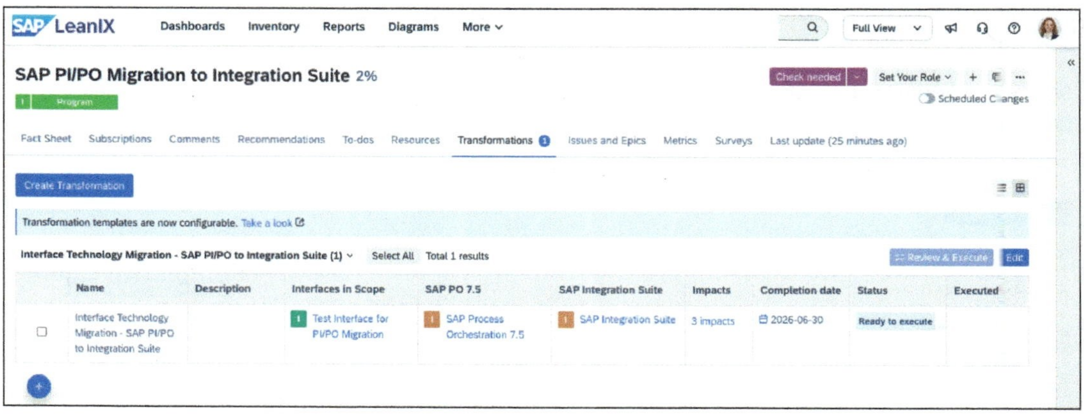

Figure 6.49: Modeled Transformation in Initiative Fact Sheet

You will also see the transformation on the impacted interface and IT component fact sheets. See Figure 6.50 for an example of the **Transformations** tab for the **Test Interface for PI/PO Migration** interface fact sheet.

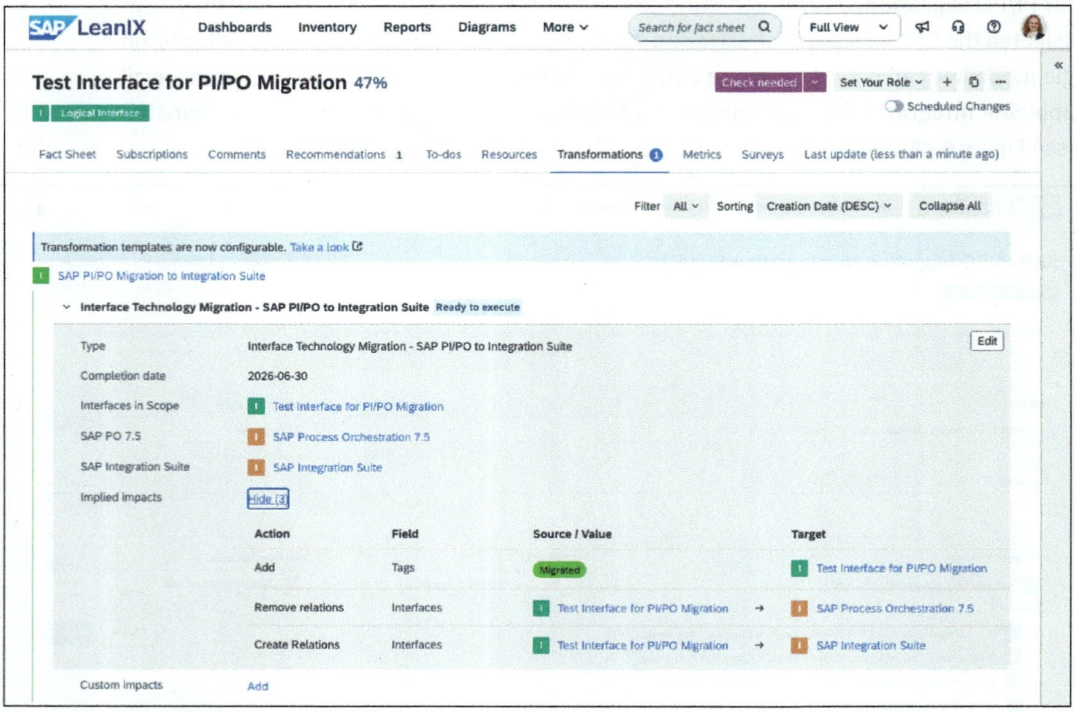

Figure 6.50: Modeled Transformation in Interface Fact Sheet

Once the transformation has been performed, navigate back to the respective initiative fact sheet and go on the **Transformations** tab. Select the executed transformation in the table, then click the blue **Review & Execute** button in the upper-right corner (see Figure 6.51).

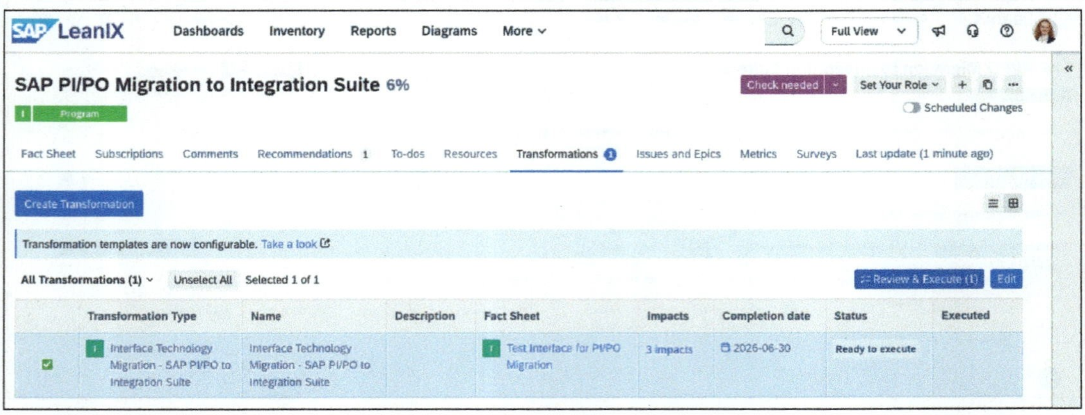

Figure 6.51: Execute Transformation for Initiative Fact Sheet

A popup window appears. Review the information, then click the blue **Execute Now** button in the lower-right corner of the window (see Figure 6.52).

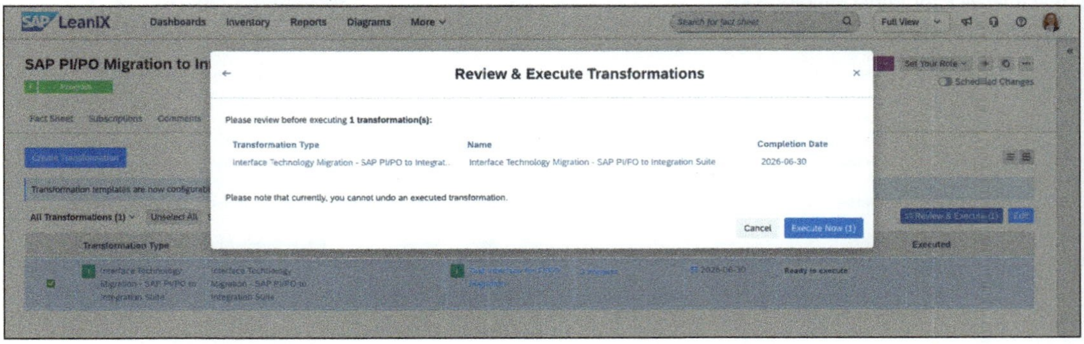

Figure 6.52: Confirming Execution of Transformation

You'll now see the executed transformation in the initiative fact sheet with the **Status** column showing **Executed** in green (see Figure 6.53).

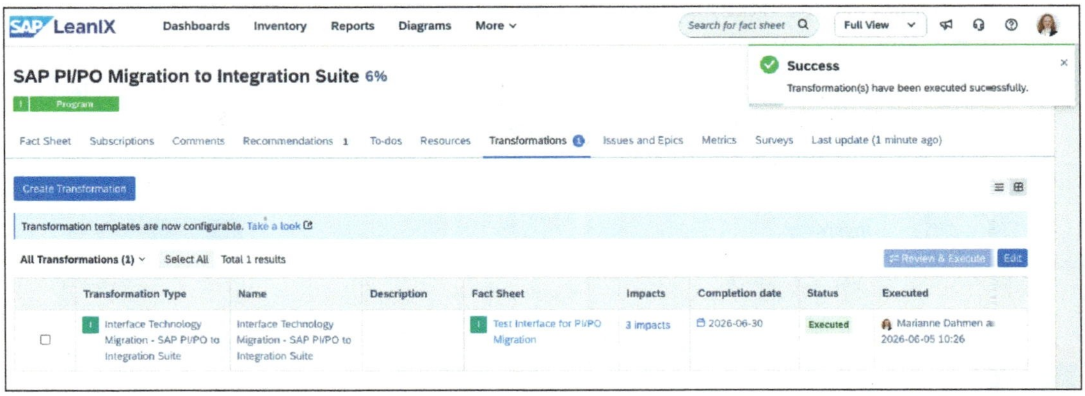

Figure 6.53: Executed Transformation in Initiative Fact Sheet

You'll see the changes performed for the interface and IT component fact sheets in the **Last update** audit log tab of the respective fact sheets. See Figure 6.54 for an example of the **Last update** tab for the **Test Interface for PI/PO Migration** interface fact sheet.

Always test your custom transformation templates with test fact sheets before rolling them out to a wider audience to confirm that the impacts are modeled correctly and the template fields offer the options that you intended.

Further Resources

For more information on custom transformation templates, visit the SAP LeanIX user documentation at *http://s-prs.co/v620656*.

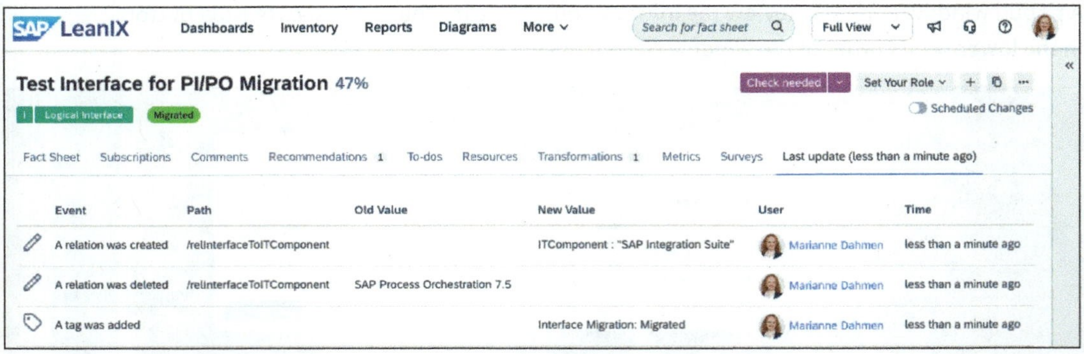

Figure 6.54: Interface Audit Log of Executed Transformation

View Transformations

Another smart way to see all created transformations in the SAP LeanIX workspace is via the transformations explorer. By default, the admin and member user roles have full create, read, update, delete (CRUD) access to the transformations explorer. This can be changed by admins if needed in the **Administration** menu under **BASIC SETTINGS • User Roles and Permissions.**

To navigate to the transformations explorer, click **More** in the top ribbon, then select the **Transformations Explorer** entry (see Figure 6.55).

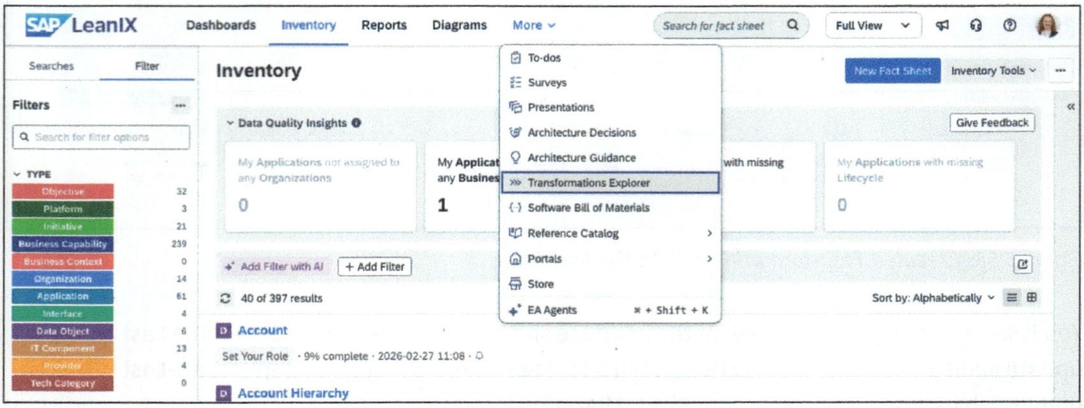

Figure 6.55: Navigating to Transformations Explorer

In the transformations explorer, you can search and filter based on various attributes (center filter pane at the top). In the left-hand menu, you can quickly filter for used transformation templates. Furthermore, you can make changes in the overview table by clicking the pen button in the upper-right corner to switch to edit mode. You can also select transformations and move them to other existing initiative fact sheets by clicking the **Move to Initiative** button in the upper-right corner. Finally, you can also **Review & Execute** selected transformations (see Figure 6.56).

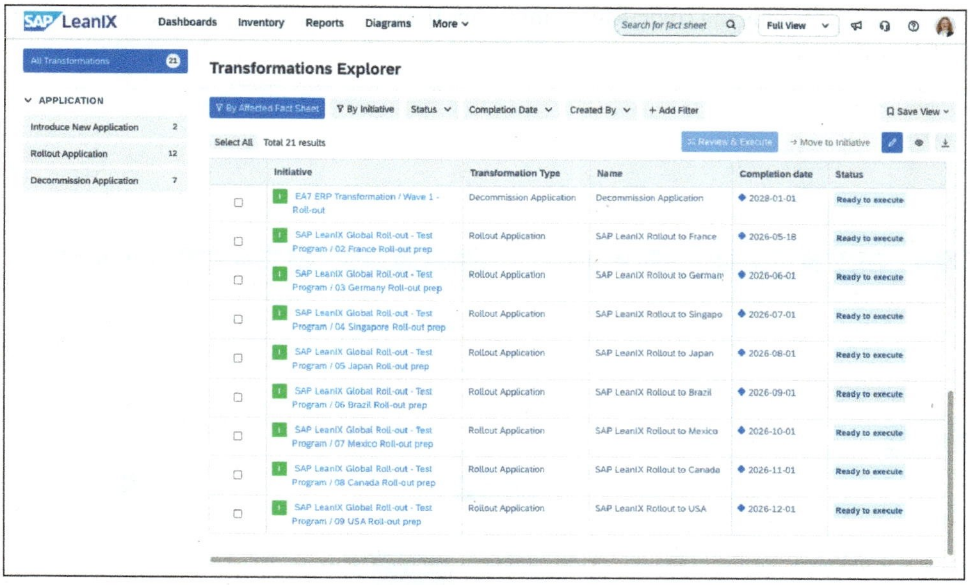

Figure 6.56: Transformation Explorer Overview

Execution means that the planned impacts to the inventory artifacts are executed/completed and will be permanently written into the referenced fact sheets as facts. The **Review & Execute** button also exists on the **Transformations** tab of the initiative fact sheet and should be used at the end of a project once the changes are implemented in order to update the SAP LeanIX inventory data accordingly.

Further Resources

For the latest information on the transformations explorer and its features, visit the SAP LeanIX user documentation online at *http://s-prs.co/v620657*.

In addition to transformation management, the SAP LeanIX Architecture and Road Map Planning product also offers an additional derived field for report color-coding in views. When transformation templates are used, you can select **Initiatives: Impact Type**, **Applications: Impact Type**, or **Interfaces: Impact Type** as a **View** option.

Click the **Reports** menu in the top ribbon and select **Landscape Reports** under **REPORT TYPES**.

Select the **Application Landscape** report, aggregate it to **Level 1** in the upper-right icon bar, and select **Initiatives: Impact Type** on the left next to **View** (see Figure 6.57). You can see various planned transformations. If you've modeled more than one impact type for an application in an initiative, the bar will still show only one color. With the **Minimum/Maximum** dropdown next to the **View** selection field, you can define which entry will overrule others on the color-coding aspect. **Minimum** means the value farthest to the right of the legend will determine the color. **Maximum** means the value farthest to the left in the legend will determine the color.

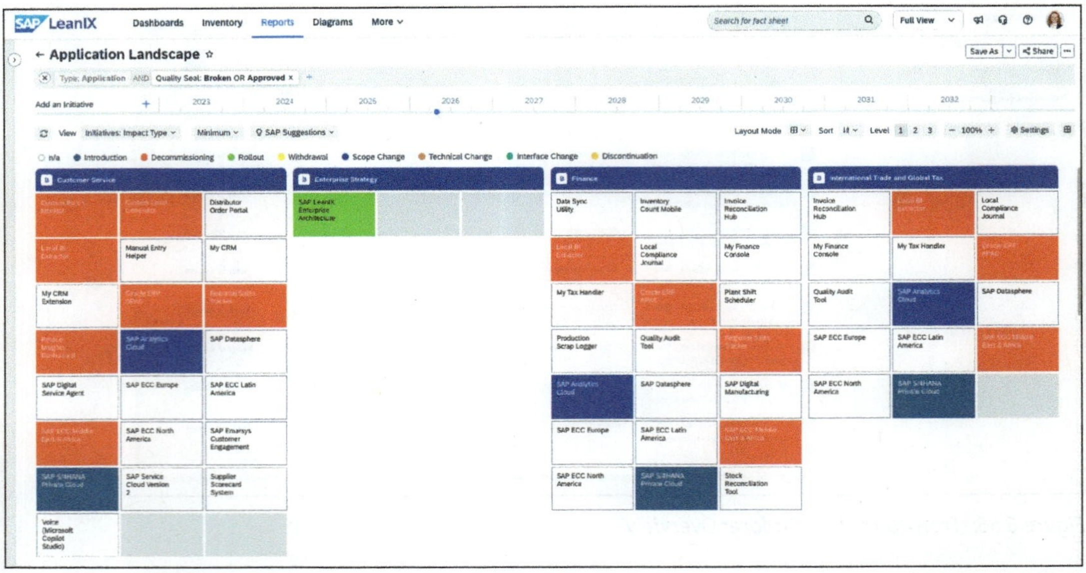

Figure 6.57: Impacts View in Reports

Caution

Do not modify the out-of-the-box **lxTransformationStatus** (see Figure 6.58) or **lxImpactType** (see Figure 6.59) fields, even though it may be tempting to include your custom status values or custom transformation impact types.

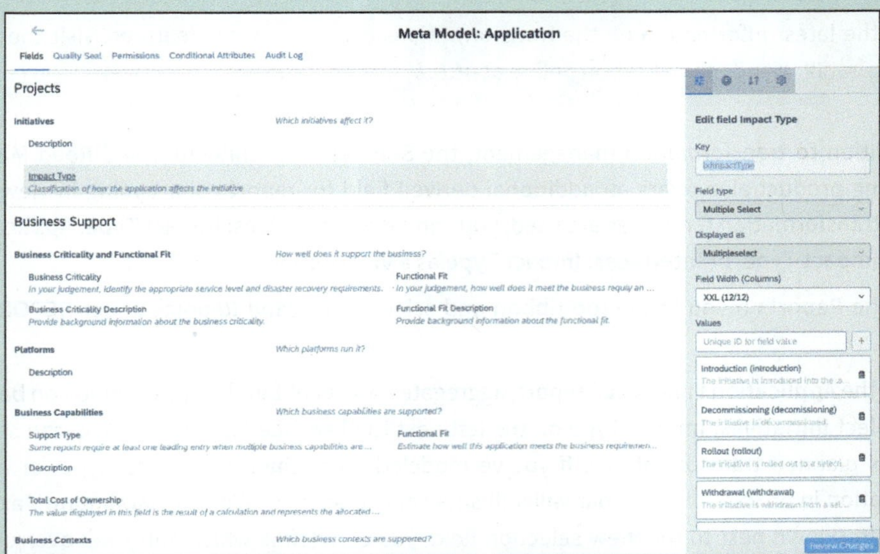

Figure 6.58: lxTransformationStatus Field in Application Meta Model

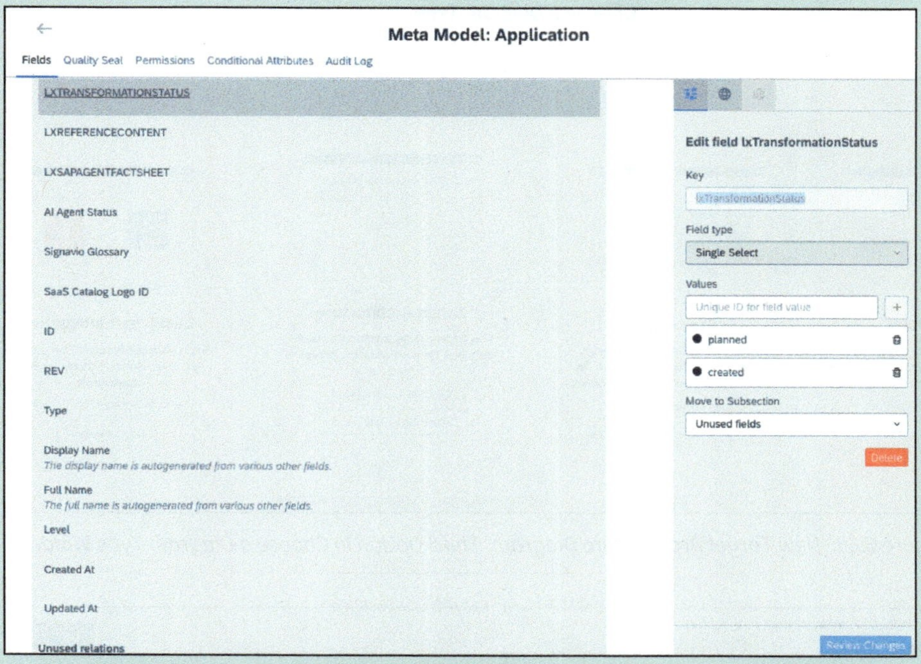

Figure 6.59: lxImpactType Field in Application Meta Model

Changing those fields will break some of the features of the transformation service in the backend.

6.4 Target Architecture Diagrams

There is one thing we know for sure in target architecture planning: Enterprise architects love to draw diagrams. Therefore, it's a natural thought to develop a feature that will help enterprise architects develop target architecture directly out of diagramming, in addition to the predefined transformations form on the initiative fact sheet.

For this reason, the SAP LeanIX product and engineering team developed a new target architecture diagram called **Target Planning**, as shown in Figure 6.60.

The initial setup is similar to the data flow and free draw diagrams (for detailed instructions, see Chapter 5, Section 5.2), with the difference that in the general information form, you can also assign initiatives to the diagram into which the transformations will be written automatically.

It's possible to import existing diagrams and develop a new target architecture based on the existing diagram. For that, click **File** in the upper-left corner once you're in the diagram editor. Hover over the **Import** menu item, and select **Diagrams** in the expanded submenu (see Figure 6.61).

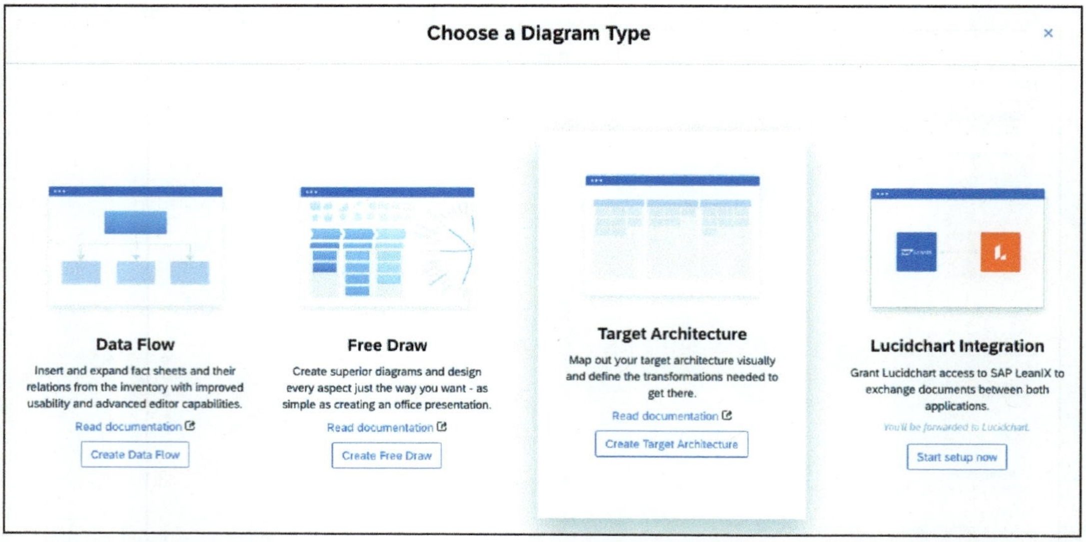

Figure 6.60: New Target Architecture Diagram: Third Option in Choose a Diagram Type Window

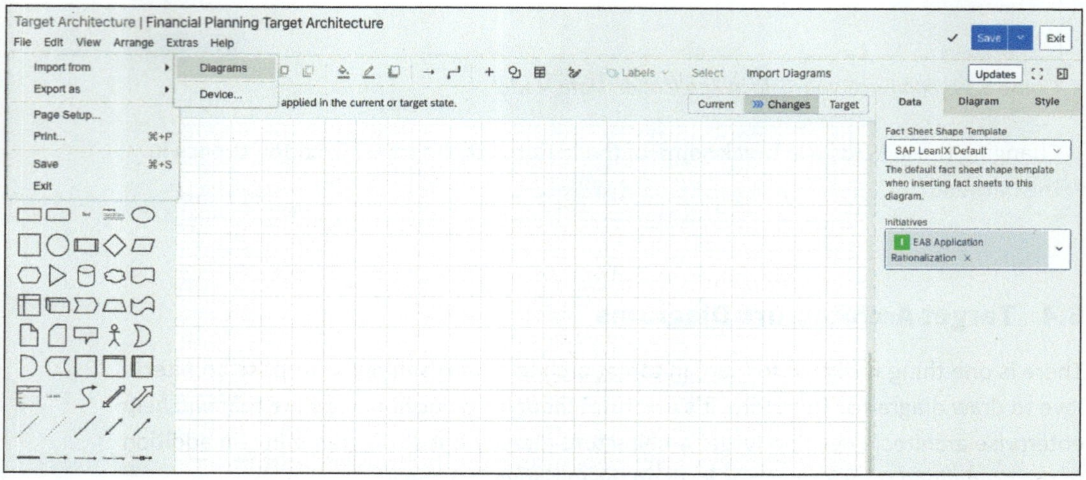

Figure 6.61: Import of Existing Diagram for Target Architecture Planning

You can filter available diagrams by searching for a title or description in the search bar or, as shown in the example, by filtering for a specific owner. Select the diagram that you want to import, then click the blue **Import Selected (1)** button (see Figure 6.62). It's also possible to import more than one diagram.

If you right-click a shape, you'll see a new line item in the context menu, called **Create Transformation**, where you can select one of the template transformations to be applied to the fact sheet connected to the shape (see Figure 6.63).

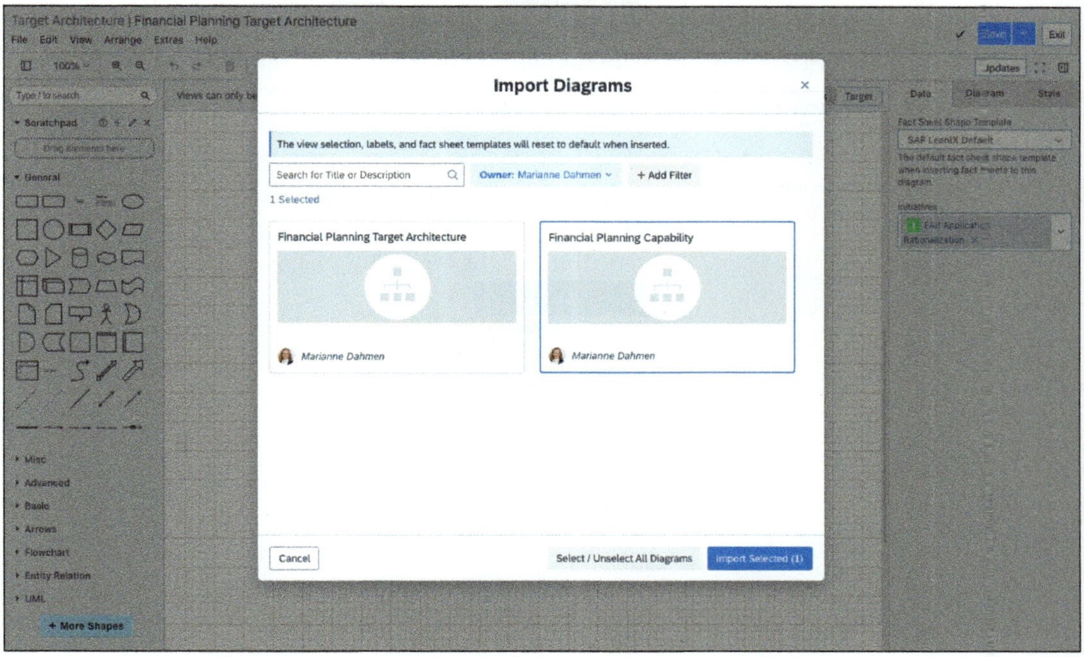

Figure 6.62: Importing Existing Diagram as Baseline for Target Architecture Planning

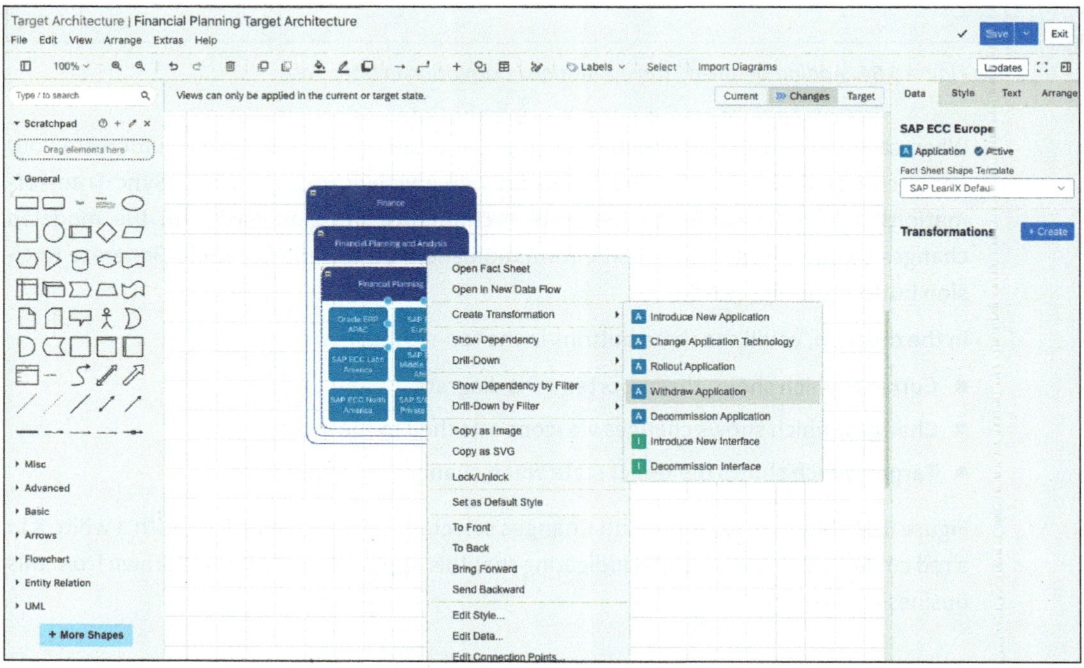

Figure 6.63: Applying Transformation to Selected Diagram Shape Linked to Application

In the right-hand side menu, the **Data** tab will update. You can specify additional information, such as related organizations or business capability fact sheets (see Figure 6.64).

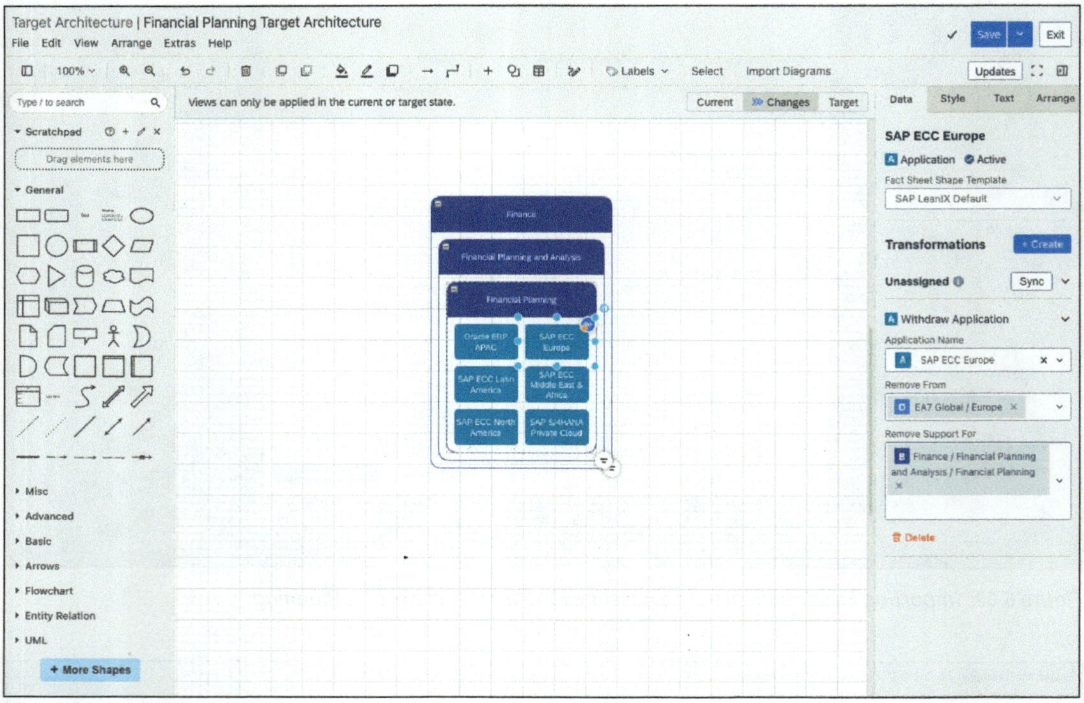

Figure 6.64: Adding Detailed Transformation Information in Data Tab

When saving the target architecture diagram, you can decide whether to autocreate your designed transformations for the connected initiative by clicking the blue **Sync Transformations** link and whether to **Generate Architecture Decision** based on the modeled changes by entering some context information and then clicking the white **Generate Decision** button (see Figure 6.65).

In the diagram, you'll see three buttons in the top-right corner:

- **Current**, which shows the imported as-is diagram
- **Changes**, which shows changes via icons attached to the shapes
- **Target**, which shows the target state when changes are applied

Figure 6.66 shows an example with **Changes** selected. You can see an icon with a white X in a red circle on **SAP ECC Europe**, indicating that this application will be withdrawn from this business capability.

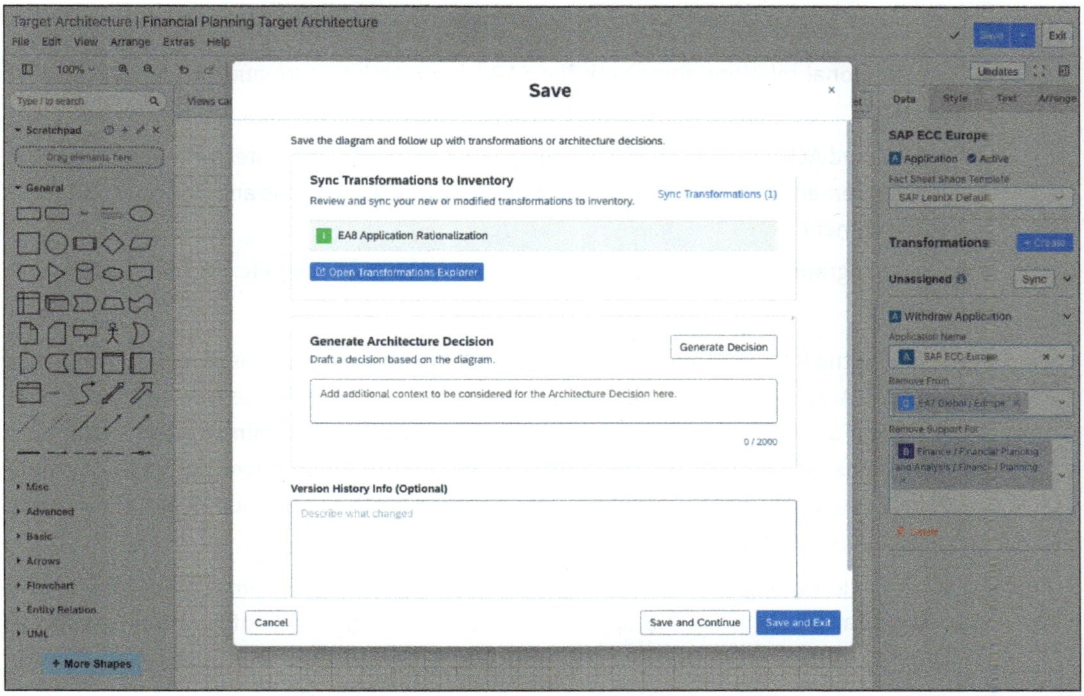

Figure 6.65: Target Architecture Diagram Save Window to Autocreate Transformations and Architecture Decisions

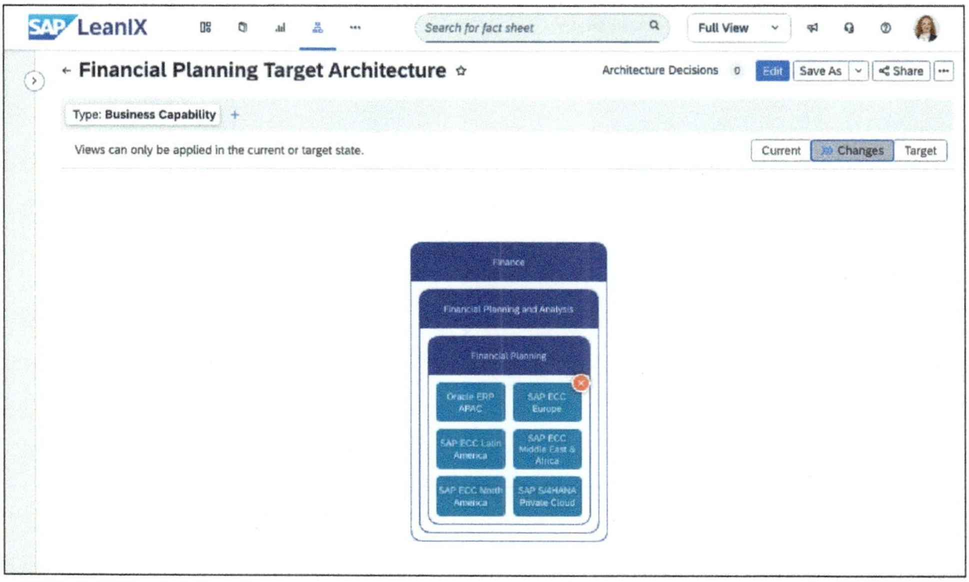

Figure 6.66: Target Architecture Diagram, Highlighting Planned Changes

6.5 Integration with Project Management Solutions

Two additional relations come with the SAP LeanIX Architecture and Road Map Planning product:

- SAP Cloud ALM integration for managing projects, which integrates with SAP Cloud ALM for implementation and is especially useful for transformations and projects in the SAP environment
- Jira integration, which links work items (epics, stories, tasks, etc.) to initiatives in SAP LeanIX

Setting up the integration of SAP Cloud ALM for managing projects is similar to setting up SAP Cloud ALM for landscape discovery integration (see Chapter 4, Section 4.2), but it's not the same. To identify the correct integration, navigate to the **Administration** menu, then **Integrations**, and click the blue **Add Integration** button in the upper-right corner. In the left-hand menu, filter for **Collaboration** and search for the **SAP Cloud ALM for Managing Projects** tile. Click the blue **Configure** button in the tile.

You can follow along with the in-tool guidance and linked documentation by clicking the **Documentation** button in the upper-left tile of the integration configuration (see Figure 6.67).

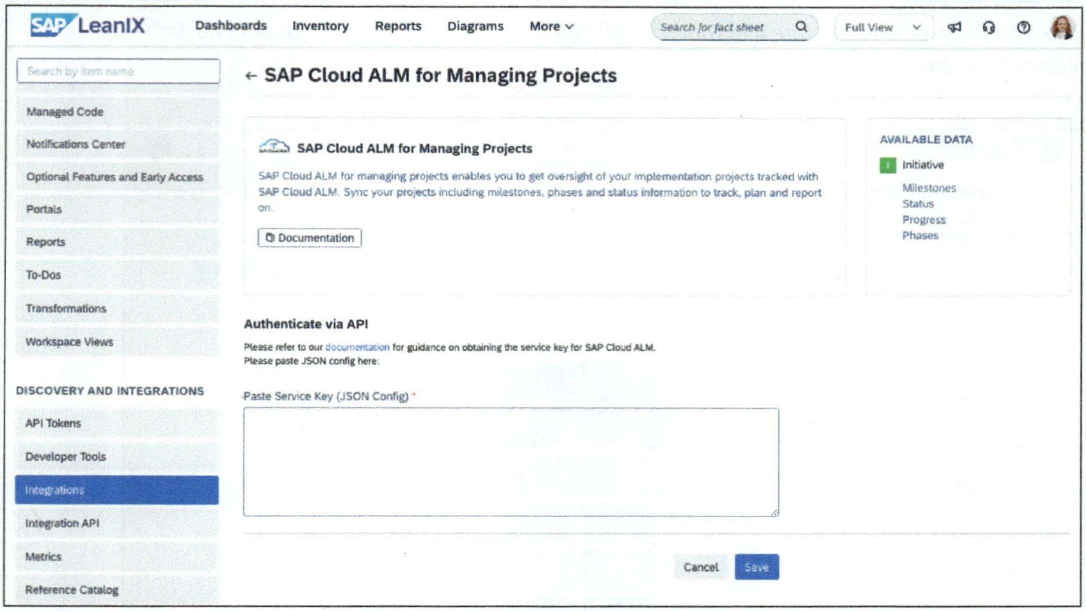

Figure 6.67: Configuring Integration of SAP Cloud ALM for Managing Projects

You can find both links, to SAP Cloud ALM and to Jira, in the **Agile Tracking** tab on the initiative fact sheet. Figure 6.68 shows an example of an initiative that was synced from SAP Cloud ALM. In the upper-right corner, you can see the grey **Open in SAP Cloud ALM** button, which will redirect you to the corresponding item in SAP Cloud ALM in a new browser window. In addition, you can also see a **Refresh** button to refresh the data sync between SAP LeanIX and SAP Cloud ALM.

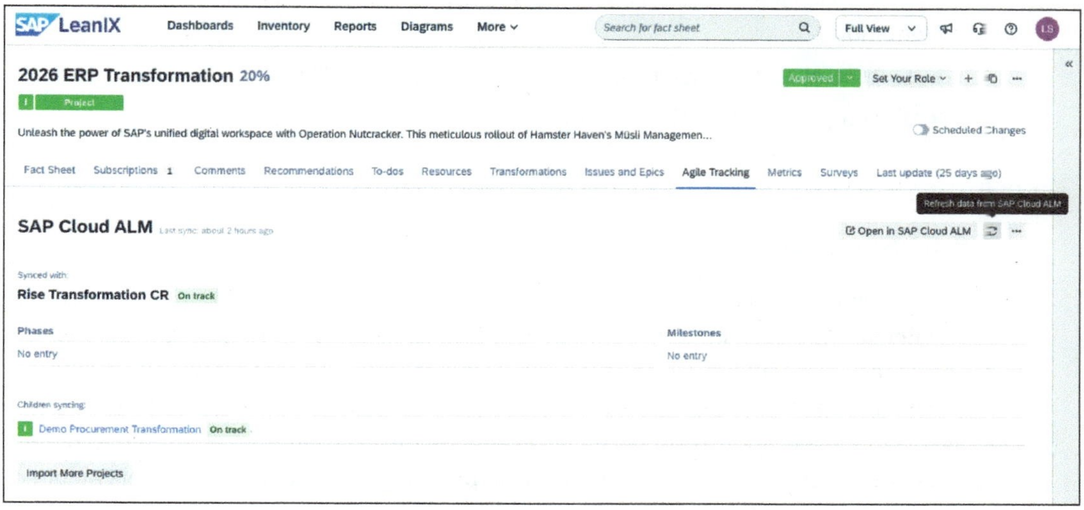

Figure 6.68: Agile Tracking Tab with Active SAP Cloud ALM for Managing Projects Integration

Figure 6.69 shows an example in which the Jira integration for project management is activated. In the upper-right corner, you can also see the icon to refresh the synced data. You can use the hyperlinks under **Key** to transfer to the corresponding **Linked Work Item** in Jira or jump to one of the **Imported Child** projects either in SAP LeanIX or Jira.

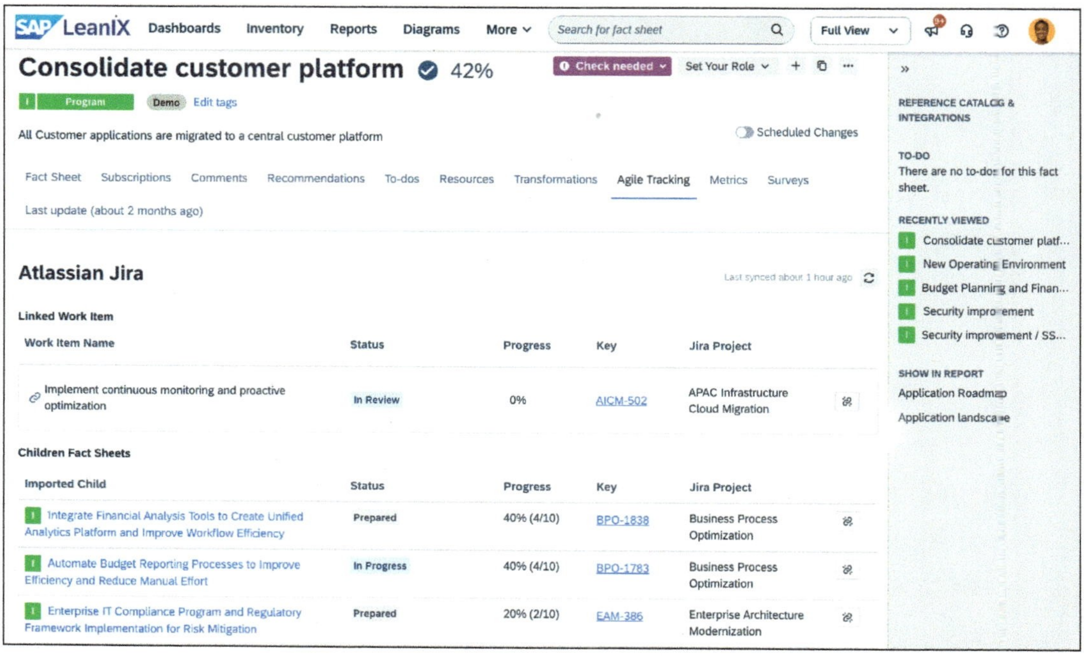

Figure 6.69: Agile Tracking Tab with Active Jira Integration

Further Resources

For the latest information on the available integrations, visit the SAP LeanIX user documentation:

- SAP Cloud ALM for managing projects: *http://s-prs.co/v620658*
- Jira for project management: *http://s-prs.co/v620659*

6.6 Summary

In this chapter, you learned how to use the initiative fact sheet type to support enterprise architecture projects and transformation initiatives. We covered modeling best practices and special fields and relations of the initiative fact sheet. We also discussed basic target architecture planning capabilities using the **Active to** and **Active from** fields available in SAP LeanIX Application Portfolio Management. We then walked through examples of how to use the milestones, transformation, and impact capabilities of the SAP LeanIX Architecture and Road Map Planning product. Next, we examined the new target architecture diagramming capability, and finally we looked at examples of project management solution integrations, for SAP Cloud ALM and Jira.

In the next chapter, we'll work through several enterprise architecture use cases, applying all the foundational knowledge you've gained in Chapter 4 through Chapter 6.

Chapter 7
Application Portfolio Management

In the first part of this book, you learned about the basics within SAP LeanIX. Now, we'll move on to leveraging the capabilities of SAP LeanIX in well-defined contexts, starting with the application modernization and rationalization use case in this chapter.

Application landscapes in modern enterprises have grown over the years through mergers, acquisitions, or quick-fix projects, while portfolios have grown out of control. The results are increased complexity, high costs to operate the application landscape, slower innovations, and increased risk.

Application portfolio management is an enterprise architecture practice that systematically assesses each application, understanding its functional and technical fit, its cost, and its strategic value. With these insights, informed decision-making in a well-defined context such as application modernization or an application rationalization use case is made transparent. Based on facts, you can decide which application needs to be modernized, rehosted, migrated, retired, and so on. However, this should not happen in isolation, with statements such as, "We need to reduce the number of applications by 10%." It should be thoughtful, with the primary focus on business and IT alignment leading to reduced risk, freeing up budget for innovation, and aiming to build an application portfolio which is fit for the future.

This chapter will dive into the topic of application portfolio management in general and the assessment, rationalization, and modernization use cases in particular in the following ways:

- Setting the stage with strategy and awareness, understanding why application portfolio management matters, what benefits it brings, and who needs to be involved.
- Moving into scoping and data enrichment, where we'll capture the functional, technical, replatforming, and cost data, including the total cost of ownership (TCO).
- Diving into evaluation from application rationalization and modernization use case perspectives, applying frameworks like the Gartner TIME model, and combining this with the TCO dimension to classify applications.
- Learning how to create an application transformation roadmap, linking decisions to initiatives, dashboards, and reports.

- Covering how to execute transformations in SAP LeanIX, modeling the impact of retiring or migrating applications.
- Focusing on tracking and reporting, using dashboards and key performance indicators (KPIs) to show progress, sustain buy-in, and continuously improve.

Note
One important aspect of the decision-making is the TCO, which we'll thoroughly explore throughout the lifecycle of a planned transformation.

7.1 Strategy and Awareness

Let's start with *why* application portfolio management matters. First, you need to understand the four primary business drivers for change:

- **Efficiency**
  Efficiency is gained through elimination of duplication. Running multiple applications for the same business capability and processes not only costs more money but also fragments processes and leads to various ways of working.
- **Transformation**
  Large programs like enterprise resource planning (ERP) modernization or cloud migration demand a simplified baseline.
- **Risk reduction**
  Legacy applications, unsupported technologies, and shadow IT increase security and compliance risks.
- **Innovation**
  By freeing up budget locked in redundant applications, organizations can invest in emerging technologies like artificial intelligence (AI), automation, or digital customer channels.

Every application portfolio initiative should define its benefit in alignment with the business drivers. The benefit is qualitative in nature and answers the question, "*Why* we are doing this?" Examples of benefits include the following:

- Reduced IT complexity
- Improved agility and innovation speed
- Lower operational risk and security exposure
- Cost savings and reallocation to digital initiatives

Benefits can be further translated into goals and objectives to address the question, "*What* do we want to achieve?" Whereas goals are broad and directional, objectives are measurable targets derived from goals, for example:

- *Goal:* Reduce application redundancy across regions
- *Objective:* Decrease the number of duplicate HR applications from eight to three by Q4 2028

To measure and track progress toward the outlined goals and target, KPIs are used to address the question, "*How* will we measure success?" The following are some examples of KPIs:

- Percentage of applications retired
- Annual cost saving amount achieved (USD)
- Percentage reduction in high-risk technologies
- Average TCO per business capability

One of the most important dimensions for measuring success in tangible terms (and for making informed decisions) is the TCO of individual applications, as this provides true cost transparency. Imagine that two applications are supporting the same business capability. Functionally and technically, they appear nearly identical. However, if one costs €200,000 per year and the other €800,000, the decision of which to retain becomes evident.

The cost dimension on the application level was recently added to SAP LeanIX, and TCO is automatically calculated based on the following three main cost levers:

- **License costs**
  Recurring subscription or license fees
- **Maintenance costs**
  Vendor-provided updates, patches, or annual maintenance agreements
- **Support costs**
  Internal or external staff time, help desk, and operations

Some organizations also include infrastructure costs, which can be linked through IT components.

> **Note**
> The standard meta model fields can either be changed or extended, which provides full flexibility to reflect a company's specific requirements for cost. Calculations can be updated accordingly.

Like with other change initiatives, our key stakeholders and their specific requirements and concerns need to be understood. Therefore, the following key stakeholders should be taken into consideration:

- **CIO and IT leadership**
  Provide the mandate and ensure the initiative aligns with corporate strategy. Without their sponsorship, application portfolio initiatives might risk traction.

- **Business stakeholders**
  Responsible for ensuring that rationalization decisions are not only strategically sound but also practical and effective in day-to-day operations.
- **Finance experts**
  Validate cost data and ensure it ties back to actual budgets. This gives credibility to the TCO figures.
- **Application owners**
  Bring firsthand knowledge about how applications are used, their value to the business, and pain points.
- **Security and compliance experts**
  Identify and assess risks arising from outdated or unsupported applications. Their insights often play a decisive role when hidden vulnerabilities could significantly impact organizational security or compliance posture.
- **Enterprise architects**
  Design the methodology, manage the inventory in SAP LeanIX, and facilitate decisions in a very coordinated manner.

One other important aspect of successful application portfolio management is the right governance process. Based on the maturity level of the organization, this could be project-driven or treated as an ongoing discipline. A well-established governance process allows organizations to capture architecture decisions in a structured way: which application is being retired, why the decision was made, who approved it, and when it will happen. Well-documented architecture decisions also ensure traceability of changes over time. For more information on the topic of enterprise architecture governance, refer to Chapter 12.

Now that you have a deeper understanding of the *why* in terms of business drivers and benefits, the *what* in the sense of goals and objectives, and the *how* defined through KPIs, you can start learning about application portfolio management initiatives in context of the stakeholder requirements and their concerns, supported by a well-structured governance process.

7.2 Choosing the Right Assessment Approach

SAP LeanIX supports the discipline of application portfolio management through multiple predefined attributes and assessment frameworks widely accepted in the industry. As described in the previous section, the *why* and *what* will determine the use cases in scope that can be applied over multiple iterations.

An initial application portfolio assessment will identify the applications in scope and their relationship to business capabilities and organizations, combined with the assessment of application-specific attributes such as the business criticality and application fit. In the following sections, we'll take a tour of the available frameworks and discuss relevant steps to activate calculations and extensions in your workspace.

7.2.1 Framework Overview

Table 7.1 provides an overview of frameworks and models that can be used as part of the SAP LeanIX standard features, including their primary motivations and individual scales.

Framework/Model	Scale	Description
Application fit	Expresses how well an application aligns with both business requirements and technical standards, indicating its overall suitability and value within the enterprise architecture. It is a dual assessment combining functional fit (business perspective) and technical fit (technology perspective) to guide rationalization, modernization, and investment decisions. Functional fit and technical fit are relevant for all use cases.	
Functional fit	Evaluates how effectively the application supports business processes, objectives, and capabilities.	
	Unreasonable	Not enough or wrong functionality
	Insufficient	Rudimentary functional support
	Appropriate	Supports all major functions
	Perfect	High number of functions available
Technical fit	Evaluates how effectively the application supports business processes, objectives, and capabilities.	
	Inappropriate	Replacement mandatory to satisfy business requirements
	Unreasonable	Replacement recommended to satisfy business requirements
	Adequate	Some parts could be optimized
	Fully appropriate	No change needed apart from regular maintenance

Table 7.1: Application Portfolio Management: Models and Scales

Framework/Model	Scale	Description
Business criticality	Represents the relative importance of an application to business operations. It indicates how severely the business would be affected if the application failed or became unavailable. Business criticality is primarily gathered as part of the application portfolio assessment and rationalization use case.	
	Mission critical	Breaks in service are intolerable and immediately significantly damaging. Availability required at almost any price. Applications are vital for business continuity. Any failure leads to immediate and severe business impact (e.g., halted operations, revenue loss).
	Business critical	Service requires continuous availability, but short breaks in service are not catastrophic. Applications are important for daily operations but can be down for a limited time without catastrophic effects.
	Business operational	Contributing to efficient business operation but out of direct line of service to customer. Applications support efficiency and productivity but are not essential for core business processes. Downtime causes limited operational disruption.
	Administrative service	Failures are undesirable, but do not affect customers and can be tolerated a little more. Applications are used for administrative or support activities, with minimal business impact if unavailable.

Table 7.1: Application Portfolio Management: Models and Scales (Cont.)

Framework/Model	Scale	Description
Gartner TIME	This framework classifies applications as *tolerate, invest, migrate*, or *eliminate*. Applications can be assessed via these four categories and strategic decisions made, such as rationalization or modernization initiatives as well as whether investments should be allowed or accepted. The TIME classification can be automatically calculated based on the architecture fit (see next section).	
	Tolerate	Applications that do not deliver outstanding business value but are still required due to their stable operation, compliance relevance, or technical quality. They remain in use with minimal investment until a better alternative is available.
	Invest	High-value applications that provide strong business differentiation and operational efficiency. These should receive continued investment for enhancement, modernization, or cloud enablement to maintain strategic advantage.
	Migrate	Applications that serve a necessary business function but currently create operational, technical, or cost issues. They should be migrated, rehosted, or replatformed to a more sustainable or modern environment to improve performance and reduce risk.
	Eliminate	Applications that no longer align with business needs, duplicate functionality, or perform poorly. They should be decommissioned, consolidated, or replaced to reduce complexity and cost.

Table 7.1: Application Portfolio Management: Models and Scales (Cont.)

Framework/Model	Scale	Description
Total cost of ownership	Quantifies the total annual and lifecycle costs of an application, helping organizations understand and optimize IT spend by linking financial data to business and technical value.	
	License cost	Fees paid to acquire the right to use a software product.
	Maintenance cost	Recurring charges that entitle you to product updates, patches, and new releases.
	Support cost	Fees for access to vendor-provided assistance such as help-desk support, trouble-shooting, incident response, and escalations.
	<custom cost field>	Additional cost elements can be added by tailoring the meta model.
AWS 6R	6R is a framework introduced by AWS used to assess the cloud readiness of an application and consists of six categories: rehost, replatform, rearchitect, repurchase, retain, and retire. The 6R framework was recently extend by a seventh R: relocation. For the sake of completion, we list this classification as well, although it is not reflected in the meta model at the time of writing this book. The AWS 6R framework is primarily used for the application modernization use case.	
	Rehosting	Move existing systems/applications to the cloud (or new infrastructure) with minimal or no change to code or architecture.
	Replatform	Move the application while introducing some modifications (e.g., to use managed services or optimize some layers) but without fundamentally redesigning the core architecture.

Table 7.1: Application Portfolio Management: Models and Scales (Cont.)

<table>
<tr><th>Framework/Model</th><th>Scale</th><th>Description</th></tr>
<tr><td rowspan="5">AWS 6R (Cont.)</td><td>Rearchitect (Refactor)</td><td>Redesign or rebuild the application architecture to fully leverage cloud-native features (microservices, serverless, PaaS) and scale/performance/improve agility.</td></tr>
<tr><td>Repurchase</td><td>Fully replace an existing application with a new software product (often SaaS) rather than migrating the current one.</td></tr>
<tr><td>Retire</td><td>Permanently decommission applications or infrastructure that are no longer useful, are redundant, or consume resources without delivering business value.</td></tr>
<tr><td>Retain</td><td>Keep applications running in their current environment (on-premises or legacy) for the time being, postponing migration.</td></tr>
<tr><td>Relocation</td><td>Move applications to the cloud without purchasing new hardware, rewriting applications, or modifying existing operations at the hypervisor level.</td></tr>
<tr><td>Pace layering</td><td colspan="2">Provides a strategic lens for capability-based planning by categorizing capabilities as innovation, differentiation, or commodity. This helps to do the following:
■ Prioritize investments based on strategic relevance and change velocity
■ Align transformation roadmaps with business intent
■ Balance stability and agility across the capability landscape
Pace layering is an attribute of the business capability and can provide value insights for the application rationalization use case.</td></tr>
</table>

Table 7.1: Application Portfolio Management: Models and Scales (Cont.)

Framework/Model	Scale	Description
Pace layering (Cont.)	Innovation	Business capabilities that enable new business models, emerging technologies, and future revenue streams. These are experimental and exploratory in nature, helping the organization test new ideas and gain early-mover advantages.
	Differentiation	Business capabilities that sustain competitive advantage through unique processes, products, or services. These reflect what makes the company distinct in its market or industry.
	Commodity	Foundational business capabilities that support standardized, transactional, and regulatory processes shared across the enterprise (e.g., finance, HR, procurement). They enable stability, efficiency, and compliance.

Table 7.1: Application Portfolio Management: Models and Scales (Cont.)

7.2.2 Preparing Your Workspace

Depending on the chosen model, you need to prepare your SAP LeanIX workspace by activating calculations or extensions. This is primarily relevant for the Gartner TIME assessment and the TCO calculation, which is delivered as a larger extension to the SAP LeanIX Application Portfolio Management product. The other frameworks do not require specific setup in the workspace. In the following sections, we'll describe the activation in more detail.

TIME Classification Calculation

Based on the combination of the functional and technical fit, you can activate a calculation to derive the TIME classification. The matrix in Figure 7.1 shows how the calculation logically works.

The calculation for the Gartner TIME assessment can be created by navigating to the **Administration** settings. Within **Advanced Settings**, go to the **Calculations** subsection. Choose **Create a Calculation** and add a calculation by selecting **Gartner TIME Assessment** as the template, as shown in Figure 7.2.

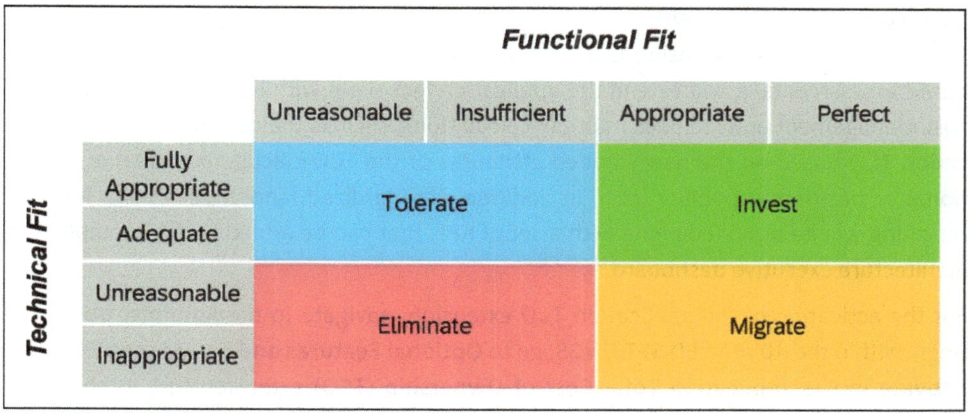

Figure 7.1: Calculation of TIME Classification Based on Application Fit

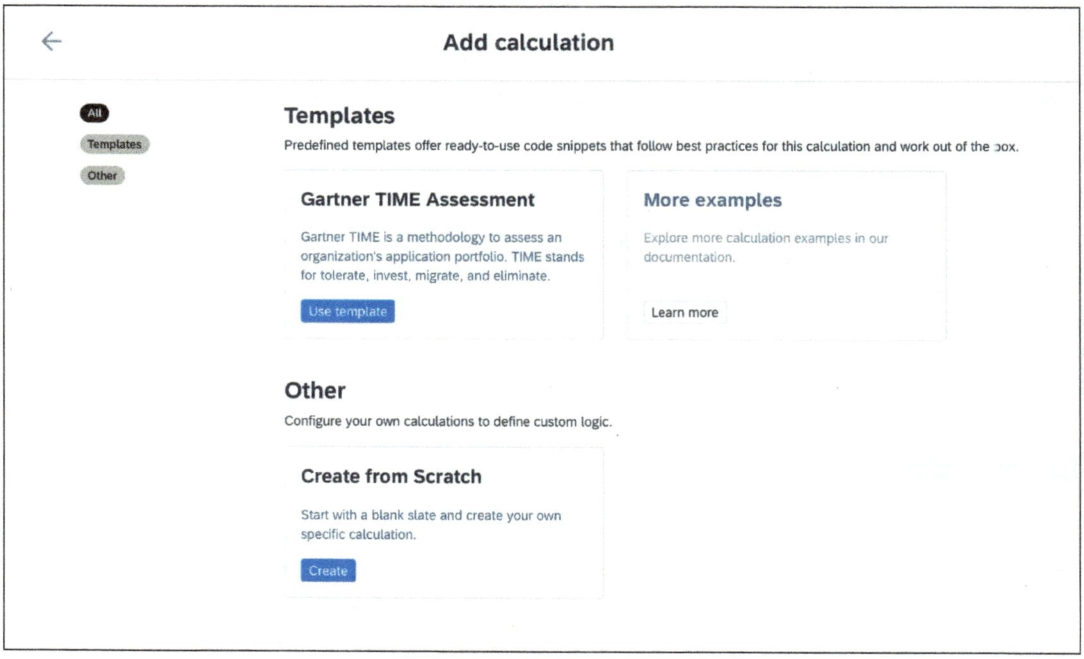

Figure 7.2: Gartner TIME Assessment: Calculation Template

The template-based Gartner TIME assessment calculation is ready to be saved and activated with the target field TIME classification in the application fact sheet type.

Note

With the activation of the calculation, the TIME classification attribute cannot be changed by a user anymore in any application fact sheets as the calculation takes place based on the functional fit and technical fit attribute of an application fact sheet.

Application TCO Extension

The application TCO extension is an optional product feature you can activate for TCO-based assessments. It will extend the application fact sheet with additional attributes for cost management and includes multiple calculations, such as the calculation of the application TCO based on the newly added attributes or the TCO calculation on an organizational or business capability level. In addition, standardized landscape reports for TCO reporting will be provided along with a set of KPIs that can be added to, for example, the architecture executive dashboard.

For the activation of the application TCO extension, navigate to the **Administration** settings. Within the **ADVANCED SETTINGS**, go to **Optional Features and Early Access**. Choose **Activate** in the **Application Total Cost of Ownership (TCO)** extension tile, as shown in Figure 7.3.

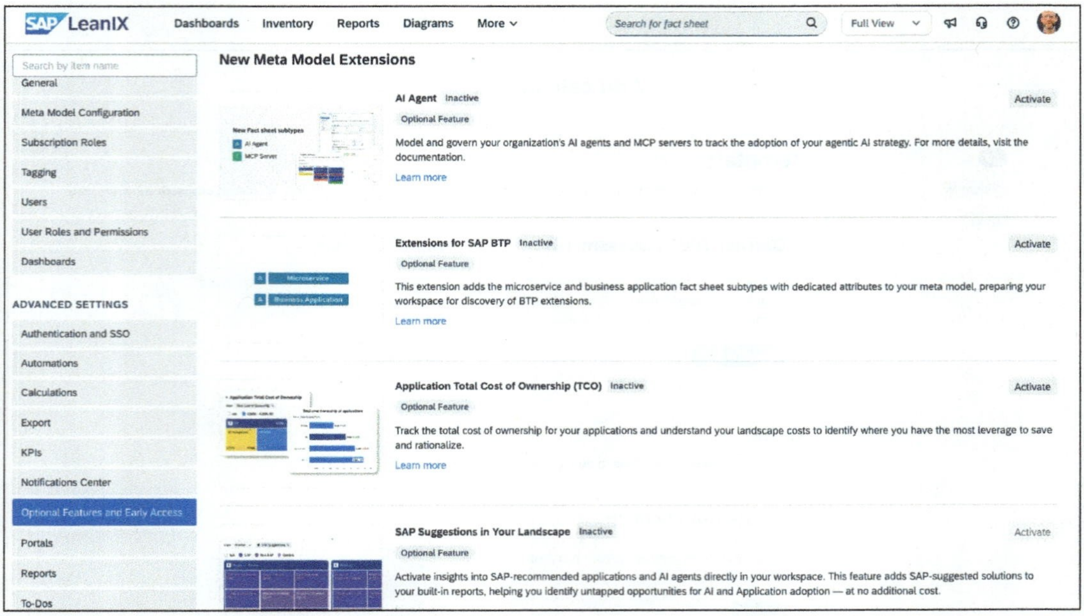

Figure 7.3: Application TCO Extension

To activate the extension in your workspace, you need to check the box indicating your agreement that the meta model will be extended. These changes can only be reverted through manual adjustments. Furthermore, the changes described in the **What Will Change** sections need to be reviewed before activating the extension in your workspace by clicking **Activate Extension**, as shown in Figure 7.4.

With the activation of the application TCO extension, the meta model is extended through various fields, which are also featured by calculation for the aggregation and allocation of costs to business capabilities and organizations.

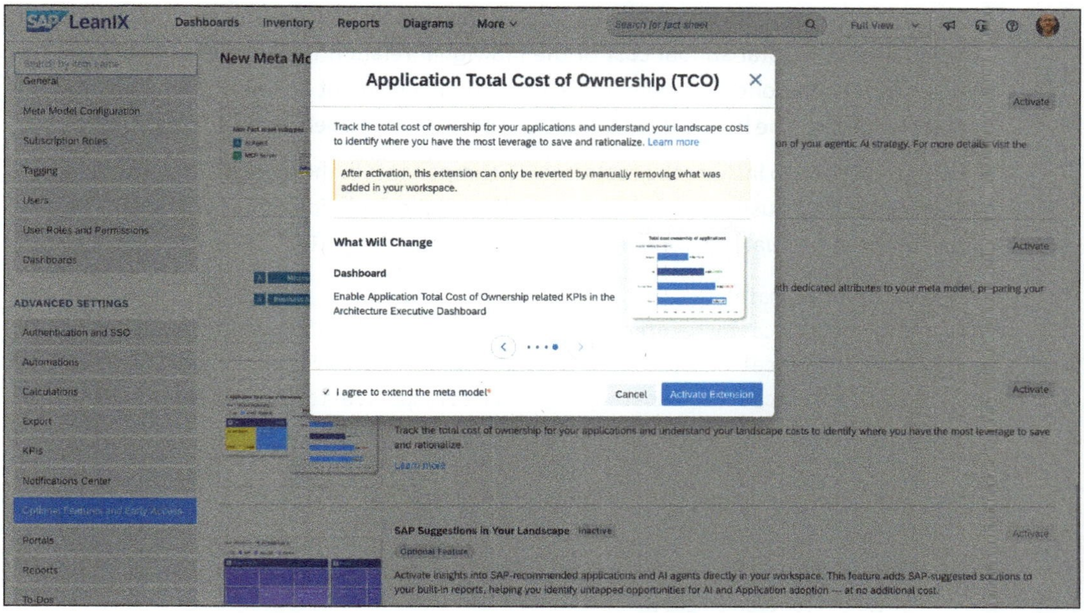

Figure 7.4: Activation of Application TCO Extension

Table 7.2 provides an overview of the metadata fields added to the application fact sheet.

Field Name	Description
Total cost of ownership (TCO)	Overall TCO automatically calculated by the sum of license cost, maintenance cost, and support cost
License cost	Fees paid to acquire the right to use a software product
Maintenance cost	Recurring charges that entitle you to product updates, patches, and new releases
Support cost	Fees for access to vendor-provided assistance such as help desk support, troubleshooting, incident response, and escalations

Table 7.2: Application Fact Sheet Metadata Extension of TCO Extension

Tip

The naming of the cost elements can be adjusted to the terminology used in your organization, and additional cost elements can be added if required. For more details, see Chapter 3, Section 3.1, or read the SAP LeanIX user documentation on the SAP Help Portal at *http://s-prs.co/v620660*.

One cost element not reflected in the TCO is the infrastructure costs. This cost element is available as the total annual cost of the individual relation between the application fact sheet and IT component. Depending on the cost management regime of your company, this cost element can be handled independently of TCO or included in the TCO calculation.

EA7 has decided to include the costs from IT components in the overall TCO calculation. We have therefore adjusted the default application TCO calculation by adding the total annual cost of the individual IT components, as reflected in Listing 7.1.

```
export function main() {
    const lxApplicationMaintenanceCosts = data.lxApplicationMaintenanceCosts || 0;
    const lxApplicationSupportCosts = data.lxApplicationSupportCosts || 0;
    const lxApplicationLicensingCosts = data.lxApplicationLicensingCosts || 0;
    let relTotalAnnualCost = 0;
    const rels = data.relApplicationToITComponent || [];
    for (const rel of rels) { if (rel.costTotalAnnual != null) { relTotalAnnualCost
+= rel.costTotalAnnual; } }
    return lxApplicationMaintenanceCosts +
        lxApplicationSupportCosts +
        lxApplicationLicensingCosts +
        relTotalAnnualCost;
}
```

Listing 7.1: Application TCO Calculation Code

For the detailed step-by-step adjustment of the calculation, refer to Chapter 10, Section 10.2. We won't go deep into the individual features of the application TCO extension at this point in time; rather, we'll show how it's applied in the context of the application rationalization use case of the EA7 case study in the next section.

7.3 Application Rationalization Use Case

In this section, we'll outline the application rationalization use case of EA7 by following this structured approach:

- Defining the context through the why, what, how, and whom
- Scoping and data enrichment
- Data evaluation and retrieval of insights
- Prioritization of candidates for rationalization
- Understanding the change impact
- Building the rationalization roadmap
- Decision-making and communication
- Monitoring with KPIs

Each section will include relevant inventory and report examples alongside the case study; these examples are also accessible in the demo workspace.

7.3.1 The Why: Strategic Drivers Behind the Transformation

EA7 entered 2026 with a clear strategic mandate to simplify its global technology landscape and accelerate digital transformation. Customer service quickly emerged as a priority domain due to its critical role in strengthening consumer trust, resolving issues efficiently, and generating the insights required for data-driven decision-making.

Three enterprise-wide business objectives made rationalization unavoidable:

- **Customer-centric growth**
  EA7 aims to deliver a unified, high-quality service experience worldwide. However, inconsistent processes and fragmented systems across regions prevent the company from offering timely, consistent, and personalized service.
- **Operational efficiency and cost optimization**
  Over time, EA7's customer service organization accumulated more than 15 regional applications: custom tools, analytics dashboards, legacy customer relationship management (CRM) extensions, and multiple SAP ERP instances. This resulted in process duplication, escalating support costs, dispersed data, and limited global reporting capabilities.
- **Digital resilience and data transparency**
  EA7's shift toward a cloud-based digital backbone, centered on SAP S/4HANA and SAP Service Cloud, required a simplified and modernized application landscape. Legacy systems hindered integration, automation, compliance reporting, and omnichannel engagement.

7.3.2 The What: The Case for Application Rationalization

The customer service capability was identified as one of EA7's most fragmented and high-impact areas, with the following key insights:

- Extensive redundancy across regions, with similar processes
- High technical debt, custom-built applications, and region-specific portals lacking integration
- Significant cost concentration, coupled with the hidden costs of maintaining numerous low-fit local applications
- Inconsistent service performance across markets due to varied tools, workarounds, and reporting gaps

These findings align the rationalization need not just with IT objectives but also with EA7's broader operational, financial, and customer experience goals.

7.3.3 The How: The Transformation Approach

The EA7 transformation approach is anchored in the principles of simplification, modernization, and global standardization. The strategic direction of the enterprise architecture and transformation leadership team defined the following goals:

- Harmonize global processes
- Adopt a unified digital platform
- Retire redundant and local applications
- Enable data transparency and analytics

Moving forward, we'll concentrate primarily on the following KPIs:

- **Application Reduction Ratio**
  Number of customer service applications retired ÷ number of applications in baseline scope
- **TCO Reduction**
  Annual cost savings from decommissioned applications
- **Redundancy Index**
  Average number of applications per customer service subcapability
- **Complexity Score**
  Total number of interfaces eliminated as part of rationalization

7.3.4 For Whom: The Stakeholders Who Benefit

The benefits of a harmonized customer service landscape extend across multiple stakeholder groups within EA7:

- **Consumers and retail partners**
  Experience faster, more consistent, and more transparent service interactions worldwide
- **Customer service agents and regional operations teams**
  Gain access to a unified platform with standardized workflows, improved usability, and integrated knowledge sources
- **Customer service leadership (CCO, regional directors)**
  Receive real-time visibility into global performance metrics, enabling data-driven decision making and cross-market improvement programs
- **IT and enterprise architecture teams**
  Reduce landscape complexity, integration overhead, operational risk, and support costs while ensuring alignment with EA7's SAP S/4HANA transformation roadmap

In the following sections, we'll walk through the individual steps of an application rationalization initiative both from general and EA7-specific perspectives. The highlighted fact sheets, reports, and dashboards can also be accessed in the demo workspace.

7.4 Scoping and Data Enrichment

Application portfolios can be broad and large depending on the size and scale of your organization. It is therefore important to understand which applications are in scope of the application portfolio initiative and who the owners of these applications are to start an initial assessment and data enrichment activity.

Once the scope is defined, you can select the relevant applications under **Inventory**, as shown in Figure 7.5, by filtering on business capabilities and the *active* lifecycle state. As per the EA7 case study, we'll focus on the customer service business capabilities.

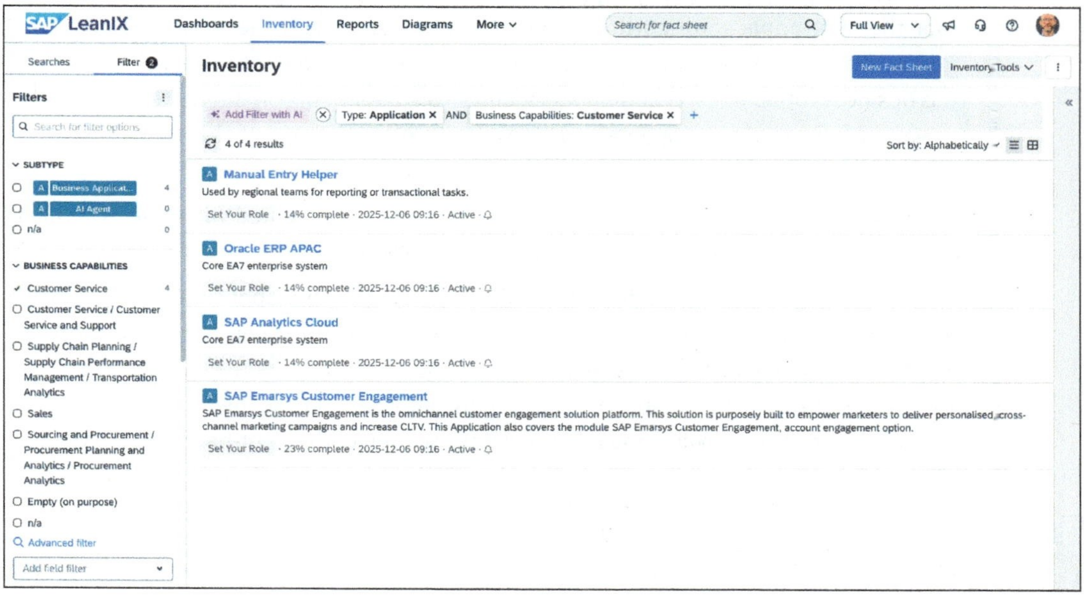

Figure 7.5: Selection of Active Applications for Business Capabilities in Scope

> **Tip**
>
> You can save your search at any time to reduce the manual effort of defining your filter criteria the next time. Open the **Inventory Tools** dropdown menu and choose **Save As New**. Define a name for the search and select the relevant permission type.

Next, verification of the applications in scope should be conducted by cross-checking the completeness of the applications in scope with the respective application and business stakeholders. Any missing application needs to be added to the SAP LeanIX inventory by following one of the approaches introduced in previous chapters:

- Manual creation of the fact sheet, a feasible approach for a small set of applications to be added. See Chapter 4, Section 4.2.7.
- Excel-based import, feasible for a larger number of applications. See Chapter 4, Section 4.2.6.

- Import through the inventory builder, a feasible approach if diagrams and/or lists of the missing applications already exist. For more information on the inventory builder, see Chapter 11, Section 11.2.
- Discovery-based integration such as the SAP landscape discovery and SaaS discovery features and import of relevant applications via the discovery inbox (refer to the respective sections in Chapter 4, Section 4.2).

> **Note**
>
> The pure import or creation of application fact sheets will most likely not be sufficient for the later application portfolio analysis. The analysis and the respective reports depend on the relations between the application fact sheet and other fact sheets such as business capabilities, IT components, and so on. Depending on the context, relevant relationships need to be maintained alongside the application fact sheets.

After ensuring that all relevant applications in scope are represented within SAP LeanIX as fact sheets, the assessment of the applications can take place.

There are multiple ways to address the individual rating of the applications. If you run a workshop meeting with your stakeholders (see previous section), you can use a landscape report (for complete instructions, see Chapter 5, Section 5.1.1) and expand the fact sheet attributes in the right-hand pane by choosing the application in scope for the assessment. The assessment options will appear in the details section of the right-hand pane, as shown in Figure 7.6.

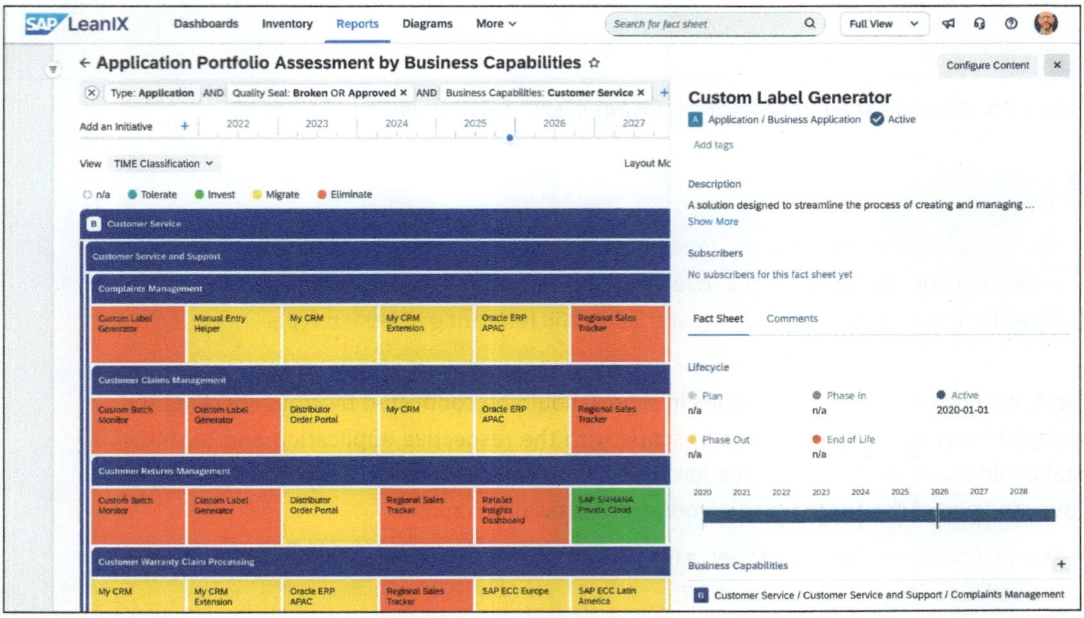

Figure 7.6: Assessment of Attributes Through Landscape Report

You have also the possibility to add multiple attributes to the detail section of the right-hand pane by navigating to **Configure Content** in that pane. This will open a screen with all relevant attributes for the application fact sheet, as shown in Figure 7.7.

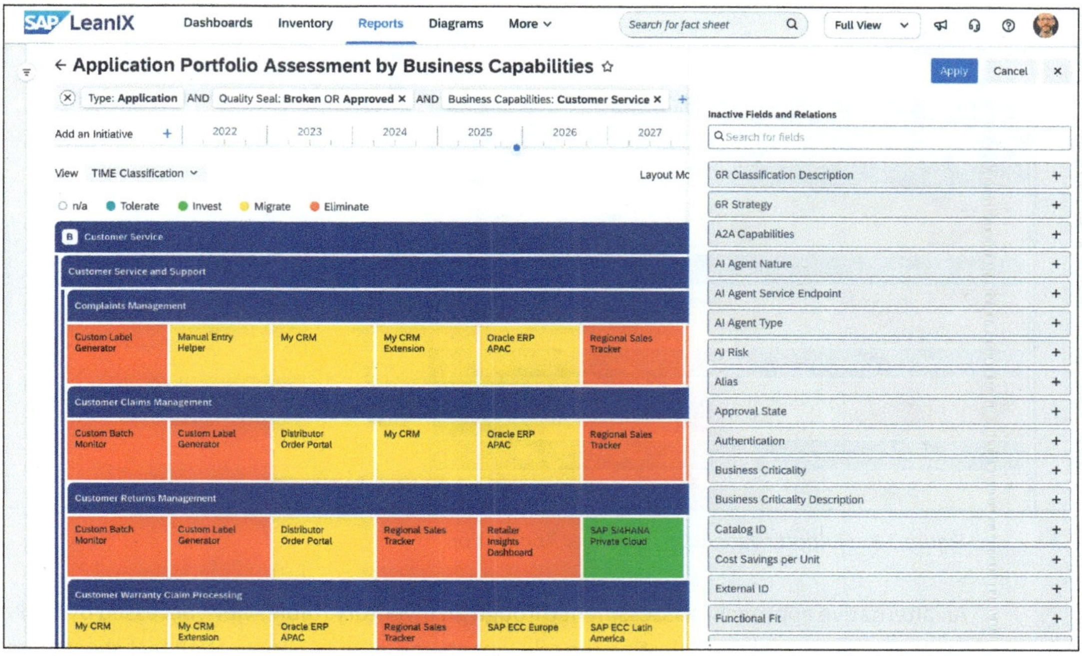

Figure 7.7: Selection of Fields and Relations

Once you have selected the relevant fields and clicked the **Apply** button, these are displayed in the **Details** section of the right-hand pane, as shown in Figure 7.8. With this, you're ready to start the assessment together with your stakeholders, based on individual applications and potentially multiple dimensions.

For the EA7 application portfolio assessment, we'll focus on three assessment categories:

1. **Functional Fit** and **Technical Fit** as inputs for the TIME classification calculation
2. **Licensing Costs, Maintenance Costs,** and **Support Costs** as inputs for the TCO calculation
3. **Business Criticality**

Note that the calculated fields are in read-only mode (not editable).

> **Note**
>
> It's important to mention that the intent of SAP LeanIX is not to replace a cost management solution. An out-of-the box integration with Apptio (*apptio.com*) is provided that can be leveraged to retrieve relevant cost information. If Apptio is not in use, an approximate cost indication is normally sufficient as you will want to identify the cost optimization potential across the enterprise landscape.

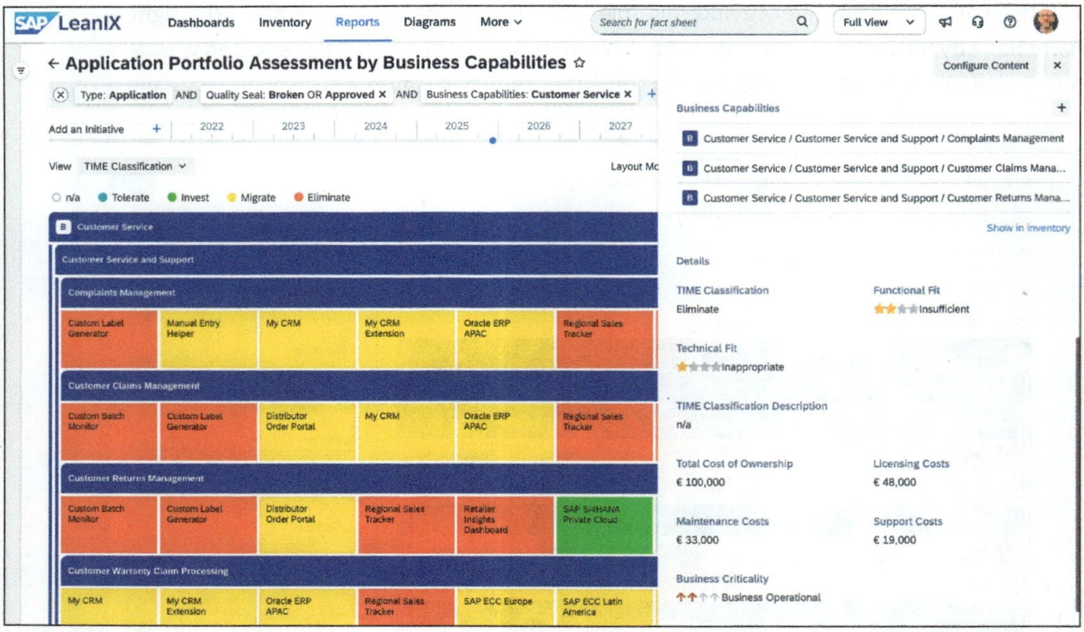

Figure 7.8: Multiattribute Assessment per Application

An alternative approach is assessment through an inventory (table) view, accessed by clicking the table symbol and selecting the relevant attributes of the application fact sheet to be displayed by clicking **Manage columns**, as shown in Figure 7.9.

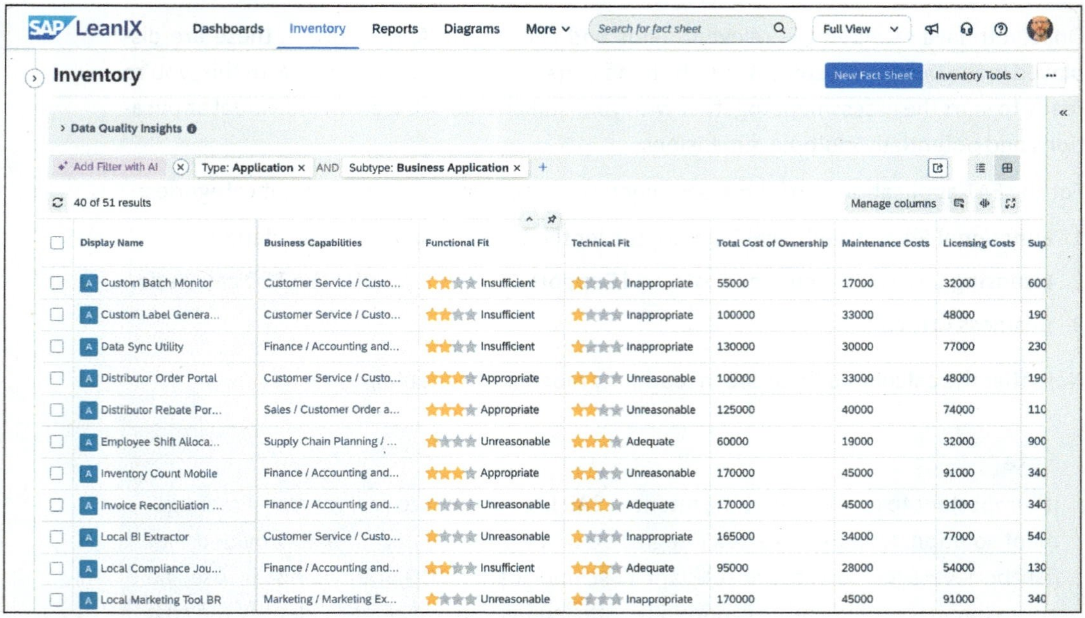

Figure 7.9: Assessment of Attributes Through Inventory Table View

If you do not have the option to directly interact with your stakeholders (for example, if your stakeholder group is far too big or distributed across multiple time zones), then multiple alternative options also exist here. If stakeholders are familiar with fact sheets within SAP LeanIX, a straightforward option might be to offer a predefined Excel sheet that needs to be filled out by your stakeholders. This prepopulated Excel sheet could be easily created through the export of the inventory table view as previously defined. Once the data for the attributes are collected by, say, the application owners in charge, the fact sheets are updated through an import process, as described in more detail in Chapter 4, Section 4.2.6.

If your stakeholders are less savvy about SAP LeanIX, we recommend creating a tailored survey to collect the relevant data. Depending on the context, the questions and descriptions within the survey might differ. Some example questions might include the following:

- How well does this application support your daily work?
- Is the technology modern and easy to maintain?
- How critical is it to your operations?
- Do you know the annual license, maintenance, or support costs?

Figure 7.10 outlines an example survey for an application rationalization TIME assessment based on a template survey provided by SAP LeanIX. For more information on the creation of surveys and the import of templates, see Chapter 2, Section 2.3.3.

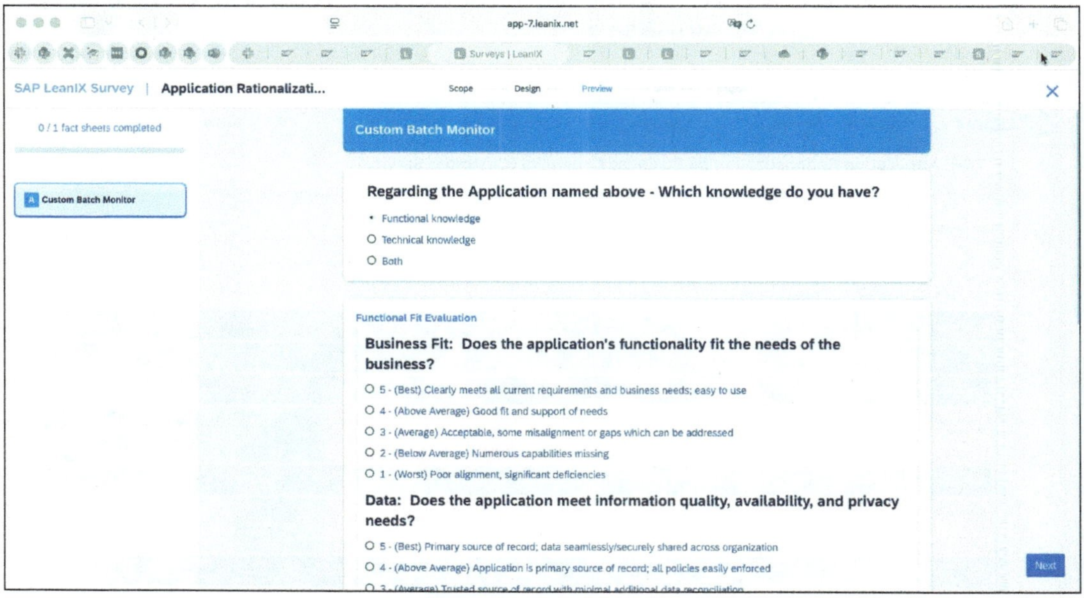

Figure 7.10: Survey for Collecting TIME Data Points

With the collection and verification of application data points in scope for the rationalization initiative, you've built a solid basis to run an impact assessment and to derive data-driven insights.

7.5 Evaluate Data and Retrieve Insights

With the help of various standard reports, you can now start to assess the application landscape of EA7 from a rationalization perspective. We recommend gaining insights from these data points, which we'll discuss in the following sections:

- TIME (individual assessment or calculated based on functional and technical fit)
- TCO
- Business criticality

7.5.1 Starting with TIME Classifications

First insights can be derived through a TIME assessment clustered by business capabilities to understand the rationalization potential of individual applications supporting similar business capabilities.

> **Note**
>
> Before creating the reports that we'll discuss in this section, you need a dedicated collection structure. For the EA7 case study, we'll create an **Application Portfolio Management** collection group; you can find each report on our demo workspace (*https://app-7.leanix.net/LeanIXPracticalGuideDemoWorkspace*). For instructions on creating collections, see Chapter 5, Section 5.1.2.

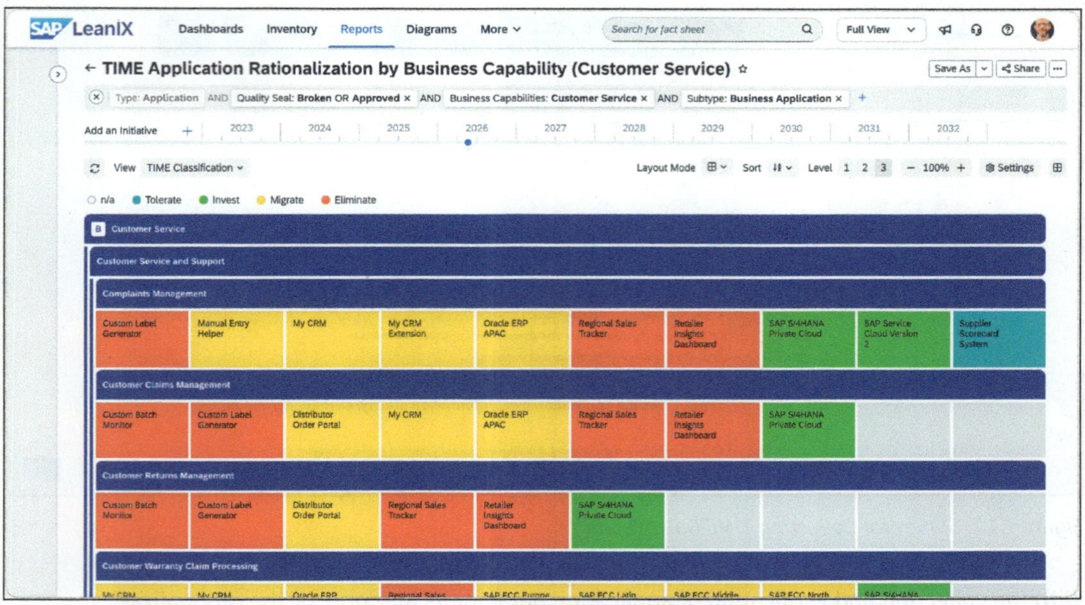

Figure 7.11: TIME Application Rationalization by Business Capability (Customer Service)

You can create the **TIME Application Rationalization by Business Capability (Customer Service)** report, as shown in Figure 7.11, by opening the default application landscape report, clicking **Save As**, and assigning the following settings:

- **View: TIME Classification**
- **Filter: Business Capability: Customer Service**
- **Subtype: Business Application**

If you now change the perspective to an organization point of view, you can identify which applications support individual versus multiple parts of the organization in combination with the differentiation of global, regional, or country-specific levels, as outlined in Figure 7.12.

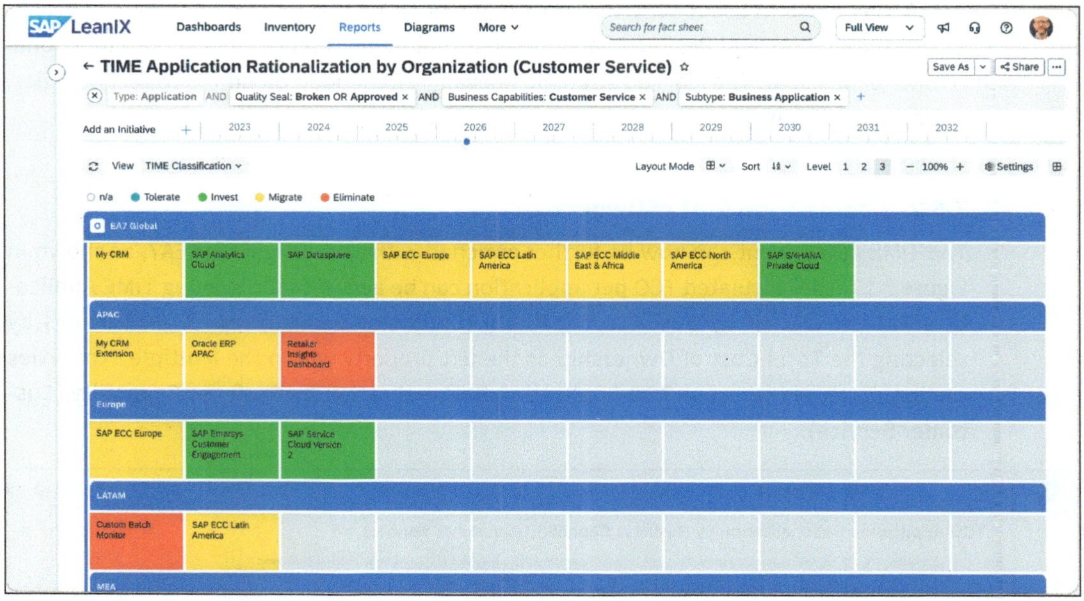

Figure 7.12: TIME Application Rationalization by Organization (Customer Service)

To create the **TIME Application Rationalization by Organization (Customer Service)** report, open the **TIME Application Rationalization by Business Capability (Customer Service)** report, go to **Settings**, and change **Cluster by** from **Business Capability** to **Organization**. Choose **Save As** to persist this report as **TIME Application Rationalization by Organization (Customer Service)**.

With the help of the TIME assessment from both business capability and organization perspectives, the following insights can be derived:

- **Tolerate**
  Several regionally specific or auxiliary applications, typically CRM extensions or lightweight analytical tools, were rated **Tolerate**. These applications deliver acceptable short-term functional value but lack strategic importance or long-term viability.

- **Invest**
  SAP S/4HANA Cloud Private Edition, SAP Service Cloud (version 2), SAP Sales Cloud (version 2), and SAP Emarsys Customer Engagement were consistently classified as **Invest** across multiple customer service subcapabilities, positioning them as future core platforms for the global service model.
- **Migrate**
  All SAP ERP instances (Europe, North America, LATAM, MEA) were rated **Migrate** due to their central process role but clear technical obsolescence within EA7's SAP S/4HANA strategy.
- **Eliminate**
  Numerous regional, locally developed applications such as Custom Batch Monitor (LATAM), Custom Label Generator (North America), Manual Entry Helper (MEA), Local BI Extractor (MEA), and Regional Sales Tracker (multiregion) scored low in both functional and technical fit. They lacked strategic relevance, depended on outdated technologies, and created fragmentation across the enterprise.

7.5.2 Adding Total Cost of Ownership

The TIME assessment can now be further enriched with cost insights for EA7, as shown in Figure 7.13. The calculated TCO per application can be added to the existing **TIME Application Rationalization by Business Capability (Customer Service)** report via **Settings** by selecting the **Total Cost of Ownership** as the left property within the **Multiple Properties** section. Save this report as **TCO Application Rationalization by Business Capability (Customer Service)**.

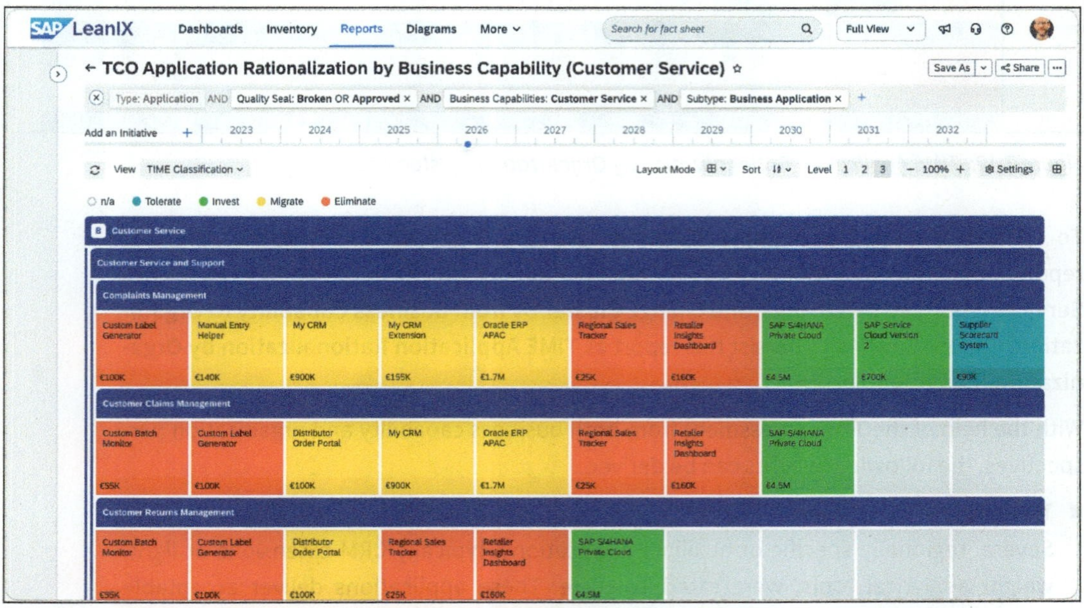

Figure 7.13: TCO Application Rationalization by Business Capability (Customer Service)

By adding the cost information, you will now gain a better understanding of, for example, potential cost reduction through the elimination of applications as well as better insights into the running costs for applications that will stay in your landscape.

Let's now assume that you do not want only to assess individual business capabilities but also to understand the cost perspective at a higher level. For this purpose, you can aggregate TCO information at individual as well as parent business capability levels. To visualize this information, change the settings by adding the **Business Capability • Total Cost of Ownership** to the right property, and activate **Sum up to cluster** for the right property. Save the changes as a new report, **TCO Application Rationalization by Business Capability (Customer Service) – aggregated.** You can now see the associated cost per application as well as the aggregated cost on various business capability levels, as outlined in Figure 7.14. With the aggregated view, you have the possibility to compare the associated application cost of individual business capabilities as well as, say, level 2 or level 1 business capabilities.

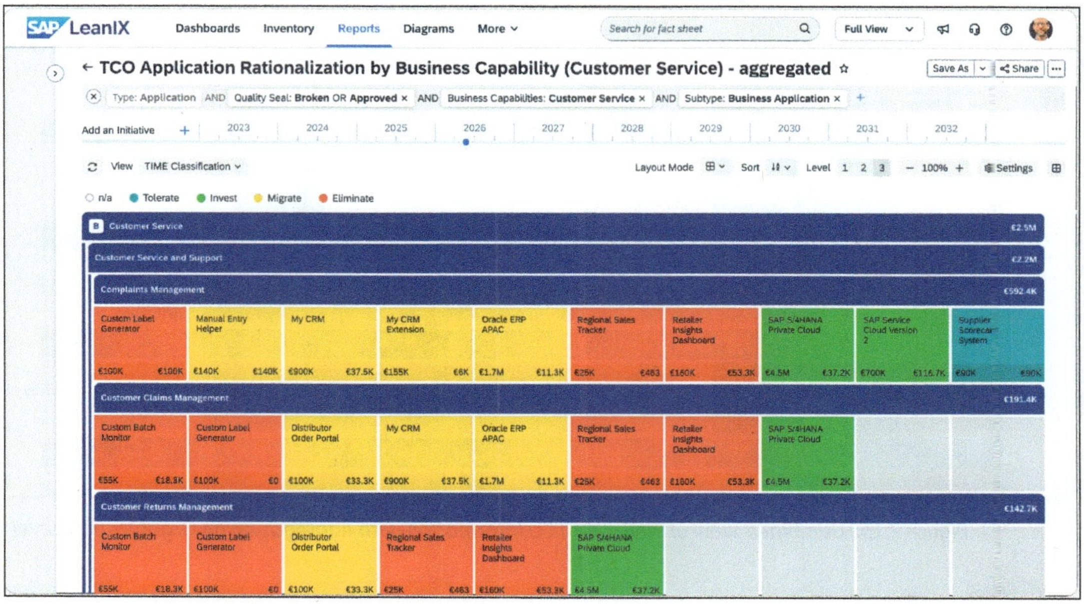

Figure 7.14: Combined View of TIME and TCO Information with Aggregation on Business Capability Level

Note

You might have recognized that in the EA7 use case, the TCO value of the left and right property might differ. This is because the TCO is split among multiple business capabilities (each getting a share of the cost). You also have the possibility to assign the full TCO to a specific business capability by setting **Support Type** to **Leading** within the **Relation-Settings** for **Application** and **Business Capability**. If you define multiple business capabilities as **Leading**, the TCO will be split among these.

A similar insight can be gained from an organization perspective by leveraging the **TCO Application Rationalization by Business Capability (Customer Service) – aggregated** report.

Within the **Settings**, change the **Cluster By** from **Business Capabilities** to **Organizations** and set the right property to **Organization Total Cost of Ownership**. Make sure that the **Sum up to cluster** is checked for the right property and save the report as **TCO Application Rationalization by Organization (Customer Service) – aggregated**.

With the cost perspective on an organizational level, as outlined in Figure 7.15, you can now get insights for EA7, in the sense of which part of the organization is causing the respective application cost. This could be of interest if you want to compare countries, plants, and so on.

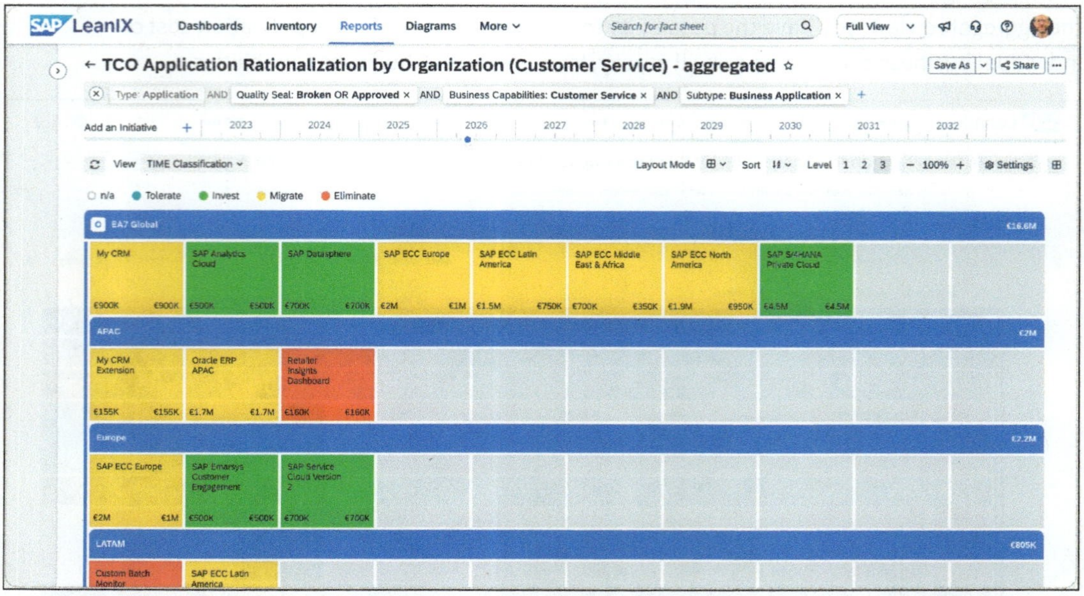

Figure 7.15: Combined View of TIME and TCO Information with Aggregation on Organization Level

> **Note**
>
> Similar to the **Support Type** of the **Application** and **Business Capability** relation, you also have the possibility to assign the full TCO to a specific part of the organization by setting the **Usage Type** to **Owner** within the **Relation-Settings** of the **Application** and **Organization**. If you define multiple organizations as **Owner**, the TCO will be split among these.

With the reports outlined, you should have a good understanding of the TIME assessment in combination with the TCO for individual applications as well as on business capability and organization levels. By combining these dimensions, valuable insights can be derived. From an EA7 perspective, you can identify the following:

- SAP ERP is the dominant cost contributor, with regional instances costing €700K to €2M each, appearing in nearly every customer service capability, and totaling ~€6–8M across regions.
- Oracle ERP APAC adds €1.7M annually, making APAC one of the highest-cost regions despite its smaller application footprint.
- Fragmented applications categorized as **Eliminate** collectively exceed €600K, including Custom Label Generator (€100K), Local BI Extractor (€165K), Manual Entry Helper (€140K), Regional Sales Tracker (€25K), and others.
- Capability hotspots drive concentrated spend, with complaints management alone reaching €525.7K and omnichannel customer engagement €295.4K, largely due to overlapping CRM and analytics tools.
- Future-state platforms already anchor major investments, including SAP S/4HANA Cloud Private Edition (€4.5M) and SAP Service Cloud Version 2 (€700K), reinforcing the business case for consolidation.

7.5.3 Adding Business Criticality

Another powerful report type in this context is the portfolio report, which combines multiple dimensions in a visually appealing format. The portfolio report in Figure 7.16 combines the TIME assessment on the x-axis with the business criticality on the y-axis, while the bubble size indicates the aggregated application TCO. By hovering over the individual bubbles, the aggregated application TCO is visualized and the basis for a business case to eliminate a certain number of applications leading to *x* amount of cost savings can be drafted.

Figure 7.16: Application Portfolio - Business Criticality by TIME by TCO

You can define the **Application Portfolio – Business Criticality by TIME by TCO** report by leveraging the default application portfolio report while changing the following under **Settings**:

- **X-Axis: TIME Classification** (with inverted x-axis order)
- **Y-Axis: Business Criticality**
- **Aggregation Fields: Total Cost of Ownership**

Also, add a filter for **Business Capabilities: Customer Service**. You can then save the report as **Application Portfolio – Business Criticality by TIME by TCO**.

Yet another useful report for the identification of application rationalization potential is a matrix report that outlines applications in use by taking the dimension of business capabilities (x-axis) as well as organizations (y-axis) into account. Applications supporting the same business capability within one or more organizations can be easily identified and, in combination with the TIME assessment, identified as rationalization candidates, as outlined in Figure 7.17.

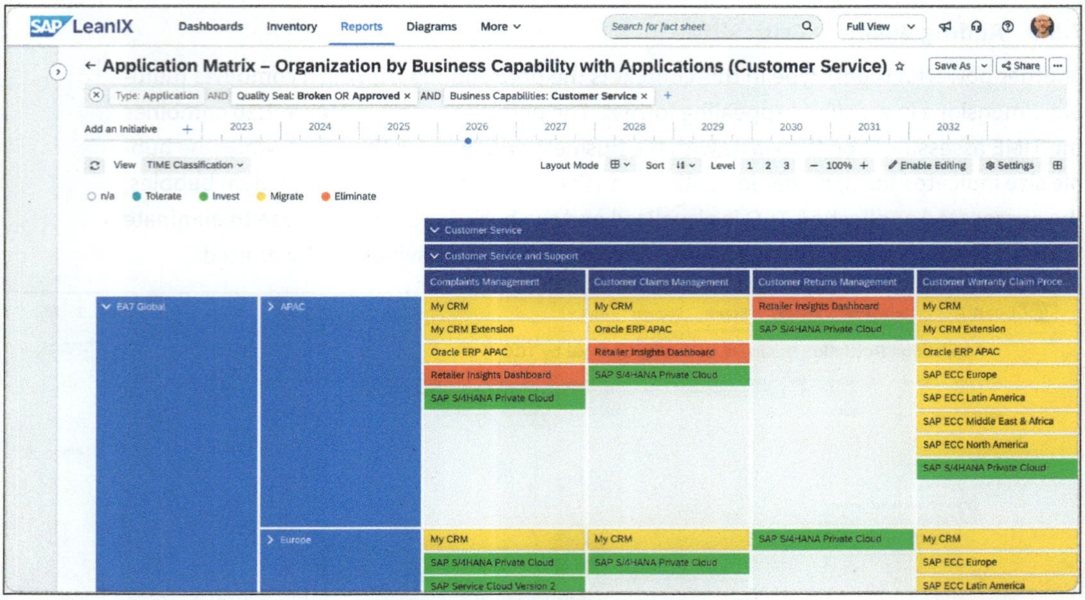

Figure 7.17: Application Matrix Report – Organization by Business Capability with Application (Customer Service)

You create the **Application Matrix – Organization by Business Capability with Applications (Customer Service)** report by leveraging the default application matrix while applying the following under **Settings**:

- Change the **View** to **TIME Classification**
- Add a filter for **Business Capabilities: Customer Service**

Save the report as **Application Matrix – Organization by Business Capability with Applications (Customer Service)**.

If the rationalization initiative is primarily driven through a cost management lens, dedicated cost reports can be used. Figure 7.18 outlines the EA7 application landscape report with the **Total Cost of Ownership** as the **View**. This viewpoint can be created by changing the **View** from the **TIME Classification** to the **Total Cost of Ownership** in our **TCO Application Rationalization by Business Capability (Customer Service) – aggregated** report. The TCO ranges can be displayed by changing **View** to **Total Cost of Ownership**.

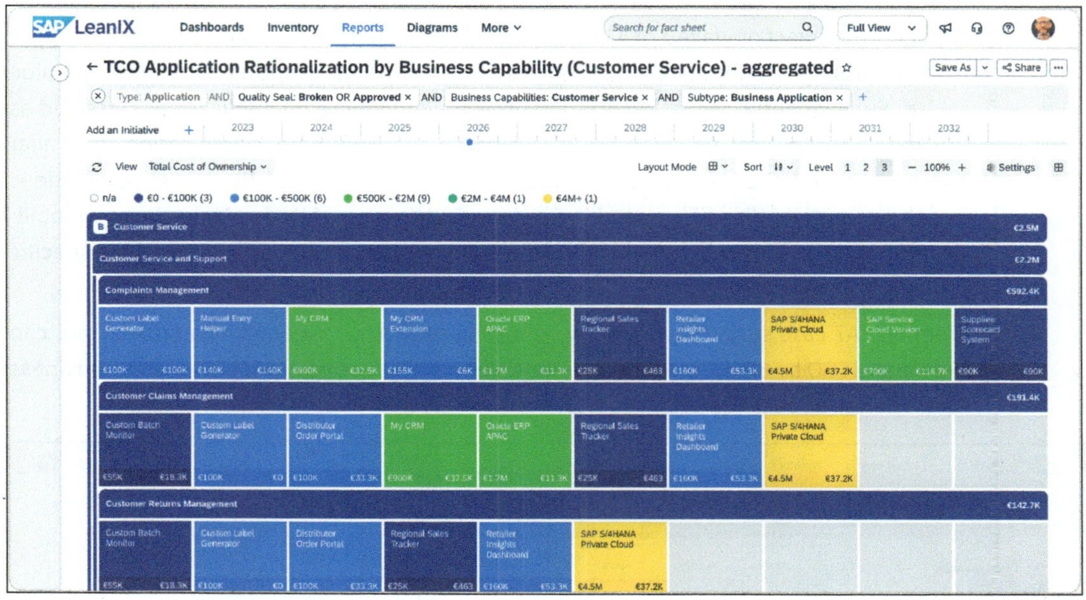

Figure 7.18: TCO Application Rationalization by Business Capability with TCO Scales

> **Note**
> The scale for the cost ranges is adjusted based on the data provided. The color-coding of the applications might therefore change as well.

With the outlined set of reports, you've been able to identify potential roadmap candidates from TIME, TCO, and business criticality perspectives, taking primarily the viewpoints of business capabilities and organizations into account. Further assessments could be run from, say, an IT component perspective to identify potential security, lifecycle, or technical fit implications, which could lead to further insights.

7.6 Prioritize Candidates and Understand Impact

In the previous section, we outlined how to retrieve potential rationalization candidates. Before adding them to a roadmap, it's important to understand the rationalization impact from cost, risk, and value perspectives.

A very valuable report to prioritize candidates is the **Application Portfolio - Business Criticality by TIME by TCO** report, as shown in Figure 7.18. Although our initial focus will be on applications classified as **Eliminate**, the business criticality classification can provide some insights from a risk perspective.

A quick-win project or wave 1 of a larger rationalization program would start with the applications classified as **Administrative Service** and **Business Operational** to realize fast value return. Subsequently, the rationalization project would focus on applications classified as **Business Critical** and finally **Mission Critical** in the **Eliminate** classification. A similar approach would apply for applications classified as **Migrate**, but the target applications and the potential rollout plan need to be clear. Furthermore, filtering by business capabilities and organizations will help with prioritization as changes normally happen in a specific part of the company.

From an EA7 case study point of view, you can see that with a quick-win project, you can reduce the TCO by €505K through the elimination of five applications classified as **Business Operational**, as outlined in Figure 7.19.

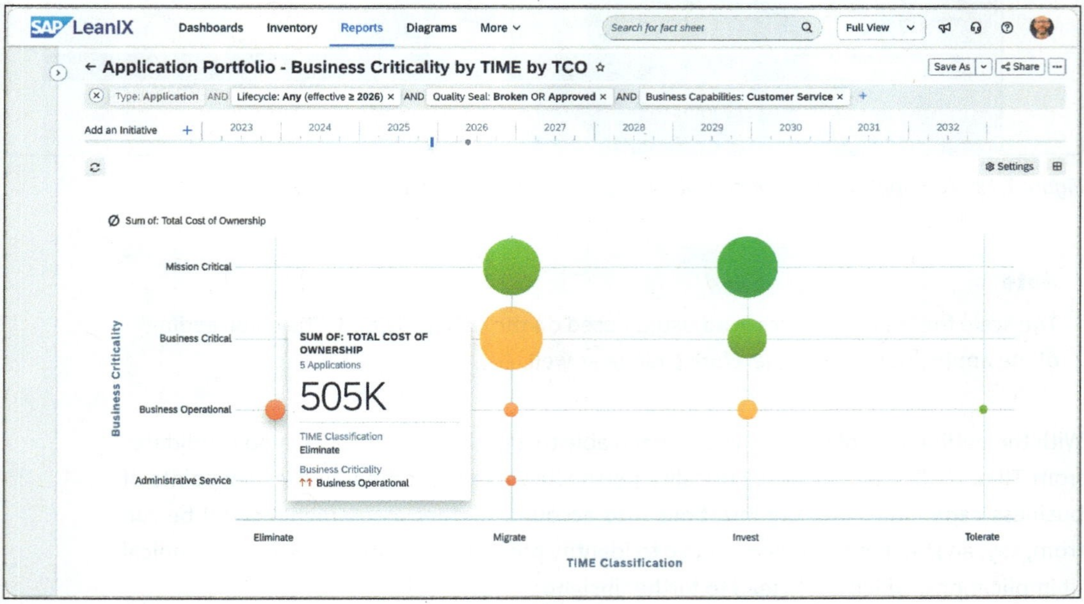

Figure 7.19: Identification of EA7 Quick-Win Potential

The elimination or migration of an application requires a fair understanding of the user population affected, interfaces that might need to be rebuilt, and the data lifecycle that needs to be kept alive. These insights, which make up the final decision-making tasks, can be supported within SAP LeanIX through interface and data reports, as outlined in Chapter 5, Section 5.1.1 and Section 5.1.2.

7.7 Building the Rationalization Roadmap

By outlining a rationalization roadmap for the applications in scope, you include the dimension of time in terms of the evolution from the baseline toward the target state. The roadmap report helps to identify potential dependencies and can simulate different options of how the evolution of the application landscape could look.

The structure of the initiatives should follow the business objectives defined from a strategic level. Depending on the type of rationalization or modernization project, you can determine a hierarchical structure of a program with subprojects.

For the EA7 application rationalization program, we have chosen the structure depicted in Table 7.3.

Wave	Initiative	Applications in Scope	Primary Action
Wave 0	Transformation office	N/A	Governance setup
Wave 1	Local tool decommissioning	Custom batch monitor; custom label generator; manual entry helper; local BI extractor; distributor order portal; regional sales tracker	Eliminate
Wave 1	Analytics rationalization	SAP Analytics Cloud; SAP Datasphere; retailer insights dashboard; BI extractor	Consolidate
Wave 2	SAP Service Cloud deployment	SAP Service Cloud v2; My CRM; My CRM extension; Oracle ERP APAC (service modules)	Invest/migrate/retire
Wave 2	Sunset CRM extensions	My CRM; My CRM extension; Oracle ERP APAC	Retire
Wave 3	SAP ERP → SAP S/4HANA migration	SAP ERP EU/NA/LATAM/MEA	Migrate
Wave 3	Interfaces and data cleanup	SAP ERP integrations; supplier scorecard; distributor portal; BI extractor	Harmonize
Wave 4	Emarsys global rollout	Emarsys; retailer insights dashboard	Invest/retire

Table 7.3: Structuring of Initiatives by Wave for EA7 Application Rationalization Program

Wave	Initiative	Applications in Scope	Primary Action
Wave 4	Unified analytics platform	SAP Analytics Cloud; SAP Datasphere; local dashboards	Standardize

Table 7.3: Structuring of Initiatives by Wave for EA7 Application Rationalization Program (Cont.)

The modeling of the application evolution is done through transformations, as introduced and explained in Chapter 6, Section 6.3.

In a first roadmap report, which is centered around initiatives, you'll provide an overview of the EA7 application rationalization initiatives, highlighting the expected business value and planned milestones, as outlined in Figure 7.20. In the example, we outline the different initiatives as defined in Table 7.3, clustered by the impacted business capabilities and a drill-down into the affected applications. For detailed instructions on creating roadmap reports, see Chapter 6, Section 6.2.

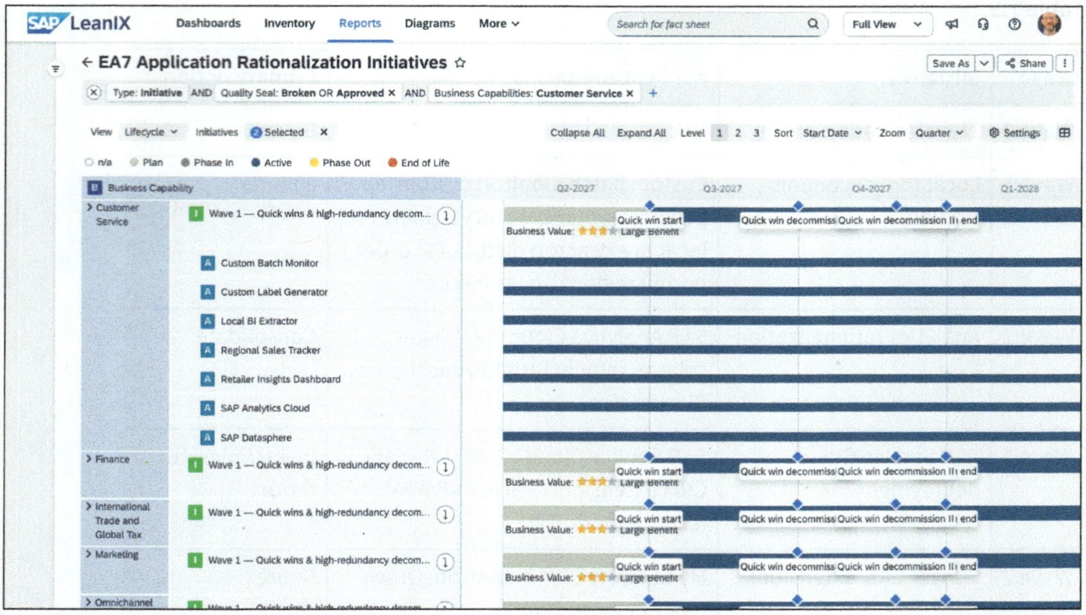

Figure 7.20: Application Rationalization Roadmap for EA7 by Initiatives

A second roadmap report, centered around applications, can provide EA7 with useful insights by taking initiatives, business capabilities, or organizations into account as clustering options. Transformations impacting the lifecycle of the application can be visualized in this roadmap report and might be a good starting point to align the proposed timelines with the stakeholders.

Figure 7.21 shows the **EA7 Application Rationalization Roadmap** clustered by initiatives. In the example, we outline the applications by initiative alongside the time dimension.

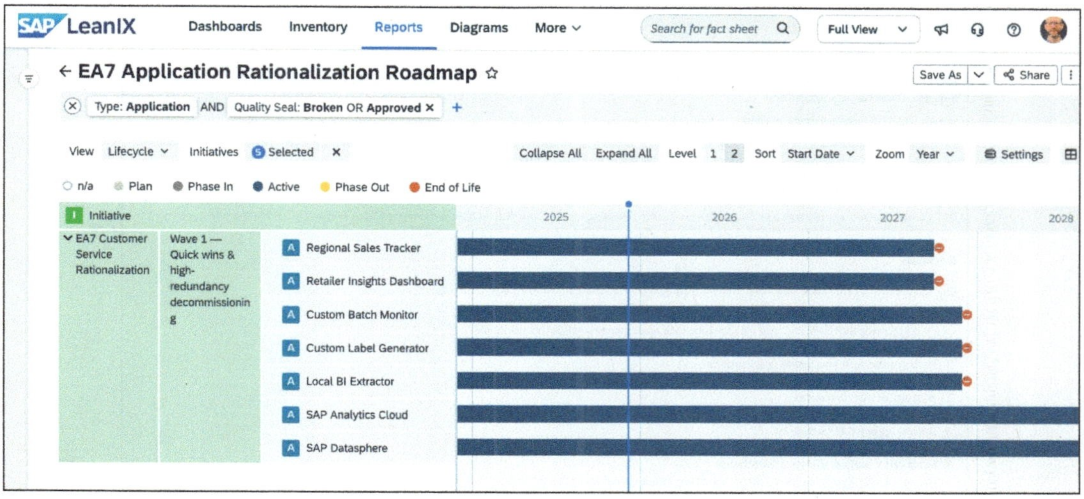

Figure 7.21: EA7 Application Rationalization Roadmap

7.8 Decision-Making and Communication

With the insights and rationales for an application rationalization project in place, a formal governance process can be initiated to retrieve the stakeholder buy-in. Individual decisions should be documented with the help of architecture decision records. We'll explain these in more detail in Chapter 12, but for now, it's important to know that they help organizations to understand the rationale for why a decision was made at a specific point in time, helping avoid reiterations of discussions in the future.

Following the structure of the context, decision, and consequences, the decision and rationale can be detailed, potential cost savings documented, and the applications in scope referenced as application fact sheets. For the endorsement of the change, the architecture decision record would undergo the company-specific enterprise architecture governance process.

Figure 7.22 outlines the EA7 architecture decision record for the application rationalization project, highlighting aspects of the context, decisions made, and consequences from positive as well as risk perspectives. In the example architecture decision record, we summarize the findings of the application rationalization use case, such as the TCO insights, and document the decisions on how to evolve the EA7 landscape by, say, adopting a global customer service platform or migrating the regional ERP systems into a global SAP S/4HANA platform.

For further details on how to create an architecture decision record, see Chapter 12, Section 12.3.

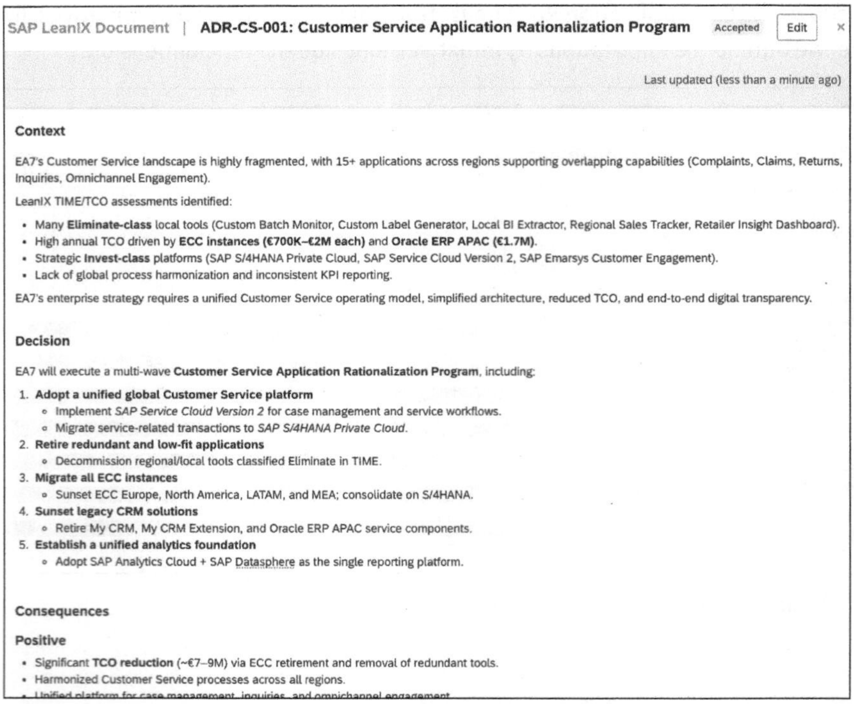

SAP LeanIX Document | **ADR-CS-001: Customer Service Application Rationalization Program** Accepted Edit ×

Last updated (less than a minute ago)

Context

EA7's Customer Service landscape is highly fragmented, with 15+ applications across regions supporting overlapping capabilities (Complaints, Claims, Returns, Inquiries, Omnichannel Engagement).

LeanIX TIME/TCO assessments identified:

- Many **Eliminate-class** local tools (Custom Batch Monitor, Custom Label Generator, Local BI Extractor, Regional Sales Tracker, Retailer Insight Dashboard).
- High annual TCO driven by **ECC instances (€700K–€2M each)** and **Oracle ERP APAC (€1.7M)**.
- Strategic **Invest-class** platforms (SAP S/4HANA Private Cloud, SAP Service Cloud Version 2, SAP Emarsys Customer Engagement).
- Lack of global process harmonization and inconsistent KPI reporting.

EA7's enterprise strategy requires a unified Customer Service operating model, simplified architecture, reduced TCO, and end-to-end digital transparency.

Decision

EA7 will execute a multi-wave **Customer Service Application Rationalization Program**, including:

1. **Adopt a unified global Customer Service platform**
   - Implement *SAP Service Cloud Version 2* for case management and service workflows.
   - Migrate service-related transactions to *SAP S/4HANA Private Cloud*.
2. **Retire redundant and low-fit applications**
   - Decommission regional/local tools classified Eliminate in TIME.
3. **Migrate all ECC instances**
   - Sunset ECC Europe, North America, LATAM, and MEA; consolidate on S/4HANA.
4. **Sunset legacy CRM solutions**
   - Retire My CRM, My CRM Extension, and Oracle ERP APAC service components.
5. **Establish a unified analytics foundation**
   - Adopt SAP Analytics Cloud + SAP Datasphere as the single reporting platform.

Consequences

Positive

- Significant **TCO reduction** (~€7–9M) via ECC retirement and removal of redundant tools.
- Harmonized Customer Service processes across all regions.
- Unified platform for case management, inquiries, and omnichannel engagement

Figure 7.22: Architecture Decision Record for EA7 Application Rationalization Project

7.9 Monitoring with KPIs

One aspect often forgotten is the actual realization of the expected change and, with this, the value realization. With the help of KPIs, you can measure the progress of your rationalization initiative over time and might get useful insights for a potential course correction.

SAP LeanIX is delivered with various out-of-the box dashboards and KPIs that can be leveraged to oversee the general application portfolio management situation as well the overall improvement progress. KPIs on the architecture executive dashboard provide senior stakeholders with an aggregated view of specific aspects, and these can be further explored through drilldowns. We'll discuss these options in detail in the following sections.

7.9.1 Application Portfolio Management Dashboard

The out-of-the box application portfolio management dashboard can be found in the **Dashboard** section within the **Predefined** subcategory. For more details on accessing dashboards, see Chapter 5, Section 5.3.

It includes the following four panels with the KPIs:

1. **Live Applications by Business Capability**
   This panel shows how many active applications an organization uses and maps them to

the top-level business capabilities. It helps identify where applications are concentrated, identify redundancies (e.g., after carve-out), and track progress in rationalization initiatives. It includes the following KPIs:

- The total number of applications
- A breakdown per top-level business capability
- Trend indicators showing changes over the selected timeframe

2. **Application Run Cost Broken Down by Business Capability**
   This panel shows the operational cost of your applications based on the total annual cost defined on each application and IT component relation. It includes the following KPIs:
   - Total run cost across the portfolio
   - Costs allocated to each business capability

   This helps you understand which capabilities drive the most IT spend and where optimization efforts could have the biggest impact.
3. **Portfolio Complexity**
   This panel provides KPIs that reflect how complex and interconnected your application landscape is. High complexity can increase the cost, risk, and effort of any change. It includes the following KPIs:
   - **Monolithic degree**
     For each application, this counts how many unique top-level business capabilities are supported. The portfolio's monolithic degree is the average across all applications. Lower values generally indicate a cleaner, more modular architecture.
   - **Average count of interfaces per application**
     Shows how many inbound and outbound interfaces applications have on average: <1 indicates likely missing data, while >10 indicates high complexity.
   - **Average count of linked business contexts/organizations/data objects/IT components**
     These KPIs show how many related elements are connected to the average application. Very low counts indicate missing data, while very high counts indicate potential complexity hotspots.
4. **Data Quality KPI**
   This panel measures whether the application data in your workspace is complete enough to support meaningful analysis and decision-making. It includes the following KPIs:
   - Overall completion of applications: Aggregates your application completion scores based on your defined completeness rules. A value above 50% indicates a higher completion rate.
   - Approved quality seal: Counts applications whose quality seal has been marked as approved.

- With lifecycle/business capability/organization: Shows how many applications have a lifecycle maintained.
- With business capability/organization: Show how many applications are linked.

Figure 7.23 outlines a snapshot of the EA7 application portfolio management dashboard. Note that the numbers and trend indication will evolve over time as historical data is taken into consideration.

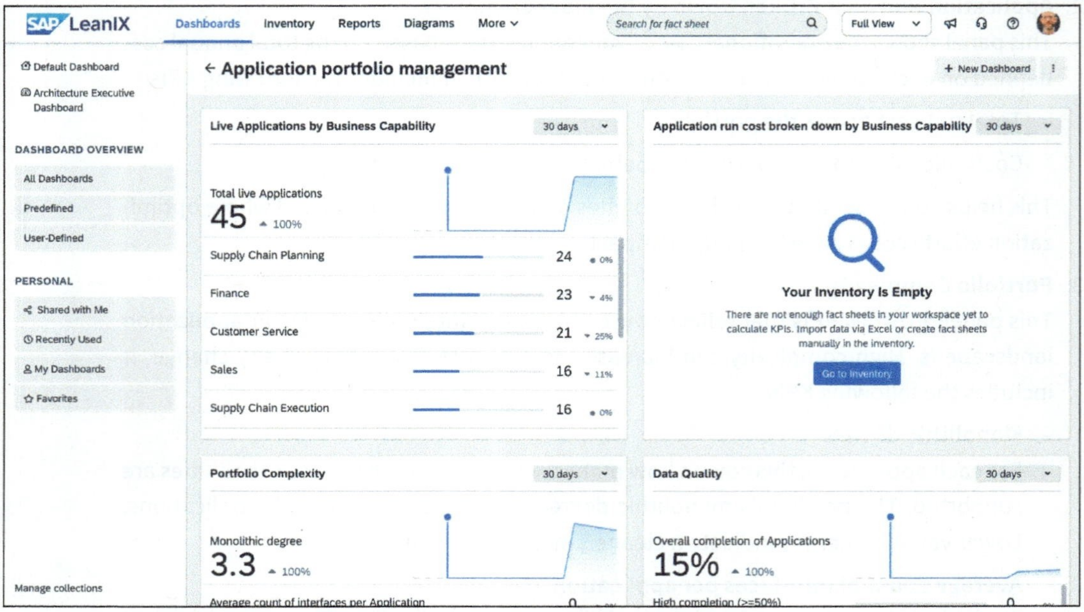

Figure 7.23: Application Portfolio Management Dashboard

7.9.2 Application Rationalization Dashboard

The second predefined dashboard is the application rationalization dashboard, which provides predefined KPIs primarily in context of cost, the Gartner TIME assessment, and general data quality. It consists of the following four panels:

1. **Optimization Candidates**
   This panel aggregates various KPIs that can indicate a potential roadmap candidate. It includes the following KPIs:
   - Monolithic degree: For each application, counts how many unique top-level business capabilities are supported (same as above)
   - Low functional and technical fit: Applications with a low functional and low technical fit
   - Legacy applications (> 10y old): Applications in use for more than 10 years
   - High cost (>500): Applications with total annual cost of more than €500K

   - Interface complexity (> 10 interfaces): Applications with more than 10 consuming and providing interfaces
2. **Cost Analysis TIME Assessment**
   This panel aggregates the annual cost attribute on the relation between IT components and applications. It includes the following KPIs:
   - Cost savings: Sum of total annual cost of all applications with eliminate TIME assessment
   - Tolerate/invest/migrate run cost: Sum of total annual cost of all applications with tolerate, invest, and migrate TIME assessments
3. **TIME Assessment**
   This panel provides insights into the TIME assessment. It includes the following KPIs:
   - Completeness of TIME assessment: Overall percentage score of applications with TIME assessment maintained
   - Tolerate/invest/eliminate/migrate: Count of applications with respective TIME assessment
4. **Data Quality**
   This panel measures whether the application data in your workspace is complete enough to support meaningful analysis and decision-making. It includes the following KPIs:
   - Applications with complete information: The following criteria must be fulfilled:
     - Functional fit filled out
     - Technical fit filled out
     - Business criticality filled out
     - At least one relation to an IT component exists and has a total cost field filled out
     - Linked to at least one business capability
     - QS approved
   - Functional fit/technical fit/business criticality: Count of applications with functional fit/technical fit/business criticality maintained
   - Applications with total annual cost: Count of applications where the related IT components have the **Total Annual Cost** field maintained
   - Applications with business capability: Count of applications with a link to any business capability

Figure 7.24 outlines a snapshot of the EA7 application rationalization dashboard. Note that the numbers and trend indication will evolve over time as historical data is taken into consideration.

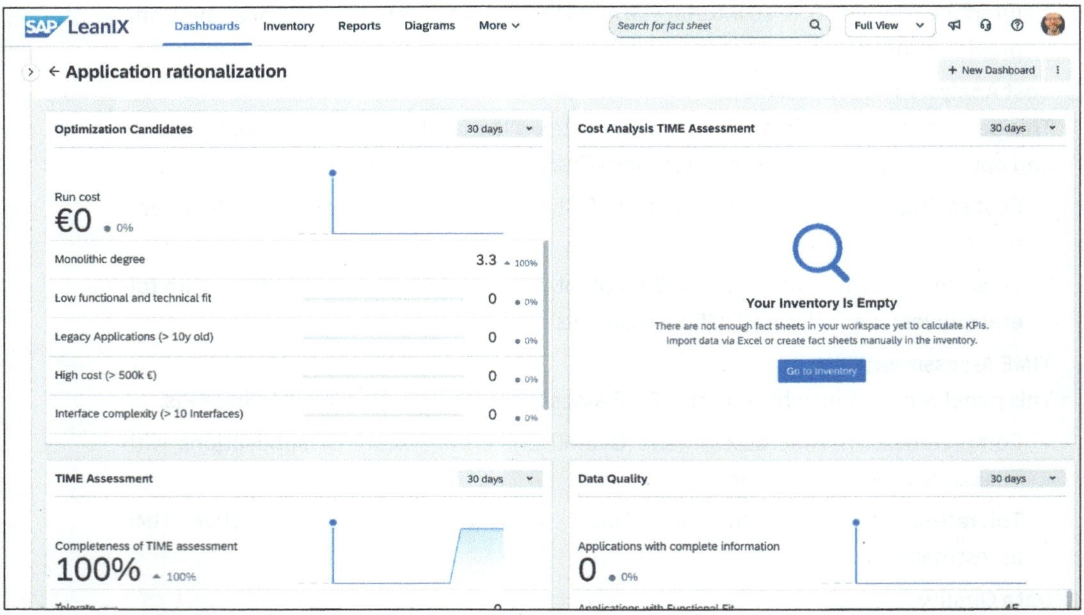

Figure 7.24: Application Rationalization Dashboard

7.9.3 Architecture Executive Dashboard

The architecture executive dashboard defines a core set of KPIs to derive data-driven insights. KPIs with aggregated information can be further drilled into to gather valuable insights. The following default KPI sections of the predefined architecture executive dashboard are relevant in context of this chapter:

- Application portfolio assessment (APA)
- Application rationalization (AR)
- Application modernization (AM)

Both the **Application Portfolio Assessment** and **Application Rationalization** KPI sections include a variety of KPIs that can be configured by choosing **Edit dashboard**. Table 7.4 provides an overview of the prebuilt KPIs and indicates in which section an individual KPI can be used.

> **Tip**
>
> We recommend using a small but relevant set of KPIs in alignment with your stakeholders.

Name	Description	APA	AR	AM
Business criticality of applications	Assess the applications within the IT landscape that are classified as either business critical or mission critical. Provides insight into how much of the application portfolio directly supports essential business operations.	X		
Application modernization (TIME)	Measures the percentage of applications classified as invest and migrate based on the Gartner TIME framework, offering insights to guide the application modernization strategy.	X	X	X
Functional fit of applications	Assesses the degree to which applications align with intended business processes and requirements, based on the percentage of applications classified as appropriate or perfect.	X		
Technical fit of applications	Assesses the degree to which applications align with the current technical standards and infrastructure, based on the percentage of applications classified as adequate or fully appropriate.	X		
Application rationalization (TIME)	Measures the percentage of applications classified as tolerate and eliminate based on the Gartner TIME framework, offering insights to guide the application rationalization strategy.	X	X	
Application portfolio	Assess number of applications categorized by their lifecycle status, supporting visibility into the health and evolution of the application portfolio, aiding in planning, modernization, and decommissioning efforts.	X		
Ownership of applications	Measures the percentage of applications that have subscribers, providing insights into the level of engagement and ownership of fact sheets.	X	X	
Lifecycle completion of applications	Assesses the ratio of existent fact sheets with lifecycle status completed, providing visibility into data completeness and supporting effective portfolio planning and transition management.	X		
Applications at end of life	Assess applications reaching their end-of-life stage, ensuring proactive decision-making for replacements, upgrades, or retirements.	X	X	

Table 7.4: Application Portfolio Management: Architecture Executive Dashboard KPIs

Name	Description	APA	AR	AM
Pace layering: innovation index	Assesses the potential of the organization's IT landscape to support and drive innovation. Evaluates the adaptability of the current architecture to new technologies and business models and, using the pace layering framework, measures the gap between the current state and market best practices.	X		
Cloud migration (business criticality plus 6R)	Assess applications already in the cloud and combines business criticality together with the 6R migration strategy framework to monitor cloud transformation progress. Reflects which workloads are suitable for cloud migration, optimization, or decommissioning, supporting targeted, data-driven cloud decisions.	X	X	
Application suitability (TIME)	Measures the completeness of TIME classification across the application portfolio. Supports faster, more accurate lifecycle decisions and ensures applications are assessed in alignment with their business value, helping identify redundancies and prioritize actions.	X		
Cloud migration (functional fit plus 6R)	Assess applications already in the cloud and use functional fit together with the 6R migration strategy framework to monitor cloud transformation progress. Reflects which workloads are suitable for cloud migration, optimization, or decommissioning, supporting targeted, data-driven cloud decisions.	X	X	
Applications-to-business capabilities	Measures the completeness of application-to-business capability mapping, ensuring that IT assets are aligned with strategic business functions. Gives clarity on how applications support key business processes, identify redundancies, and assess the impact of changes.	X		
Applications per organization	Assess number of applications across the organization, grouped by business unit or organizational entity. Helps visualize application distribution and support rationalization or ownership accountability.	X		

Table 7.4: Application Portfolio Management: Architecture Executive Dashboard KPIs (Cont.)

Name	Description	APA	AR	AM
TCO of applications	Measures the full cost associated with owning and operating a specific application within your IT structure. Used with the TIME framework to leverage processes of application rationalization or modernization.		X	
Application lifecycle TCO	Measures the full cost associated with owning and operating a specific application within your IT structure at a certain lifecycle stage.		X	
Maintenance cost of applications	Reflects the total expenditure associated with the upkeep, support, and ongoing operation of applications.		X	
Share of support cost	Measures the percentage of application TCO attributed specifically to support activities, allocated to a specific business capability.		X	
Share of maintenance cost	Measures the percentage of application TCO attributed specifically to maintenance activities, allocated to support a specific business capability.		X	
Licensing cost of applications	Refers to the total expenditure associated with software licensing fees paid for applications.		X	
Support cost of applications	Refers to the total expenditure associated with providing technical, functional, and end user support for applications.		X	
Share of licensing cost	Measures the percentage of application TCO attributed specifically to licensing, allocated to support a specific business capability.		X	
Unsuitable applications	Visibility into applications that are in phase-out or end-of-life lifecycle status, enabling timely actions for retirement, replacement, or migration.		X	

Table 7.4: Application Portfolio Management: Architecture Executive Dashboard KPIs (Cont.)

The configuration of views will help to derive more detailed insights through drilldowns and should be configured in alignment with the stakeholders, such as by region, capabilities, and so on.

> **Further Resources**
>
> For more information, refer to the SAP LeanIX help documentation at *http://s-prs.co/v620661.*

Figure 7.25 provides an example of the EA7 architecture executive dashboard from a **Europe** view perspective. This helps stakeholders see information relevant to their specific context.

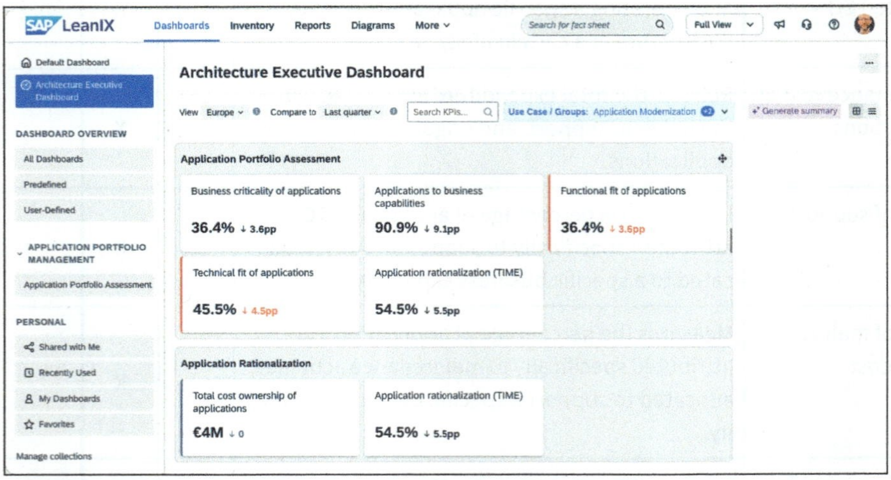

Figure 7.25: Architecture Executive Dashboard for EA7: Europe View

The individual KPIs can now be further assessed through drilldowns. Depending on the views defined, the pivot can be flexibly changed to derive relevant insights. Figure 7.26 outlines the drilldown of the **Functional fit of applications** KPI.

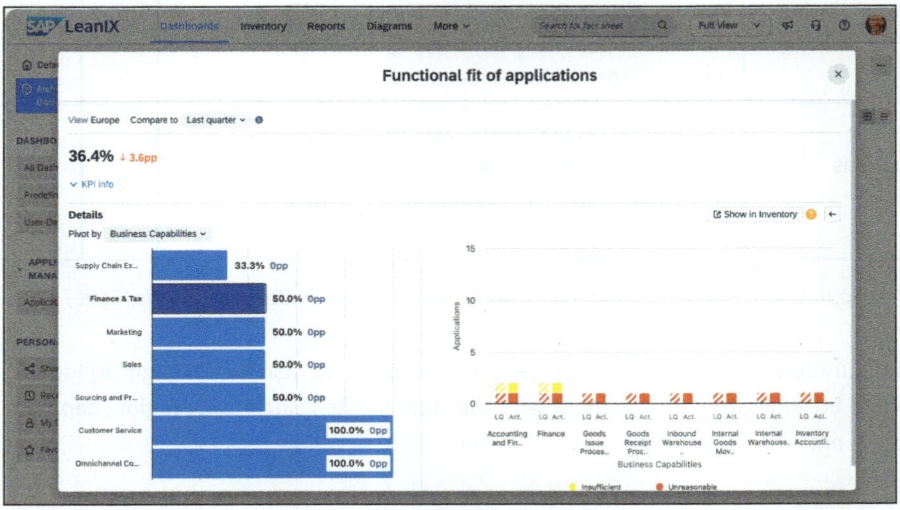

Figure 7.26: Drilldown into Functional Fit KPI

The predefined business capability view clusters the applications accordingly. By selecting one of the business capability categories, further details in context of the functional fit can be derived.

7.10 Summary

This chapter introduced the principles and practices of application portfolio management and demonstrated how SAP LeanIX enables organizations to evaluate, rationalize, and modernize their application landscapes. By establishing clear strategic objectives and activating the right assessment frameworks such as functional and technical fit, business criticality, TIME, and TCO, enterprises can build a reliable foundation for decision-making.

Through the EA7 case study, the chapter illustrated how fragmented, costly environments can be transformed using data-driven insights. Visualizations revealed rationalization opportunities across customer service, highlighted high-cost legacy systems, and identified *invest* platforms supporting EA7's target architecture. These insights were translated into a structured, multiwave roadmap and supported by governance mechanisms such as architecture decision records and executive dashboards.

In essence, this chapter showed how application portfolio management creates transparency, reduces complexity, and enables organizations to align their application portfolios with business strategy.

In the next chapter, we'll explore enterprise-wide transformation with the support of SAP Enterprise Architecture Framework, a structured approach to run a full architecture development cycle.

Chapter 8

Enterprise-Wide Planning with SAP Enterprise Architecture Framework

An enterprise architecture management solution provides many capabilities. An enterprise architecture framework combines these capabilities with a structured approach to implementing an architecture development cycle.

This chapter will provide an overview of the SAP Enterprise Architecture Framework with the focus of its applicability in context of SAP LeanIX. With support of the EA7 case study, we'll explore an entire architecture development cycle, taking a subset of the artifacts of SAP Enterprise Architecture Framework into account. The focus of this exploration will be the SAP-specific features and extensions for discovery and the usage of reference content.

8.1 Overview of SAP Enterprise Architecture Framework

The purpose of SAP Enterprise Architecture Framework is to align business strategies with IT roadmaps and implementation projects via a structured approach and well-defined deliverables featured in reference content to support sophisticated modeling tools.

SAP Enterprise Architecture Framework was initially introduced by a cross-functional team within SAP to align the methodology and approach of how enterprise architecture development should take place, independent of the roles (consultants, partners, customers, etc.) involved. Furthermore, the acceleration of architecture work through standardized and curated reference content as well as the usage of sophisticated modeling tools supported the evolution of SAP Enterprise Architecture Framework over time.

Figure 8.1 outlines the five key pillars of SAP Enterprise Architecture Framework:

- **Methodology**
  The enterprise architecture methodology is largely based on The Open Group Architecture Framework (TOGAF) in its latest form, version 10. It also leverages multiple other well-established standards such as American Productivity & Quality Center (APQC, *www.apqc.org*), Business Process Model and Notation (BPMN, *www.bpmn.org*) and Unified Modeling Language (UML, *www.omg.org/uml/*). A large amount of enterprise architecture project experience was woven into the methodology by including techniques and concepts to make it not a theoretical construct but rather a pragmatic approach to support outcome-based architecture work.

- **Reference architecture content**
  The reference architecture content consists of two parts: the solution-agnostic SAP Reference Business Architecture and the solution-specific SAP Reference Solution Architecture. These cornerstones of SAP Enterprise Architecture Framework provide unique value to customers and partners, as they literally breathe life into the methodology by providing curated content such as business capability models and target application information. We'll discuss these solutions in more detail in Section 8.3.
- **Enterprise architecture tool**
  A sophisticated enterprise architecture tool was brought into the SAP Enterprise Architecture Framework equation with the acquisition of SAP LeanIX in 2023. Enterprise architecture work started to be performed via a much more collaborative and scalable approach.
- **Enterprise architecture practice**
  The enterprise architecture practice outlines the organizational implementation of an enterprise architecture function in an organization, considering different organizational setups and company cultures. This cornerstone was added in one of the later versions of SAP Enterprise Architecture Framework. Refer to Chapter 12 to gain further insights into the governance of enterprise architecture.
- **Services**
  Services provide an overview of planning and implementation offerings for multiple aspects of an enterprise architecture project as well as an enterprise architecture function. These services are in alignment with SAP Enterprise Architecture Framework.

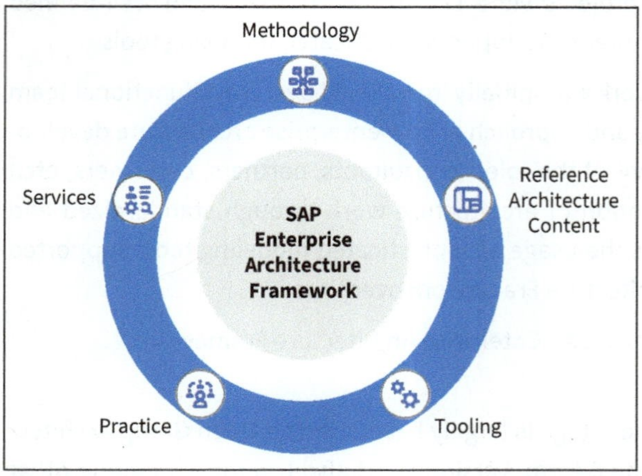

Figure 8.1: SAP Enterprise Architecture Framework Overview

Further Resources

In this chapter, we'll focus on the combination of the methodology, the reference architecture content, and the tooling with SAP LeanIX. For more details on SAP Enterprise Architecture Framework, refer to *http://s-prs.co/v620662*.

For a complete guide, we recommend *Enterprise Architecture with SAP: Planning, Management, and Transformation* by Anup Das, Peter Klee, and Johannes Reichel (SAP PRESS 2025, *www.sap-press.com/5863*).

8.2 Structure of SAP Enterprise Architecture Framework

SAP Enterprise Architecture Framework is centered around seven domains that structure the methodology from vision to roadmap. Table 8.1 provides an overview and description of the seven domains.

Domain	Description
Architecture vision	Defines the overall outcome of architecture work, scope, and stakeholder needs
Strategy and motivation	Translates business strategy into objectives and motivates stakeholders
Business architecture	Describes business aspects (capabilities, processes, data, structure) for communication and alignment
Solution architecture	Designs solutions that bridge business goals with technology needs
Technology architecture	Documents IT components (systems, networks, hardware) and their deployment
Roadmap and transition	Plans migrations and transitions to achieve project objectives efficiently
Requirements and governance	Manages governance, principles, risks, and architecture decisions across all domains

Table 8.1: SAP Enterprise Architecture Framework: Domain Definitions

Each domain contains a set of artifacts, including guidance for creating them, a template for each artifact to jump-start the creation in your architecture work, and tangible examples to get a first glimpse of what "good" looks like. Figure 8.2 provides an overview of the seven domains with the associated artifacts.

As SAP Enterprise Architecture Framework is based on TOGAF's Architecture Development Method (ADM), Figure 8.3 depicts the previously introduced domains and artifacts alongside the TOGAF ADM phases.

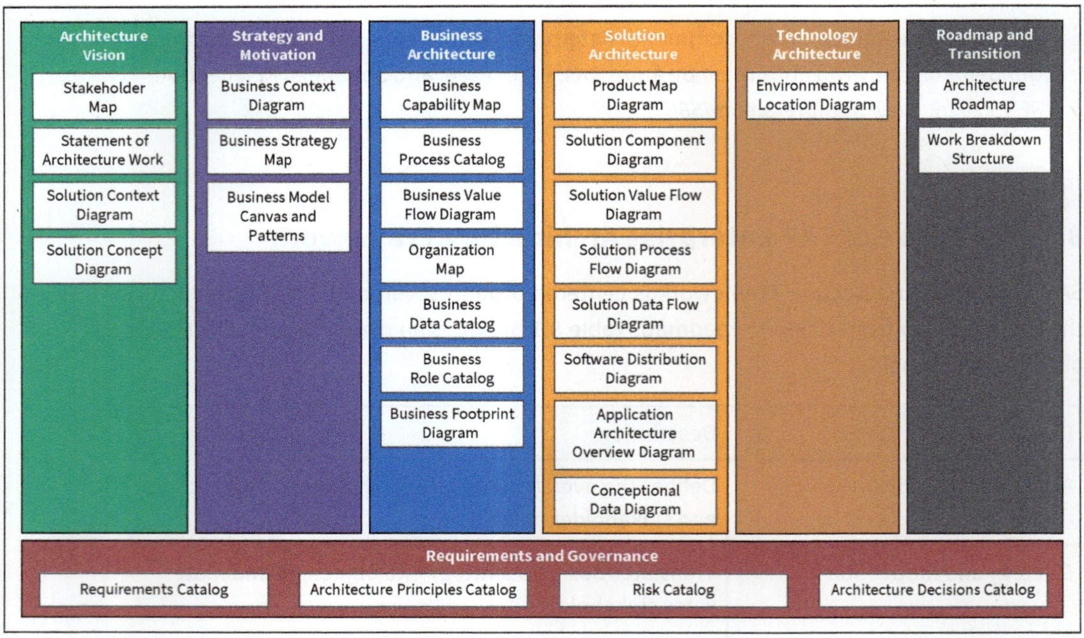

Figure 8.2: SAP Enterprise Architecture Framework: Domain Overview with Artifacts

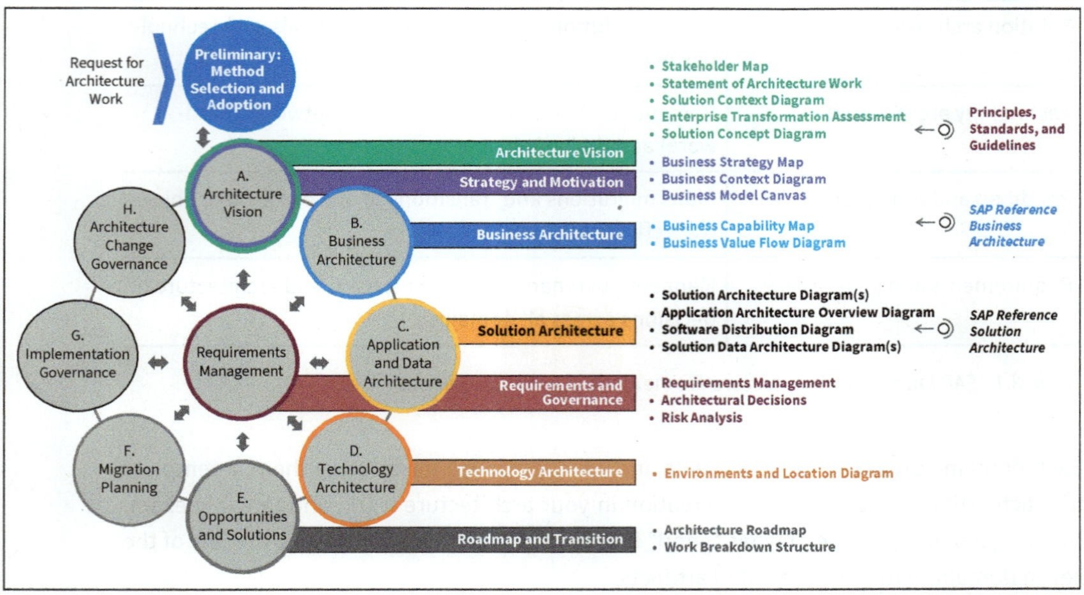

Figure 8.3: TOGAF ADM with SAP Enterprise Architecture Framework Domains and Artifacts

Readers who are familiar with the TOGAF standard will recognize that phase C has been renamed, from the TOGAF definition of *information systems architecture* to *application and data architecture*. Furthermore, the focus of architecture work within SAP Enterprise Architecture Framework is primarily covered in phases A, *architecture vision*, to E, *opportunities and solutions*.

All phases are connected to *requirements management*, centered in the middle of the ADM wheel, which advocates for a highly requirement- and thus stakeholder-driven architecture design. At any phase of the ADM, new requirements might be identified, which leads to the conclusion that multiple iterations throughout the ADM will be required to address stakeholder requirements.

One important aspect to point out is the *preliminary phase*, which is initiated with the *request for architecture work*. During this phase, the following activities are executed to prepare for architecture engagement:

- Tailoring of the taxonomy
- Selection of reference content to be used
- Definition of the architecture approach (e.g., phases in scope, iterations)
- Selection of artifacts to be produced to address stakeholder needs and concerns

Depending on the type of request for architecture work, architecture engagement can be more business- or technology-focused, or a combination of both in larger, business-led transformation programs.

A very handy and useful view for the tailoring of artifacts and activities throughout the architecture engagement is the SAP Enterprise Architecture Framework metro map, as shown in Figure 8.4.

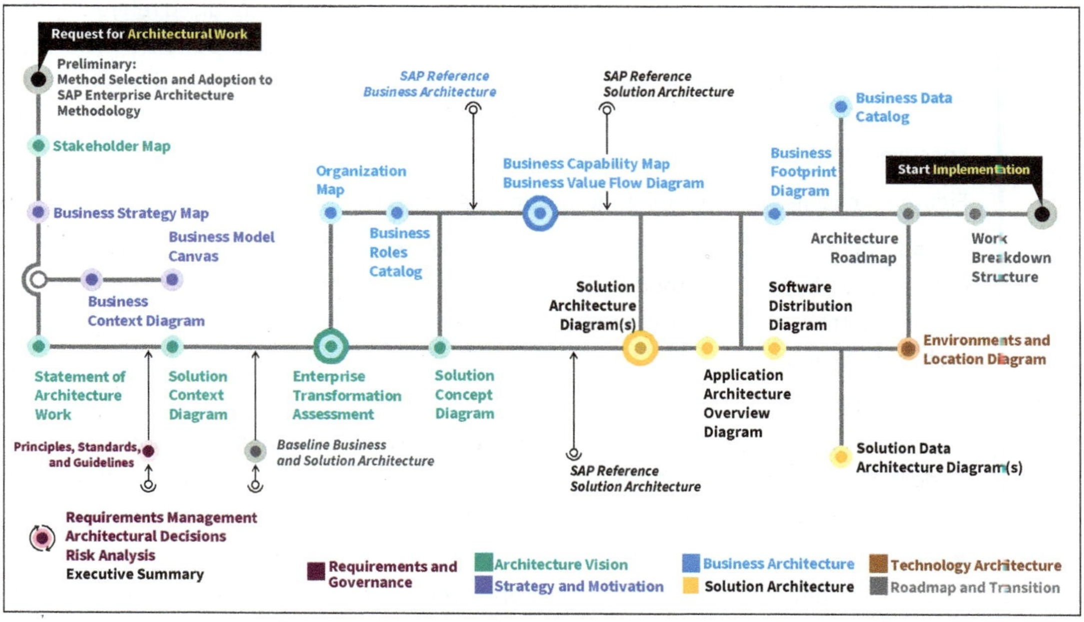

Figure 8.4: SAP Enterprise Architecture Framework: Metro Map

The metro map depicts the previously introduced domains and artifacts in a logical sequence, starting on the left with the vision and strategy and moving toward the right, focusing on the architecture roadmap and preparation of the implementation for the defined target architecture.

In various SAP Enterprise Architecture Framework trainings, the following questions are raised: How would the approach look if an enterprise architect only gets involved later on, or if a project is IT- rather than business-driven? The metro map offers the possibility to enter the architecture development process across the domains at any point in time and to work potentially from right to left, with the overarching aim of enabling business and IT alignment. Let's imagine the following two entry points:

- A business capability model was defined, but the link to the overarching strategy is missing. The next step could be to create a business strategy map and to link the drivers, goals, and objectives a company is envisioning in order to enable you as enterprise architect to tell a consistent storyline from the company's strategy down to the implementation roadmap.
- If the entry point is at a very late point in time, and you become engaged while defining the target architecture, you can make a shift to the left by introducing a business capability model (maybe complementing an existing process model) in order to provide a contextual business perspective with regard to the planned IT changes.

8.3 SAP Reference Architecture Content

A cornerstone of SAP Enterprise Architecture Framework is the standardized reference content provided by SAP in the form of SAP Reference Business Architecture and SAP Reference Solution Architecture. These components help organizations accelerate business transformation by providing ready-to-use architectural models that bridge business and IT perspectives.

Figure 8.5 outlines the SAP Reference Business Architecture content.

As you can see, SAP Reference Business Architecture supports multiple artifacts across the enterprise architecture methodology, especially those outlined in the SAP enterprise architecture metro map. The structured reference content is delivered through four primary architecture views:

- **Capability view**
  Provides a structured, outcome-oriented map of what an organization does, independent of how it does it. On the business side, it organizes capabilities hierarchically. On the solution side, it maps those capabilities to SAP product and solution capabilities.
- **Process view**
  Describes how work flows through the organization. The business process model documents end-to-end business processes. The solution process model maps those processes to SAP solution-level process flows and transactions.

- **Data view**
  Defines the key business objects and data entities relevant to the enterprise, and aligns them with the corresponding data structures within SAP solutions. It provides traceability from business-level data concepts to the technical data model in the solution landscape.
- **Organization view**
  Captures the roles, responsibilities, and organizational structures involved in executing business processes. The business role model defines business roles and the activities they perform. The application role model maps those business roles to the corresponding authorization roles and personas within SAP applications.

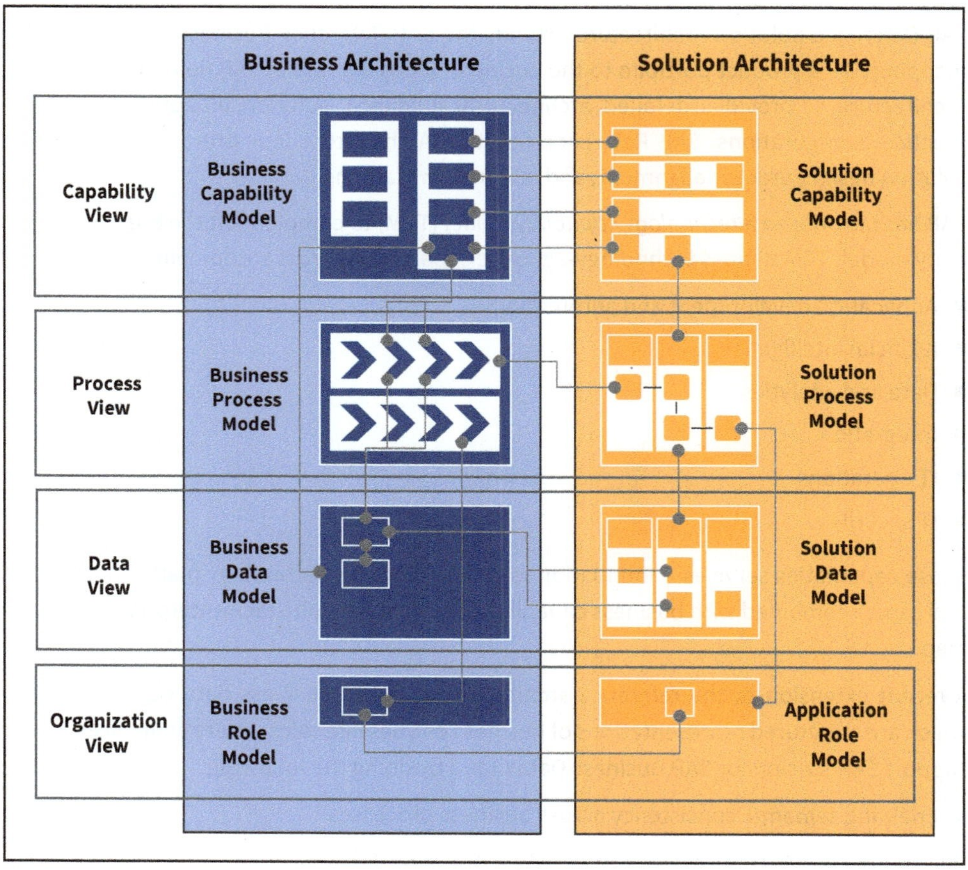

Figure 8.5: SAP Reference Business and Solution Architecture Framework

These views span both business and IT perspectives, enabling traceability and consistency across capability models, process flows, data usage, and organizational structures.

SAP Reference Business Architecture is designed to describe enterprises from a business-centric, product-agnostic viewpoint. It is based on SAP's 50-plus years of industry experience and aligned with standards such as APQC. The model is composed of the following:

- **Enterprise domains**
  Products and services, supply, customer, corporate
- **Business domains and business areas**
  Logical groupings of business capabilities
- **Business capabilities**
  Descriptions of what an enterprise can do to create value

SAP Reference Business Architecture also defines eight generic, end-to-end business processes, such as idea-to-market, lead-to-cash, and source-to-pay. These are supplemented with process variants for different business models and industries (e.g., B2B commerce, subscription models).

SAP Reference Solution Architecture extends the SAP Reference Business Architecture by mapping SAP's product portfolio to the business needs defined in SAP Reference Business Architecture. It provides detailed architectural guidance, solution blueprints, and best practice configurations. SAP Reference Solution Architecture is instrumental in guiding solution design, process alignment, and integration planning.

SAP has introduced a *technology capability model* (TCM) to complement the business capability model. TCM defines technology-agnostic capabilities across six domains:

- Application development and automation
- Artificial intelligence
- Data and analytics
- Integration
- IT operations
- IT security

These capabilities serve as a foundation for mapping SAP's technology platform offerings and can be combined with business capabilities to provide a full-stack enterprise capability map.

A recent extension to the reference architecture includes business data objects (BDOs), which are structured representations of business entities like sales orders or invoices. BDOs support SAP's vision for SAP Business Data Cloud by doing the following:

- Enabling semantic consistency across business processes
- Supporting cross-system data modeling and integration
- Facilitating data lineage, transparency, and governance

BDOs are on the roadmap to eventually be natively available in SAP LeanIX and SAP Signavio, where they will provide integrated modeling and data flow analysis.

Together, SAP Reference Business Architecture (including TCM and BDOs) and SAP Reference Solution Architecture enable a comprehensive, multiperspective understanding of enterprise structure and transformation strategy. These assets form the foundational content layer of SAP Enterprise Architecture Framework and are embedded directly into SAP LeanIX for active use in transformation projects.

8.4 Applying SAP Enterprise Architecture Framework in SAP LeanIX

The tailoring of SAP Enterprise Architecture Framework and the definition of the artifacts in scope for your architecture engagement are of crucial importance and depend very much on the type of engagement and the environment you are operating in. For the sake of this chapter, we've tailored SAP Enterprise Architecture Framework alongside the following criteria to provide the best value for money:

- The why, what, and how of the strategy to implement the roadmap need to be well articulated.
- The use and application of SAP Reference Business Architecture and SAP Reference Solution Architecture content must be demonstrated to their full extent.
- SAP-specific features within SAP LeanIX accelerating architecture work must be outlined and demonstrated.

Taking these key criteria into account, Table 8.2 outlines the short-listed artifacts, including the rationales for selection.

Artifact	Rationale
Requirements management	We are architecting not for the sake of architecture but for the sake of stakeholder needs and requirements.
Architecture principles	Architecture principles build our guardrails for any target architecture decision and help to govern the architecture over time.
Business strategy map	A breakdown of the business strategy into manageable pieces helps to define the context in which we are operating.
Enterprise transformation assessment	We must have a general understanding of the baseline architecture as well as the readiness of the organization for change.
Organization map	An understanding of responsibilities and ownership across the organization is important.
Solution architecture diagram(s)	Application architecture design should be understood in terms of target applications.
Architecture roadmap	Planning the evolution of our architecture should occur from a business outcome as well as an application architecture perspective.

Table 8.2: Scoping of Architecture Engagement: Rationale for Each Artifact in Scope

For a more visual overview of the artifacts, refer back to the metro map in Figure 8.4. With this scope in mind, we'll ensure that you get the best experience by structuring the subsequent sections of this chapter as follows:

1. Introduction of the artifact as defined by SAP Enterprise Architecture Framework
2. General guidance and approach of how to model the artifact with SAP LeanIX, where relevant
3. Exemplary modeling of the artifact in the context of the EA7 case study
4. Sharing of best practices and insights based on project experience

8.4.1 Requirements Management

The business requirements of EA7 in context of the planned ERP transformation program have been defined as follows:

- Define a unified, enterprise-wide architecture vision that reflects EA7's strategic direction, including direct-to-consumer and product-as-a-service capabilities.
- Harmonize global business processes across regions to support scalable operations and a consistent customer experience.
- Enable the SAP S/4HANA global template program by providing a capability-based view of the future operating model.
- Support organizational alignment by establishing a shared understanding of business capabilities and outcomes across business and IT.
- Support clean core principles by distinguishing in-core standard processes from out-of-core innovations delivered via SAP Business Technology Platform (SAP BTP).

Over the next sections, we'll address these requirements via a structured approach by applying SAP Enterprise Architecture Framework artifacts in scope.

8.4.2 Architecture Principles

The purpose of an *architecture principle* is to act as a guardrail for enterprise design and transformation. Principles set the fundamental rules and boundaries that must be considered when developing an architecture. They define the constraints and non-negotiable factors that guide decision-making, ensuring architecture remains aligned with strategic objectives while adaptable to change.

The benefits of embracing and governing architecture principles can be summarized as follows:

- **Ensure consistency and coherence**
  Architecture principles enforce a common direction across initiatives, ensuring decisions remain aligned with enterprise goals. They reduce conflicting approaches, foster uniformity across projects, and create a coherent enterprise-wide architecture.

- **Enable clear communication**
  By providing a shared and recognizable language, principles make architectural intent explicit. They help bridge the gap between business and IT stakeholders, reducing misunderstandings and supporting effective collaboration across teams.
- **Guide strategic decision-making**
  Principles serve as a reference point for evaluating alternatives, weighing trade-offs, and prioritizing investments. They help ensure that choices remain aligned with both business objectives and technical standards, supporting sustainable growth.
- **Accelerate transformation and innovation**
  By clarifying which aspects are fixed and which are flexible, principles enable teams to innovate within safe boundaries. This balance fosters agility, speeds up decision making, and allows organizations to embrace change while protecting enterprise integrity.
- **Strengthen governance and accountability**
  Architecture principles establish traceability and the rationales behind architectural choices. They increase accountability, provide a foundation for governance, and ensure compliance with agreed-upon standards, building trust and transparency among stakeholders.

SAP Enterprise Architecture Framework proposes the following three-step approach for the management of architecture principles:

1. **Identify principles**
   In the given organization and architectural context, a set of architecture principles is defined. We recommend defining between 10 and 12 architecture principles, with a recognizable *name* and clear *statement* for each to keep them manageable.
2. **Understand benefits**
   The *rationale* for why you're defining a specific architecture principle is of crucial importance for later communication and needs to reflect the business or IT strategy.
3. **Identify impact**
   The impact of each architecture principle can be translated into an *implication* that states what potentially needs to be stopped or needs to be done in a more precise manner.

As we're focusing on SAP Enterprise Architecture Framework in this chapter and on modeling of individual artifacts with SAP LeanIX, we recommend referring to Chapter 12, Section 12.3 for more detailed information about the governance of architecture principles.

Our EA7 case study highlights the importance of a *clean core*, which is centered on five dimensions: processes, extensibility, data, integration, and operations.

Table 8.3 defines the five clean core principles by name, with a statement, rationale, and implications for each. Feel free to use these principles as templates for your next clean core–driven project.

Principle #1	Business Process Standardization
Statement	Design, implement, and manage business processes that align with SAP standards to reduce complexity, foster innovation, and enable scalable transformation. Process deviations are limited to clearly differentiating needs with measurable value.
Rationale	▪ **Business agility** Fit-to-standard processes allow faster adaptation to change and easier integration of innovations. ▪ **Reduced TCO** Lower effort in customization, upgrades, and long-term maintenance. ▪ **Accelerated innovation** Easier adoption of SAP-delivered updates and innovations. ▪ **Efficiency** Leverage modular, end-to-end, and integrated SAP best-practice processes.
Implication	▪ Adopt a fit-to-standard approach for future process design using SAP best practices. ▪ Classify business processes into standard, differentiating, and innovating groups, based on business value and capabilities. ▪ Minimize customizations by aligning to SAP reference architecture and limiting deviations to well-justified business needs. ▪ Run process diagnostics and fit-gap analysis to identify standardization opportunities. ▪ Govern business requirements centrally, using a solution standardization board and clean core KPIs. ▪ Link process changes to business capabilities and maintain traceability of deviations and their impacts.
Principle #2	**Extensibility**
Statement	Design and implement extensions that are decoupled from the digital core, leveraging released APIs and the SAP extensibility model. Choose the cleanest available option, either on-stack or side-by-side, based on clear business value and architectural fit.
Rationale	▪ **Upgrade stability** Decoupled extensions ensure SAP updates won't break custom code and vice versa. ▪ **Reduced technical debt** Clean extensions are transparent, traceable, and easier to manage.

Table 8.3: Clean Core Principles

Rationale (Cont.)	■ **Flexibility and scalability** Supports a full spectrum from low-code (SAP Build) to pro-code (ABAP Cloud, SAP BTP). ■ **Governed differentiation** Extensions only where necessary, and where they deliver competitive advantage.
Implication	■ Use released APIs by default to ensure upgrade-safe extensions and reduce future maintenance overhead. ■ Apply the full SAP extensibility model (on-stack with ABAP Cloud or side-by-side with SAP BTP), choosing the cleanest option. ■ Promote clean development practices through design guidance, developer upskilling, and enforcement of clean core-compatible tooling. ■ Manage technical debt transparently, documenting deviations, setting thresholds, and reviewing them regularly. ■ Measure and improve using KPIs like clean core share, technical debt score, and adherence to tier 1 extension practices.
Principle #3	**Data Integrity and Ownership**
Statement	Ensure data quality, accountability, and lifecycle integrity across the SAP landscape through clearly defined ownership, governance, and compliance structures. Treat data as a strategic asset to enable trusted, clean, and scalable enterprise operations.
Rationale	■ **Trusted decision-making** Reliable, consistent data is the foundation of every insight-driven business decision. ■ **Compliance and security** Proper handling of sensitive data safeguards against breaches and regulatory risks. ■ **Operational efficiency** Clean data reduces effort, increases automation, and simplifies transformation. ■ **Upgrade readiness** Clear ownership and harmonized models improve upgrade agility and reduce rework.
Implication	■ Protection of sensitive data with encryption, access control, and response protocols that align with standards like General Data Protection Regulation (GDPR). ■ Quality controls and cleansing routines using SAP tools and defining standards for harmonization across systems.

Table 8.3: Clean Core Principles (Cont.)

Implication (Cont.)	■ Data volume management via retention policies, archiving, and tiered storage to ensure performance, cost control, and legal compliance. ■ Operational monitoring and automation through SAP tools and continuous data validation workflows. ■ Data ownership and roles through a data strategy that aligns business and IT, supported by skill-building and stewardship. ■ Data governance framework including policies, technical standards, a data board, and lifecycle tools (e.g., catalog, audits).
Principle #4	**Integration**
Statement	Integrations must be designed to be modular, standardized, and decoupled from the ERP core, using modern API-based approaches and central governance frameworks.
Rationale	■ **Future-ready architecture** Decoupled integrations increase agility, enabling faster adoption of new systems and process innovation. ■ **Operational efficiency** Reusable, standardized interfaces reduce maintenance effort and lower TCO. ■ **Transparency and control** Centralized visibility into integration flows improves governance, impact analysis, and compliance. ■ **Sustainable modernization** API-first and platform-based integration approaches reduce legacy complexity and support clean core alignment.
Implication	■ Use standard APIs wherever possible to maintain upgrade compatibility and reduce the need for custom code. ■ Avoid point-to-point connections by adopting a central integration platform to orchestrate and manage all integration flows. ■ Adopt the side-by-side extension model to run custom logic and data transformations outside the ERP core, ensuring upgrade safety and flexibility. ■ Document integration dependencies in your architecture repository to ensure transparency, impact analysis, and better governance. ■ Design for change by building flexible interfaces that can adapt to evolving systems, data formats, and business processes. ■ Ensure monitoring and observability through real-time tracking, error handling, and diagnostics to maintain integration reliability and performance.

Table 8.3: Clean Core Principles (Cont.)

Principle #5	Operations and Lifecycle Management
Statement	Design and operate systems with lifecycle agility and operational transparency in mind. Align architecture, operations, and compliance to ensure reliability, observability, and adaptability across the entire solution lifecycle.
Rationale	▪ **Sustainable agility** Embedded operational readiness enables fast, stable delivery cycles. ▪ **Lifecycle control** Governance across design, deployment, and decommissioning supports resilience and clean core alignment. ▪ **Observability** Transparent system monitoring supports proactive operations and reduces downtime. ▪ **Compliance and auditability** Structured operational processes ensure secure, compliant enterprise landscapes.
Implication	▪ Operational design is embedded into architecture design processes, covering monitoring, logging, and compliance controls. ▪ Lifecycle policies are defined for extensions, interfaces, and applications, including archiving, retirement, and refactoring. ▪ Observability and transparency are gained through end-to-end monitoring and alerting across hybrid and cloud environments. ▪ DevOps practices are employed, using continuous integration/continuous delivery (CI/CD) pipelines, transport governance, and automated testing. ▪ Lifecycle tools allow for tracking, managing, and optimizing across solution lifecycles. ▪ Defined operational responsibilities offer clear ownership models, handover processes, and SLO/KPI tracking frameworks.

Table 8.3: Clean Core Principles (Cont.)

For the detailed modeling of architecture principles, see Chapter 12, Section 12.3. In addition, the five clean core principles are modeled in the demo workspace (*https://app-7.leanix.net/LeanIXPracticalGuideDemoWorkspace*): Go to **More** and select **Architecture Decisions**.

8.4.3 Business Strategy Map

The business strategy map is a core artifact within the strategy and motivation domain with the purpose of formally structuring and breaking down a company's strategy. A given strategy will be structured by *strategic priorities (drivers)*, *goals*, and *value drivers (objectives)*.

Note

The terminology used is a combination of terms from the reference architecture as well as the TOGAF terms driver, goals, and objectives.

The business strategy map is of crucial importance to define the business context in which the architecture engagement takes place. Without a proper business context, you run the risk of investing in changes and transformations not in alignment with the overall business strategy.

The following three-step approach is recommended by SAP Enterprise Architecture Framework to define a sound business strategy map:

1. **Understand the strategy**
   With the help of a business strategy assessment, a core enterprise architecture technique, you can achieve a holistic understanding of external and internal drivers, plus identification of areas of transformation and opportunities and risks. These are translated into strategic priorities (drivers), which fuel and motivate change in the organization.
2. **Understand the goals**
   In a second step, one or more goals for each of the strategic priorities (drivers) are identified. A goal is defined by an end state the organization wants to achieve and is therefore specific to the organization.
3. **Understand the value drivers**
   Each goal is further broken down into one or more value drivers (objectives): precise and actionable targets guiding transformations and decisions in each business context. Value drivers (objectives) ideally follow the SMART principle; that is, they are defined in a specific, measurable, achievable, relevant, and time-bound way.

With the definition of the strategic priorities (drivers), goals, and value drivers (objectives), the business strategy map is defined, and a formal stakeholder review should be conducted to get the artifact approved.

The real value of the business strategy map kicks in when you link initiatives in the sense of *containers* for change and business capabilities against the value drivers (objectives). The business strategy map acts as the business context in which you're operating and informs the scope and extent of change for an initiative or informs the required level of maturity for a set of business capabilities in each context.

The creation of the business strategy map requires some thorough work. Items such as publicly available annual and financial reports, internal strategy narratives, and interviews with business and IT stakeholders can help to gather the required information.

> **Note**
>
> The expectation for an enterprise architect is not to define the company strategy but to formally structure the information that is often scattered across multiple documents or even in the heads of key stakeholders.

Figure 8.6 outlines the template for the business strategy map. For the sake of completeness, a vision statement was added, which is often used by organizations. Furthermore, the relations to subsequent architecture information such as business capabilities or initiatives are outlined. Keep in mind that this artifact will be created over potentially multiple iterations and therefore complemented over time.

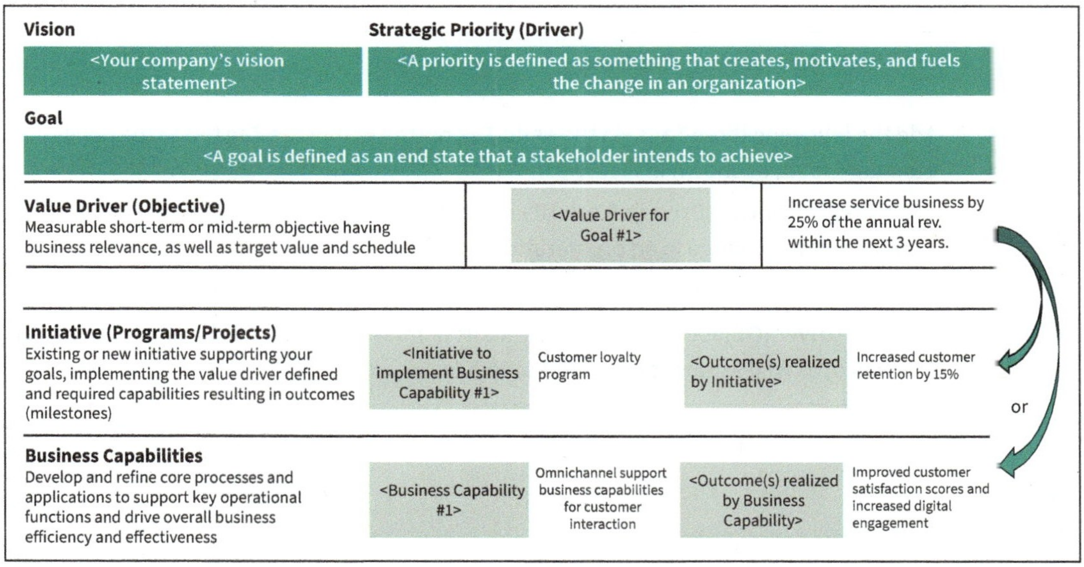

Figure 8.6: Business Strategy Map: Template

For modeling the business strategy map in SAP LeanIX, we'll use the objective fact sheet, as outlined in the SAP LeanIX meta model (see Chapter 2, Section 2.2). Although the objectives fact sheet in the standard meta model V4 does not contain any subtypes, we recommend applying a hierarchical (parent-child) modeling approach in combination with tagging fact sheets in terms of strategic priorities (drivers), goals, and value drivers (objectives).

Follow these steps to create a new tag group for modeling the business strategy map:

1. Open the SAP LeanIX **Administration** menu and select the **Tagging** section in the left menu pane.
2. Go to the **Tag Group** tab, as shown in Figure 8.7, and create a new single-mode tag group. In the **Name** field, enter "Business Strategy," and in the **Description** field, state that this is relevant for the business strategy map. Choose **Objective** as the **Fact Sheet Type**.

Figure 8.7: Business Strategy Tag Group

3. Add the following three **Tags** via the **+Add Tag** button within the **Tag** tab to the newly created tag group, as shown in Figure 8.8:
   - **Strategic Priority (Driver)**: A priority is defined as something that creates, motivates, and fuels the change in an organization
   - **Goal**: A goal is defined as an end state that a stakeholder intends to achieve
   - **Value Driver (Objective)**: Measurable short or mid-term objective having business relevance, as well as target value and schedule

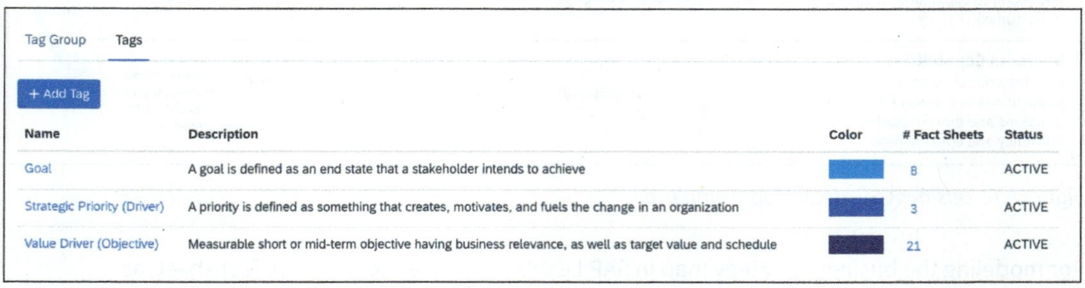

Figure 8.8: Business Strategy Tags

Once the tag group is created, you can start to create individual objectives fact sheets reflecting the company's business strategy by setting the fact sheets up with their respective parent-child relations and by tagging individual fact sheets with the newly created tags.

A well-defined business strategy map enables the opportunity to create relations to initiatives and business capabilities in scope. This will help to craft a consistent storyline from business strategy to implementation roadmap.

Note

Before creating the reports that we'll discuss in this section, you need a dedicated collection structure. For the EA7 case study, we'll create an SAP Enterprise Architecture Framework collection group in which you can later find each report on our demo workspace. For instructions on creating collections, see Chapter 5, Section 5.1.2.

For the creation of the **EA7 Business Strategy Map** report, you need to do the following:

1. Create a new landscape report with the objective as the base fact sheet type.
2. Select the **Business Strategy** tag for **View**.

Strategic priorities (drivers), goals, and value drivers (objectives) will now be visualized in a hierarchical structure, indicated by the color-coding defined from the tag definition.

The business strategy map of the EA7 case study looks as shown in Figure 8.9. For the sake of this chapter, we've filtered to **Operational Resilience & Efficiency** as **Strategic Priority (Driver)**.

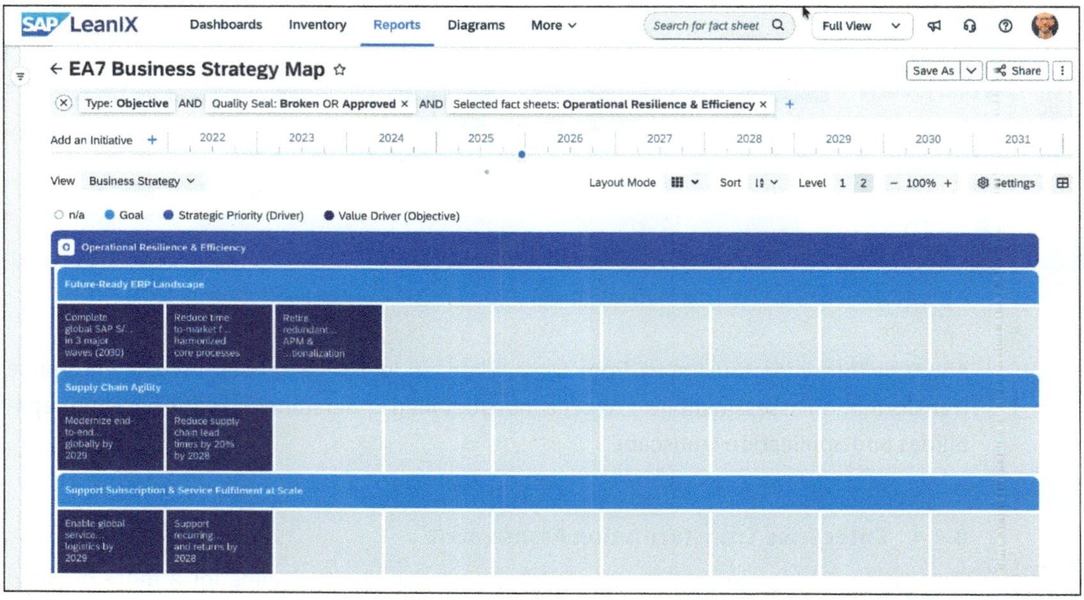

Figure 8.9: Business Strategy Map: EA7 Case Study

The following objectives are of primary importance for the EA7 transformation program:

- Complete global SAP S/4HANA clean core rollout in three major waves (2030).
- Reduce time-to-market for new business initiatives by 40% through harmonized core processes.
- Achieve 90%+ data quality in customer, product, and supplier master data by 2029.
- Modernize end-to-end planning processes globally by 2029.

By focusing on the first objective, you can create a "container" in an early stage for the changes to come. Based on what you learned in Chapter 6, you're creating an EA7 ERP transformation initiative as a program with multiple subinitiatives as projects, starting with the definition of a global template and followed by three rollout waves. The program and projects include their respective milestones and lifecycle phases, inspired by a first, high-level planning perspective.

To create the **EA7 ERP Transformation Initiative Roadmap** report, as shown in Figure 8.10, you need to do the following:

1. Create a new roadmap report with initiative as the base fact sheet type.
2. Select the **EA7 ERP Transformation Initiative** as parent filter.

For more details and insights on initiatives, milestones, and roadmaps, refer to Chapter 6.

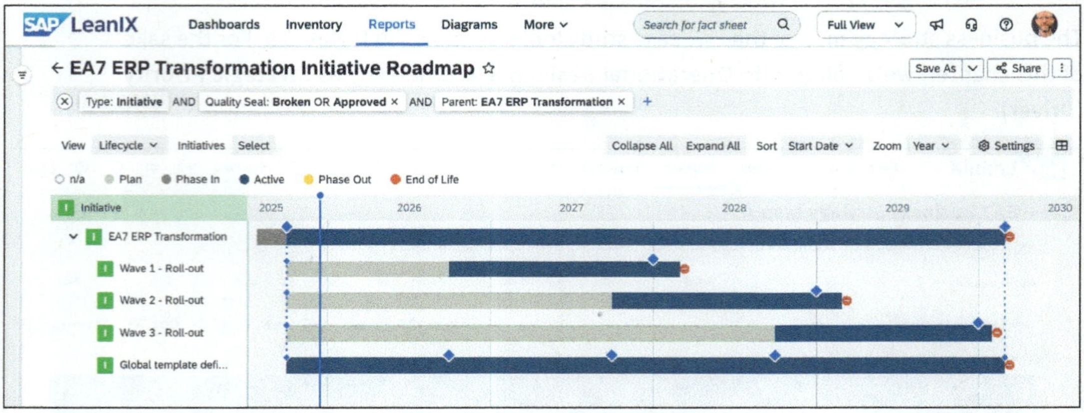

Figure 8.10: EA7 ERP Transformation Initiative Roadmap

Any transformation must start from somewhere, so let's dive deeper into the enterprise transformation assessment next to get a more detailed understanding of the baseline capabilities and application landscape.

8.4.4 Enterprise Transformation Assessment

The enterprise transformation assessment is a technique that aims for a more detailed understanding of the readiness of an organization for the change to come. It entails the following three activities:

- **Enterprise architecture practice capability assessment**
  A maturity level-based assessment to gain insight into the readiness and potential shortages, such as in skills, governance process, and the like, of an enterprise architecture practice itself

- **Preliminary business and technology capability assessment**
  A high-level assessment of the baseline business and technology capabilities to enable a better understanding of when the transformation will start and which major gaps might arise
- **Business and technology transformation readiness assessment**
  A high-level assessment of the readiness of the organization with regards to the foreseen change, focused highly on the *people* aspects, such as mindset, skills, talent, accountability, and so on

The focus of the EA7 case study is on the preliminary business and technology assessment, which is subdivided into two more detailed assessments: the baseline capability assessment and the application portfolio assessment.

Business Capability Assessment

The purpose of the *baseline capability assessment* is to gain a first understanding of the impact the strategic change will have on the business and technology capabilities a company operates. A high-level reflection of the current capabilities is taken into consideration, centered on the strategic context previously defined by the business strategy map. With the evaluation of the capabilities in scope based on well-defined criteria, the strategic change impact can be identified and communicated. Furthermore, potential risks can be highlighted in an early stage of a project or transformation program and potential mitigation activities initiated.

> **Note**
> The baseline capability assessment is a preliminary assessment to gain first insights. It does not replace the more detailed business capability assessment within the business architecture domain. We recommend focusing on level 2 during the preliminary assessment and further detail out the business capability map as described in Section 8.4.6.

The preliminary business capability assessment leverages the maturity and strategic importance view and is based on the attribute values of the business capability fact sheet, which can be found in the maturity section in Table 8.4.

Strategic Importance	■ Commodity: Maintains commodity and operational needs, but does not provide any competitive advantage ■ Differentiation: Is important to achieve excellence in order to compete with and differentiate from others ■ Innovation: Is critical to competitive advantage

Table 8.4: Strategic Importance and Maturity

Current/Target Maturity	▪ Level 1 (ad hoc): Processes are largely manual and inconsistent. Success often depends on individual heroics. ▪ Level 2 (repeatable): Basic and consistent processes are established and can be repeated for similar projects. ▪ Level 3 (defined): Processes are automated and integrated across all processes for the organization. ▪ Level 4 (managed): Processes are automated and measured for success.

Table 8.4: Strategic Importance and Maturity (Cont.)

The context for the high-level assessment builds the previously defined business strategy map. This context will indicate the strategic importance and maturity of each of the business capabilities for EA7 (see Chapter 4 for the importance of business capabilities).

Figure 8.11 outlines the preliminary business capability assessment for the EA7 case study, classifying the business capabilities in level 1 and 2 into commodity, differentiation, or innovation business capabilities from a strategic importance perspective. The current maturity level is indicated by the left property while the target maturity is outlined by the right property, both indicating the maturity level through the number of circles indicating the maturity scale (see Section 8.4.6).

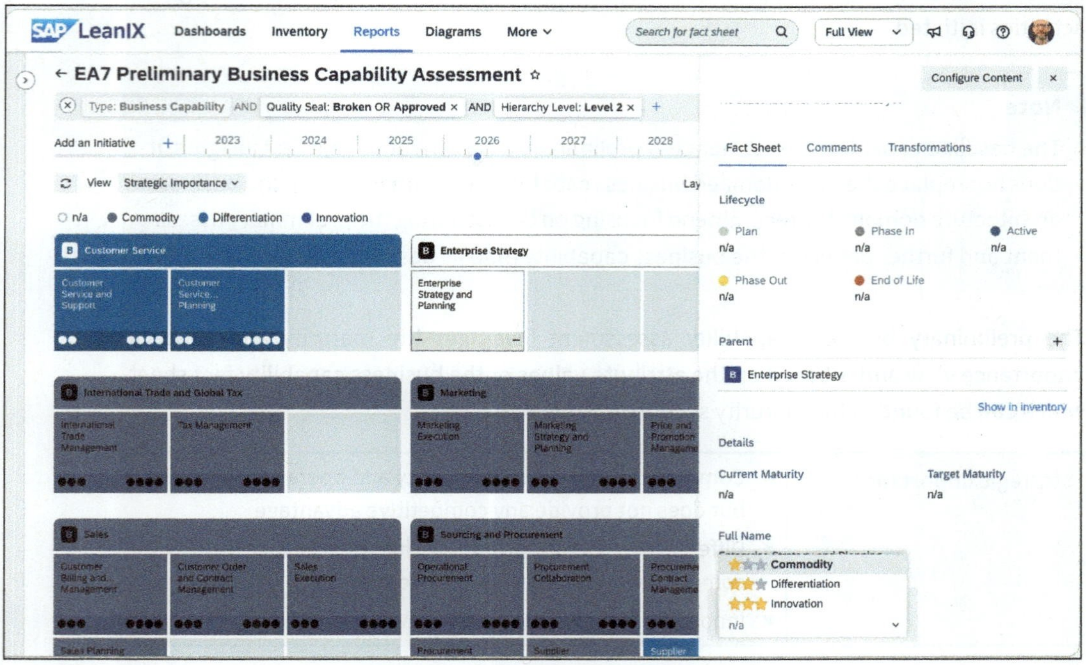

Figure 8.11: Preliminary Business Capability Assessment: EA7 Case Study

For the creation of the **EA7 Preliminary Business Capability Assessment** report shown in Figure 8.11, follow these steps:

1. Create a new landscape report with business capability as the base fact sheet type.
2. Set a hierarchy level filter on level 2 of the business capabilities.

For an interactive assessment of the strategic importance and maturity levels, we recommend using the side panel as shown in Figure 8.11.

An alternative approach for the assessment would be using the inventory view with the relevant fields in a tabular format, as outlined in Figure 8.12. You can create this **EA7 Preliminary Business Capability Assessment** view by doing the following:

1. Open the inventory and select **Business Capabilities.**
2. Show the business capabilities as a table.
3. Select **Strategic Importance, Current Maturity,** and **Target Maturity** as columns (eye symbol).
4. Turn to edit mode to maintain the individual values.

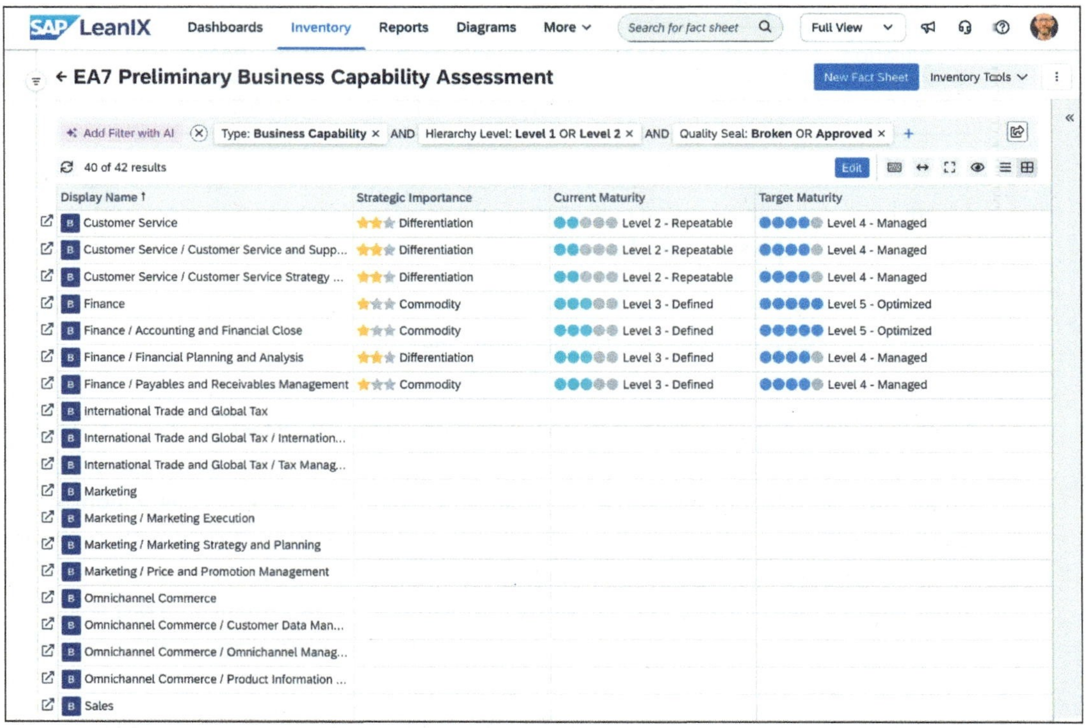

Figure 8.12: Preliminary Business Capability Assessment: Saved Search for EA7 Case Study

Application Portfolio Assessment

Similarly to the baseline capability assessment, the understanding of the baseline application landscape from the perspective of business and technical fit as well as a TCO perspective is of importance for any change or transformation to come. SAP Enterprise Architecture Framework leverages here the concepts of the TIME and TCO assessments, as introduced in Chapter 7, and reuses these within the application portfolio assessment.

The first step of the application portfolio assessment is building an application inventory. SAP LeanIX offers the *SAP landscape discovery* feature to automatically discover your installed or provisioned applications. This ranges from on-premise to private cloud to SaaS applications. The discovered applications will be made visible in the *discovery inbox*, in which you have the option to choose the applications that you want to import into your inventory by either linking a discovered application to an existing application fact sheet or by creating a new one. SAP landscape discovery identifies additional related fact sheets, as outlined in Table 8.5. For more information on these discovery capabilities, see Chapter 4, Section 4.2.2 (discovery inbox) and Section 4.2.3 (landscape discovery).

Fact Sheet	Description
Application	SaaS, private cloud, or on-premise business applications
IT component	The IT component the application is based on, such as SAP S/4HANA 2025
Provider	Provider of the IT component
System	An extension to the SAP LeanIX V4 meta model; can be used to capture system-specific information such as installation number, system ID, or system type (DEV, TST, PRD, etc.)

Table 8.5: Fact Sheet Types Provided by SAP Landscape Discovery

Once the system landscape information is retrieved, all identified fact sheets will be visible in the discovery inbox. The discovery inbox can be accessed in the SAP landscape discovery configuration itself or by navigating to **Inventory • Inventory Tools • Discovery Inbox • SAP Landscape Discovery**. Figure 8.13 provides an example of the discovery inbox. Through various filters, such as classification, product, system role, customer, free text, or a self-defined filter, you can focus on a subset of information. Figure 8.13 filters on the free test **IP** as well as **On-Prem ERP** as a classification, with the result of listing all IPx discovery items, such as IPP, IPD, and so on.

Each discovered item consists of multiple fact sheet types, which are visualized in the **Fact Sheet Link** column. Figure 8.14 provides an overview of the fact sheets that have been discovered for the IPP discovery item. The following logic applies:

1. **SAP ERP IPP** is the leading application fact sheet and represents the business application.

2. Two IT components are identified: one for the SAP HANA database version and a second for the enhancement package version of the ERP application. Based on these IT components, potential risks from a lifecycle perspective can be identified moving forward.
3. SAP as a provider is identified.
4. Two system fact sheets are identified: one productive system for SAP ERP and one productive system for the SAP HANA database; both include their system number.

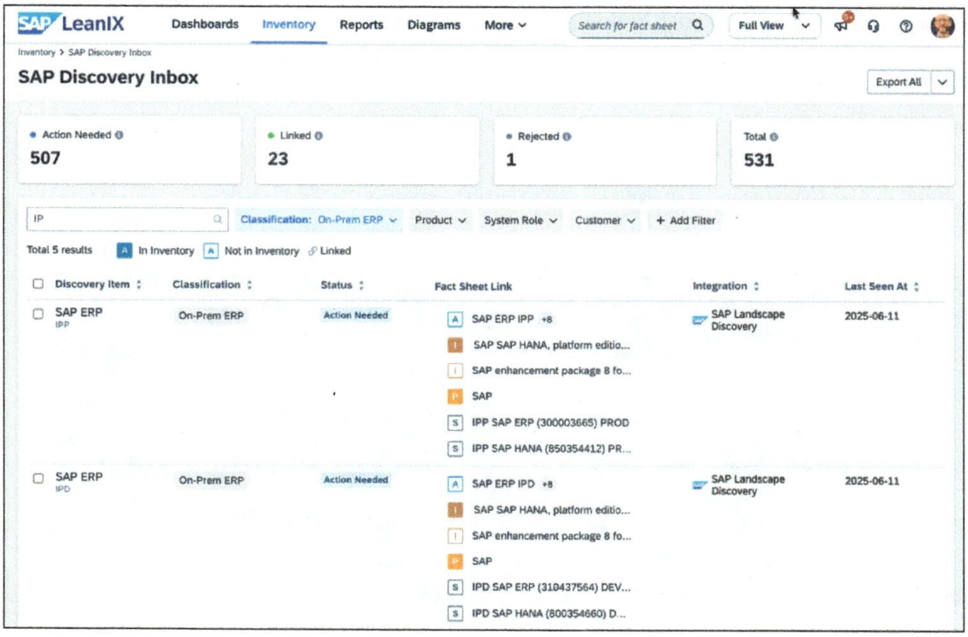

Figure 8.13: SAP Landscape Discovery: Discovery Inbox

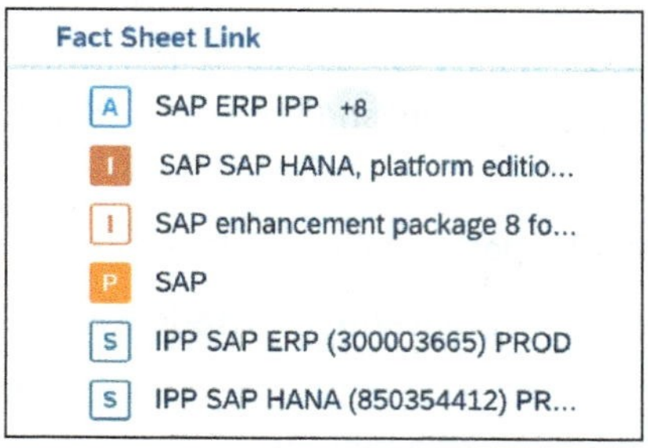

Figure 8.14: SAP Landscape Discovery: Fact Sheet Types

An alternative to the SAP landscape discovery approach is the import of the baseline application architecture with the help of the inventory builder, as explained in Chapter 11.

With the automatically discovered, imported, or manually created application fact sheets, you can build a first overview of the application landscape in context of the architecture engagement, as shown in Figure 8.15 in the context of the EA7 engagement. A review with application owners will help to identify any missing applications.

You can create the **EA7 Application Landscape** report by following these steps:

1. Create a new landscape report with the application fact sheet type as the base fact sheet.
2. In the settings, choose **Lifecycle** for **Cluster by**.
3. Optionally set the **Lifecycle** filter to **Active** (to only see active applications).
4. Make sure that the lifecycle information is maintained.

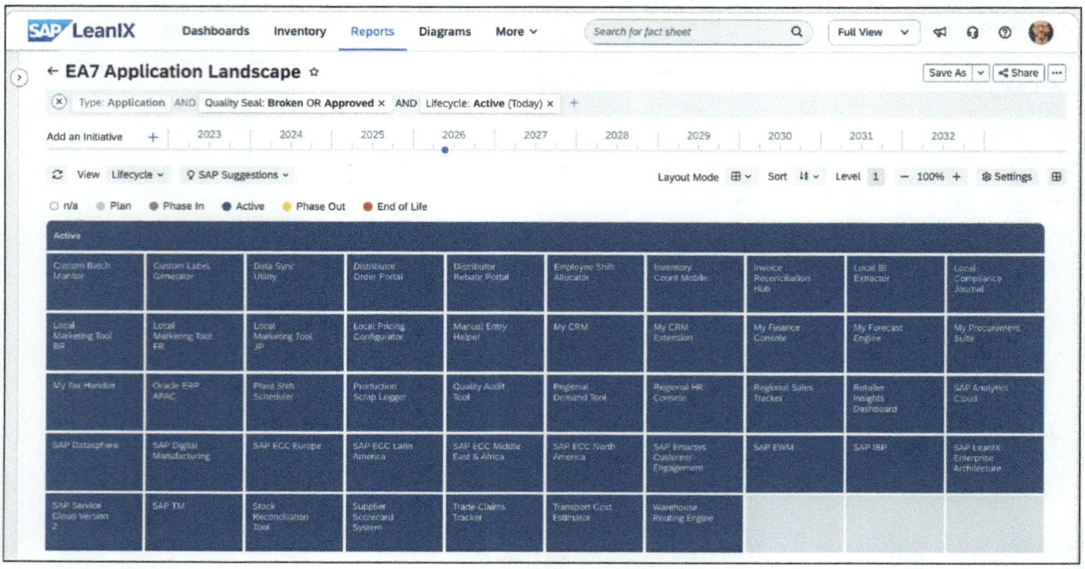

Figure 8.15: EA7 Application Landscape

Once you have a good understanding of the baseline application portfolio, you can combine the application view with the preliminary business capability assessment to reflect the strategic importance of the business capabilities in the application landscape, as outlined in Figure 8.16.

You can create the **EA7 Application Portfolio Assessment** report by following these steps:

1. Save the **EA7 Application Landscape** report as the **EA7 Application Portfolio Assessment** report.
2. In the **Settings**, change **Cluster by** from **Lifecycle** to **Business Capability**.
3. Change the report **View** from **Lifecycle** to **Business Capability: Strategic Importance.**
4. Set the report level to **2**.

With the overlay of the strategic importance rating from the preliminary business capability assessment onto individual applications, you can gain some first insights into the strategic requirements for the applications in your current portfolio. This is a valuable insight for target planning.

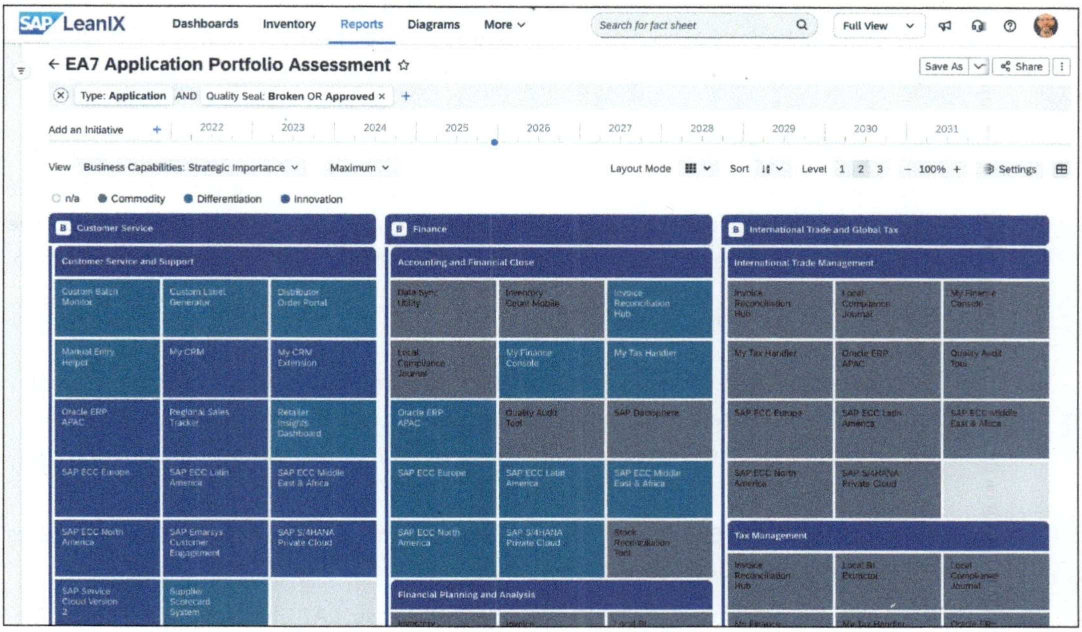

Figure 8.16: EA7 Application Portfolio Assessment

8.4.5 Organization Map

The purpose of the organization map is to model the existing and/or future organization of a company by applying a hierarchical structure to outline individual business units, divisions, geographies, countries, and so on. Its nature and level of detail will be defined by the scope of the architecture engagement. SAP Enterprise Architecture Framework proposes the following three-step approach:

1. **Identify the context**
   Identify the relevant organization elements in context of the enterprise architecture engagement.
2. **Identify relationships**
   Identify relationships between the different organizational elements.
3. **Visualize the organization map**
   Visualize the organization map as a hierarchical structure, network, or web diagram and outline relevant relationships.

Tip

From an SAP LeanIX perspective, we recommend structuring the organization alongside the ownership of, for example, applications or initiatives as well as the planned transformations. If you plan an SAP S/4HANA greenfield implementation with a phased rollout across multiple countries, for example, countries should be the lowest level in your organization map as this can be used for detailed rollout planning during the architecture roadmap definition (see Section 8.4.8).

Within SAP LeanIX, the modeling of the organization map is centered around the organization fact sheet. The respective level of an organization element is identified through the parent-child relationship, similar to the organization chart used in various companies. The organization chart serves also as a valid source to start the creation of the organization map.

Tip

We recommend structuring the organization map around the entities such as region, countries, and business units, which will be subject to the changes to come. For the rollout of a global platform, for example, the lowest hierarchy level would be a country.

For the creation of the **EA7 Organization Map** report, the required organization fact sheets need to be created. Follow these steps:

1. Start with the creation of the highest entity, in this case, **EA7 Global**.
2. For the respective regions and countries, create individual organization fact sheets and relate these in parent-child relationships.
3. Maintain the lifecycle information of the individual fact sheets.
4. Create a new landscape report with the organization as the base fact sheet type to visualize the EA7 organization structure, as shown in Figure 8.17.

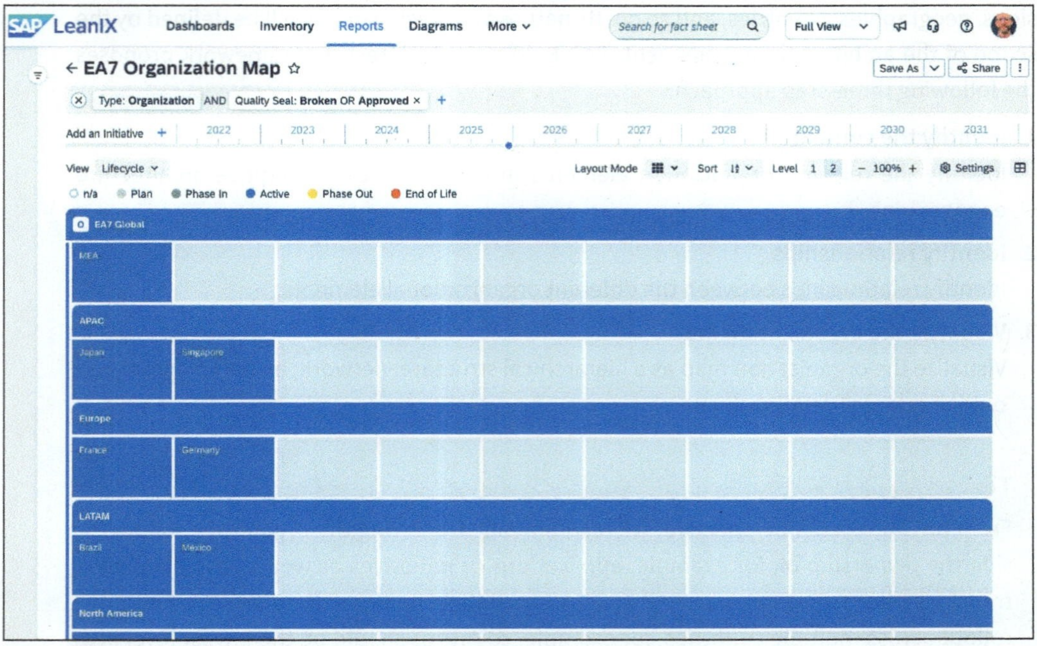

Figure 8.17: EA7 Organization Map

With the organization map, you get a detailed insight into the *who* of an organization. The next artifact, the business capability map, will focus on the *what*.

8.4.6 Business Capability Map

As introduced in Chapter 4, Section 4.2.1, business capabilities describe what an organization does, not how or by whom. By focusing on these stable constructs, business capability maps do the following:

- Provide a shared language between business and IT
- Enable capability-based planning and portfolio analysis
- Support transformation planning by linking capabilities to applications, data, and initiatives

For the creation and application of the business capability map, SAP Enterprise Architecture Framework proposes the following three-step approach:

1. **Identify business capabilities**
   For the given business context defined with the help of the business strategy map, the relevant business capabilities are defined. This can either happen by revising an existing business capability map that was assessed during the enterprise transformation assessment or by augmenting the same with the help of reference business capability models. This could be particularly relevant if an extension to new business models takes place.
2. **Heat mapping of business capabilities**
   Once the business capability map is agreed upon and endorsed by the organization, the foundation is set and active usage can be initiated by applying heat mapping techniques. Business capabilities are assessed in a well-defined heat mapping context, such as current and target maturity level, strategic importance, or other relevant contexts.
3. **Visualization of business capabilities**
   Stakeholder engagement and communication are of crucial importance for any enterprise architecture engagement. The communication of the business capability map and the results of heat mapping exercises ensure that stakeholders stay engaged or endorse the insights retrieved.

SAP provides predefined, hierarchical business capability maps as part of SAP Reference Business Architecture. These maps serve as the foundational structure for modeling an enterprise's business architecture and can be consumed directly within SAP LeanIX. Each capability represents a stable, strategic function of the business, abstracted from the organizational structure, processes, or systems.

The reference business capability map is classified into *enterprise domains*, with a top-level view centered on the customer, supply, corporate, and products and services domains and organized on three levels:

- **Business domains (BD)**
  Level 1 grouping of business capabilities according to functional similarities
- **Business area (BA)**
  Level 2 grouping of business capabilities along functional similarities
- **Business capabilities (BC)**
  Define what the business does, constituting a particular ability or capacity that a business may possess or exchange to achieve a specific purpose or outcome

An important aspect of the SAP Reference Business Architecture meta model we want to highlight here is the relation between the business capability and the business activity, the lowest level of the business process model. This relationship provides insights into which business activities and thus processes are enabled by a specific business capability.

Furthermore, you can see that the business capability holds an "is realized by" relation to the solution capability of SAP Reference Solution Architecture. In Section 8.4.7, we'll demonstrate in detail how this relation can be used to derive a target application architecture suggestion.

Figure 8.18 provides an overview of the SAP reference architecture content meta model, with the business capability linking to business activities and solution capabilities.

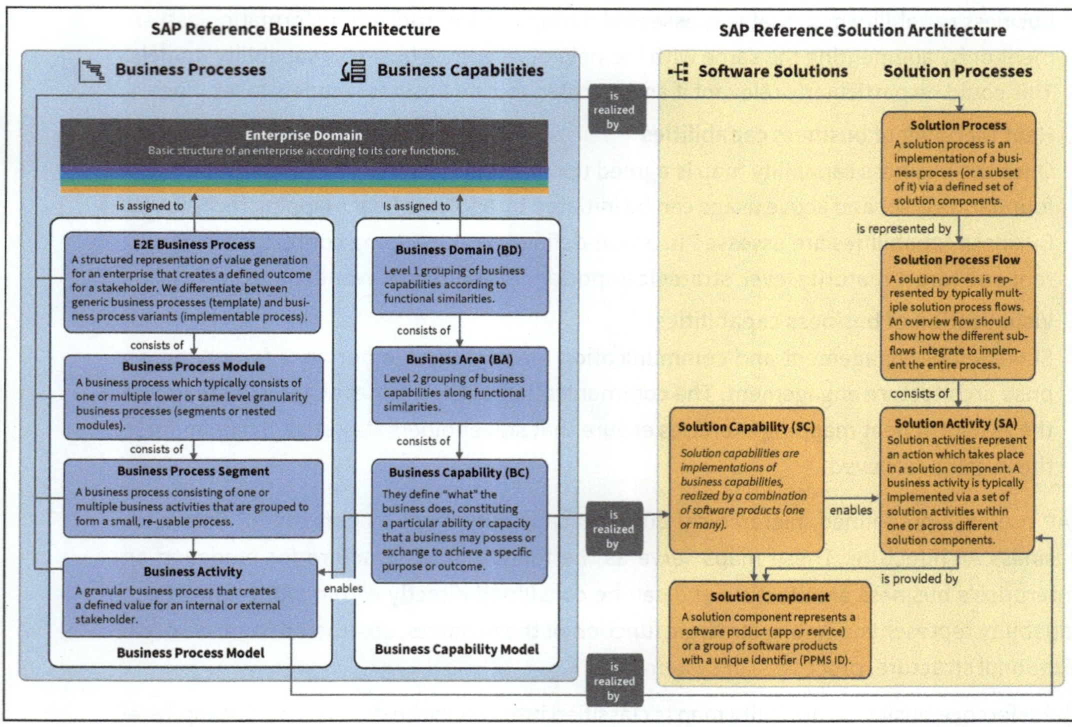

Figure 8.18: SAP Enterprise Architecture Framework, SAP Reference Architecture Content Meta Model

In SAP LeanIX, these capability maps can be leveraged and tailored to the enterprise's context, creating a robust, business-driven baseline for architecture work. SAP's reference business capabilities can be newly imported as business capability fact sheets, or existing fact sheets can be linked with reference business capabilities.

To access the reference business capabilities, navigate to **More** and select the **Explore Business Capability** option within the **Reference Catalog** section.

Figure 8.19 shows an import of reference business capabilities structured around the described three levels.

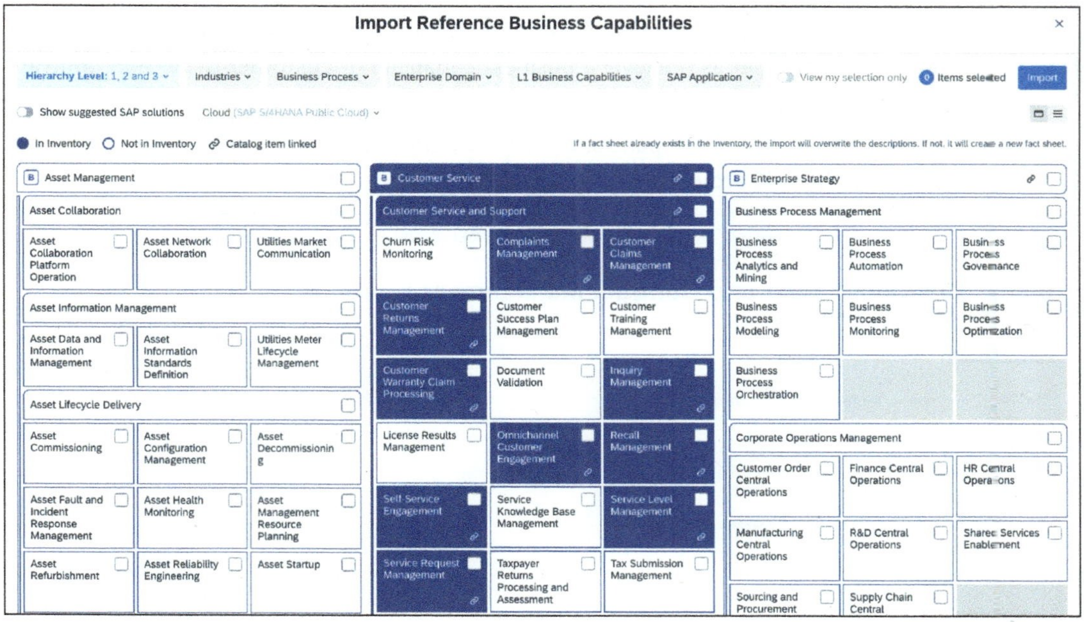

Figure 8.19: Import of Reference Business Capabilities

Tip

Reference business capabilities can be filtered by various filter options:

- Hierarchy levels
- Industries
- Business processes including business process variants
- Enterprise domains
- Level 1 business capabilities (business domains)
- SAP applications

Applying a filter for SAP applications will enable a "reverse engineering" approach to identify which business capabilities are supported through which applications.

Furthermore, a preview of suggested SAP solutions for either cloud or hybrid deployments can be enabled, helping you gain a better understanding of available solution options.

With the import of reference business capabilities, you have the possibility either to create a business capability map from scratch or to extend an existing business capability model. Often, there might be a business capability map that already exists, and you might want to leverage the reference architecture information provided by SAP. In this case, you can link existing business capabilities to the reference business capabilities at any level of the business capability model. To link an existing business capability fact sheet with a reference catalog entry, follow these steps:

1. Open the business capability fact sheet you want to link. In this example, **Subscription Management** is set as the level 3 business capability of **Customer Order** and **Contract Management** in the **Sales** domain.
2. Navigate to the right-side pane and click **Start Linking** within the **REFERENCE CATALOG & INTEGRATIONS** section, as shown in Figure 8.20.

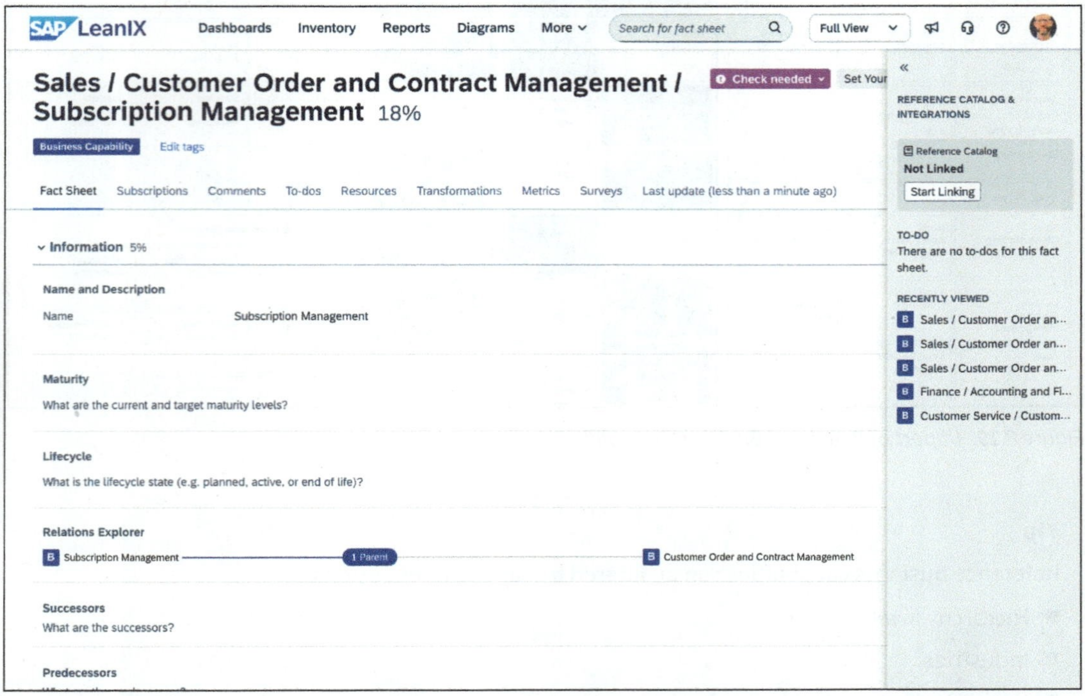

Figure 8.20: Linking of Reference Catalog

3. On the next screen, as shown in Figure 8.21, either choose a recommended business capability or enter a search term. Note that recommendations will only be provided if the name of the existing business capability can be found in the reference catalog. Prior to linking the respective business capability, you can review the details of the reference business capability, including the suggested SAP solutions. Figure 8.21 shows an example of linking the **Subscription Management** business capability. The respective business capability is recommended, and detailed information such as the description or relevance for industries can be reviewed next to the suggested SAP solutions.

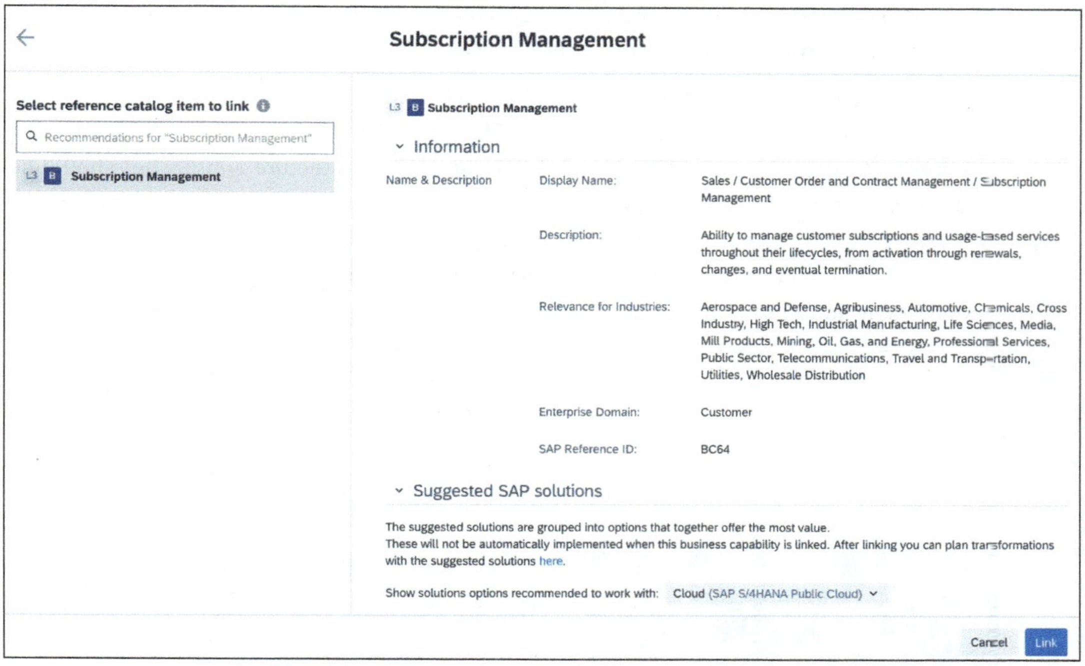

Figure 8.21: Selection of Reference Content Item

4. The existing business capability will be enriched with reference information and the suggested SAP solutions when you click the **Link** button.

> **Note**
>
> We recommend linking existing business capabilities to reference business capabilities on similar levels. If you link a business capability on level 2, the subsequent level 3 business capabilities can be imported either via the **Import of Reference Business Capabilities** function as described before or with the help of the transformations planning feature, which will be described in detail in the next section.

With the definition of a holistic business capability map reflecting all relevant aspects of an enterprise, you build a solid baseline for interactions with various stakeholder groups in your organization. The business capabilities build a common language for business and IT alignment and can be used for specific heat mapping exercises. As part of the preliminary business capability assessment, we concentrated here on defining the strategic importance and maturity in a level 2 business capability model. These heat maps will now be further detailed and expanded to level 3 by taking the information in the strategy map into account.

Figure 8.22 provides an overview of the EA7 business capability map with a detailed (level 3) maturity and strategic importance assessment. Create the **EA7 Business Capability Map** report by saving a variant of the previously created **EA7 Preliminary Business Capability Assessment** report with the filter for business capabilities at level 2 turned off.

With the help of this report, you can identify and highlight the major gaps between the current and target maturity level of each individual level 3 business capability. The strategic importance provides insight for decisions about what architecture approaches should be used for specific business capabilities. Business capabilities rated as commodities will leverage highly standardized processes and solutions, while business capabilities rated as innovations might be implemented in an agile environment subject to short learning cycles and timely adoption of changes.

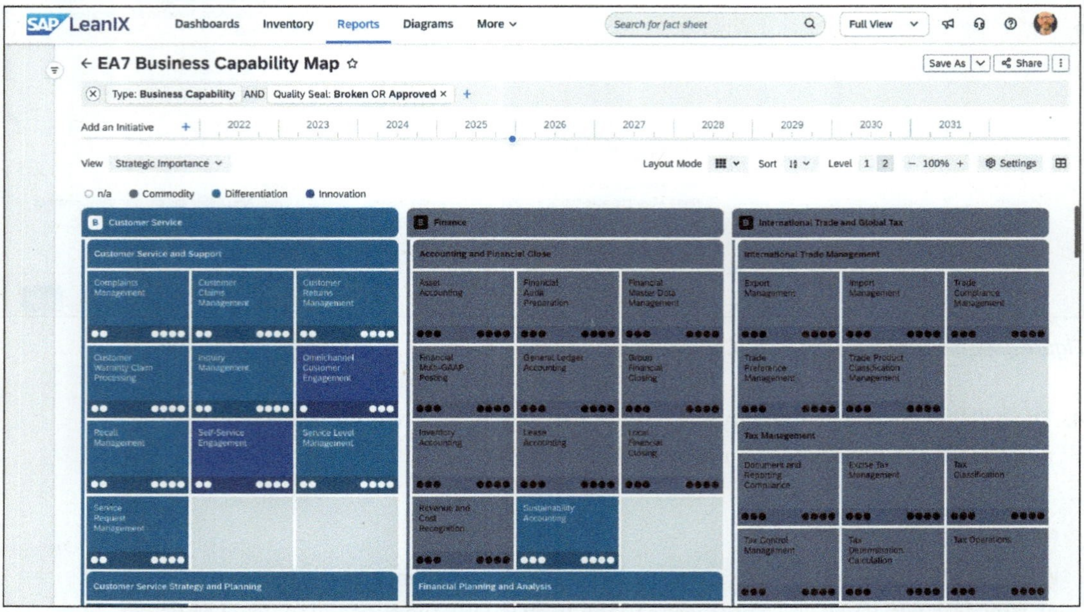

Figure 8.22: EA7 Business Capability Map

By integrating reference business capability maps into SAP LeanIX, organizations gain a strategic anchor for their enterprise architecture work and ensure that transformation initiatives are always rooted in what truly drives business value.

8.4.7 Application Architecture Overview Diagram

With the definition of the business capability map, you should have a solid understanding of the strategic relevance of business capabilities in your business context as well as insight into which areas you need to mature to address company objectives and goals. One core artifact of each and every architecture engagement is the target application architecture. The business capability map is owned by business stakeholders, with the enterprise architect acting as a catalyst to enable the alignment between business and IT. However, the enterprise architect is primarily accountable for the target architecture. A solid and sound target application architecture needs to be created in context of the given business strategy and requirements while taking the baseline application architecture, potential constraints, and readiness of the organization into account. Once the target architecture has been defined and endorsed by the key stakeholders, the documentation of decisions leading to the target architecture and the communication with the broader ecosystem is of crucial importance to align everybody with the future direction.

SAP Enterprise Architecture Framework foresees the application architecture overview diagram as one of the key artifacts for communication with the ecosystem. The purpose of the application architecture overview diagram is to sketch in an abstract way all relevant applications (solution components or solution building blocks) required for the proposed application architecture. Optionally, it also shows the relationships between the defined applications (solution components).

The following three-step approach for the creation of the application architecture overview diagram is recommended:

1. **Identify solution components**
   Define and select the relevant solution building blocks.
2. **Understand dependencies**
   Capture and analyze relationships between the solution components.
3. **Visualize the diagram**
   Present the components and dependencies in a diagram for better communication and understanding.

The identification of the appropriate applications (solution components) in the first step has been a major endeavor in the past. Although the business context was maybe clear with the help of the business capability and requirements, the potential target candidates were defined based on past experience, research, or even requests for proposals (RFPs). This is where the SAP reference architecture content as outlined in Figure 8.5 provides a solution recommendation.

SAP LeanIX hits the sweet spot of combining SAP Reference Business Architecture content with SAP Reference Solution Architecture suggestions to support the definition of the target architecture. The starting point for the target architecture planning and definition can be based on level 3 or level 2 of the business capability model. Both approaches will be demonstrated in this section.

As defined by the SAP reference content meta model (see Figure 8.18), business capabilities on level 3 are linked to solution capabilities and solution components. This information is integrated and leveraged within the business capability fact sheet under the **Transformations** tab. The business capability first must have been either imported from the reference catalog or linked to it as described in the previous business capability map section.

Figure 8.23 shows an example of the suggested SAP solutions for the general ledger accounting business capability, with one solution option proposed. Note that the solution options might differ in an SAP S/4HANA implementation context as a differentiation between the following two options takes place:

- Cloud (SAP S/4HANA Cloud Public Edition)
- Hybrid (SAP S/4HANA Cloud Private Edition or on-premise)

Each option defines mandatory and optional applications and indicates if an application is already in the inventory or not, as outlined in the legend of Figure 8.23.

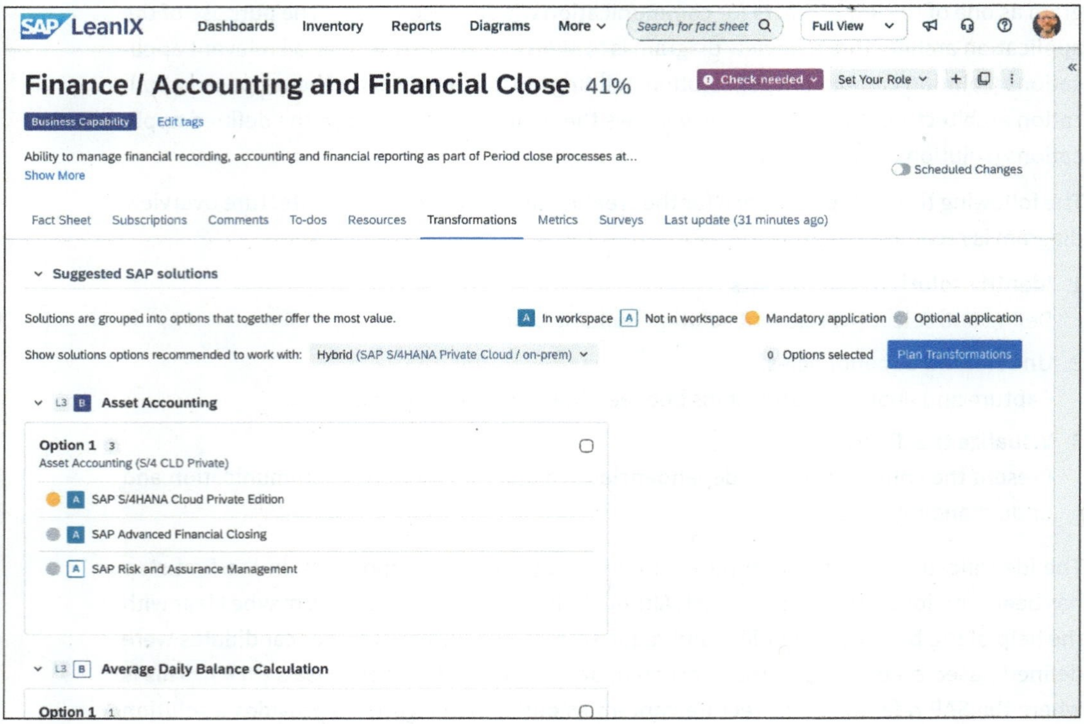

Figure 8.23: Suggested SAP Solutions

Note

Not all business applications of SAP Reference Business Architecture have an option assigned. This is because SAP Reference Business Architecture is solution-agnostic and broader than the SAP product portfolio.

If a business capability offers multiple options, we advise collaborating with subject-matter experts (SMEs) to see which option best addresses the functional and nonfunctional requirements.

Once you decide on a relevant option, you can start the transformation planning by clicking **Plan Transformations**. For further information on transformations, see Chapter 6, Section 6.3.2.

On the **Plan Transformations** screen, the relevant applications for the rollout in context of the given business capability can be selected, as shown in Figure 8.24. For this example, both **SAP Advanced Financial Closing** and **SAP S/4HANA Cloud Private Edition** are mandatory applications for the **General Ledger Accounting** business capability. By setting the cross-check on the right side, SAP Advanced Financial Close (an application that does not exist in our inventory) will be created.

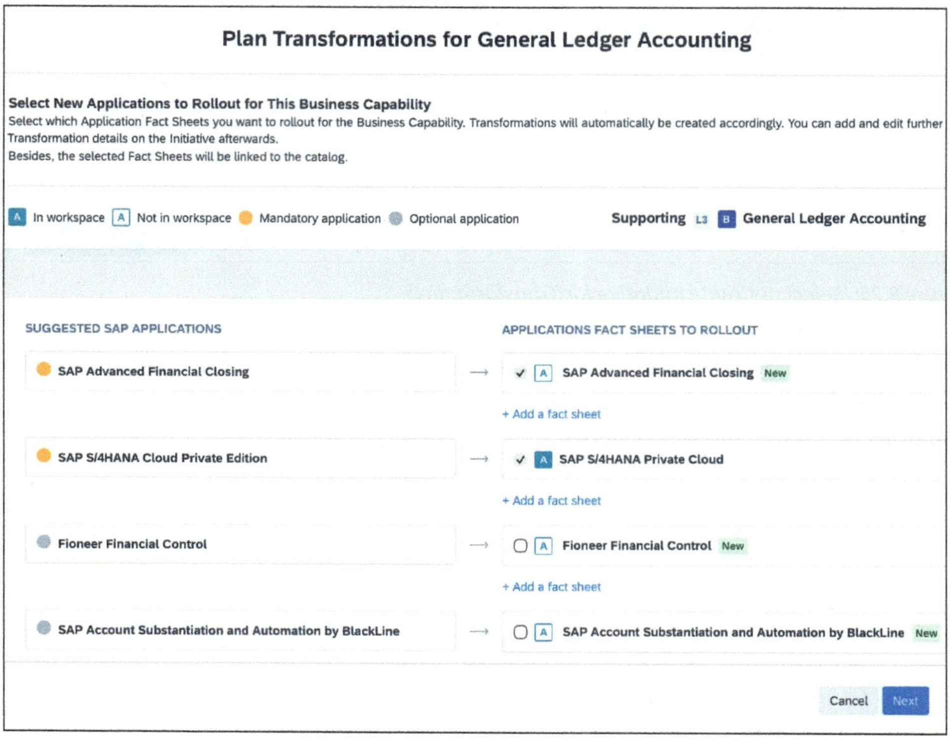

Figure 8.24: Plan Transformation: Select Relevant Applications

Once the relevant applications are selected, the following occurs:

- The planned transformation can be assigned to an existing initiative, or a new initiative can be created.
- A milestone or target date is assigned.
- The transformation can be created by choosing **Create Transformations**, as shown in Figure 8.25.

> **Note**
>
> We recommend using milestones for transformation planning as they can be visualized in roadmap or landscape reports. Following this recommendation requires the creation of initiatives with milestones prior to transformation planning. For more information on milestones, see Chapter 6, Section 6.3.1.

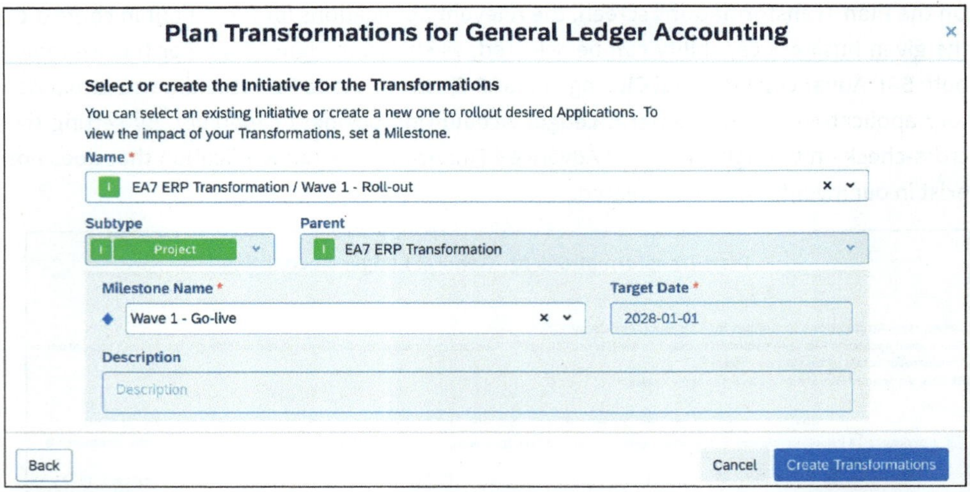

Figure 8.25: Select or Create Initiative for Transformation

For each application, a transformation is created and will be listed in the **Transformations** tab in the **Planned transformations** section, which can be found below the **Suggested solutions** section, as shown in Figure 8.26.

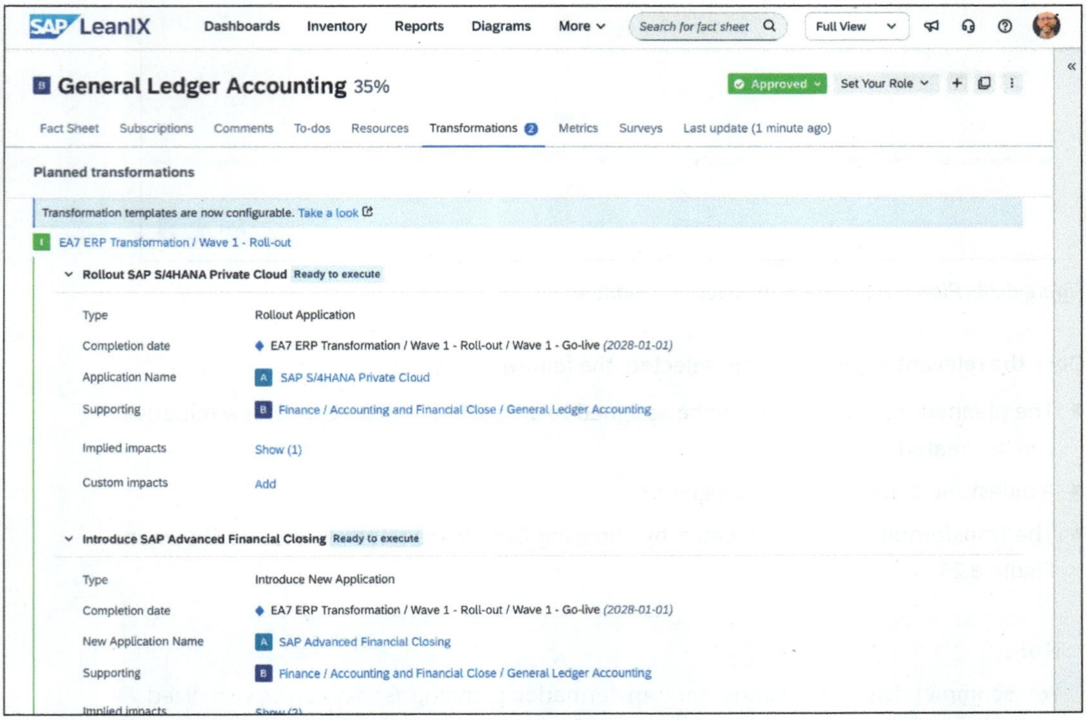

Figure 8.26: List of Planned Transformations

For each combination of application and business capability, two implied impacts are created:

1. The setting of the lifecycle phase of the application to **Active** at the point in time of the milestone or date defined.
2. The creation of a relationship between the business capability and the application.

Note

The transformation planning for individual level 3 business capabilities can be very time-consuming, especially if a larger number of business capabilities is in scope for the architecture engagement. We therefore recommend planning transformations on level 2 business capabilities. This approach should also be followed if a business capability model is only defined on levels 1 and 2.

Like the level 3 business capability transformation planning, the planning for level 2 business capabilities takes place within the **Transformations** tab, as shown in Figure 8.27.

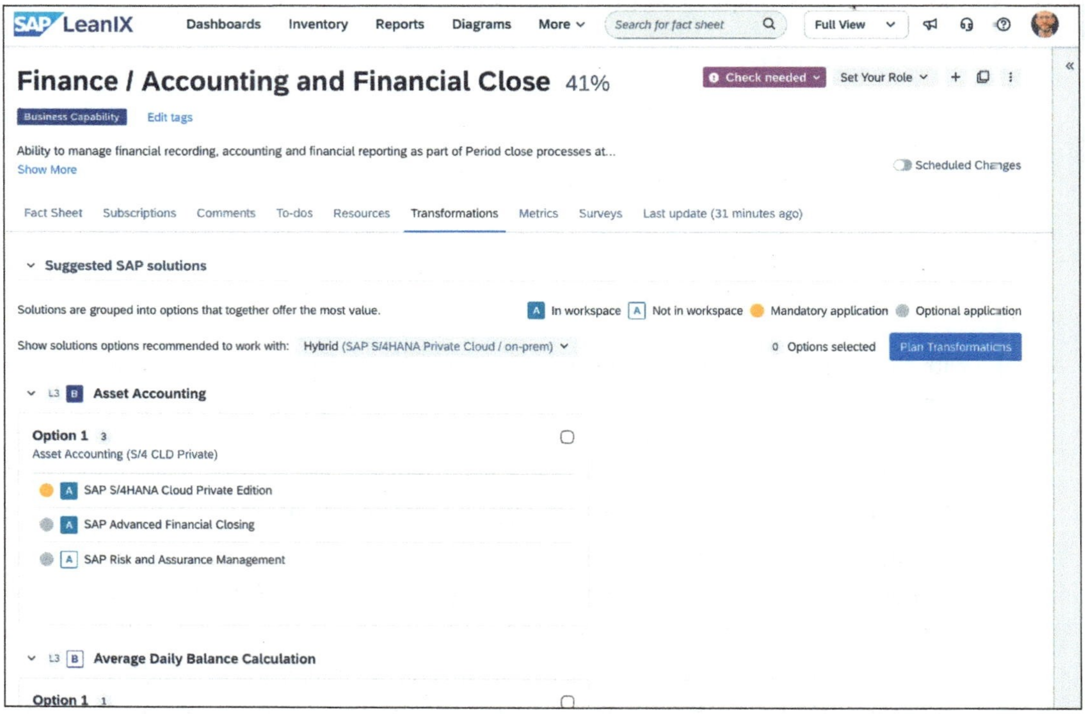

Figure 8.27: Transformation Planning for Level 2 Business Capabilities

Within the **Suggested SAP solutions** section of the level 2 business capability, all level 3 business capabilities and their options are listed. This is independent of the existence of the level 3 business capability fact sheet in the inventory. If an option will be selected for a level

3 business capability that is not in the inventory, an import of the respective level 3 business capability will be initiated.

To review the planned transformations, you can either check the **Transformations** tab of the individual level 3 business capability or refer to the initiative fact sheet for which the transformation was created. An alternative to this is to use the transformations explorer, which can be accessed via the **Transformations Explorer** option in the **Inventory Tools** dropdown in the **Inventory** section. Multiple filters exist to find the relevant information, as outlined for the EA7 case study in Figure 8.28.

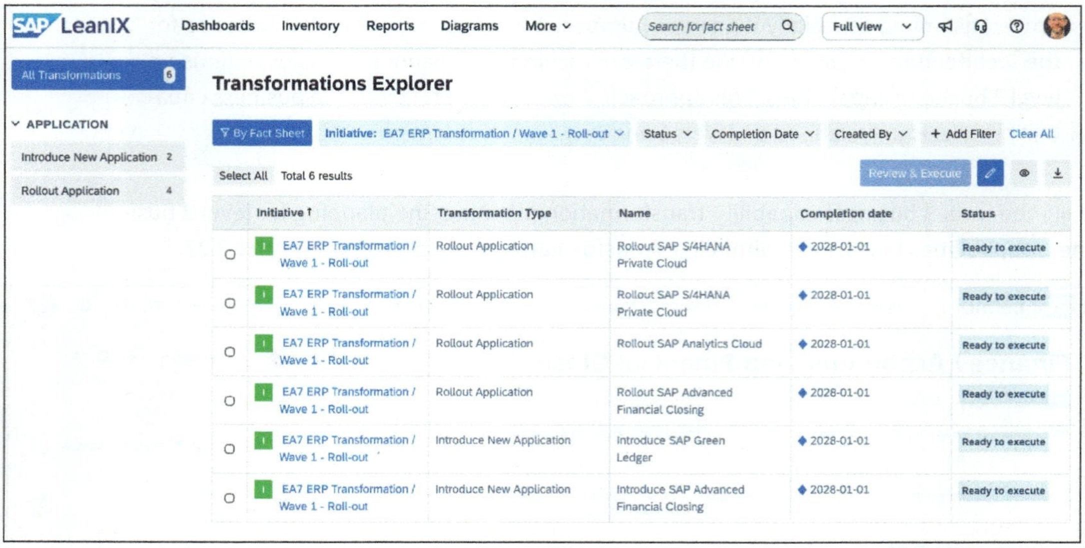

Figure 8.28: Overview of Planned Transformation of Initiative

With the target architecture planning of business capabilities on level 2 and level 3, the curated content of the SAP reference content can be leveraged. As previously mentioned, not all business capabilities might have a suggestion, and, of course, there are other software vendors beyond SAP. Once the target application for a non-SAP application is known, a transformation can be planned accordingly, as described in Chapter 6, Section 6.3.2.

In addition to modeling the target architecture by creating transformations for new, to-be-introduced applications or rollouts of existing applications, it is equally important to model the decommissioning of applications.

With the help of the application landscape report, you can visualize the baseline and target application architecture by leveraging one or more initiatives that contain detailed information about the transformations. With the help of the time slider, transition architectures can be reviewed at a given point in time and comparisons between different points in time can be visualized, as shown in Figure 8.29.

For the creation of the **EA7 Target Architecture** report, the following steps need to be performed:

1. Create a new landscape report with the application as the base fact sheet type. Name the report **EA7 Target Architecture**. The default **Cluster By** option in the settings is **Business Capabilities**, which works fine in this case.
2. Select the initiative for which you want to visualize the target architecture. In this example, it is the **Wave 1 Roll-out** of the **EA7 ERP Transformation** program.

The report is now showing the timeline of the selected initiative(s) and the change for the relevant applications is visualized as a small amendment to the application fact sheet boxes, indicating the lifecycle status at a given point in time or if an application is added, such as SAP Advanced Financial Close in this transformation planning example. The point in time can be chosen in the time slider either by selecting a milestone or by choosing any point in time.

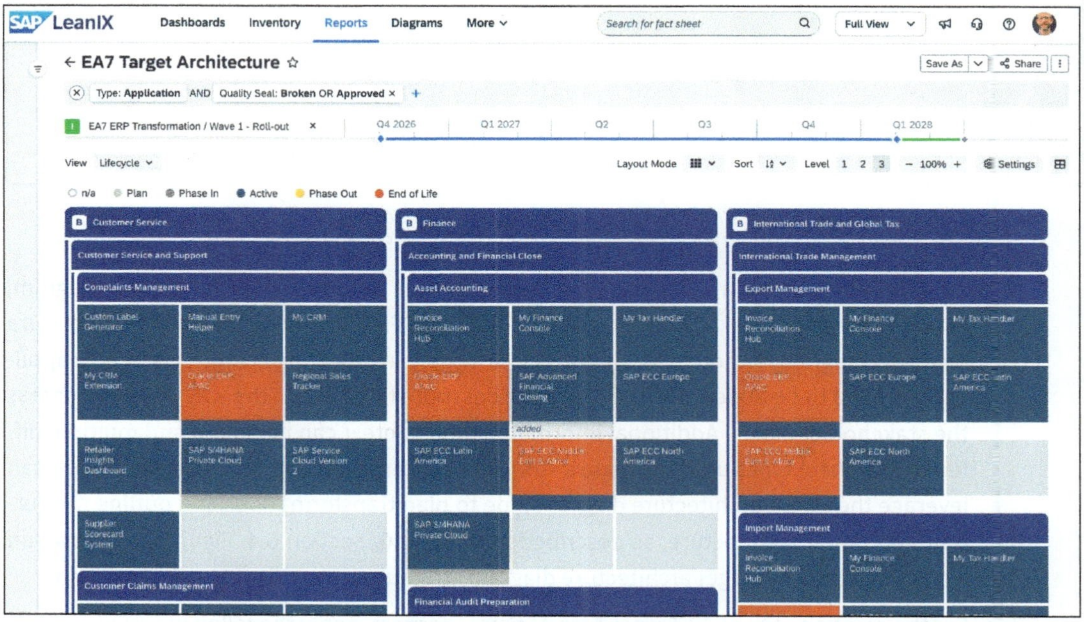

Figure 8.29: EA7 Target Application Architecture

An alternative to the reference architecture–based target planning is to use the *SAP suggestions in the reference catalog* feature, an optional feature that augments existing application fact sheets within an application landscape report with potential target SAP solutions, as shown in Figure 8.30.

> **Note**
>
> This feature needs to be activated in the **Administration • ADVANCED SETTINGS • Optional Features and Early Access** section, and it relies on the AI agent feature. For more details on its activation and usage, refer to Chapter 11.

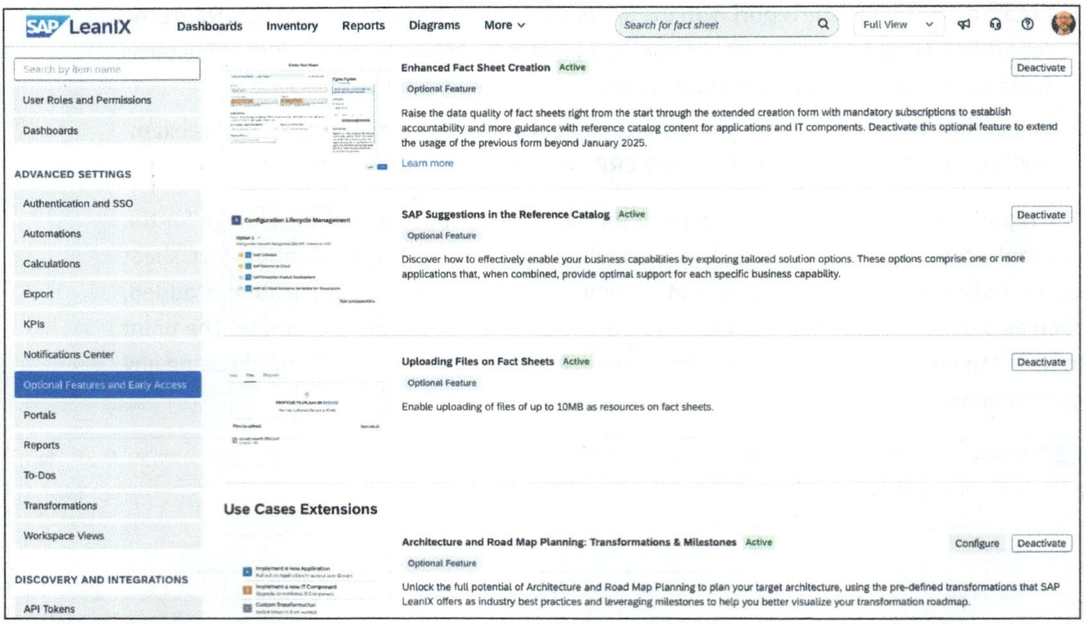

Figure 8.30: SAP Suggestions in Reference Catalog (Optional Feature)

An alternative to the application landscape report is an application landscape diagram, which represents the SAP Enterprise Architecture Framework application architecture overview diagram. Although a diagram is less flexible in reflecting changes (e.g., new applications need to be added manually), it provides more flexibility in the way you can address the stakeholder needs. Additional information and context can be added and multiple different fact sheet types used to create a story for your target audience. Furthermore, you can leverage the target architecture diagram type to plan transformations and outline the current and target architecture, as described in Chapter 6, Section 6.4. Figure 8.31 shows an example of the EA7 target architecture diagram structure alongside functional domains.

For the creation of the **EA7 Target Architecture Diagram**, apply the following steps:

1. Create a new **EA7 Target Architecture Diagram** by selecting **Target Architecture** as the diagram type.
2. Import from diagrams the **Template Architecture Overview Diagram** (this can be downloaded from the demo workspace) for the import of the functional domain visualization.
3. Add **EA7 ERP Transformation Program** and **Wave 1 - Roll-out** in the **Data** section.
4. Add the relevant applications to the diagram. Any transformation that already exists can be visualized by expanding the blue arrow.

In this example, we focused on the applications subject to the wave 1 rollout. The arrows indicate the changes to happen; for example, Oracle ERP APAC is replaced by SAP S/4HANA Cloud Private Edition in multiple functional domains. You can also leverage the capabilities of the target architecture diagram type to create transformations.

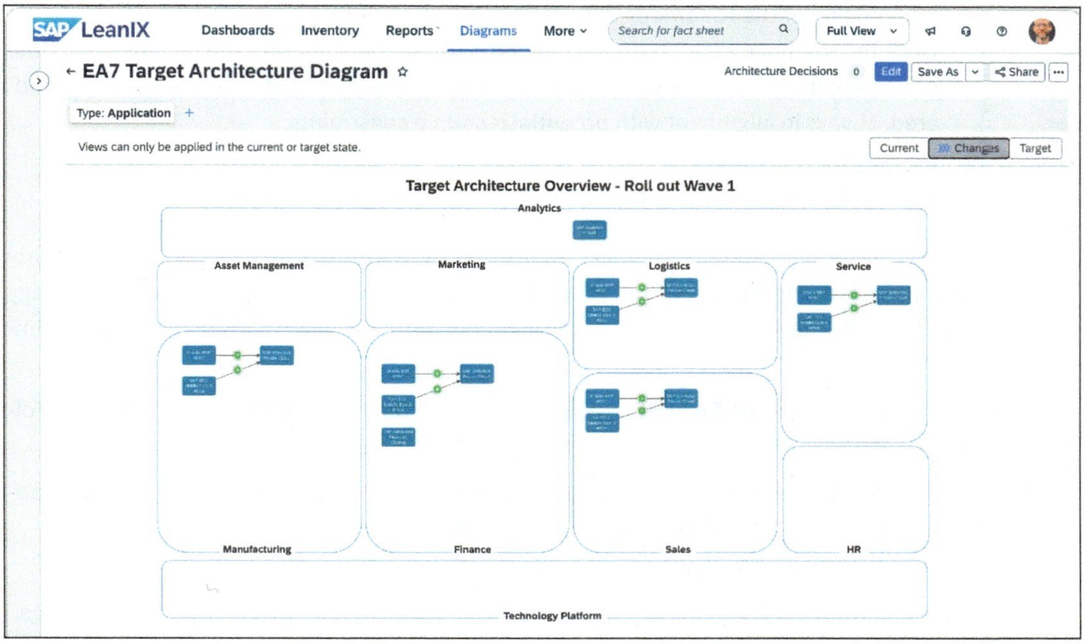

Figure 8.31: EA7 Target Architecture Diagram

With the definition of the target architecture, you've achieved a solid understanding of the destination your architecture engagement is heading toward. For an alternative approach using SAP Build and SAP BTP, see Section 8.5. For more information on in-depth target architecture modeling, see Chapter 6, Section 6.4.

Further Resources

For the discovery of Kyma runtime–based applications, refer to *http://s-prs.co/v620663*.

8.4.8 Architecture Roadmap

Although the architecture roadmap looks like a single stop on the SAP Enterprise Architecture Framework metro map, two variants have been defined: the *business architecture roadmap* and the *application architecture roadmap*. Both variants go together and can be used in different stakeholder situations.

SAP Enterprise Architecture Framework proposed the following approach to create the architecture roadmap artifacts:

1. **Identification of the roadmap candidates**
   For both architecture roadmap variants, it is important to understand which roadmap candidates are relevant. This could be in the context of either the business objectives or the target architecture to be achieved. Every company has limited resources, so the scope, budget, timeline, and so on need to be considered to create a realistic roadmap.

2. **Sequencing of the roadmap**
   Once the roadmap candidates are defined, logic sequencing with the dimension of time can be applied. Dependencies need to be identified and parallel activities can be considered, always in alignment with potential resource constraints.
3. **Visualization of the roadmap**
   The results will be visualized and socialized with the stakeholders to retrieve their buy-in.

The business architecture roadmap focuses on business outcomes structured around functional clusters and a time horizon of about one to three years, linking the business strategy and context. Figure 8.32 depicts the business outcomes as milestones on the program initiative level.

For the creation of the **EA7 ERP Transformation Initiative Roadmap** report, apply the following steps:

1. Create a new roadmap report with the name **EA7 ERP Transformation Initiative Roadmap** and select **Initiative** as the base fact sheet type.
2. Set the **EA7 Transformation Program Initiative** as a parent filter.
3. In the **Settings**, check the **Show Labels of Milestones** option within **Advanced Settings.**
4. Apply the settings.
5. Expand the **EA7 ERP Transformation** initiative to visualize the child initiatives with their milestones and lifecycle information.

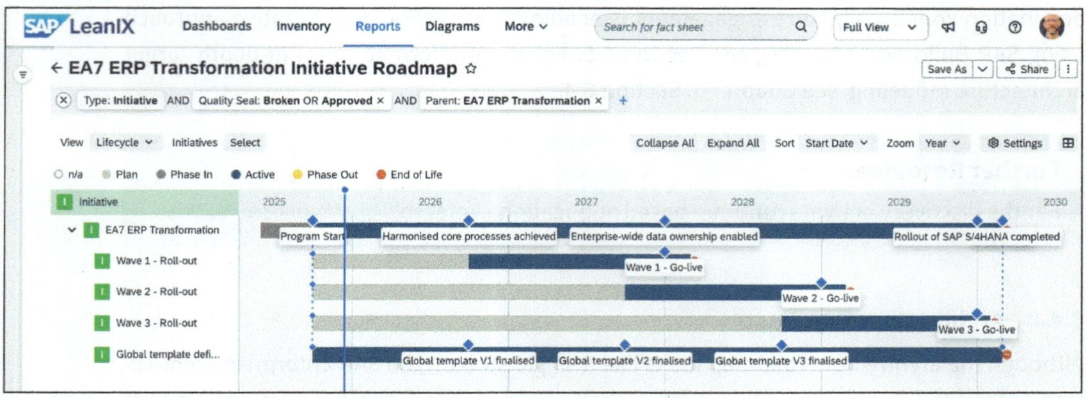

Figure 8.32: Business Outcome-Based Roadmap

The focus of the application architecture roadmap is on applications, as the name suggests, and it should be structured around the same dimensions as the business architecture roadmap.

With the fact sheets and relationships modeled in SAP LeanIX, you have the option to structure the application architecture roadmap along multiple dimensions. Depending on the stakeholder perspective, you could organize the application roadmap report via business capabilities, organizations, initiatives, and so on. Figure 8.33 outlines the applications in scope of the ERP transformation, clustered by organization.

For the creation of the **EA7 Application Architecture Roadmap** report, apply the following steps:

1. Create a new roadmap report with the name **EA7 Application Architecture Roadmap** and select **Application** as the base fact sheet type.
2. Select the **EA7 ERP Transformation Program** and **Wave 1 – Roll Out** as initiatives.
3. In **Settings**, check **Show only affected fact sheets**.
4. Apply the settings.
5. Expand **EA7 Global Organization** to visualize the child organizations.

Figure 8.33: EA7 Application Architecture Roadmap

8.5 Integration with SAP Build and SAP BTP

The architecture overview diagram that we discussed in Section 8.4.7 provides a holistic picture of the target architecture, but often you'll find that for a subset of applications, a more detailed modeling approach is required. This is true especially when modeling an SAP S/4HANA application that entails multiple solution capabilities and thus supports many business capabilities. With the promotion of clean core principles and the side-by-side extension of, for example, SAP S/4HANA, the modeling of SAP BTP–based applications gains momentum. Therefore, let's look at the discovery of SAP BTP–based applications and services plus best practices for how to model SAP S/4HANA in more detail.

Because out-of-the box integrations, such as the previously introduced SAP landscape discovery, provide valuable insights into how modeling best practices should look, we'll start by exploring additional out-of-the box integrations, namely:

- SAP Build integration to discover applications based on SAP Build Apps, SAP Build Code, and SAP Build Process Automation
- SAP BTP, Cloud Foundry environment integration for the discovery of multitarget applications (MTAs)
- SAP BTP, Kyma runtime integration for the discovery of self-built applications

Similar to SAP landscape discovery, these integrations can be configured individually and leverage the concept of the inbox to enable a staged discovery and import of new fact sheets or linking to existing fact sheets.

For the configuration of the SAP Build integration, the following two prerequisites need to be fulfilled:

1. Creation of an SAP Build Process Automation instance within your SAP BTP account and creation of a service key
2. Creation of a cloud management and service instance and creation of a service key

Further Resources

For more information about these prerequisites, visit *http://s-prs.co/v620664*.

Within SAP LeanIX, you set up the SAP BTP integration by first navigating to the **Integration** section of the **Administration** area, which can be found under **Discovery and Integration**.

After choosing **Add integration**, you can narrow down the integration options by filtering on **SAP Discovery**. The integration options shown in Figure 8.34 will be presented.

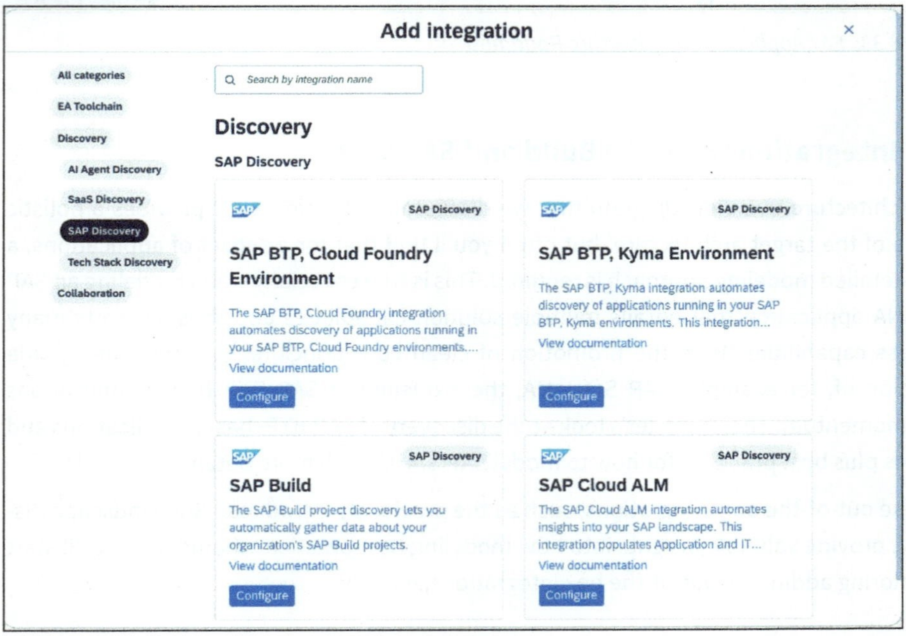

Figure 8.34: SAP Discovery Integration Options

Choose to configure a new SAP Build integration and provide an integration name, as shown in Figure 8.35.

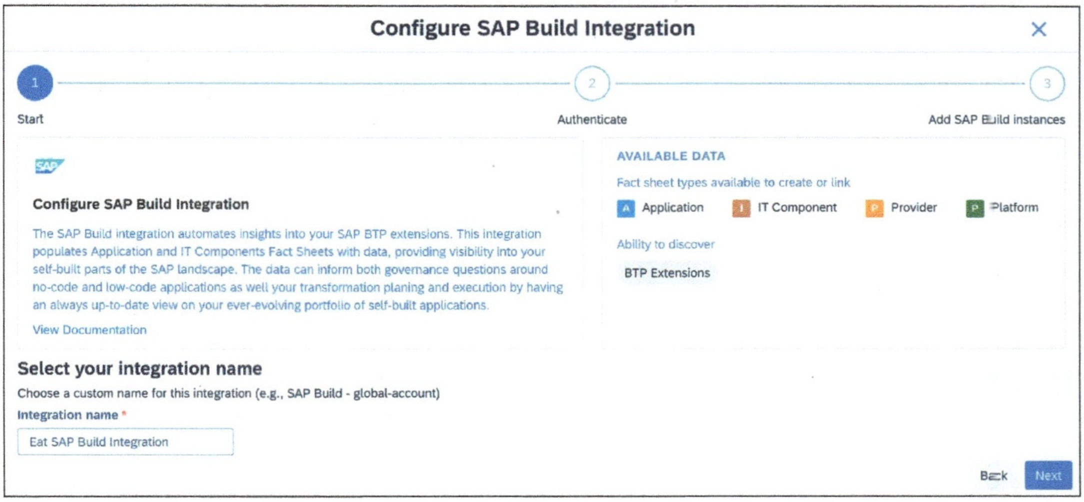

Figure 8.35: Definition of SAP Build Integration Name

In the next step, in **Cloud Management Service Key**, paste your key, as outlined in Figure 8.36.

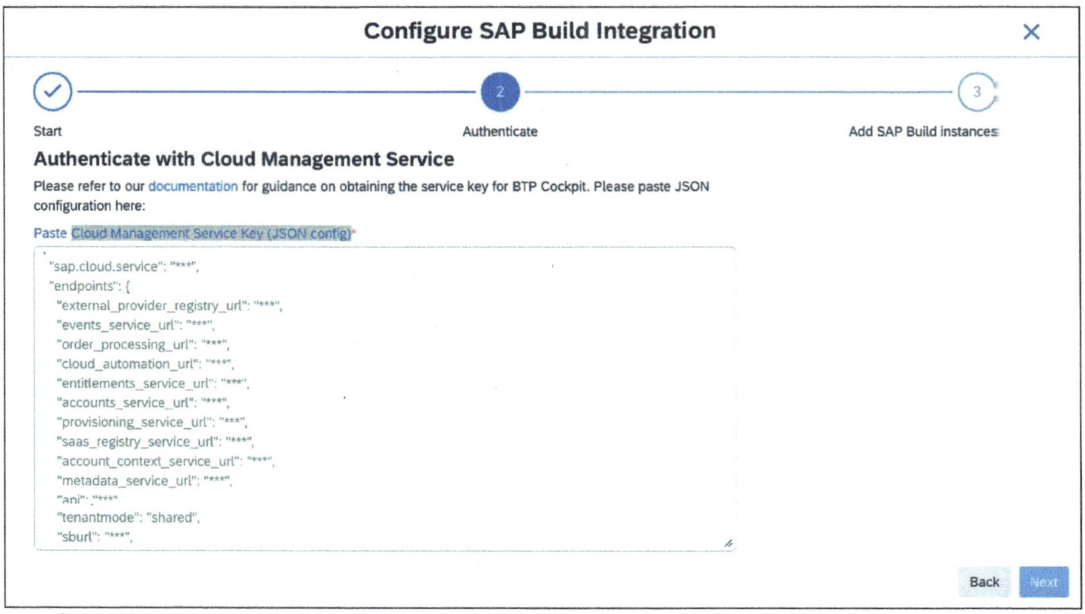

Figure 8.36: Cloud Management Service Key for SAP BTP Integration

In the last configuration step, you must define the SAP Build instance by pasting in the SAP Build instance service key, as shown in Figure 8.37.

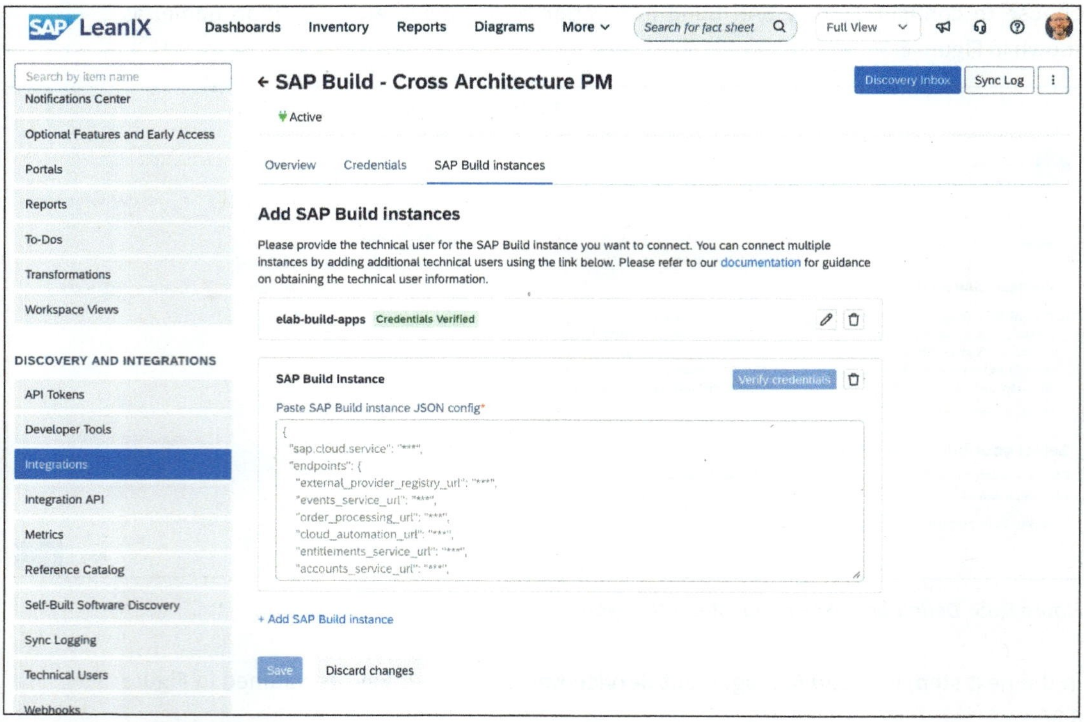

Figure 8.37: SAP Build Instance Service Key for SAP Build Integration

With the activation of the SAP Build integration, information will be retrieved from the SAP Build instance and made available in the discovery inbox.

For the configuration of the SAP BTP, Cloud Foundry environment integration, the following two prerequisites need to be fulfilled:

1. Creation of a platform user in the Identity Authentication service tenant and creation of a Cloud Foundry certificate
2. Creation of a cloud management service instance and creation of a service key

Further Resources

Refer to this help page to meet the prerequisites: *http://s-prs.co/v620665.*

For the setup of an SAP BTP, Cloud Foundry environment integration, choose the respective integration options as shown previously in Figure 8.34. Enter the name of the SAP BTP, Cloud Foundry environment integration as shown in Figure 8.38.

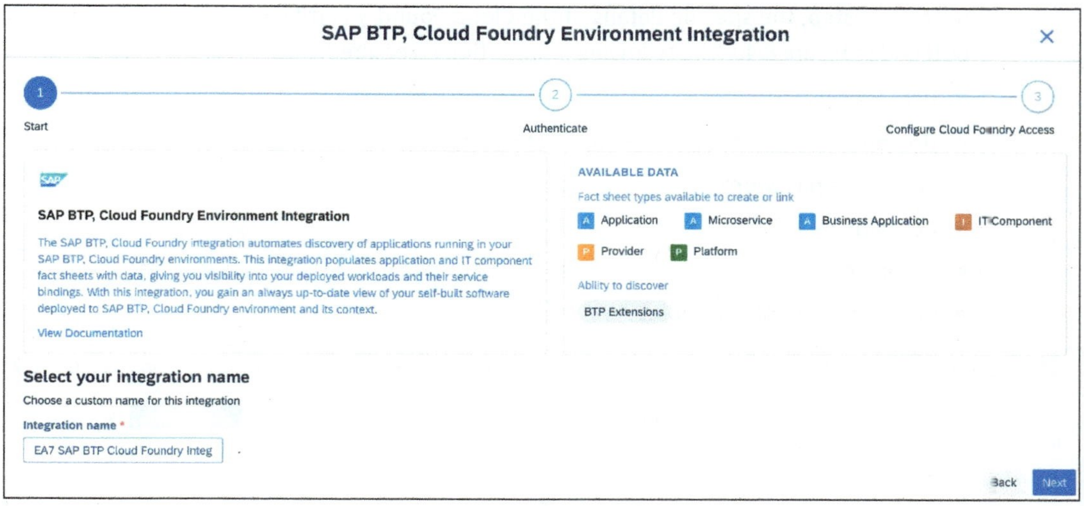

Figure 8.38: Definition of SAP BTP Cloud Foundry Environment Integration Name

In the next step, paste in your **Cloud Management Service Key**, as shown in Figure 8.39.

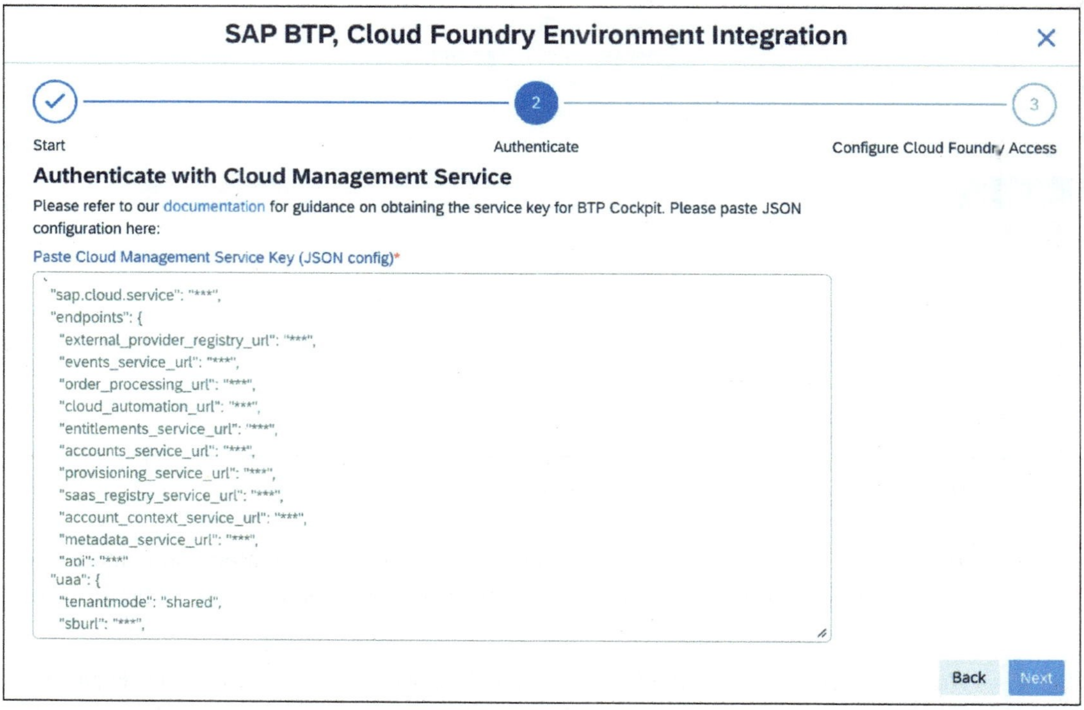

Figure 8.39: Cloud Management Service Key for SAP BTP Cloud Foundry Environment Integration

In the last step, the specific details of the Cloud Foundry certificate need to be added, as outlined in Figure 8.40. The following information is required:

- User email
- Upload of the certificate in PFX or P12 format
- Certificate password
- Domain
- Origin key
- Cloud Foundry API endpoint

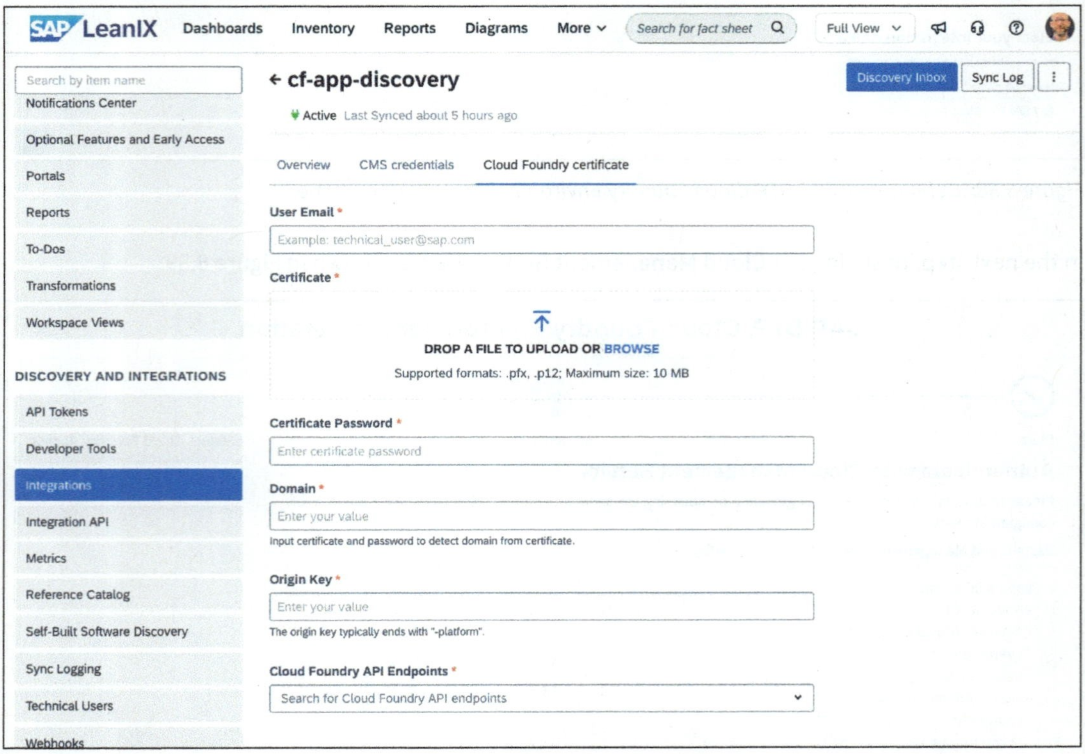

Figure 8.40: Configuration of Cloud Foundry Certificate

With the activation of the SAP BTP, Cloud Foundry environment integration, information will be retrieved from the Cloud Foundry instance and made available in the discovery inbox.

We recommend adding a filter for the **Type** of the discovery item, as outlined in Figure 8.41, to allow a better potential selection of various discovery items. In this example, we selected **SAP Build Events** as a type to identify the **Customer Return Event** application.

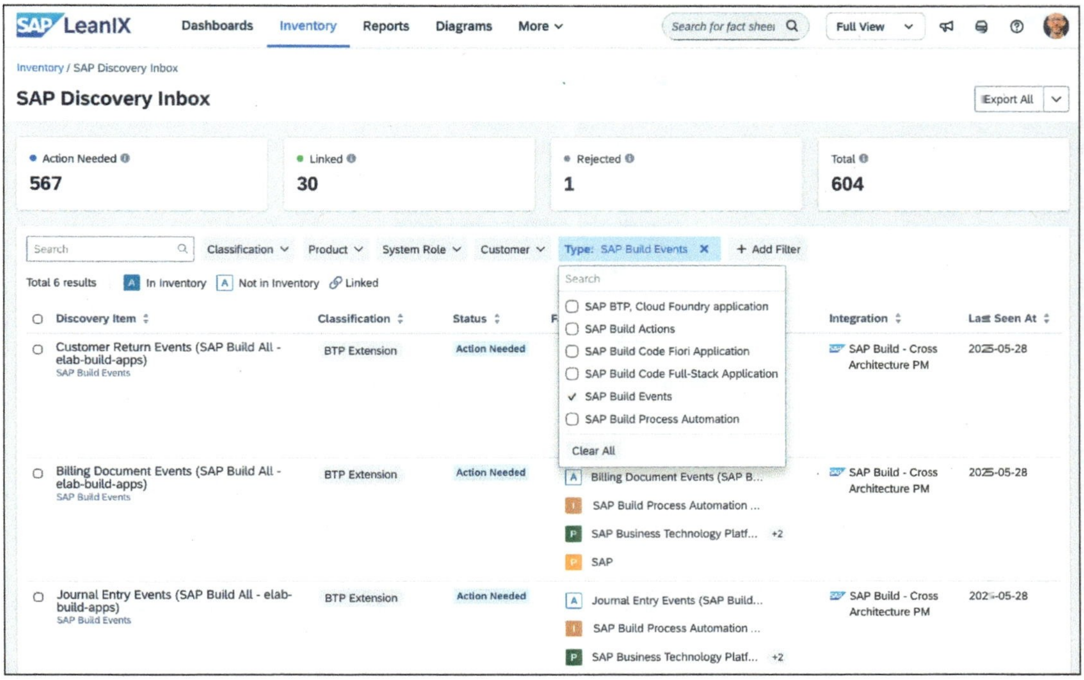

Figure 8.41: Filtering of SAP Discovery Inbox Items Based on Type

Each discovery item provides an overview of the discovered fact sheets of the application, IT component, platform, and provider types, in addition to general information such as name, description, creation time, and so on. If the fact sheet does not exist in the inventory yet, it will be highlighted without a solid fact sheet-specific color; existing fact sheets can be easily identified by the solid, fact sheet–specific color-coding. Before the actual linking and import into the inventory takes place, you can edit individual fact sheets to ensure correct assignments while avoiding duplications.

Figure 8.42 provides an example of an SAP BTP, Cloud Foundry environment application and the discovered fact sheets. With the linking of the discovery item, existing fact sheet types will be reused and related while new ones will be created as new inventory entries.

With the insights gained from the out-of-the box SAP BTP integrations, you can derive the SAP BTP modeling best practices outlined in Figure 8.43. The platform fact sheet plays an important role here, as it is used to reflect the global and subaccount setup in a hierarchical structure.

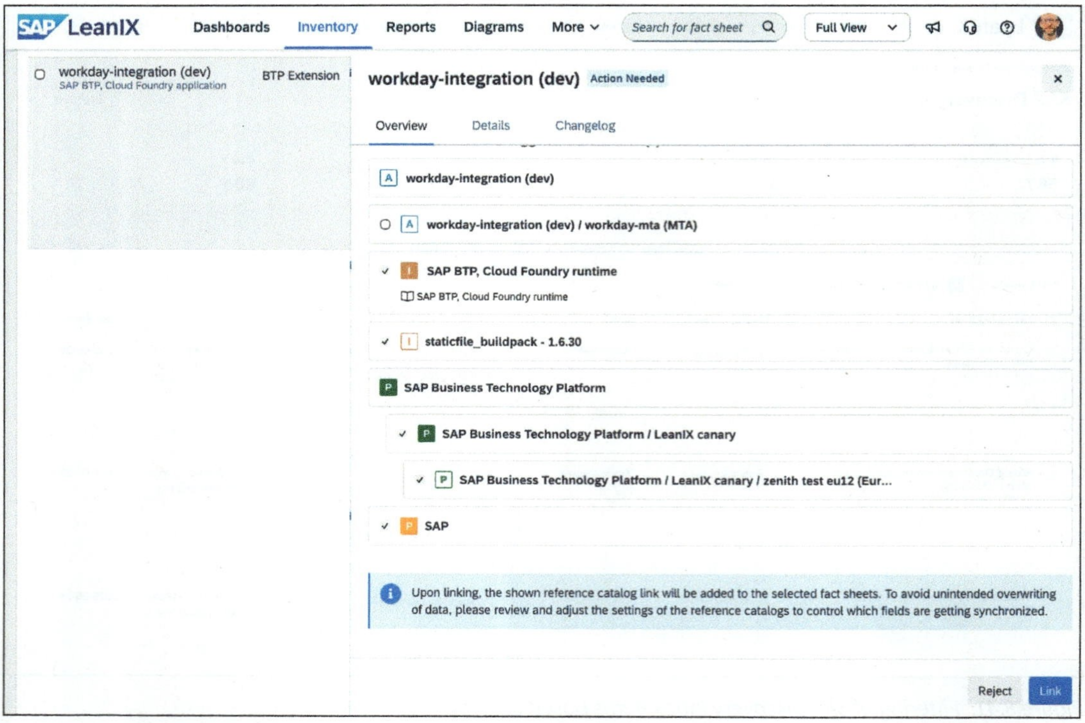

Figure 8.42: Example of SAP BTP, Cloud Foundry Environment Application in Discovery Mode

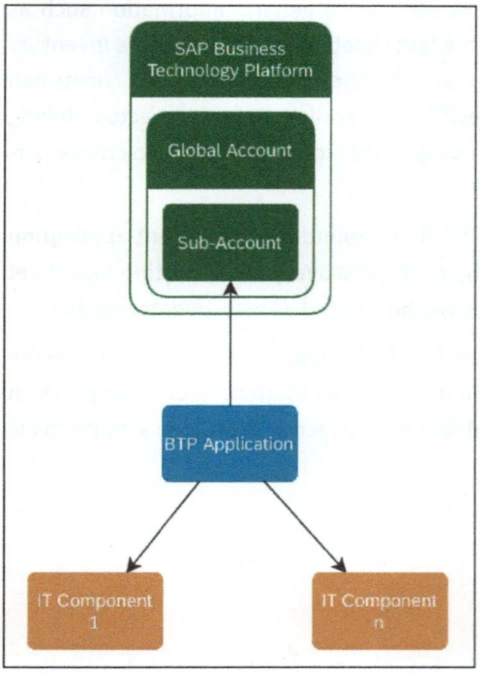

Figure 8.43: SAP BTP Modeling Best Practices

New SAP BTP–based applications to be added should apply these modeling best practices for the definition of their target architecture.

8.6 Summary

Throughout this chapter, we outlined how SAP LeanIX acts as an integral part of SAP Enterprise Architecture Framework by creating relevant artifacts and leveraging advanced features such as SAP landscape discovery or SAP reference content. The EA7 case study demonstrated how the scoped artifacts look in an architecture engagement and can provide a reference for your next project.

In the next chapter, we'll discuss additional use cases such as technology risk and compliance and mergers and acquisitions, as well as a regulatory use case within the finance and insurance industry.

Chapter 9
Additional Use Cases and Extensions

Now that you've learned a lot about "classic" enterprise architecture use cases, we want to spark ideas for different questions and less obvious use cases that can also be supported by SAP LeanIX and the enterprise architecture discipline. Are you ready? Let's dive into new and maybe unfamiliar challenges.

In this chapter, we'll discuss additional capabilities and features of the SAP LeanIX Technology Risk and Compliance product, specifically, those for self-built software discovery and management. Furthermore, we'll look at mergers and acquisitions (M&A) and how SAP LeanIX can be used in these endeavors. Finally, we'll look at highly regulated industries, like finance and insurance, and how to prepare for the Digital Operational Resilience Act (DORA) and the open finance regulations using SAP LeanIX and selected in-tool capabilities.

SAP LeanIX Extension Hub

At the time of writing, SAP LeanIX product and engineering just switched from the SAP LeanIX store and the **Optional Features** submenu in the **Administration** menu to the SAP LeanIX Extension Hub. In the future, use case-specific meta model extensions will be available for activation via the Extension Hub. In the interim, some meta model extensions are still in the **Administration** menu under **Optional Features and Early Access**, or are even connected to integration activations.

9.1 Self-Built Software Management and Product Architecture

As you might remember from Chapter 2, before the acquisition by SAP, LeanIX once offered a product called LeanIX Value Stream Management, which was a solution targeting internal developers to manage self-built software. After the acquisition, the product was sunsetted, but its core functionality was moved into the SAP LeanIX Technology Risk and Compliance product.

The following are a few key reasons you should monitor your self-built software:

- **Differentiation**
  You have built something to protect your intellectual property, but this makes your organization vulnerable to cyberattacks or obsolescence risk during ongoing maintenance.

- **Customization**
  You have enriched off-the-shelf solutions to extend critical capabilities, which might pose a risk during the next forced update cycle.
- **AI agents**
  You have increasing numbers of agents that built something for you quickly with libraries and frameworks you have never heard of, and you don't know whether they come from a legit source.
- **Mergers and acquisitions**
  You have acquired a new company and are taking over their booming products, but also their decades-old, self-built technical debt.

In the following sections, we'll briefly look at the core features that will help you to discover, analyze, and manage self-built software in SAP LeanIX.

Further Resources

For more information on self-built software management, visit the SAP LeanIX user documentation at *http://s-prs.co/v620666*.

To learn about the out-of-the-box GitHub integration, visit *http://s-prs.co/v620667*.

For more on the SAP LeanIX OpenAPI Explorer to set up your custom integrations for microservice discovery and SBOM intake, see the following resources:

- *http://s-prs.co/v620668*
- *http://s-prs.co/v620669*

You can also rewatch a webinar from June 18, 2025, titled "Discover and Govern Your Self-Built Software with the new GitHub Enterprise Integration" at *http://s-prs.co/v620670*. Just enter your business email address in the **Watch Now** box in the upper-right corner of the screen and click the **Proceed** button.

9.1.1 Self-Built Software Discovery

If you have purchased the SAP LeanIX Technology Risk and Compliance product, you will see the additional **Self-Built Software Discovery** submenu in the **DISCOVERY AND INTEGRATIONS** section under the **Administration** menu (see Figure 9.1).

During the development of these features, the SAP LeanIX product and engineering team discussed how to best model this use case with various customers. It was decided that self-built software discovery would use an additional subtype of the application fact sheet, called *microservice*. A microservice is a component or service for a specific function or task, often self-developed, that can be reused by one or more business applications. The microservice fact sheet subtype is not provisioned in the out-of-the-box SAP LeanIX workspace and needs to be provisioned via a meta model extension. You can see an overview of the newly created subtypes and relations in Figure 9.2.

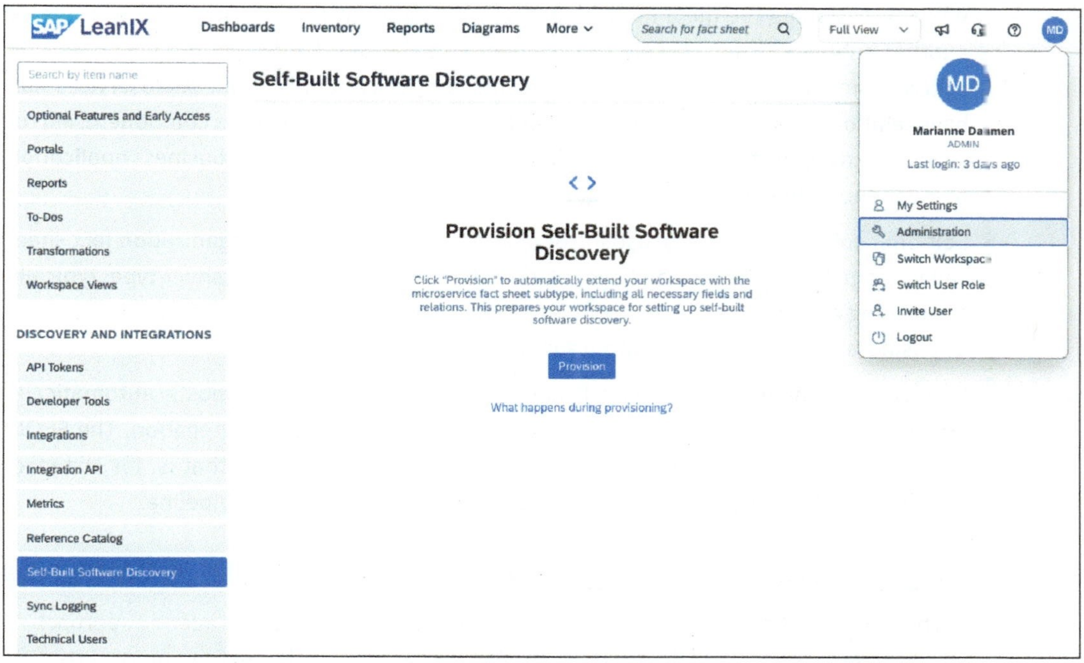

Figure 9.1: Additional Self-Built Software Discovery Submenu in Administration Menu

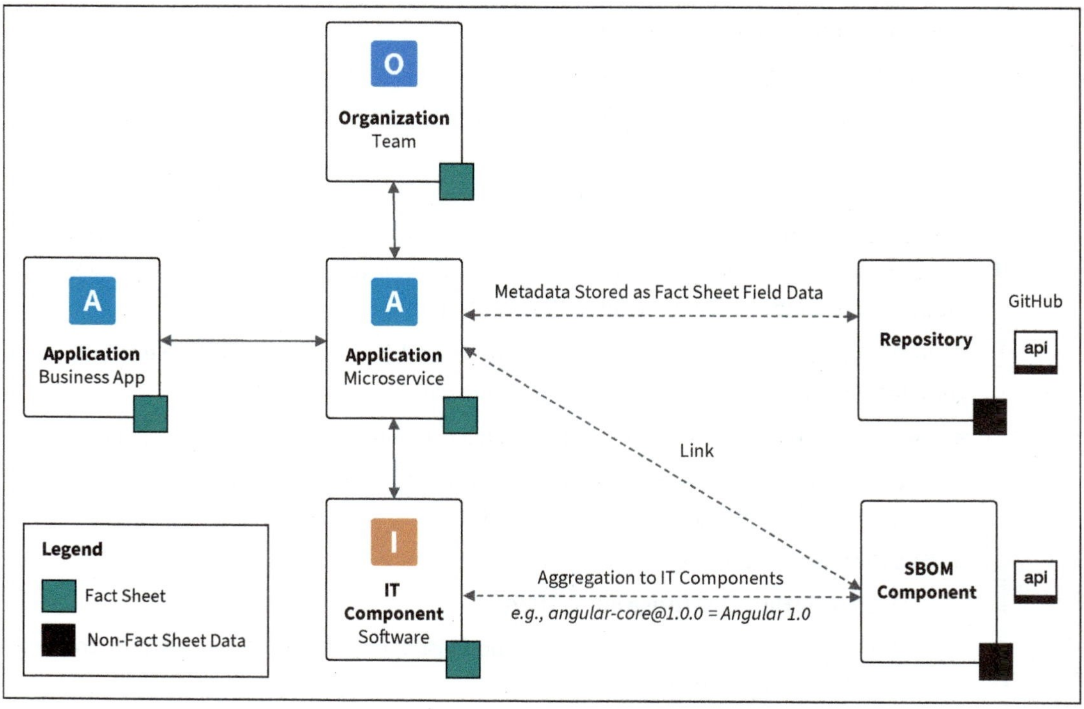

Figure 9.2: Meta Model Extension for Self-Built Software Discovery

When provisioning the meta model extension, the application fact sheet type will receive another subtype, if not already provisioned with the SAP landscape discovery feature, called *business application*, which represents the logical application. Microservices can have relations to one or more business applications, of which they are a component. Microservices can support a business capability through its connection to a business application or independently as its own entity.

Self-built software discovery also introduces a new subtype in the organization fact sheet called *team*. Microservices can have relations to the organization fact sheet type, typically the owning developer team. However, they can also be connected to other subtypes, such as business unit, region, or a custom subtype.

Microservices can have relations to IT components, which ideally are mostly automatically created via the software bill of materials (SBOM) attachment and aggregation. The SBOM can be attached to the microservice fact sheet via a custom script that is, for instance, embedded in the continuous integration/continuous delivery (CI/CD) pipeline.

SAP LeanIX Self-Built Software Agent

The SAP LeanIX product and engineering team is also working on an agentic approach to facilitate the automatic discovery, analysis, and information extraction of GitHub repositories. They have developed an open source command line interface tool and library.

For more information, visit the following pages:

- *http://s-prs.co/v620671*
- *https://github.com/SAP/leanix-self-built-software-agent/*

To add the meta model extension to your workspace, first click the blue **Provision** button in the **Self-Built Software Discovery** submenu under **Administration**. A new popup window will appear, in which you must confirm that you want to provision the workspace meta model extension by clicking the respective button.

Once the meta model extensions have been applied (see Figure 9.3), the **Self-Built Software Discovery** submenu under **Administration** will show the following:

- A grey button with a link to the SAP LeanIX user **Documentation** in the upper-right corner
- An overview of provisioned meta model changes to the inventory and an option to configure the default relations via the **Edit** button
- Two tiles to set up the self-built software discovery:
  - One for **GitHub Enterprise** integration
  - One for a custom CI/CD discovery setup (**Discovery via CI/CD**), which calls the Self-Built Software Discovery API

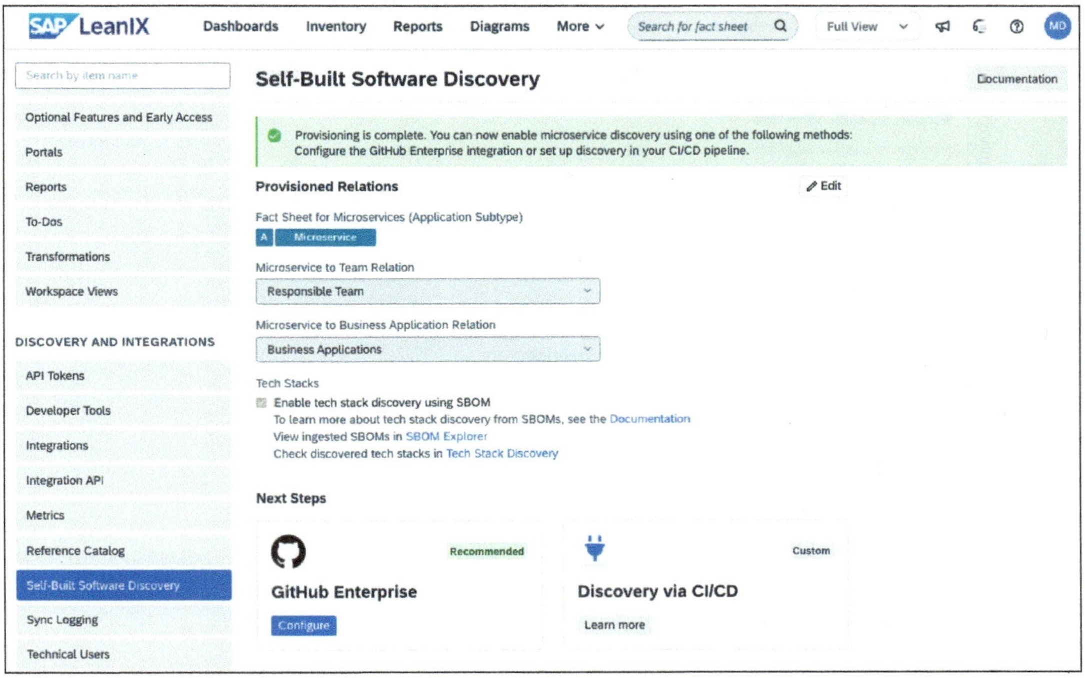

Figure 9.3: Self-Built Software Discovery Submenu After Meta Model Extension Provisioning

For both variants, GitHub Enterprise integration and custom integration, you need a YAML manifest file with basic information about the microservice, which is placed in the root directory of the respective repository. You can have multiple manifest files in one mono-repository, but each manifest file represents only one microservice fact sheet and should be placed in structured subdirectories.

Further Resources

The latest schema and structure of the needed SAP LeanIX manifest file can be found in the SAP LeanIX user documentation at *http://s-prs.co/v620672*.

Figure 9.4 shows an example of a detected and enriched microservice fact sheet. It was detected in GitHub via the *leanix.yaml* file placed in the root directory and was enriched with an SBOM via a GitHub workflow that triggered the SBOM push to the Self-Built Software Discovery API.

Figure 9.5 shows an example report, in this case, an **Application Landscape** report clustered by business application and filtered for microservices. **Layout Mode Rolled-up** was selected, and **SBOM Status** was selected for **View**, which means that color-coding indicates how many microservices have an SBOM attached. This provides a good holistic overview of your self-built software to product mapping and the data completeness in your workspace.

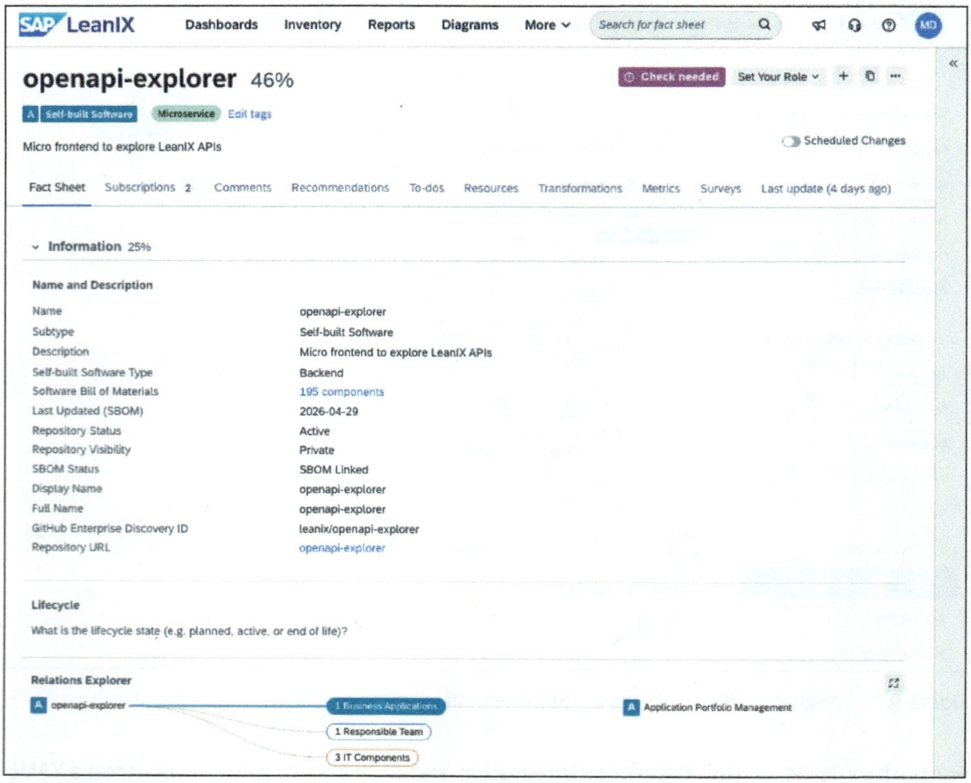

Figure 9.4: Example Microservice from SAP LeanIX Internal Test Workspace

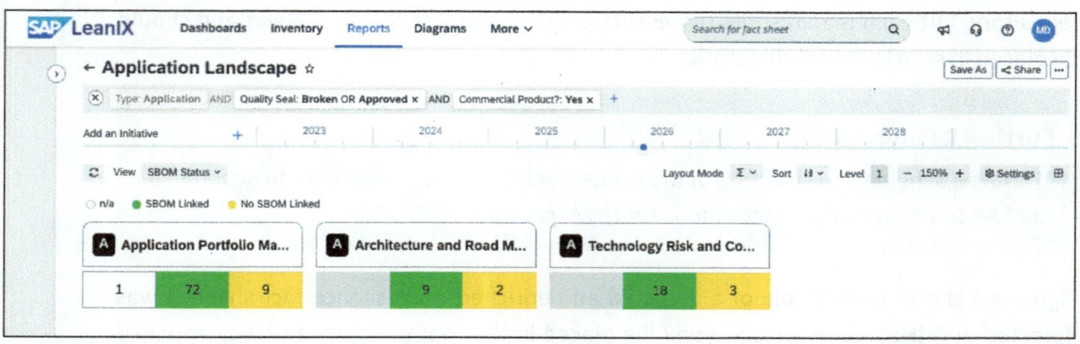

Figure 9.5: Example Landscape Report Showing Number of Microservices per Business Application

9.1.2 Software Bill of Materials Analysis

You cannot generate an SBOM with SAP LeanIX, but you can attach an SBOM in the Cyclone Data Exchange (CycloneDX) or Software Package Data Exchange (SPDX) format to a microservice via the Self-Built Software Discovery API or manually via the right-hand side menu in the user interface.

> **SBOM Standards**
>
> Increasingly, governments mandate that software companies publish a product-specific list of used components, packages, libraries, dependencies, and so on embedded in their application (software supply chain). To structure and standardize those inventory lists, and to make them comparable as well as machine-readable, SBOM standards were developed that define a unified structure and format, which can be read and reused by other tools.
>
> SBOMs not only help to quickly identify security vulnerabilities but also ensure license compliance.
>
> Two frequently used standards, which are supported by SAP LeanIX, are CycloneDX (see *https://cyclonedx.org/*) and SPDX (see *https://spdx.dev/*).

The automated upload process can be embedded into various steps in your software development lifecycle, such as during container registries or security scans. Make sure that the SBOMs contain good quality package URLs as this info is needed to derive aggregated IT components from the SBOM line items.

> **Further Resources**
>
> For more information on how to create an SBOM and which quality criteria need to be met for a successful SBOM ingestion into SAP LeanIX, visit the online user documentation at *http://s-prs.co/v620673*.

An additional feature in SAP LeanIX Technology Risk and Compliance with self-built software discovery is the SBOM viewer. In the top ribbon, click **More** and select **Software Bill of Materials** (see Figure 9.6). The SBOM viewer by default is accessible for users with an admin or member role. To change that, navigate to the **Administration** menu and then to **User Roles and Permissions**, and adjust the permissions for the respective role.

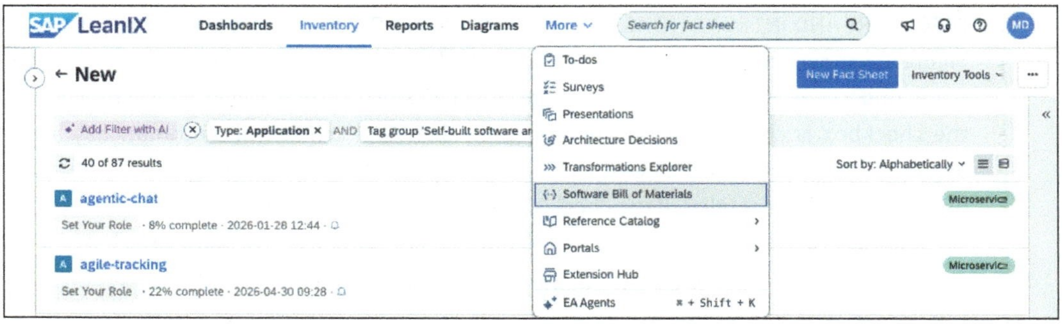

Figure 9.6: Navigating to SBOM Viewer

In the SBOM viewer, you can see all uploaded SBOM components, and you can see their versions when you expand their line items, plus additional information such as package manager, license, number of microservice SBOMs this component is part of, number of teams that are using this component, and number of business applications that are connected to this SBOM component via a microservice relation (see Figure 9.7).

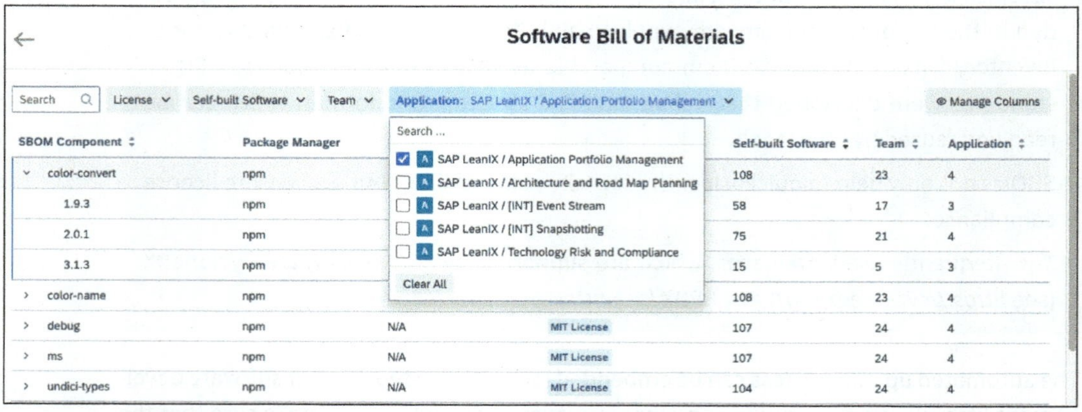

Figure 9.7: SBOM Viewer Example

This overview is especially crucial during security incidents such as Log4j or Log4Shell, which happened in 2021 and were categorized as the highest-level cybersecurity threat across the world. Enterprise architects can quickly filter for vulnerable components and get an overview of where in their self-built software landscape the component is used, which teams to approach, and which logical applications might be impacted. The SBOM viewer also provides technical users with detailed insights into open-source usage during tech stack migration and modernization initiatives.

In addition to this detailed SBOM component view, SAP LeanIX Technology Risk and Compliance also includes an SBOM component aggregation feature that detects technology versions from SBOM data, such as Bootstrap 3.X.

To use this feature, make sure that you have it enabled when configuring self-built software discovery in the administration area. Navigate to the **Administration** menu and then to the **DISCOVERY AND INTEGRATIONS** section. Click the **Self-Built Software Discovery** submenu. Figure 9.8 shows that we successfully provisioned the meta model extension. In the center pane, underneath the configured relations, you can see the **Tech Stacks** subsection. When the checkbox is checked (as in our example), then SAP LeanIX will automatically scan, aggregate, and create SBOM items as IT components based on the configurable tech stack discovery mapping catalog. If the checkbox is not checked, click the **Edit** button next to the **Provisioned Relations** header, then check the checkbox and click the blue **Save** button.

To navigate to the **Tech Stack Discovery** mapping catalog, which already includes curated entries from the SAP LeanIX catalog research team, navigate to the top ribbon **Inventory** menu and click the grey **Inventory Tools** button in the upper-right corner. Select **Tech Stack Discovery** from the expanded menu (see Figure 9.9).

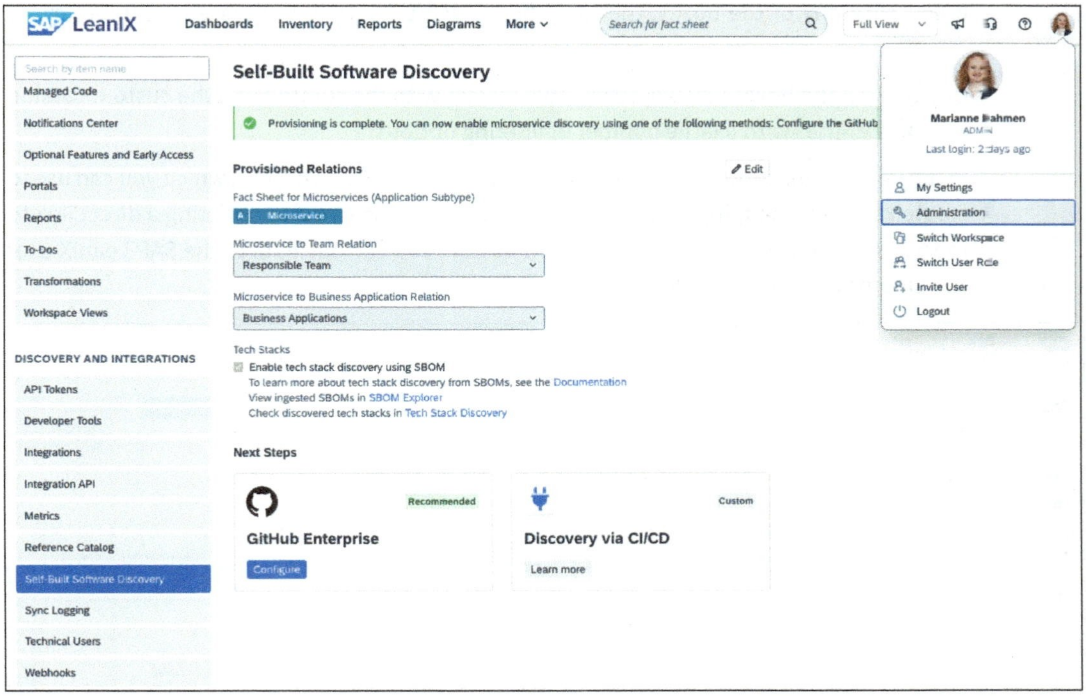

Figure 9.8: Enabling Tech Stack Discovery from SBOM Items

Figure 9.9: Navigating to Tech Stack Discovery Mapping Catalog

You will now see the **Tech Stack Discovery** catalog (see Figure 9.10). In the top row, you can see three boxes, showing you the number of items in each of the following categories:

- **Discovered**
  Total number of discovered, created, and linked IT component fact sheets based on SBOM scans
- **Discovered and End of Life**
  Tech stack IT components that have a known and due end-of-life risk
- **Custom Tech Stacks**
  IT components that were created based on custom tech stack mappings that you created

If you click any of the boxes, the table below will be autofiltered based on the relevant criteria.

Underneath the boxes you will see a table format with the curated and the custom catalog mapping entries with a large number of filtering options.

In the upper-right corner, you can see the blue **Add Tech Stack** button, which you can use to create your own mapping catalog entries. You can also see a button offering a direct link to the **SBOM Explorer**, and if you click the **...** button, you can navigate to the SAP LeanIX user **Documentation** for more information.

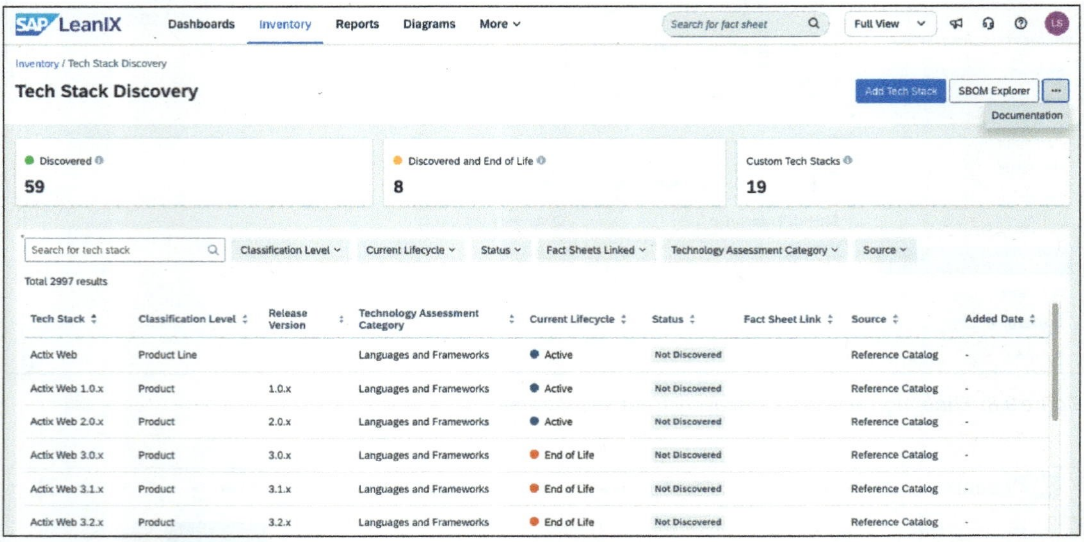

Tech Stack	Classification Level	Release Version	Technology Assessment Category	Current Lifecycle	Status	Fact Sheet Link	Source	Added Date
Actix Web	Product Line		Languages and Frameworks	Active	Not Discovered		Reference Catalog	-
Actix Web 1.0.x	Product	1.0.x	Languages and Frameworks	Active	Not Discovered		Reference Catalog	-
Actix Web 2.0.x	Product	2.0.x	Languages and Frameworks	Active	Not Discovered		Reference Catalog	-
Actix Web 3.0.x	Product	3.0.x	Languages and Frameworks	End of Life	Not Discovered		Reference Catalog	-
Actix Web 3.1.x	Product	3.1.x	Languages and Frameworks	End of Life	Not Discovered		Reference Catalog	-
Actix Web 3.2.x	Product	3.2.x	Languages and Frameworks	End of Life	Not Discovered		Reference Catalog	-

Figure 9.10: Tech Stack Discovery Mapping Catalog

The detected technology versions are automatically created as IT component fact sheets and linked to the IT component lifecycle catalog, and a relation to the relevant microservice is created. This relation can then be used for active IT component obsolescence risk and lifecycle management. Figure 9.11 shows an **IT Component Landscape** report, clustered by **Tech Category** and using a color-coded **Vendor Lifecycle** view.

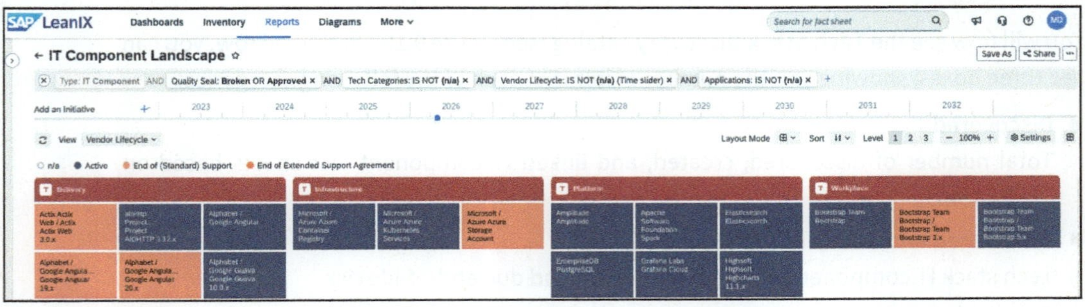

Figure 9.11: IT Component Landscape Report with Vendor Lifecycle Color-Coding

9.1.3 Tech Standards Management

Now that you've autodetected IT components via the SBOM aggregation, you should ask your developer teams to also maintain nondetectable IT components of their microservices, such as infrastructure or data management components. In the next step, for instance, a dedicated committee can perform a technology assessment and enter the information into the SAP LeanIX IT component fact sheets (see Figure 9.12).

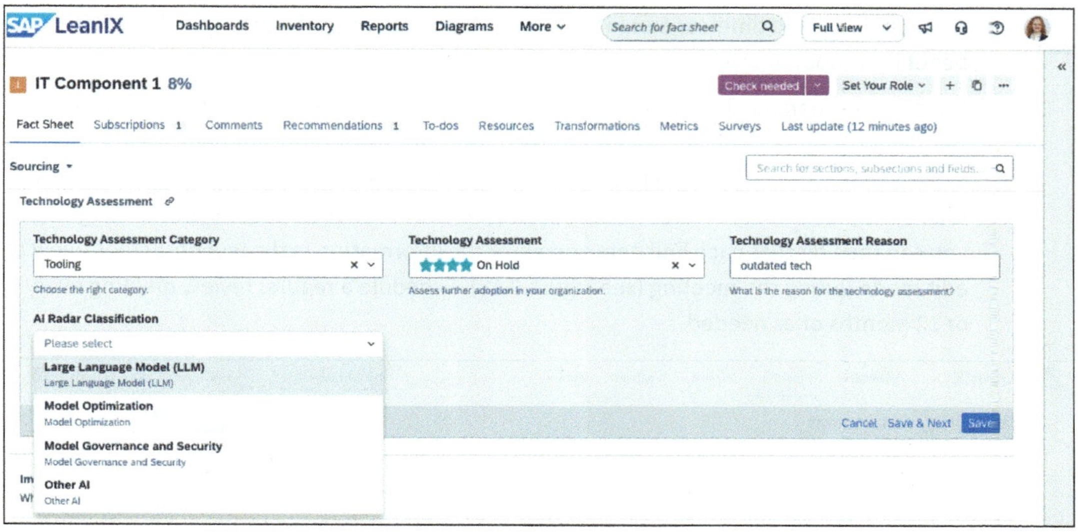

Figure 9.12: Example of Technology Assessment on IT Component Fact Sheet

The out-of-the-box select values include the following (these can be adjusted in the meta model configuration as needed):

- **Technology Assessment Category**
  - **Tooling**
  - **Languages and Frameworks**
  - **Data Management and Storage**
  - **Infrastructure**
- **Technology Assessment**
  - **Adopt**
  - **Trial**
  - **Assess**
  - **On Hold**
- **Technology Assessment Reason** (text field)
- **AI Radar Classification** (for AI-related IT components)
  - **Large Language Model (LLM)**
  - **Model Optimization**

- **Model Governance and Security**
- **Other AI**

There are multiple ways to gather the technology assessment information. You can update the data manually on each individual IT component fact sheet (see Chapter 4, Section 4.3.2 for how to do that, and then adapt your knowledge to enter information on the IT component fact sheet's **Fact Sheet** tab, under the **Technology Assessment** section). This process is very manual and time-consuming, so we advise using it only if you have only a small number of IT components in your inventory.

If you have a manageable number of IT components, it might be sufficient to create a saved search of IT components in a table view with relevant supporting columns (revisit Chapter 4, Section 4.3.3 to check how to create an editable table view). In a working meeting with the IT component responsible persons or subject matter experts like staff or lead engineers, you can then discuss each line item and enter the information in the inventory table view in edit mode during the meeting (see Figure 9.13). Schedule a regular review meeting every 6 or 12 months or as needed.

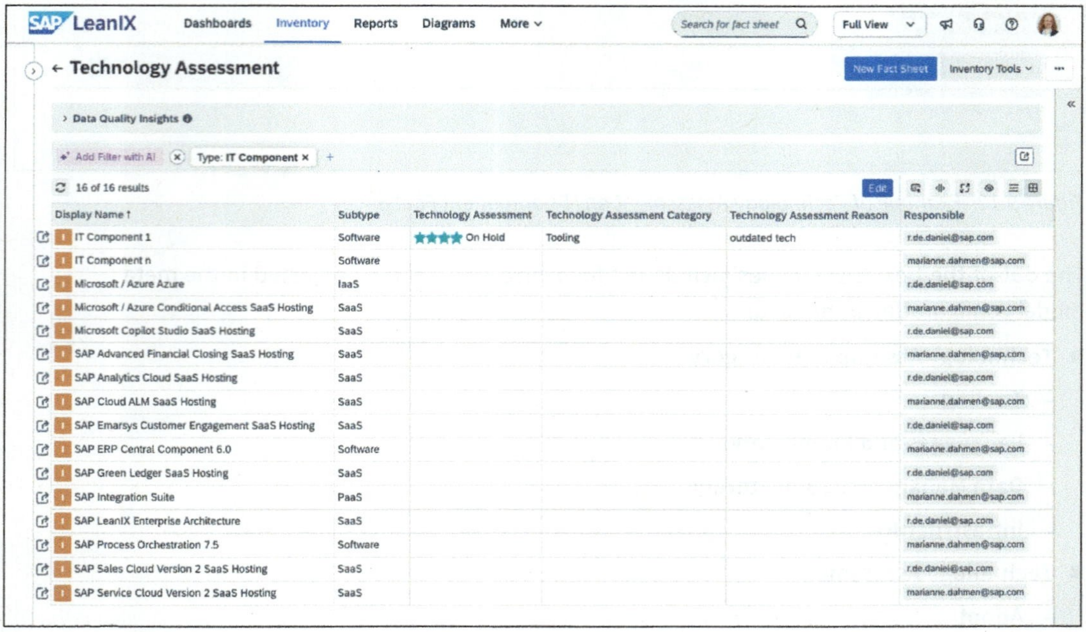

Figure 9.13: Prepared Saved Search for Technology Assessment Working Session

You can also use the survey capability to get the initial technology assessment data and then schedule recurring surveys for data reviews. Check Chapter 2, Section 2.3.3 to refresh your knowledge of surveys.

Once you have the data collected, open the technology radar report by navigating to the top ribbon **Reports** menu. In the left-hand menu under **REPORT TYPES**, click the **Radar** submenu. Click the **Technology Radar** tile (see Figure 9.14).

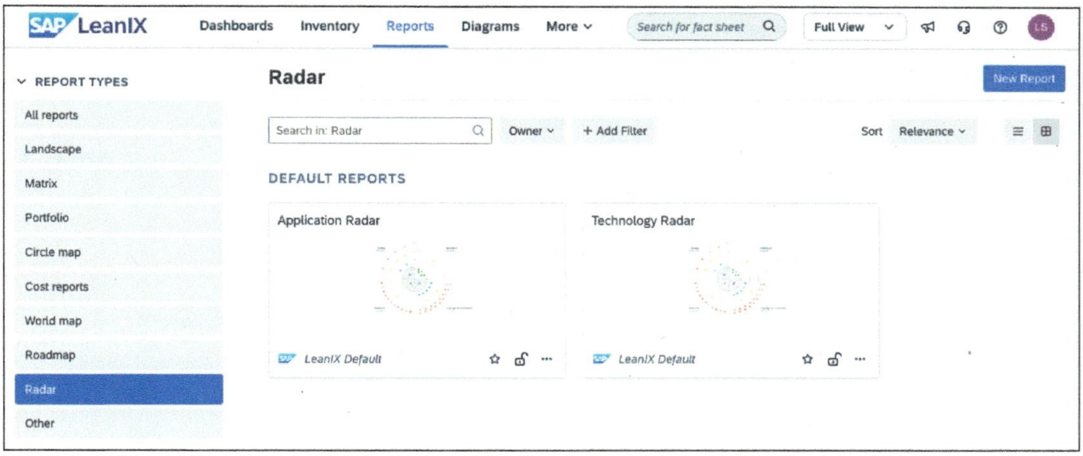

Figure 9.14: Navigating to Technology Radar Report

By default, the radar report settings (accessed by clicking **Settings**; see Figure 9.15) show **Technology Assessment** under **Ring Category** (similar to tree rings in a tree trunk) and **Technology Assessment Category** under **Sector Category** (similar to pizza slices of a round pizza). **View** is set to **Technology Assessment**, which is an information redundancy. Feel free to adjust either the ring category or the color-coded view for an additional data point as needed.

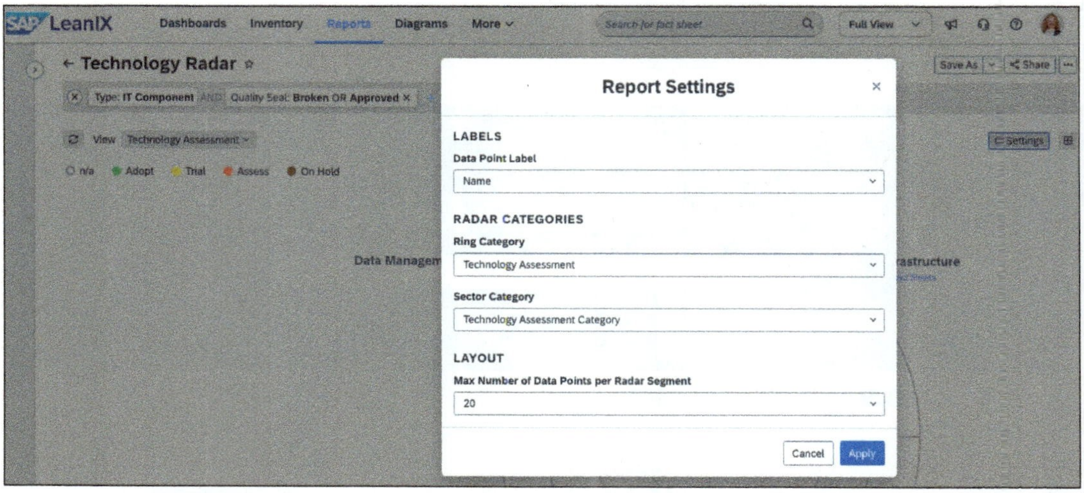

Figure 9.15: Technology Radar Report Default Settings

Share the filled **Technology Radar** report (see Figure 9.16) with your internal developer community (e.g., via a dashboard or a portal; see Figure 9.17) to provide a simple overview of approved and nonapproved tech stacks.

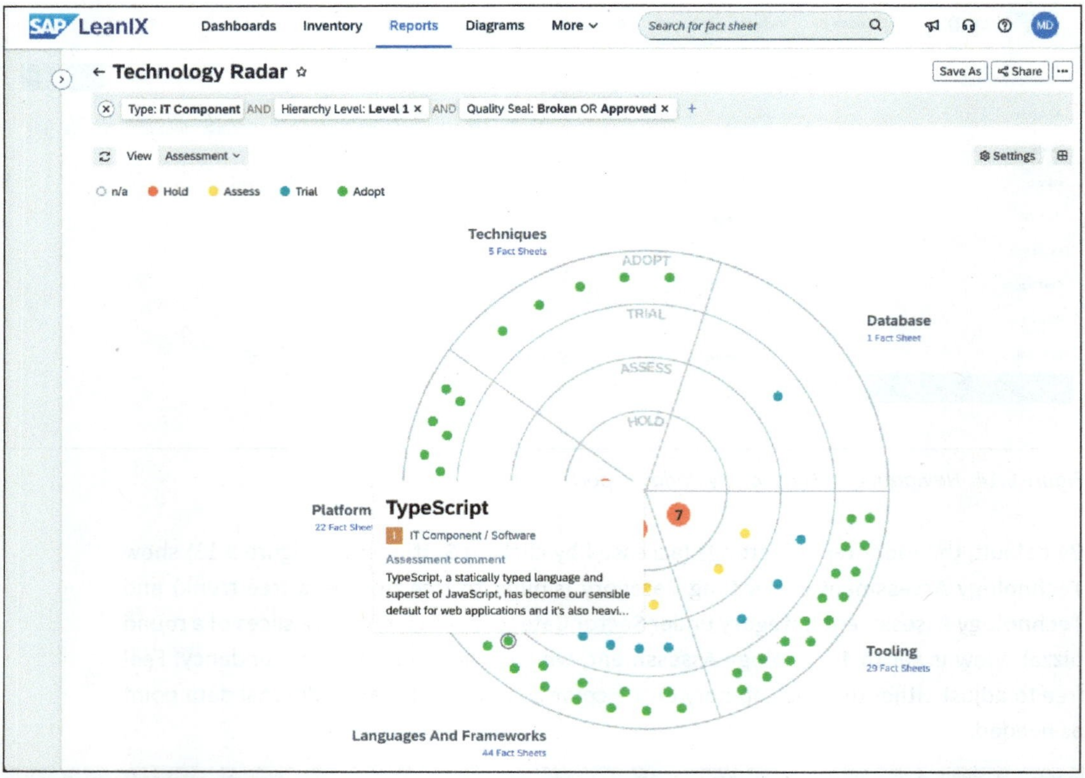

Figure 9.16: Example of Technology Radar Report

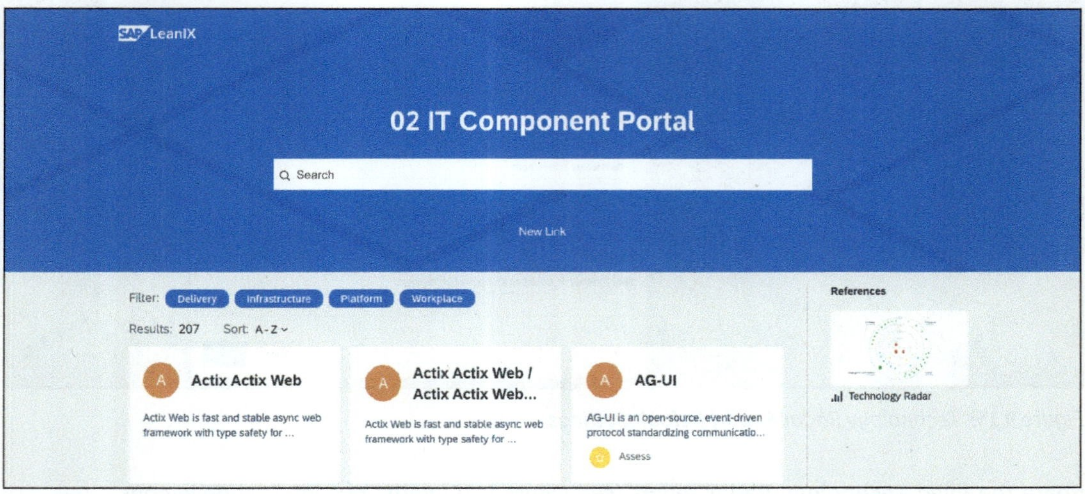

Figure 9.17: Technology Radar Report in Configured IT Component Portal

9.2 Mergers and Acquisitions

Another interesting use case that SAP LeanIX supports is M&A. You may be familiar with the scenario in which you must make critical decisions under time pressure, with incomplete or inaccurate data, and with legal and/or monetary implications. SAP LeanIX can help improve transparency in the following M&A-relevant areas:

- IT due diligence
- Postmerger integration planning and execution
- Portfolio optimization
- IT modernization

Various features in SAP LeanIX can make those projects easier to handle, more efficient in their execution, and easier to document. It is important to bring enterprise architecture into the M&A project from the beginning to align terminology and make use of synergies and aligned target architecture early on.

Table 9.1 offers an overview of activities and features that can support you and your M&A teams during pre- and postmerger/postacquisition phases.

Strategy Alignment	■ Deal scope, diligence, assumptions, transitional service agreement ■ Leadership alignment and separation strategy ■ Enterprise architecture governing body and governance model
Capability Analysis	■ Reference catalog functionality ■ Current and target business capability map ■ Capability maturity and strategic assessment
Current-State Enterprise Architecture Analysis	■ Discovery functionality for enterprise architecture landscape ■ Reporting on focus areas and value creators ■ Ownership (functional and technical) and assessment
Target-State Enterprise Architecture Analysis	■ Target architecture with reporting structure ■ Data capturing with visualization capabilities ■ Plan for app rationalization and tool harmonization
Data Migration Strategy	■ Gap analysis and configuration needs ■ Data/interface mapping and data migration ■ Best practice modeling guidelines
Roadmap Execution	■ Lifecycle management capabilities ■ Transformation roadmap and progress tracking ■ Resource and budget management
Handover to Organization	■ Transitional service agreement exit and handover to organization ■ Harmonized and transparent data model ■ Solid foundation for continuous improvement

Table 9.1: SAP LeanIX Capabilities and Where They Can Support the M&A Activity Chain

Let's break down these phases in more detail:

1. **Strategy alignment**
   The first phase, strategy alignment, can be supported by agreeing to use SAP LeanIX for IT due diligence and by defining the objectives, decision bodies, and capabilities to be used.
2. **Capability analysis**
   For each assessment, a separate SAP LeanIX workspace should be used. We recommend starting with a capability analysis. For that, define the capability map of the target company and compare it with the existing capability map of the acquiring company. Ideally, you can use the business capability reference catalog to ensure a comparable baseline. Perform a maturity assessment and check for overlaps, synergies, and mismatches.
3. **Current state enterprise architecture analysis**
   Ideally, the target company is already a user of SAP LeanIX and is following best practice modeling guidelines. If so, then the transfer of relevant inventory items for the current state enterprise architecture analysis will be very easy. Define the scope of fact sheets of interest and exchange the relevant information via Excel export/import. It is not necessary to copy all custom fields or fact sheet types. Focus on the relevant SAP LeanIX out-of-the-box fact sheets, like applications, IT components, interfaces, data objects, and organizations, if maintained. In the due diligence workspace, perform a new functional and technical fit assessment of the applications and IT components of the target company.
4. **Target state enterprise architecture analysis**
   Next, compare the current-state enterprise architecture of the target company with your own architecture and perform a target-state enterprise architecture analysis including app rationalization or app modernization potentials. Use the initiative fact sheet to capture different plans and scenarios.
5. **Data migration strategy**
   Following the high-level target planning, use SAP LeanIX for insight generation for the data migration strategy. This is where you should model high-level interface fact sheets with exchanged data objects. This is also your last chance to modify the enterprise architecture frameworks and modeling best practices to reach a harmonized meta model, including the fact sheet types and fields needed.
6. **Roadmap execution**
   During roadmap execution, which is usually postacquisition, you can use the SAP LeanIX Transformation and Project Management integration capabilities as well as high-level enterprise architecture initiative monitoring to track the post-merger integration progress.
7. **Handover to organization**
   During handover to the organization, you either remove the old assessment workspace once everything is merged into a joint SAP LeanIX workspace and the new colleagues are onboarded to the new workspace and processes, or you keep the old workspace with a harmonized meta model, if your units operate as independent entities with own workspaces.

Further Resources

For more information, visit the following SAP LeanIX wiki article: *http://s-prs.co/v620674*.

You can also rewatch the webinar titled "How Enterprise Architecture Management Supported Post Merger Integrations at ZF" from March 10, 2021, at *http://s-prs.co/v620675*. Just enter your business email address in the **Watch Now** box in the upper-right corner and click the **Proceed** button.

Be aware that the tool has evolved a lot; ZF was using the old meta model version 3 at the time of recording the webinar. The reports shown might look different in your SAP LeanIX workspace today. Nonetheless, the insights from a project perspective are still very valuable.

9.3 Digital Operational Resilience Act

You might wonder: What is DORA? It is an EU regulation in effect since January 2025 for financial institutions operating in the EU and is meant to assess/manage technology risks in order to remain resilient to Information and Communication Technology (ICT) disruptions.

Experts from SAP LeanIX, SAP Signavio, and consulting partner Deloitte worked on a concept to support financial institutions to prepare for DORA using SAP LeanIX and SAP Signavio. In this section, we will focus on the SAP LeanIX part of the story.

Further Resources

For more information on DORA management in SAP LeanIX, visit the SAP LeanIX user documentation at *http://s-prs.co/v620676*.

You can also rewatch the webinar titled "Navigating DORA Compliance - Achieving Digital Resilience with SAP LeanIX and SAP Signavio" from February 13, 2025, at *http://s-prs.co/v620677*. Just enter your business email address in the **Watch Now** box in the upper-right corner and click the **Proceed** button.

The DORA meta model extension provides curated fields for various fact sheet types, which helps customers in the financial industry assess and document DORA-relevant details and procedures. Additional fields and relations will affect the application, IT component, business context, and provider fact sheet types.

You can find this extension in the SAP LeanIX Extension Hub. Navigate there by clicking **More** in the top ribbon and selecting **Extension Hub** from the menu. In the Extension Hub, filter **Extension Type** to **Meta Model**, then click the **Digital Operational Resilience Act (DORA)** tile (see Figure 9.18).

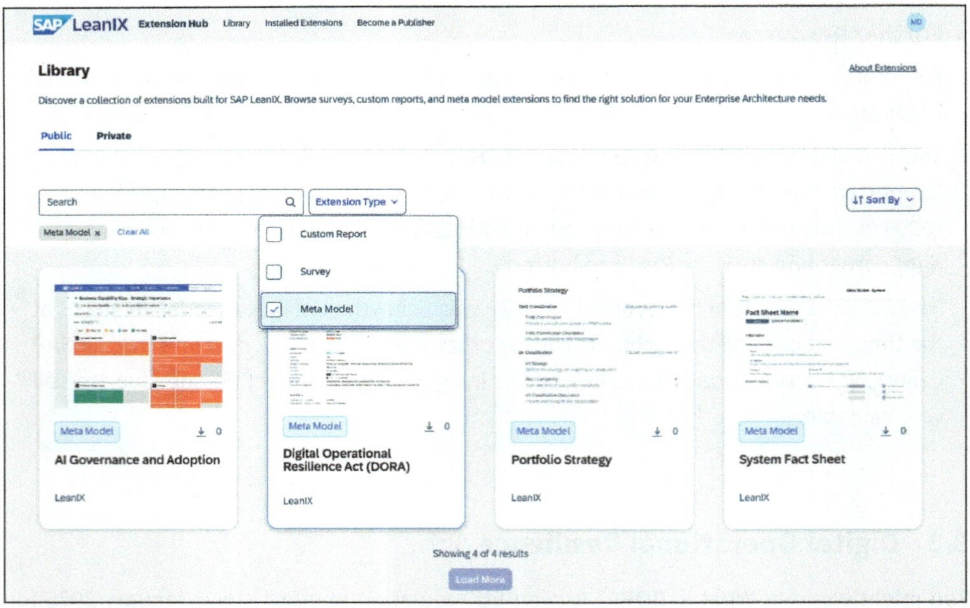

Figure 9.18: Searching for DORA Meta Model Extension in Extension Hub

On the next screen, you will find more information about the meta model extension. To load it into your workspace, click the blue button in the upper-right corner that says **Add to <Workspace Name>** (see Figure 9.19).

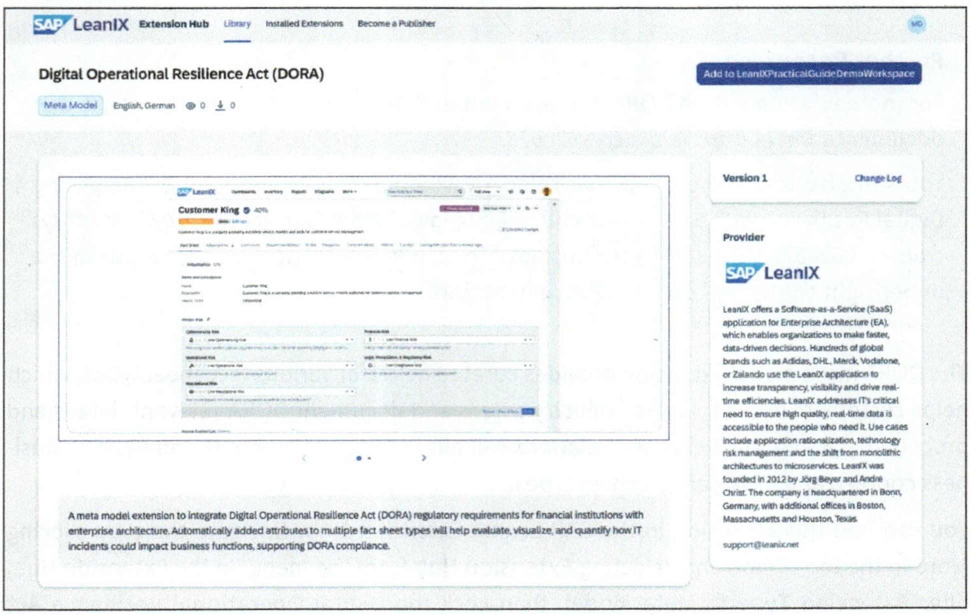

Figure 9.19: Adding Meta Model Extension to Workspace

Make sure that you are logged into the correct workspace before activating the meta model extension provisioning. After clicking the button, you will be prompted to confirm the meta model extension action, as this step can't be undone automatically. Click the blue **Continue** button to confirm.

Documenting DORA-relevant information in SAP LeanIX can help the enterprise architect to increase regulatory awareness among users and support cross-departmental alignment with powerful visuals.

First, you need to identify which applications, IT components, and providers (vendors) are relevant and in scope for the DORA assessment. This can be done by filling out the simple yes/no single-select **DORA Critical** field in the general **Information** section under **Name and Description** (see Figure 9.20).

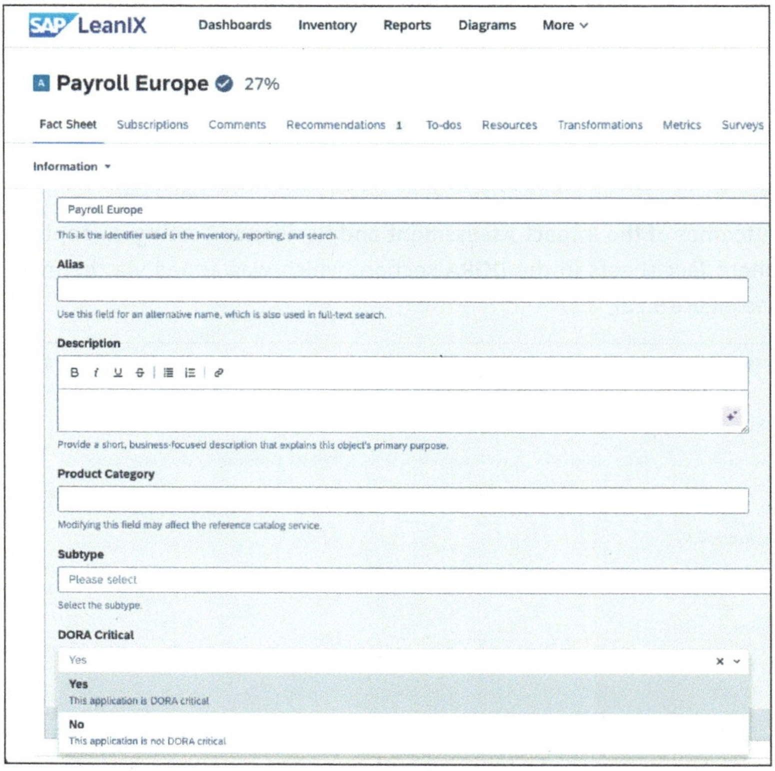

Figure 9.20: Marking DORA Criticality in Information Section of Application Fact Sheet

Second, you need to assess the third-party vendors on **Criticality and Quality**. This includes their **Financial Stability**, **Integration Capabilities**, **Reputation**, **Operational Resilience**, **Sourcing Risk**, **NPS**, and **Exit Criteria**. Those fields can be found in the provider fact sheet under the **Evaluation and Cost** section, in the **Criticality and Quality** subsection (see Figure 9.21).

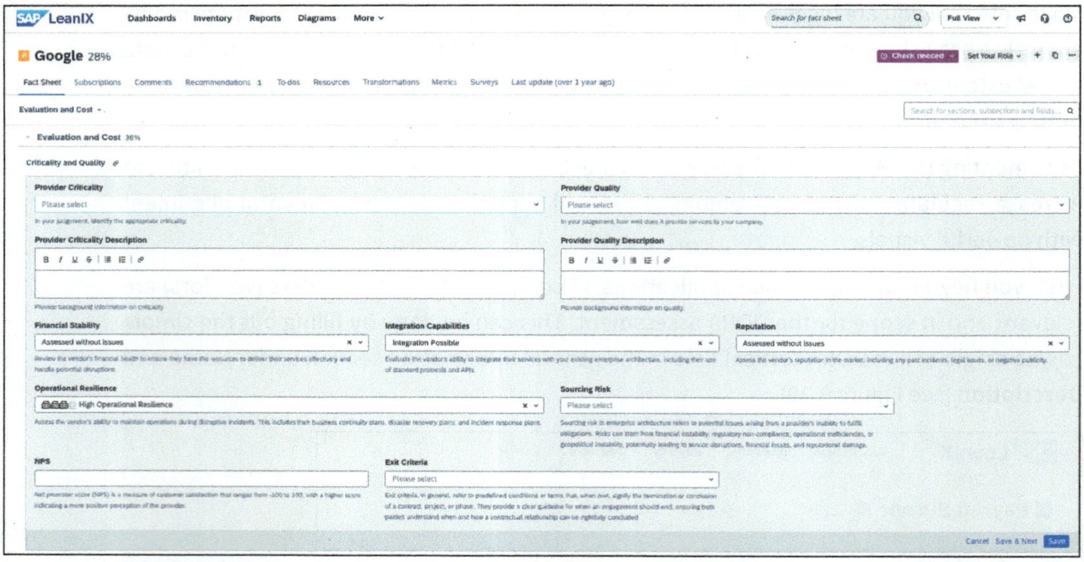

Figure 9.21: Example of DORA Provider/Vendor Assessment

Third, document outcomes of the **Impact Assessment** and **Resilience Testing** on application and IT component fact sheets in the **DORA** section, which was added via the meta model extension (see Figure 9.22).

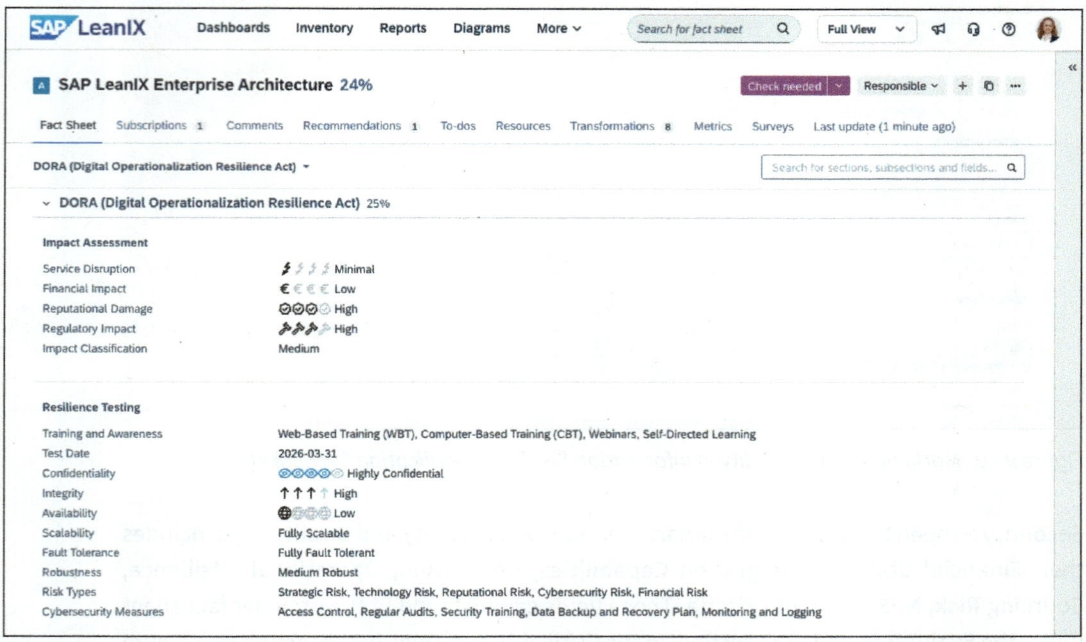

Figure 9.22: DORA Impact Assessment and Resilience Testing Attribute Example

The fields should be filled out by the owners/responsible persons for the respective application and IT component team in collaboration with the IT security team. The impact analysis values will help with prioritizing mitigation strategies during a global incident, to quickly identify the most critical applications/IT components (those that will cause not only service disruptions but also carry financial, reputational, or regulatory risks). The resilience testing, on the other hand, gives an indication of how robustly the application and IT component are set up and how well the workforce is trained in regard to risks and countermeasures to prevent exposures or other cybersecurity issues.

To add the information most efficiently, make use of the SAP LeanIX collaboration features, such as surveys, to-dos, and comments, plus automations and calculations where it makes sense to crowdsource information.

Finally, capture your recovery plans and training needs in the business context fact sheets under the new **Continuity Plans** and **Training and Awareness** subtypes and connect them to the relevant applications and IT components.

Remember that this is not a one-time only exercise; it requires regular updates in a defined cadence. You can use the quality seal feature and scheduled surveys for that.

9.4 Preparation for Open Finance

Around the globe, open finance or open banking regulations are being developed. In the past, banks had closed banking systems with in-person service for their customers. With the shift to online and mobile banking, more and more service providers and fintech companies entered the market with higher demands for customer data sharing. Bank systems nowadays must interact with third-party market players other than the customer and need to provide secure technology for sensitive data exchange (see Figure 9.23). New regulations for secure and fair data exchange are touching not only the finance industry but also insurance and adjacent areas.

Experts from SAP LeanIX, Collibra, and consulting partner EY-Parthenon worked on a concept to support financial and insurance customers preparing for open finance via SAP LeanIX and *Collibra*, a data intelligence platform focused on data governance, data cataloging, and data lineage.

Note

This use case primarily deals with current and future EA landscapes, including their high-level interface connections and data products exchanged. It is important to note that enterprise architects must work very closely with business users and data domain owners, as well as the data stewards of the company to be successful in this use case.

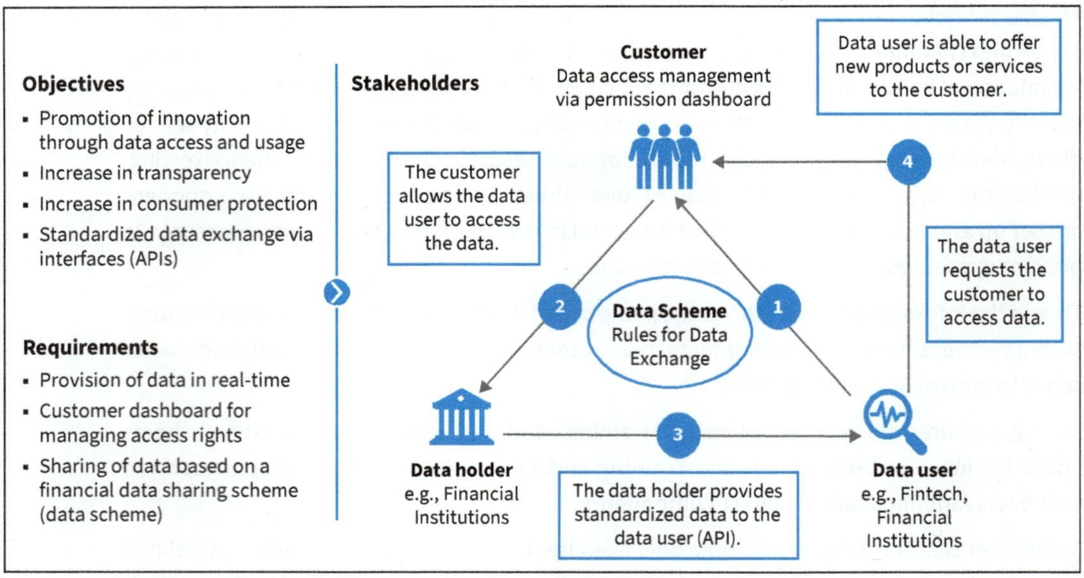

Figure 9.23: Open Finance Scenario (Source: EY-Parthenon)

SAP LeanIX delivers architectural visibility, mapping systems, APIs, and data flows across the IT landscape. On the other side, Collibra provides the data governance layer for managing lineage, classification, consent, and audit trails. The business logic comes from the business units involved in the day-to-day operations.

For this use case, we work with the out-of-the-box integration between SAP LeanIX and Collibra, available with the SAP LeanIX Application Portfolio Management product.

Further Resources

Be aware that data objects will get an overhaul with the introduction of new SAP data products. The SAP LeanIX and Collibra integration also will see some facelifts and changes to its functionality.

Therefore, we highly recommend checking the latest information about the SAP LeanIX and Collibra integration in the SAP LeanIX user documentation at *http://s-prs.co/v620678*.

On a high level, SAP LeanIX is the source of truth for applications, business capabilities, and interfaces and their relations, and it syncs this information to Collibra. Collibra is the source of truth for data objects and, in the future, *data products*, which are a type of business asset in Collibra synced to SAP LeanIX (see Figure 9.24).

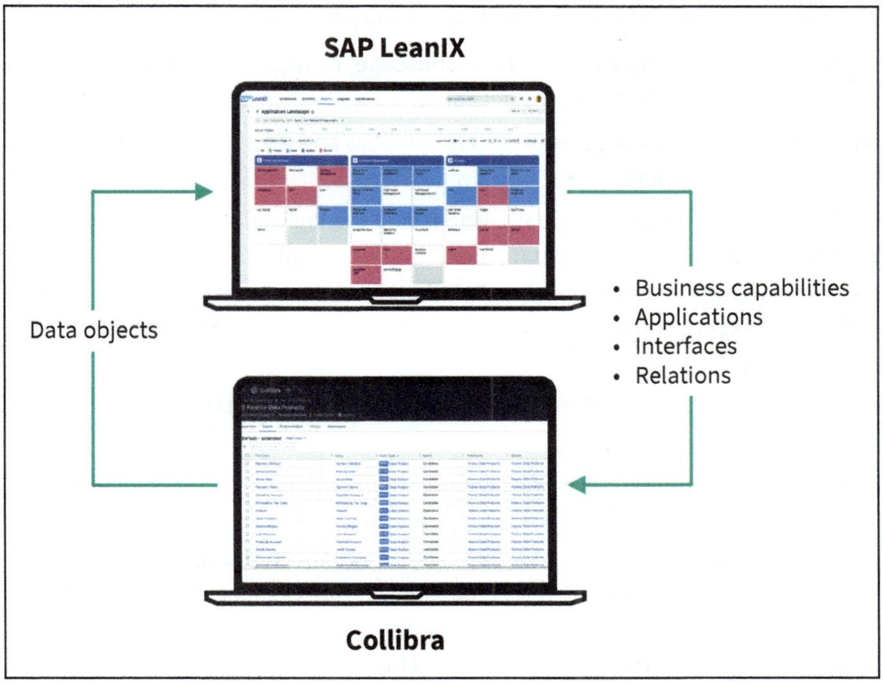

Figure 9.24: Simplified Asset Exchange Between SAP LeanIX and Collibra

The out-of-the-box meta model is slightly enriched with the following fields, within the business capability meta model configuration, in the **Fields** tab, under the **Open Finance** section (see Figure 9.25):

- **Open Finance Relevance** field (single-select, yes/no)
- **Open Finance Criticality** field (single-select field, with low/medium/high/critical as values)

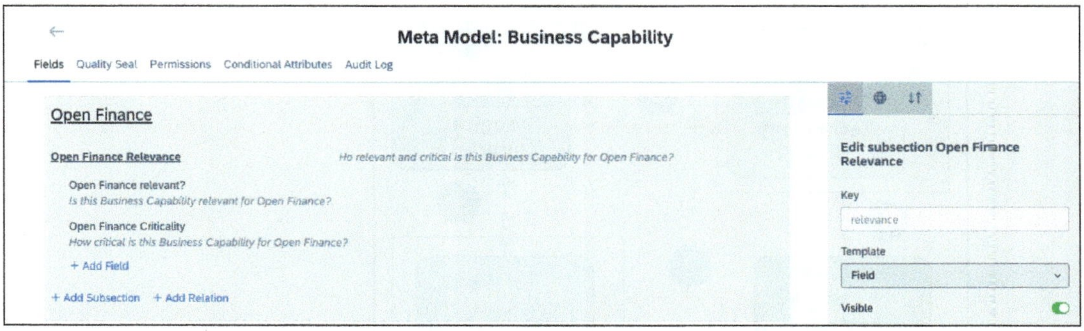

Figure 9.25: Open Finance Meta Model Modifications in Business Capability Data Model

For this use case, the flow of high-level steps is as follows (see Figure 9.26):

1. Gather your current enterprise architecture landscape, including business capabilities, applications, and logical interfaces in SAP LeanIX.
2. Maintain your data objects as business assets in Collibra and sync them to SAP LeanIX as data objects or data products.
3. Connect the high-level data objects/data products to the logical interfaces and applications (see Figure 9.27 for a simplified data model). Here the enterprise architecture team should work closely with business experts. Relations can be created either within the fact sheets, via Excel export/import, or visually in diagrams (see *http://s-prs.co/v620679* for more information).
4. Sync all maintained relations in SAP LeanIX back to Collibra for full transparency across systems.
5. If needed, enrich the information in SAP LeanIX with underlying IT component data and modeled interface subtype API (see Figure 9.28 for an extended data model). For open banking, it is not only important to know that there is a connection between two or more business applications, and which high-level data objects are exchanged; it is also important to know what interface technology is used and how secure this IT component is against security threats. The same holds true for APIs and endpoints used, which can be modeled using the optional API subtype of the interface fact sheet in SAP LeanIX.
6. Plan your target architecture in SAP LeanIX (not depicted in Figure 9.26).

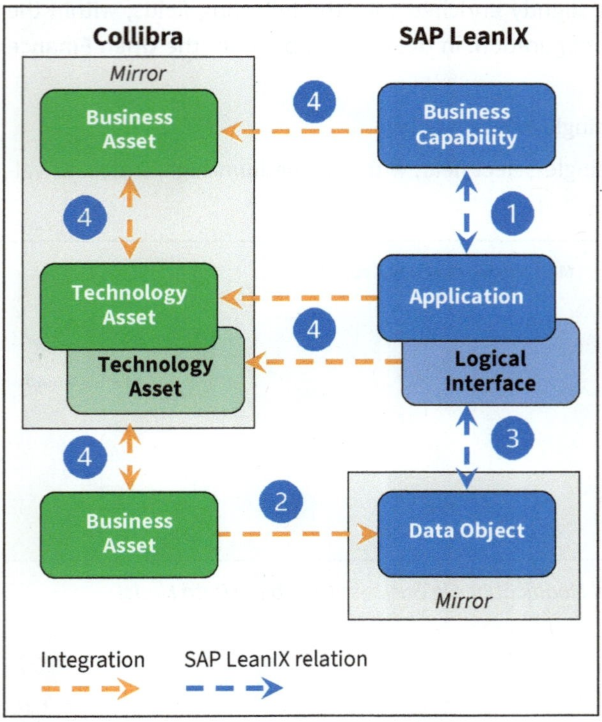

Figure 9.26: Open Finance: High-Level Modeling Concept Between SAP LeanIX and Collibra

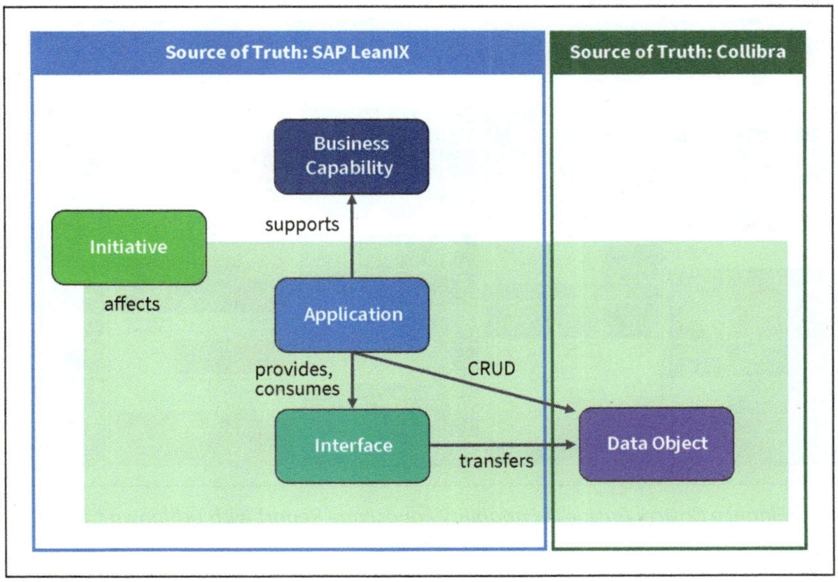

Figure 9.27: Simplified Data Model for Open Finance Modeling

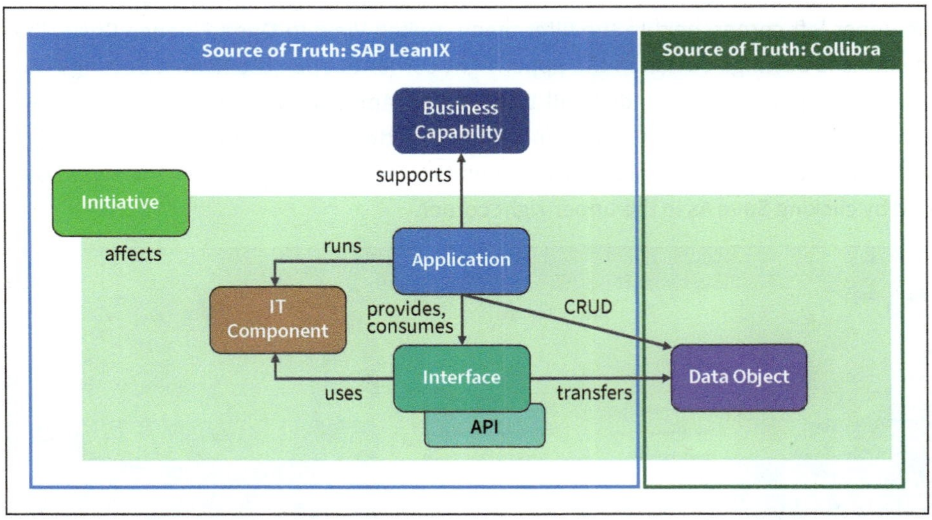

Figure 9.28: Extended Meta Model for Open Finance with IT Components and APIs

In SAP LeanIX, we recommend starting with a business capability–driven analysis to identify potential *white spots*; that is, business capabilities that are needed but missing in the current business capability map. Next, assess your business capabilities regarding their open finance criticality (custom field) and review which active applications currently support those capabilities.

In Figure 9.29, we created a **Business Capability Landscape** report, with the color-coded view set for the custom **Open Finance Criticality** field and a drilldown into applications.

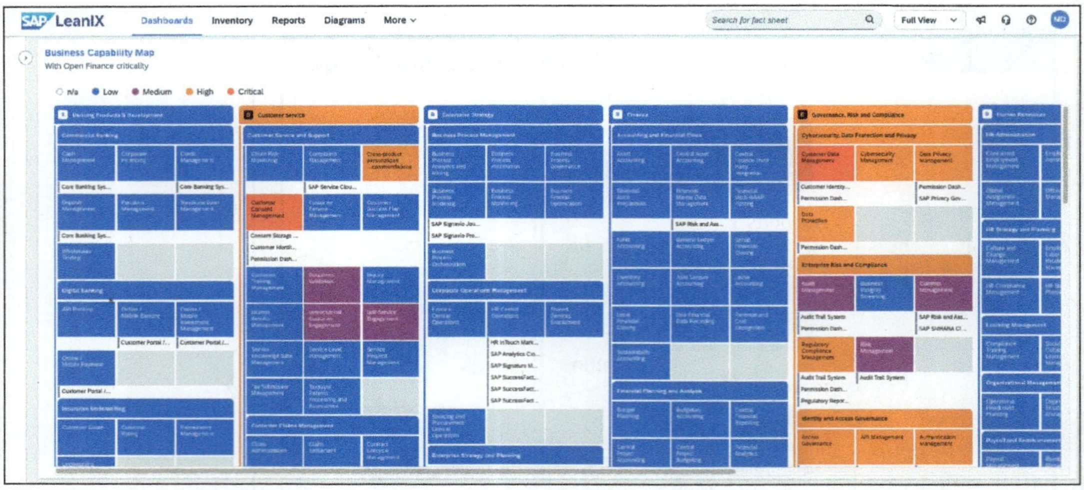

Figure 9.29: Open Finance Critical Business Capability Landscape Report with Drilldown to Supporting Applications

To create such a report, click **Reports** in the top ribbon menu and select **Landscape** in the left-hand side collection overview. Click the tile with the default **Business Capability Map**. In the upper-left corner, next to the filter menu, switch **View** to **Open Finance Criticality**. Then click the **Settings** button in the upper-right corner to open the **Report Settings** window. In the dropdown menu under **Drill-Down**, select **Applications** to be able to drill down into the related applications. Click the blue **Apply** button in the lower-right corner of the **Report Settings** window to finalize the configuration (see Figure 9.30). Save this report as a variant by clicking **Save As** in the upper-right corner.

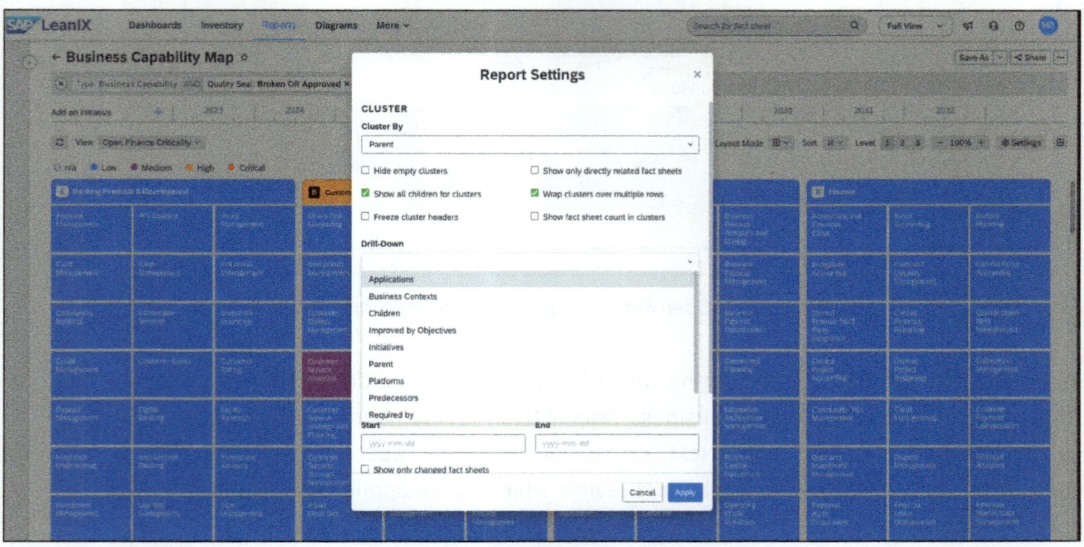

Figure 9.30: Configuring Application Drilldown into Business Capability Landscape

After the capability-driven preanalysis, you can take a closer look at the supporting applications and their interface ecosystem. For that, create a high-level current state diagram as a free draw diagram with logical interface connections between applications. You can also map the synced data objects from Collibra to the respective logical interfaces (see Figure 9.31). Revisit Chapter 5, Section 5.2.4 to refresh your knowledge of how to design free draw diagrams in the diagram editor.

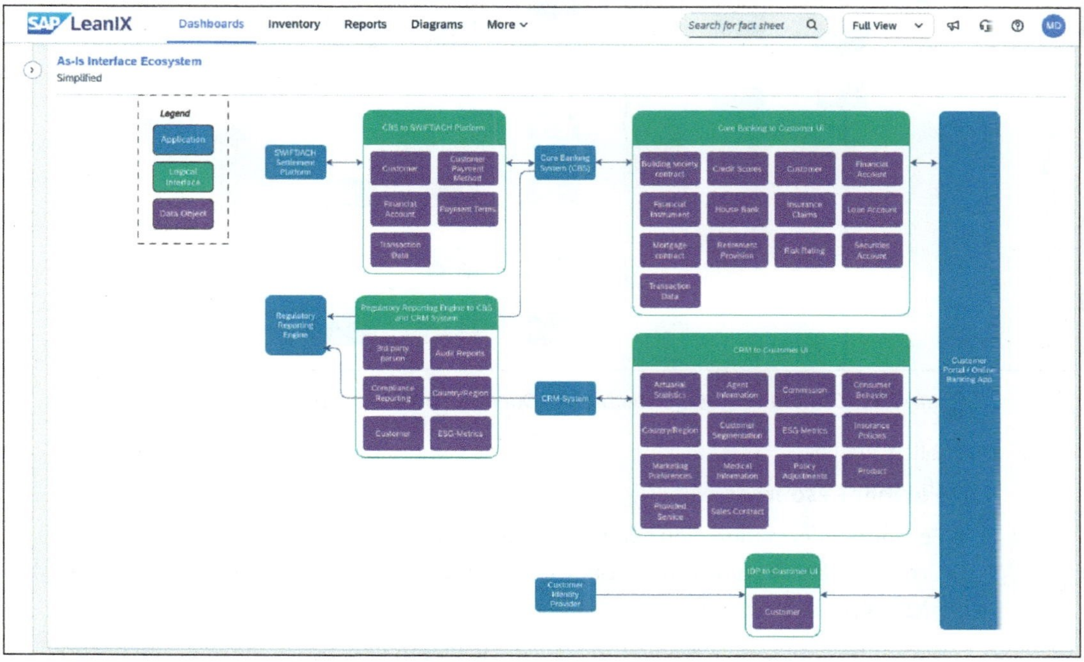

Figure 9.31: Simplified Current Integration Landscape Modeled in Free Draw Diagram

Now that the as-is state is captured, you can start to develop a target state architecture. In this example, we don't want a greenfield implementation; instead, we'll build on what we have with a brownfield approach. Therefore, start from the as-is diagram and enrich it with the open finance–relevant extensions and additions (see Figure 9.32). Ideally, you'll use the target architecture planning diagram for this exercise. Revisit Chapter 6, Section 6.4 to refresh your knowledge of target architecture diagrams.

In the diagram, the lines between applications each represent an interface fact sheet. The arrow direction indicates the data flow direction. A solid line means that the interface exists and is active. A dotted line means that the interface is still in a planned lifecycle phase.

Finally, use the target architecture and roadmap planning features of the SAP LeanIX Architecture and Road Map Planning product to detail the open finance migration project in initiative fact sheets using transformations. For more information, see Chapter 6.

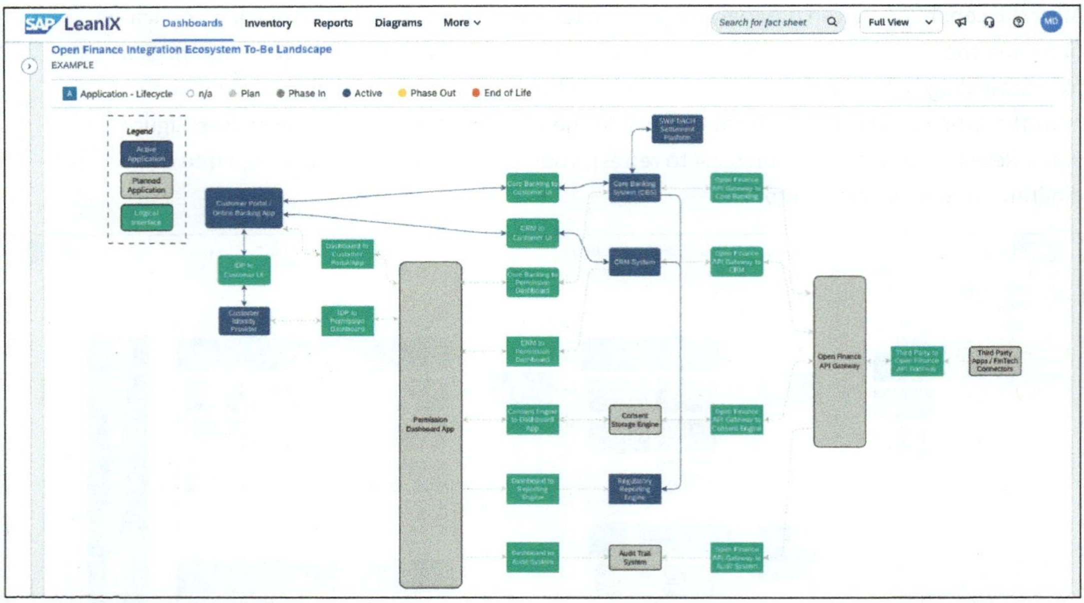

Figure 9.32: Drafted Target Architecture for Open Finance, Including As-Is Landscape

Further Resources

For more information, you can rewatch the webinar titled “Open Finance in Practice: Orchestrating Architecture & Data Governance at Scale” from March 25, 2026, at *http://s-prs.co/v620680*.

Just enter your business email address in the **Watch Now** box in the upper-right corner and click the **Proceed** button.

9.5 Summary

In this chapter, you learned about additional use cases for and additional capabilities and features of the SAP LeanIX Technology Risk and Compliance product, specifically, for self-built software discovery and management.

Using SAP LeanIX in M&A, for preparation for DORA, or for open finance regulations are just a few of the many possible use cases showing how SAP LeanIX and a well-maintained IT landscape can benefit your organization as a whole, and senior executives specifically, when making strategic decisions or planning significant changes to the enterprise architecture landscape. We are curious to see how you will use SAP LeanIX. If you have a great new use case idea that significantly added value for your organization, share it with us and maybe you will be featured in our customer success stories, or we may even present your story as a lighthouse customer example at one of our signature events on stage.

Further Resources

Check the SAP LeanIX user documentation on typical enterprise architecture use cases at *http://s-prs.co/v620681*.

For more inspiration, check out our customer success stories at *http://s-prs.co/v620682*.

In the next chapter, we'll explore some very useful features from the advanced settings menu of the administration area, calculations and automations. Both techniques will make your life as an SAP LeanIX admin much easier and will help you to increase data integrity, completeness, and accuracy.

Chapter 10
Advanced Techniques: Calculations and Automations

Every enterprise architecture team will at some point add custom calculations to their framework. Similarly, every enterprise architecture team has too much to do and too few colleagues. This is where SAP LeanIX calculations and automations can take over some heavy lifting for smooth and efficient enterprise architecture operations.

In this chapter, we will dive deeper into some advanced techniques that can be set up by admins in the SAP LeanIX workspace: calculations and automations. First, you'll learn the differences and similarities between the two features and when to use them. Next, we'll explore how to work with calculations in the workspace with a hands-on example. In addition to discussing the **ADVANCED SETTINGS** portion of the **Administration** menu, we'll also discuss example code for templates and individual use cases. Then, we'll explore the automations setup and the various configuration options that can be chosen to automate for instances, data quality checks, or workflows for data governance. Finally, we'll look at the managed code feature, in which you'll find an overview of all custom code hosted in your SAP LeanIX workspace.

Further Resources

For latest information on the calculations and automations features, check out the SAP LeanIX user documentation:

- Calculations: *https://help.sap.com/docs/leanix/ea/calculations*
- Automations: *https://help.sap.com/docs/leanix/ea/automations*
- Comparison of both features: *http://s-prs.co/v620683*

10.1 Overview

Calculation literally means computing a new value for a defined attribute field based on a defined algorithm. A calculation field is read-only and can be a select field, a text field, or a numerical field. Each calculation can only update one field. It is not possible to have a calculation update multiple fields; to accomplish that task, the calculations have to be chained, such as by automations or a trigger field.

Automations, on the other hand, support rule-based automation of a wide range of actions (add or remove a tag, set a field value, run a script, etc.) based on a defined triggering event (changes in fact sheet relations, fact sheet creation, changes in field values, etc.).

Both features can also be used in combination for complex scenarios.

Automations offer a user interface in the **Administration** menu in which an admin can design and create an automation without any special skills (guided setup).

For script-based automations, a basic understanding of JavaScript is needed. Calculations are based on JavaScript as well. There are lots of examples and templates available in the **Administration** menu and the SAP LeanIX online user documentation; you can modify or adjust sample code to fit your needs. In today's world of AI coding helpers, it should even be possible for nontechnical users to adjust templates to their needs using AI coding support. It is important to test the script, such as in a sandbox, before implementing it in the production workspace as automations especially can quickly perform mass updates that are difficult and time-consuming to reverse if something goes wrong.

Keep in mind that there are no rate limits on calculations, but there are quotas for automations as they are more complex and require more computing power. Therefore, demo and sandbox workspaces have a limit of 10,000 automation executions per month, and production workspaces have limits based on the purchased app quota.

10.2 Calculations

Creating calculations is only possible for SAP LeanIX admins. In this section, we'll walk through the steps to add a calculation and activate a calculation template. We'll also walk through custom calculation syntax and considerations for error handling and deactivation.

Tip

Before opening the tools, read these quick tips. In this chapter, we'll revisit these points in more detail:

- **Use correct syntax**
  Every calculation must start with `export function main() { ... }`.
- **Write in modern JavaScript**
  Use ECMAScript 2024, but no external libraries.
- **Optimize for performance**
  Avoid complex loops or unnecessary processing.
- **Test and validate thoroughly**
  Handle nulls and undefined and empty fields as well as edge cases; test your logic with sample data.
- **Monitor and refine continuously**
  Watch how values behave in real life; adjust logic if output diverges from expectations.

- **Start with templates**
  Don't reinvent the wheel; use SAP LeanIX templates to get started.
- **One calculation per target field**
  Plan dedicated output/target fields per metric that you want to calculate.
- **Make calculated fields read-only**
  Prevent accidental edits.
- **Use technical keys**
  Always reference attributes and relations by their technical key, not the display name.

10.2.1 Adding a Calculation

To navigate to the calculations management page, first enter the administration menu by clicking your user icon in the upper-right corner and selecting **Administration**. In the left-hand menu, scroll to the **ADVANCED SETTINGS** area and select **Calculations**. The **Calculations** page will be empty initially except for a blue button that says **Create a Calculation** and a link to the SAP LeanIX user documentation that reads **Learn more about calculations** (see Figure 10.1).

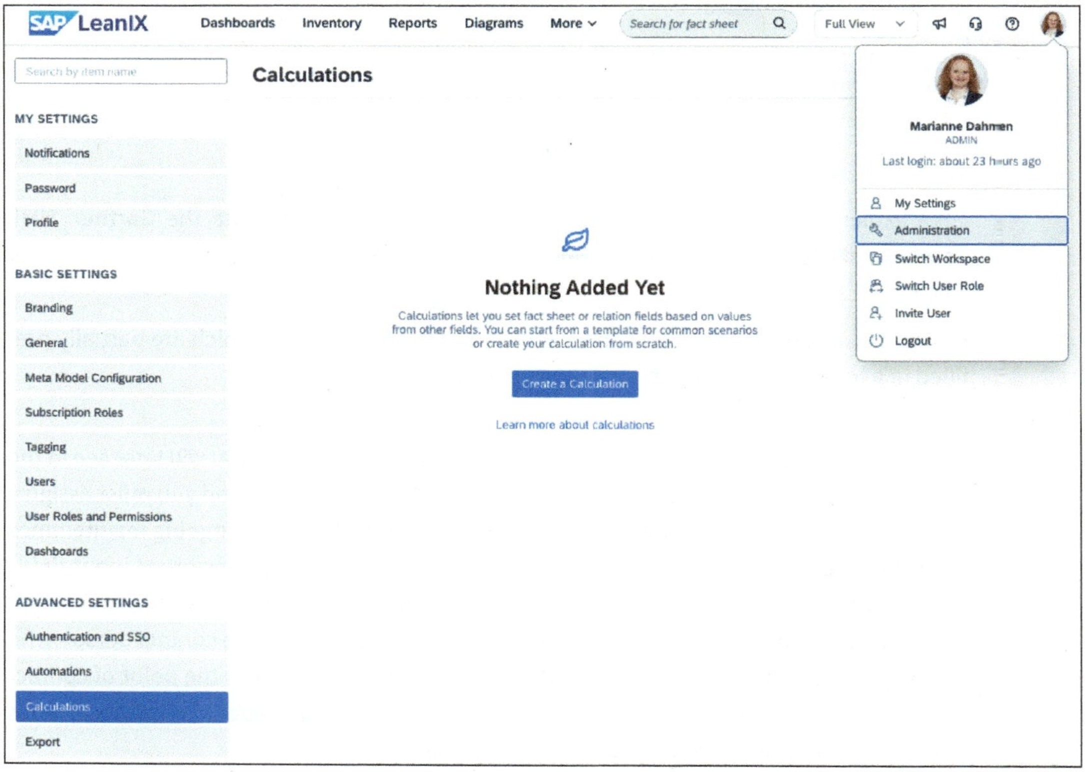

Figure 10.1: Navigating to Calculations in Administration Menu

After clicking the **Create a Calculation** button, you are brought to a new screen on which you will see the available template calculations under the heading **Templates** and the tile to create a custom calculation under the heading **Other** with the **Create from Scratch** tile and a blue **Create** button (see Figure 10.2).

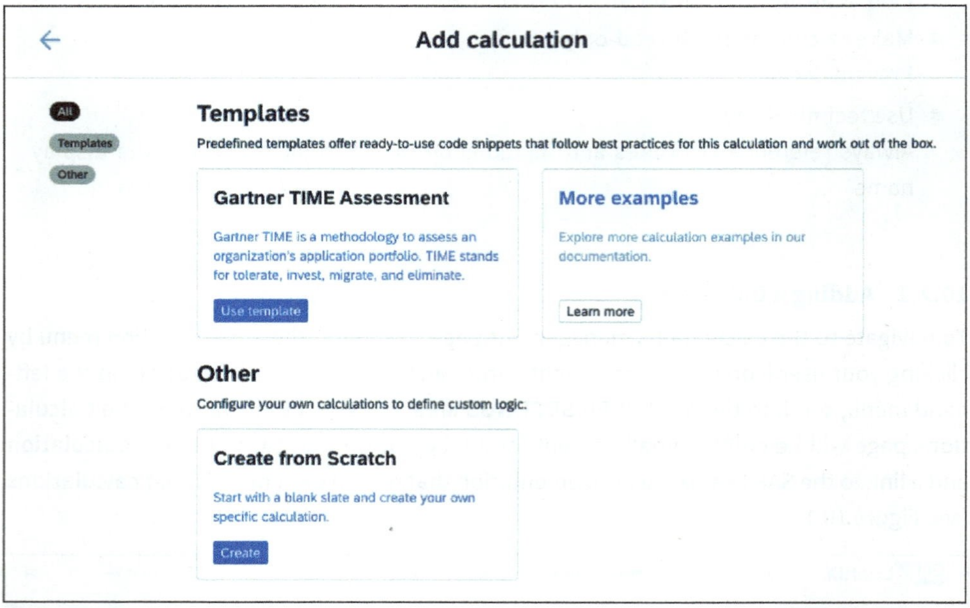

Figure 10.2: Create Calculation Screen

10.2.2 Activating a Template Calculation

Let's explore a template first. One available template is offered via the **Gartner TIME Assessment** tile. Click the blue **Use template** button to open the calculations configurator (see Figure 10.3).

As you can see, the calculation configurator has different sections, which are partially pre-filled in a template calculation:

- **Details**
  In the **Details** section, you can define a **Calculation Name**, which you will later see in the calculations management page; you can enter a **Description** text; and you must assign a calculation **Owner**. The owner is a user who is responsible for the ongoing maintenance of the calculation. For instance, if the calculation logic changes, or other input fields should be used, the owner must ensure that the changes are made in the calculation configurator. The owner does not need to be an admin. The owner is the contact person who reaches out to the admins if something needs to be changed and is the point of contact for the admins if they have a question regarding a specific calculation, for instance, if this calculated field is still needed on the business side.

- **Target Field**
  In the **Target Field** section, the calculation creator must select a value for the **Target Fact Sheet Type** dropdown, in which they select the fact sheet type containing the field that should get filled by this calculation. In addition, the precise location must be selected in the **Write to** dropdown, indicating whether the field is in the **Fact Sheet** or a **Relation**. Finally, the exact **Target Field** must be selected from a dropdown menu. Certain fields are excluded from being a target field, such as base fields like **Fact Sheet ID**, **Lifecycle Date**, and **Hierarchy Level**, or fields like the **Project Status** in the initiative fact sheet type.
- **Code**
  The **Code** area is an embedded terminal that can be used to develop and store the JavaScript code. It is securely saved in SAP LeanIX and hosted on SAP LeanIX infrastructure without additional third-party involvement. Figure 10.4 shows the structure of the **Code** section. In the upper-left corner above the terminal, you can find a link in blue font reading **View Calculation Examples**, which directs to the SAP LeanIX user documentation. Below the terminal, you will find a white **Test run** button and the blue **Save and Activate** button. Clicking the arrow next to the **Save and Activate** button will allow you to select the **Save in Inactive state** option.

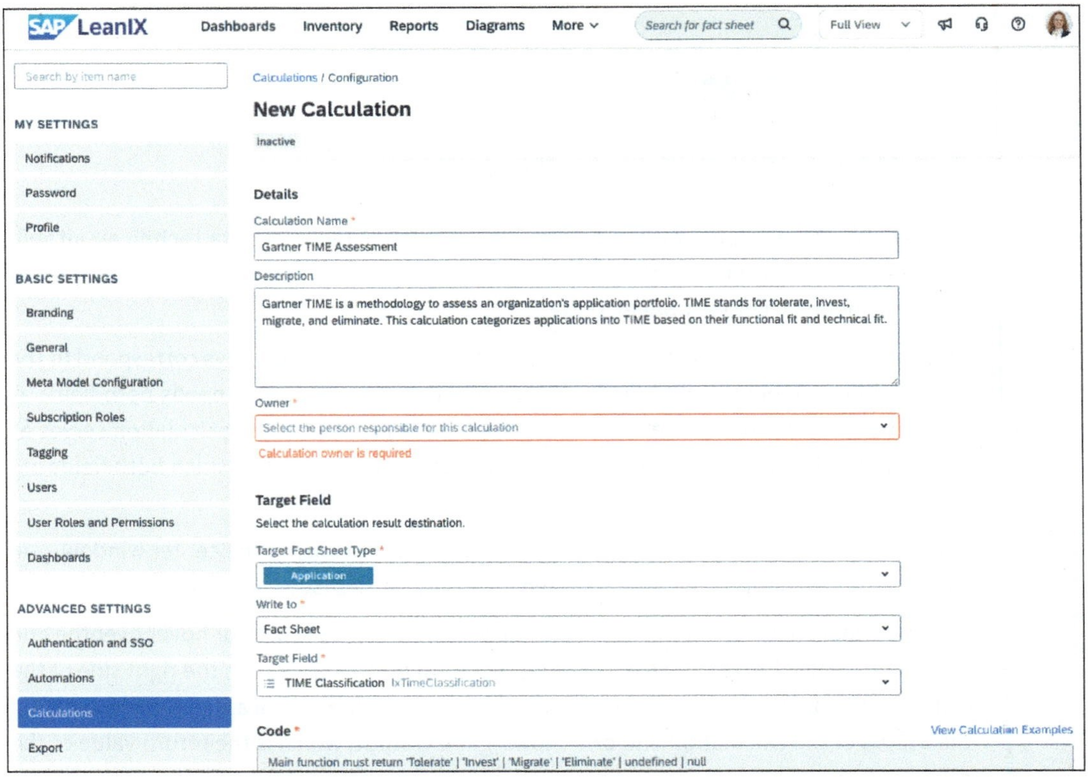

Figure 10.3: New Calculation Form with Prefilled Fields from Calculation Template

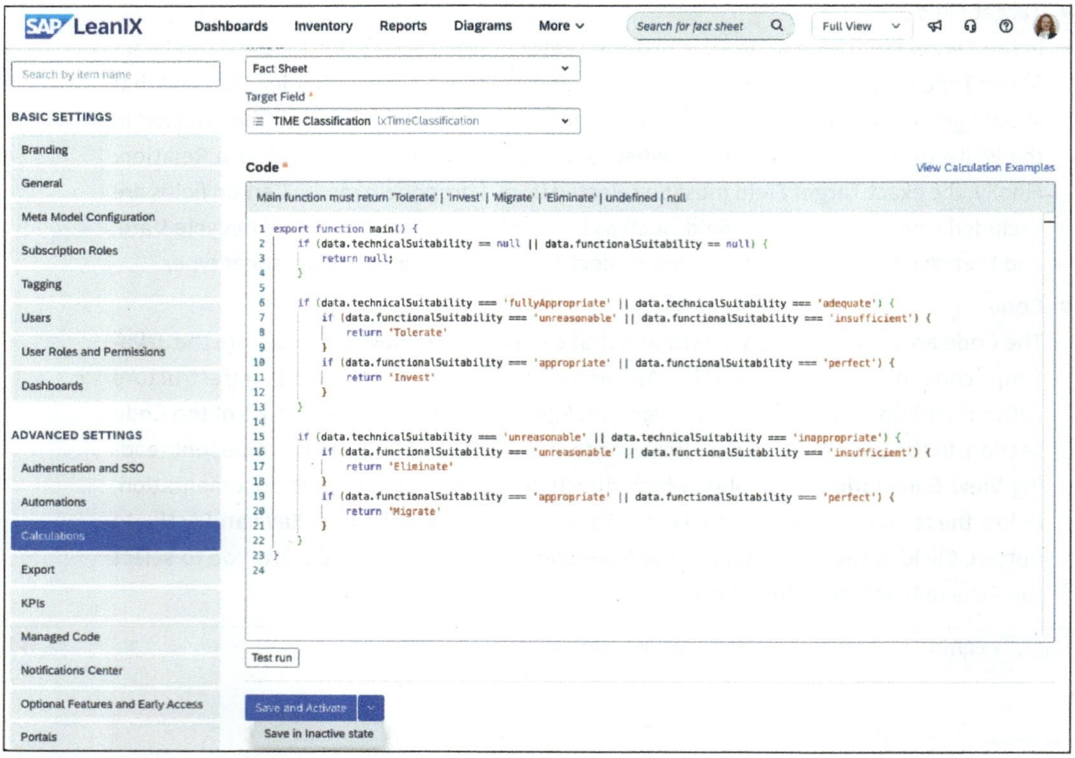

Figure 10.4: Code Terminal in Calculations Form

In this template example, you can see that the calculation uses the **data.technicalSuitability** and **data.functionalSuitability** fields in the application fact sheet (**Application** is selected under **Target Fact Sheet Type**).

An expert in the SAP LeanIX meta model knows that those technical keys correspond to the technical fit and functional fit fields. This means that the calculation needs both fields as inputs to calculate the Gartner TIME assessment. The first line of the calculation serves as the error handler, ensuring that only those entries are considered in which both fields are not empty.

When you click the grey **Test run** button in the lower-left corner of the center window, you can test the calculation logic with selected inventory entries (see Figure 10.5).

You can either enter text in the **Search for Applications** bar to search or hover over the line item of an application to make a play (**Request Run**) button appear on the right side of the cell. Click the button to run a calculation test. Two new columns will appear, one indicating the **Status** of the calculation and one showing the **Output**, which is the return value of the calculation.

The **Status** can be **Succeeded** in green or **Failed** in red. For a **Failed** calculation, the **Output** column will indicate what the error is (e.g., a certain "text" cannot be entered into the target field, which indicates that you have used the wrong technical key name).

If the status shows **Succeeded** and the **Output** holds the value you would expect, then the calculation is ready for activation.

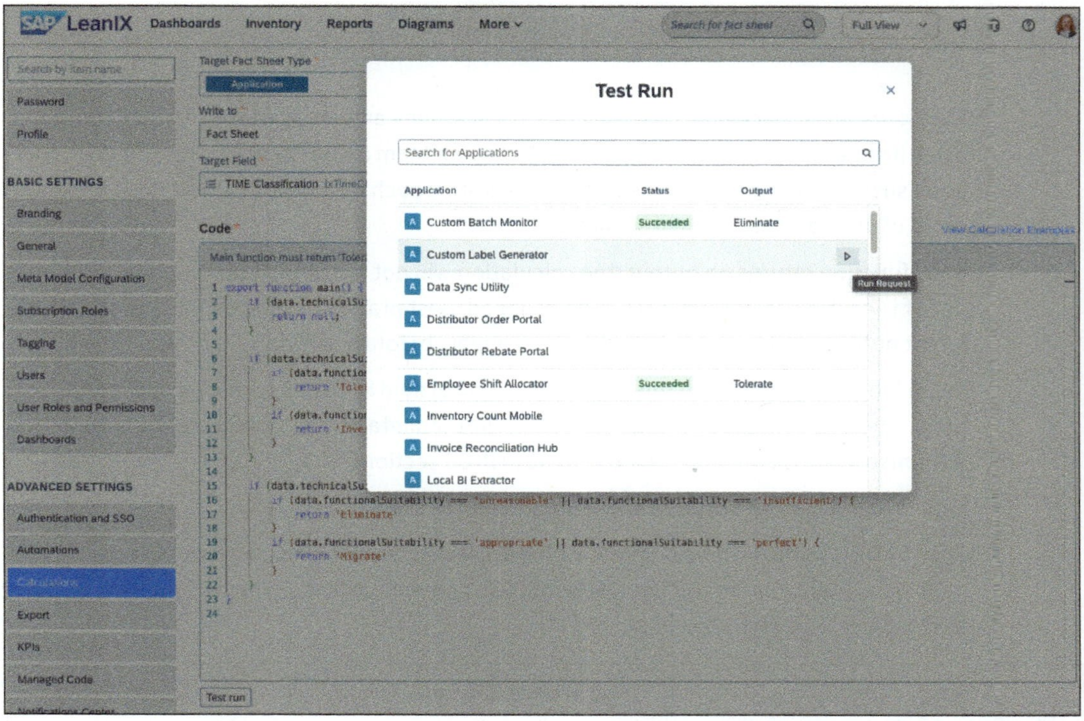

Figure 10.5: Test Run Window, to Individually Test Calculations on Fact Sheets Before Activation

At first activation, the calculation will calculate the field on all fact sheets of the defined type in the inventory. Therefore, it may take a few minutes until everything is calculated and updated depending on the number of fact sheets and complexity of the calculation.

After the initial setup, the recalculation is triggered whenever there is a change in one of the input fields/relations used for the calculation. The recalculation is performed instantly on the affected fact sheet; however, even though the calculation is done instantly, it does not instantly update the field in the UI without a browser refresh.

10.2.3 Syntax Examples for Custom Calculations

When creating a new calculation from scratch, the fields in the different sections are blank and must be filled manually. To learn about the different sections and the purpose of the various fields, revisit Section 10.2.2. Only the code terminal at the bottom has the wrapper function syntax prepopulated.

Let's now look at some syntax examples that will help you and your AI helper develop the calculation your team needs.

Calculation codes start with the following syntax:

```
export function main() {
< your code >
}
```

Always include checks for `null` or `undefined` values and account for them in your code using if statements to avoid errors. The SAP LeanIX team recommends using nonstrict operators such as `==` or `!=`. When using strict operators such as `===` or `!==`, you need to handle cases of `null` and `undefined` separately.

If the function returns no output, the calculation will not run. If the output of the function is `undefined`, the target field will not be changed. Best practice is to account for possible different outputs and to test calculations thoroughly before activation.

In the function, you need to reference the technical field key and technical name of the relations. The technical keys can be looked up in the **Meta Model Configuration** area of the **Administration** menu under the **BASIC SETTINGS** section.

Figure 10.6 shows an example of the application fact sheet type.

The **TIME Classification** field under the **TIME Classification** subsection was selected. In the right-hand menu, under the **Overview** tab, you can see the first characteristic, **Key**, which is the technical key of the selected field.

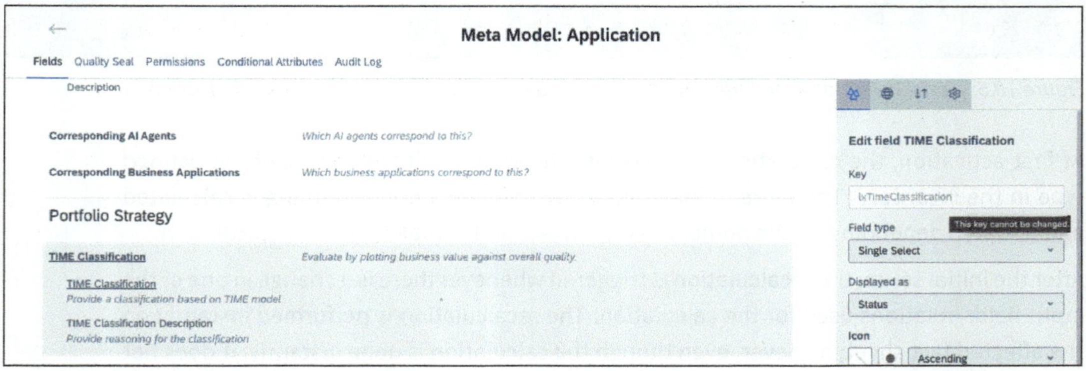

Figure 10.6: Application Meta Model Configuration with TIME Classification Field Key

Figure 10.7 shows where to find the technical keys of defined field values with dropdowns. For that, switch from the **Overview** tab to the **Translation** tab with the globe icon. Under the **Field Values** header, you can see a value name and, underneath that, a white box, which is editable. The value name above the box is the defined key that was entered when creating the select values, and the name in the white box is the defined display name that users see in the UI of the fact sheet.

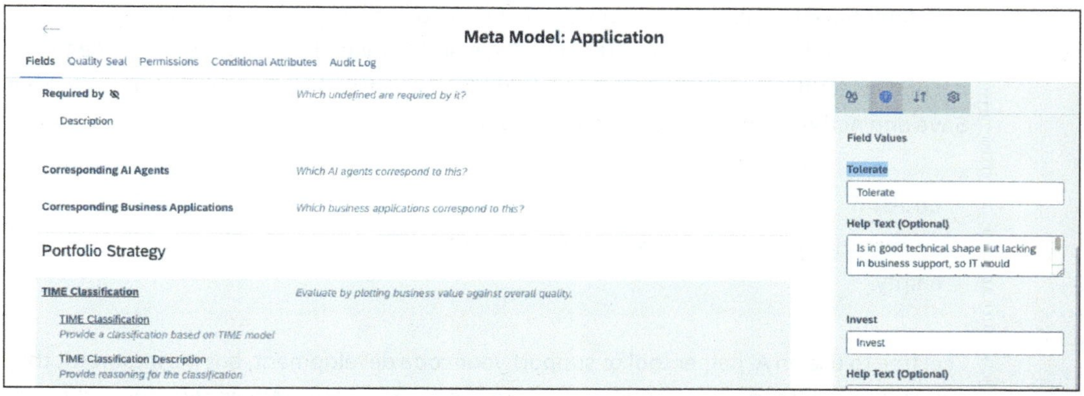

Figure 10.7: Meta Model Configuration Translation Tab with Field Value Technical Key and Display Name

To reference a field name on the same fact sheet type as the target field and when **Write to Fact Sheet** was selected, use the following syntax:

```
data.FieldName
```

If **Write to Relation** was selected, use the following syntax:

```
data.fromFactSheet.fieldName
```

To reference a lifecycle date field, use the following syntax:

```
data.lifecycleFieldName.phaseName
```

Example: `data.lifecycle.active`

Calculations that read data from a field of type `String - Date` run daily at 12 a.m. UTC.

Table 10.1 offers a few more syntax examples for referencing relations.

Reference	Syntax
Relations	`data.relationName`
Count of relations	`data.relationName.length`
Fields on relations	`data.relationName.fieldOnRelation`
Looping through fields on relations	`for (const relation of data.relationName) { … relation.fieldOnRelation…}`
Fields on related fact sheets	`data.relationName.factsheet.fieldName`

Table 10.1: Calculation Syntax Overview for Relation-Related Attributes

You don't have to finish the custom calculation in one sitting. To save your progress, click the down-pointing arrow next to the blue **Save and Activate** button under the code terminal and select the grey **Save in Inactive state** button. After thorough testing, use the blue **Save and Activate** button to start your calculation.

> **Caution**
> Be sure to always end with a return statement; otherwise, the target field will remain empty.

Feel free to use an AI helper tool to support your code development, but it's important that you still understand the calculation logic and algorithm. With AI support, think through edge cases, for instance, what the return value would be if an input field value was empty. Let the AI helper explain what certain parts of the code do and how to perform proper error handling. Don't skip thinking through the problem statement and possible solutions yourself. AI is fast, but AI can also make mistakes.

> **Further Resources**
> For the latest information regarding supported attributes, visit the SAP LeanIX user documentation at *http://s-prs.co/v620684*.

10.2.4 Error Handling, Error Log, and Deactivating a Calculation

Once you have multiple calculations up and running, SAP LeanIX offers multiple features to monitor the calculation operations.

First, you will see a status in the **Calculations** tab overview tiles (see Figure 10.8). In the upper-left corner, you will see a green **Active** indicator if everything is working fine, or a red **Active - invalid** indicator if there are errors. In case of errors, the calculation execution is paused until the error is resolved. To investigate the cause, you can switch to the **Error Log** tab next to the **Calculations** tab in the top row of the center **Calculations** overview window.

Common error sources include meta model changes, such as a source field/relation or target field being deleted or changed. This error will be listed as a configuration error. After meta model changes, the system will validate the active calculations and indicate whether a configuration error was found in the respective calculation tile.

The **Managed Code** area in the **Administration** menu will provide an overview of all deployed custom code in the workspace, not only the code in the calculations, but also, for instance, the code of automations. To learn more about that feature, see Section 10.4.

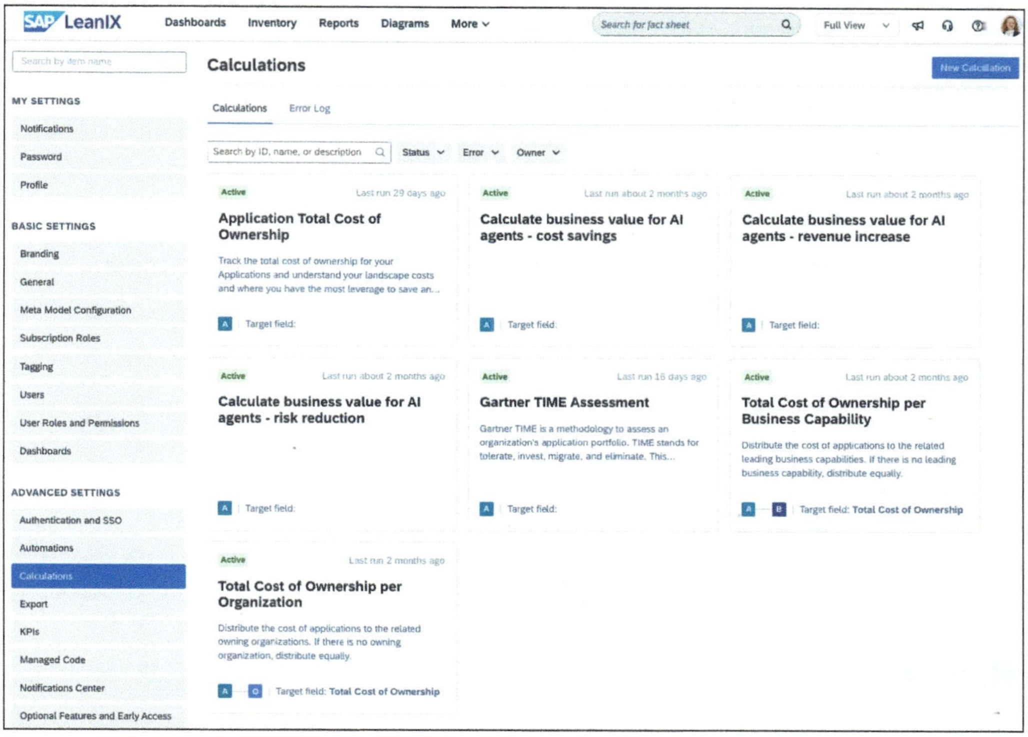

Figure 10.8: Calculation Tiles with Status and Last Run Information

To deactivate or delete a calculation, open the respective calculation tile and click the **...** button in the upper-right corner. There you will see different options (see Figure 10.9). Use the grey **Disable** selection to deactivate and pause the calculation. Instead of a green **Active** status, the calculation will then show a grey **Inactive** status in the calculation overview. To turn the calculation back on, click the blue **Save and Activate** button under the code terminal. If you won't need the calculation anymore, you can also click the red **Delete** selection in the **...** menu. This action is nonreversable, so we recommend deactivating a calculation first and only deleting it once you are certain that the calculation will no longer be needed.

Further Resources

If you want to create more advanced calculations that also include data outside of SAP LeanIX, the best way is to host the calculation outside of SAP LeanIX and use event-triggered automations via webhooks that leverage SAP LeanIX APIs to update data in an automated way.

For more information, visit the SAP LeanIX developer guide user documentation:

- *http://s-prs.co/v620685*
- *http://s-prs.co/v620686*
- *http://s-prs.co/v620687*

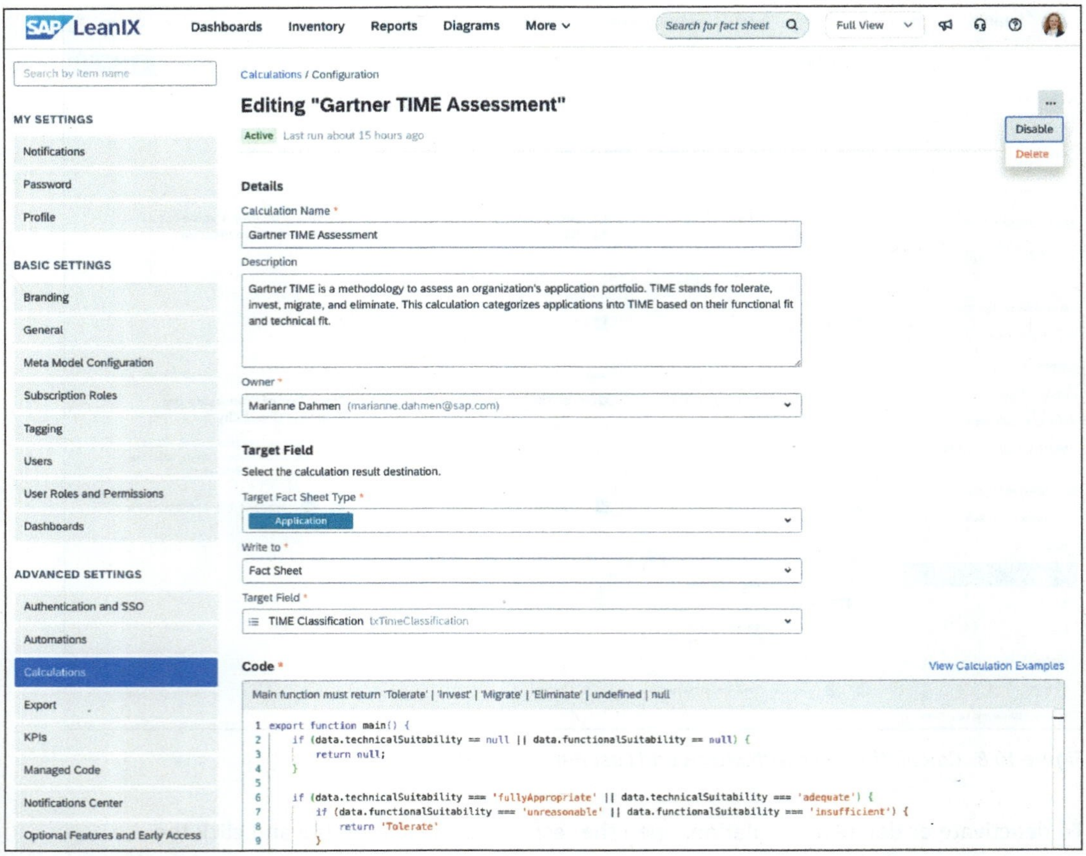

Figure 10.9: Disabling and Deleting Calculation in Calculation Configuration Form

10.2.5 Calculation Examples

In the following sections, you will find some example calculations that can spark ideas for your own custom calculations.

Cost Aggregation and Calculation

The following calculation example can be used to sum budget fields on the initiative fact sheet type. We will use the out-of-the-box fields for operating expenses (OpEx; technical key: `budgetOpEx`) and capital expenditure (CapEx; technical key: `budgetCapEx`) from the initiative data model and will need an additional custom total budget field (technical key: `totalBudget`).

To create the custom field, open the **Meta Model Configuration** area in the **Administration** menu and select the **Initiative** fact sheet (see Figure 10.10).

In the **Fields** tab, scroll to the **Initiative Setup** section and click the **Budget Costs** subsection. Click **+ Add Field** and enter **totalBudget** in the **Key** field. Select **Double** in the **Field type** dropdown and change **Displayed as** to **Costs**. Adjust the **Field Width (Columns)** as needed. The double format allows for decimal numbers to be entered, and the **Costs** setting will show the decimal with the currency defined in the **General** settings in the **Administration** menu. When you're done, click the blue **Create** button (see Figure 10.11).

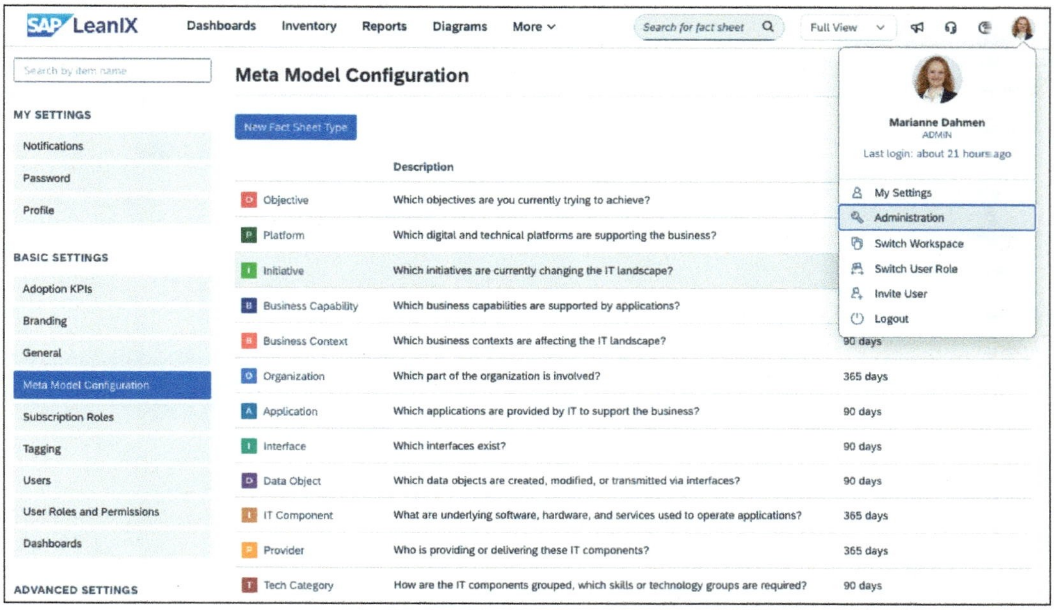

Figure 10.10: Opening Initiative Meta Model Configuration

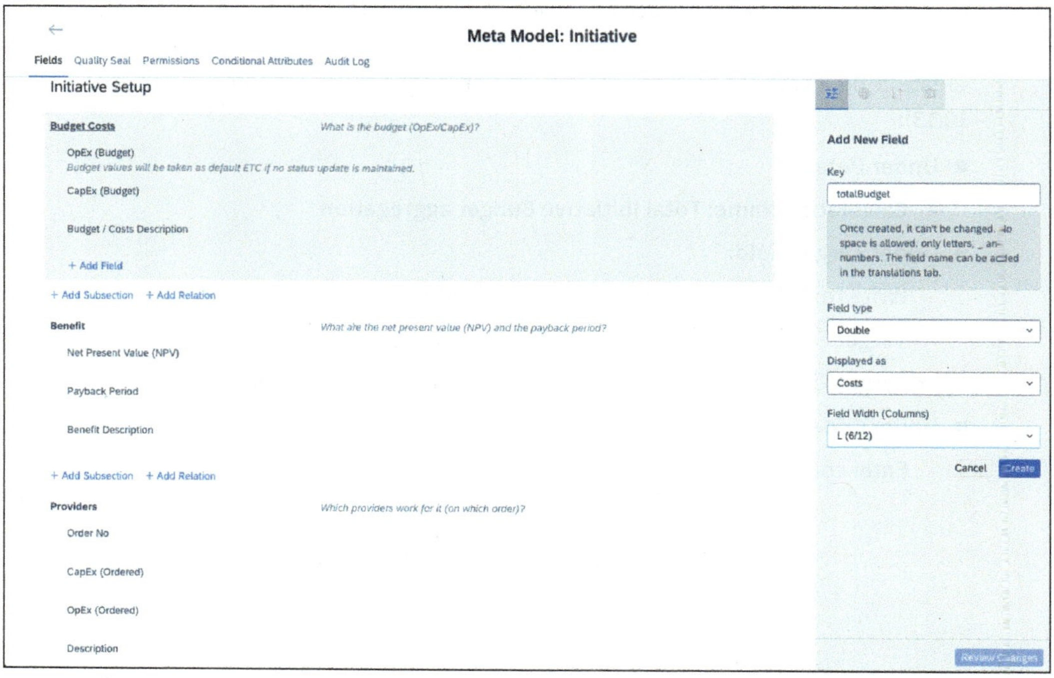

Figure 10.11: Creating Total Budget Field in Initiative Data Model

Next enter the **Labels** and **Display Names** in the **Translations** (globe icon) tab, then switch to the **Settings** (gear icon) tab to toggle on the setting for **Include in views** so that you can color-code reports based on this field (see Figure 10.12).

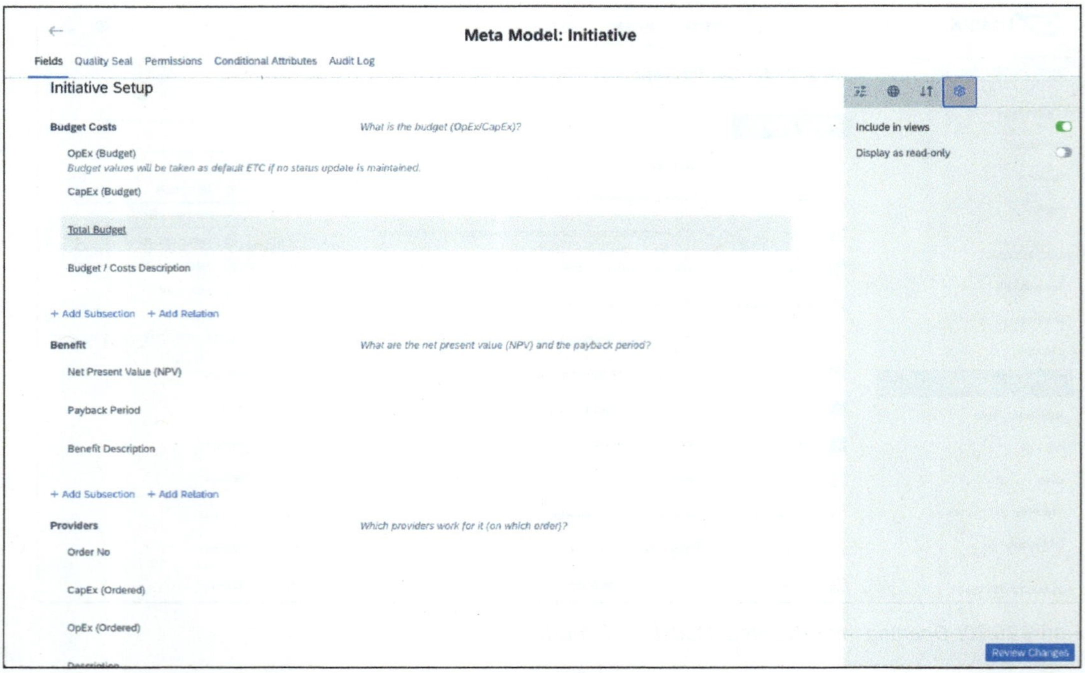

Figure 10.12: Making Total Budget Available as View in Reports

Create a new calculation from scratch and enter the following information (see Figure 10.13):

- **Under Details:**
  - **Calculation Name: Total Initiative Budget aggregation**
- **Under Target Field:**
  - **Write to**: Select **Fact Sheet Field** from dropdown
  - **Target Fact Sheet Type**: Select **Initiative** from dropdown
  - **Target Field**: Select **Total Budget totalBudget** from dropdown
- **Under Code:**
  - Enter code from Listing 10.1.

```
export function main() {
    if (data.budgetCapEx == null || data.budgetOpEx == null) {
        return null;
    }
```

```
        const totalBudget = data.budgetCapEx + data.budgetOpEx;
        return totalBudget;
    }
```

Listing 10.1: Example Code for Initiative Budget Aggregation

Click the blue **Save and Activate** button after testing a few example calculations.

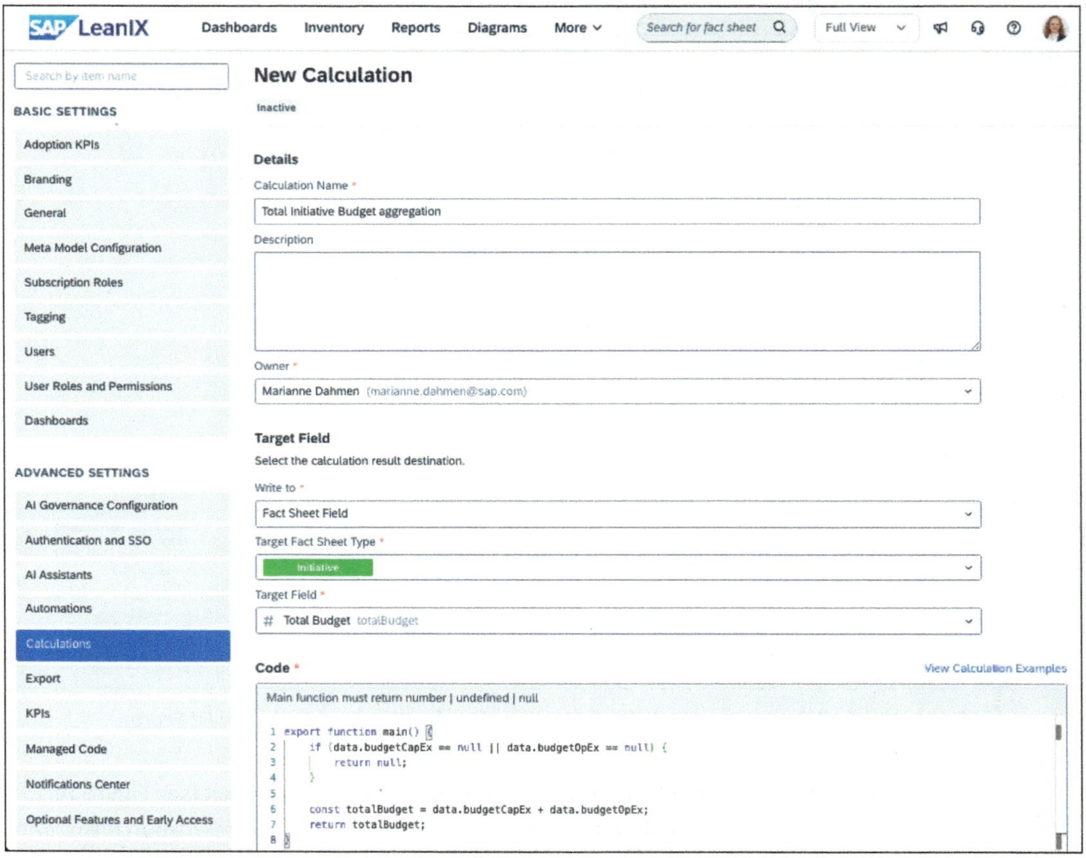

Figure 10.13: Calculation Form for Initiative Total Budget Calculation

Weighted Technical Fit Score

This calculation creates a weighted technical fit numerical score in the IT component fact sheet type derived from three custom, single-select fields that capture information on stability, maintainability, and security.

To use this example, you must create the following custom fields in the IT component meta model configuration (see Figure 10.14):

- Target custom field: **Technical Fit Score** (technical key: `technicalFitScore`); numeric field

- Source custom fields formatted as single selects:
  - **Stability** (technical key: `stability`)
  - **Maintainability** (technical key: `maintainability`)
  - **Security** (technical key: `security`)

  The technical key values of the source fields' single-select values are `best`, `aboveAverage`, `average`, `belowAverage`, and `worst`.

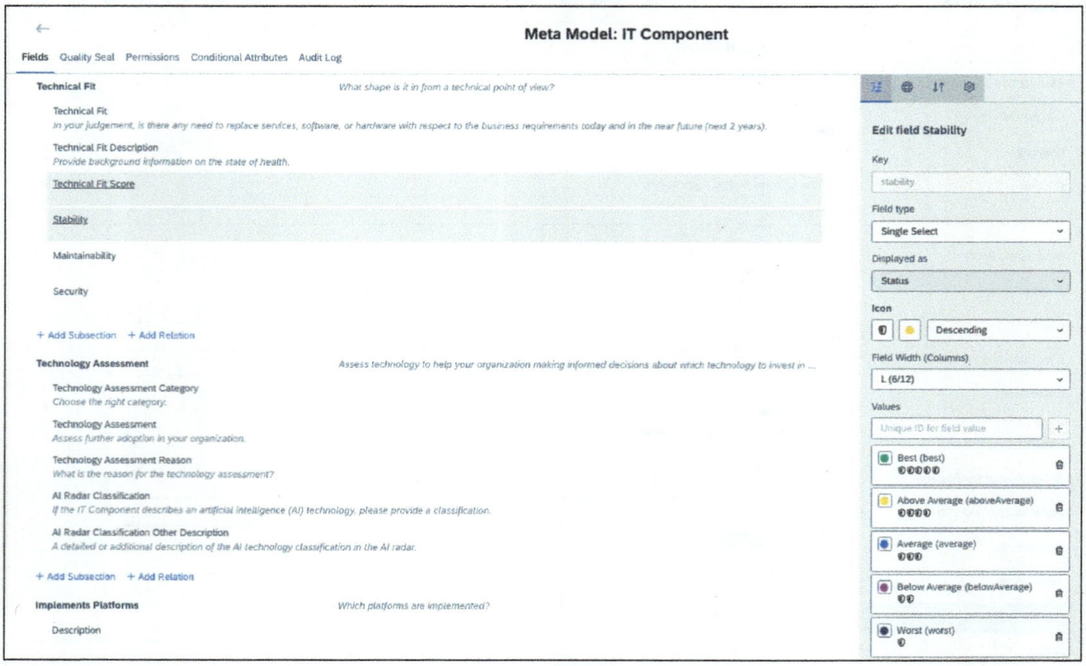

Figure 10.14: IT Component Meta Model Modification for Weighted Technical Fit Score

Ensure that the IT components then have the new custom fields filled out (see Figure 10.15). You can use SAP LeanIX collaboration features such as surveys to crowdsource the data from experts.

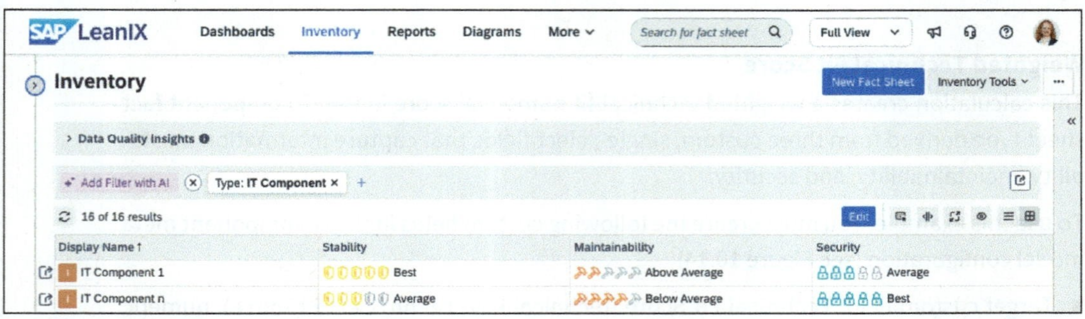

Figure 10.15: Filled Out Custom Fields on IT Component Fact Sheets

Create a new calculation from scratch and enter the following information (see Figure 10.16):

- **Under Details:**
  - **Calculation Name: Weighted Technical Fit Score on IT Component**
- **Under Target Field:**
  - **Write to**: Select **Fact Sheet Field** from dropdown
  - **Target Fact Sheet Type**: Select **IT Component** from dropdown
  - **Target Field**: Select **technicalFitScore** from dropdown
- **Under Code:**
  - Enter code from Listing 10.2.

```
export function main() {
    if (data.stability == null || data.maintainability == null ||
        data.security == null) {
        return null;
    }

    const scoring = {
        best: 100,
        aboveAverage: 75,
        average: 50,
        belowAverage: 25,
        worst: 0,
    };
    const weighting = {
        stability: 1,
        maintainability: 1,
        security: 2
    };
    const totalWeights = weighting["stability"] +
      weighting["maintainability"] + weighting["security"];
    const technicalFitScore = (weighting["stability"] *
      scoring[data.stability] + weighting["maintainability"] *
      scoring[data.maintainability] + weighting["security"] *
      scoring[data.security])/totalWeights;
    return technicalFitScore;
}
```

Listing 10.2: Example Code for IT Component Weighted Technical Fit Score

Click the blue **Save and Activate** button after testing a few example calculations.

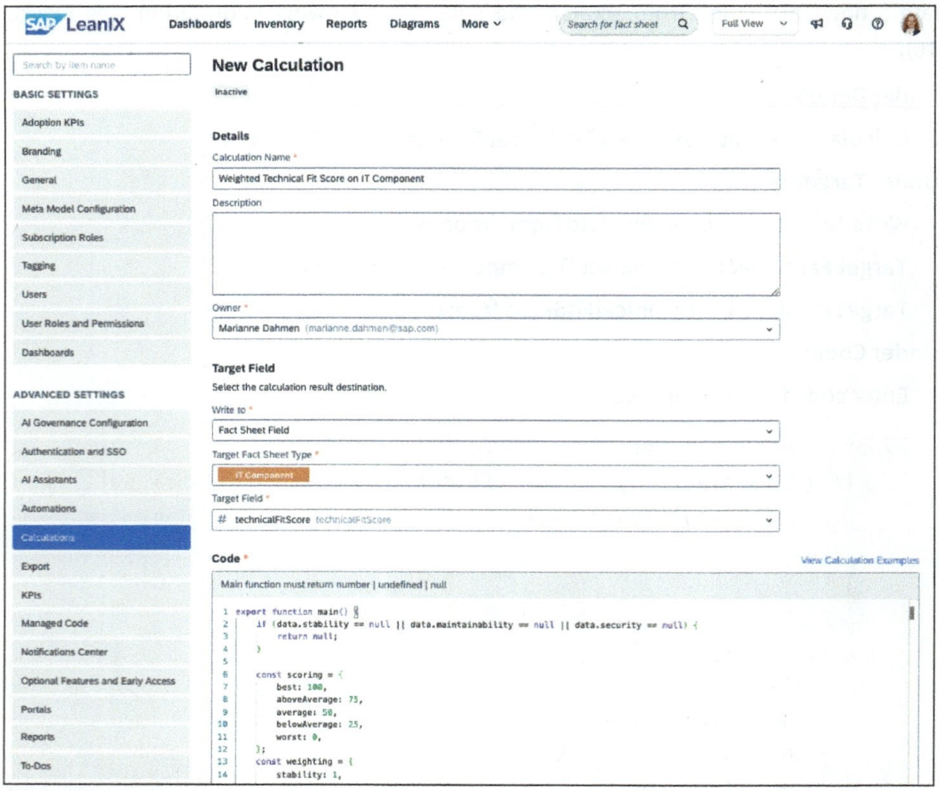

Figure 10.16: Calculation Form for IT Component Weighted Technical Fit Score

Total Number of Related Fact Sheets

In the next example, we will count how many interface fact sheets are connected to a related application fact sheet. We will add both provided and consumed interfaces.

The target fact sheet type for this calculation is application, and the **Write To** value is **Fact Sheet**. In the application meta model configuration, add a custom numeric integer field to hold the result of the calculation. You can call it, for instance, **Number of Connected Interfaces** (technical key: `countInterfaces`), which will be the **Target Field**.

The code for the calculation is shown in Listing 10.3.

```
export function main() {
    const numberConsumedInterfaces =
      data.relConsumerApplicationToInterface.length;
    const numberProvidedInterfaces =
      data.relProviderApplicationToInterface.length;
    return numberConsumedInterfaces + numberProvidedInterfaces;
}
```

Listing 10.3: Example Code for Count of Total Application Interface Relations (Provider and Consumer)

Automated Data Classification

In the next example, we will aggregate the highest data object data classification up to the related interfaces and ultimately to the related applications of those interfaces. This ensures that each application is classified with the highest data classification that it touches in any documented integration scenario.

For this example to work, you must replicate the **Data Classification** field of the data object meta model configuration to the interface and application fact sheet types. Figure 10.17 shows which field has which settings and values; you can recreate this in the interface and application meta model configuration.

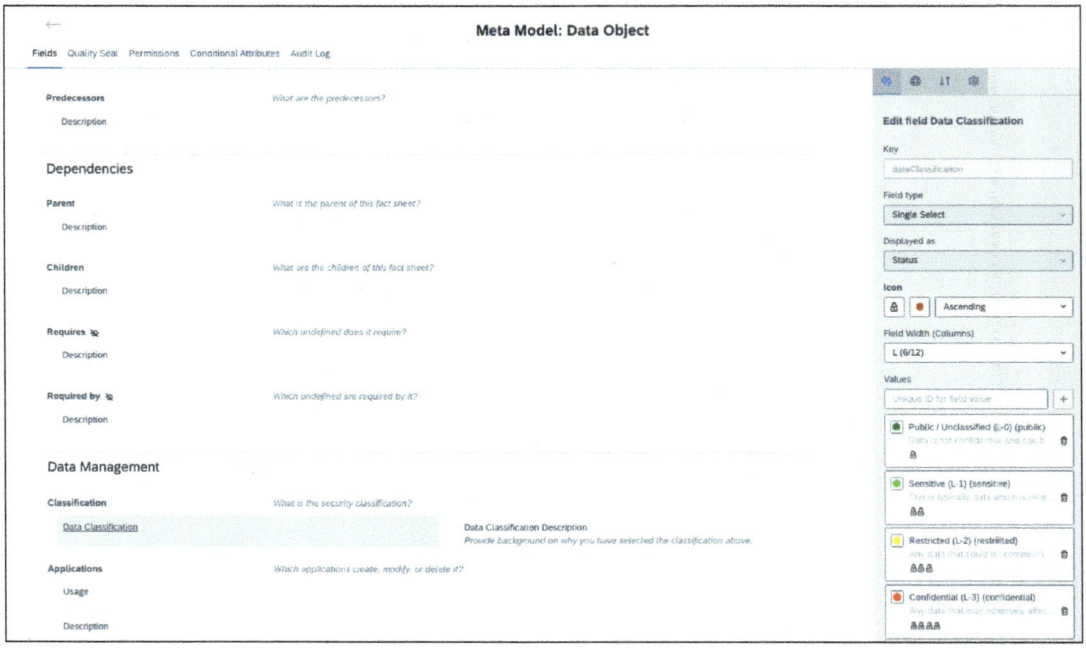

Figure 10.17: Data Classification Field in Data Object Meta Model Configuration to Be Recreated in Interface and Application Configuration

The first calculation aggregated the highest data classification to the related interface fact sheets.

For **Target Fact Sheet Type**, select **Interface**, and set **Write To** to **Fact Sheet**. The **Target Field** is the newly created data classification field (technical key: `dataClassification`) on the interface fact sheet type.

The code is shown in Listing 10.4.

```
export function main() {
    // 1) Define the comparison hierarchy (higher number = more sensitive)
    const classificationRank = {
        "public": 0,
```

```
        "sensitive": 1,
        "restricted": 2,
        "confidential": 3
    };

    // 2) Initialize defaults (used when no classified objects are found)
    let maxRank = 0;
    let maxLabel = "public";

    // 3) Loop over related Data Objects
    for (const dataObject of data.relInterfaceToDataObject) {
        // Access the classification stored on the related
        // Data Object fact sheet
        let classification = dataObject.factSheet.dataClassification;

        // 4) Only consider values that exist in hierarchy map
        if (classification != null &&
          classificationRank.hasOwnProperty(classification)) {
            const currentRank = classificationRank[classification];

            // 5) Keep the highest rank seen so far and remember its label
            if (currentRank > maxRank) {
                maxRank = currentRank;
                maxLabel = classification;
            }
        }
        // Unknown / empty classifications are ignored to keep result stable
    }

    // 6) Return the label of the most sensitive classification encountered
    return maxLabel;
}
```

Listing 10.4: Example Code for Data Classification Aggregation on Interface Fact Sheets Based on Relation to Data Objects

In a second calculation, the application fact sheet aggregates the maximum data classification of all provided and consumed interfaces.

For **Target Fact Sheet Type**, select **Application**, and set **Write To** to **Fact Sheet**. The **Target Field** is the newly created data classification field (technical key: `dataClassification`) for the application fact sheet type.

The code is shown in Listing 10.5.

```
export function main() {
    // 1) Define the comparison hierarchy (higher number = more sensitive)
    const classificationRank = {
        "public": 0,
        "sensitive": 1,
        "restricted": 2,
        "confidential": 3
    };

    // 2) Initialize defaults (used when no classified interfaces are found)
    let maxRank = 0;
    let maxLabel = "public";

    // 3) Loop over consumer interfaces
    for (const rel of data.relConsumerApplicationToInterface) {
        // Access the classification stored on the related Interface fact sheet
        let classification = rel.factSheet.DataClassification

        // 4) Only consider values that exist in hierarchy map
        if (classification != null &&
          classificationRank.hasOwnProperty(classification)) {
            const currentRank = classificationRank[classification];

            // 5) Keep the highest rank seen so far and remember its label
            if (currentRank > maxRank) {
                maxRank = currentRank;
                maxLabel = classification;
            }
        }
        // Unknown / empty classifications are ignored to keep the result stable
    }

    // 6) Loop over Provider Interfaces (same logic)
    for (const rel of data.relProviderApplicationToInterface) {
       let classification = rel.factSheet.DataClassification

        if (classification != null &&
          classificationRank.hasOwnProperty(classification)) {
            const currentRank = classificationRank[classification];
            if (currentRank > maxRank) {
                maxRank = currentRank;
                maxLabel = classification;
            }
        }
```

```
    }

    // 7) Return the label of the most sensitive classification encountered
    return maxLabel;
}
```

Listing 10.5: Example Code for Data Classification Aggregation on Application Fact Sheets Based on Relation to Consumed Interface Fact Sheets

10.3 Automations

Automations are no-code features that help you create in-tool workflows triggered by certain events on fact sheets.

Only admins can create and manage automations. To navigate to the **Automations** area, first click your user icon in the upper-right corner and select **Administration**. Then in the left-hand menu, scroll down to the **ADVANCED SETTINGS** area and click the **Automations** menu item (see Figure 10.18).

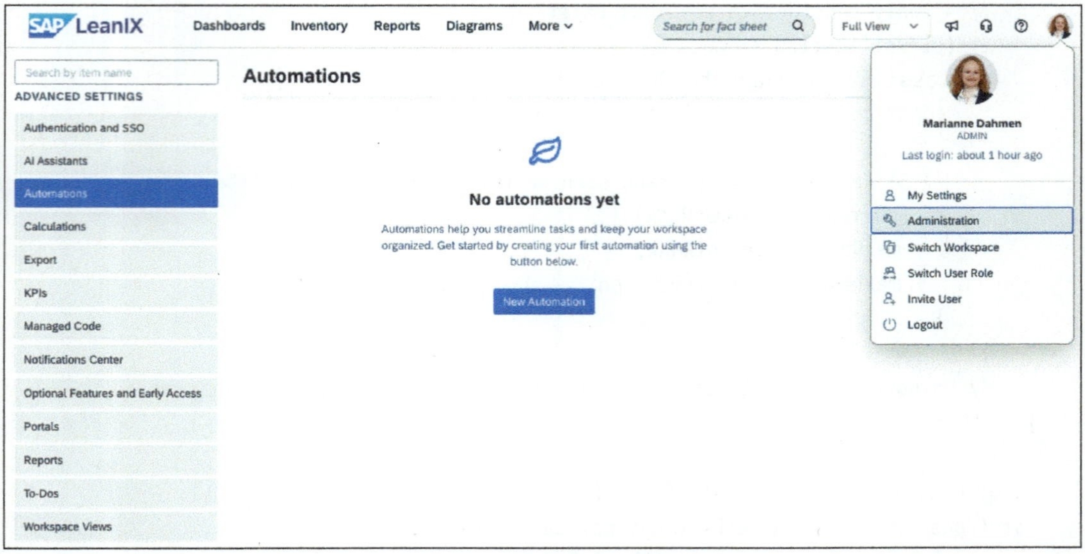

Figure 10.18: Navigating to Automations in Administration Menu

When you click the **New Automation** button, you will enter the **Create a New Automation** guided tour (see Figure 10.19). In the center pane, you will see the automations workflow, which will be designed from top to bottom. Click the + sign under the **Trigger** box to add new elements (more on that later). On the right-hand side, you can see a menu with two tabs:

- **Details**, to enter general information such as **Name, Description**, and **Owner**
- **Builder**, a configuration guide for the selected tile in the automations workflow

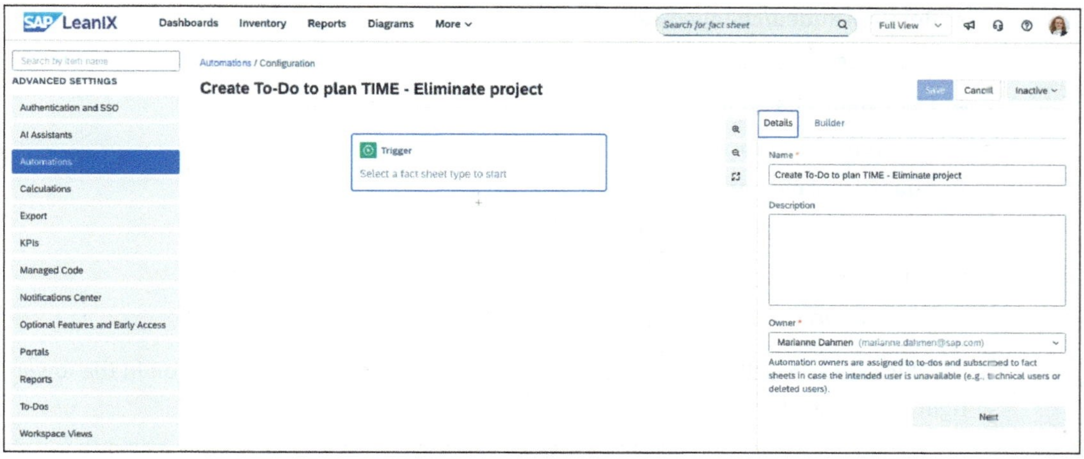

Figure 10.19: New Automation Configuration Guide

We have already entered the general information, as shown in Figure 10.19.

Click the **Trigger** box and switch to the **Builder** tab. You will now be prompted to select a **Fact Sheet Type** in which the trigger will originate. In this example, select the **Application** fact sheet type. Click the blue **Next** button to continue with the next configuration step (see Figure 10.20).

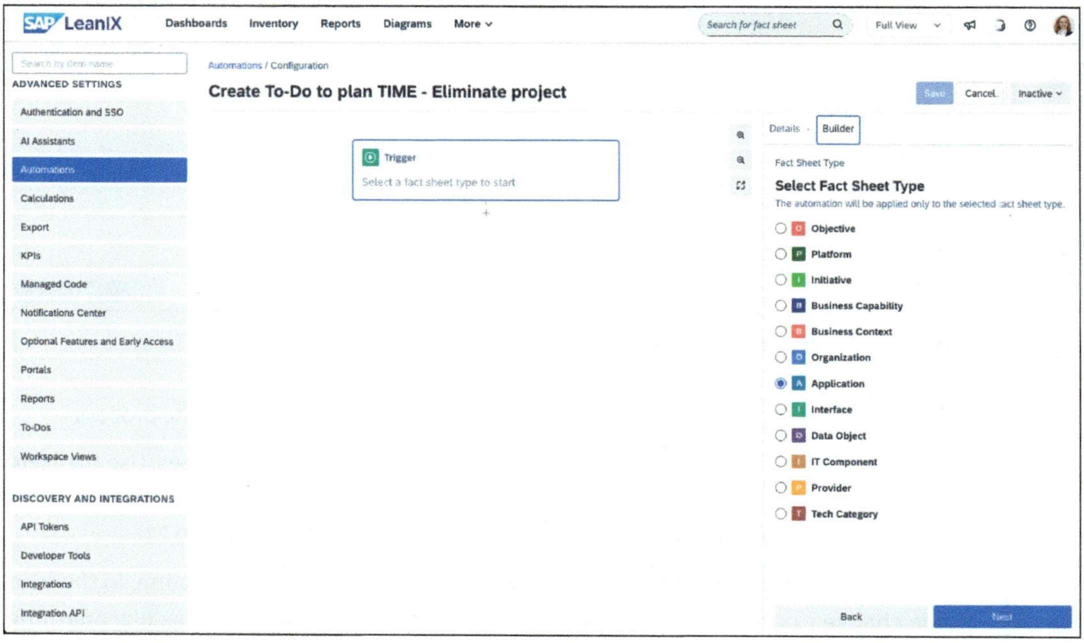

Figure 10.20: Selecting Fact Sheet Type for Trigger

On the next overview, in the right-hand menu, the **Select Trigger** area offers various options. Possible trigger events you can choose include the following:

- **Relation added/changed/removed**
- **Fact sheet created**
- **Field value changed**
- **Lifecycle state reached**
- **Quality seal changed**
- **Subscription added/removed**
- **Tag added/removed**
- **Completion score changed**

For this example, select **Field value changed** and click the blue **Next** button in the lower-right corner (see Figure 10.21).

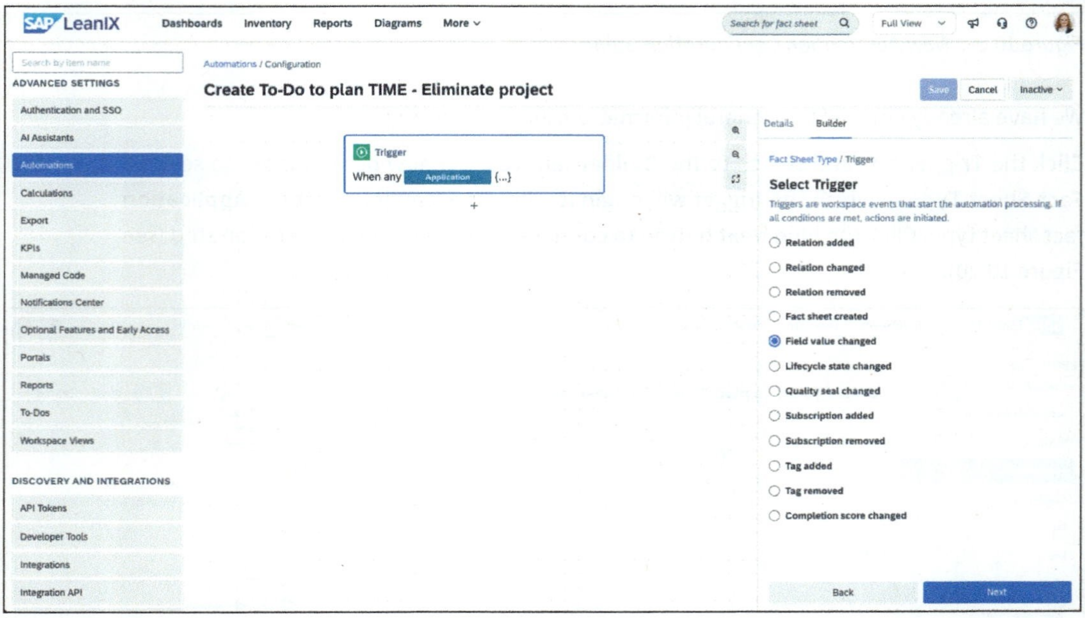

Figure 10.21: Select Trigger Event

You can only select one trigger, but you can add more conditions later in the process.

Now, select the field that will trigger the automation when its value changes. Use the **Field** dropdown to search for available fields. In general, select and text fields are available in this scenario. In this example, select the **TIME Classification** field (see Figure 10.22).

After selecting the triggering field, specify which value changes are in scope. In the field value change example, you can choose in the dropdowns between status like **Anything** or **Empty**, or select values, in this example **Tolerate**, **Invest**, **Migrate**, or **Eliminate**. In this example, define that in-scope changes from **Anything** to **Eliminate** (see Figure 10.23). Click the blue **Confirm** button in the lower-right corner of the right-hand menu to finalize the trigger configuration.

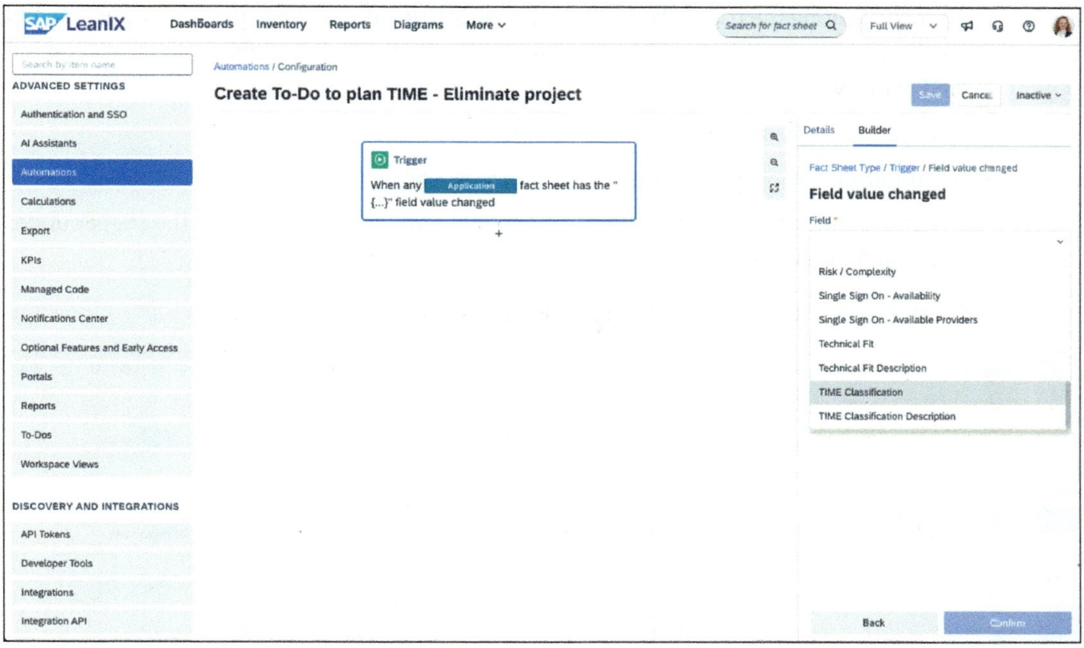

Figure 10.22: Selecting Field to Trigger Automation Based on Value Changes

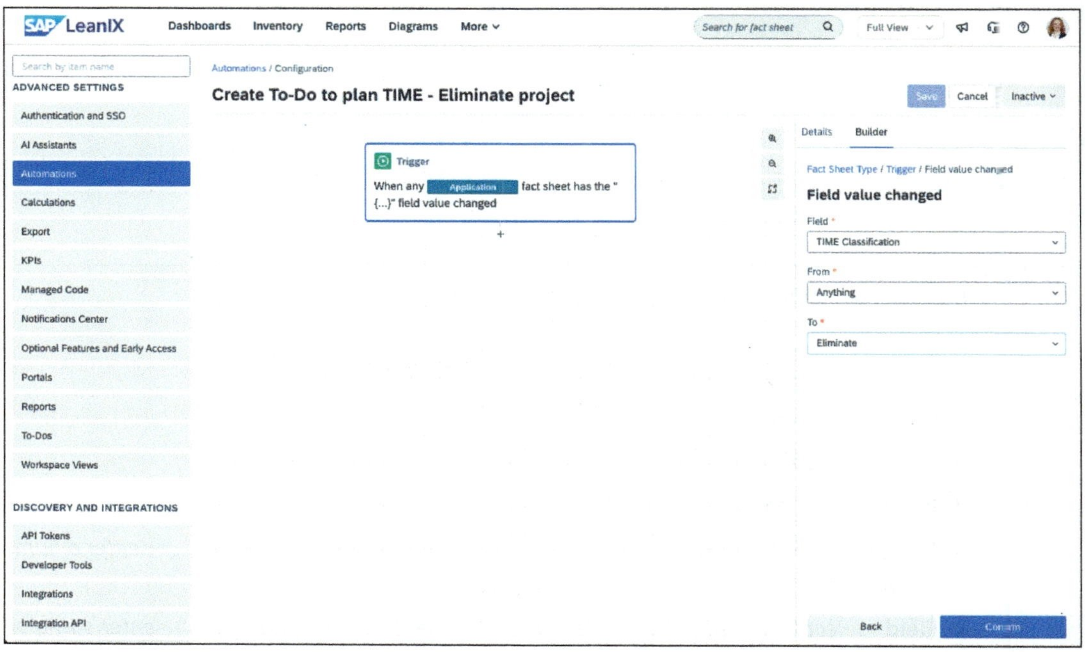

Figure 10.23: Defined Field Value Changes to Trigger Automation

After configuring the trigger box, click the **+** button under the box in the center pane to add a new workflow box. In the right-hand menu, under the **Builder** tab, you can select what kind of box you want to add.

The options are as follows:

- **Condition**, which adds additional criteria to the triggering event
- **Action**, which defines the action that should happen at occurrence of the trigger under the defined conditions

For this example, select **Condition** (see Figure 10.24).

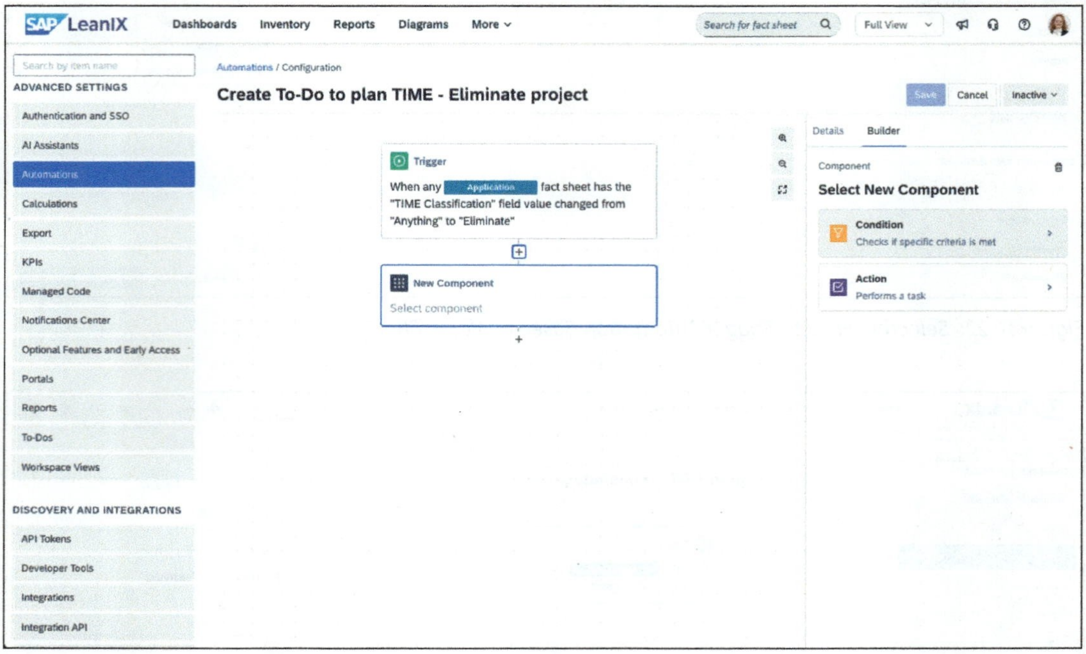

Figure 10.24: Adding Another Automation Workflow Box

As conditions, you have the following options to choose from:

- **Category** *<available subtypes>*
- **Tags** *<available tags>*
- **Ignore Technical User** *included / ignored*
- **Completion Score** *<select operator like equals, greater than, less than, …> <enter a number value>%*
- **Field** *<select fact sheet field> <select operator like is, is not, contains, …> <enter string or select single select value>*

In this example, we select **Ignore Technical User** and click the blue **Next** button to get to the next screen (see Figure 10.25).

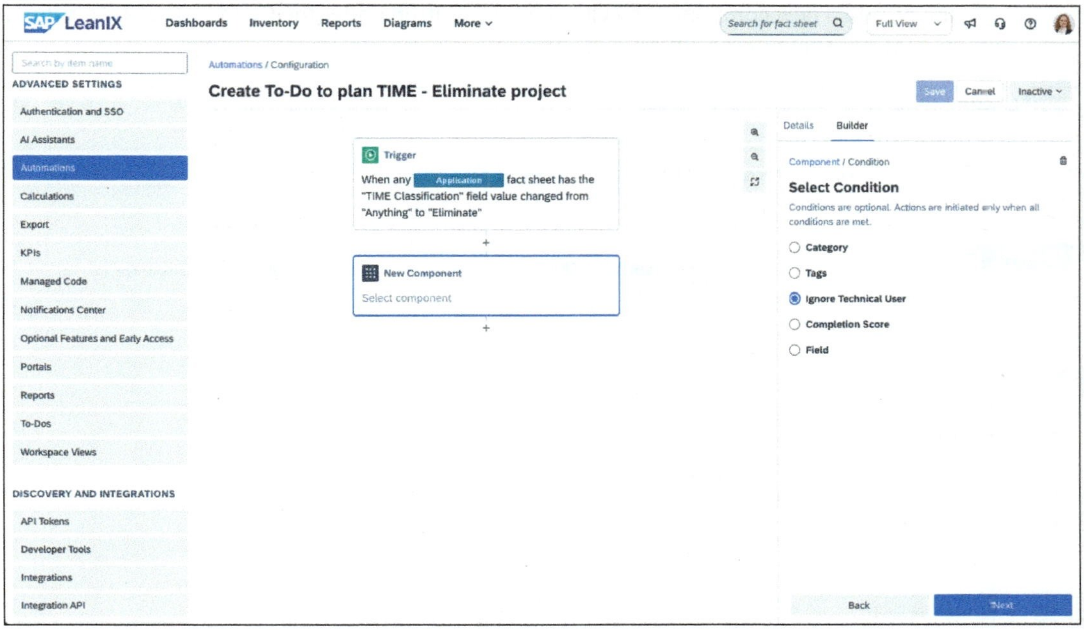

Figure 10.25: Possible Condition Criteria

In the next step, you can toggle on a condition stating that events triggered by technical users should be ignored. In this example, we will skip that change as it is not relevant here. Click the blue **Confirm** button to finalize the condition configuration (see Figure 10.26).

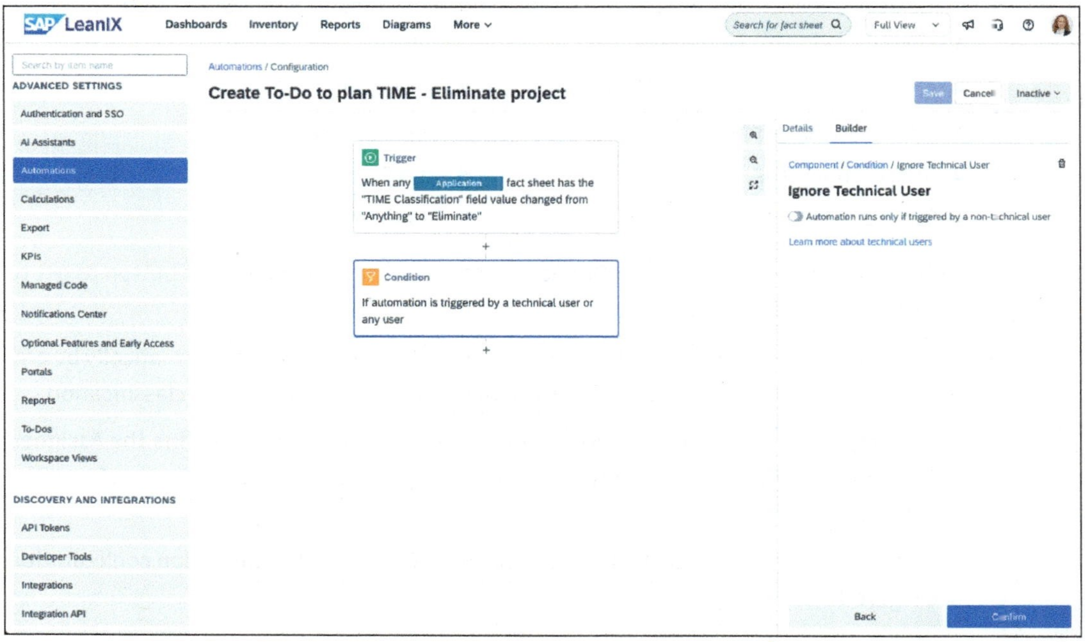

Figure 10.26: Configuring Ignore Technical User Condition

Finally, add a new component by clicking the **+** button under the **Condition** box. This time, select the **Action** type (see Figure 10.27).

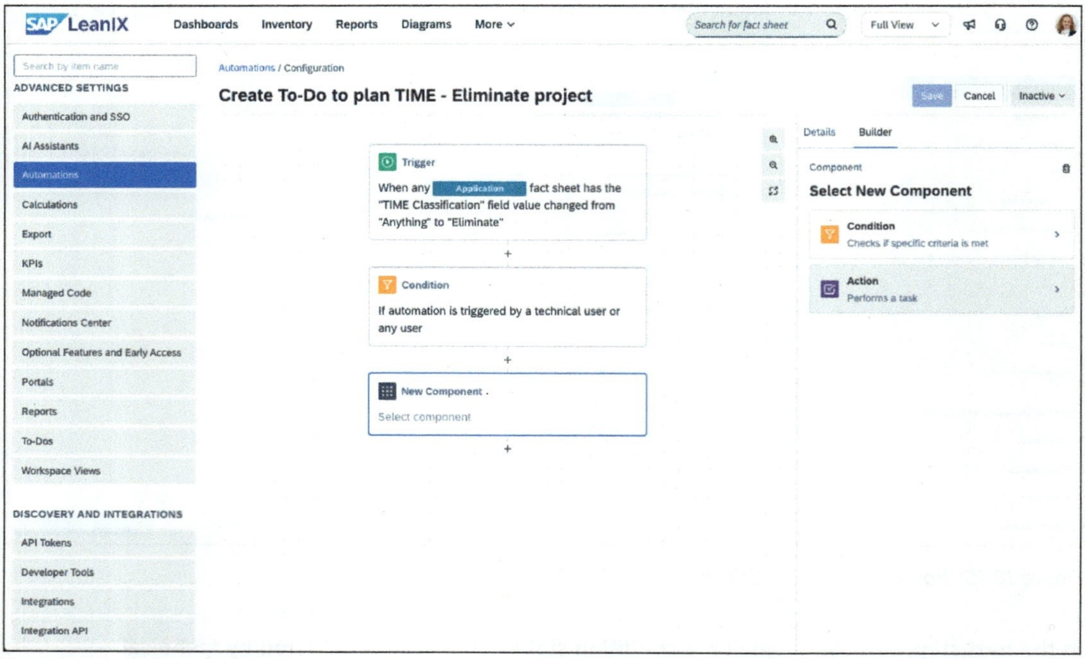

Figure 10.27: Adding Action Component to Workflow

In the next screen, in the **Builder** tab of the **Action** component, under **Select Action**, choose among the following options:

- **Create To-Do: Action Item/Approval**
- **Add/Set Subscription**
- **Add/Remove Tag**
- **Set Field**
- **Set Quality State**
- **Send Webhook**
- **Run Script**
- **Send Email**

In this example, select **Create To-Do: Action Item** and click the blue **Next** button to get to the next screen (see Figure 10.28).

In the next screen, configure the to-do of type action item that should automatically be created once an application switches from any value to *eliminate* in the TIME classification.

In the form, you can define an action item **Title**, enter a **Description**, define the **Assignee** and the relevant **Fact Sheet Subscription Role and Type**, and establish the timeframe for **Days until due**.

When done with the entry, click the blue **Confirm** button to finalize the action configuration (see Figure 10.29).

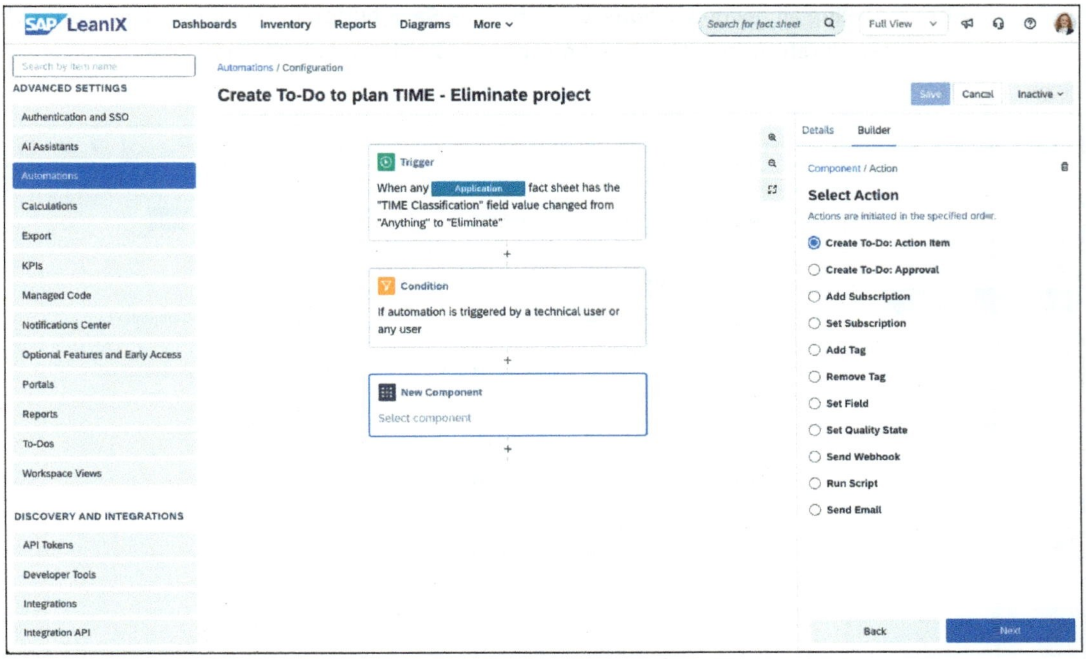

Figure 10.28: Selecting Action: Create To-Do: Action Item

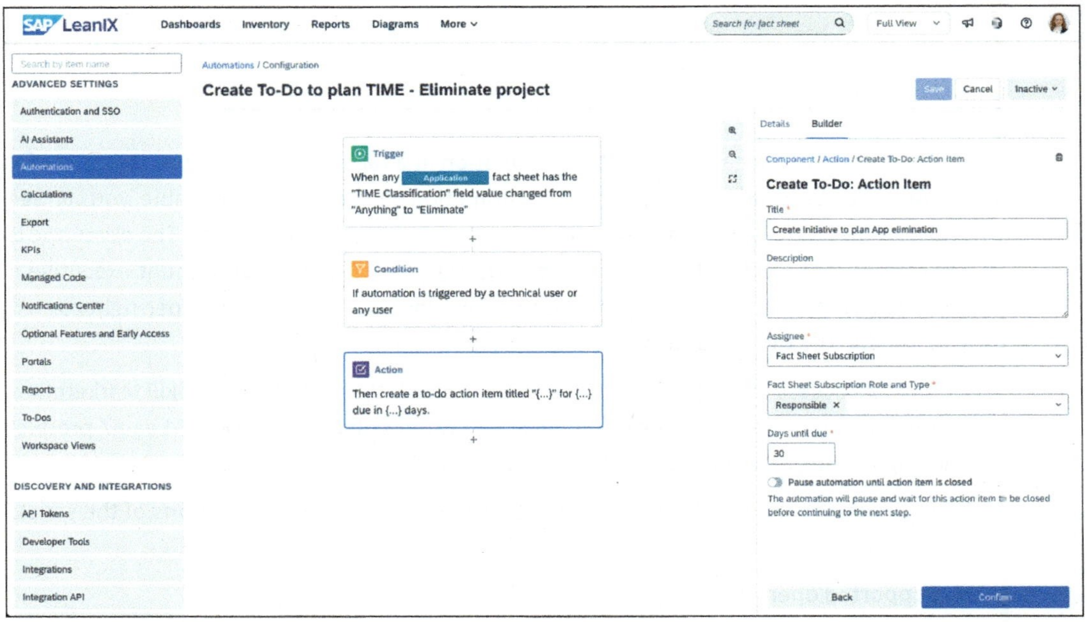

Figure 10.29: Configuring Action: Create To-Do: Action Item

You can define multiple actions that should run with an automation. They are performed sequentially in the order that they were created.

After modeling all needed actions, click the blue **Save** button in the upper-right corner of the screen to return to the automations overview in the administration menu (see Figure 10.30).

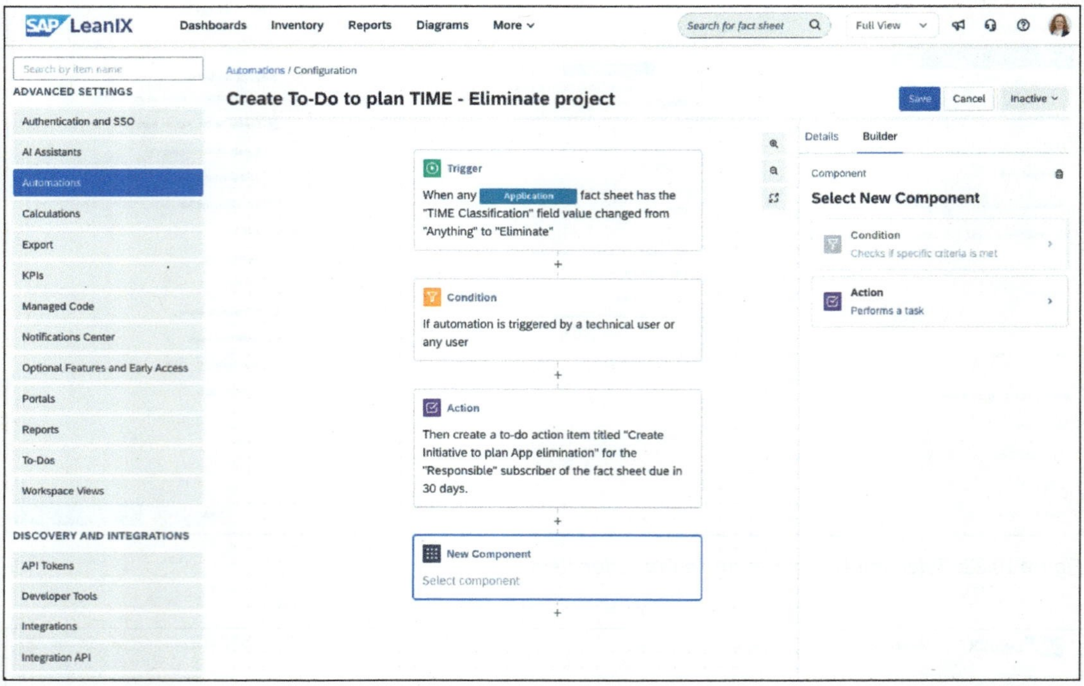

Figure 10.30: Saving Automation Workflow

In the overview, you can now see the automation in a table view (see Figure 10.31). Under the **Active** column, admins can toggle an automation on or off. Above the table, you can see two new tiles. One is the overview of the number of created **Automations**. The other one is the overview of the **Monthly Automations Quota** used. Talk to your account executive if you need a higher quota than 10,000 automation runs per month. In the upper-right corner, you can find the **New Automation** button to create a new automation.

If you have access to an AI assistant, you can also visit the SAP LeanIX AI Skill Marketplace on GitHub and reuse any of the designed and curated skills offered by SAP LeanIX product and engineering experts, such as the automations toolkit.

When using AI helper tools, always verify that they adhere to the limitations of the script-based approach:

- **Supported operations**
  - Read a field value
  - Update a field value
  - Delete a field value, which will result in empty

- **Available target attributes**
  - All fact sheet fields, with a few base field exceptions
- **Operation on base field exceptions**
  - Fields not supported for operation `Update`: `id`, `type`, `rev`, `level`, `completion`, `updatedAt`, `createdAt`
  - Fields not supported for operation `Delete`: `name`
- **Unsupported attributes, references in code**
  - Relations
  - Fields on relations
  - Subscriptions
  - Fields of type location

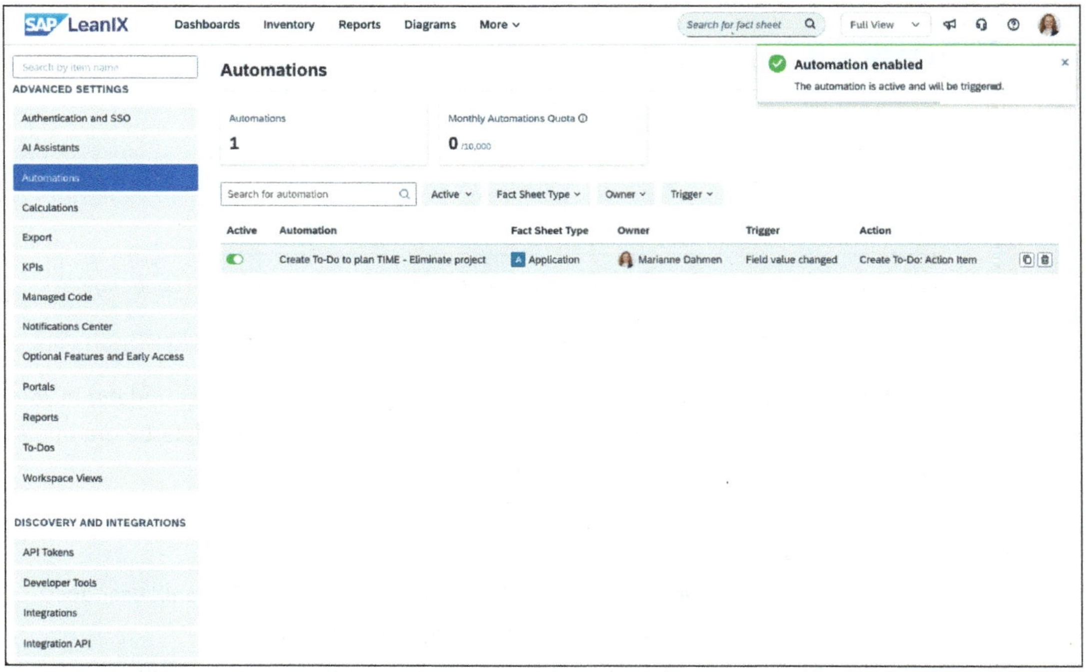

Figure 10.31: Automation Overview After Creating Automation

Beta Version and Future In-Tool AI Support

At the time of writing, the SAP LeanIX AI Skill Marketplace was still in development and not released.

Check for the latest announcements and updates at *https://updates.leanix.net/*.

For tools on GitHub, visit *https://github.com/SAP/leanix-ai-plugins*.

The SAP LeanIX product and engineering team is diligently developing new in-tool AI features. For automations, the team is planning to embed in-tool AI support to generate automations that include scripts with custom code based on natural language instructions (see Figure 10.32).

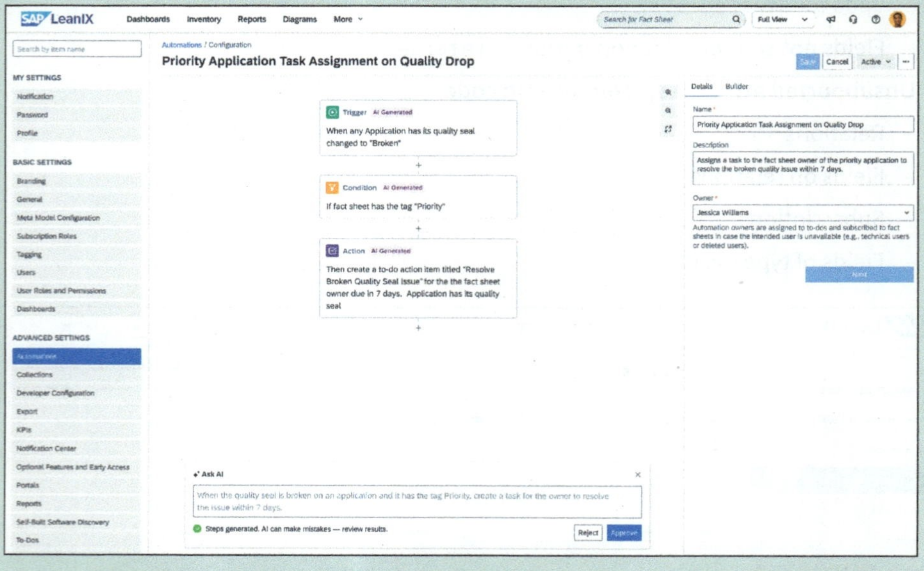

Figure 10.32: Preview: In-Tool AI Support to Create Automations from Natural Language Instructions

Always test your script! An automation is very powerful and can cause mass updates in your workspace. If you have a sandbox workspace, always test it with test data before implementation.

Further Resources

For the latest information on the automations feature, visit the SAP LeanIX user documentation at *https://help.sap.com/docs/leanix/ea/automations*.

Visit the SAP LeanIX user documentation for lots of examples and best practices:

- Automation best practices: *http://s-prs.co/v620688*
- Example automations with scripts: *http://s-prs.co/v620689*
- Example automations with webhooks: *http://s-prs.co/v620690*

10.4 Managed Code

The AI assistant and agent era will most likely increase the number of customizations and automations within the workspace. SAP LeanIX workspace admins need to have an overview of scripts with custom code deployed in automations, calculations, and webhooks to gauge impacts on existing workflows when discussing any potential meta model or data model changes. It is also important to prevent duplicated or redundant efforts that use up calculation capacity or, for automations, the monthly quota. This is why the SAP LeanIX product and engineering team introduced a new administration submenu: **Managed Code**.

In the **Administration** menu under the **ADVANCED SETTINGS** section, you will find the **Managed Code** submenu item (see Figure 10.33). There you will see an overview of not only the active calculations but any other deployed code, such as the code from automation scripts.

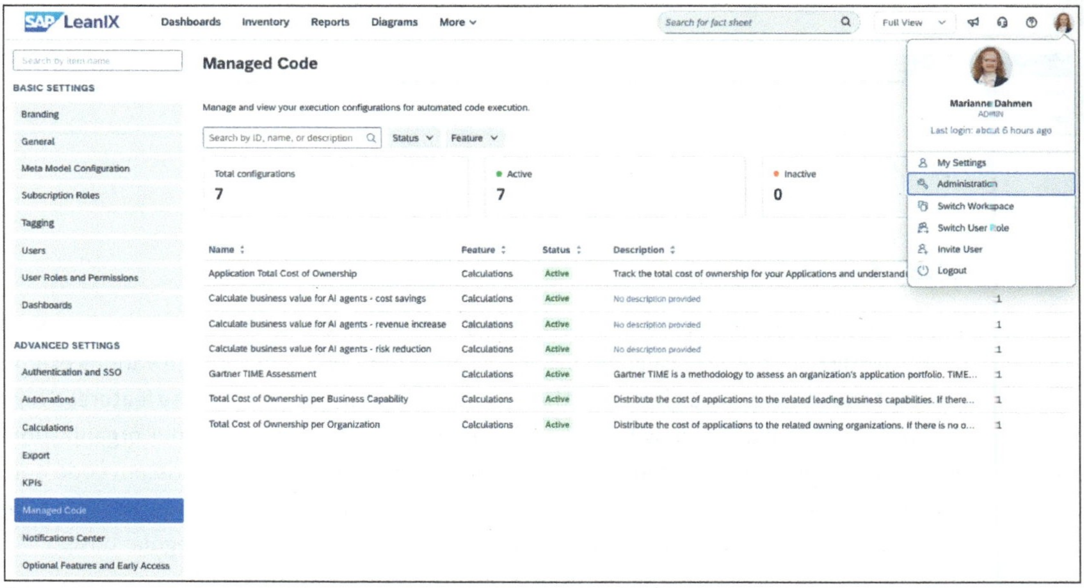

Figure 10.33: Navigating to Managed Code in Administration Menu

In this overview, you can monitor which scripts and codes are active, and by clicking their names, you can check whether the last runs were performed without errors in the **Execution Log** tab (see Figure 10.34).

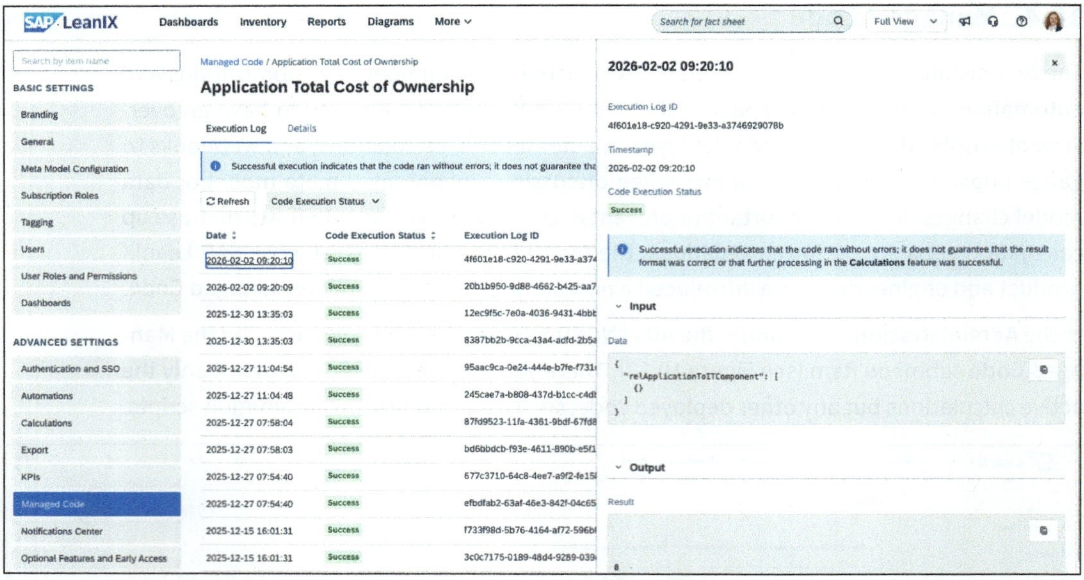

Figure 10.34: Code Insights with Execution Log and Details Tab

10.5 Summary

In this chapter, we introduced some advanced techniques that are only available for SAP LeanIX workspace admins: calculations and automations. We went through various examples that should give you a glimpse of the potential of and use cases for these features. Now it's up to you to take this knowledge further and create added value for your organization using SAP LeanIX. To make sure you keep oversight over your deployed custom code and scripts in calculations, automations, and webhook connections, be sure to establish a routine to monitor the **Managed Code** area in the administration menu on a regular basis.

Remember that you can combine your newly gained knowledge with AI-supported tools, either within SAP LeanIX or outside of it. Just remember: Never get rid of the human in the loop for review and final decisions.

In the next chapter, you'll learn more about the new AI capabilities in SAP LeanIX.

Chapter 11
AI Capabilities in SAP LeanIX

There are few presentations today that don't include the term AI. AI is sometimes overhyped and other times underestimated. AI will change the way we collaborate and work today. In this chapter, we want to identify in a very pragmatic way the different AI capabilities SAP LeanIX offers currently, and a glimpse into what the future holds.

Over the last several months, SAP LeanIX has evolved rapidly in terms of AI capabilities, from simple embedded intelligence to the foundation of standalone AI applications and endpoints for interaction in an agentic enterprise.

AI is not replacing enterprise architects, but it's clear that with the productivity gains and insights it offers, the focus and awareness of enterprise architects will shift to more value-adding activities. A lot of time today is allocated to keeping the inventory data quality high, reports in a consistent state, and diagrams up to date, but these routine tasks will in time be automated via AI capabilities, leading to a "living" inventory (sometimes called a *digital twin* of your organization). Enterprise architects will collaborate with AI on an increased level of sophistication, which offers a great opportunity to reposition the role of an enterprise architect as a catalyst and strategic advisor for change.

Depending on the maturity of your organization, this reality may be already lived, while others are currently evolving into this space that is rapidly evolving on an almost daily basis. For us as authors, it's therefore impossible for this chapter to be completely up to date; we therefore ask you to leverage the various channels available for updates, such as the product updates and roadmaps, or reach out to us on platforms such as LinkedIn for our latest updates and insights.

In this chapter, we will outline and provide insights into the various AI capabilities SAP LeanIX offers. We have classified the AI capabilities via a five-level collaboration model, as shown in Table 11.1, to abstract the product features from the human interaction perspective in order to make it easier for you as a reader to understand which capability might help you in a given situation.

Collaboration Mode	AI Capability
AI assists me	Embedded services that handle routine tasks on demand: natural language filters, standard prompts, semantic analysis of descriptions and relations

Table 11.1: AI Capabilities in Context of Collaboration Modes

Collaboration Mode	AI Capability
AI works alongside me	Inventory builder and enterprise architecture assistants that actively process documents, research the web, and build the repository together with the enterprise architect
AI advises me proactively	Architecture guidance that continuously scans the repository and pushes improvement recommendations without being asked
AI acts on my behalf	External agents connected via the Model Context Protocol (MCP) server that query, create, and update architecture data across tools and platforms and provide insights
AI collaborates with others	Joule as the universal interface to bring architecture intelligence to the entire organization for custom AI-driven reports and standalone applications

Table 11.1: AI Capabilities in Context of Collaboration Modes (Cont.)

Note

Note that for the usage of AI capabilities in the context of SAP LeanIX, agreeing to the *SAP AI terms* is a prerequisite. For more information on AI terms, user data handling, and data privacy and security, refer to the *responsible AI* page at *http://s-prs.co/v620691*.

11.1 Embedded Services

SAP LeanIX offers multiple embedded services featured through AI capabilities to support interactions, automate routine tasks, and increase productivity. The following sections outline the various services and how they can be consumed.

Prerequisite

For the usage of the following AI features in this section, the *base AI capabilities* extension must be activated. This can be done in the **Optional Features and Early Access** area within the **ADVANCED SETTINGS** in the **Administration** options (see Figure 11.1).

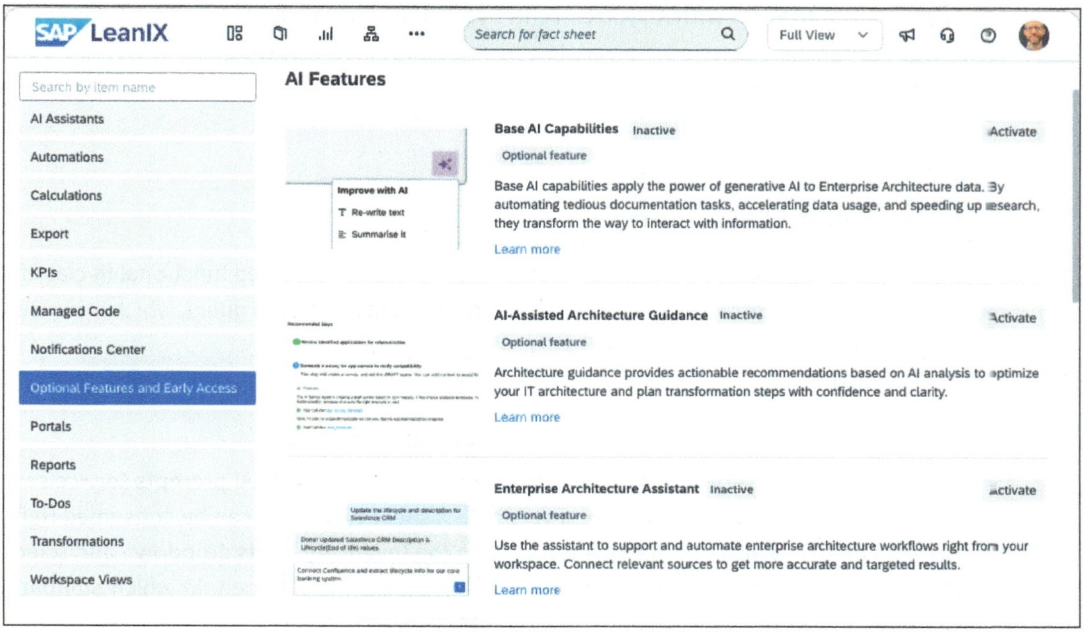

Figure 11.1: Activation of AI Features

11.1.1 Natural Language Filters

Within the **Inventory** tab, you have the option to add a filter with the support of AI, as shown in Figure 11.2.

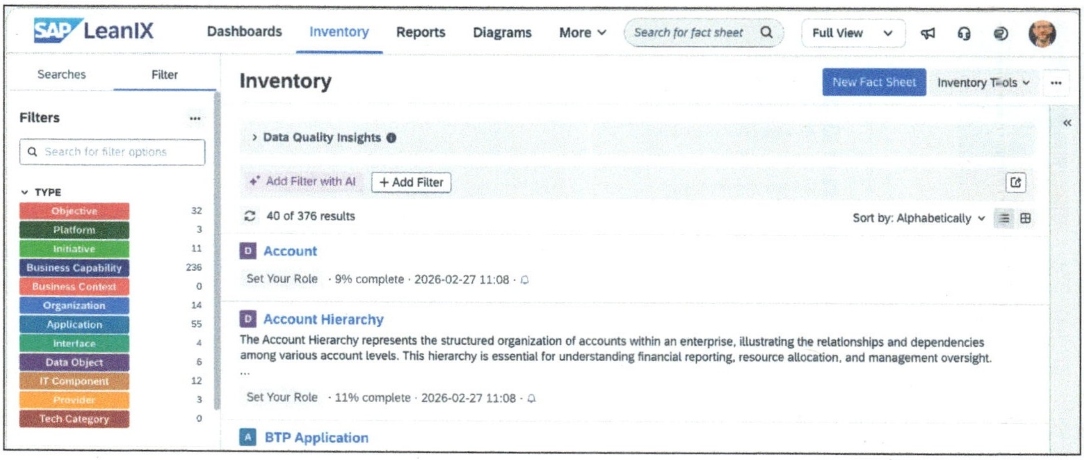

Figure 11.2: Define Filter with Support of AI

By choosing **Add Filter with AI**, you can enter your query in plain English, such as "Show me all applications in France with a very low business fit." After applying the query, AI would translate the query into filters such as the following:

- Type: Application
- Functional fit: Unreasonable
- Organization: EA7/Global/France

Infrequent users and business stakeholders not aware of the detailed functional fit classification can benefit from the translation of a natural language–based query, which increases adoption and reduces search time.

11.1.2 Standard Prompts

Another embedded service in the **Inventory** tab is AI prompts. The AI prompts service sets up commonly used prompts in the context of fact sheet types that can be consumed right away. To do so, navigate in the **Inventory** tab to the **Inventory Tools** dropdown and select **Start AI Prompt**. The dedicated **Inventory AI Prompt** view will open, in which standard prompts can be selected as outlined in Figure 11.3.

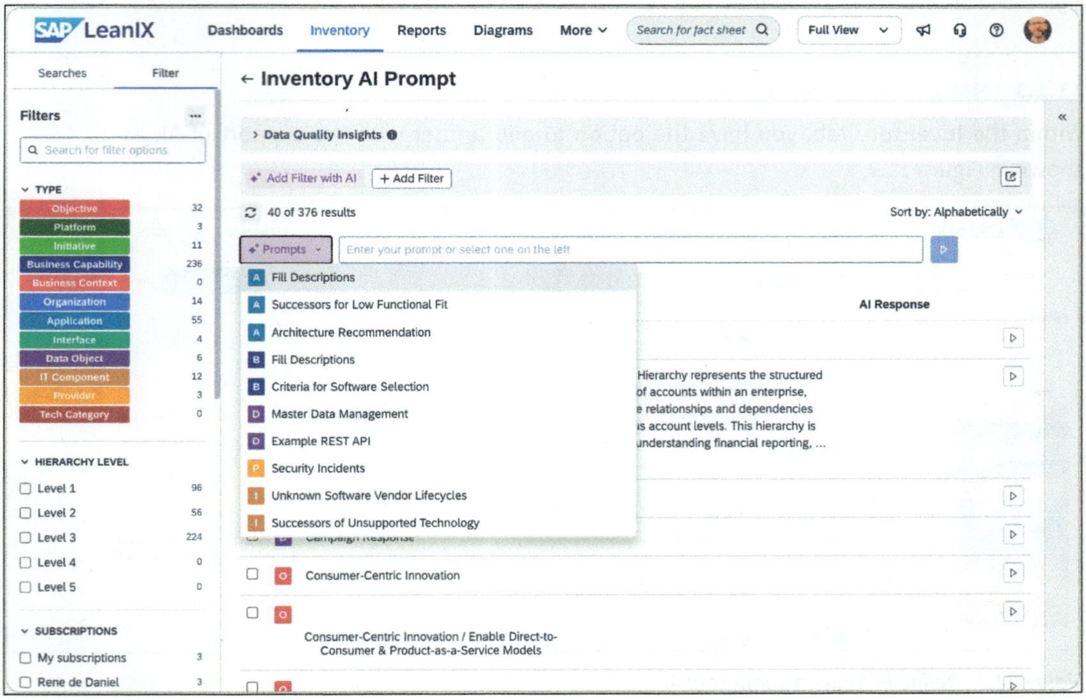

Figure 11.3: Inventory AI Prompts

Choosing one of the prebuilt prompts means that relevant inventory fact sheets will be selected, as outlined in Figure 11.4.

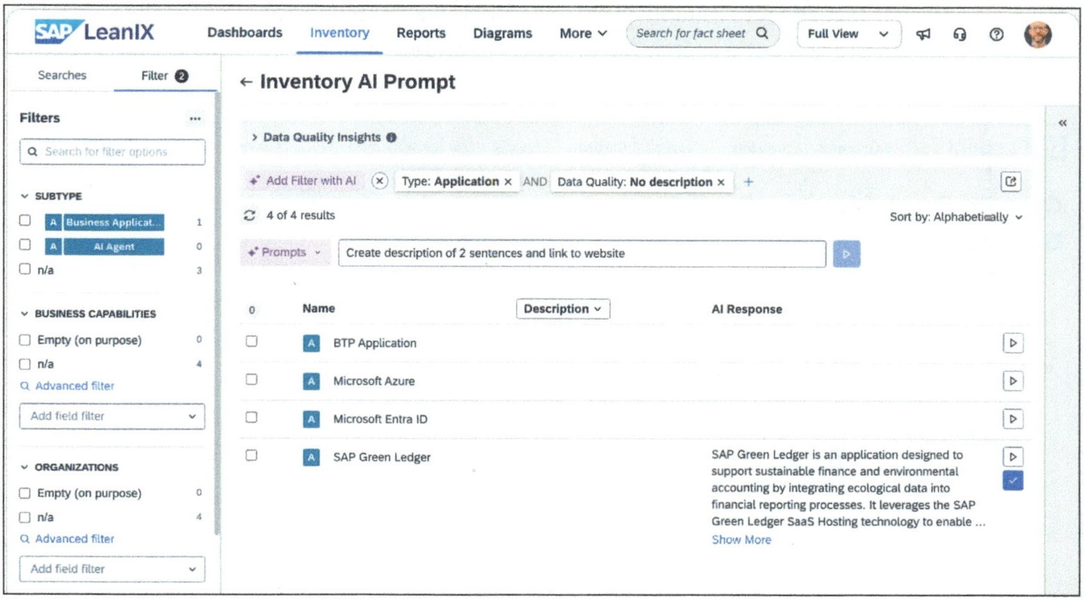

Figure 11.4: Execution of AI Prompt on Inventory Fact Sheets

On this screen, you have multiple options to collaborate with the AI and to augment fact sheet data in the following ways:

- Adjusting the AI prompt in the prompt field
- Defining which target field of the fact sheet should be populated with the AI response
- Running the AI prompt via the **Play** button
- Accepting the results, after potential multiple iterations, and saving these in the defined target field

You have also the option to select multiple fact sheets and run the AI prompt in bulk. The human in the loop is built in by design as confirmation must be provided before saving the generated content.

11.1.3 Semantic Analysis

The descriptions and relationships of individual fact sheets are important to ensure general understanding and to show how fact sheets relate to each other, which builds the basis for creating reports and insights.

Within the **Recommendations** tab of an individual fact sheet, a semantic analysis capability is available to create a context-aware description of the fact sheet by analyzing the fact sheet's name, type, relationships, and attributes to propose an accurate summary.

Depending on the fact sheet type and availability of the description, a second semantic analysis is offered to propose relations between the fact sheet in scope and relevant adjacent fact sheets available in the inventory in a context-aware approach. Figure 11.5 outlines the semantic search for the recommended business capability relations of an application.

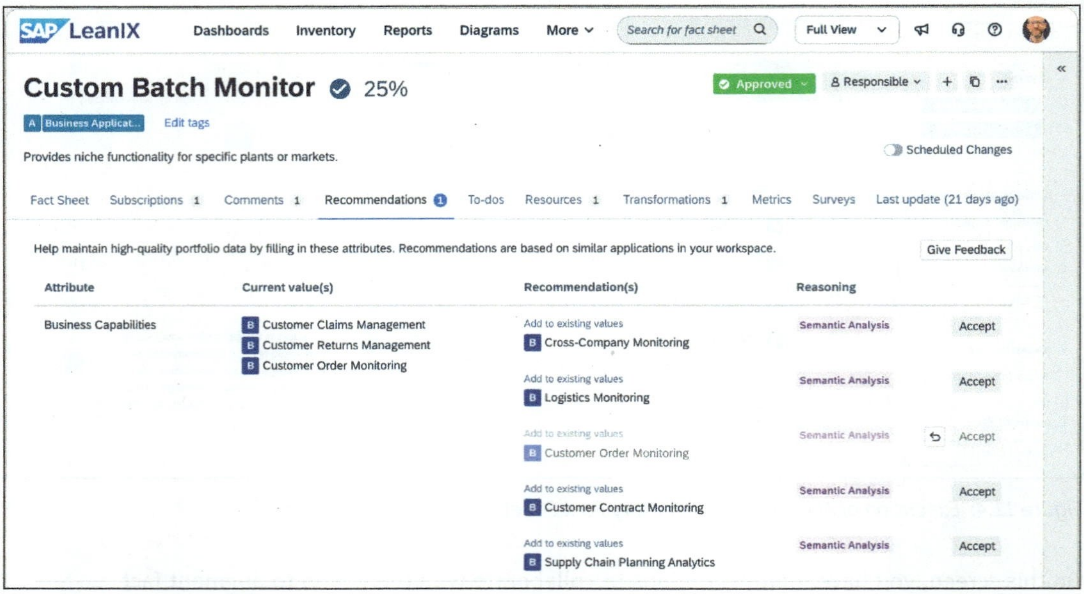

Figure 11.5: Semantic Analysis for Application to Business Capability Recommendations

In this context-aware approach, the name and description of the fact sheet are considered to recommend a set of business capabilities that make sense to link semantically. Individual recommendations can be accepted with the option to reverted later, as outlined in Figure 11.5.

Relation recommendations have proven very successful in daily architecture work. These often-cumbersome activities can be accomplished in a very timely and user-friendly manner.

> **Tip**
>
> The **Recommendations** tab of an individual application fact sheet indicates if a recommendation is available by displaying the number of available recommendations within the tab title.

11.1.4 Additional Embedded AI Services

The following additional embedded AI services are available for more occasional use:

- **AI-supported translation**
  As part of the meta model configuration, as outlined in Chapter 2, the translation of labels, help texts, and attribute descriptions might be required. This activity is supported by AI across all supported languages.

- **AI-generated field mapping**
  When configuring integrations (e.g., ServiceNow), AI suggests field mappings to streamline the initial setup.

The outlined AI capabilities in this section do not require the enabling of the AI terms. The *confirm before save* design ensures that architects review AI output first, while occasional users can interact through natural language.

11.2 AI-Assisted Inventory Builder

Imagine you are starting a new architecture engagement, and you find out that for this particular function or organization, the architecture information is not available in the SAP LeanIX workspace. After the first stakeholder interactions, you gain the insight that the baseline architecture has been documented in PowerPoint slides and Excel sheets. This is your window of opportunity to save hours of work that would otherwise be caused by either manually modeling the required fact sheets or massaging the data into importable formats.

The inventory builder helps you speed up your work and save time by taking the following actions:

- Uploading document, diagram, PDF, image, CSV, JSON, or XML files and discovering and extracting architectural elements and their relations
- Transforming unstructured artifacts into structured fact sheets and relations defined by your meta model (including custom fact sheets, fields, and relations)

> **Note**
> To use the inventory builder, the *AI-assisted inventory builder* extension must be activated. This can be done in the **Optional Features and Early Access** options within the **ADVANCED SETTINGS** of the **Administration** menu.
>
> The AI-assisted inventory builder is a premium AI feature subject to additional licensing.

With the activation of the AI-assisted inventory builder, the **Inventory Tools** dropdown is extended to include the **Inventory Builder** option, as shown in Figure 11.6.

You have here the option to either start a new import or open an existing import. If you start a new import, the inventory builder's import screen will appear as shown in Figure 11.7.

On the left pane of the screen, you have the option to either drag and drop or browse to upload a file that should be imported. The screen also informs you about the file formats supported and maximum file size.

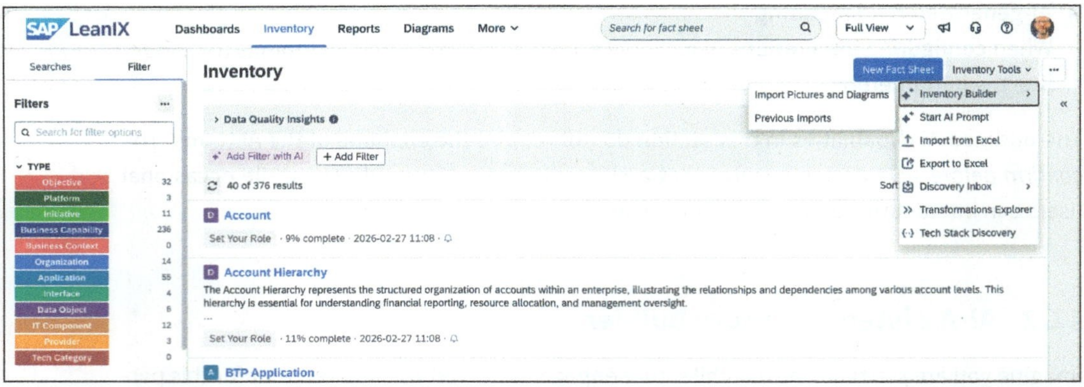

Figure 11.6: Inventory Builder

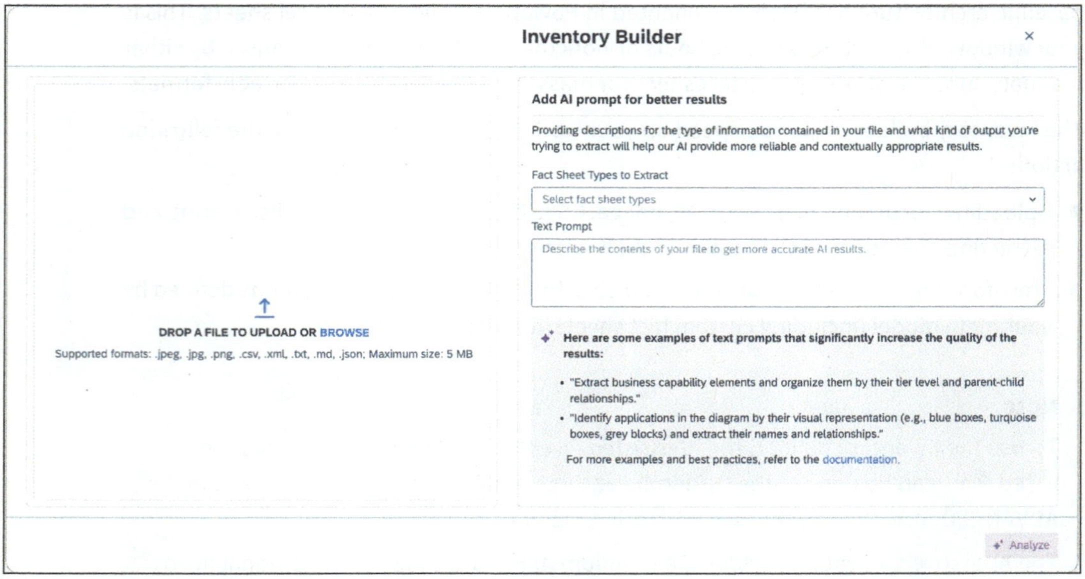

Figure 11.7: Inventory Builder: Import Screen

Once you upload a file, the inventory builder starts to analyze potential fact sheet types, which appear in the **Fact Sheet Types to Extract** section. Make sure to review and potentially correct the proposed fact sheet types as this will set the context for the subsequent matching.

Within the **Text Prompt** section, you can enter a specific prompt to enable a higher-quality outcome from the AI in terms of the identification and structuring of fact sheets. The link to the documentation is provided in the subsequent description, which helps to jump-start the prompt engineering.

Tip

You can be very prescriptive in describing what you see and how, for example, colors, shapes, or line styles should be interpreted. Feel free also to define how certain results are expected to appear in terms of relations between fact sheets or even naming conventions of fact sheets.

Our recommendation is to take a closer look at the sample prompts provided in the SAP LeanIX documentation, which helps to jump-start the prompt engineering. The examples and best practices are accessible either via the embedded link in the inventory builder import screen or the following link: *http://s-prs.co/v620692*.

Select **Analyze** to start a first inventory builder run; this might take a few moments.

Once the analysis is done, the screen will be updated by outlining a summary of the number of fact sheets and relations identified. The source file, sheets in scope, and text prompts will be displayed in the left pane, as shown in Figure 11.8.

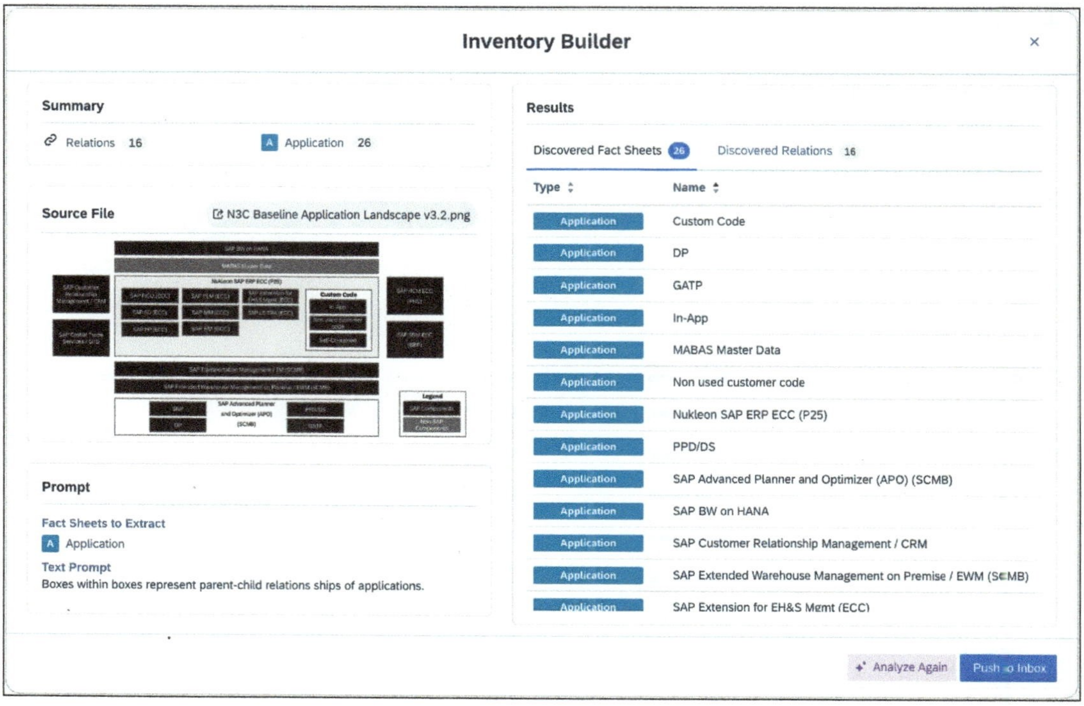

Figure 11.8: Inventory Builder: AI-Assisted Results

In the right pane, you'll see all the discovered fact sheets and relations.

If the discovered fact sheets and relations are not meeting your expectations, you can run a new analysis via the **Analyze Again** option. This will bring you back to the previous screen, where you can revise the text prompt.

By selecting the **Push to Inbox** option, the discovered fact sheets and information about the relations are moved into the import inbox, as shown in Figure 11.9. The inbox indicates how many discovery items are in this inbox and classifies them in terms of their status:

- **Action Needed**
  A suggested fact sheet or relation between fact sheets needs to be confirmed
- **In Progress**
  Inbox items that haven't been reviewed fully
- **Reviewed**
  Number of reviewed items
- **Rejected**
  Number of rejected items

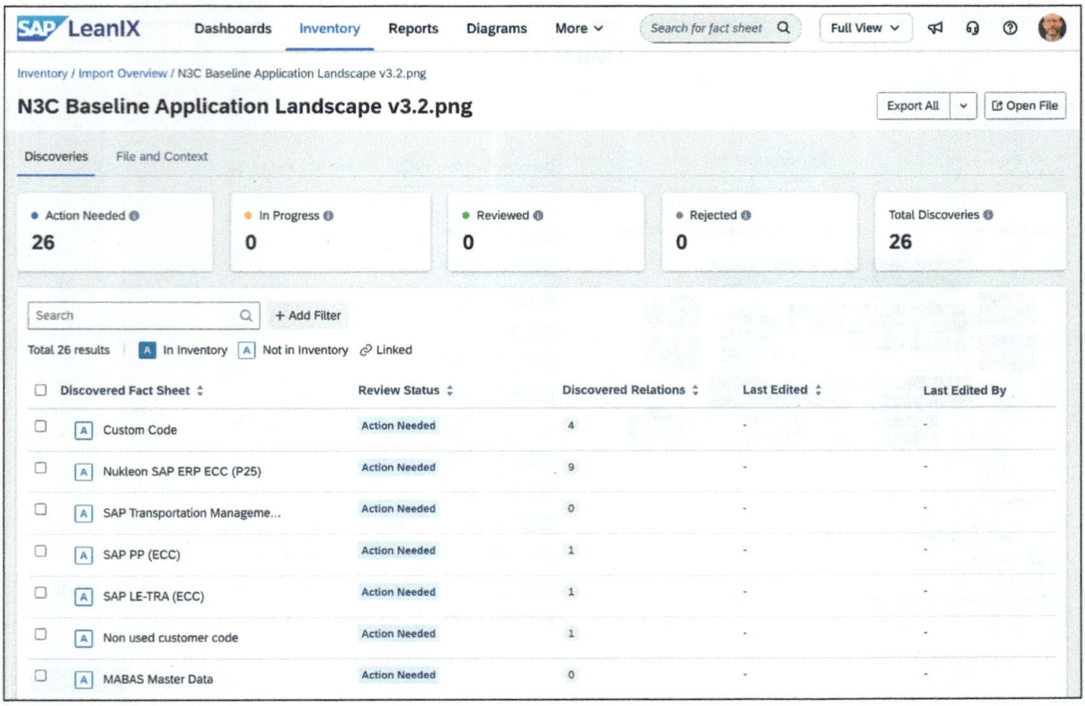

Figure 11.9: Inventory Builder: Import Inbox

The confirmation or rejection must be done on two levels: the fact sheet level and the relations level.

Fact sheets can be linked or rejected in the inbox screen as shown in Figure 11.9 by selecting the relevant fact sheets (the option to link or reject will appear).

Relations between facts sheets are confirmed or rejected in a detailed screen, which is opened by selecting any of the inbox fact sheets, as shown in Figure 11.10.

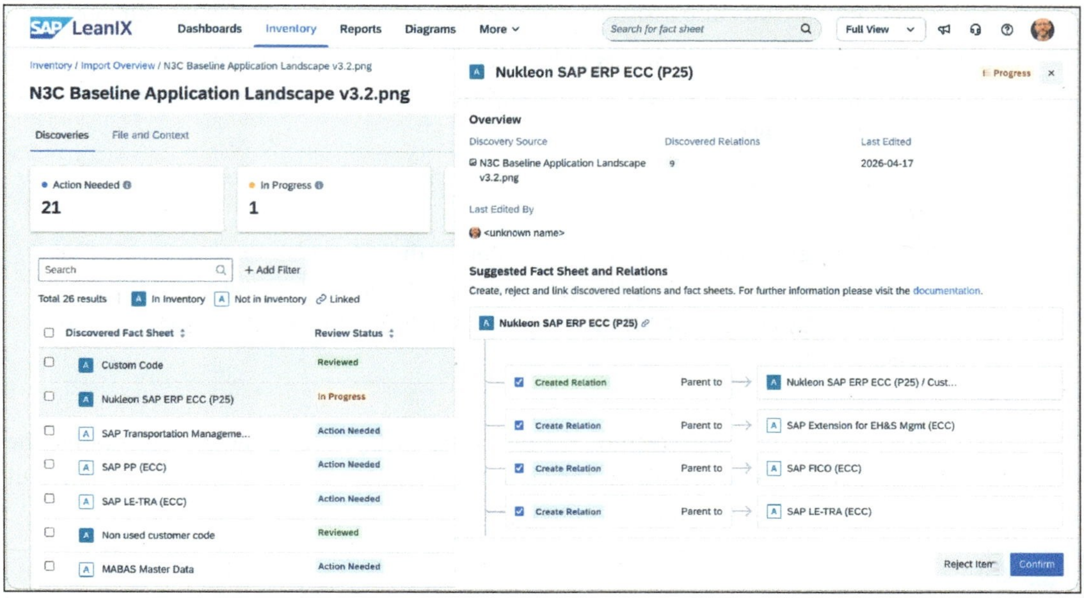

Figure 11.10: Rejection and Confirmation of Fact Sheet Relations

You have also the possibility to edit the fact sheets that have been identified as targets. This is very important if fact sheets have been identified that exist already in your inventory under another name.

Fact sheets already in the inventory are indicated by the opaque color of the respective fact sheet.

> **Tip**
>
> After linking a fact sheet or confirming a relation, the **Inventory** tab is updated immediately. This is also shown by the color of the fact sheet, which turns from transparent, for a fact sheet that does not already exist, to opaque.

With the help of the inventory builder, you can ingest existing architecture artifacts in an accelerated way. Fact sheets of any type as well as relations between fact sheets are identified and suggested, and the text prompt provides you an opportunity to instruct the AI on an individual level. The human-in-the-loop design ensures that only relevant information is entered into the inventory.

11.3 Enterprise Architecture Assistant

The enterprise architecture assistant is an AI-powered chat assistant embedded in SAP LeanIX that acts in a workspace-aware context consisting of multiple agents. It is connected to the SAP LeanIX MCP server and can therefore leverage the available toolset. In addition, the enterprise architecture assistant is also aware of the SAP LeanIX documentation.

The enterprise architecture agents fulfill a certain purpose and can be seen as assistance for your daily work, with access to generally available and unstructured information relevant to your architecture work.

Note

To use the enterprise architecture assistants, the *enterprise architecture assistant* extension must be activated. This can be done in **Optional Features and Early Access** within the **ADVANCED SETTINGS** area of the **Administration** section, as shown earlier in Figure 11.1.

To open the enterprise architecture assistant side panel, you can either navigate within the **More** section to **EA Assistant**, as shown in Figure 11.11, or use the keyboard shortcut `Ctrl`+`Shift`+`K` for Windows or `⌘`+`Shift`+`K` for Mac.

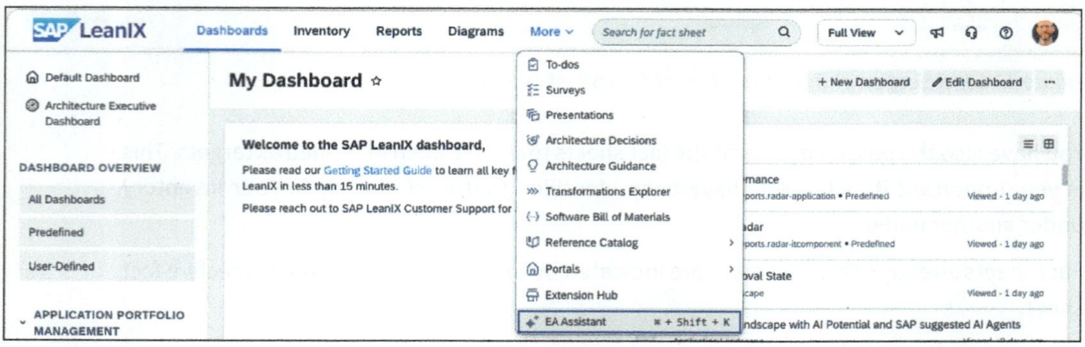

Figure 11.11: Start of Enterprise Architecture Assistant

The enterprise architecture assistant side panel will appear on the right-hand side of the screen, as shown in Figure 11.12.

Note that at the time of writing, the enterprise content research agent was just released.

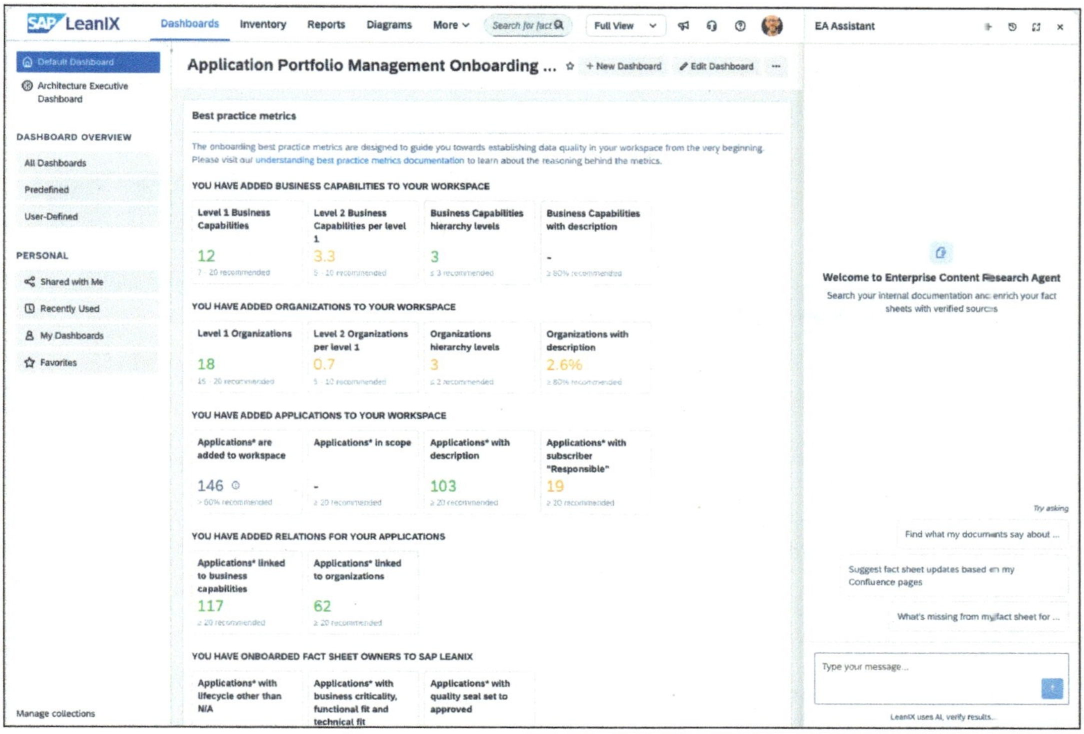

Figure 11.12: Enterprise Architecture Assistant: Side Panel

The enterprise content research agent has an AI-powered capability to read your internal documentation, such as uploaded PDFs, Word documents, wikis, Confluence pages, and so on. These sources of unstructured data can be used to gain insights and enrich your fact sheets or to increase their data quality. Manual copying of information can be avoided, and the heavy lifting of finding the information you are looking for can be handed over to the agent.

The enterprise content research agent is based on the conceptual architecture outlined in Figure 11.13.

The enterprise content research agent indexes uploaded and connected data and retrieves relevant content at query time. Files on the fact sheet level (to be found in the **Resource** tab) also are indexed automatically.

File-based content can be uploaded in the **Administration** area. You can find a dedicated **AI Assistants** section under **ADVANCED SETTINGS**, where you can upload and manage the content for your workspace as outlined in Figure 11.14.

An alternative to uploading content is to set up connectors to data sources in your ecosystem. At the time of writing, a connector to Atlassian was offered, with additional connectors on the roadmap. Connectors can be configured via the **Connectors** tab, as shown in Figure 11.15.

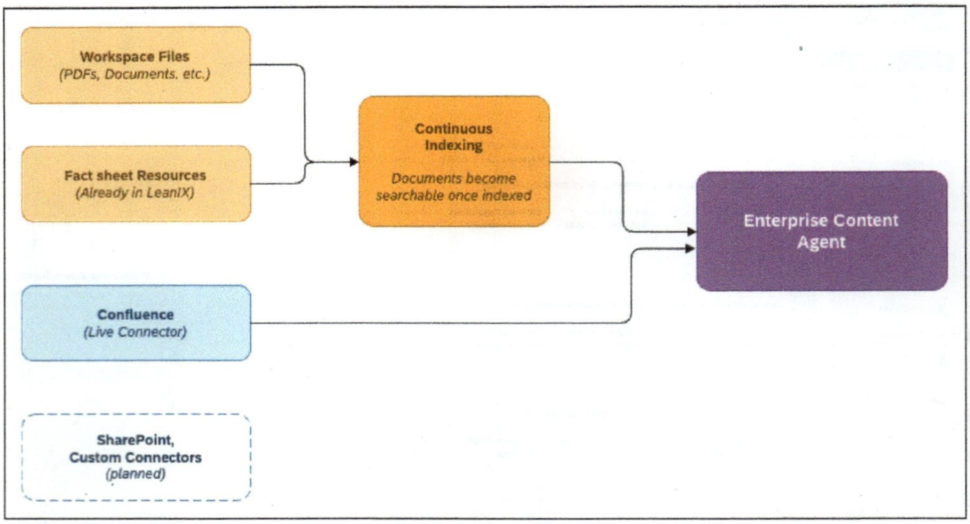

Figure 11.13: Enterprise Content Research Agent: Conceptual Architecture

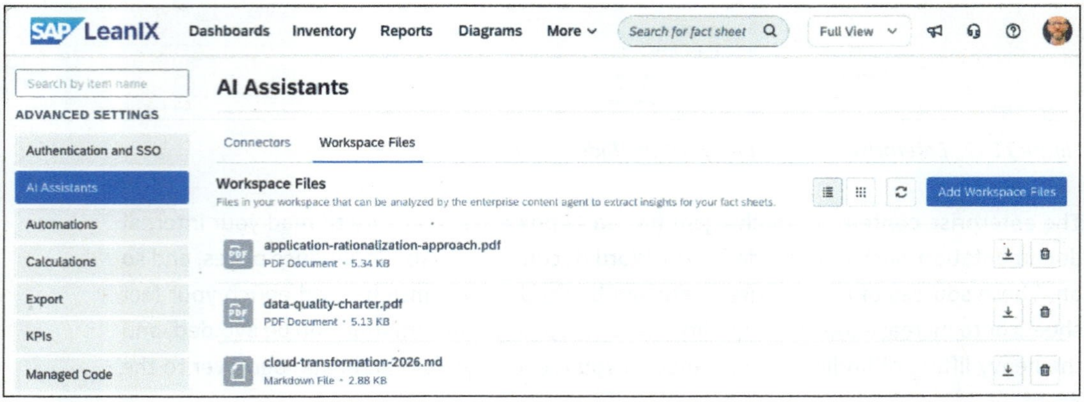

Figure 11.14: Content Upload for AI Assistants

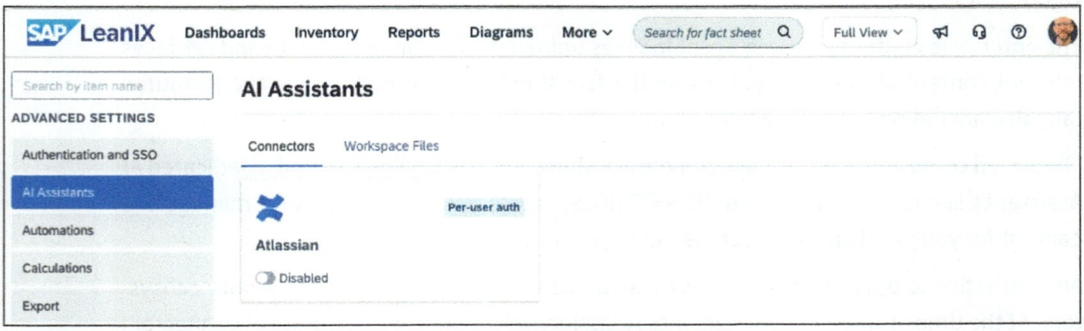

Figure 11.15: Connector Configuration for AI Assistants

One use case is the analysis of project documentation for RFP responses to discover undocumented applications, technologies, interfaces, and so on. Like the web research agent, any change to the workspace would require approval.

Outlook

Additional subagents are planned for the future, with, for example, the capability of conducting web research. Technically, these would use a dedicated Perplexity MCP integration to retrieve up-to-date, web-sourced information in combination with a workspace-aware context. The following primary use cases are imaginable (this is not an exhaustive list):

- **Market research**
  Open-ended questions about vendors, technologies, customer portfolios, and so on can be posed to gain insights into vendor comparisons, roadmap information, portfolio optimization recommendations, and more.
- **Market news**
  Recent news can be retrieved to, for example, understand the potential impact on your application landscape. This could include security vulnerabilities, acquisitions, supply chain incidents, and updated end-of-life announcements.

The use cases very much depend on the prompt you are entering and can expand across these primary use cases. Any result you retrieve through your web research can be considered as an option to update your fact sheets by the agent. Prior to any change of data in the workspace, the agent would ask for approval thanks to the human-in-the-loop design.

11.4 AI-Assisted Architecture Guidance

The AI-assisted architecture guidance introduces the paradigm shift from pull to push. The previously described AI capabilities were initiated by the SAP LeanIX user, but the architecture guidance changes the dynamic. It runs continuously in the background, proactively scans the inventory, and provides insights and recommendations without being asked. Users can review the surfaced insights and decide if the proposed recommendations, such as creation of surveys, architecture decision records, or transformations, should be followed or not.

Note

To use the architecture guidance, the *AI-assisted architecture guidance* extension must be activated. This can be done in **Optional Features and Early Access** within the **ADVANCED SETTINGS** area of the **Administration** menu, as shown in Figure 11.1.

The **More** menu will be extended by the activation of the architecture guidance extension as outlined in Figure 11.16.

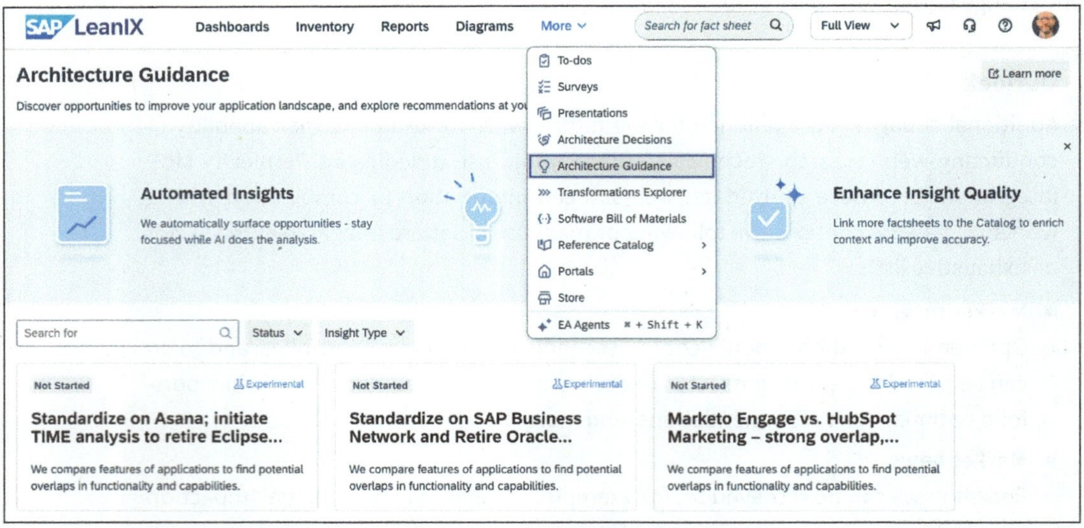

Figure 11.16: Architecture Guidance Navigation and Landing Page

> **Note**
> With the activation of the architecture guidance, insights and recommendations will not be immediately available. Inventory scans run autonomously in the background once the extension is activated, and insights rely on information and the data quality available in your workspace.

The architecture guidance provides the following types of insights (at the time of writing):

- Applications with overlapping features that are consolidation or retirement candidates
- Applications with low technical or functional fit compared to market standards
- Recommendations to adjust Gartner TIME classifications (tolerate, invest, migrate, eliminate)
- Suggestions to autopopulate fit attributes using industry benchmarks

The proactively provided insights should be reviewed on a regular basis and acted upon. The status of the insights helps to organize the work, such as by following up on recommendations for individual insights or by dismissing insights that are not relevant. The insights gained from your inventory fact sheet are supported by the following intelligence:

- **Industry best practices**
  Experience leveraged from 1,000+ enterprise customers
- **Peer benchmarking**
  Comparisons against anonymized best practice blueprints of similar organizations

- **TIME framework**
  Alignment with TIME classification

Once you open an insight, you will get a more detailed description of the rationales and recommendations for it, as highlighted in Figure 11.17.

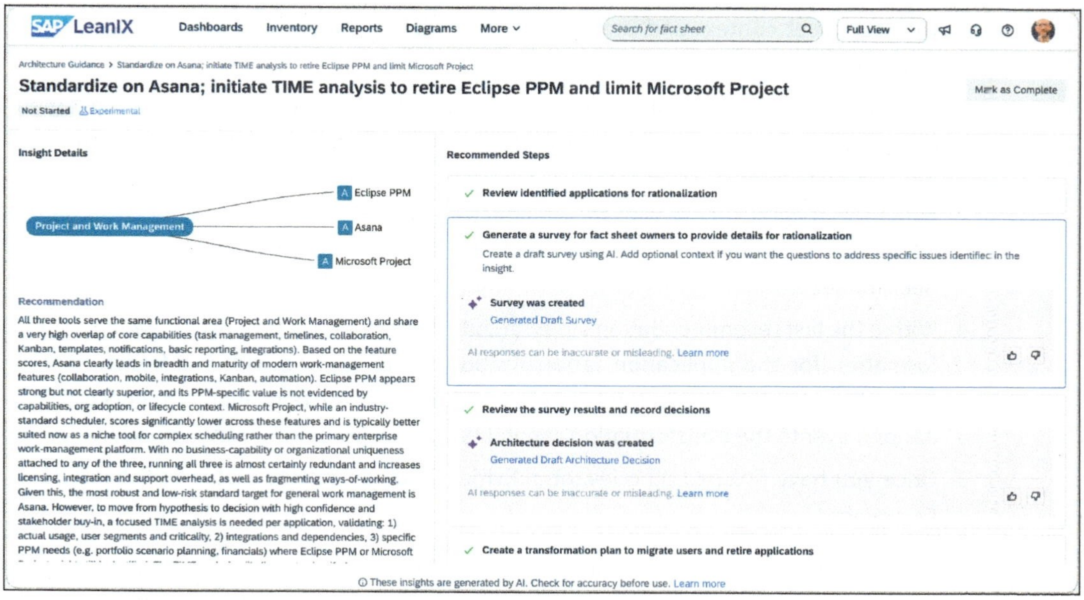

Figure 11.17: Architecture Guidance Insight Example

In this example, the architecture guidance is recommending the standardization of various project management tools into one standardized solution. Various recommended steps are proposed in the right-hand pane to realize these insights:

1. The first recommended step is to review the identified applications for rationalization. Within this step, a more detailed view is provided (see Figure 11.18).

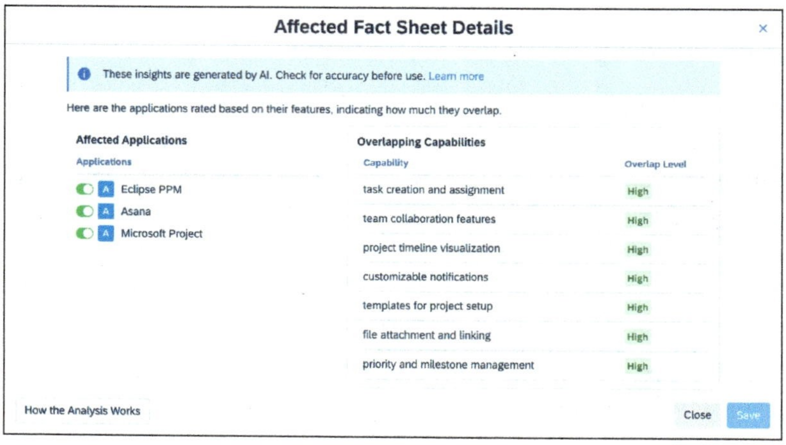

Figure 11.18: Affected Fact Sheets in Insight Example

The insight is based on a comparison of overlapping business capabilities in the context of the three applications in scope, which leads to the rationalization proposal.

2. The second step in the recommendation proposes to run a survey to collect further information in the context of the proposed rationalization. Architecture guidance integrates here with the survey capability introduced in Chapter 2, Section 2.3.3 by handing over the fact sheet context and drafting a survey. You as a user can decide if a survey is required.
3. The rationalization of applications (in this example, standardization of a project management solution) has an impact on the organization. Therefore, we recommend creating an architecture decision record as the next step. Like the survey feature, architecture guidance integrates here with the architecture decision record capability (see Chapter 12, Section 12.3.1) by handing over the fact sheet context and drafting an architecture decision record.
4. Within the last recommendation steps, architecture guidance proposes creating a transformation for the application rationalization by integrating with the transformation capability introduced in Chapter 6, Section 6.3.2. Again, the relevant fact sheet context is handed over to the transformation capability by creating a draft transformation.
5. Once you have finished all relevant activities, you can set the status of the insight to **Completed**.

As the architecture guidance is only as good as the underlying data, we strongly recommend investing in data quality first to achieve meaningful insights. Furthermore, an insight should be seen as a hypothesis in its first instance, which needs to be validated against the organization context. A good practice is to review the insights on a weekly basis.

Note

Your data is not used for AI training nor shared with other businesses.

11.5 External Agent Integration via MCP Server

With the release of Model Context Protocol (MCP) by Anthropic in November 2024 as an open-source standard, AI agents were enabled to connect to any external data sources and tools, avoiding nonreusable and costly custom integrations. The standard, which was adopted by all major providers—Anthropic, OpenAI, Microsoft, AWS, Google, and so on—is the most quickly adopted AI infrastructure standard, moving from 2 million downloads in November 2024 to 97 million downloads in March 2026. MCP is seen as the universal connector for the agentic enterprise; once built, any compliant agent can connect.

The SAP LeanIX MCP server has been available since November 2025 to all organizations at no additional cost. The MCP server offers a secure and standardized API gateway for agents and LLMs to connect to SAP LeanIX workspaces.

The true value of the MCP server is centered on the information available within SAP LeanIX, which can be seen like an architecture knowledge graph (the combination of fact sheets and their relations). This architecture knowledge graph is exposed via the MCP server to AI agents built on, for example, Claude Desktop, Microsoft Copilot Studio, Cline, a custom LLM, or any MCP-compliant agent, allowing these to query, create, or update live data as outlined in Figure 11.19. Agents in turn can interact with users via natural language (without the need to understand the technicalities) or autonomously. The release of the SAP LeanIX MCP server can therefore be seen not only as a means of technical integration but as making SAP LeanIX a participant in the agentic enterprise.

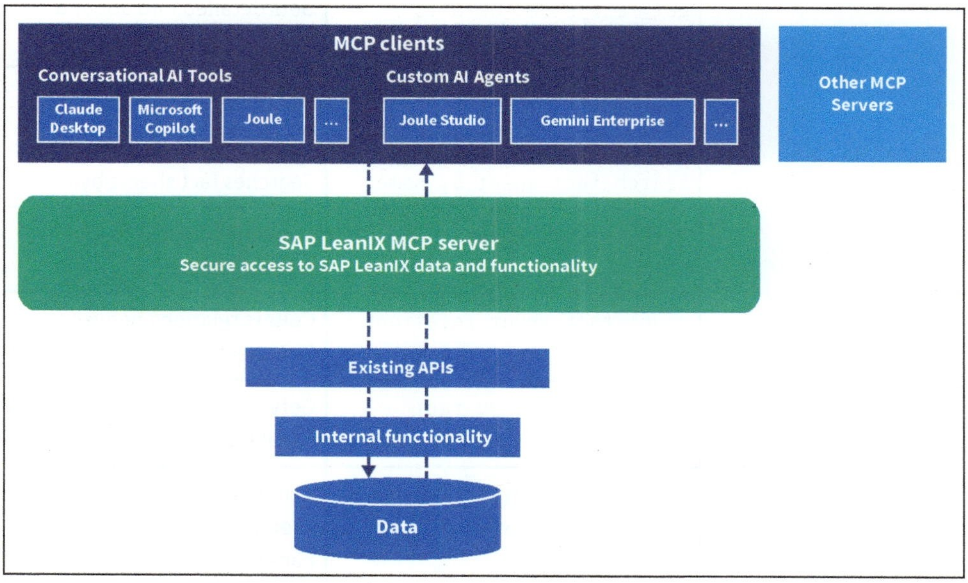

Figure 11.19: SAP LeanIX MCP Server: Conceptual Architecture

The SAP LeanIX MCP server is characterized by the following design principles:

- **Secure integration**
  Workspace-level permissions for agent access
- **Permission management**
  Role-based access controls extended to agents
- **Context-rich data**
  Structured data to enhance AI reasoning quality

We'll take a closer look at the MCP server in the following sections, including how it fits into SAP LeanIX, important dos and don'ts, and a practical example of building an external AI agent with Claude.

11.5.1 MCP Server Tools and Components

Although the MCP standard defines the protocol for how an agent interacts with the MCP server, the essence of the MCP server is defined by the core tools it exposes. Table 11.2 provides an overview of the tools offered by the SAP LeanIX MCP server.

Category	Tool	Description
Fact sheet management	get_fact_sheet_types	Lists all available fact sheet types
	get_fact_sheet_details	Gets detailed info about a specific fact sheet
	get_fact_sheet_subscriptions	Retrieves subscriptions for a fact sheet
	search_fact_sheet_by_name	Searches fact sheets by name
	text_to_fact_sheets	Converts free text to matching fact sheets
	count_fact_sheet_relations	Counts relations between fact sheets
	get_milestones_by_fact_sheet_ids	Gets milestones for given fact sheet IDs
	list_fact_sheets	Renders a fixed list of fact sheets (by ID) in the inventory panel
	single_fact_sheet	Renders a single fact sheet's detail page (attributes, lifecycle, subscriptions)
	get_overview	Workspace summary with fact sheet counts per type
Applications	get_applications	Lists applications in the workspace
	get_application_components	Gets components linked to an application
	get_initiatives	Lists initiatives/projects in the workspace

Table 11.2: SAP LeanIX MCP Server Tools

Category	Tool	Description
Tech stack/component analysis	get_all_tech_stacks	Lists all tech stacks
	get_tech_stack_details	Gets details for a specific tech stack
	create_tech_stack	Creates a new tech stack
	preview_tech_stack_matches	Previews component matches for a tech stack
	search_components_by_purl	Searches components using a package URL (PURL)
	batch_search_components_by_purls	Batch-searches components using multiple PURLs
	search_components_by_license	Searches components filtered by license type
Architecture	get_architecture_decisions	Retrieves architecture decisions
	get_architecture_decision_as_markdown	Retrieves an architecture decision formatted as markdown
	create_architecture_decision	Creates a new architecture decision
	update_architecture_decision	Updates an existing architecture decision
	delete_architecture_decision	Deletes an architecture decision
	get_architecture_templates	Lists architecture templates
	get_architecture_template_components	Gets components within an architecture template
Diagrams	get_diagrams	Lists diagrams in the workspace
	get_diagram_by_id	Retrieves a specific diagram by ID

Table 11.2: SAP LeanIX MCP Server Tools (Cont.)

Category	Tool	Description
Surveys	get_survey_templates	Lists available survey templates
	create_survey	Creates a new survey
	search_survey_runs	Searches for survey runs
	get_survey_recipients	Gets recipients of a survey
	get_survey_results	Retrieves results from a survey
	change_survey_status	Changes the status of a survey
	create_survey_reminder	Creates a reminder for a survey
	get_reminders	Lists existing reminders
Reports	get_reports	Lists available reports
	get_insights	Retrieves workspace insights and analytics
	get_custom_report_guide	Gets guidance for building custom reports
GraphQL API introspection	get_graphql_type_definitions	Returns GraphQL type definitions for the API
	list_graphql_types	Lists all available GraphQL types
User management	search_users	Searches for users in the workspace
	get_user_subscriptions	Gets subscriptions for a user
Transformations	get_transformation_types_tool	Lists available transformation types
	search_transformations_tool	Searches for transformations
Relation explorer	open_relations_explorer	Opens the relations explorer UI for a fact sheet

Table 11.2: SAP LeanIX MCP Server Tools (Cont.)

Tip

The tools offered by the SAP LeanIX MCP server will evolve at such a speed that the documentation might not reflect a full overview. A best practice is to ask the SAP LeanIX MCP server about the tools on offer by applying the following prompt in your AI client, such as Claude Code: "Provide me a summary table of SAP LeanIX MCP tools available by category."

If we use a cooking analogy for a moment, the tools offered by the MCP server can be seen as individual ingredients. If you ask an LLM to cook a specific meal, it might use different ingredients depending on the model version or conversation length. A vague query most likely will end in a vague result, and the outcome of a similar query in another session most likely will differ. Therefore, skills are of primary importance to allow a guided, step-by-step approach, including the ingredients to be used and in which order: basically, a recipe. A full dinner in turn resonates with the analogy of an agent, able to combine multiple skills while considering a given context. Table 11.3 outlines the concepts of tools, skills, and agents by leveraging the analogy of cooking.

Concept	Definition	Analogy
Tool	A single MCP API call exposed to the LLM	The ingredient
Skill	A guided, step-by-step workflow that tells the LLM which tools to use, in which order, including instructions for handling responses and formatting output	The recipe, or how to process the ingredients
Agent	A combination of multiple skills, with additional context/system prompt, focused on delivering a specific outcome	The full dinner: cooking a specific meal by using multiple recipes

Table 11.3: Cooking Analogy

Let's now take this analogy and translate it into a mental model for the best usage of the MCP server by building agents with reliable outcomes that combine multiple skills with reliable workflows and guided tool usage, provided via the MCP server, as outlined in Figure 11.20.

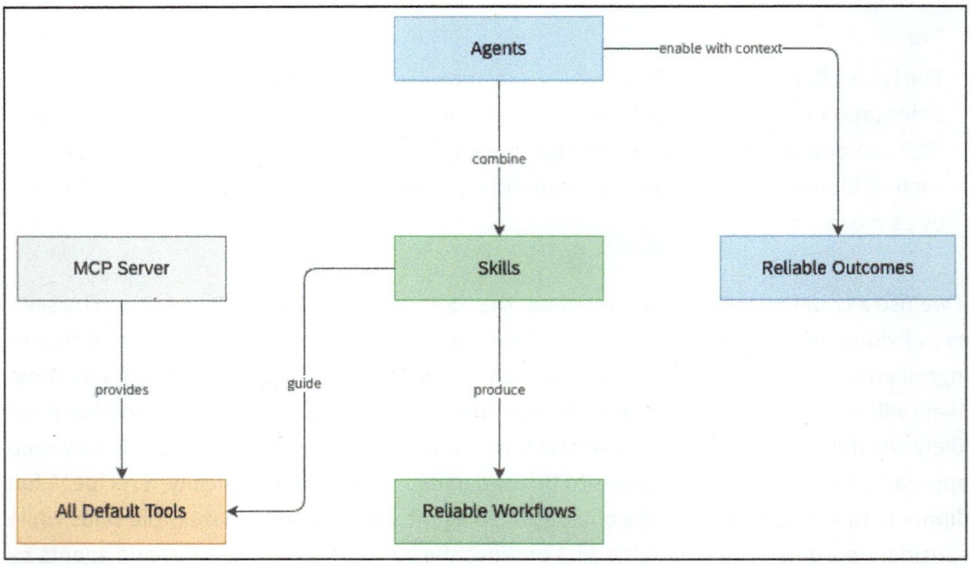

Figure 11.20: Mental Model for MCP Server

From a technical architecture perspective, the following characteristics are worth mentioning:

- **Protocol layer**
  The MCP server communicates via SSE and HTTP streaming, enabling real-time, low-latency interactions between AI agents and the SAP LeanIX repository.

  The SAP LeanIX GraphQL API is exposed as a set of MCP-compatible tools, with the server acting as a translation layer between agent requests and repository operations.

- **Authentication**
  Access is authenticated via API tokens, issued to a dedicated technical user provisioned specifically for agent interactions.

  We strongly recommend that personal or administrative accounts never be used for agent authentication. Furthermore, the technical user's permissions should be scoped precisely to the operations required while applying the principle of least privilege.

- **Deployment**
  The SAP LeanIX MCP server supports two deployment modes: Node.js for lightweight local or server-based deployments, and Docker for containerized, enterprise-grade environments.

  Both options require configuration rather than custom development, making deployment accessible to architects with foundational technical literacy. Therefore, no bespoke engineering is required to establish a functional agent connection.

11.5.2 Dos and Don'ts

From a practitioner perspective, we would like to share the dos and don'ts outlined in Table 11.4. These should be considered while using the MCP server.

Do	Why?
Build skills before using the MCP	Raw tool access without guidance leads to unreliable, nonrepeatable outcomes.
Use tool sets, scope down the tools	Fewer tools = smaller context window = cheaper + faster + more reliable. Only expose the tools the skill needs.
Start small, then combine	Begin with a single focused skill with a complete, small outcome. Build complexity incrementally.
Make questions specific and scoped	Tell the LLM exactly what you want: country filter, lifecycle filter, business criticality; don't leave it open-ended.
Iterate manually first, then encode as a skill	Walk through the workflow manually (multiple prompts, filtering), then ask the LLM to "create an agentic skill for this whole approach" to capture your learnings.
Reuse the marketplace	Check the marketplace, GitHub, and the like for existing skills before building from scratch.
Use OAuth, user-based authentication	This provides a proper audit log and simplifies setup versus technical user tokens.
Allow tool confirmations selectively	Keep confirmation on in engineering/development environments to stay in control; preapprove tools in production skills.
Don't	**Why?**
Expose all tools to every agent	This bloats the context window, increases costs, and confuses the LLM into picking the wrong tools.
Expect out-of-the-box agents from the MCP server alone	The MCP server provides ingredients, not a finished dish; you must add skills.
Rely only on persona prompts	"You are an EA expert" is not sufficient to narrow down tool selection or produce reliable output.

Table 11.4: Dos and Don'ts

Don't	Why?
Ask the same vague question louder	Rephrasing "tell me about my landscape" to "tell me about the landscape in the US" without structured guidance doesn't solve the problem.
Give AI-generated workspace reports directly to your stakeholders	Reports need human judgment; AI output is a starting point, not a finished deliverable.

Table 11.4: Dos and Don'ts (Cont.)

11.5.3 Build Your First Claude Agent Skill

Now, let's build our first external agent skill with Claude Code as an example. (Note that the instruction will be in the context of a Mac computer.)

Installation and Setup

The following steps must be performed to install Claude Code and set up the MCP server connection:

1. Open a terminal window and enter the following command to install Claude Code:

   ```
   npm install -g @anthropic-ai/claude-code
   ```

2. Open the SAP LeanIX **Administration** menu and create a new technical user under the **DISCOVERY AND INTEGRATIONS** section. Make sure to assign the relevant permissions role. By saving the technical user settings, the access token will be shared as shown in Figure 11.21. Copy the access token (make sure not to share it with anyone else).

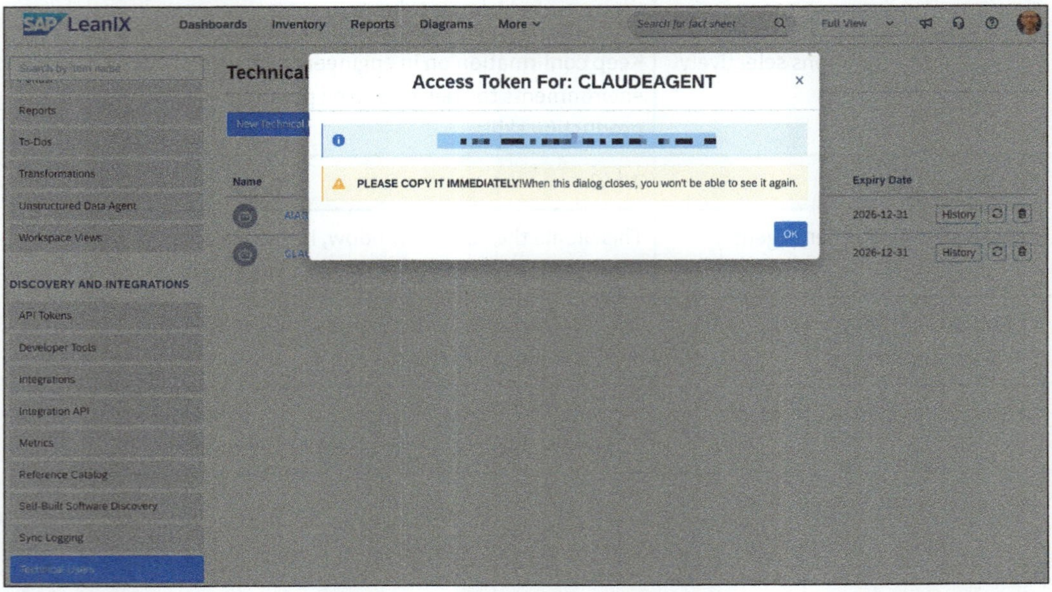

Figure 11.21: Access Token for Technical User

3. Define your SAP LeanIX workspace-specific MCP endpoint by entering your workspace domain into the following SAP LeanIX MCP server endpoint:

   ```
   https://{SUBDOMAIN}.leanix.net/services/mcp-server/v1/mcp
   ```

   Make sure to enter your specific SAP LeanIX workspace subdomain by extracting the subdomain from your SAP LeanIX workspace URL.
4. For the initial configuration of the SAP LeanIX MCP server, enter the following command in a terminal window, replacing the placeholders with your own info:

   ```
   claude mcp add --transport http leanix-mcp https://[your-leanix-mcp-
   endpoint] \ --header "Authorization: Token [your-access-token]"
   ```

5. Verify the MCP server connection by running the following command:

   ```
   claude mcp list
   ```

   The status of the MCP server connection should be **Connected**.
6. Start Claude by running the following command: `Claude`

   Claude is now ready for interaction with the SAP LeanIX MCP server of your workspace.
7. Create your first prompts in natural language and see what information Claude is returing:
   - *"List all applications in our portfolio with end-of-life dates before 2027."*
   - *"Show me all fact sheets tagged as business critical with low technical fit."*
   - *"What applications are owned by the finance business unit?"*
   - *"Generate an architecture decision record for adopting the SAP LeanIX MCP server."*

Figure 11.22 shows the result of the *"Show me all fact sheets tagged as business critical with low technical fit"* against the MCP server of the SAP LeanIX demo account. A subsequent interaction could bring more information to the surface, or more detailed queries could be combined with the initial search result, such as including a TCO perspective, for example.

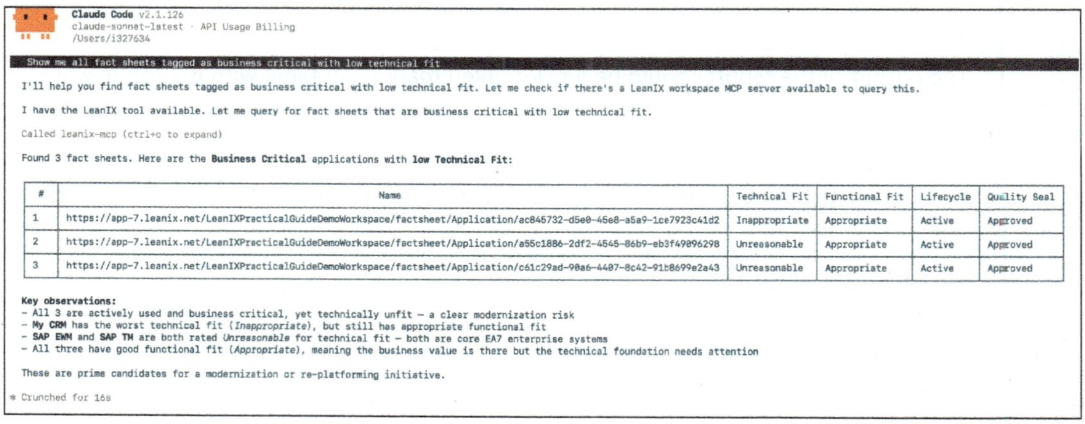

Figure 11.22: Sample Prompt Response

The subsequent interaction will most likely be tailored to your context and the output; eventually, the tools used and the format might differ.

Create a Skill

To ensure a consistent outcome, you can create a skill by following a reliable workflow structure, as outlined in Figure 11.20. Before you start to write a skill, you need to answer the following questions:

- What does the user say to trigger this skill?
- What data does the skill need, and from which tools?
- What does it produce?

By answering these three questions, you can define the description, the fetch sequence for the tool, and the output format and destination path.

Let's assume you want to create an application risk report. The user query for this report would be "Show me application risks for [domain]" or "risk report." By taking the tools outlined in Table 11.2 into account, the fetch sequence looks as outlined in Table 11.5.

MCP Tool	Purpose
`get_overview`	Establish the workspace baseline
`text_to_fact_sheets("[domain] apps")`	Resolve scope to IDs
`get_fact_sheet_details(id) × N`	Retrieve lifecycle, fit, and TIME tag per app
`get_application_components(ids)`	Retrieve IT component relations and end-of-life status
`get_insights()`	Get insights into workspace-level risk flags

Table 11.5: Fetch Sequence

The output of the report should be a risk-ranked markdown table with the top five action recommendations and should be saved to `~/Documents/Claude/LeanIX-Reports/app-risk-[domain]-[date].md`.

First, create the directory structure for the skill by using the following command in the terminal:

```
mkdir -p ~/.claude/plugins/local/app-risk-reporter/1.0.0/skills/ \
app-risk-reporter
```

The following folder structure and file will be created:

```
~/.claude/plugins/local/app-risk-reporter/
└── 1.0.0/
  └── skills/
    └── app-risk-reporter/
      └── SKILL.md
```

Note

Claude Code discovers skills by scanning *~/.claude/plugins/*. The version directory (*1.0.0/*) is required for the loader to find the skill.

In our example, *SKILL.md* is simultaneously the agent specification and the system prompt overlay Claude receives when the skill is invoked. Precision in the definition of the individual sections of *SKILL.md* directly determines reliability. The generic *SKILL.md* structure looks like Listing 11.1.

```
---
name: [machine-readable kebab-case name (lowercase and separated by
hyphens)]
description: [natural language trigger conditions – what user says to
invoke this]
---
# [Skill Title]
[One sentence: what this skill does and what it produces]
## Workflow
### Step 1 – [name]
[what to do, which tools to call, in what order]
### Step n – ...
## Safety Rules
[what requires user confirmation, what to do on errors]
## Output Style
[format, sort order, file path, summary format]
```

Listing 11.1: SKILL.md Structure

Note

SKILL.md builds the core of the skill. We therefore demonstrate the structure in full.

Feel free to use Anthropic's skill creator skill (see *http://s-prs.co/v620693*) to build and register a new skill.

The full example for the *app-risk-reporter/SKILL.md* looks as shown in Listing 11.2.

```
---
name: app-risk-reporter
description: >
  Generate an application risk report for SAP LeanIX. Use when the user
asks for
  application risks, portfolio health, technology risk analysis, or a
risk-ranked
  application summary. Trigger phrases: "risk report," "application
risks,"
  "portfolio health," "tech risk," "which apps are at risk," "show me
risks for."
---
# Application Risk Reporter
Produces a risk-ranked application summary from SAP LeanIX live data using
the LeanIX MCP server.
## Workflow
### Step 1 – Clarify scope
Ask the user the following before fetching any data:
- Which business domain or tag filter? (leave blank to include all
applications)
- Include applications in end-of-life or phase-out lifecycle? (default:
yes – these are typically highest risk)
- Preferred output format: table / full markdown report / CSV for LeanIX
import?
If the user provides a domain in the trigger phrase (e.g. "risks for
Finance"), skip this step and proceed with that domain as the filter.
### Step 2 – Fetch data
Run these tool calls in order. Parallelize where inputs are independent.
1. `get_overview` – establish workspace baseline (total application count,
workspace health score)
2. `text_to_fact_sheets("[domain] applications")` – resolve the scope to a
list of fact sheet IDs
   - If no domain filter: use `get_applications` to retrieve all
application fact sheets
3. For each Application fact sheet ID (run in parallel batches):
   - `get_fact_sheet_details(id)` – extract: name, lifecycle phase,
functional_fit, technical_fit,
     business_criticality, TIME classification tag, last_updated_at
4. `get_application_components(ids)` – retrieve IT Component relations for
all applications;
   check the lifecycle phase of each linked component to identify end-of-
life dependencies
```

```
5. `get_insights()` - pull workspace-level risk signals and data quality
flags
**Scope guard:** If step 2 returns more than 50 fact sheets, pause and ask
the user to narrow the domain or tag filter before continuing. Do not
silently process large sets.
### Step 3 - Score risk
Apply the following model to every application.
**Factor weights:**
| Factor | Weight | High Risk Condition |
|---|---|---|
| Technical fit | 30% | fit = unreasonable or insufficient |
| Functional fit | 20% | fit = unreasonable or insufficient |
| IT Component EOL | 25% | any linked component in end_of_life lifecycle
phase |
| Lifecycle phase | 15% | application phase = end_of_life or phase_out |
| Business criticality | 10% | mission_critical amplifies base score by
×1.2 (capped at 1.0) |
**Fit score mapping:**
| LeanIX Value | Numeric Score |
|---|---|
| unreasonable | 1.0 |
| insufficient | 0.75 |
| adequate | 0.4 |
| good | 0.15 |
| perfect | 0.0 |
**Lifecycle phase risk:**
| Phase | Risk Contribution |
|---|---|
| end_of_life | 1.0 |
| phase_out | 0.7 |
| active / phase_in / plan | 0.0 |
**Business criticality multiplier:**
| Criticality | Multiplier |
|---|---|
| mission_critical | ×1.2 (capped at 1.0) |
| important | ×1.0 |
| administrative | ×0.8 |
**Risk tier thresholds:**
- Score ≥ 0.65 → **HIGH** ?
- Score 0.35-0.64 → **MEDIUM** ?
- Score < 0.35 → **LOW** ?
```

````
**Priority labels:**
| Tier | Label |
|---|---|
| HIGH | Immediate |
| MEDIUM | Plan this quarter |
| LOW | Monitor |
Assign each application a tier. If fit score data is missing, mark the field `[no data]`
and exclude that factor from the weighted calculation rather than defaulting to any value.
### Step 4 – Produce output
Lead with a one-line summary:
> "N HIGH, N MEDIUM, N LOW risk applications found in [domain or 'the workspace']."
**Markdown table (default):**
| App | Criticality | Tech Fit | Func Fit | Lifecycle | EOL Components | Risk | Priority |
|---|---|---|---|---|---|---|---|
| App Name | Mission Critical | Insufficient | Adequate | Active | 2 | ? HIGH | Immediate |
Sort order: HIGH → MEDIUM → LOW. Within each tier, sort by business_criticality descending
(Mission Critical first).
Use these emoji indicators:
- ? HIGH
- ? MEDIUM
- ? LOW
Follow the table with a **Top 5 Actions** section: one concrete, actionable next step per HIGH-risk application. Frame each action as a specific recommendation (e.g. "Schedule technical assessment for [App] – linked IT Component [X] reaches end-of-life in Q3 2025").
**CSV format (if requested):**
```csv
name,business_criticality,technical_fit,functional_fit,lifecycle_phase,eol_components,risk_tier,priority
```

### Step 5 – Save output
Save the report to:
`~/Documents/Claude/LeanIX-Reports/app-risk-[domain]-[YYYY-MM-DD].md`
Create the directory if it does not exist.
Print confirmation: `Report saved to [full path]`
````

```
## Safety Rules
- **Never call any write tool** (`create_*`, `change_*`) without first
stating the intended action and receiving explicit confirmation from the
user.
- If any `get_fact_sheet_details` call fails, log the failed ID in a
"Fetch errors" section at the end of the report – do not silently skip or
halt the run.
- If the MCP server returns an authentication error, stop immediately and
report it clearly rather than returning partial results without a warning.
- Do not make assumptions about missing fit scores – mark them as `[no
data]` in the table and exclude them from the risk calculation rather than
defaulting to a value.
## Output Style
- Executive summary: 3 sentences maximum
- Table sorted HIGH → MEDIUM → LOW, mission-critical applications first
within each tier
- Actions section: one bullet per HIGH-risk app, phrased as a specific
recommendation
- Tone: direct, practitioner-level – omit explanations of basic EA
concepts
- Dates: use ISO 8601 (YYYY-MM-DD)
```

Listing 11.2: SKILL.md for Application Risk Reporter

> **Note**
>
> We registered the skill as a local plugin in this example. For the sake of completeness, we will outline the additional steps to be taken if you manually register the skill.
>
> A recommended alternative is building the skill via Anthropic's skill creator skill (see *http://s-prs.co/v620693*).

The new skill requires a *plugin.json* metadata file within a new subfolder of the previously created folder structure:

1. Create the *.claude-plugin* folder via the following command:

   ```
   mkdir -p ~/.claude/plugins/local/app-risk-reporter/1.0.0/ \
   .claude-plugin
   ```

2. Place *plugin.json* into this newly created folder. In this example, the file looks like Listing 11.3.

   ```
   {
     "name": "app-risk-reporter",
     "description": "Generate an application risk report for SAP
   ```

```
LeanIX. Produces a risk-ranked application summary using the LeanIX
MCP server.",
    "version": "1.0.0",
    "author": {
      "name": "Local"
    },
    "keywords": ["leanix", "risk", "application", "portfolio", "tech-
risk"]
  }
```

Listing 11.3: Plugin.json for Application Risk Reporter

3. A registration entry needs to be added into *.claude/plugins/local/.claude-plugin/marketplace.json*, as shown in Listing 11.4.

```
{
       "name": "app-risk-reporter",
       "description": "Generate an application risk report for SAP
LeanIX",
       "version": "1.0.0",
       "source": "./app-risk-reporter"
}
```

Listing 11.4: Marketplace.json for Application Risk Reporter

4. Finally, register the newly created plugin within *.claude/settings.json* by adding the `"app-risk-reporter@local": true` entry within the `enabledPlugins` section.
5. After restarting Claude, the app-risk-reporter plugin is loaded. You can also verify this by entering `/plugins list` in Claude Code; the plugin should show the status **Enabled**.
6. You can trigger the skill by using a natural language phrase like "Show me application risks for the supply chain domain." The output for the demo workspace would look like Figure 11.23.

MCP in general, as well as in the context of SAP LeanIX, is still evolving. We therefore foresee ongoing improvements in terms of security, scalability, and collaboration. One important aspect in this context is the sharing of skills. With the launch of the SAP LeanIX GitHub skill library, shown in Figure 11.24, you have the opportunity to install predefined skills via the options at *https://github.com/SAP/leanix-ai-plugins*.

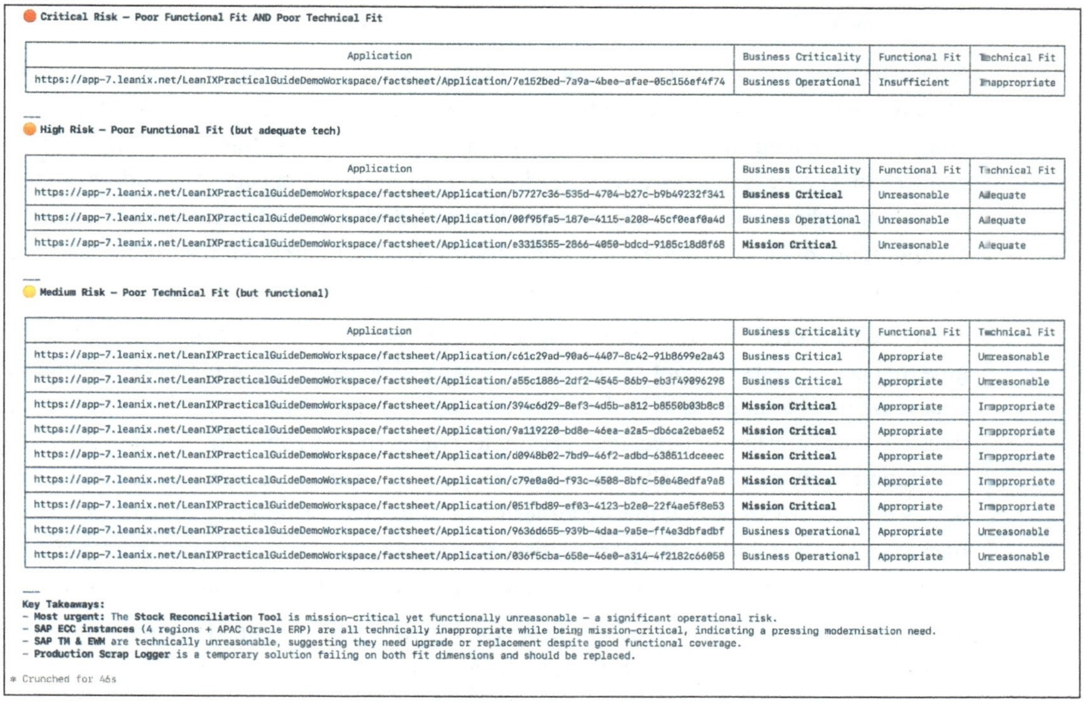

Critical Risk – Poor Functional Fit AND Poor Technical Fit

Application	Business Criticality	Functional Fit	Technical Fit
https://app-7.leanix.net/LeanIXPracticalGuideDemoWorkspace/factsheet/Application/7e152bed-7a9a-4bee-afae-05c156ef4f74	Business Operational	Insufficient	Inappropriate

High Risk – Poor Functional Fit (but adequate tech)

Application	Business Criticality	Functional Fit	Technical Fit
https://app-7.leanix.net/LeanIXPracticalGuideDemoWorkspace/factsheet/Application/b7727c36-535d-4704-b27c-b9b49232f341	**Business Critical**	Unreasonable	Adequate
https://app-7.leanix.net/LeanIXPracticalGuideDemoWorkspace/factsheet/Application/00f95fa5-187e-4115-a208-45cf0eaf0a4d	Business Operational	Unreasonable	Adequate
https://app-7.leanix.net/LeanIXPracticalGuideDemoWorkspace/factsheet/Application/e3315355-2866-4050-bdcd-9185c18d8f68	**Mission Critical**	Unreasonable	Adequate

Medium Risk – Poor Technical Fit (but functional)

Application	Business Criticality	Functional Fit	Technical Fit
https://app-7.leanix.net/LeanIXPracticalGuideDemoWorkspace/factsheet/Application/c61c29ad-90a6-4407-8c42-91b8699e2a43	Business Critical	Appropriate	Unreasonable
https://app-7.leanix.net/LeanIXPracticalGuideDemoWorkspace/factsheet/Application/a55c1886-2df2-4545-86b9-eb3f49096298	Business Critical	Appropriate	Unreasonable
https://app-7.leanix.net/LeanIXPracticalGuideDemoWorkspace/factsheet/Application/394c6d29-8ef3-4d5b-a812-b8550b03b8c8	**Mission Critical**	Appropriate	Inappropriate
https://app-7.leanix.net/LeanIXPracticalGuideDemoWorkspace/factsheet/Application/9a119220-bd8e-46ea-a2a5-db6ca2ebae52	**Mission Critical**	Appropriate	Inappropriate
https://app-7.leanix.net/LeanIXPracticalGuideDemoWorkspace/factsheet/Application/d0948b02-7bd9-46f2-adbd-638511dceeec	**Mission Critical**	Appropriate	Inappropriate
https://app-7.leanix.net/LeanIXPracticalGuideDemoWorkspace/factsheet/Application/c79e0a0d-f93c-4508-8bfc-50e48edfa9a8	**Mission Critical**	Appropriate	Inappropriate
https://app-7.leanix.net/LeanIXPracticalGuideDemoWorkspace/factsheet/Application/051fbd89-ef03-4123-b2e0-22f4ae5f8e53	**Mission Critical**	Appropriate	Inappropriate
https://app-7.leanix.net/LeanIXPracticalGuideDemoWorkspace/factsheet/Application/9636d655-939b-4daa-9a5e-ff4e3dbfadbf	Business Operational	Appropriate	Unreasonable
https://app-7.leanix.net/LeanIXPracticalGuideDemoWorkspace/factsheet/Application/036f5cba-658e-46e0-a314-4f2182c66058	Business Operational	Appropriate	Unreasonable

Key Takeaways:
- **Most urgent:** The **Stock Reconciliation Tool** is mission-critical yet functionally unreasonable – a significant operational risk.
- **SAP ECC instances** (4 regions + APAC Oracle ERP) are all technically inappropriate while being mission-critical, indicating a pressing modernisation need.
- **SAP TM & EWM** are technically unreasonable, suggesting they need upgrade or replacement despite good functional coverage.
- **Production Scrap Logger** is a temporary solution failing on both fit dimensions and should be replaced.

* Crunched for 46s

Figure 11.23: Skill-Based Risk Report

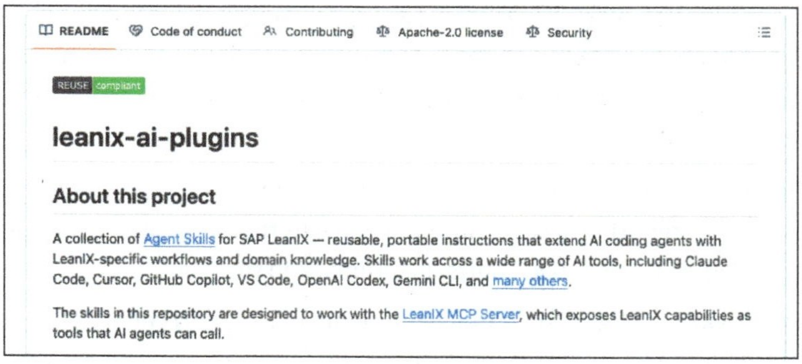

README Code of conduct Contributing Apache-2.0 license Security

REUSE compliant

leanix-ai-plugins

About this project

A collection of Agent Skills for SAP LeanIX — reusable, portable instructions that extend AI coding agents with LeanIX-specific workflows and domain knowledge. Skills work across a wide range of AI tools, including Claude Code, Cursor, GitHub Copilot, VS Code, OpenAI Codex, Gemini CLI, and many others.

The skills in this repository are designed to work with the LeanIX MCP Server, which exposes LeanIX capabilities as tools that AI agents can call.

Figure 11.24: SAP LeanIX GitHub Skill Library

Note

With the introduction of write access to the SAP LeanIX workspace, the governance of permission management is crucial. We recommend using OAuth as an authentication mechanism, since agent interactions will be traced in audit logs, and the overarching governance of agents should happen in SAP AI Agent Hub, as outlined in Chapter 13.

We recommend applying the principle of starting small and thinking big by allowing read-only access in the first instance and opening further up once trust and maturity are established.

11.6 Custom AI Solutions

AI-assisted development of code, often called *vibe coding*, may be best described as a style of software development in which the developer describes what they want in natural language, the so-called vibe of the outcome, and an AI generates the code iteratively. The real game changer here is that the developer does not write any line of code. The code is refined further through interaction and feedback with the AI, during which the outcome of a screen, an operation, or so on is initially defined and fine-tuned or corrected over time. Even the deployment of the code itself is managed by the interaction with the AI.

What does this mean for us as enterprise architects? Some of you see the huge numbers of governance requests in front of us, but let's table this aspect for a moment (though of course we need to address it) and explore the opportunities. If you argue that a developer does not write any line of code anymore but leverages the outcome-driven (vibe) software development style—that is, all of us are developers—then software development becomes democratized. We know that this hypothesis can cause many emotions, but let's dial them back for a moment and see the possibilities opening. As we are only at the beginning of that journey, we have selected two use cases evolving these days for vibe coded solutions.

In the following sections, we'll outline how vibe coding can be leveraged to create your custom reports or standalone applications with SAP LeanIX as the source of data.

11.6.1 Vibe Coding of AI-Driven Reports

By leveraging the previously introduced technologies of AI-assisted development (vibe coding) and access to the architecture knowledge graph in combination with other external and internal sources, we have the possibility to create any custom report and to deploy this report in SAP LeanIX.

Think of custom reports such as compliance dashboards, cost optimization reports, transformation trackers, or any other report leveraging internal and external sources while visualizing the results in appealing and interactive charts, diagrams, and the like.

To get a first impression of what the AI-assisted development of a custom report in SAP LeanIX looks like, we'll use Claude Code as the coding assistant. In the example, we'll provide a screenshot of a report that will be used as input instead of providing the instructions for how the custom report should look.

The following prompt was used to create a report like that shown in Figure 11.25 (the prompt's Image #1):

```
I want to build a custom report in my SAP LeanIX workspace based on the
following example [Image #1]
```

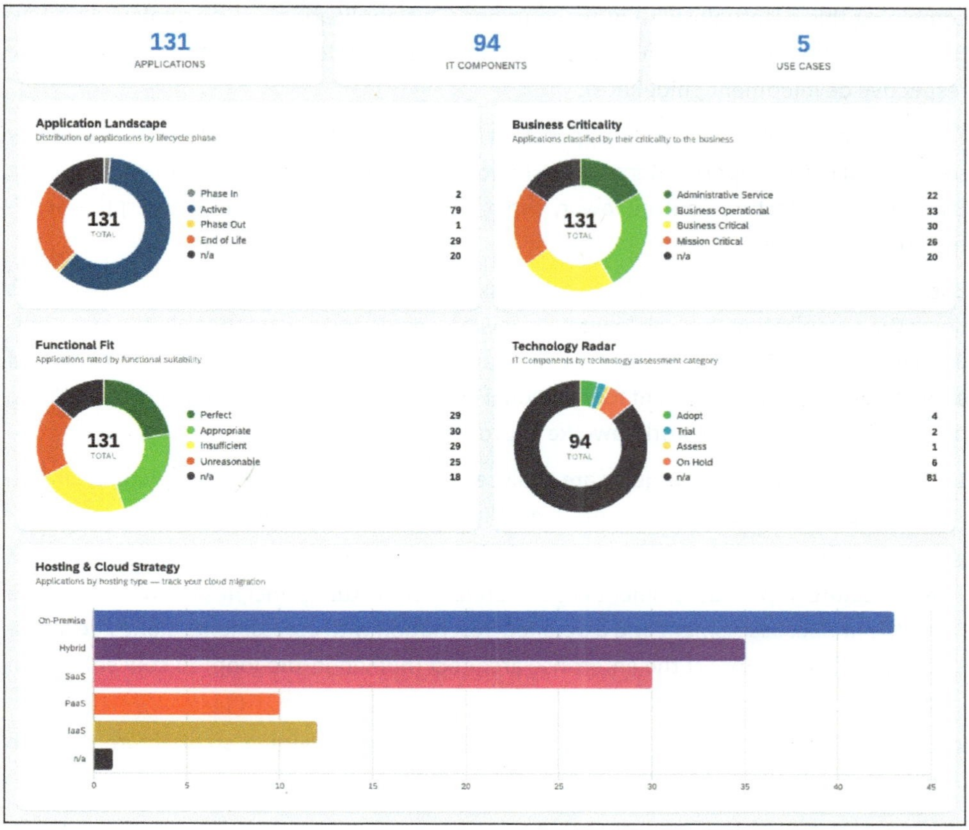

Figure 11.25: Custom Report Template

While scaffolding the custom report project and defining the content of the report, Claude will ask for the confirmation of specific change activities and for the API token, as described in the MCP server sections. We recommend using OAuth-based authentication for reporting purposes.

Over multiple iterations, the custom report will be defined and deployed within your SAP LeanIX workspace. Simulations of the report, all orchestrated by Claude, will verify if the expected output is achieved.

> **Further Resources**
>
> For more information and demonstrations of how to set up and create custom reports with AI, refer to *https://www.youtube.com/playlist?list=PLLo70LNRp9UE*.

11.6.2 Building Standalone Applications

While custom reports operate in the SAP LeanIX environment, we could go one step further and build a standalone application by extending or complementing SAP LeanIX and combining this eventually with other solutions. This application can be deployed on SAP Business Technology Platform (SAP BTP) or as an independent application (following the respective development guidelines).

To get a first idea of what such an application looks like, we will demonstrate an exemplary solution built to support enterprise architecture governance activities. We'll call it the Enterprise Architecture Control Tower, and it was inspired by a common enterprise architecture governance use case.

Enterprise architecture practices run the risk of being perceived as an entity that slows down the organization through lengthy review processes and governance obstacles, as outlined in Chapter 12. The enterprise architecture control tower addresses the bottleneck of architecture impact assessment for individual enterprise architecture engagement with the help of AI capabilities in a workflow-like approach:

1. Enterprise architecture projects are analyzed to understand the broader picture, and triage is applied to determine if a more detailed impact assessment is required.
2. From the information provided within the enterprise architecture project, an AI-driven architecture impact assessment is generated. Architecture principles build the baseline for the impact assessment and a detailed description with assessments alongside multiple dimensions support the enterprise architect (human-in-the-loop) to concentrate on the most imminent deviations.
3. Based on the architecture impact assessment, a corresponding architecture decision record is generated, leveraging the insights gained.

The enterprise architecture control tower solution can be described from a conceptual architecture perspective as shown in Figure 11.26.

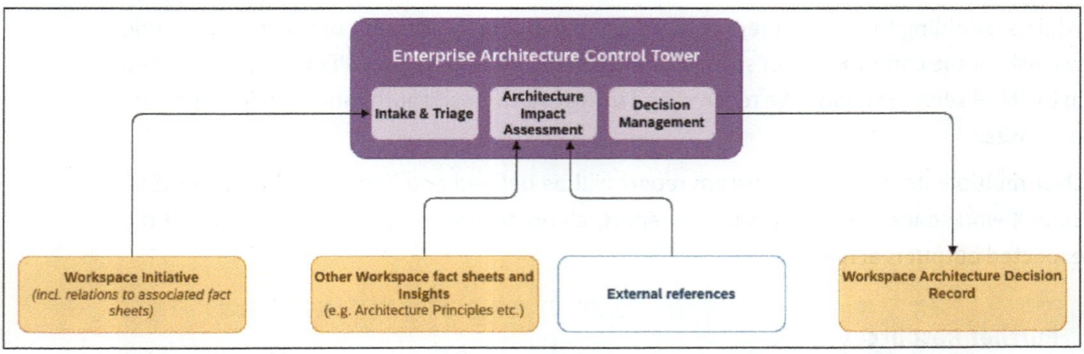

Figure 11.26: Conceptual Architecture of Enterprise Architecture Control Tower Solution

The standalone solution was built as a vibe coding solution with Claude Code and leverages various AI agents to support architecture decision-making. The SAP LeanIX workspace acts as the system of record and is complemented with a governance workflow to allow timely decision-making, based on information modeled in the workspace. The enterprise architecture control tower leverages the MCP server capability, and two AI agents for the generation of the architecture impact assessment and architecture decision record have been defined with their specific skills. The starting point of the standalone solution builds the landing page outlined in Figure 11.27.

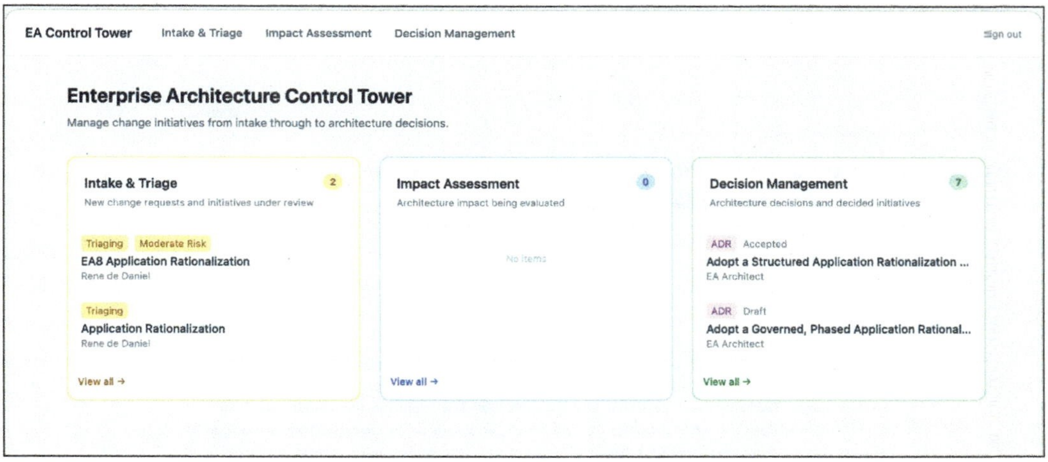

Figure 11.27: Enterprise Architecture Control Tower Landing Page

The solution is designed in a workflow-like approach to allow the evolution of enterprise architecture projects via an individualized impact assessment toward an architecture decision record. To begin, SAP LeanIX projects can be imported with their related fact sheets into the standalone solution, and the creation of the architecture impact assessment can be requested from an agent, as outlined in Figure 11.28. This agent builds the core value of the standalone solution as references such as architecture principles are used to derive a conclusion about the impact.

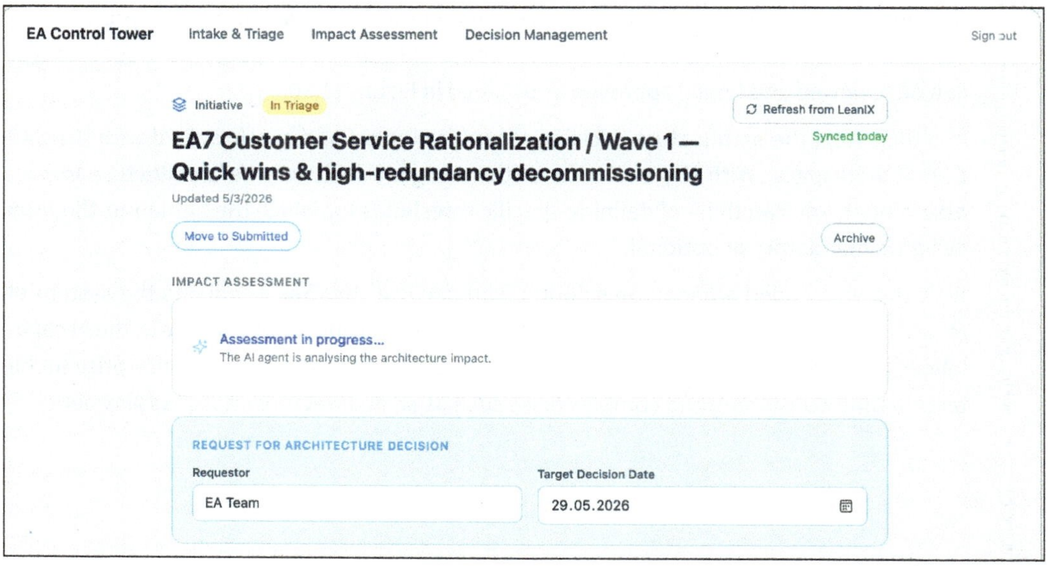

Figure 11.28: AI Agent-Driven Architecture Impact Assessment Creation

With the architecture impact assessment, the project moves into the second phase, in which the architecture impact assessment can be reviewed and refined, as shown in Figure 11.29. Alternatively, the project might be waived if the impact is rather limited.

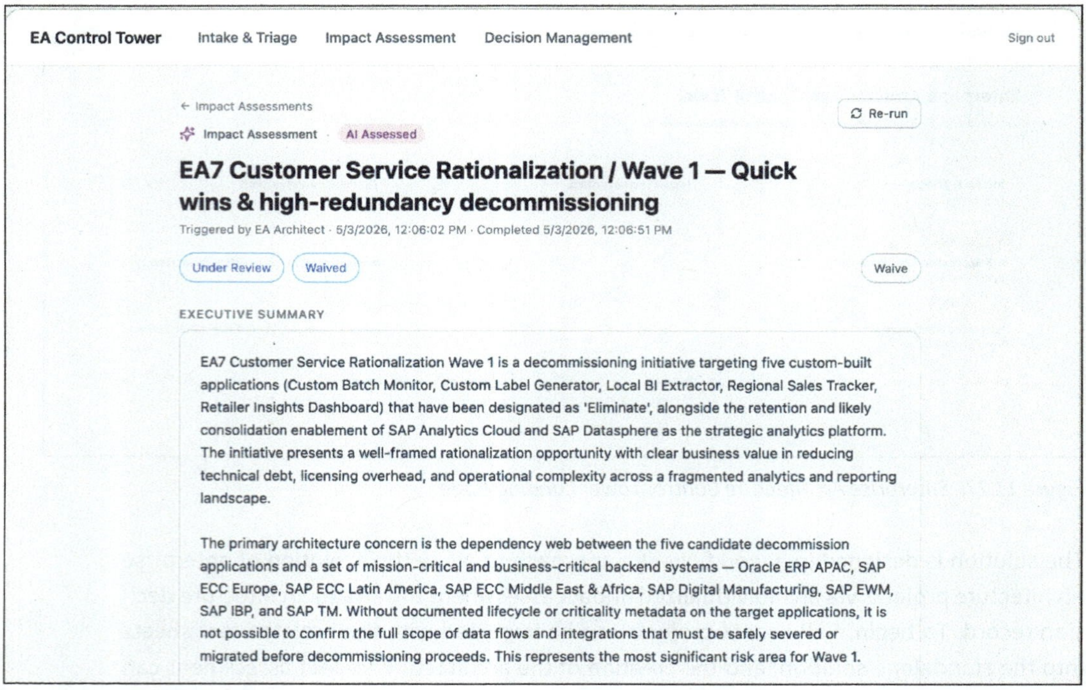

Figure 11.29: Review of Architecture Impact Assessment Proposal

In preparation for the third phase, an AI agent can be requested to draft the architecture decision record based on the reviewed architecture impact assessment. In sync with the status of an architecture decision record in the SAP LeanIX workspace, individual records can be reviewed and finally approved as outlined in Figure 11.30.

In a final step, the architecture decision record is created and synchronized with the SAP LeanIX workspace. With the evolution of the AI agent skills for the architecture impact assessment, you can think of defining specific thresholds for which the human-in-the-loop design is mandatory or optional.

With this vibe coded standalone solution, you can see that SAP LeanIX as the system of record can be extended for the purposes of differentiation and innovation. With the AI capabilities at hand, we cannot only create new applications but also support enterprise architects in time-consuming and complex tasks such as an architecture impact assessment.

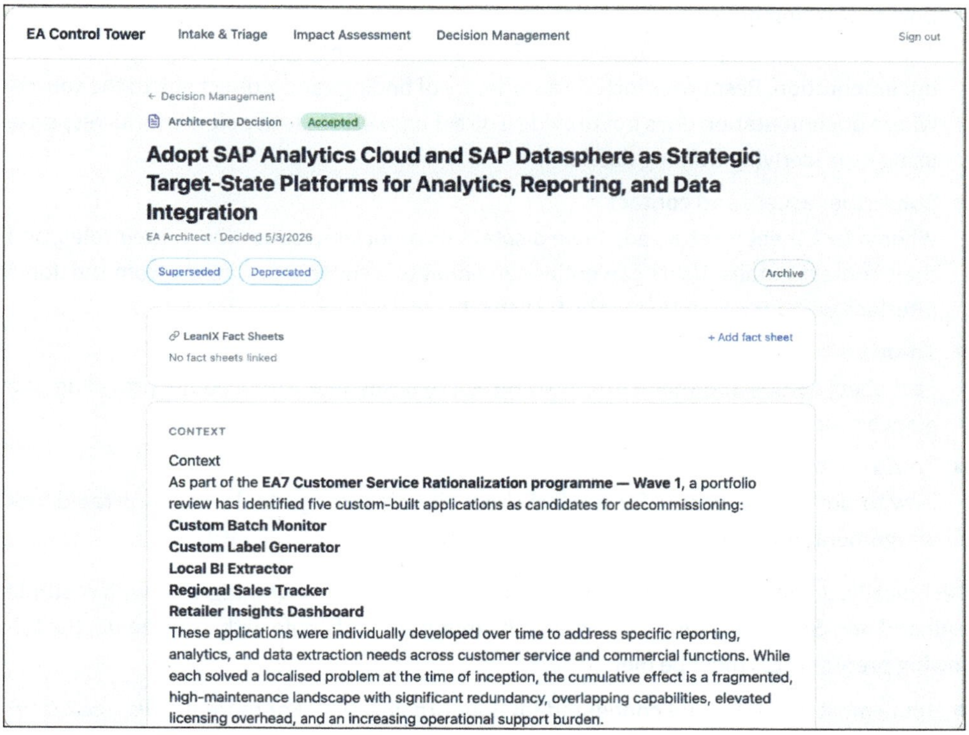

Figure 11.30: Architecture Decision Record

11.7 Joule

Joule is SAP's generative AI copilot, which is embedded across SAP Business Suite. It integrates with SAP LeanIX, SAP Signavio, and many other SAP as well as non-SAP applications. With the integration of SAP LeanIX and Joule, not only enterprise architects but anyone in the organization can uncover architecture insights through natural language. With the bidirectional integration of Joule and Microsoft 365 Copilot, architecture data and insights can be democratized even further.

The conversational interface access to SAP LeanIX offers the following capabilities:

- **Diagram and report search**
  Users can search for diagrams and reports using natural language queries. Results can be previewed directly within the Joule interface or opened in full within the inventory or diagram editor, depending on availability.
- **Fact sheet search**
  The full SAP LeanIX inventory is accessible via conversational search. Users can retrieve fact sheet summaries within Joule or navigate directly to the selected fact sheet in the inventory for detailed review.

- **Documentation query**
  Joule responds to questions about SAP LeanIX functionality by leveraging the official documentation. Responses include a summary of findings and a direct link to the source. Where documentation does not provide a direct answer, Joule supplements its response using its underlying training data.
- **Subscriber access and contact**
  When a fact sheet is retrieved, Joule displays its associated subscribers, their roles, and their contact details. Users can initiate an email to a subscriber directly from the Joule interface without navigating to the fact sheet.
- **To-do visibility**
  Fact sheet to-dos and their current statuses are accessible within Joule, providing at a glance a view of outstanding actions.
- **To-do creation**
  New to-dos can be created for any fact sheet directly through Joule, enabling rapid task assignment from within the conversational interface.

Technically, Joule is offered as a service running on SAP BTP that integrates with multiple SAP and non-SAP applications. To enable the integration of Joule with SAP LeanIX, the following prerequisites must be met:

- Joule must be set up and configured for your organization. For more details, see *http://s-prs.co/v620694*.
- The Identity Authentication service must be configured as the identity provider (IdP) for SAP LeanIX.
- The integration between Joule and SAP LeanIX must be configured.

Figure 11.31 shows Joule integrated into the SAP LeanIX workspace.

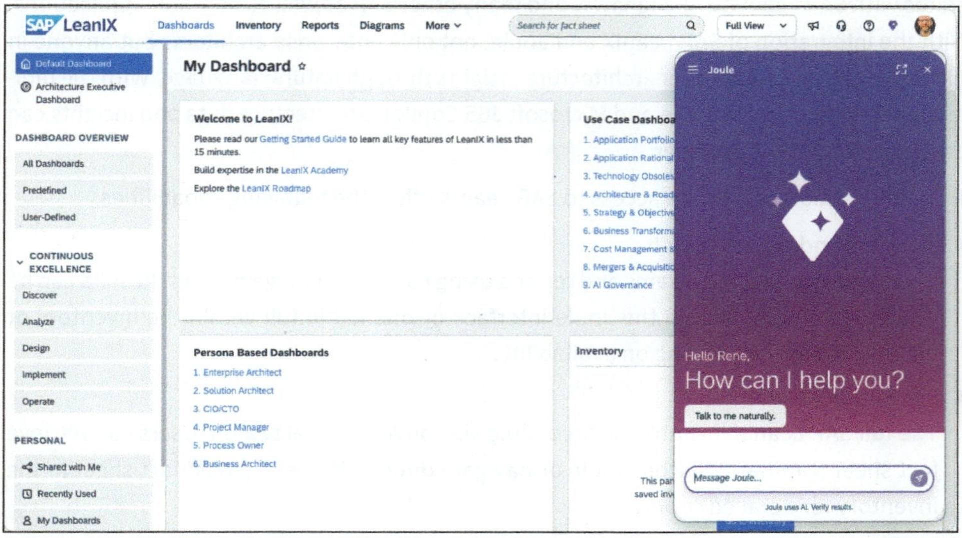

Figure 11.31: Joule in SAP LeanIX

11.8 Risks and Limitations

In the previous section, we outlined the various AI capabilities available in the context of SAP LeanIX. However, like in most AI-specific documentation or articles, we also want to highlight an overview of the risks, limitations, and mitigation actions you should be aware of:

- **Data quality dependency**
  All AI features are only as accurate as the underlying data. Establish ownership, use the inventory builder for enrichment, and set up quality seals as governance mechanisms.
- **AI hallucination**
  AI-generated insights may be incorrect. Maintain confirm-or-reject workflows; never autoexecute recommendations for critical decisions.
- **Agent governance**
  As agents gain write access via MCP, unauthorized changes become a risk. Use scoped technical users, log interactions, and govern agents in SAP AI Agent Hub.
- **Overreliance and skill decay**
  Maintain periodic "AI-free" portfolio reviews. Use AI insights as input, not substitutes for architectural judgment and decisions.
- **Shadow AI risk**
  As architecture intelligence becomes broadly accessible via Joule and custom apps, ensure governance policies extend to all AI touchpoints.
- **Change management**
  Start with low-risk quick wins; share success stories; and position AI as a time-saving assistant, not a replacement.
- **Privacy and compliance**
  SAP LeanIX's data privacy commitment (no data is used for AI training) is of primary importance, leveraging workspace-level permissions and standard authentication mechanisms.

11.9 Summary

In this chapter, we have shown how AI capabilities boost the productivity of enterprise architects, enabling them to move away from repetitive and non-value-adding activities to become a catalyst for change and decision-making. SAP LeanIX as a living repository, providing the architecture knowledge graph, builds the foundation for sophisticated AI capabilities, such as architecture guidance, MCP, vibe coded reports, and agentic solutions. Architecture information and insights are democratized across the organization through natural language conversation tools embedded within the stakeholder's working environment, such as Joule or Copilot.

We cannot predict the future, but we assume that the advisory nature of agents will mature into autonomously acting agents, performing tasks such as monitoring compliance with and adherence to architecture standards and guardrails. Agents could enable a living digital twin of the enterprise, enabling simulations of transformation options, what-if scenarios, and in vitro, synthetic architecture evaluations in complex environments.

Further Resources

To stay updated with future use cases and product developments, you're invited to subscribe to the "Enterprise Architecture in Practice" YouTube channel, which also includes a playlist focusing on AI-native enterprise architecture: *https://www.youtube.com/@Rene.de.Daniel*.

In the next chapter, we'll focus on the topic of governance from the broader enterprise architecture perspective, which is a crucial discipline for each and every enterprise architecture practice.

Chapter 12

Enterprise Architecture Governance

Enterprise architecture governance is essential to steer companies' investment in the right architecture choices. Successful enterprise architecture governance setups focus on outcomes rather than on stage gates and act as a catalyst for change.

In today's dynamic enterprise environments, organizations face rising complexity, fast-paced change, and growing pressure to align IT investments with strategic goals. Governance provides the structure, processes, and transparency needed to steer architecture decisions, reduce risk, and ensure long-term coherence across business and IT.

This chapter explores architecture governance from both theoretical and practical perspectives. Building on foundational frameworks like TOGAF, we outline the principles, governance outcomes, and core foundations required for effective control. We also present real-world governance implementations using SAP LeanIX, highlighting patterns in both agile and traditional organizations. Finally, a case will provide insights into how governance can be realized in a large-scale organization.

12.1 The Building Blocks of Enterprise Architecture Governance

Architecture governance is the practice of directing and controlling architecture decisions to ensure alignment with business goals, compliance with standards, and effective risk management. It encompasses roles, responsibilities, processes, and metrics that guide how architecture is developed, reviewed, and enforced.

We'll explore the basics of enterprise architecture governance in more detail in the following sections, including the guiding principles, outcomes, and core concepts.

12.1.1 Principles

While issuing, following, and enforcing architecture principles, an enterprise architecture governance practice itself should follow the principles of good governance:

- **Transparency**
  Decisions and their rationales should be visible to all stakeholders.
- **Accountability**
  Roles and responsibilities must be clearly defined.

- **Agility**
  Governance must enable timely decisions, not hinder them.
- **Consistency**
  Standards and principles should guide architecture design.

12.1.2 Outcomes

Enterprise architecture governance ensures that an organization's architecture practices are controlled and aligned with overall business needs. An enterprise architecture governance practice should aim for the following outcomes, which should be considered while defining the enterprise architecture governance practice:

- **Business alignment**
  Business alignment is the practice of ensuring that all architectural plans, projects, and solutions directly support the business's goals and objectives. Under effective enterprise architecture governance, IT initiatives and architecture decisions are tightly aligned with the organization's overarching business strategy. This means that every technology investment or architectural change is evaluated for how well it contributes to strategic business outcomes. By maintaining this alignment, enterprise architecture governance helps bridge the gap between business and IT, ensuring technology efforts deliver clear business value and improving overall performance. In essence, business alignment keeps the architecture focused on enabling the company's mission and competitive strategy, rather than pursuing IT for its own sake.
- **Decision clarity**
  Decision clarity refers to clarifying and formalizing who has the authority to make, review, and approve architecture-related decisions. Enterprise architecture governance establishes a clear decision-making framework that defines roles and responsibilities for architecture decisions (e.g., creating an architecture review board or assigning a chief architect). By outlining who can decide what, the organization avoids confusion or overlap in decision-making. A well-defined governance model will specify how architectural decisions are proposed, reviewed, approved, and communicated throughout the organization. This clarity ensures accountability and consistency. The right stakeholders (such as business leaders for strategic priorities or architects for technical design choices) have the final say in their domain, and decisions align with agreed-upon standards and strategy. Clearly defined decision rights lead to faster, more coherent decision-making and prevent ad hoc or conflicting architecture changes.
- **Risk and compliance**
  Risk and compliance in enterprise architecture governance involves embedding risk management practices and compliance controls into the architecture process. This dimension ensures that architectural decisions do not expose the organization to undue risk and that they adhere to required standards and regulations. Effective enterprise architecture governance will implement policies, procedures, and checkpoints to identify, assess, and mitigate risks associated with technology choices, for

example, risks related to security, data privacy, or system failure. At the same time, governance enforces compliance with both internal architecture standards and external regulatory requirements. This means architecture designs are reviewed for alignment with corporate IT standards, and solutions must meet industry or legal regulations (e.g., compliance with GDPR, good practice [GxP] guidelines, or other relevant policies). By integrating risk assessment and compliance checks into every stage of architecture development, organizations can avoid costly noncompliance penalties and reduce the likelihood of architecture-related failures. In practice, this might involve formal architecture compliance reviews and audits to ensure standards are followed and risks are under control.

- **Controlled evolution**
  As an enterprise architecture governance outcome, controlled evolution focuses on controlling and guiding the impact of architectural changes. Because enterprise architectures evolve over time, governance establishes structured processes for how changes are proposed, assessed, approved, and implemented. This typically includes formal change review boards or workflows that evaluate the impact, feasibility, and risk of proposed changes before they are executed. A robust architecture change management process will ensure that modifications to the architecture (such as introducing a new system, modifying an integration, or decommissioning a component) are handled in a predictable, low-risk manner. For example, governance may require impact analysis and stakeholder approval for significant changes. By having well-defined change management processes in place, an organization can minimize disruptions. In practice, this means that any architectural change is vetted to ensure it doesn't inadvertently introduce security vulnerabilities, performance issues, or compatibility problems. Overall, the controlled evolution outcome of enterprise architecture governance helps the enterprise adapt its architecture in a controlled way, maintaining stability while allowing necessary evolution.
- **Asset stewardship**
  Asset stewardship involves overseeing the entire lifecycle of architecture assets, from their introduction through maintenance and to eventual retirement. In an enterprise, architecture assets (such as applications, technologies, or IT components) should not be deployed in an uncontrolled way or without oversight. Enterprise architecture governance therefore puts controls in place for introducing new technologies or systems in line with strategy and standards, maintaining and updating existing assets, and retiring or sunsetting outdated ones. This lifecycle perspective ensures the architecture remains up-to-date and efficient. Concretely, governance defines processes for the evolution of architectures, including adopting innovations (bringing in new solutions), modernizing or optimizing current systems, and decommissioning legacy systems when they become obsolete (rationalization). By managing the introduction, maintenance, and retirement of architecture components, the organization prevents technology sprawl and minimizes technical debt. Asset stewardship makes sure that as business needs and technologies

change, the enterprise architecture is continuously adjusted and shaped to stay aligned with business objectives and to use resources optimally.

- **Stakeholder visibility**
  Stakeholder visibility means making the architecture, and the governance process around it, visible and understandable to all relevant stakeholders. In enterprise architecture governance, transparency is achieved by clearly documenting architecture decisions, standards, models, and roadmaps, and by openly communicating these to business and IT stakeholders. The goal is that anyone who needs to can understand the current and planned state of the architecture and the rationale behind decisions. This openness demystifies the IT architecture for nontechnical team members and promotes trust. When architecture governance is transparent, business leaders, IT staff, and other stakeholders can see how IT initiatives support organizational goals. It encourages feedback and collaboration, as everyone has a clear view of the architecture's direction and constraints. Typical practices to enhance transparency include maintaining an accessible enterprise architecture repository, using dashboards or visualizations to show architecture landscapes, and communicating decisions and guidelines in a common language. The stakeholder visibility outcome ensures that enterprise architecture is not a black box: It's a well-understood, shared concern across the enterprise.
- **Governance maturity**
  Governance maturity is the practice of regularly evaluating and refining the enterprise architecture governance processes themselves based on feedback and performance metrics. A good enterprise architecture governance model is not static; it evolves as the organization learns what works best and as business and technology environments change. This dimension involves establishing feedback loops to gather input from stakeholders and lessons learned from past projects, then using that information to improve policies, standards, and decision-making processes over time. For example, after an architecture initiative or major change, the governance team might conduct a retrospective to identify what governance practices were effective and what needs adjustment. Organizations also use metrics and key performance indicators (KPIs) to measure the effectiveness of enterprise architecture governance and then refine their approach to address any weaknesses. By embracing governance maturity, the enterprise architecture governance capability itself becomes more mature and effective over time, ensuring it remains relevant and valuable as the enterprise grows and changes.

12.1.3 Core Foundations

Taking the governance outcomes into account, an enterprise architecture governance practice can be established, executed, and evolved by defining the seven core foundations described in Table 12.1. These are essentially the key ingredients of an enterprise architecture governance practice.

Foundation	Description	Aspects
Standards and principles	Standards and principles provide the guardrails for solution design.	▪ Architecture principles derived from strategic objectives ▪ Technology standards and solution patterns
Boards and roles	Boards and roles define the stakeholder responsibilities.	▪ Role profiles for chief architect, domain architect, solution architect, and so on ▪ Scope and ownership of architecture board, domain councils, and so on ▪ RACI model for decision types
Operating model	The operating model defines the structure for decision-making and oversight.	▪ Architecture boards and review councils ▪ Escalation paths and authority levels ▪ Integration with the broader ecosystem: project management office (PMO), agile release trains, information security, and so on
Decision-making model	The decision-making model establishes how architecture decisions are made, documented, and enforced.	▪ Use of architecture decision records ▪ Structured workflows for review and approval ▪ Delegated authority and escalation rules
Processes and workflows	Processes and workflows operationalize governance throughout the architecture lifecycle.	▪ Definition of the architecture review process ▪ Workflows for technology onboarding and retirement ▪ Processes for exception and change handling
Tools and repositories	Tools and repositories support transparency and traceability.	▪ A central platform for enterprise architecture management, such as SAP LeanIX ▪ Definition of documenting standards, decisions, and lifecycles
Metrics and KPIs	Metrics and KPIs enable the measurement of governance effectiveness over time.	▪ Percentage of reviewed projects ▪ Decision cycle time ▪ Standards compliance rate ▪ Lifecycle risk exposure

Table 12.1: Seven Core Foundations of Enterprise Architecture Governance

12.2 Enterprise Architecture Governance Case

As the previously introduced outcomes and foundations of an enterprise architecture governance practice are rather theoretical in nature, we want to provide some tangible insights in the context of a real case: that of a leading pharmaceuticals company operating a sophisticated enterprise architecture governance practice. For the sake of readability, this section was structured to match the enterprise architecture governance foundations introduced in Section 12.1.

12.2.1 Context

The organization is at a strategic inflection point where healthcare and digital are converging. To lead in this new era, the organization is undergoing one of the most comprehensive business and digital transformations in the healthcare industry. This is not just an IT project; it's a fundamental business transformation program enabled by technology, reshaping how the organization works to deliver products and innovations to patients with greater speed and efficiency.

The primary driver for this change is the need to address inefficiencies from a fragmented systems landscape, focus time on value-creating activities, and harness new technologies like AI and big data, as outlined in Figure 12.1. Previously, the organization operated with 11 different core SAP systems, leading to inconsistent processes and poor data quality.

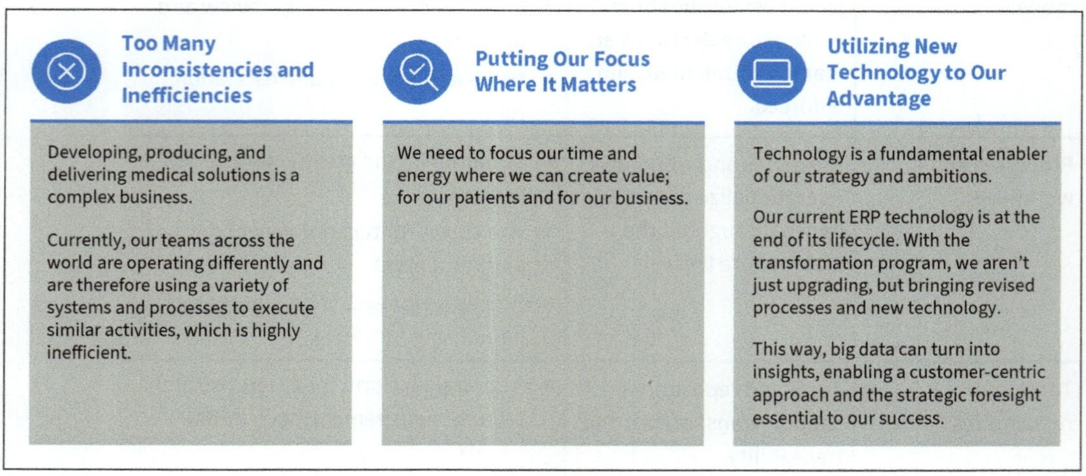

Figure 12.1: Three Main Drivers for Change

The transformation program aims to create a unified digital backbone for the organization. Its vision is to implement a single, harmonized process house across the enterprise, supported by an integrated and simplified IT landscape with SAP S/4HANA at its core.

This transformation will enable the following:

- **One solution**
  Moving from 11 core SAP systems to an integrated system landscape.
- **Consistent, high-quality data**
  Establishing a single source of truth by capturing data once at the source, enabling data-driven decision-making.
- **Scalability and future-readiness**
  Building a strong foundation for advanced digital solutions, automation, and AI by harmonizing underlying processes and systems.

This massive undertaking is one of the largest SAP S/4HANA projects worldwide, impacting approximately 100,000 end users across about 260 legal entities. By harmonizing processes and systems, the organization is not just improving efficiency today but also laying the groundwork for innovative, scalable solutions for the future.

Starting from the current landscape and moving to the target architecture requires a lot of architecture work (scenario planning, transitions scenarios definition, decisions, governance, etc.).

12.2.2 Structuring the Use Case

The global transformation program is centered around a set of harmonized business processes described in the structure of a process house, as shown in Figure 12.2.

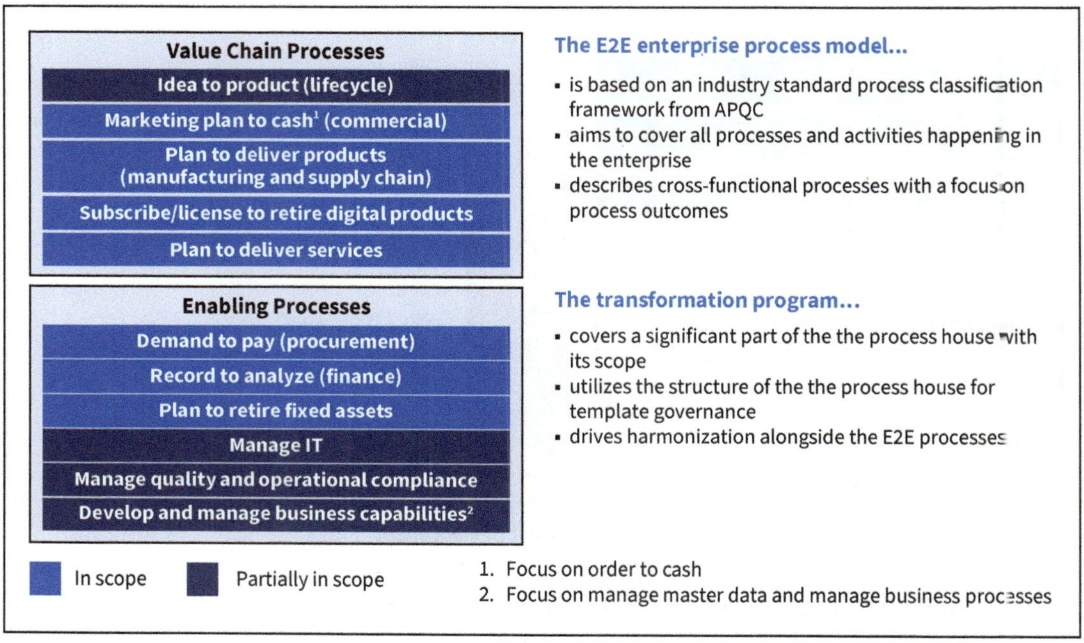

Figure 12.2: Process House

The *process house* consists of value chain and enabling processes, which are based on the American Productivity & Quality Center (APQC) standard. The process house and the scope of the individual processes build the foundation for the development and governance of a process template to be deployed across the world in multiple iterations.

The development of the template takes place in a product-centric and agile approach. Each product area is structured around product development, solution trains, agile release trains (ARTs), and agile teams. The highest levels of authority are the sponsor boards and executive sponsors, as illustrated in Figure 12.3.

Further Resources

For more information on agile frameworks and methodologies, refer to the Scaled Agile Framework (SAFe) at *https://framework.scaledagile.com/* as well as Scrum (*scrum.org*) and Kanban.

The product development as defined in Figure 12.3 is governed through the following:

- The *design authority*, focusing on globally harmonizing business processes
- The *cross-architecture authority*, focusing on globally harmonizing the technical solution approach

While there are overlaps and close collaboration, the organization intentionally separates business process discussions from technical solution approaches to ensure the proper focus is maintained.

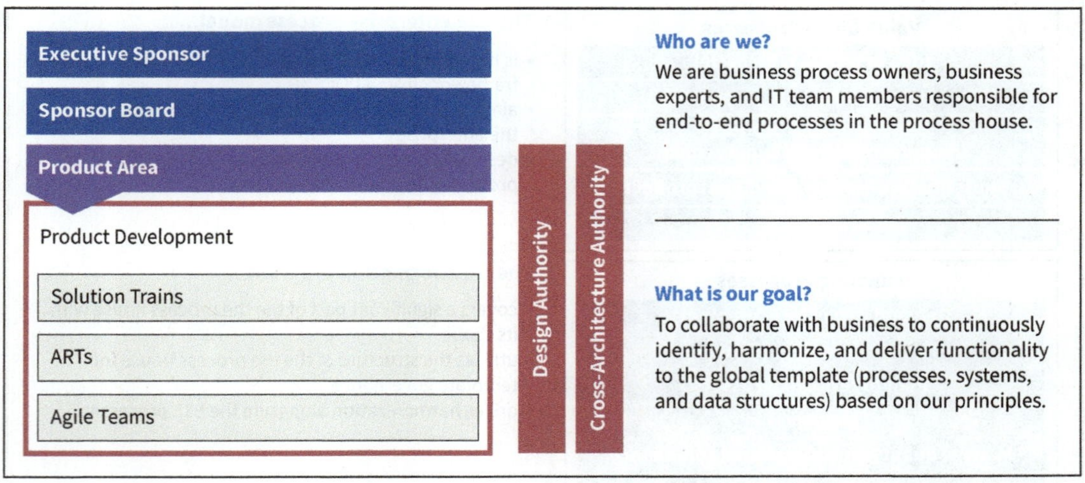

Figure 12.3: Transformation Program: Product Development

To steer this complex transformation and manage the harmonization of the template for global processes and solutions (the so-called digital backbone) effectively, the organization has established a robust architecture governance framework. The size of the endeavor requires amplification into the organization based on standards and principles and empowered architects. In the following sections, we'll apply the core foundations introduced in Section 12.1.3 to this use case.

Standards and Principles

Standards and principles are the guardrails that guide the solution design, ensuring strategic alignment, reducing risk, and maximizing return on investment (ROI).

The customer's principles are derived from business and IT strategy and provide high-level direction for all ERP-related architectures. The key principles are as follows:

- Keep the digital core lean and clean. Limit extensions to ensure cloud readiness.
- Use packaged standard solutions. Buy rather than build and reuse wherever possible.
- Apply a greenfield approach. Rethink processes instead of replicating old ones.
- Evolve one enterprise-wide template. Secure a standardized and harmonized template using best practices.

Enterprise architects, in collaboration with product teams, are responsible for identifying gaps and defining new technology standards. These standards are mandatory, and adherence is the accountability of every solution architect. Noncompliance can risk stopping an implementation.

Roles and Governance Bodies

Clear roles and responsibilities are fundamental to the organization's governance model. The following architect roles have been defined:

- **ERP enterprise architects**
  Accountable for technology standards and the function-specific architecture roadmap. ERP enterprise architects identify gaps, define standards, provide strategic guidance, and can escalate cases of noncompliance to the overall IT architecture authority (an IT architecture body encompassing the whole enterprise).
- **ERP system and solution architects**
  Every application must have an identified solution architect. Solution architects are accountable for shaping and documenting the solution architecture (with the help of a solution outline and architecture decision records) and ensuring it adheres to technology standards and principles.

Transformation program-specific governance bodies and boards include the following:

- **Design authority (DA)**
  Safeguards the process house and global template, making key design decisions in order to drive business process harmonization.

- **Cross-architecture authority (CAA)**
  A strategic body for ERP and the transformation program that sets architectural direction, ensures clean core integrity, and promotes innovation. The CAA board reviews and approves major architectural decisions.

Architecture Governance Framework

The organization's architecture governance framework is designed to ensure consistent, standardized architecture practices across all teams. It defines the structure for decision-making and oversight, with a clear vision to architect a clean-core, cloud-enabled, and intelligent ERP system, as shown in Figure 12.4.

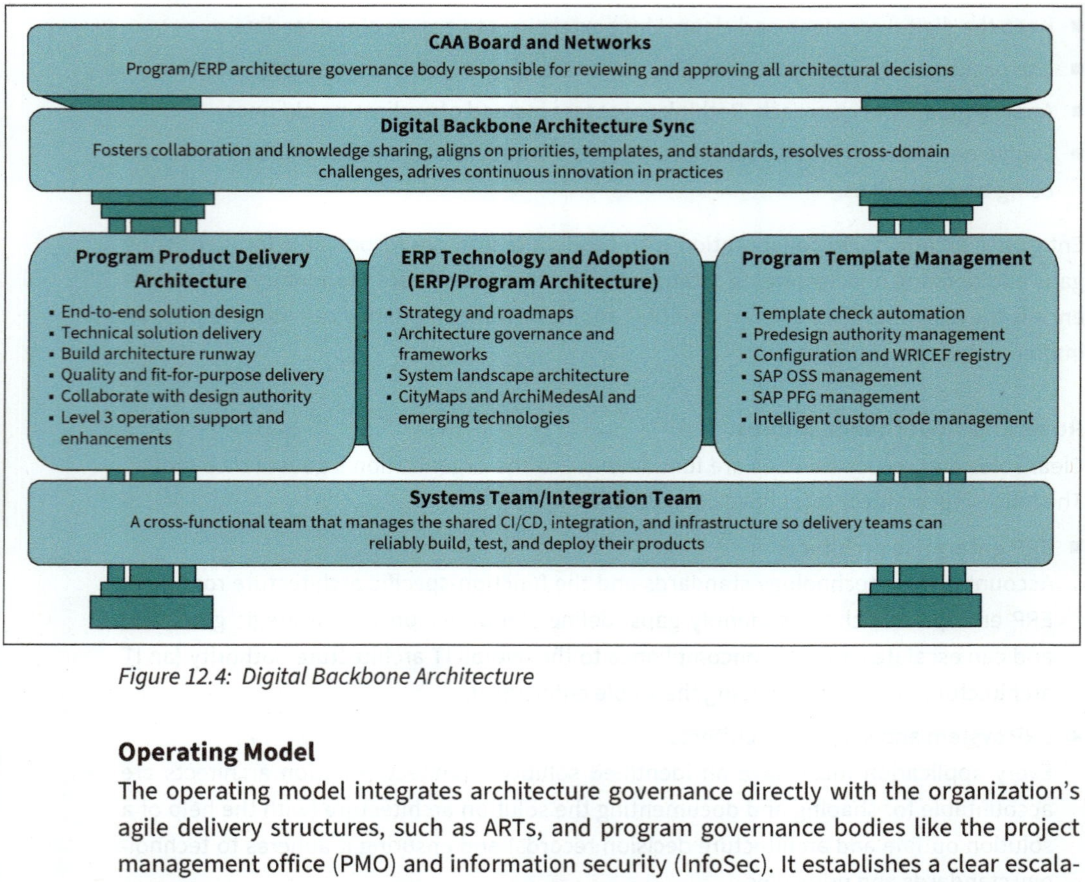

Figure 12.4: Digital Backbone Architecture

Operating Model

The operating model integrates architecture governance directly with the organization's agile delivery structures, such as ARTs, and program governance bodies like the project management office (PMO) and information security (InfoSec). It establishes a clear escalation process for noncompliance with technology standards, ensuring accountability up to the chief digital technology officer. In the organization's ERP world, architects on the program, network, and ART levels, which are structured around end-to-end processes, systems, and technologies, focus on architecture deliverables as depicted in Figure 12.5. These deliverables depend on the nature of the architecture work and metaphorically can be linked to various layers of the architecture elevator, reaching from enterprise roadmaps and principles on the penthouse level to component diagrams or integration specifications on the lowest level.

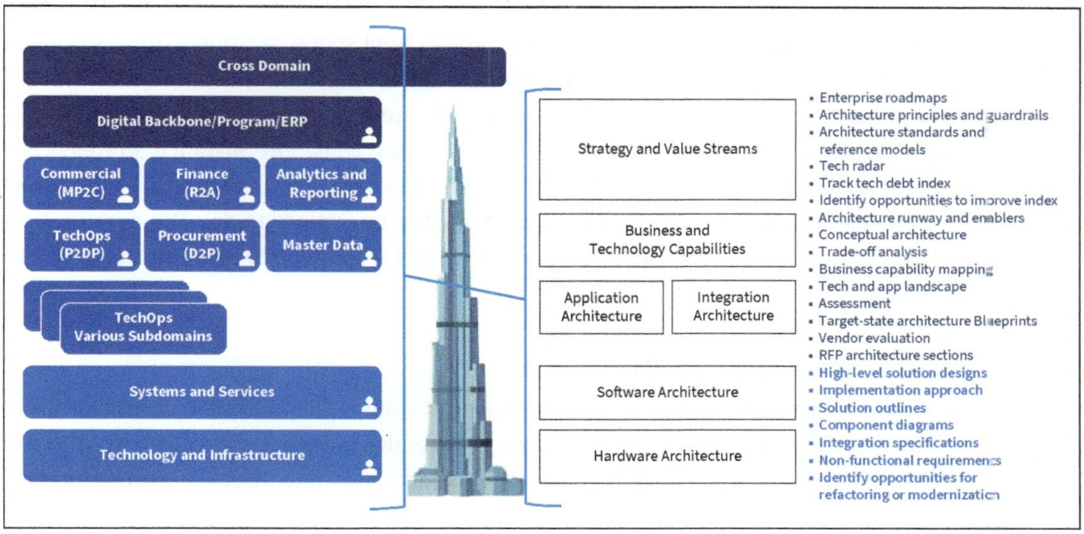

Figure 12.5: Organization's ERP World

A key outcome of the ERP organization are the decisions generated based on the operating model, as outlined in Figure 12.6.

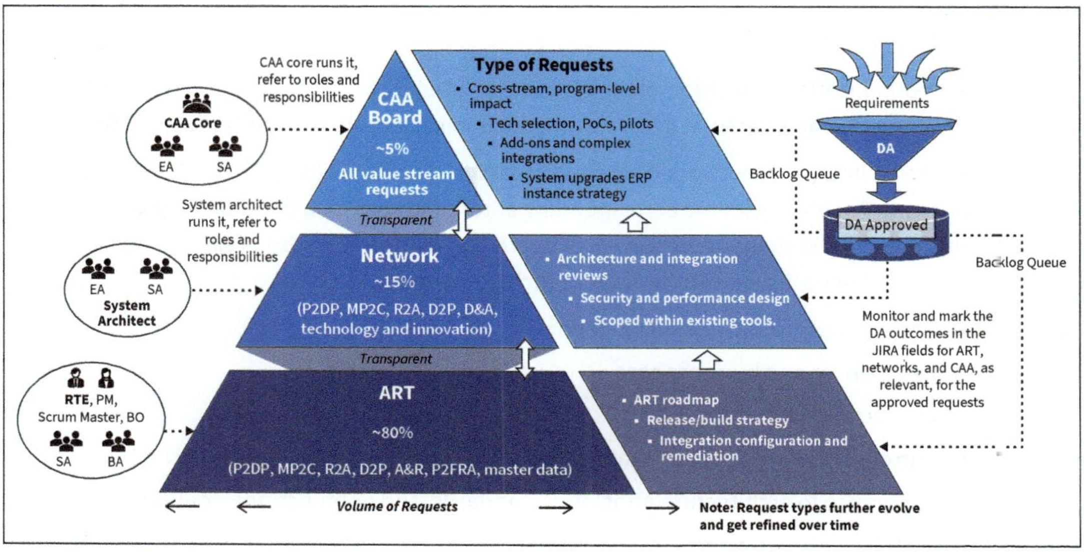

Figure 12.6: ERP Operating Model

The operating model defines how empowered architects are making architecture decisions based on the ART (~80% of volume) and network (~15%) levels, in close collaboration with the DA. The bulk of decisions are made directly by system architects at the ART level, covering the ART roadmap, release/build strategy, and integration configuration and remediation, guided by ERP architecture principles. Cross-ART concerns (e.g., architecture and integration reviews, security and performance designs, tool scope decisions) are resolved at the network level. Decisions that are cross-network or have a foundational impact (e.g., cross-stream or program-level impact, technology selection, proofs of concept (PoCs)/pilots, add-ons and complex integrations, system upgrades, ERP instance strategy) are brought to the CAA board (~5%) to drive alignment, diligence, and decision-making, backed by detailed architecture decision records. With the operating model, the organization scales and delivers the architecture intention across a community of approximately 100 architects.

Decision-Making Model and Processes

Architects are empowered to make architecture decisions within their ART and network levels. Certain topics with bigger impacts require higher due diligence and alignment with different parts of the organization, and these will be discussed and decided upon by the CAA board, for example:

- Cross-stream, program-level impacts
- Tech selection, PoCs, pilots
- Add-ons and complex integrations
- System upgrades
- ERP instance strategy

The organization's decision-making model is structured to be both rigorous and agile, ensuring that architectural decisions are documented, reviewed, and enforced effectively.

If a topic in an ART/network has cross-impact (see previous list), it's labeled as *CAA board relevant*. Together with the corresponding enterprise architect, the solution architect starts due diligence in preparation for a CAA board review. This includes prealignment and gathering further background information, such as the following:

- Project background and introduction
- Value driver and business case
- Organizational impact
- Current state versus future state
- Adherence to architecture principles, standards, and nonfunctional requirements
- Operational documentation and references

Once a topic is prepared, presented, discussed, and decided, an architecture decision record is created, and the decision is documented with the help of a Google form, as shown in Figure 12.7.

ERP Cross Architecture Authority - Architecture Decision Record (ADR)

Decision Title	tbd			
Architecture Status	☐ Approved under previous CAA	☐ Review not Applicable	☐ ADR superseded	☐ Not Approved
	☐ Approved	☐ Approved with Exception(s)	☐ Approved with Action Item(s)	☐ Further Review Needed
Decision Context	tbd			
Decision	tbd			
Decision Rationale	tbd			
Alternatives Considered	tbd			
Implications & Impact	tbd			
Consequences	tbd			
LeanIX - IT Landscape Fit				
LeanIX - Technology Fit				
LeanIX - Financial Impact				
LeanIX - Security Relevance				
LeanIX - Overall Decision Impact				
LeanIX - Organisation in Scope	tbd			
LeanIX - Initiative	ASPIRE			
LeanIX - Applications	tbd			
LeanIX - Digital Product (Ix Product)	tbd			
LeanIX - Capabilities	tbd			
LeanIX - Business Context (ERP Functional Area)	tbd			
ERP Enterprise Architect(s)	tbd			
System / Solution Architect	tbd			
Requestors / Presenters	tbd			
Date	tbd			
Project Code (e.g. Jira Issue)	tbd			
Modelling Information updated (CityMap, LeanIX)	☐ Yes	☐ No	☐ N/A	
Solution Architecture Document Complete	☐ Yes	☐ No	☐ N/A	
Related Documents				
Supporting slides as pdf Link to ADR as pdf				

Action Items

#	Action	Assigned To	Due Date	Completion Status
1				
2				

Recommendations	
Comments	
Date for Additional Reviews	

Figure 12.7: Architecture Decision Record Template

The decision documentation leads to an architecture decision record in SAP LeanIX, as shown in Figure 12.8. The architecture decision record in SAP LeanIX is captured manually as a dedicated fact sheet type, since the architecture decision feature was not ready at the point of introducing the tool-supported governance.

In this example, the decision to use SAP S/4HANA as a core solution is captured. In addition to general information, the relations to the governance body, affected applications, initiatives, and architecture principles are outlined.

Tools and Repositories

To support transparency and traceability, the organization leverages specific tools and repositories in the context of governance activities.

SAP LeanIX is the organization's central enterprise architecture platform for documenting standards and decisions. Solution outlines and architecture decision records serve as key artifacts for documenting solution designs and decisions.

For the transformation program, Jira is used to manage the workflow for the DA and CAA, providing a transparent dashboard of all decisions.

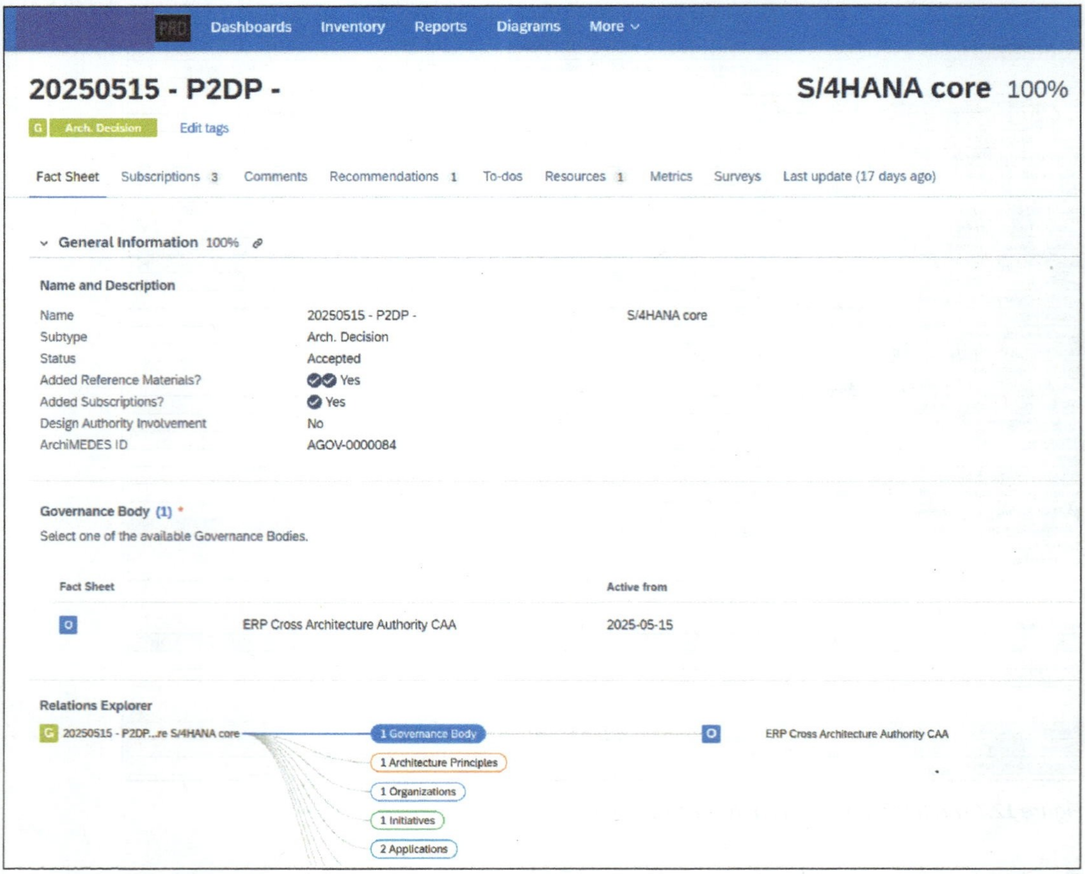

Figure 12.8: Architecture Decision Record

Metrics and KPIs

Measuring the effectiveness of governance is crucial for continuous improvement. The organization plans to measure and report on the adoption of and compliance with the defined technology standards. Establishing these metrics is a key part of maturing the governance practice. The benefits the organization tracks, such as reduced costs, improved agility, and lower risk, serve as indicators of the governance effectiveness.

12.3 Enterprise Architecture Governance in Practice with SAP LeanIX

Now that you understand the seven foundations of enterprise architecture governance, let's take a close look at the capabilities SAP LeanIX offers in the context of enterprise architecture governance. The core capability is architecture decisions, which we'll explain at the beginning of this section, as they can be used across multiple foundations.

In the following sections, we will outline how each foundation can be supported through SAP LeanIX–specific features. We have explained many of these features in previous chapters, and we will provide specific references or links to additional resources here.

12.3.1 Architecture Decisions

Architecture decisions are a core capability of SAP LeanIX that enables enterprise architecture governance. Unlike a fact sheet, architecture decisions are handled as a document type that has the ability to integrate with fact sheets and diagrams. To access architecture decisions, select **More** in the top ribbon and then choose **Architecture Decisions** as shown in Figure 12.9.

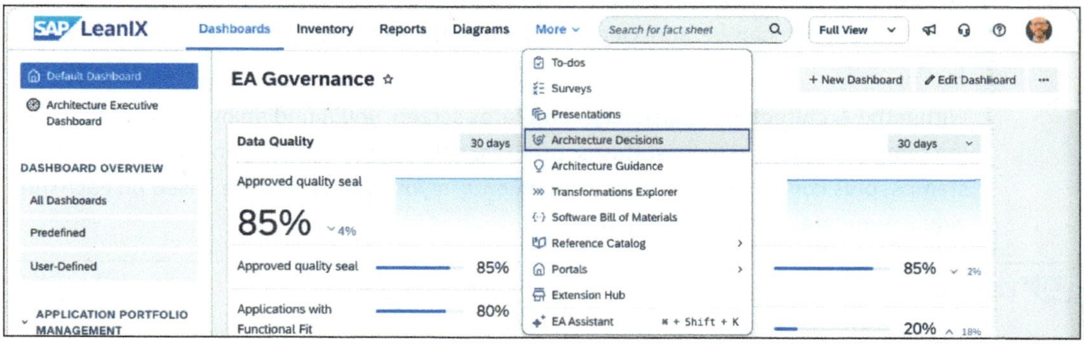

Figure 12.9: Architecture Decisions Navigation

You will now see an overview of all architecture decisions, including their IDs, titles, statuses, creators, and last update timestamps, as shown in Figure 12.10.

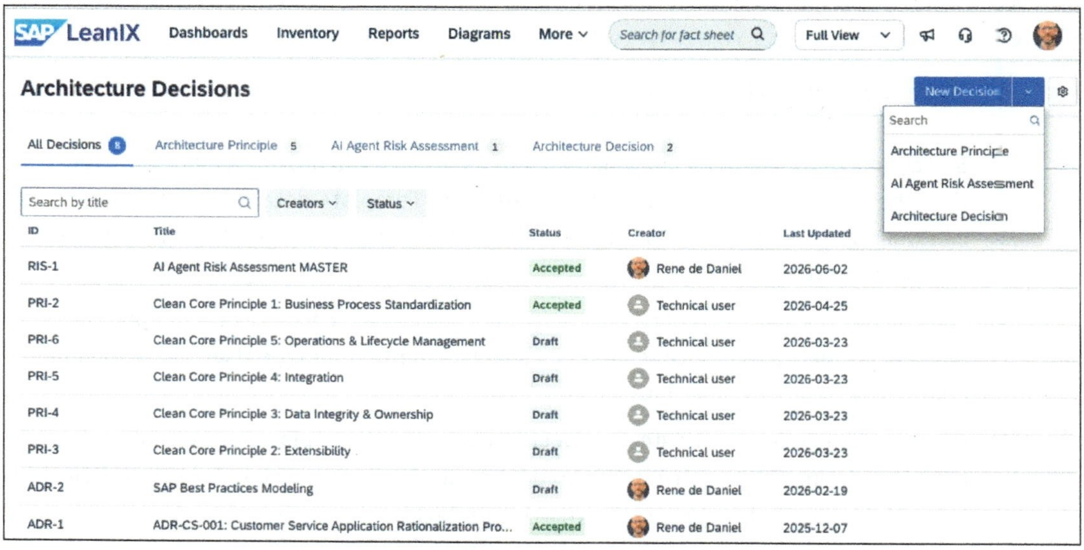

Figure 12.10: Architecture Decision Overview

You also have the possibility to create a new decision by clicking the **New Decision** dropdown and selecting an architecture decision template; in this example, the options are **Architecture Principle, AI Agent Risk Assessment** (discussed in Chapter 13), or **Architecture Decision**, as shown in Figure 12.10.

> **Note**
> Note that the features of architecture decisions can be used in different contexts. With the help of specific templates, you can define the content of an architecture decision, which could be a principle or risk assessment.

To customize and categorize the architecture decisions, you leverage architecture decision templates. These can be accessed by selecting the **Settings** (gear) icon shown in Figure 12.10.

Within the **Architecture Decisions / Templates** screen, you'll find an overview of the existing templates, their activity statuses, prefixes, names, descriptions, and last update timestamps, plus counters of how many documents in your workspace are based on each individual template, as shown in Figure 12.11.

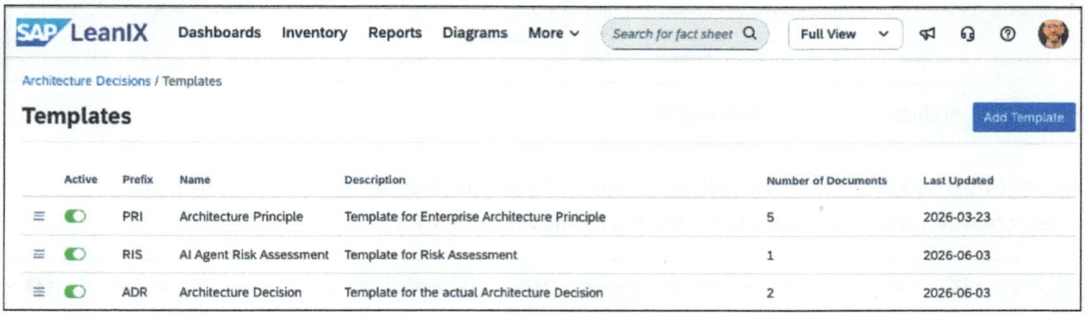

Figure 12.11: Architecture Decision Template

Select **Add Template** to create your own template. Once you provide a name and a description, the template editor opens with a predefined **Template details** section in which you can maintain the **ID Prefix** and **Description**, as well as a configurable content area, as shown in Figure 12.12.

You can flexibly add fields of type **Date** to, for example, indicate a review or approval date; type **Fact Sheet** to reference related fact sheets; type **Text**; or type **User**. Once you finalize editing your template, save it by clicking **Save**.

Once you create a new architecture decision based on a template, you can also maintain the lifecycle information, which is predefined as **Draft**, **Proposed**, **Accepted**, **Rejected**, **Superseded**, or **Deprecated**, as shown in Figure 12.13.

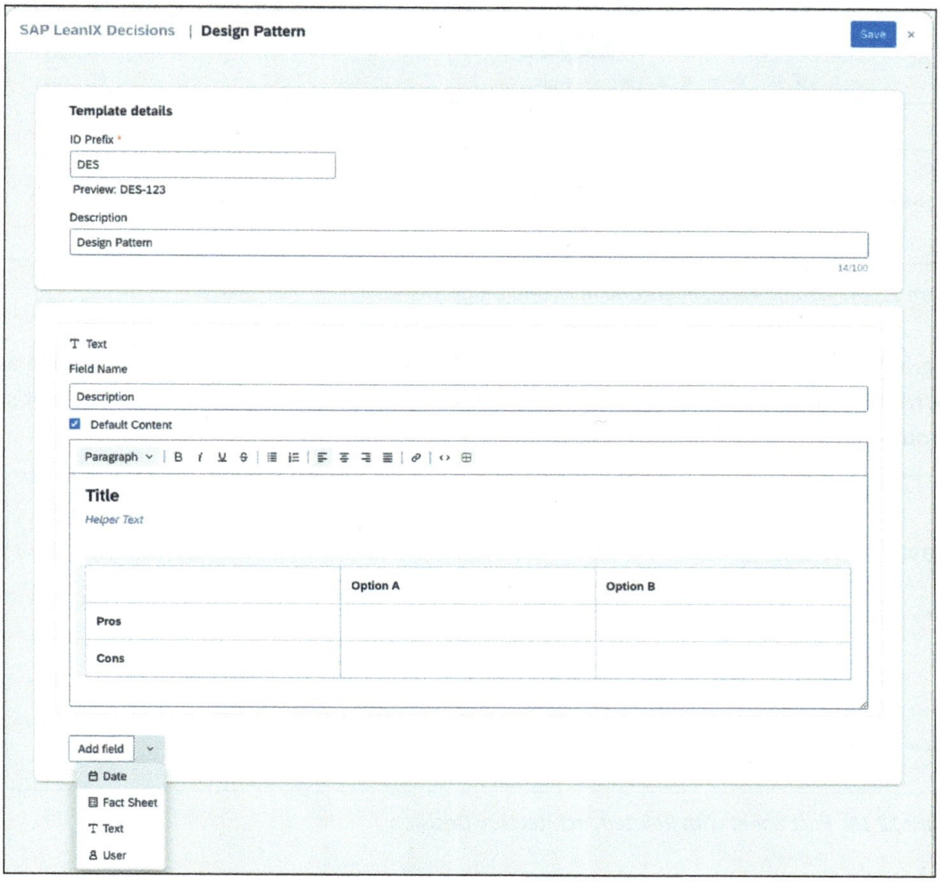

Figure 12.12: Architecture Decision Template Editor

Figure 12.13: Architecture Decision Lifecycle Status

SAP LeanIX adds value by linking architecture decisions to fact sheets and diagrams. Whereas fact sheets can be added within the **Impacted Fact Sheets** area by choosing **+Add fact sheet**, diagrams can be linked within text fields by choosing the **Insert Diagram** symbol, as shown in Figure 12.14.

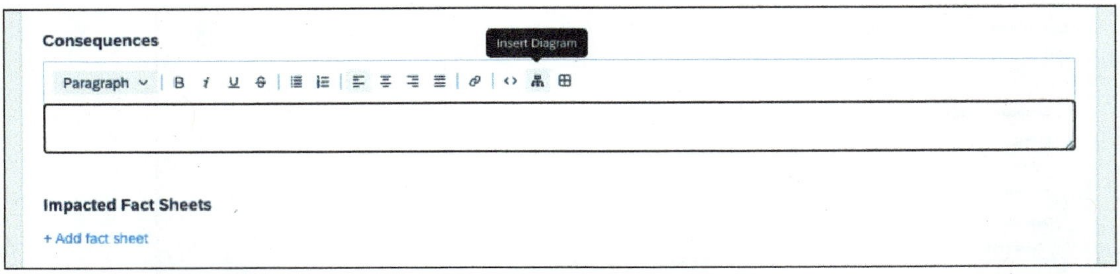

Figure 12.14: Adding Fact Sheets and Inserting Diagrams

Diagrams and fact sheets that have been linked to an architecture decision will include a link to the architecture decision in turn. This link can be found in an individual fact sheet's **Resources** tab, as shown in Figure 12.15.

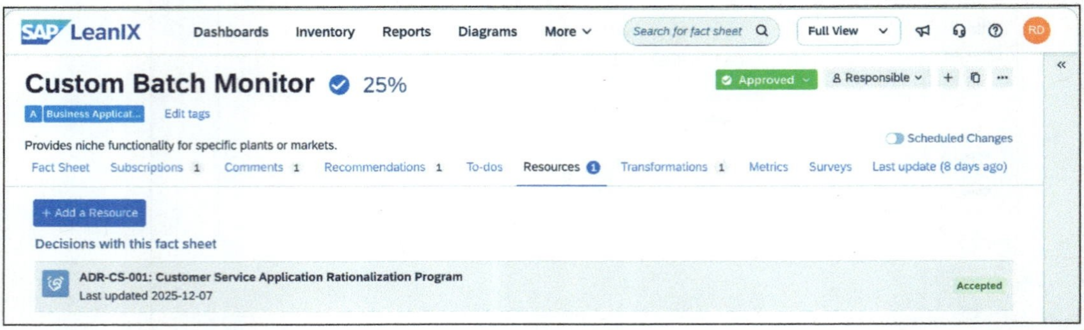

Figure 12.15: Fact Sheet with Related Architecture Decision

For diagrams, the link to the architecture decision is visible on the diagram header level, as shown in Figure 12.16.

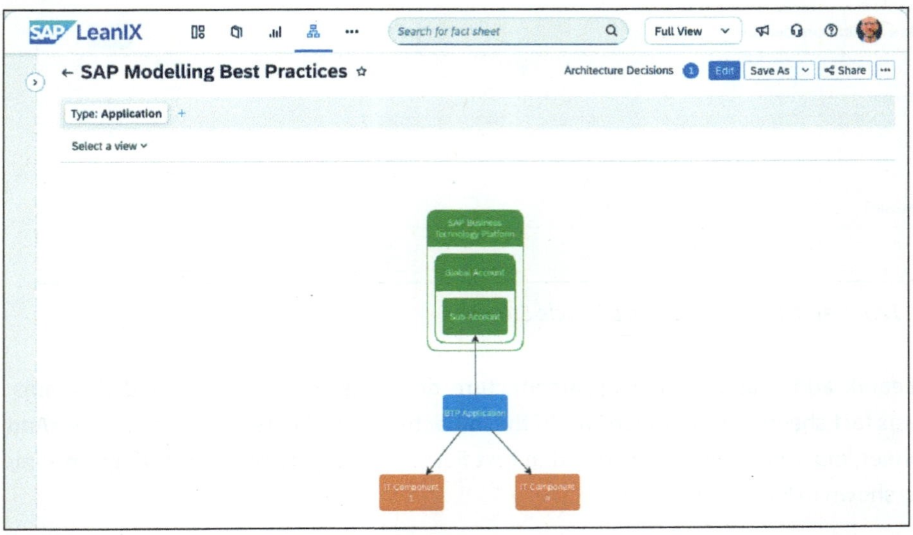

Figure 12.16: Diagram with Related Architecture Decisions

12.3.2 Standards and Principles

Architecture principles can be managed by leveraging the architecture decision capability, which you do by defining a dedicated template as outlined in Section 12.3.1. Figure 12.17 provides an overview of the first clean core principle in the demo workspace. The template was structured by the definition of the principle artifact, including the name, description, rationale, and implication, as well as the approver and date, as described in Chapter 8, Section 8.4.2.

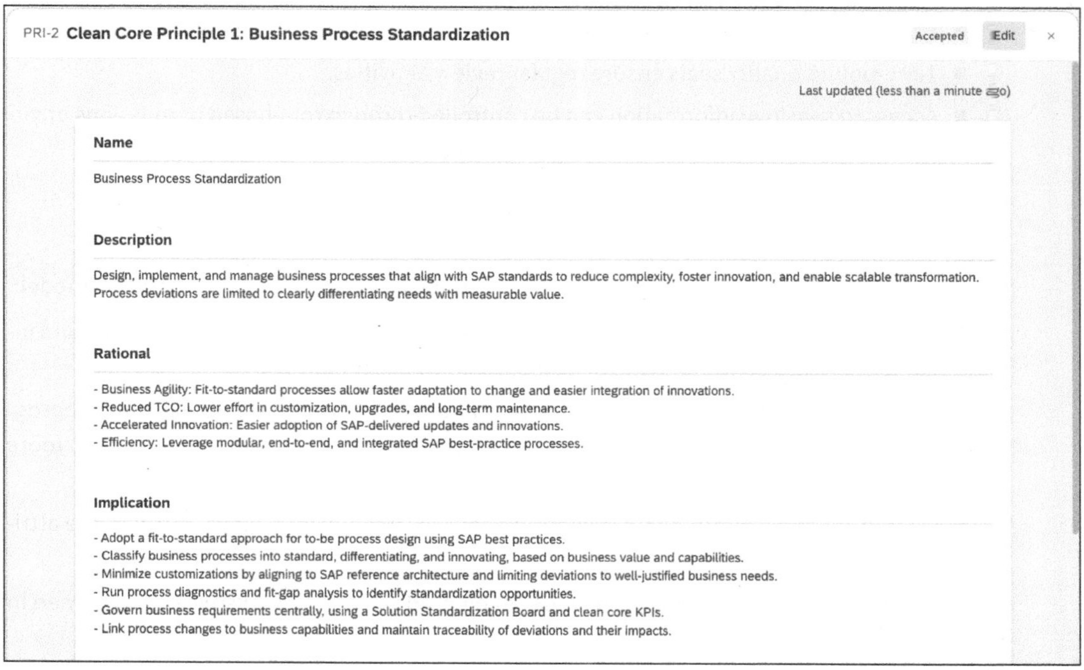

Figure 12.17: Architecture Principle

The governance of standards can be enabled through the following features:

- Technology standards governance can be enabled through IT component fact sheet attributes such as the vendor lifecycle (based on built-in lifecycle catalog), tags, or dedicated attributes (e.g., for clean core compliance).
- Technology radar reports provide an overview of which standards are accepted, rejected, and so on in an organization. For more details, see Chapter 5, Section 5.1.1.
- Portals can be used to ensure the adoption of standards in an organization. For more details, see Chapter 2, Section 2.3.4.
- Heatmaps based on standard tags or attributes provide an overview of the standardization adoption across the architecture landscape.

12.3.3 Boards and Roles

SAP LeanIX offers the following capabilities in the context of boards and roles:

- The accountability model foresees the assignment of stakeholders with different degrees of ownership, such as responsible person, observer, or accountable person.
- Fact sheet maintenance in the sense of accuracy and completeness is achieved through sign-off of the accountable person through quality seals. For more details, see Chapter 2, Section 2.3.1.
- Fact sheet subscribers are informed if changes are applied to individual fact sheets.
- Time-bound quality seals ensure regular review activities.
- Access to sensitive information can be controlled through role-based permissions or virtual workspaces, as outlined in Chapter 2, Section 2.3.4.

12.3.4 Operating Model

SAP LeanIX offers the following capabilities in context of the governance operation model:

- Architecture boards and review councils are supported through a living inventory and in-context discussion in the style of *presentation-proposal-approval*.
- Governance dashboards with key artifacts and KPIs provide a status overview of progress and identify potential deviations that will lead to follow-up activities. For more details, see Chapter 5, Section 5.1.3.
- Quality dashboards provide information about the general data quality on, say, the attribute or relation level.
- Access to sensitive information can be segmented via virtual workspaces, as outlined in Chapter 2, Section 2.3.4.

12.3.5 Decision-Making Model

The decision-making model builds the core of the enterprise architecture governance and is featured in the architecture decision, as introduced in the previous section. A minimum set of information such as a context, the decision, consequences, approver, review date, and impacted fact sheets and diagrams should be included here. As you have seen in the case presented in Section 12.2, more detailed information can be added if required. The architecture decision template offers full flexibility.

You can find an example of an architecture decision record in the demo workspace, as shown in Figure 12.10. Alternatively, you can find the architecture decision record for the EA7 Application Rationalization in Chapter 7, Figure 7.22.

12.3.6 Processes and Workflows

SAP LeanIX offers the following capabilities in the context of processes and workflows:

- Surveys can be used to gather data across the organization (e.g., migration strategy survey for cloud governance) and as a mechanism to approach stakeholders for ongoing maintenance and risk assessment, as discussed in Chapter 2, Section 2.3.3.
- Automations can be leveraged to enable governance controls, for example, fact sheets can be auto approved under specific conditions, and actions can be created if mission-critical data quality is compromised or notifications can be sent in case data quality is low. For more details, see Chapter 10, Section 10.3.

12.3.7 Tools and Repositories

The enterprise architecture governance foundation of tools and the repository can be seen as the home turf of SAP LeanIX.

SAP LeanIX is the single source of truth for a data-driven governance process. Fact sheets act as the system of record to capture information relevant in the context of enterprise architecture governance. With the help of the configurable meta model, missing attributes can be added and mandatory versus optional fields defined. Based on this information, the completeness of a fact sheet can be calculated with a metric that can also include a weighting beside applicability and status controls.

Data quality governance is achieved through formal approval and reapproval mechanisms based on the quality seal. Responsible and accountable users approve the quality of a fact sheet, while nonresponsible users will break the quality seal, which leads to an email notification. This event-driven approach can be combined with the definition of a time period for reapproval set by an administrator. Any changes to a fact sheet will be traceable in the audit log.

Traceability is gained through relations and dependencies, which can be used for impact assessments and (multidimensional) visualizations. Relevant governance information can be tracked on the relation level.

12.3.8 Metrics and KPIs

SAP LeanIX offers the following capabilities in the context of metrics and KPIs:

- Dashboards with reports and KPIs act as an entry point for governance-specific views. For more details, see Chapter 5, Section 5.1.3.
- Standard and custom KPIs provide insights that include historical trending, for example:
  - Enterprise architecture governance effectiveness
  - Architecture decision quality and speed
  - Standards and lifecycle compliance
  - Risk and technology obsolescence exposure

  - Portfolio rationalization and cost transparency
  - Data quality and stewardship maturity
  - Governance process efficiency
- Calculations feature KPIs through transformation or aggregation of information, assignment on the relation level, and so on, as discussed in Chapter 10, Section 10.2.

Figure 12.18 shows an exemplary enterprise architecture governance dashboard, accessed by navigating to the **Dashboards** section. In this example, we included the **Data Quality** and **Relations Quality** KPIs as well as the **Best practice metrics** in the dashboard.

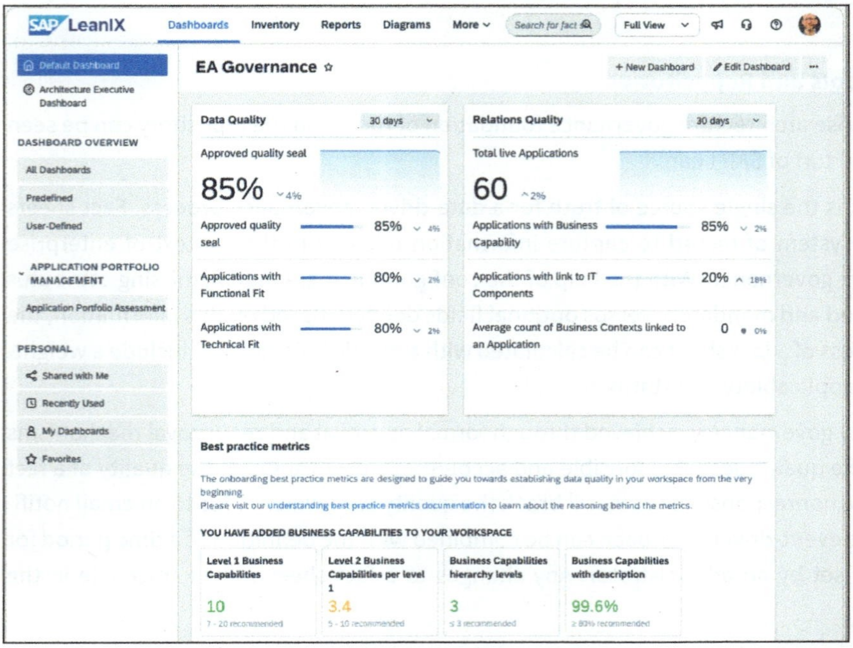

Figure 12.18: Enterprise Architecture Governance Dashboard

12.4 Summary

This chapter introduced the outcomes of a successful enterprise architecture governance practice and described the seven foundations that need to be considered and defined to operationalize it. A tangible case outlined how enterprise architecture governance is lived in practice. Each of the foundational elements can be supported through the features and capabilities of SAP LeanIX, the majority of which have been described in previous chapters. The focus of this chapter was on the creation and management of architecture decision records and architecture principles. By leveraging these capabilities, transparency and insights can be brought to an organization.

In the next chapter, we'll focus on the topic of governance in context of AI, which is a core discipline of enterprise architects in the era of AI.

Chapter 13
AI Governance

Welcome to the era of AI! Enterprise architects will play a crucial role in enabling the adoption of AI at scale in enterprises. This will require new skills and approaches, such as risk management across multiple dimensions or business value assessments. If done right, enterprise architects will be seen as catalysts for innovation.

An *AI agent* is an intelligent software entity that can perceive its environment, reason about how to achieve goals, and autonomously execute actions to meet those goals. Enterprise leaders see AI agents as critical enablers of business transformations. Key strategic drivers for using AI agents include the following:

- **Business agility and speed**
  AI agents provide automation and the agility to scale and respond in real time.
- **Operational efficiency and cost reduction**
  AI agents can offload labor-intensive tasks and run routine processes 24/7.
- **Innovations and new value creation**
  AI agents augment human capabilities and proactively identify patterns or opportunities in data that humans might miss.
- **Enhanced decision support and alignment**
  AI Agents can analyze vast numbers of datasets, detect trends, or even run simulations for decision-making.
- **Improved customer and employee experience**
  AI agents can handle common inquiries instantly, personalize responses, and escalate complex issues to humans while improving satisfaction levels.
- **Competitive necessity**
  Leaders recognize that failing to invest in AI capabilities could leave their organization at a disadvantage.

These strategic imperatives are driving enterprises to introduce AI agents into their enterprise architecture to achieve greater agility, efficiency, and competitive differentiation.

In the following sections, we'll outline how SAP LeanIX supports the AI governance use case in multiple dimensions:

1. We'll start with an introduction to the AI landscape and how it impacts your enterprise architecture.

2. We'll then prepare the SAP LeanIX workspace for the era of AI by activating the following:
   - The *AI governance* extension
   - The *SAP suggestions in your landscape* extension
3. To establish a sound AI governance practice, you need to create transparency in your baseline architecture and assess potential risk in combination with a more detailed architecture impact assessment around technical debt and data architecture requirements.
4. From strategic alignment and value points of view, you should identify the potential for AI use cases from a business capability perspective and assess which AI agents already exist with the help of a reference catalog.
5. Once a viable AI use case is identified, the value contribution needs to be articulated.
6. A broader communication of the lifecycle stage of an AI use case will provide transparency into reusability, adoption potential, and broader organizational change.
7. The measurement and oversight of the AI use case landscape will provide valuable insights to steer the AI adoption.

13.1 Why AI Governance Has Become an Enterprise Architecture Responsibility

The introduction of AI agents offers huge potential in productivity gains and new business models. However, each new technology being introduced in the past also introduced a new responsibility. With the introduction of AI-based technologies, the responsibility requirements are reaching new levels and the requirements for the governance capability are increasing, primarily through the following drivers:

- **Cost, value, reuse, and scale (avoiding agent sprawl)**
  The introduction of AI technologies comes at a cost; we are all aware of the huge demand for new data centers, which currently is manifesting itself in almost weekly news about new centers to be built or partnerships between hyperscalers and AI chip manufacturers. The investment must enable a return (which often still needs to be proven). However, time is too limited to focus on each and every individual AI use case, so reuse and scalability are of importance to enable timely adoption at scale. At the same time, AI use cases that do not hold a return on investment (ROI) promise need to be stopped in a controlled way to ensure that investment streams are redirected into more promising opportunities.
- **Risk, trust, and brand protection**
  With new technologies there comes also the risk of false insights, biased recommendations, and wrong decisions, which could accumulate in an overarching lack of trust or even an impact on the brand value at larger scale. The quality of data and the maturity of large language models (LLMs) are just two possible root causes for these situations, and these need to be understood in detail.

- **Regulatory pressure (e.g., the EU's Artificial Intelligence Act)**
  While new technologies are being introduced, the number of regulatory controls is steadily increasing, with a time gap of a few months between the issue of a new regulation and the necessary implementation in the tool. This can lead to situations in which AI use cases in operation today might not be compliant anymore tomorrow. A continuous assessment of new regulations followed by a thorough impact assessment of existing and new AI use cases will be the norm.

The enterprise architecture function is the ideal candidate to govern the company-wide evolution and adoption of AI as AI cuts across the business, application, data, and technology domains and requires at the same time the capability to integrate with strategy, risk, and execution.

In the era of AI, enterprise architects must evolve from technology custodians to ethical stewards, strategic designers, and change leaders with the aim of enabling scalable innovation and sustainable value creation. Table 13.1 outlines the required capabilities of enterprise architects in the era of AI.

Capability	Description
Ethical AI governance	■ Define and enforce ethical guardrails for AI usage (fairness, explainability, accountability) ■ Establish transparency across AI models, agents, and decision logic ■ Ensure human-in-the-loop and accountability mechanisms are architecturally embedded
AI regulatory translation and compliance enablement	■ Translate regulatory frameworks (e.g., the EU's AI Act) into concrete architecture requirements ■ Classify AI solutions by risk and ensure traceability across architecture artifacts ■ Enable auditability and evidence-based compliance through architecture data
AI transparency and inventory management	■ Maintain a complete inventory of AI agents, models, and embedded AI features ■ Establish ownership, lifecycle, and usage context for AI assets ■ Detect and control AI sprawl across the enterprise landscape
AI-ready data and platform architecture	■ Design scalable, reusable data architectures that support AI workloads ■ Ensure data quality, lineage, and governance for AI consumption ■ Align AI platform decisions with long-term enterprise architecture strategy

Table 13.1: Enterprise Architecture Capabilities in AI Era

Capability	Description
AI-driven technical debt management	■ Identify AI agents and models as a new form of technical debt ■ Assess sustainability, reuse, and long-term cost of AI solutions ■ Drive standardization and rationalization of AI capabilities
AI value and strategic alignment	■ Link AI initiatives to business capabilities and strategic objectives ■ Distinguish between differentiating, commodity, and experimental AI use cases ■ Guide investment decisions across build, buy, and embed options
Organizational change and leadership enablement	■ Enable federated decision-making for AI across teams ■ Act as facilitator and coach rather than gatekeeper ■ Support cultural change to move toward responsible and transparent AI adoption
Architecture decision and governance	■ Structure AI-related architecture decisions using architecture decision records ■ Ensure decisions are transparent, traceable, and revisitable ■ Embed governance into everyday ways of working

Table 13.1: Enterprise Architecture Capabilities in AI Era (Cont.)

13.2 Preparing Your Workspace

The AI governance use case is based on SAP AI Agent Hub, a vendor-agnostic command center with the ability to discover and govern AI agents, LLMs, and Model Context Protocol (MCP) servers across a company's architecture landscape. Technically, it is based on two extensions of the meta model, as well as a set of technical integrations, reports, reference content, saved searches, and predefined key performance indicators (KPIs).

Note

Two versions of the SAP AI Agent Hub exist: a standalone version and a version serving as an integral part of SAP LeanIX. The latter will be the focus in this chapter.

13.2.1 Activating the AI Governance Extension

The AI governance extension is delivered as part of the SAP LeanIX Application Portfolio Management product and provides new capabilities for the discovery of AI agents across your landscape through an extension of the meta model. Furthermore, it activates a set of reports, saved searches, calculations, and KPIs.

> **Note**
>
> The availability of the AI governance extension depends on the creation date of your workspace. If the workspace was created before May 11, 2026, the steps detailed ahead need to be applied. Otherwise, the extension is activated by default.

For the activation of the extension, log into your workspace as an administrator and open the **Administration** section. From there, follow these steps:

1. Select **Optional Features and Early Access** within **ADVANCED SETTINGS.**
2. Scroll down to the **New Meta Model Extensions** section, as shown in Figure 13.1, and select **Activate** for the **AI Governance** extension.

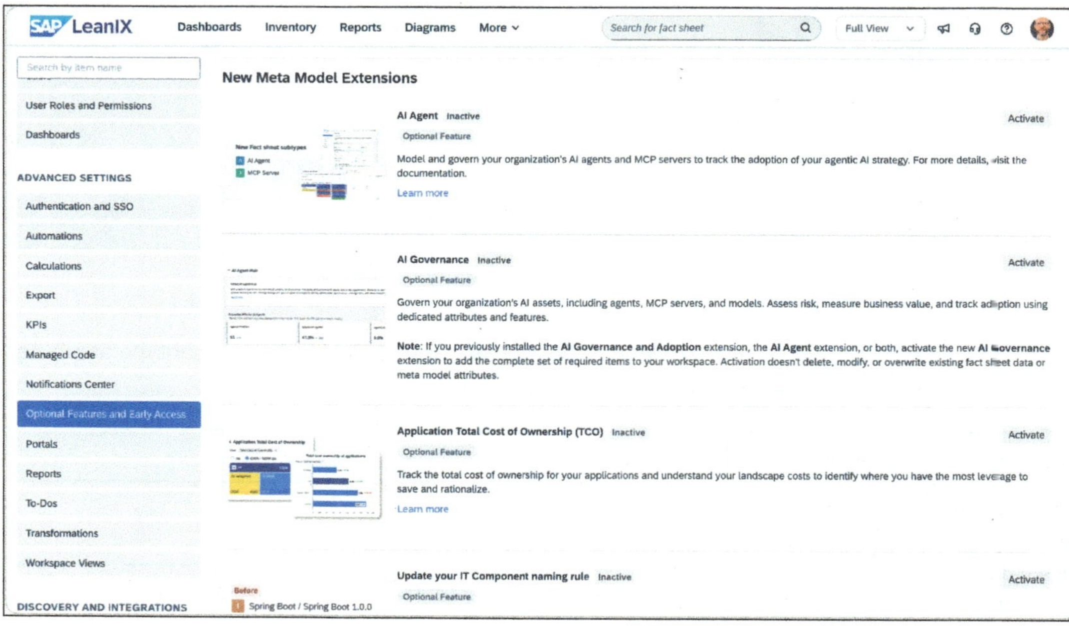

Figure 13.1: AI Governance Extension Activation

In the popup window, make sure that you check the **I agree to extend the meta model** box prior to clicking **Activate Extension**, as shown in Figure 13.2.

With the activation of the meta model extension, two subtypes for the application fact sheet are introduced:

- **Business application**
  This subtype should be used for the existing application fact sheets to allow a better differentiation from the newly introduced AI agent subtype.
- **AI agent**
  This is a new subtype for modeling and autodiscovery of AI agents.

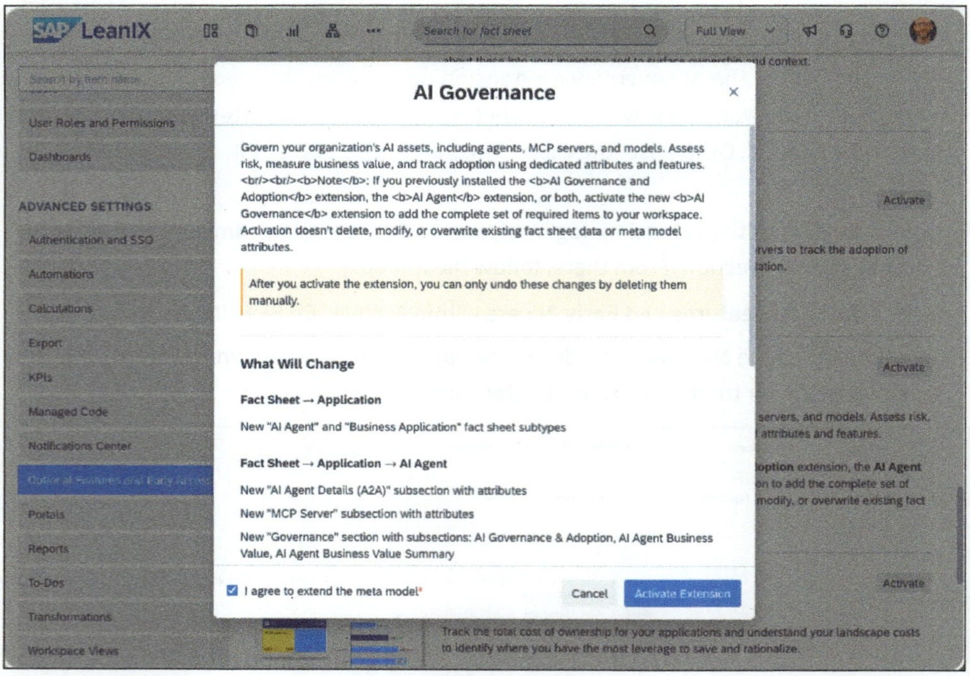

Figure 13.2: Agreement for AI Governance Extension

Table 13.2 provides an overview of the AI agent application fact sheet subtype attributes and the field values relevant for the AI governance use case.

Section	Attribute	Field Value
AI Agent Details (A2A): Provide agent details based on the agent-to-agent (A2A) protocol specification.	**AI Agent Service Endpoint**	URL to reach the AI agent.
	Authentication	Authentication requirements for accessing the agent's endpoint: ■ **OAuth 2.0** ■ **Basic Auth** ■ **API Key** ■ **OpenID Connect** ■ **Other**
	Skills	A specific capability, function, or area of expertise the agent can perform or address.
	A2A Capabilities	Specify optional supported A2A protocol features like streaming or push notifications.

Table 13.2: AI Agent Application Fact Sheet Subtype Attributes

Section	Attribute	Field Value
MCP Server: MCP server connected to the agent	**MCP Server Support**	Indicates whether the agent or application can be accessed via an MCP server: ■ **Built-in** ■ **Custom** ■ **No**
	MCP Server Endpoint	URL to reach the MCP server.
	MCP Server Classification	Indicates whether the MCP server queries data or has writing permissions: ■ **Read** ■ **Write**
AI Governance and Adoption: Defines how AI is used, evaluates associated risks and benefits, and ensures alignment with legal and business frameworks	**AI Usage**	Describes whether the application is using an AI-type algorithm: ■ **Unknown** ■ **No AI Usage** ■ **AI Available** ■ **AI Enabled**
	Approval State	Indicate the agent's approval state: ■ **Draft** ■ **In Preview** ■ **Accepted** ■ **Rejected**
	Types of Artificial Intelligence	Indicates the type of algorithmic solution or approach being used by this application: ■ **Neural Net** Computing model inspired by the human brain that consists of interconnected layers of artificial neurons capable of processing and learning from complex, multidimensional data. ■ **Large Language Model (LLM)** An LLM is a type of AI model trained on a vast amount of text data, capable of generating human-like text by predicting subsequent words or sentences in a given piece of writing. ■ **Machine Learning (ML)** Machine learning is a subfield of AI that uses statistical techniques to enable computers to improve performance and make data-driven decisions automatically without being explicitly programmed. ■ **Other** Please specify in the text field what other type of AI is employed

Table 13.2: AI Agent Application Fact Sheet Subtype Attributes (Cont.)

Section	Attribute	Field Value
	AI Potential	▪ **Low** Hardly any potential ▪ **Medium** Limited potential ▪ **High** Good opportunity to improve business value ▪ **Very High** Mandatory to unlock business AI potential, e.g., by enabling available features or investing in development
	Autonomy Level	Categorizes agents based on how they interact with users, whether through direct conversation or by performing tasks autonomously in the background: ▪ **Autonomous** ▪ **Conversational**
	Build Type	Defines the configuration mode of the agent, whether it was custom, prebuilt, or prebuilt expanded: ▪ **Prebuilt** ▪ **Custom** ▪ **Prebuilt expanded**
	AI Risk	Assessment of threats to the safety, livelihoods, and rights of people according to the EU's AI Act: ▪ **Minimal or No Risk** The AI Act allows the free use of minimal-risk AI. This includes applications such as AI-enabled video games or spam filters. ▪ **Limited** Limited risk refers to risks associated with lack of transparency in AI usage. ▪ **High** High-risk AI systems will be subject to strict obligations before they can be put on the market. ▪ **Unacceptable** AI systems considered a clear threat to the safety, livelihoods, and rights of people will be banned.

Table 13.2: AI Agent Application Fact Sheet Subtype Attributes (Cont.)

Section	Attribute	Field Value
AI Agent Business Value: Defines the type and use of the agent to calculate its business value	**Type of Business Value**	Business value type that best describes the agent's primary benefit. ■ **Cost Savings** ■ **Revenue Increase** ■ **Risk Reduction**
	Usage	Defines the usage of the agents, based on users, session runs, sent messages, and so on, and is used for the calculation.
	Cost Savings per Unit	Base unit for cost saving calculation.
	Revenue Increase per Unit	Base unit for revenue increase calculation.
	Risk Reduction per Unit	Base unit for risk reduction calculation.
AI Agent Business Value Summary: Calculated fields based on the AI agent's **Business Value Input** section	**Total Expected Cost Savings**	Value of the field is determined by the calculation: calculate business value for AI agents—cost savings Formula: *Usage x Cost Savings per Unit*
	Total Expected Revenue Increase	Value of the field is determined by the calculation: calculate business value for AI agents—revenue increase Formula: *Usage x Revenue Increase per Unit*
	Total Expected Risk Reduction	Value of the field is determined by the calculation: calculate business value for AI agents—risk reduction Formula: *Usage x Risk Reduction per Unit*

Table 13.2: AI Agent Application Fact Sheet Subtype Attributes (Cont.)

In addition to the application fact sheet subtypes, the interface and IT component fact sheets were extended by dedicated subtypes. The interface fact sheet is extended by the new *MCP server* subtype, as outlined in Table 13.3.

Section	Attribute	Field Value
Name and description	**Authentication**	Authentication requirements for accessing the agent's endpoint: ■ **OAuth 2.0** ■ **Basic Auth** ■ **API Key** ■ **OpenID Connect** ■ **Other**
MCP server details	**MCP Server Endpoint**	URL to reach the MCP server.
	MCP Server Classification	Indicates whether the MCP server queries data or has write permissions: ■ **Read** ■ **Write**
	Available Tools	Defines the tools the MCP server is authorized to access.
	MCP Server Type	Mode of operation for the communication between the AI client and MCP server: ■ **Remote** ■ **Local**

Table 13.3: MCP Server Interface Subtype Attributes

In addition to the AI agent and MCP server fact sheet subtypes, the IT component fact sheet is extended by a new *AI model* subtype. This subtype is also featured by the IT component references catalog to speed up the creation of AI model subtype fact sheets and to increase data quality. If you create a new IT component (e.g., Gemini), you can leverage the reference information, including the creation of the provider factsheet.

These three newly added subtypes—AI agents, MCP server, and AI models—form *AI assets* within SAP LeanIX. Although the governance of AI assets is of primary importance, a new verification mechanism is introduced for the AI assets: the verification badge. The verification badge is located next to the quality seals on the header level of an AI asset fact sheet, as shown in Figure 13.3.

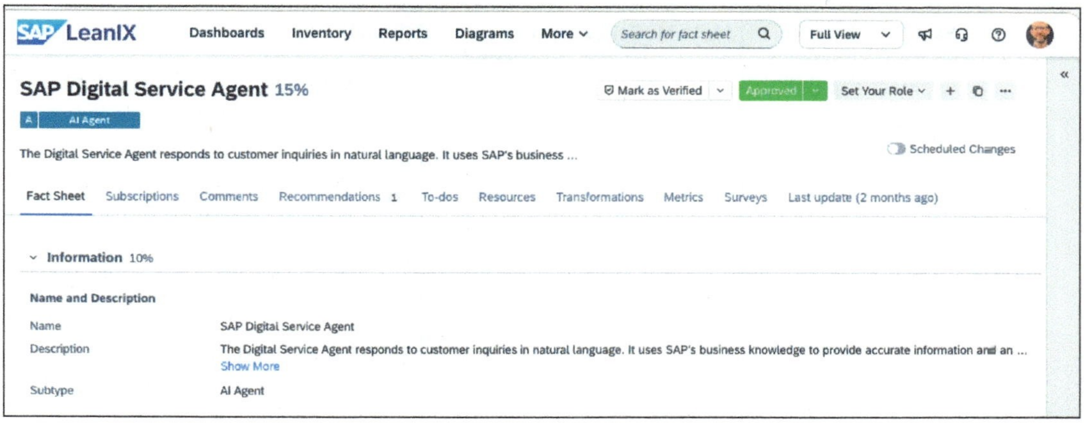

Figure 13.3: AI Asset Verification Badge

> **Tip**
> For each individual AI asset subtype, the verification badge can be configured within the **AI Governance Settings** of the **Administration** menu.

In addition to the newly introduced subtypes, Table 13.4 shows attributes that were added to the business capability, interface, and IT component fact sheet types to support the AI governance use case.

Fact Sheet	Meta Model Extension	Field Value
Business capability	**AI Potential** within new **AI Potential Assessment** section	■ **No Potential** There is no potential for LLM AI in this capability at this time (or it should be deliberately avoided). ■ **Efficiency Improvement** Evaluate if and how LLM AI can streamline processes, automate tasks, and improve overall efficiency. This could involve speeding up decision-making, reducing errors, or automating manual tasks such as document classification or customer support.

Table 13.4: Meta Model Additions for AI Governance Extension

Fact Sheet	Meta Model Extension	Field Value
Business capability (Cont.)		■ **User Interaction Enhancement** Consider how LLM AI could improve interactions with users, be they customers, employees, or partners. This could involve personalizing user experiences, improving accessibility, or facilitating multilingual communication. ■ **Process Transformation** Assess the potential of LLM AI to disrupt and transform existing processes. This could involve changing how tasks are performed, redefining roles, or introducing new ways of working, such as using sentiment analysis for market research instead of traditional methods.
Interface (API subtype)	**Authentication** as a new field in the **Name and Description** section	Authentication requirements for accessing the agent's endpoint: ■ **OAuth 2.0** ■ **Basic Auth** ■ **API Key** ■ **OpenID Connect** ■ **Other**
IT component	**AI Radar Classification** as a new field in the **Technology Assessment** subsection	■ **Large Language Model (LLM)** AI models designed to understand, generate, and reason over human language, typically trained on large volumes of text data. ■ **Model Optimization** Capabilities and techniques focused on improving AI model performance, efficiency, scalability, and cost-effectiveness, such as fine-tuning, compression, or inference optimization.

Table 13.4: Meta Model Additions for AI Governance Extension (Cont.)

Fact Sheet	Meta Model Extension	Field Value
IT component (Cont.)		■ **Model Governance and Security** Capabilities that ensure AI models are used responsibly and securely, including access control, compliance, risk management, monitoring, and protection against misuse or vulnerabilities. ■ **Other AI** Any other AI capability not covered by the preceding categories. Details should be specified separately.

Table 13.4: Meta Model Additions for AI Governance Extension (Cont.)

13.2.2 Activation of SAP Suggestions in Your Landscape

The second extension in the context of AI is the SAP suggestions in your landscape extension, which provides the capability to overlay existing application landscape reports with reference fact sheets for applications and AI agents, which we will explain in more detail in the subsequent sections.

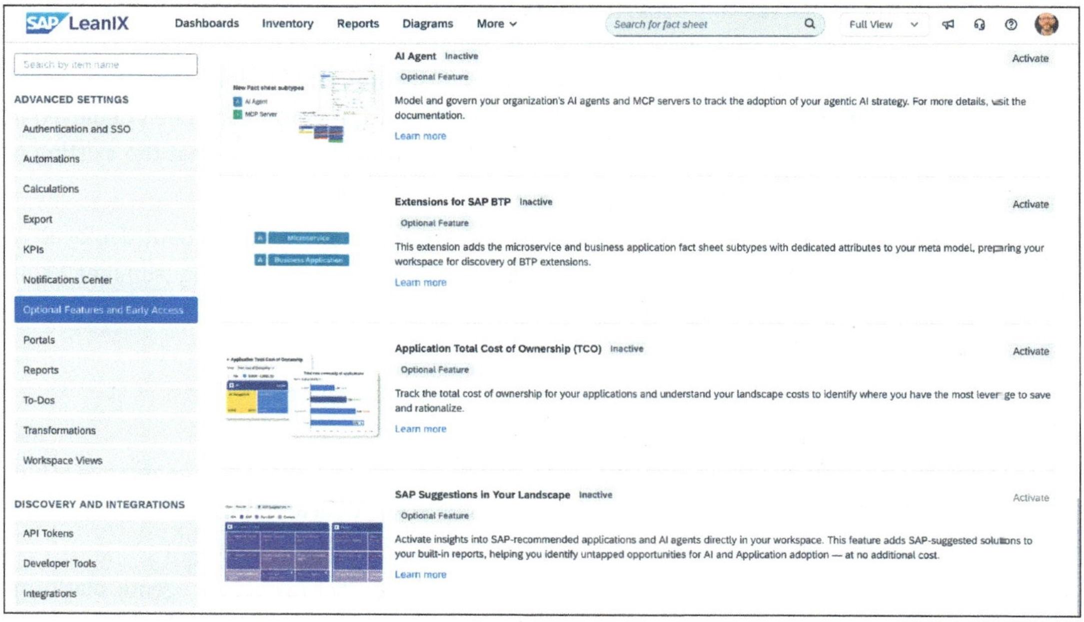

Figure 13.4: SAP Suggestions in Your Landscape Meta Model Extension Activation

The activation of the SAP suggestions in your landscape extension follows the same workflow. Log into your workspace as administrator, open the **Administration** section, and follow these steps:

1. Select **Optional Features and Early Access** within **ADVANCED SETTINGS**.
2. Scroll down to the **New Meta Model Extensions** section, as shown in Figure 13.4, and select **Activate** for **SAP Suggestions in Your Landscape**.

Note

The SAP suggestions in your landscape extension can only be activated if the AI governance extension has already been activated.

13.3 Governing Your Agents in the Landscape

Now that you understand the meta model extensions and features, let's define how you can govern your AI agents in your landscape. This includes building the baseline architecture, planning the target architecture, and applying governance techniques.

13.3.1 Baseline Architecture Assessment

By the time you read this book, the likelihood that your company has already deployed a set of AI agents is rather high. We therefore want to start the AI governance use case with the assessment of your baseline architecture in the context of AI agents. The purpose here is to gain a holistic understanding of and insight into what has already been deployed in your company context and to apply a first risk assessment to ensure that the existing AI agents comply with ethical, regulatory, and compliance standards.

SAP AI Agent Hub offers multiple possibilities to retrieve information about deployed AI agents that can be configured through integrations. At the time of writing (summer 2026), the following integrations were offered:

- **AI Agent API**
  REST API for ingesting AI agent cards based on the A2A protocol.
- **AI Model Discovery via SAP AI Core**
  Automatic discovery of AI model deployments managed through SAP's generative AI hub across SAP Business Technology Platform (SAP BTP) global accounts.
- **AWS Bedrock AgentCore**
  Automatic discovery of AI agents from the Amazon Bedrock AgentCore agentic platform.
- **AWS Bedrock Agents**
  Automatic discovery of AI agents built and configured with Amazon Bedrock agents.
- **Google Agent Engine**
  Automatic discovery of AI agents from Google Workspace.

- **Microsoft Azure API Center**
  Automatic discovery of AI agents through MCP servers managed within Azure API Management.
- **Microsoft Entra ID**
  Automatic discovery of AI agents registered through the Microsoft Entra Agent ID framework.
- **SAP AI Agents**
  Automatic discovery of AI agents deployed in your SAP landscape.
- **ServiceNow (AI Agent Studio)**
  Automatic discovery of AI agents from ServiceNow AI Agent Studio.

13.3.2 Prebuilt Integration of AI Agents

To configure a new integration, log into your workspace as administrator and open the **Administration** section. Then, follow these steps:

1. Select **Integrations** within **Discovery and Integrations** and choose **Add Integration**.
2. Select **AI Agent Discovery** under **Discovery** to select any of the available AI agent integrations, as shown in Figure 13.5.

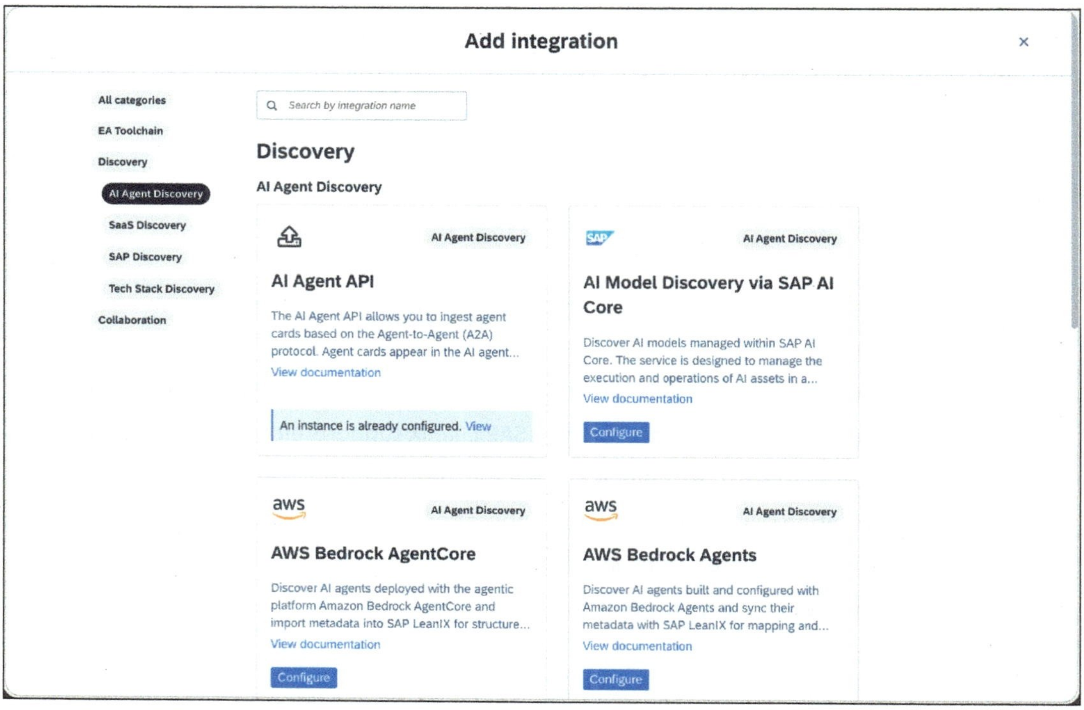

Figure 13.5: Integrations for Discovery of AI Agents

To showcase an example, we'll outline a step-by-step approach to configure a Microsoft Entra ID integration in the demo landscape.

Note

Each integration might require specific prerequisites, such as credentials or access to service endpoints, and these might differ from integration to integration. More detailed information can be found via the **View documentation** link, as shown in Figure 13.6.

To create a new Microsoft Entra ID integration, follow these steps:

1. Select **Configure** within the **Microsoft Entra ID AI Agent Discovery** tile.
2. The screen shown in Figure 13.6 will open, in which you can define a **Name** for the new integration to be configured and choose **Next**.

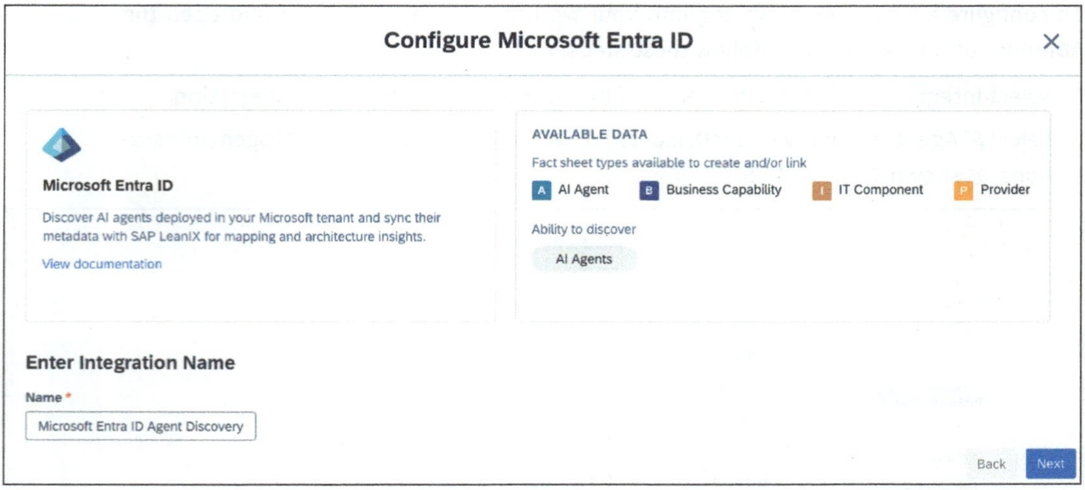

Figure 13.6: Configuration of Microsoft Entra ID Agent Discovery

3. You'll arrive at the screen shown in Figure 13.7, where you will fill in the following fields:
   - **Region**: Choose a region. Currently, SAP LeanIX supports either **Global** or **China**.
   - **Directory (tenant) ID**: Enter your directory (tenant) ID.
   - **Application (client) ID**: Enter your application (client) ID.
   - **Client Secret**: Enter your client secret.

   When you're done, click the **Finish** button.

Note

For the retrieval of the required credentials, SAP LeanIX needs to be registered in the Microsoft Entra ID tenant and permissions need to be granted. For more details and prerequisites, refer to *http://s-prs.co/v620695*.

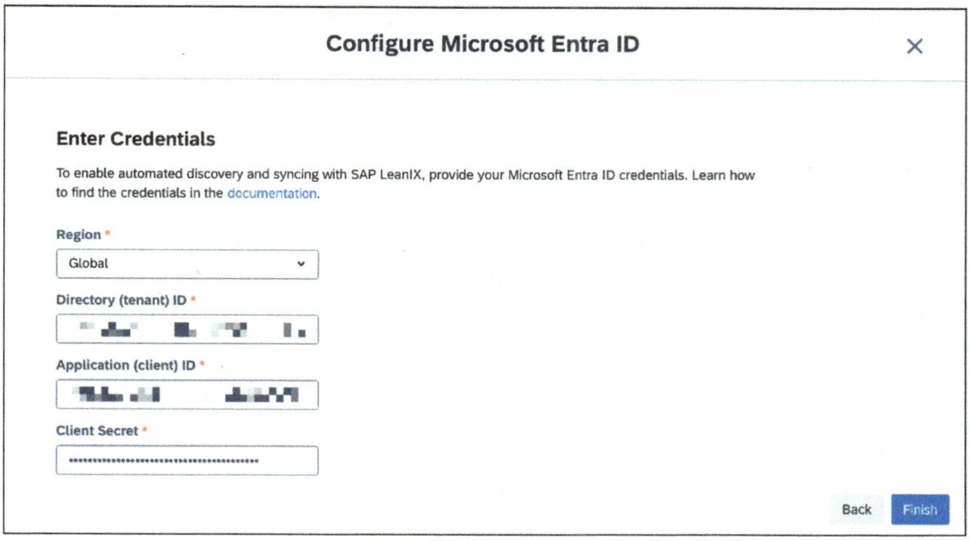

Figure 13.7: Credentials for Microsoft Entra ID Agent Discovery Configuration

The newly created Microsoft Entra ID integration will now be activated and the credential and AI agent discovery status provided if you navigate to the **Integrations** menu option under **DISCOVERY AND INTEGRATIONS**, as shown in Figure 13.8.

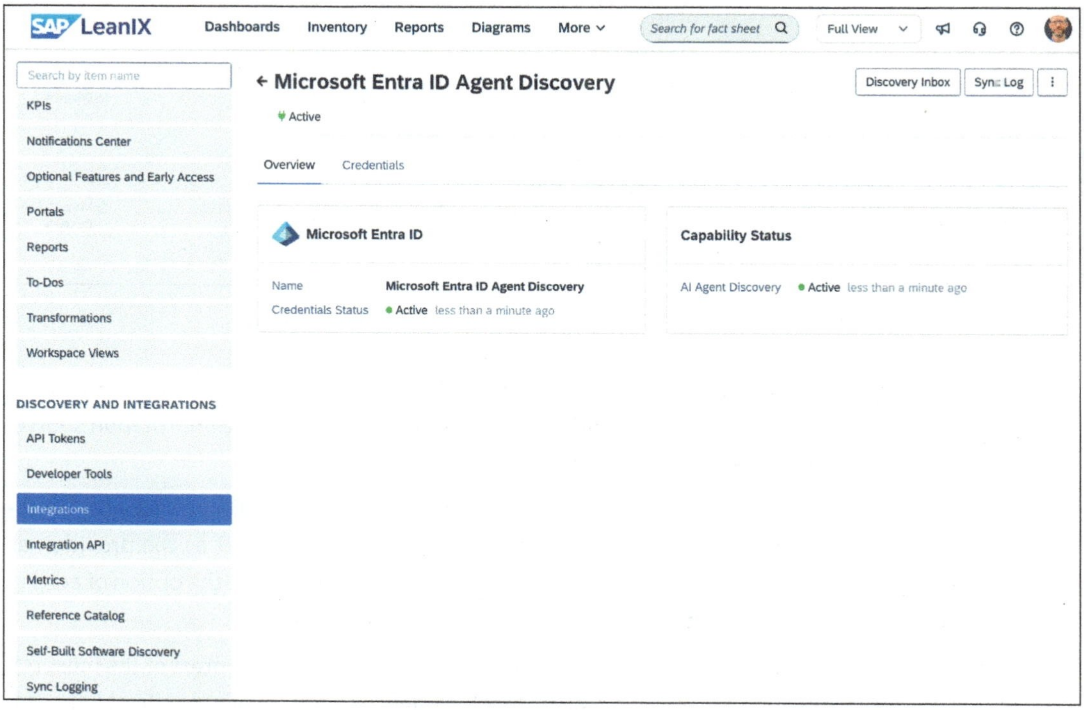

Figure 13.8: Microsoft Entra ID Agent Discovery Activation

Metadata information for AI agents will now be retrieved and added to the AI agent discovery inbox, a similar concept to the discovery inbox (refer to Chapter 4, Section 4.2.2). The discovery inbox can be accessed directly by clicking **Discovery Inbox**, as shown in Figure 13.8.

Alternatively, the AI agent discovery results are also accessible within the **Discovery Inbox** option under **Inventory Tools** in the **Inventory** tab. All discovered AI agents will be shown in the AI agent discovery inbox, and various filters can help to narrow down the results by selecting the relevant **Status, Business Capability, Integration**, and so on, as shown in Figure 13.9.

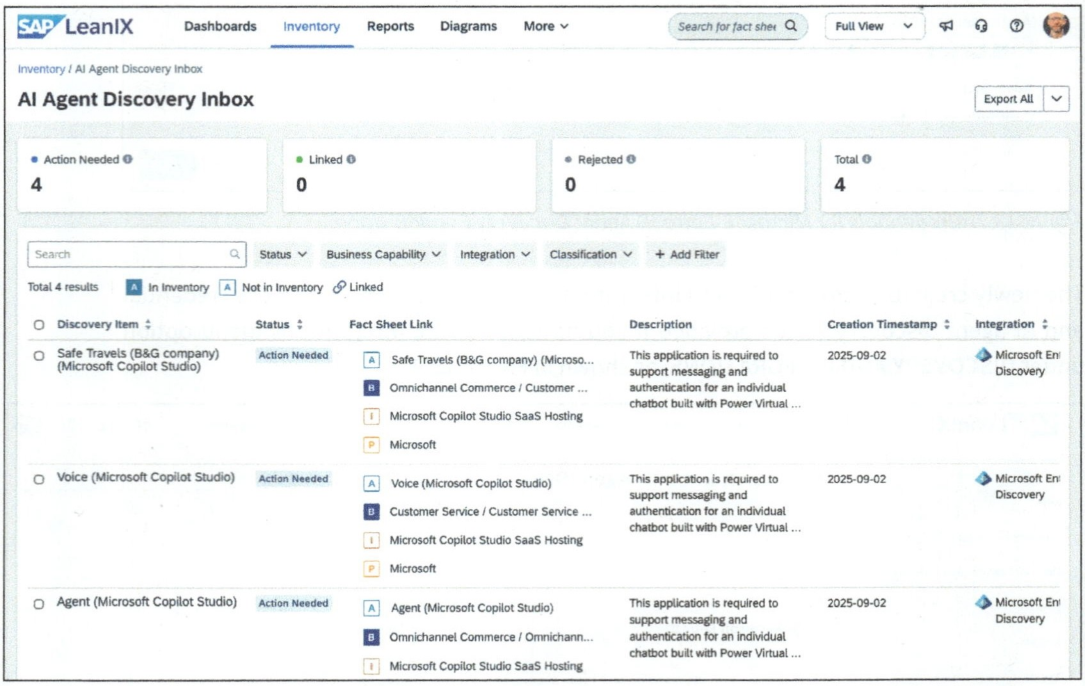

Figure 13.9: AI Agent Discovery Inbox

The AI agent subtype fact sheets have been discovered, including their relationships to IT components and providers. Depending on the metadata shared by the information source, relevant details will be leveraged.

In our example, the **Voice (Microsoft Copilot Studio)** AI agent was discovered with **Microsoft Copilot Studio SaaS Hosting** as the IT component and **Microsoft** as the provider, as shown in Figure 13.10. For each of the discovered fact sheets, you have the option of editing these details and assigning existing fact sheets from your inventory.

Based on the information shared about the AI agent, a suggestion in context of the most relevant business capability is made. SAP LeanIX leverages an internal AI capability to suggest the best match with existing business capabilities in your inventory, as shown in Figure 13.11. You can change the suggestion via the **Edit** option.

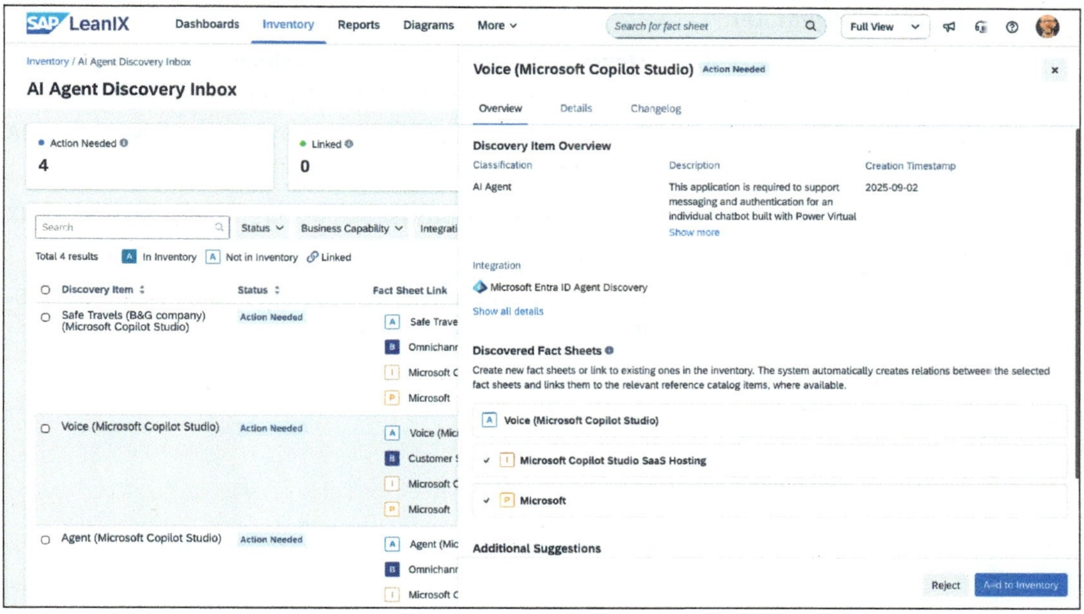

Figure 13.10: Example of Discovered Voice AI Agent

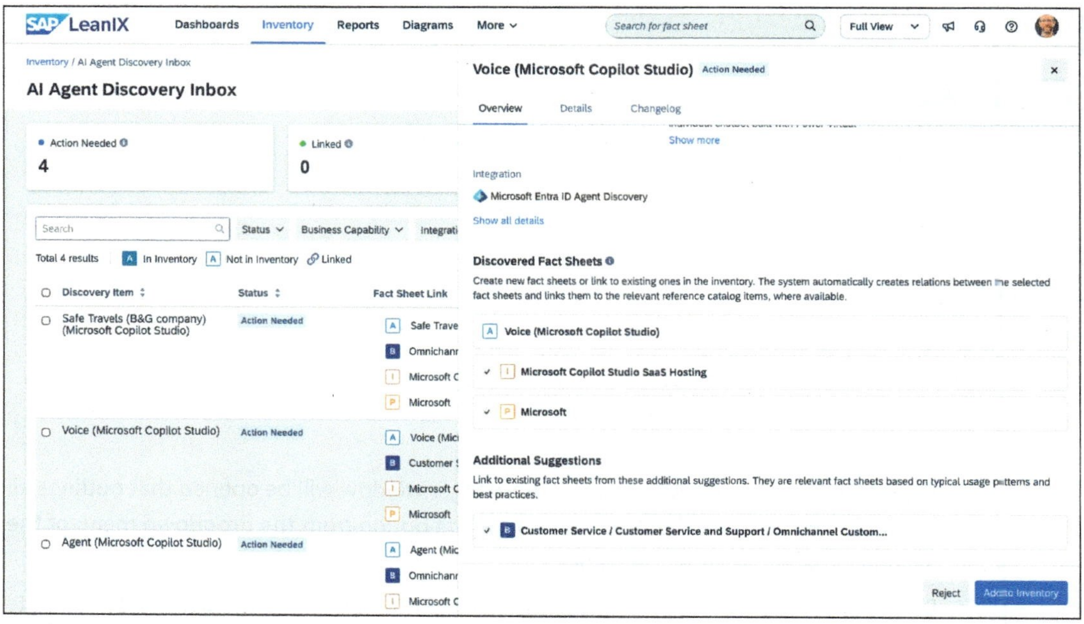

Figure 13.11: Business Capability Suggestion for AI Agent

Once you select **Add to Inventory**, new fact sheets and relations will be created. If reference catalog information exists (e.g., for IT components), the newly created fact sheets will be enriched accordingly.

13.3.3 AI Agent API

An alternative to the prebuilt integrations is the AI Agent API, as shown previously in Figure 13.5. With the activation of this API, information about the AI agents in the style of agent cards can be ingested and made available in the AI agent discovery inbox, following the A2A protocol.

You can test the AI Agent API by leveraging the **Browse API** feature within the SAP LeanIX **Developer Tools** section. Follow these steps:

1. Open the SAP LeanIX **Administration** view and select **Developer Tools** within the **DISCOVERY AND INTEGRATIONS** section, as shown in Figure 13.12.
2. Select **Browse API**.

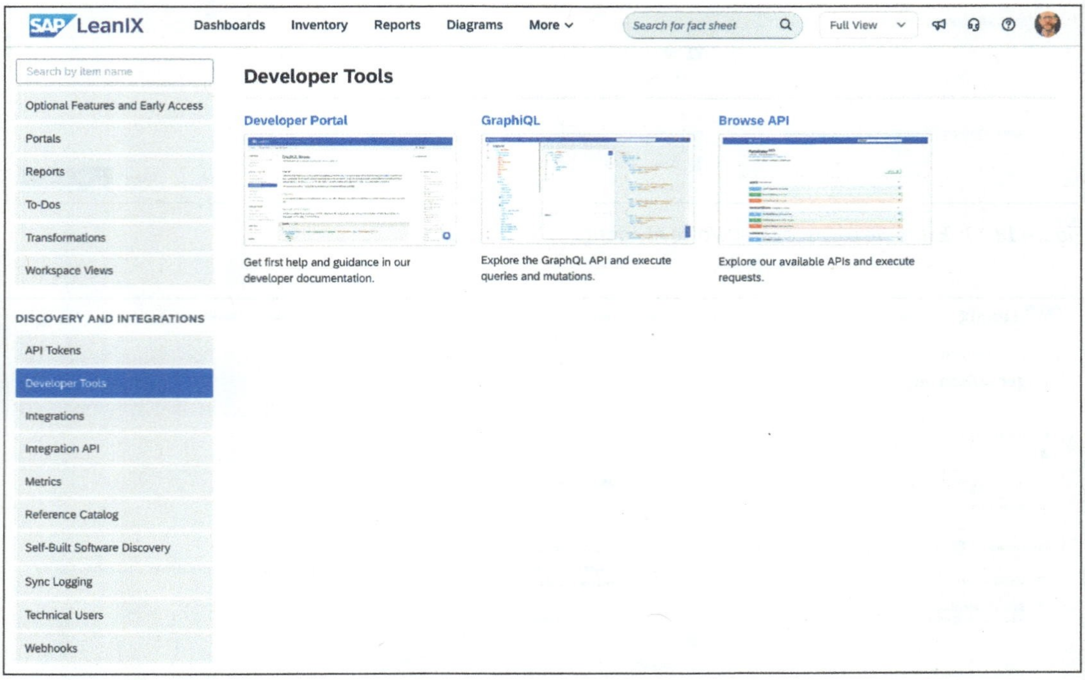

Figure 13.12: SAP LeanIX Developer Tools

3. By opening the **Browse API**, a separate Swagger window will be opened that outlines all available APIs. Choose the **Discovery AI Agents** option from the dropdown menu of the Swagger header, as shown in Figure 13.13.
4. Drop down the **POST** section of the AI Agent API to review the **Request body** example provided.
5. Choose **Try it out** to generate a new API call, as shown in Figure 13.14.

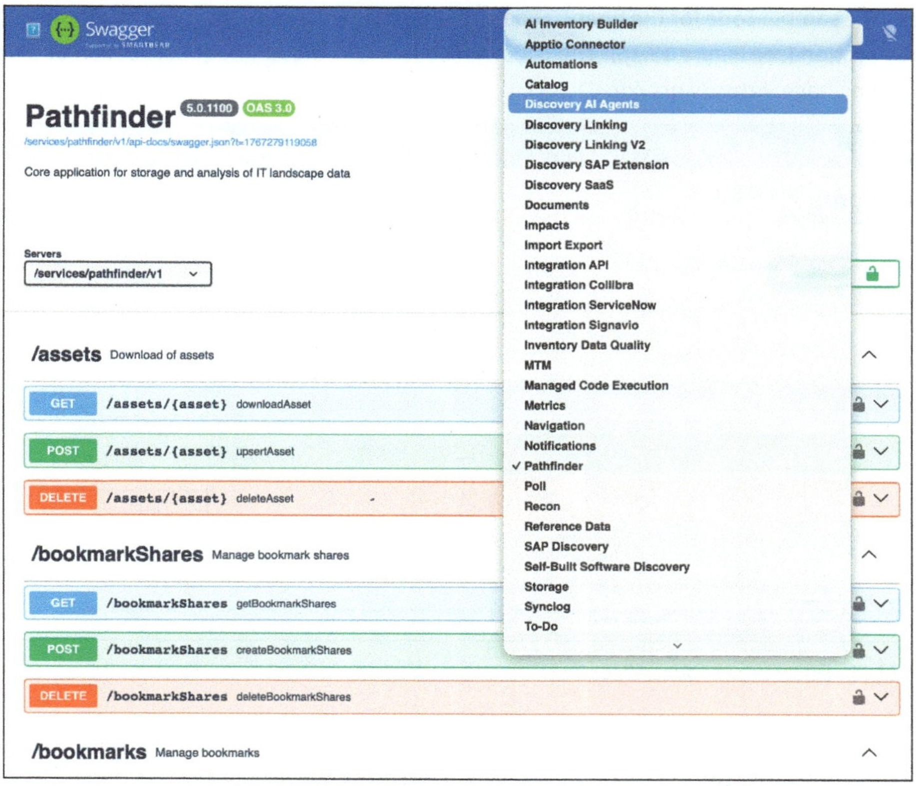

Figure 13.13: Selection of Discovery AI Agent API

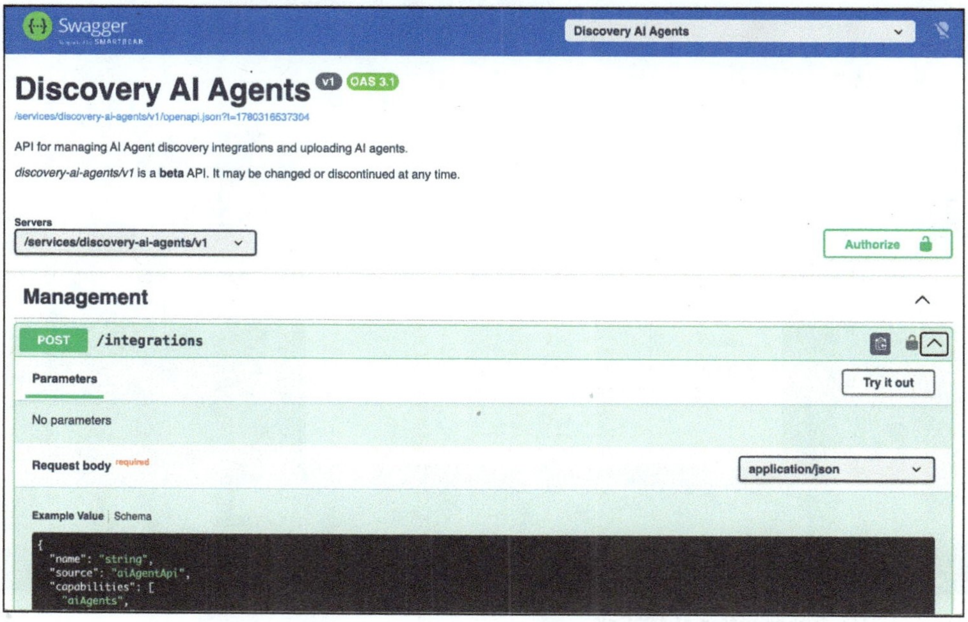

Figure 13.14: Try It Out

6. Copy the sample code for the financial data analytics agent and paste it into the **Request body** section (see Figure 13.15). The code can be found in the **API Request** section at the following page: *http://s-prs.co/v620696*.

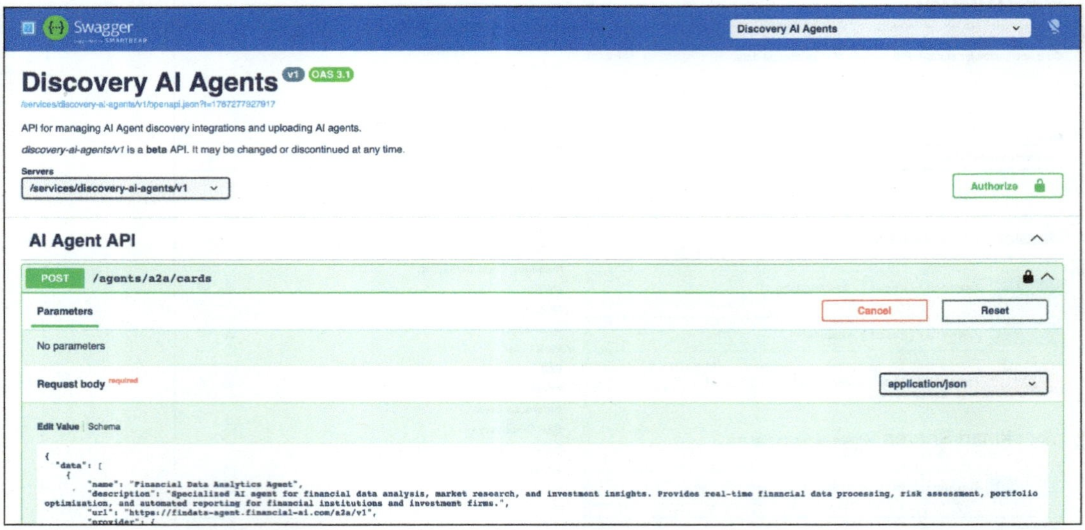

Figure 13.15: Discovery AI Agent API Call

7. To test the API call, a technical user needs to be created and the security token copied. Click the **Authorize** button to authorize the API call, as shown in Figure 13.16. Enter "api-token" in the **client_id** field and provide the copied security token in **client_secret**.

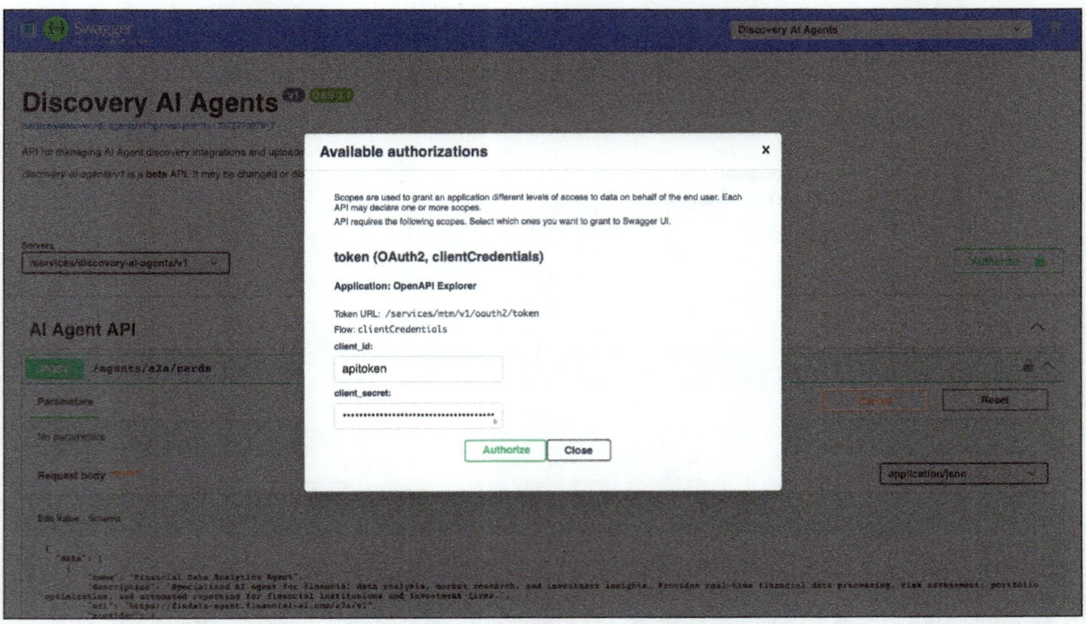

Figure 13.16: Authorization of API Call

8. Once you execute the API call by clicking **Execute**, a positive response message should be provided, and the AI agent information will be available in the AI agent discovery inbox, as shown in Figure 13.17 (in this case, for the financial data analytics agent).

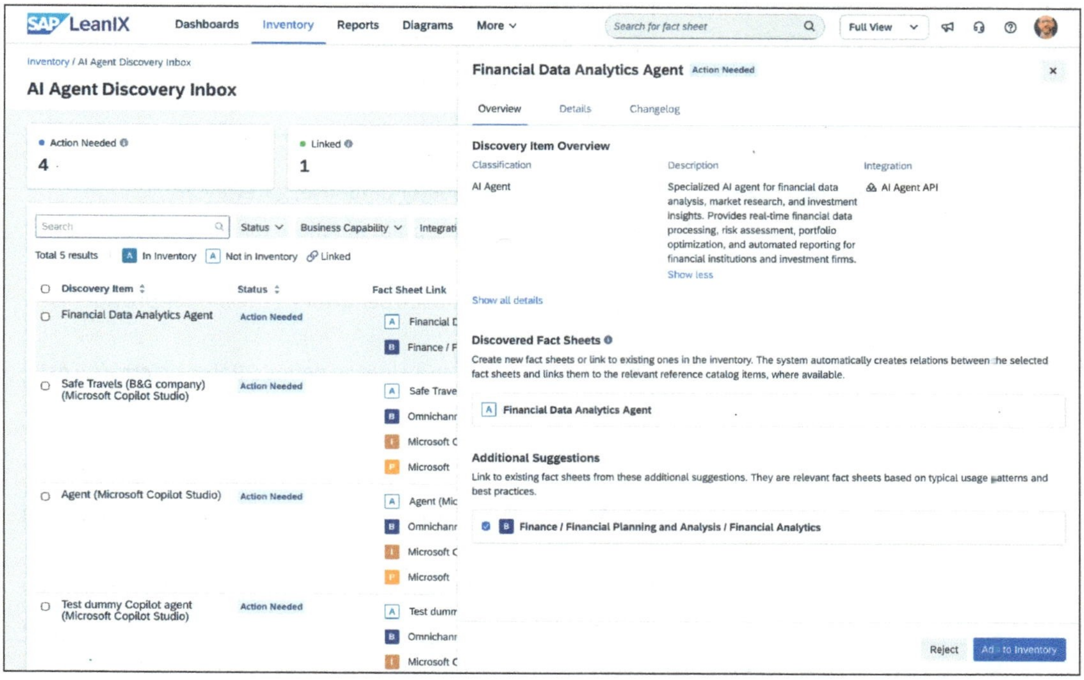

Figure 13.17: Financial Data Analytics Agent Information in AI Agent Discovery Inbox

13.3.4 Risk Assessment

With the different vendor-specific discovery options and the AI Agent API, you can gain a holistic understanding of the AI agents deployed in your landscape. This will be the basis of a more detailed architecture impact assessment that takes several risk dimensions and key questions into account, as outlined in Table 13.5.

Risk Dimension and Scope	Key Assessment Questions
Data risk assessment: sensitivity, exposure, residency, retention, and control of data used/created by the AI agent	■ What data types are accessed/generated (public, internal, confidential, personal, regulated)? ■ Does the agent send data to external providers (prompts, context, logs)? ■ Is personally identifiable information (PII) or sensitive enterprise data used in prompts or retrieved context? ■ Where is data processed/stored (region, cloud, subprocessors)?

Table 13.5: AI Agent Risk Assessment

Risk Dimension and Scope	Key Assessment Questions
Data risk assessment: sensitivity, exposure, residency, retention, and control of data used/created by the AI agent (Cont.)	■ What is the retention policy for prompts, responses, and logs? ■ Are anonymization/masking and access controls enforced? ■ Are data lineage and ownership defined for every data object used?
Technology risk assessment: robustness, dependency risk, resilience, observability, lifecycle, and operational readiness	■ Which IT components are involved (model, runtime, orchestration, vector database, gateways)? ■ Is there any single point of failure (one model, one provider, one API)? ■ What are the fallback behaviors if the model/provider is unavailable? ■ Are monitoring and alerting defined (latency, errors, cost, throttling)? ■ Are model/prompt/tool changes versioned and ready for rollback? ■ Are security controls in place (secrets, encryption, network boundaries)? ■ Are service-level agreements (SLAs) or service-level objectives (SLOs) defined for availability and response time?
Ethical risk assessment: bias, fairness, transparency, explainability, human oversight, and potential harm	■ Does the agent influence decisions about people (hiring, pricing, eligibility, performance)? ■ Could outputs lead to discrimination or exclusion? ■ Is a human in the loop required for critical outcomes? ■ Are users informed they are interacting with an AI agent? ■ Is explainability required and supported (traceability to sources)? ■ Is there a defined escalation path for harmful/incorrect outputs? ■ Are bias tests performed and repeated over time?
Regulatory risk assessment: external legal exposure (General Data Protection Regulation [GDPR], sector rules, AI regulations) and readiness for audit/controls	■ Which regulations apply given the users, regions, and data processed? ■ Is the agent classified as low, limited, or high risk under AI regulation frameworks? ■ Are consent and lawful bases documented (for personal data)? ■ Are data processing agreements or subprocessor terms in place?

Table 13.5: AI Agent Risk Assessment (Cont.)

Risk Dimension and Scope	Key Assessment Questions
Regulatory risk assessment: external legal exposure (General Data Protection Regulation [GDPR], sector rules, AI regulations) and readiness for audit/controls (Cont.)	▪ Can you produce audit evidence (logs, model provenance, decision trace)? ▪ Are cross-border transfers compliant with residency requirements?
Compliance risk assessment: adherence to internal enterprise architecture standards, security policies, architecture principles, and governance gates	▪ Does the agent comply with internal enterprise architecture principles (approved platforms, integrations, reuse)? ▪ Has a security/privacy review been completed and recorded? ▪ Are ownership roles assigned (business, technical, risk)? ▪ Are operational controls defined (incident management, change management, periodic review)? ▪ Is there a lifecycle plan (decommissioning, model retirement, data cleanup)? ▪ Are governance states defined (draft → reviewed → approved → monitored)?

Table 13.5: AI Agent Risk Assessment (Cont.)

To make this AI agent risk assessment tangible and applicable in your project, we have created an architecture decision record template in the demo workspace and created a master decision record that includes the key questions listed in Table 13.5 as well as an executive summary and the AI agent context.

Note

For more details on the management of architecture decisions, refer to Chapter 12.

You can find the link to the **AI Agent Risk Assessment MASTER** form, as shown in Figure 13.18, within the **Architecture Decisions** section under **More**.

Within the architecture decision record, you can link the affected fact sheet and open it by clicking the fact sheet name. With the insights gained through the risk assessment, you can now initiate the verification of the AI agent. By clicking the verification badge, you can mark the AI agent as either verified or rejected. If you select the **Mark as Verified** option, a popup will appear that provides an overview of the key details, as shown in Figure 13.19.

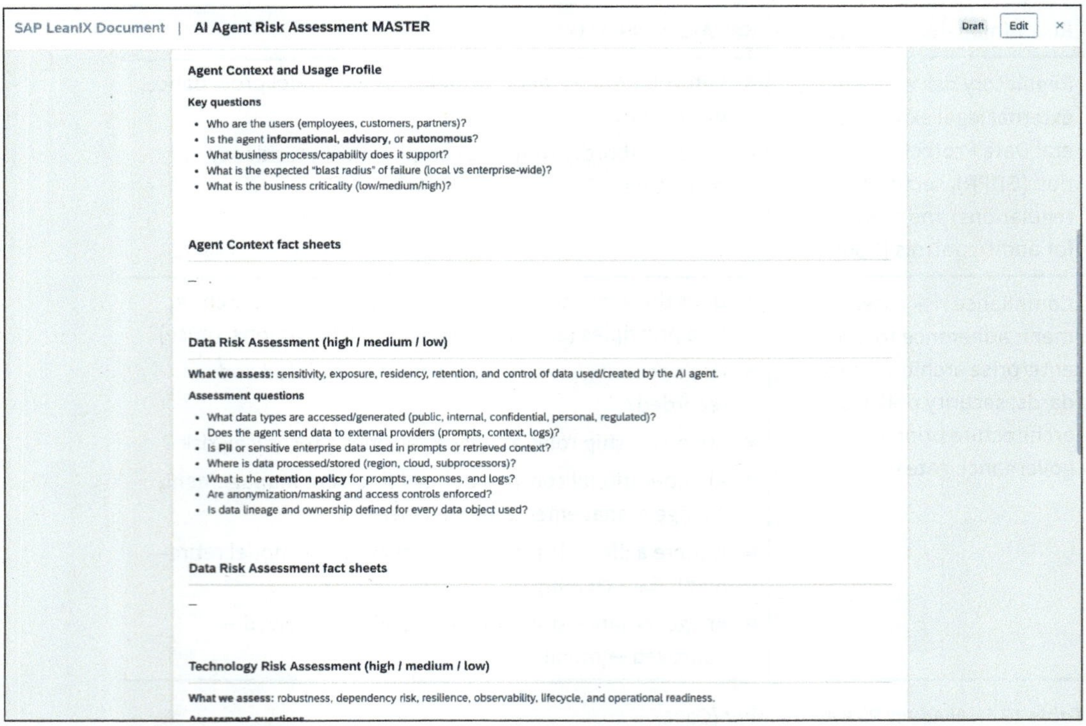

Figure 13.18: AI Agent Architecture Decision Record Master with Risk Assessment Questions

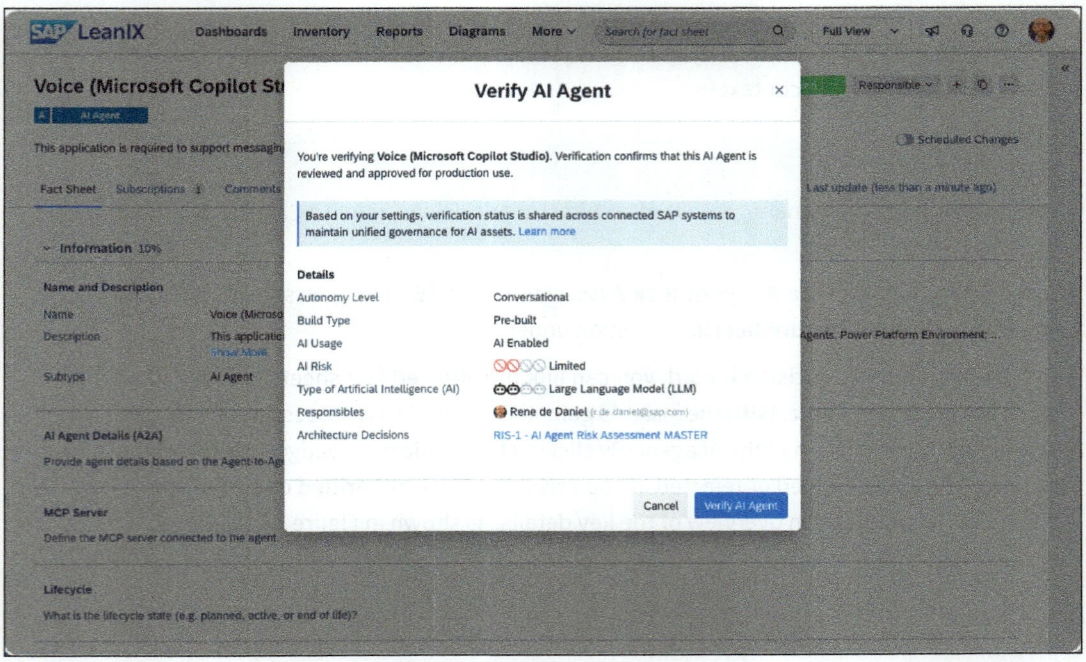

Figure 13.19: Verification of AI Agent

If you confirm by clicking **Verify AI Agent**, the verification badge of the AI agent will be set to **Agent Verified**.

> **Note**
> The status of an AI agent verification badge goes beyond SAP LeanIX and will be propagated to solutions in the ecosystem for governance controls. At the time of writing, additional features were not yet available.

We will further explore the reporting and communication capabilities for the risk assessment results in Section 13.6.

13.4 AI Target Planning

With the holistic understanding of the baseline AI agent landscape and its associated risks, you can introduce a level of control that provides space for the intentional design of a future AI landscape. Enterprise architects will play an important role in linking the business strategy with technological capabilities and guiding the organization through a value-driven innovation conversation at scale.

For an assessment of the AI innovation potential alongside a given business strategy, we will leverage the business capability map in combination with the new AI potential attribute, as introduced earlier in Table 13.4. Figure 13.20 shows the *business capability map—AI potential* report for EA7.

> **Note**
> Before creating the reports that we'll discuss in this section, you need a dedicated collection structure. For the EA7 case study, we'll create an AI governance collection group in which you can find each report. For instructions on creating collections, see Chapter 5, Section 5.1.2.

To create the report, apply the following steps:

1. Create a new landscape report with **Business Capability** selected as the base fact sheet type.
2. Select **AI Potential** for the **View** option.

We recommend conducting the assessment of the AI potential via an interactive approach with the business stakeholders by updating the AI potential rating in real time. With the insights gained, you can take the next step in understanding which AI capabilities the supporting applications are already offering or which whitespaces need to be closed by implementing a custom AI capability, as shown in Figure 13.20.

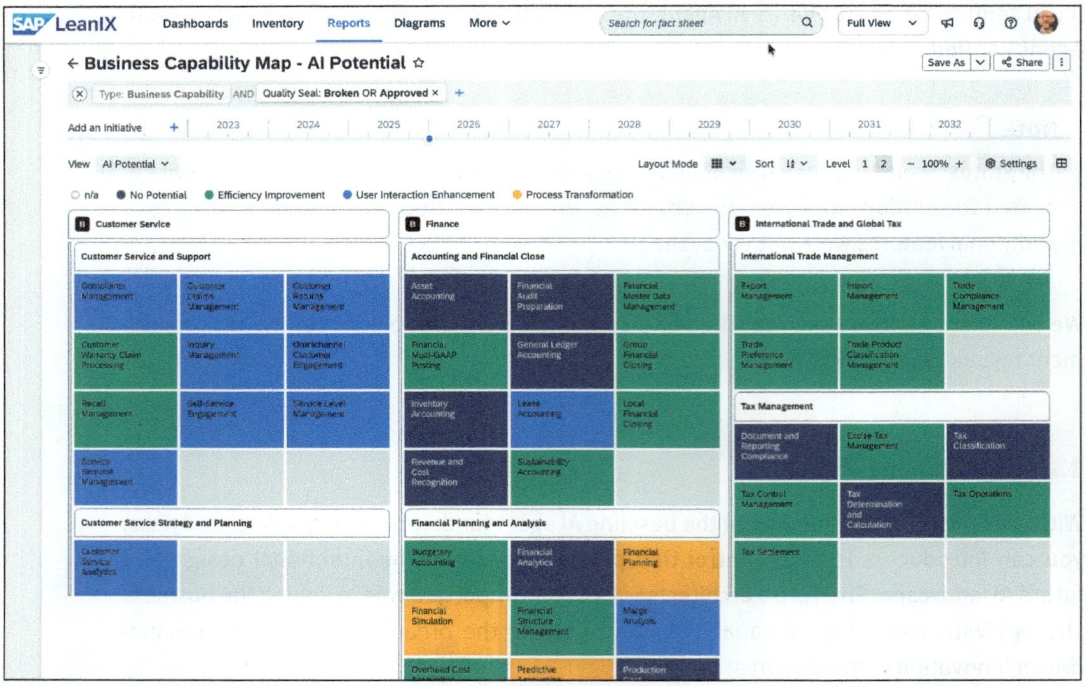

Figure 13.20: Business Capability Map with AI Potential

For the former case, you can use an application landscape report that highlights the previously assessed AI potential in combination with SAP-suggested AI agents, as shown in Figure 13.21. This report combines existing fact sheets with reference fact sheets highlighted by the light bulb symbol and the asterisk sign (*) at the end of the reference fact sheet name.

These fact sheets are not active in your inventory but are visible as suggestions. The different types of reference fact sheets can be visualized with the help of the **SAP Suggestions** dropdown (which shows **AI Agents** and **Applications** options).

To create the application landscape with AI potential and SAP-suggested AI agents report, apply the following steps:

1. Create a new landscape report with **Application** as the base fact sheet type.
2. Set **View** to **Business Capabilities: AI Potential.**
3. Set a filter for **Subtype: AI Agent**.
4. Within **Settings**, navigate to **Suggestions** and check the **Enable SAP suggestions** option.

The combined view of actual and reference fact sheets (marked by the light bulb and asterisk) eases the identification of AI capabilities in a given context. Next, you'll leverage the reference fact sheet to turn it into an actual fact sheet as part of your target architecture.

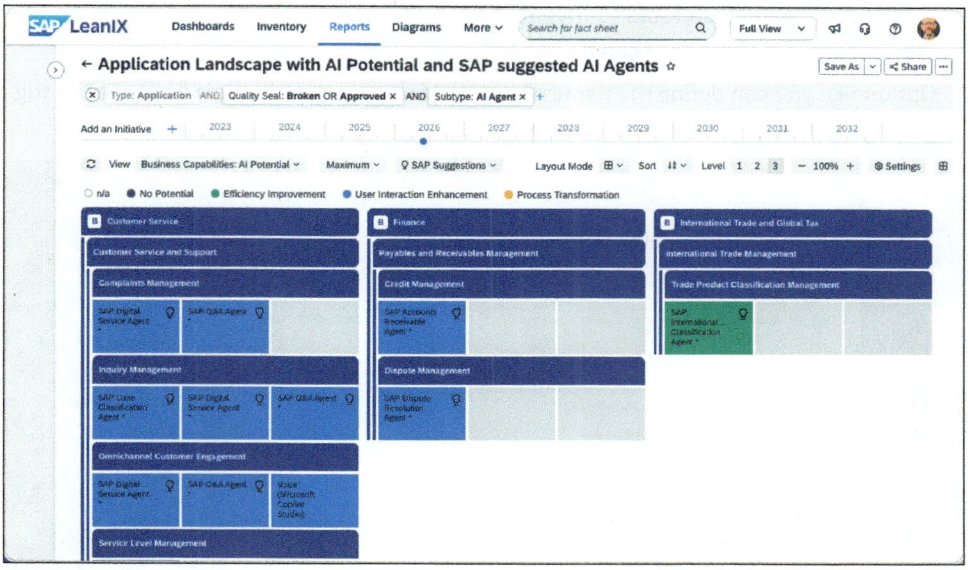

Figure 13.21: Application Landscape with AI Potential and SAP Suggestions

As customer interaction is a primary business objective of EA7, it was decided to introduce SAP's Digital Service Agent, which is able to interact with customers through natural language while leveraging SAP's business knowledge. The following steps describe how an AI agent for the reference content can be added to the inventory:

1. Select the reference fact sheet. More detailed information is provided in the right panel, as shown in Figure 13.22.

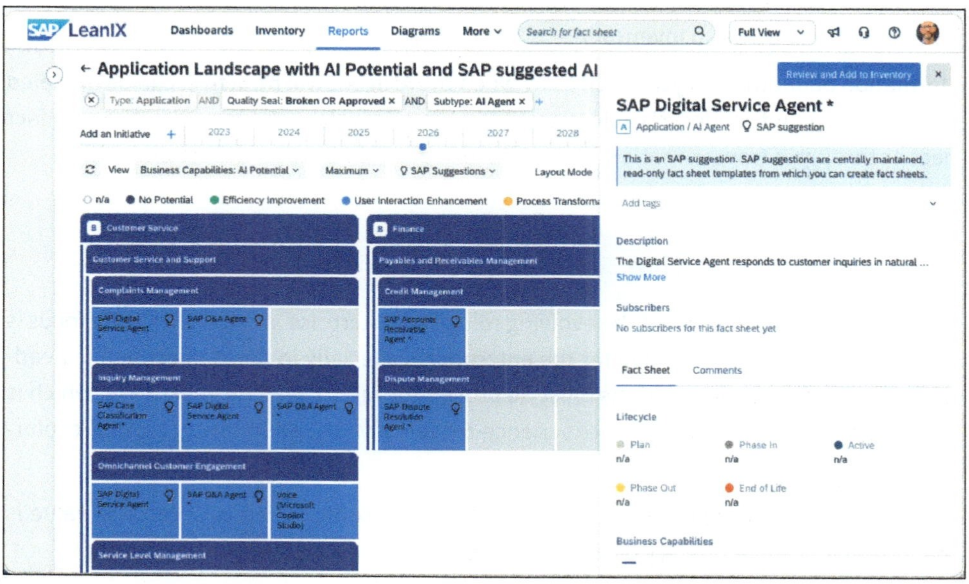

Figure 13.22: SAP Digital Service Agent Reference Fact Sheet Details

2. By selecting **Review and Add to Inventory**, relationships that will be created alongside the AI agent fact sheet itself can be defined.
3. Optionally, you can define the name of the new fact sheet to be created in case the suggested name does not resonate well, as indicated in Figure 13.23.

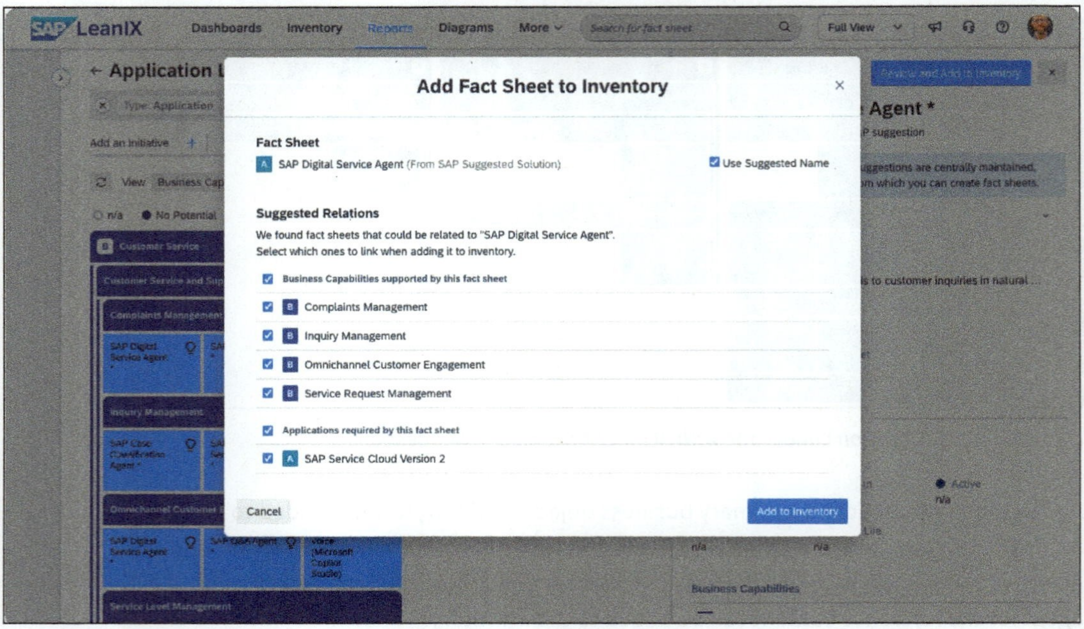

Figure 13.23: New AI Agent Fact Sheet with Relations

The new fact sheet and the relations to business capabilities and applications will be created by confirming **Add to Inventory**.

Information about the lifecycle, such as by when the AI agent will be active, can be modeled either by defining the lifecycle information via the **Lifecycle** settings of the AI agent fact sheet or through a transformation, as explained in detail in Chapter 6, Section 6.3.2.

13.5 Value Perspective

As outlined when we discussed the evolving role of an enterprise architect, a strong focus is on demonstrating tangible value for the enterprise. Especially in times of economic headwinds, the funding for innovations must be proven through a valid business case, which is often difficult to build, as proven experience is typically limited in emerging and exploratory AI domains.

To assess and justify the investment in AI agents, their contribution to enterprise value is categorized into three primary business value types:

- **Cost savings**
  Reducing operational effort and expenses through automation, efficiency gains, and the elimination of manual or redundant activities
- **Revenue increase**
  Enabling new revenue streams or accelerating existing ones by improving customer experience, shortening cycle times, or enhancing decision-making
- **Risk reduction**
  Mitigating operational, compliance, security, or reputational risks by increasing consistency, transparency, monitoring, and control

To estimate scale effects, validate assumptions, and continuously reassess the realized value of AI agents over time, the type of business value, the expected value per unit, and the usage need to be defined.

The *usage indicator* describes to what extent an AI agent is consumed in an enterprise. Depending on the nature of the AI agent, a usage indicator could be, for example, the number of active users, session runs, executed tasks, processed transactions, or exchanged messages.

By multiplying the usage by the expected value per unit (depending on the business value type, this could be expressed through the revenue increase per unit, cost savings per unit, or cost reduction per unit), the overarching value of an AI agent can be calculated.

Figure 13.24 shows an example for the EA7 use case. Here you can see the **Type of Business Value** (**Revenue Increase**), the **Usage** (**500**), and the expected **Revenue Increase per Unit** (**€ 200.00**), which are factored into the calculation for the **Total Expected Revenue Increase** field (**€ 100,000.00**), found in the **Governance** section of the AI agent fact sheet.

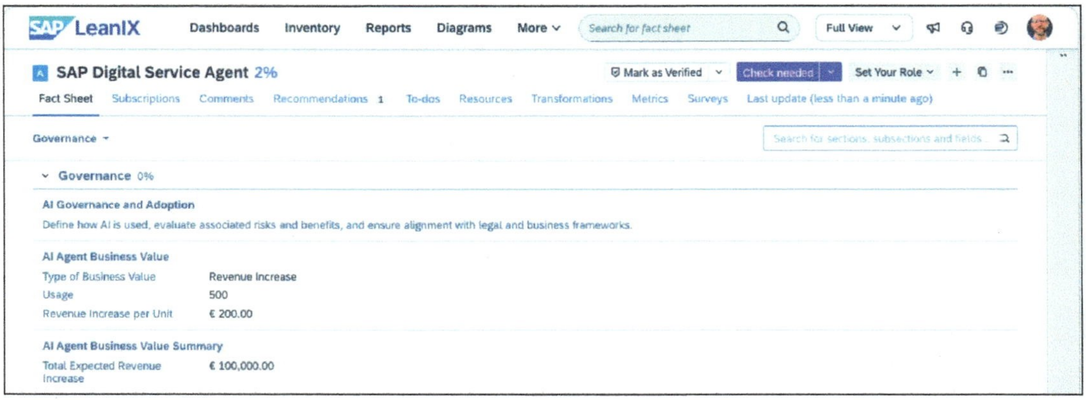

Figure 13.24: AI Agent Business Value and Usage Indicator

The **AI Agent Business Value Summary** information can be used to build reports and heatmaps to track and course-correct the AI value realization over time, which we will explain in more detail in Section 13.7.

Similar to the risk assessment and verification of existing AI agents in our landscape, the same procedure applies for the planned AI agent, as explained in Section 13.3.4.

13.6 Communication to Stakeholders

In the previous section on the risk assessment of the baseline AI landscape, we demonstrated how architecture decisions are made by following a risk-based approach. From a general AI governance perspective, architecture decisions will be supported by enterprise architecture principles relevant in the context of AI, which act as guardrails for the assessment and rationalization of decisions.

Once an architecture decision is made, communication with stakeholders is of primary importance. This is especially true in the context of AI capabilities, as other parts of the organization may want to leverage outcomes from, for example, pilot projects or proofs of concept (PoCs) by adopting proven AI capabilities.

For an overview of the status of AI agents in your landscape, SAP LeanIX offers the predefined AI agent governance report, as shown in Figure 13.25. This radar report provides an overview by categorizing AI agents by their nature (**Prebuilt**, **Prebuilt Expanded**, or **Custom**) and **AI Agent Status**.

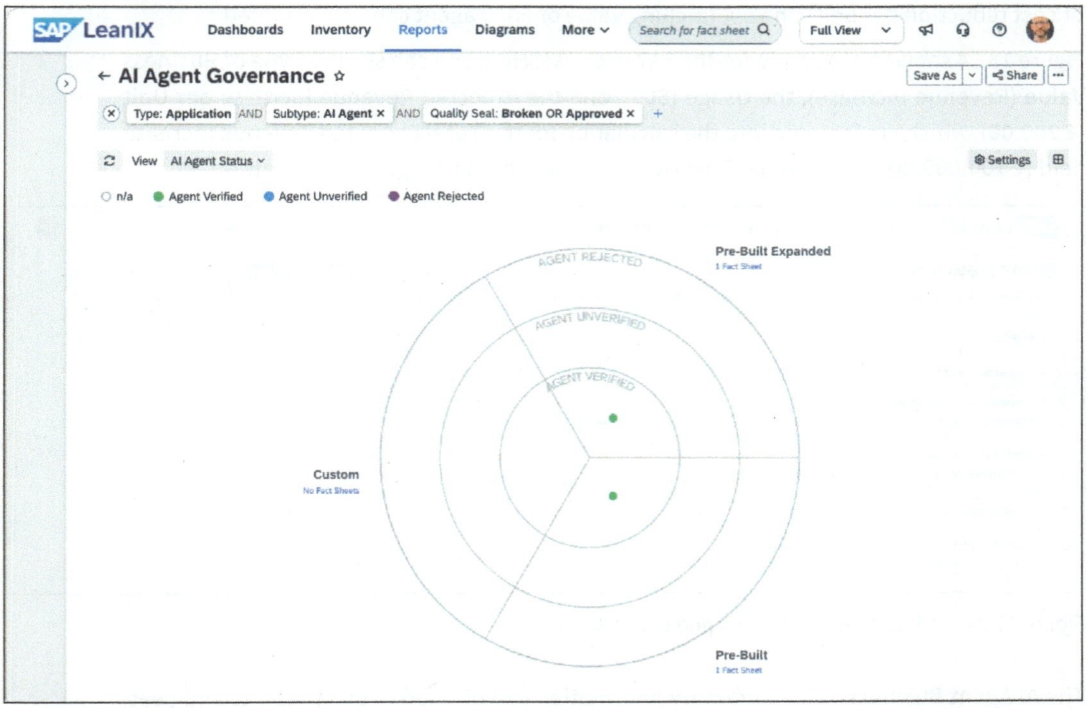

Figure 13.25: AI Agent Governance Radar

With the help of the radar report, stakeholders can immediately see which AI agents are accepted or rejected in a given company context.

For broader communication, we recommend that you set up an AI agent portal that can be shared with a broader community. To configure an AI agent portal, proceed with the following steps:

1. Log into SAP LeanIX as an administrator and open the **Administration** menu.
2. Open **Portals** within **ADVANCED SETTINGS** and select **New Portal** to arrive at the form shown in Figure 13.26.

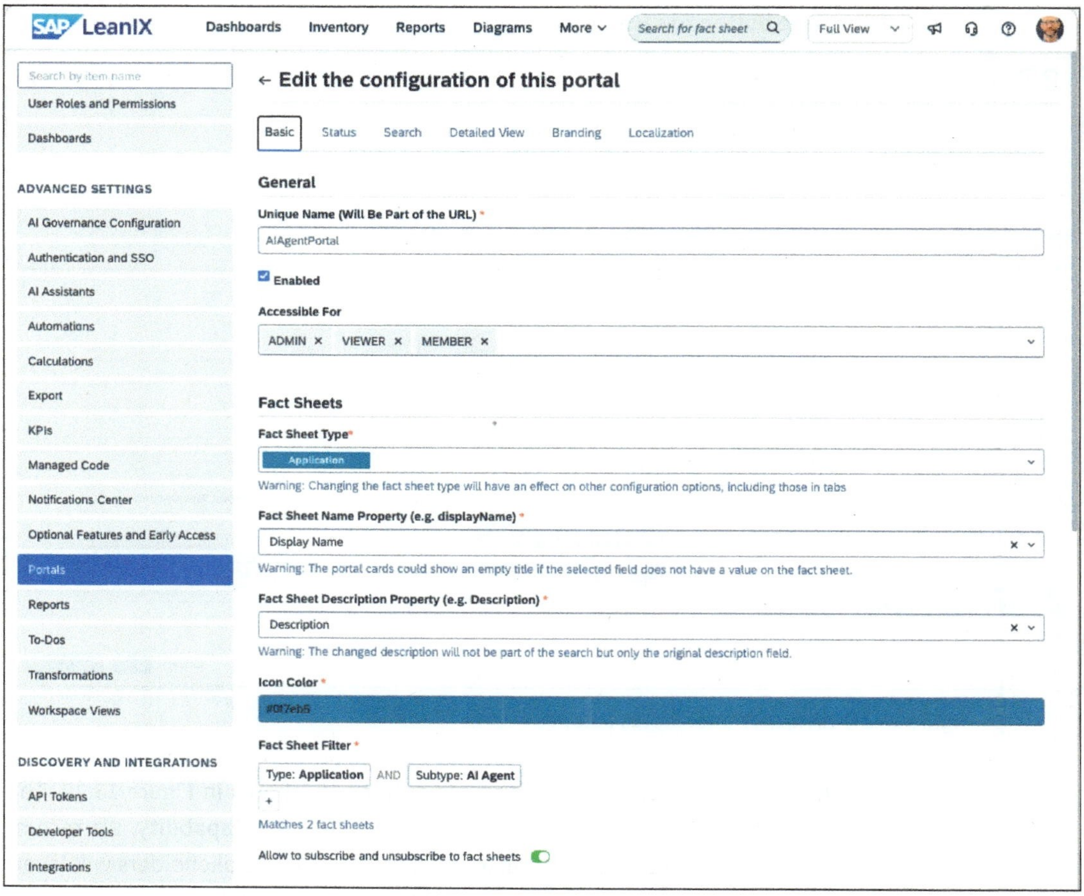

Figure 13.26: AI Agent Portal Configuration

3. Define a unique name (e.g., "AIAgentPortal").
4. Within the **Accessible For** field, add the **VIEWER** and **MEMBER** roles to allow everyone in the organization to see the portal content.

5. Change the **Fact Sheet Filter** settings by removing the **Lifecycle** filter and adding the **AI Agent** and **Approval State: Active** subtypes.
6. Change the **Portal Header** field to "AI Agent Portal".
7. Change the **Global Action Button Label** field to "Request New AI Agent".
8. Click the **Save** button to save the settings.
9. Move to the **Status** tab, as shown in Figure 13.27, and change the **Fact sheet field** value to **AI Agent Status**.
10. Feel free to add icons and icon colors for the different values of the AI agent status.
11. Click the **Save** button to save the settings.

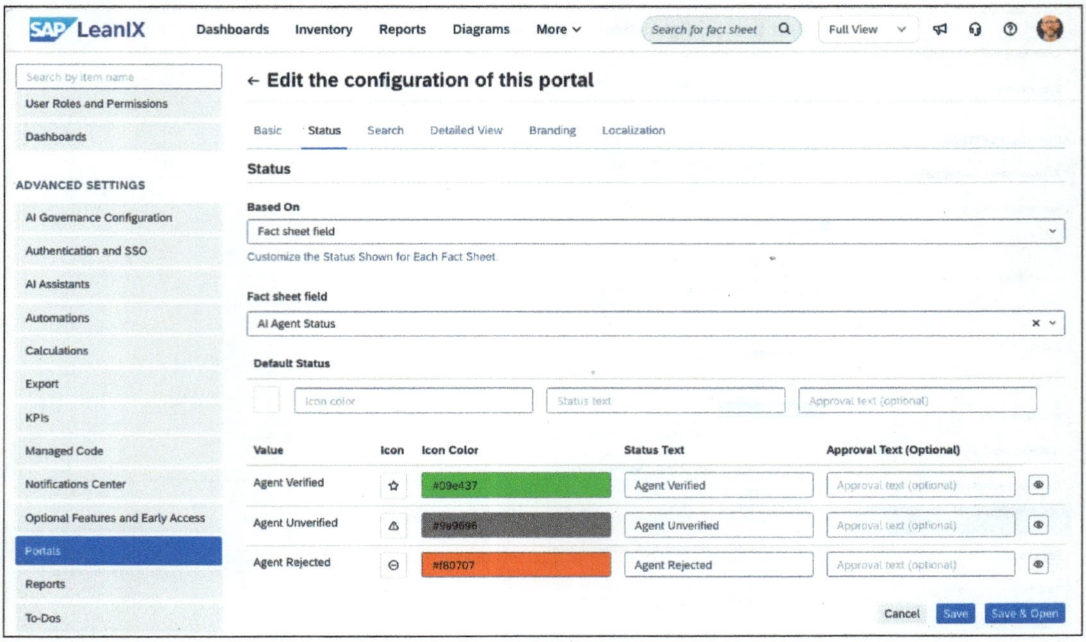

Figure 13.27: AI Agent Portal Status Settings

12. Move to the **Search** tab and review the filter options, as shown in Figure 13.28. The default setting for **Related Fact Sheet Type** is set to **Business Capability**. We recommend using the default setting, keeping in mind that business stakeholders will leverage the information provided in your newly created AI agent portal.

The newly created AI agent portal can now be accessed via the **More** menu option. Within the **Portals** section, **AI Agent Portal** will be listed next to **Application Portal**. Note that you may need to refresh your browser to update the menu options. An initial AI agent portal might appear as shown in Figure 13.29.

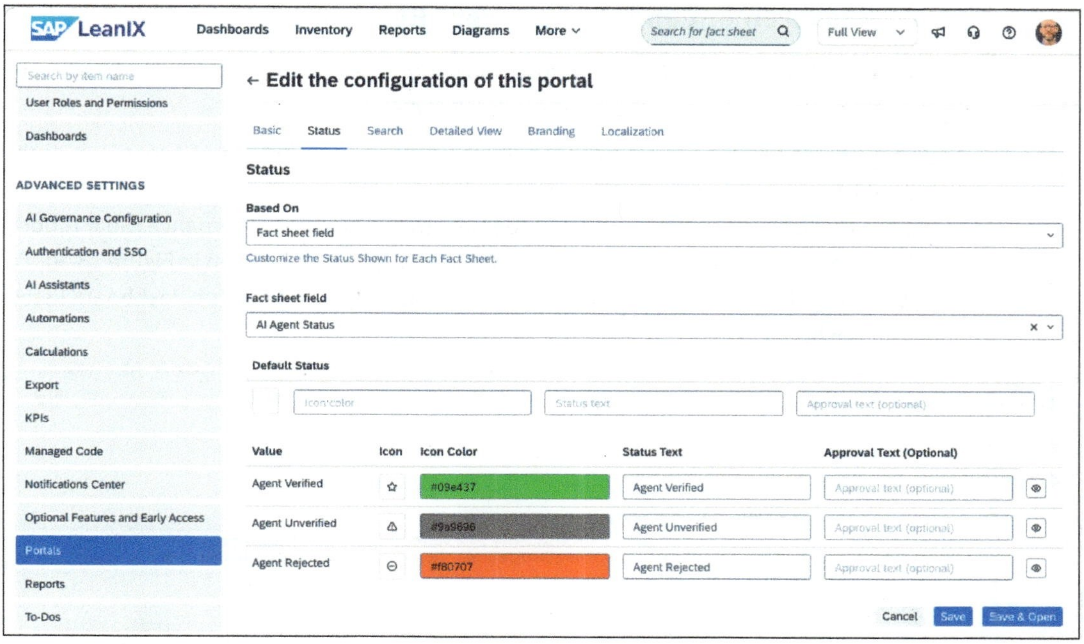

Figure 13.28: AI Agent Portal Filter Options

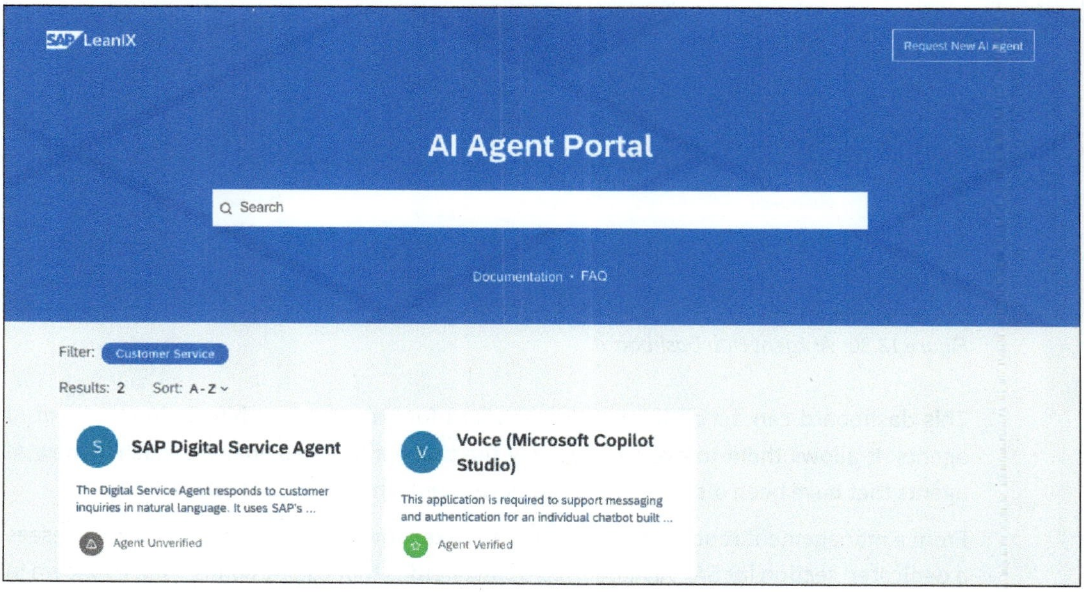

Figure 13.29: AI Agent Portal

With the help of the free text search and options to set filters on business capabilities, approved AI agents will be shown to the stakeholders.

13.7 Measurement and Management Reporting

To round out this chapter, let's take a closer look at the measurement- and KPI-based management reporting of AI capabilities in your landscape.

As part of the AI governance extension, SAP LeanIX provides a default AI Agent Hub dashboard, which includes executive KPIs, insights into the AI agent discovery inbox, inventory searches, a list of AI agents, the previously demonstrated AI agent governance radar report, and the AI agent landscape report to visualize the AI potential, as shown in Figure 13.30. To access it, navigate to the **Dashboards** section and select the **AI Agent Hub** dashboard listed under **All Dashboards**.

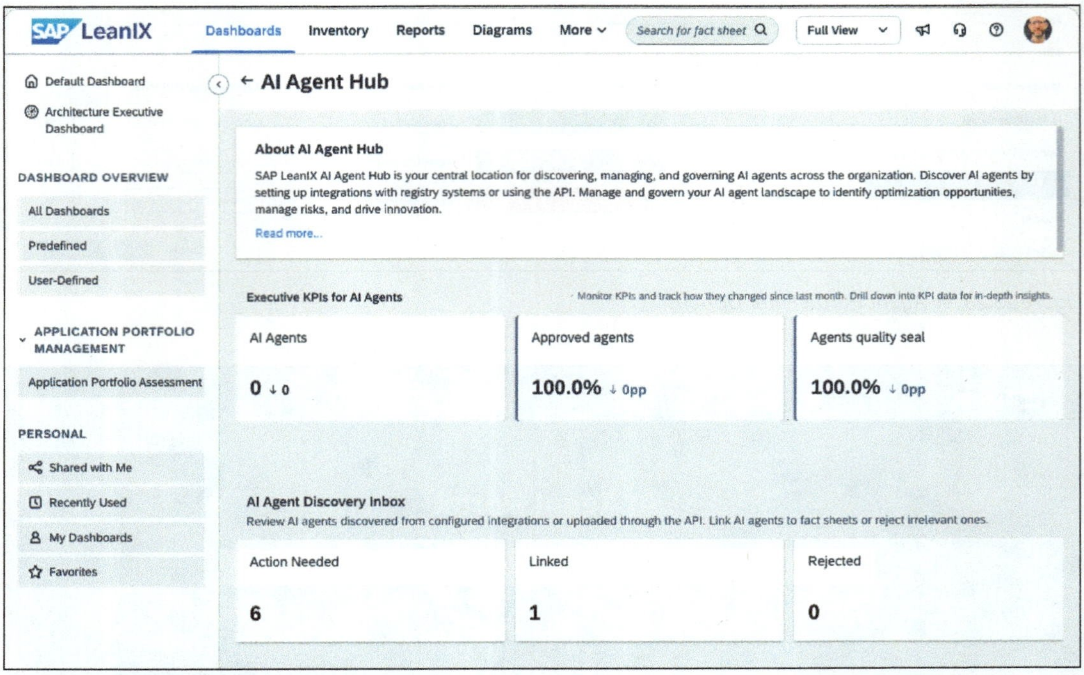

Figure 13.30: AI Agent Hub Dashboard

This dashboard can act as the go-to dashboard for enterprise architects in charge of AI agents. It allows them to oversee not only the process but also changes, such as new AI agents that have been discovered and made available in the AI agent inbox.

From a management reporting perspective, the executive architecture dashboard foresees a dedicated section for SAP AI Agent Hub–relevant KPIs that can be configured, as shown in Figure 13.31. To access this option, navigate to **Administration** and select **KPIs** within the **ADVANCED SETTINGS** section. For ease of use, select **AI Agent Hub** under **Use Case / Groups.**

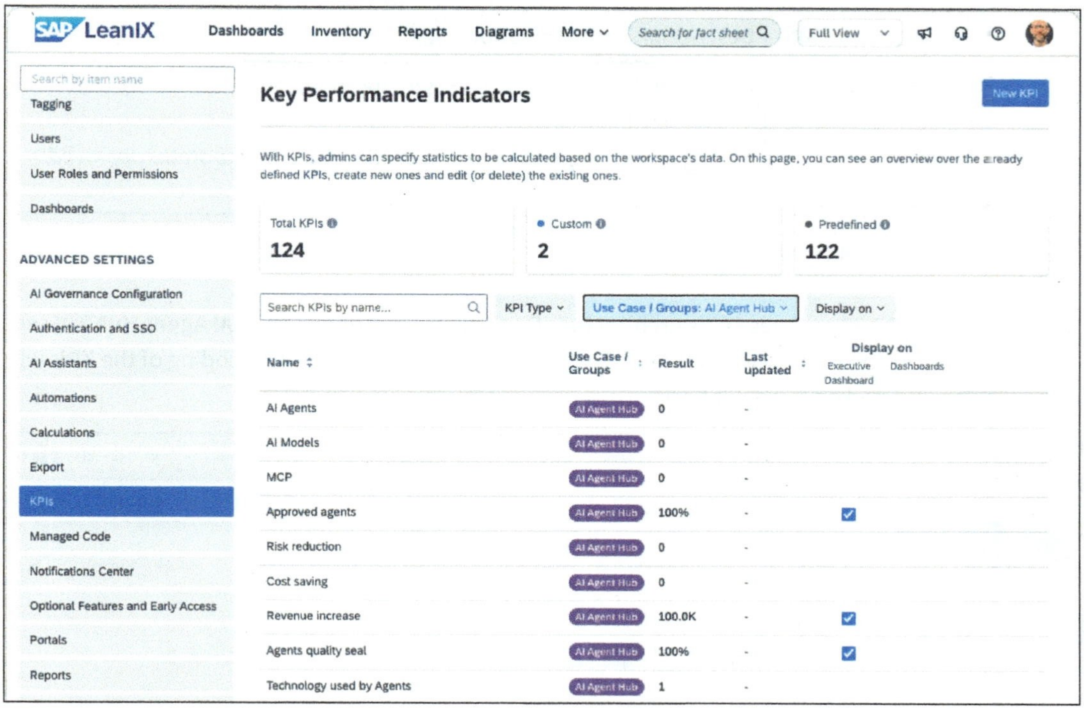

Figure 13.31: AI Agent Hub KPIs

Table 13.6 describes the individual AI Agent Hub KPIs and their calculation logic. You can activate up to four KPIs at the same time.

AI Agent Hub KPI	Description and Calculation Logic
Agents portfolio	■ The total number of AI agents that have been formally approved ■ *Σ(Agents, except rejected)*
Approved agents	■ Percentage of AI agents that have been formally approved relative to the total number of AI agents ■ *Σ(Agents approved) ÷ Σ(Agents) × 100*
Risk reduction	■ Total estimated reduction of risk achieved by AI agents ■ *Σ(Total estimated risk reduction of approved agents)*
Cost saving	■ Total estimated cost savings generated by AI agents ■ *Σ(Total estimated cost savings of approved agents)*
Revenue increase	■ Total estimated revenue increase attributed to AI agents ■ *Σ(Total estimated revenue increase of approved agents)*

Table 13.6: AI Agent Hub KPIs

AI Agent Hub KPI	Description and Calculation Logic
Agent quality seal	■ The total number of AI agents in the approved status and with an approved quality seal ■ *∑(Agents with approved quality seal) ÷ ∑(Agents with status approved) × 100*

Table 13.6: AI Agent Hub KPIs (Cont.)

Figure 13.32 outlines EA7's **Architecture Executive Dashboard** and its **AI Agent Hub** KPI section. As the dashboard includes a trend analysis over time, the color-coding of the KPIs will be updated on a regular basis.

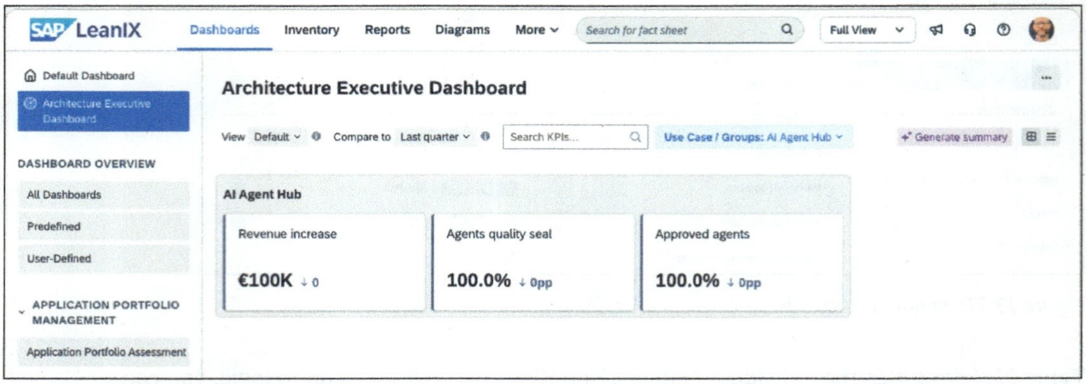

Figure 13.32: Architecture Executive Dashboard with AI Agent Hub KPIs

Each KPI can be further drilled down into to gain more granular information. For EA7, we have introduced views for business capabilities, organization, and TIME classification in Chapter 7, which can be leveraged for the AI Agent Hub. Taking the revenue increase KPI as an example, you can assess the KPI in more detail by understanding from a business capability perspective how AI agents support the revenue increase, as shown in Figure 13.33. For example, here you can see that with the introduction of AI agents, a revenue increase of 100k was achieved in comparison to last year across the customer service business capabilities.

With the help of the architecture executive dashboard and the outlined AI Agent Hub KPIs, senior stakeholders get an up-to-date status for the evolution of AI capabilities across the organization. With the drilldown into individual KPIs, insights can be extracted that increase adoption over time.

Figure 13.33: Drilldown into AI Agent Hub Revenue Increase KPI

13.8 Summary

Throughout this chapter, you've seen how the capabilities of enterprise architects will evolve in the era of AI. Through automated discovery, you can get control of AI capabilities deployed in your organization and assess and manage the risk associated with them. Alignment with the business strategy and the value assessment of potential AI capabilities are of crucial importance to invest in the "right" options. Finally, continuous measurement and governance are key to enable innovation at scale.

Outlook

At the time of writing, the roadmap for SAP AI Agent Hub was shared at the annual SAP Sapphire conference, outlining the broader scope of SAP AI Agent Hub capabilities and integration points with the SAP ecosystem. Table 13.7 provides a high-level overview of the planned capabilities.

Capability	Ecosystem Integration
Agent evaluation and assessment	Joule Studio
Identity and access control	SAP Cloud Identity Services

Table 13.7: Future AI Agent Hub Capabilities

Capability	Ecosystem Integration
Agent observability through telemetry	SAP Cloud ALM
Agent mining	SAP Signavio
Agent organization and skills mapping	SAP SuccessFactors
MCP runtime enforcement	SAP Integration Suite

Table 13.7: Future AI Agent Hub Capabilities (Cont.)

The Authors

René de Daniel has more than two decades of experience in enterprise architecture, having worked for several years at the intersection of strategy, transformation, and technology. His career spans both customer-side leadership roles, where he managed architecture teams and drove enterprise-wide change, and consulting roles, where he contributed to some of the most complex transformation programs across industries.

As a principal enterprise architect at SAP, René plays a pivotal role in enabling product success across the SAP LeanIX and integrated toolchain landscape. He is widely recognized for his pragmatic, hands-on insights into SAP LeanIX, shared through social media content and community enablement. His work bridges frameworks with real-world applications—always with a clear focus on delivering business value.

Certified as a Solution-Focused Coach (Erickson International) and Life Trust Coach, René integrates coaching principles into his approach to enterprise architecture, supporting individuals and teams to grow in clarity, confidence, and strategic alignment.

Marianne Dahmen is a principal internal consultant for strategic projects and initiatives and part of the SAP LeanIX Product & Engineering department. She works closely with the SAP LeanIX product management team and selected strategic customers to test and validate new ideas for how SAP LeanIX can deliver value to enterprise architects and other roles in a corporate environment.

Prior to her role in Product & Engineering, Marianne worked as a senior professional services consultant and supported the onboarding and implementation of SAP LeanIX at various Fortune 500 and DAX 40 companies. In her work on projects, she benefits greatly from her experience in management consulting and financial auditing. Customers value her hands-on mentality paired with a deep passion for the product, and her perseverance to think through use cases from high-level strategic perspectives down to very technical details and their pragmatic implementation in SAP LeanIX.

Index

B

C

D

E

F

G

I

J

K

L

M

N

O

P

Q

R

S

V

W

A Hands-On Guide to Designing Embedded Systems

For a listing of recent titles in the
Artech House Integrated Microsystems Library,
turn to the back of this book.

A Hands-On Guide to Designing Embedded Systems

Adam Taylor
Dan Binnun
Saket Srivastava

ARTECH HOUSE
BOSTON | LONDON
artechhouse.com

Library of Congress Cataloging-in-Publication Data
A catalog record for this book is available from the U.S. Library of Congress.

British Library Cataloguing in Publication Data
A catalog record for this book is available from the British Library.

ISBN-13: 978-1-63081-683-4

Cover design by Charlene Stevens

685 Canton Street
Norwood, MA 02062

10 9 8 7 6 5 4 3 2

Contents

Acknowledgments

I have been truly lucky in my career to be able to work with some amazing engineers and companies on a range of exciting projects, many of which are literally out of this world. In recent years, I have been extremely fortunate to be able to run my own engineering consultancy focusing on embedded system development. The opportunity to run my own business came from a blog that I wrote called the MicroZed Chronicles (www.microzedchronicles) over 8 years. Now I write weekly blogs on embedded system development.

However, I have often wanted to go a little further than allowed in the blog or series of blogs and really walk new and younger engineers through the life cycle of an engineering project from concept to delivery. Hopefully, this book will act as a guide and aide to those who are starting out in what I consider to be one of the most rewarding careers there is. Unlike many engineering texts, my coauthors and I have set out to tell the story of an embedded development project, in parallel with the writing of the book, that we have designed, manufactured, and commissioned for the board.

All design information will be made available, along with additional technical content, on the book's website, looking further into deep technical areas than several books possibly could.

The creation of a book is a challenge and there are several people to thank, most especially my coauthors Dan and Saket. It has been truly a pleasure to work on the creation of this book with such talented engineers.

I would like to thank Steve Leibson, Mike Santarini, and Max Maxfield, who first published my articles; Rebecca To and the wider Xilinx community, who have been so supportive over the years as I work with their products to create blogs, projects, and real applications for clients; and the Avnet team of Kristine Hatfield, Bryan Fletcher, Kevin Keryk, Fred Kellerman, Dan Rozwood, and the wider Avnet/Hackster and Element 14 communities.

No man is an island, and I must thank my wife and son Jo and Dan for understanding the long hours that I spend both working for clients and creating content.

—Adam Taylor

It's with genuine humbleness that I write these acknowledgments, as I never thought I would be writing a book on how to do this kind of work. This is a professional accomplishment that I am very proud of and I would not be writing these words without the help of many people.

First and foremost, I want to thank my family for pushing me to go to college and pursue engineering. As a young man I didn't really know what I wanted to do—and with some pushing and prodding from my family, I ended up in a career that I truly find engaging and interesting. For that alone I consider myself very fortunate.

I also want to thank my wife, Katie. She gave me the courage to jump headfirst (after years of merely sticking a toe in the water) into running my own consulting company, E3 Designers. Without her push, I would not have taken the plunge mostly out of fear of failure. She believed that I could do it and be successful before I did.

In the context of this book, I have to start by thanking my coauthor Adam Taylor. Adam approached me about joining him as a coauthor after we had worked together on a project for a customer that he introduced me to. Without the invitation, I would certainly not have gone down the path of writing a book on engineering, trying to share the lessons I have been fortunate enough to be given by my more seasoned peers over my career.

Along the way I had a lot of help from wonderful people and companies in creating the design to support our story, I'd like to thank them all as well.

My colleagues Brandon Smith and Dave Chisholm, whom I consider among the best I've worked with in my career, helped an incredible amount with the PCB design, reviews, and layout. I truly couldn't have done it without their help and skillset. I have been fortunate enough to work with them professionally and consider them both friends as well. Their technical expertise and high standards are always valued, and I lean on them frequently in my consulting ventures as well.

The companies that helped me throughout this process also deserve a heap of thanks—Advanced Assembly, Altium, Analog Devices, Ansys, FLIR Systems, Mentor Graphics, Sparx Systems, and Trilogic all personally worked with my in some form or fashion in this process. I am grateful for all they did because none of it was owed to me, but was given out of their good will.

Lastly, I do want to thank my first mentor in this industry, Patrick Joyce. Pat was a senior engineer in the engineering team at my first job in Keene, NH at Markem (which would become Markem-Imaje shortly after I joined). Not only was he an extremely talented engineer and designer, but he dedicated time to teaching all of us in the group as well. It was because Pat was generous with his knowledge that I was able to grow professionally. My hope is that over the course of my career I can do that for others and pay that generosity forward.

—Dan Binnun

As an academic, it was always my dream to write a book that made a real difference in the life of everyday engineers. Writing a technical paper is bread and butter for academics, but the audience for technical papers is often limited to a particular research area and hardly reaches the general public. So when my dear friend Adam Taylor approached me to coauthor a book on this topic, my first question to him was what was going to be the flavor of this book, as it was always my desire to write something that can help the readers not only solve real-life challenges but also is easy enough to understand and follow. When we had our preliminary discussions, it was exciting to see that all of us were aiming for the same goals, which really motivated me to join the team. That was the easy part; writing a book from scratch and, most importantly, making sure that it is technically sound but still easy enough to read and follow are the most challenging part. During the journey of writing this book, we all learned a lot from each other. Having never worked in the industry, personally it was enlightening for me to see the point of view of my coauthors, both of whom have extensive knowledge of what embedded design engineers in the industry are really like.

I would like to thank my wife, parents, and other family members, who have been encouraging me throughout my life to achieve higher academic goals and to write a book for as long as I could remember. I especially thank my wife Nupur, who helped me keep my focus even during the trying times of the Covid-19 pandemic, when major portions of this book were written and did not let me lose track when I was getting lost without new ideas.

I again thank my wonderful coauthors for the patience that they showed throughout this journey, even when I was missing deadlines, and to Artech House, for keeping faith in this project despite the delays. Without the understanding and commitment of the entire team, this book project may not have been successful.

—Saket Srivastava

Introduction

Designing products is hard. A product is born out of a series of smaller developments done with a common goal in mind, and each of those smaller developments on their own is incredibly challenging.

Whether you are designing packaging for a product that needs to ship around the world in a cargo plane or shipping container or designing the mechanical enclosure of the product that goes inside that packaging, you are going to be faced with challenges from the very first day of the project.

When we decided to write this book on designing embedded systems, we did so because we felt that there was a need to address the challenges involved in designing products with a more holistic view than is often presented in technical books and publications.

There is certainly no shortage of wonderful references on the nitty-gritty technical information and nuance involved in electronics, firmware, software, systems, and reliability engineering; we reference many of them in this book. What we felt was missing was a framework for people to use that tries to explain how a product begins with an idea; that idea then becomes a set of requirements, and those requirements evolve into a focused product. Contrasted against an evaluation kit that can be overwhelmed with peripherals and features, a true embedded system is often a bare-bones collection of hardware required to perform a very specific set of tasks. Adding in superfluous interfaces, functionality, or features can cause further complexity to be introduced. This almost always results in additional technical risks and cost risks and, perhaps most significantly, can extend development timelines to the point of being untenable for a company to deal with.

The goal of this book is to illustrate the development cycle more holistically to provide better context as to why certain features are present on certain systems. Understanding this process holistically is critical to being a productive embedded systems engineer and technical contributor.

That is not to say that the why is more or less important than the how, merely that we wanted to bring the why into the picture as well.

In trying to come up with a concept for how to show this process in a realistic light, we decided that following the journey of a team of imaginary engineers at an imaginary company (SensorsThink) as they developed a new platform would be a fresh take on what can be a stale topic.

In addition to following along with this team of imaginary engineers, we also decided to practice what we preached and actually design the heart of the system to be available as reference material.

We started with a set of basic requirements, created a system architecture, and on down the line to designing a custom printed circuit board (PCB) and writing firmware, field programmable gate array (FPGA) code, and application code as well.

Our hope is that, in sharing those design artifacts and finished product in addition to this book, a more complete picture will be available and it will be easier to see how we got from the starting line to the finish line.

Lastly, we intend to make a wealth of other information available for reference as part of a larger ecosystem of information and content: things like simulations, test plans and results, design calculations and notes (as we took them in the course of the design work), and the schematics, layout, and source code.

We sincerely hope that you all find this book and ecosystem around it useful, no matter your experience level or where you are in your career.

CHAPTER 1

Programmatic and System-Level Considerations

Before we can design, we must have a clear and frozen technical baseline, which allows for progressive assurance and risk retirement.

1.1 Introduction: SensorsThink

Another day begins bright and early at SensorsThink, a product design company that specializes in sensing technology solutions for all types and sizes of markets. A typical day for an engineering team member here is similar to a typical day for most engineering team members: chock-full of engineering change orders (ECOs), project status meetings, and other interesting, but mostly turn-the-crank engineering work. Today will be different and more exciting. There is a buzz in the office, a new product platform has been talked about for months, and the day may finally have arrived when management and marketing are ready to unveil the idea that will become the next great SensorsThink product. After months of research and speaking with potential customers, they have concluded that one product platform can be developed to suit the needs of several potentially large and growing markets:

- Smart buildings/smart infrastructure;
- Factory automation;
- Environmental monitoring.

In situations like this, there will often be a project kick-off meeting to generate some momentum for the work. This may take the form of a company-wide meeting or an email to the project team or engineers may be pulled aside by the project team leader and given an overview. So if you were to find yourself as part of a new product development team, perhaps pulled into a meeting with representatives from management and marketing, what would come next? How would you approach the problem of creating a product from what may be as vague as a bullet-point list of must-have features?

This is a situation in which almost every design engineer will be put at some point in their careers, and it can seem daunting, overwhelming, and intimidating. By the time you have read through this book, the task at hand will seem more manageable and you will understand how to take that list of features and turn them into aspects of a well-engineered and well-designed product.

1.2 Product Development Stages

In order to set the stage for further understanding of what we consider best practices for doing product development work, it is vitally important that you understand the stages (or phases) of product design and development. It is equally important to note that there is a clear distinction between methods used in product design and development (Agile, for example, is an extremely popular method in the software engineering world) and the generalized stages that a development life cycle has. The focus here is not on the methods applied, but rather the stages that should be worked through.

For the purpose of this book, a classic stage-gate style approach will be discussed. However, before exploring that approach, it should be noted that there are many other approaches that are used in industry. These approaches are regularly tailored on a project-by-project basis for any number of reasons: project complexity, project timeline, product volumes, and cost target, for example. Any of these factors can change the approach or tailor the approach that a company uses.

Another key concept to understand is that the length of the entire process, or the length of a stage within the overall process, can vary wildly from project to project. Often, this is directly proportional to product complexity and anticipated product volumes.

A critical takeaway to have is that while the process may be documented in a certain way by your company, it may not always be followed to the letter of the law. Adaptability is a critical characteristic to have in the modern development world with aggressive timelines and aggressive requirements often seeming to be in direct opposition with each other. Remember the spirit of the process, but try not to be paralyzed from making progress by feeling restricted by it.

A classical stage-gate type product development typically consists of several stages of works, followed by checkpoints in the process referred to as gates. It is at these checkpoints where stakeholders in a project determine whether the project and/or product are still viable both from a business and technical perspective. One of several items that is evaluated at these gates is risk. Risk can be from any number of areas: financial risk, market risk, technical risk, and staffing risk, to name a few. In general, the best-performing projects tend to reduce risk early in this stage-gate process so that as you near the product launch date, the remaining risk is minimal, and no major risk items are left for the final hours. This book is not intended to serve as a complete reference for a stage-gate process; there are many excellent resources available both online and in reference texts for learning more about these types of processes and how they can be tailored to a specific company, product, or project.

For the purpose of this development journey, we will assume that SensorsThink has chosen to follow a 5-stage development path, shown as a simple flow diagram in Figure 1.1 [1].

A brief overview of each stage in the process is given here to provide further understanding and context.

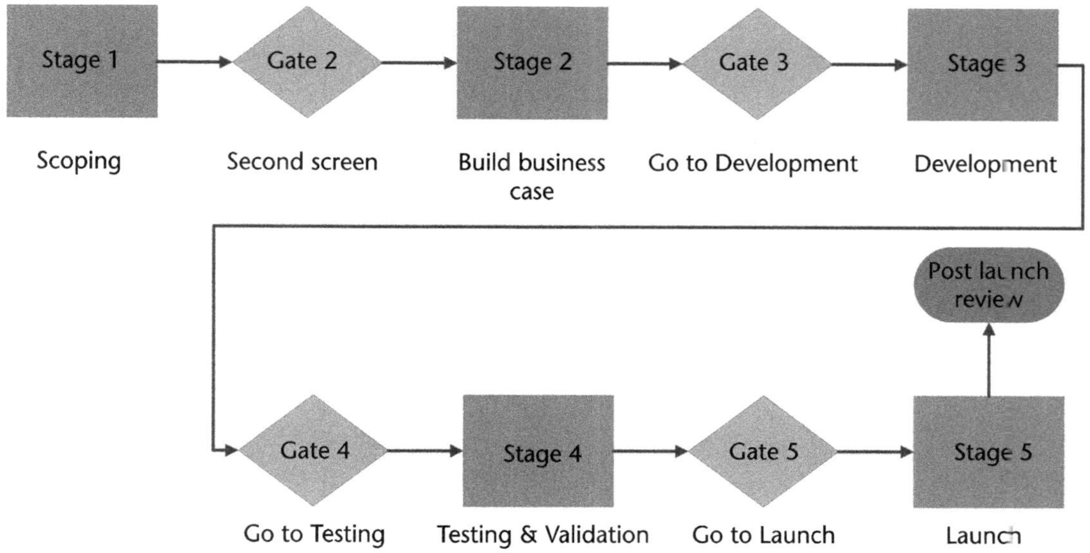

Figure 1.1 A 5-stage development process.

- *Stage 1:* During this stage of the process, the main goals are essentially to evaluate the product and project from all angles. What size is the market? What is your competition doing? What competition already exists? What will differentiate your product from those of your competitors? Does your company possess the skills, experience, and know-how to deliver?
- *Stage 2:* Building a business case is exactly as it sounds; this is where the company must determine the following:
  - What is the product? What precisely is going to be offered and what technical capabilities will it have? What level of reliability is required by the consumers of the product?
  - What legal and regulatory requirements will be imposed upon the new product prior to it entering the marketplace? In safety-critical applications, the level of rigor and documentation can be quite daunting and is very resource-intensive. Knowing this well ahead of time can go a long way towards having a viable case for developing and launching that new product.
  - What risks are present in the product? This includes business risks, financial risks, and technical risks.
  - What is the plan for the project? This can vary in level of detail greatly, but most often will include key milestones, required staffing, and required financial resources (including any capital expenditures for new equipment or for expensive time at qualification labs at the end of the project).
  - Is this feasible? After spending a great deal of time and energy in analyzing the project and product, the company then decides if it is worth doing. If they feel it is worth doing, stage 3 will occur; this is typically when most design engineers will be involved in the process.

- *Stage 3:* This is the stage in the product development process with which most engineers are familiar. This is the execution of the (hopefully) detailed and rigorous analysis performed in the prior stages of work. This can include deployment of prototypes to customers during the process to get valuable customer feedback on the product as it is being developed. Ultimately, the product development team is tasked with delivering a prototype product ready for the next stage of the process, testing and validation.
- *Stage 4:* Testing and validation take on many forms, depending on in which department you work.
    - As a design engineer doing hardware, you will likely be rigorously testing your hardware against requirements and specifications to ensure performance.
    - As a software developer, you will be checking your code for coverage, perhaps compliance with coding standards, and regression testing releases.
    - In a regulatory and compliance role, you will be testing the product against worldwide standards for things such as electromagnetic interference/electromagnetic compatibility (EMI/EMC) and general product safety. Depending on what type of product you have designed and the markets in which you will be selling your product, there are different harmonized standards and directives against which your product will be evaluated.
    - There is also field testing, perhaps redeploying updated or further refined products to your early adopters from stage 3.
    - Lastly, there is market testing, where a business may tease the product and gauge interest prior to moving to the final stage of development, product launch.
- *Stage 5:* This stage is the culmination of the prior 4 stages of work (there are actually a few additional stages sometimes) to which the company has devoted countless resources. This is a very exciting time for all involved in the development process, from marketing, engineering, sales, and field support and service. This is also a very intense time for all involved in the process, with all involved hoping that the product will be successful.

In addition to understanding these 5 stages, it is equally important to understand what an effective gate can do for a development process. A gate must have "teeth," and it must be empowered to determine that a product is no longer feasible and "kill" the development life cycle. It can be for any number of reasons: too much residual risk, key staff departures, and changes in market conditions. This can be a disheartening thing for team members, as months (sometimes years) of hard work and effort are placed on a shelf and sometimes never continued. Healthy companies kill projects; it is part of the natural course of running a business and developing products. As a design engineer (or in whatever role you may have), it is crucial to remember that, if done for the right reasons, this is absolutely a healthy and viable choice for a business to make.

1.3 Product Development Stages: Tailoring

Before we proceed further into our journey, let's touch on some of the other aspects of the development process. As mentioned previously, this is not an in-depth guide to the product development process, just an overview. However, some additional discussion on the topic is useful to provide a broader understanding for anyone not familiar with a stage-gate process. The other aspects that are perhaps most important to touch on are:

- Ways that this process can be tailored based on project complexity;
- Ways that ideas can be fed into the process and turned into development projects.

A full 5-stage development process is not always necessary, even when considering the intent of always doing best-practice engineering development. This can be true if a product development is starting from ground zero. For example, consider the case of a value engineering (also known as cost reduction) or a customizable product family that may need a minor modification to complete a sale, sometimes referred to as a design-in-win. These are very different cases from the new idea to new product development cycle, and as such, the process can be tailored accordingly. This does not mean that steps should be skipped randomly and caution thrown to the wind. Generally speaking, the same deliverables are going to be required, but stages can be combined. We will examine these two cases in a bit more detail.

For the case of a value engineering project, it might be prudent to apply a slightly condensed version of this process; refer to Figure 1.2.

In this example, assume that the changes being made are to change the enclosure paint and hardware of an existing product. These changes may have an impact on performance in terms of regulatory or safety issues, but hopefully no impact to functionality. After scoping the work, the business chooses to go directly to development and test in one step, knowing that they intend to replace the current product offering with the new and lower-cost product. Once the engineering and product management teams are satisfied with the test results of the changed

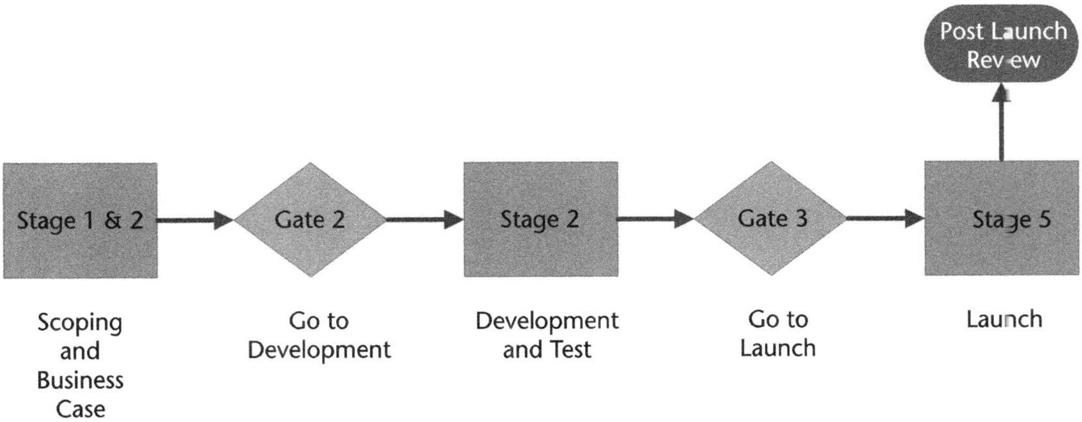

Figure 1.2 A 5-stage development process condensed to 3 stages.

product (having skipped the gate to proceed to testing), the product simply "cuts in" to production (meaning that, as old inventory is depleted, the new, lower-cost model replaces it).

For the case of a design-in-win scenario, it might be prudent to condense the process even further; refer to Figure 1.3.

In this example, assume that an existing product needs only to have the paint color changed and a prospective customer will purchase 1,000 units. In a case like this, the necessity for complete engineering rigor does not have justification. After a quick scoping of the necessary changes and product management impact, this product can be launched as a new orderable part number for the customer in a rapid fashion. The same deliverables can be required of the team, but most can be recycled from the original product.

Being able to adapt to these conditions and differences in product development undertakings as a design engineer is very important in today's landscape. A truly brand-new product development is far less common than a derivative product; think, for example, of mobile phones or televisions or even cars and other vehicles. Perhaps there are a few new features, but upon close examination, the similarities almost always outweigh the differences.

Lastly, it is important to understand that this process of development must be fed ideas; this idea buffet can be filled from many different places: fundamental research (true research and development (R&D)), internal idea capture from employees, and voice of customer (which can be anything from direct interaction with potential customers to focus groups and even collaborative design efforts with the customer).

In this development adventure on which we are about to embark, we are assuming that this is a true ground-up development project for a new platform. We are also under the assumption that marketing and management have done their homework properly, and this idea has been vetted in terms of the marketplace and in terms of the available skills the company has to deliver upon it. This places the bulk of the content of our journey in what was previously described as stage 3 (development) and stage 4 (testing and validation).

1.4 Product Development: After Launch

As a final note, once a product is released, it enters a new life cycle of sorts, often divided into four distinct stages:

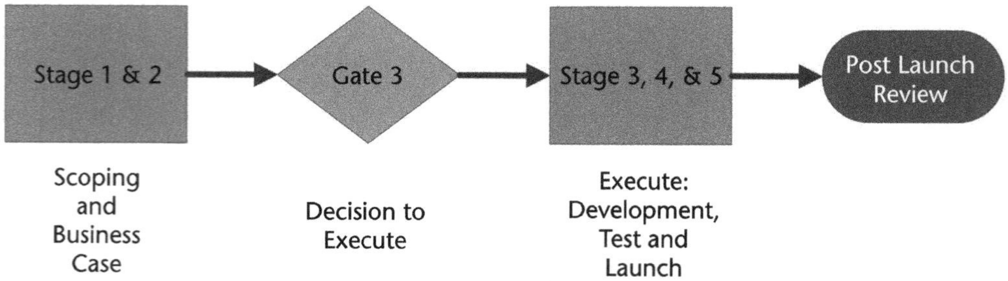

Figure 1.3 A 5-stage development process condensed to 2 stages.

- *Product introduction:* This is where the development process that we have just discussed ends, and the product is introduced to the marketplace and is ready to be sold.
- *Growth stage:* This is hopefully where the product starts to sell better as the hard work and groundwork laid by marketing and sales starts to translate into sales. Demand and distribution hopefully both increase, and the product becomes a profitable one for the business.
- *Maturity stage:* At this point, the product may be considered to be in a maintenance phase; perhaps there are fewer sales because the market has become saturated or competition has increased. A business may opt for a value engineering project here to really try to squeeze the value out of the product, knowing it needs to be maintained for existing customers.
- *Product decline:* This is the point in time when the business case for support of the product no longer stands up to critique and analysis. It is time to end the product's active life and hopefully introduce its successor. This is also referred to as the end of life (EOL) or obsolescence stage.

Throughout this post-launch product life cycle, engineering is still needed to support products in the field, perhaps fix any bugs or defects that escaped initial launch, or even introduce new features. The launch of a new product is certainly not where the engineering work ends; the difference is in the mindset of the engineers and other contributors and the context of the work being done.

This hopefully has oriented us as we begin to follow the journey of SensorsThink as they start the development of their new platform and give some context as to where this part of the journey sits in the overall life cycle of both a design project and a product's own life cycle. What comes next is understanding precisely what must be delivered and capturing that in a clear and concise manner. Remember our key message: before we can design, we must have a clear and frozen technical baseline, which allows for progressive assurance and risk retirement.

1.5 Requirements

At this point in our journey, SensorsThink has announced their intention to develop and launch a new product platform; they have done their homework through the first two stages of a typical 5-stage development process. Marketing and management sit the broader engineering team down in a room (yourself included) and present a set of requirements to help the team start to better comprehend the task at hand (see Figure 1.4).

Figure 1.4 is a Systems Modeling Language (SysML) requirements diagram, meant to be a visual representation of the high-level requirements of the new product platform; it does not contain textual information when presented this way, but that textual information can be viewed in a number of ways depending on the systems engineering toolset that is in use. For SensorsThink, Enterprise Architect 14 (created by Sparx Systems) [2] is the in-house tool of choice. The diagram presented in Figure 1.4 is an abstracted view of the full set of textual requirements through which we will be working during our development process. (As an aside,

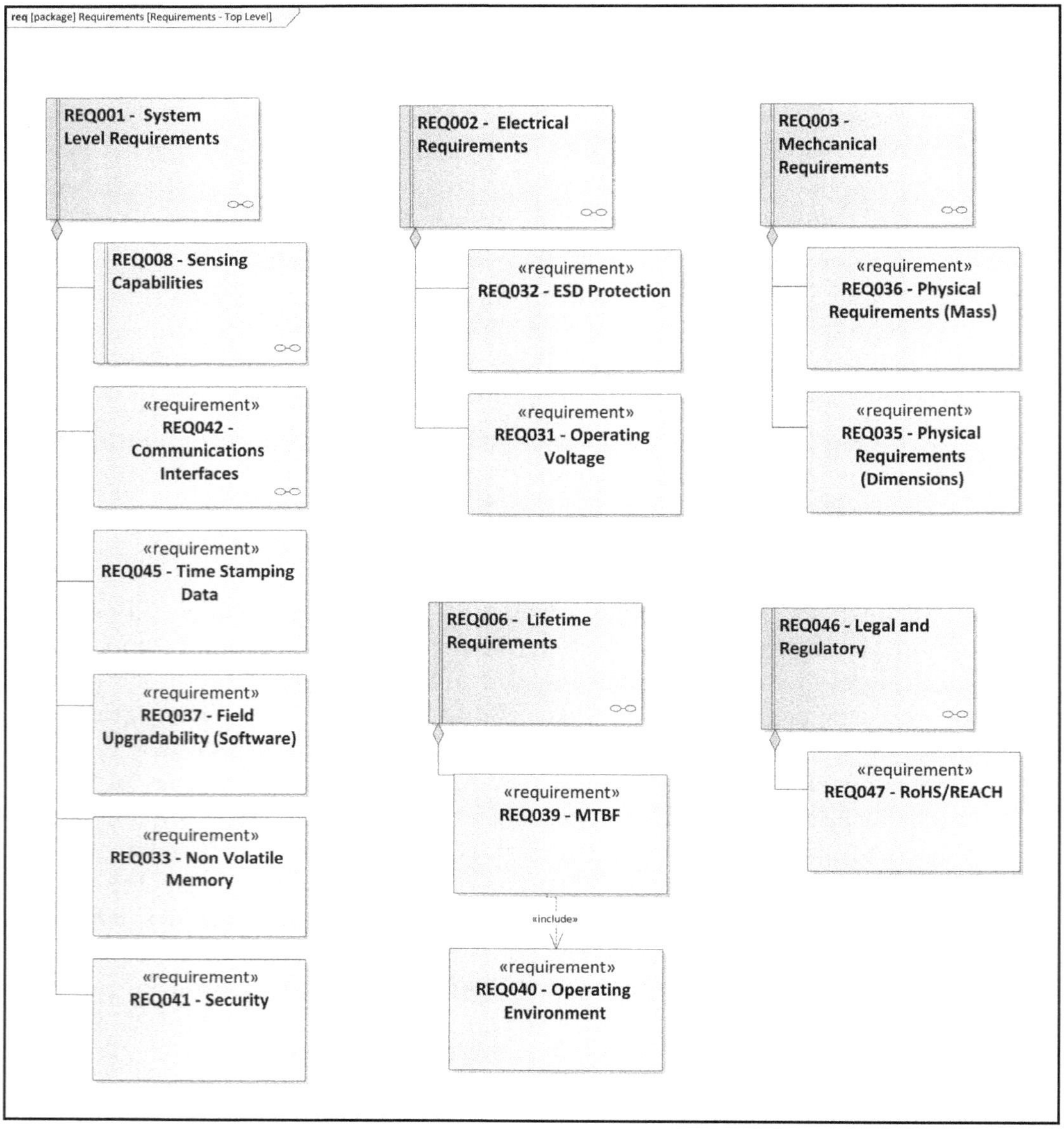

Figure 1.4 Requirements tree as presented by marketing and management.

if marketing and management ever present something that looks like this, a celebration is in order. Starting out with requirements entered in a modeling tool is a fantastic starting point and often is not the norm.)

A modeling tool that supports SysML is extremely useful during product development. It makes the application of many modern engineering best practices much easier and, in some cases, may be the only way to implement them in a satisfactory way for a critical or safety relevant system. Some of these best practices supported by SysML tools such as Enterprise Architect include:

- The ability to trace multiple SysML element types to and from each other; requirements to other requirements, requirements to test cases, requirements to use cases, and requirements to design elements.
- The ability to monitor the status of a requirement (or any other element) as reviewed or approved with built-in attributes for modeled elements. This can be vitally important in a safety-critical system environment. Even better is the ability to make these things visible in diagrams.
- Traceability through the use of relationship matrices to view exiting traceability between element types and identify gaps in traceability. Refer to Figure 1.5.
- Checking for unrealized requirements; as an example, consider a large system with thousands of requirements. Manually checking that each requirement

	Requirements::Checklist	Requirements::REQ001 - Syster	Requirements::REQ002 - Electri	Requirements::REQ003 - Mech	Requirements::REQ006 - Lifetin	Requirements::REQ007 - Introdu	Requirements::REQ008 - Sensin	Requirements::REQ011 - Infrare	Requirements::REQ012 - Infrare	Requirements::REQ013 - Infrare
Requirements::REQ027 - Local Temperature Sensing Ra...										
Requirements::REQ029 - Local %RH Sensing Range										
Requirements::REQ030 - Remote %RH Sensing Range										
Requirements::REQ031 - Operating Voltage			↑							
Requirements::REQ032 - ESD Protection			↑							
Requirements::REQ033 - Non Volatile Memory		↑								
Requirements::REQ034 - Non Volatile Memory Contents										
Requirements::REQ035 - Physical Requirements (Dimen...				↑						
Requirements::REQ036 - Physical Requirements (Mass)				↑						
Requirements::REQ037 - Field Upgradability (Software)		↑								
Requirements::REQ039 - MTBF					↑					
Requirements::REQ040 - Operating Environment										
Requirements::REQ041 - Security		↑								
Requirements::REQ042 - Communications Interfaces		↑								
Requirements::REQ043 - Ethernet										
Requirements::REQ044 - Bluetooth										
Requirements::REQ045 - Time Stamping Data		↑								
Requirements::REQ046 - Legal and Regulatory										
Requirements::REQ047 - RoHS/REACH										
Requirements::REQ049 - Remote Temperature Sensing ...										

Figure 1.5 Example of a traceability matrix for SensorsThink.

has been realized through a use case would be extremely cumbersome. Model validation in Enterprise Architect makes this task much more manageable.

These are just a few examples of the kinds of tasks that must be considered and properly managed throughout the development process and through a product's life cycle.

It is important to note that what is shown here is not intended to represent a full set of product requirements, if you are familiar with the requirements development process, you probably already know that a product of even moderate complexity can have thousands of requirements. A full elicitation and decomposition of requirements is not the focus of this journey. The requirements that are shown here will be addressed throughout the later parts of this text, so that the full process of requirement creation to requirement validation can be shown.

1.5.1 The V-Model

Before we dive any deeper into the requirements for our new product, a discussion on the V-model and what constitutes properly written requirements is needed in order to provide context and clarity for the next steps and how the pieces fit together.

Shown in Figure 1.6 is a diagram with which hopefully most readers are familiar. This development approach is a tried-and-true method for developing robust products and systems. The model emphasizes requirement-driven design and test. In the modern development, there are certainly detractors of this approach; many of these detractors are from the software engineering domain, and their objections to use of the model are certainly valid for their work. Some of the more common objections can be [3]:

- The model is too simple to reflect the realities of software development and is more suited towards management than developers or users.
- It is inflexible and has no inherent capability to respond well to changes.

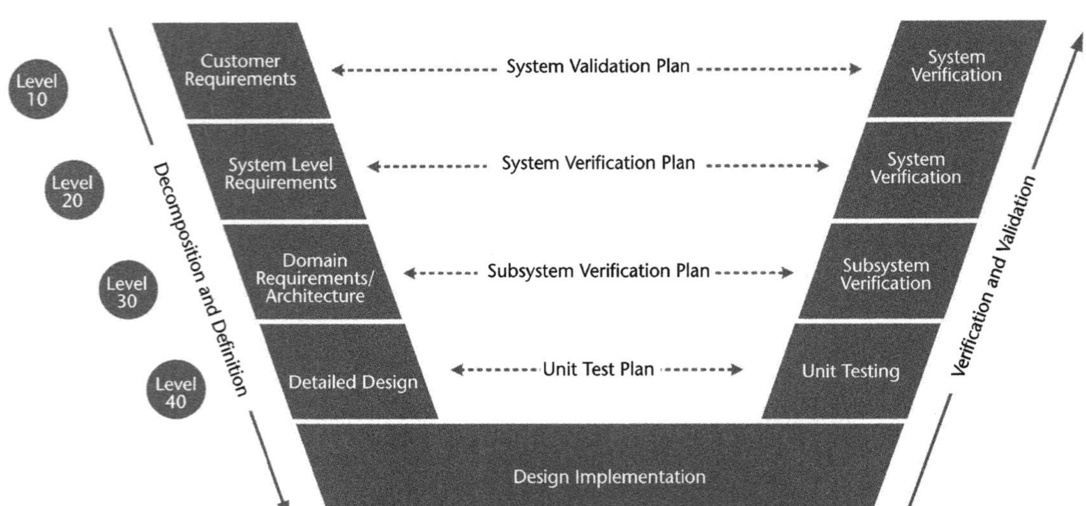

Figure 1.6 The classical V-model.

- The approach often forces testing to be fit into a small window at the end of the development cycle as earlier stages take longer than anticipated, but the launch date remains fixed [4].

That said, there is certainly merit to the use of the V-model. It has been widely adopted by both the German government and U.S. government. The model provides assistance on how to implement an activity and the steps to take during that implementation. It also provides instructions and recommendations for those activities. Starting with the V-model (or any established model) is almost always better than trying to reinvent the wheel. A framework like this can be tailored and adapted to a company's culture or a specific project's needs quite easily, and most mature models allow for such tailoring, providing guidance on how to approach that task.

Taking a cursory look at the V-model, it is evident that there are both different sides and different levels, with links traversing the model. It is important to note that from a purely theoretical perspective, as you move down the left side, across the bottom, and up the right side of the model, you are always moving forward in time. This makes the V-model an extension of sorts of waterfall product development and project management.

1.5.1.1 The Left Side: Decomposition and Definition (D&D)

The left side of the V is the D&D side of the model. During this stage of the project, customer requirements (which we will refer to as level 10) are decomposed into system-level requirements (which we will refer to as level 20), which are then decomposed into domain requirements and architecture (which we will refer to as level 30). For SensorsThink's project, domains are areas of engineering: electrical, mechanical, firmware, software, testing, and reliability.

As requirements are decomposed, they are always traced back through the levels such that they can point directly to a customer requirement that they fulfill or address.

Once level 30 is reached, the design implementation can begin, with execution on a firm set of requirements now the task at hand, the bottom of the V. Any documentation about the design implementation (engineering calculations, theory of operation-type documents for hardware, firmware, and software, for example) can be considered to be at level 40. These documents and calculations serve an important purpose in the overall life cycle of the product and should not be forgotten during the development process. They are often the basis for knowledge transfer internally and can be key artifacts for critical or safety-related systems.

1.5.1.2 The Right Side: Verification and Validation (V&V)

The right side of the V is the V&V side of the model. During this stage of the project, the physical hardware and written code are tested against ascending levels of requirements. The further the ascension up the V, the more integrated the system is becoming. The first tests performed at level 40 are often deemed unit tests where individual components of software and hardware are tested.

As an example for software or firmware, consider a program that consists of several mathematical functions that ultimately integrate into a calculator. The developer would unit test the addition function, subtraction function, and so on, individually, prior to attempting to integrate all functions into a larger framework.

As an example for hardware, consider a printed circuit board (PCB) that has a microprocessor, a power supply, and a serial port. The designer would verify that the power supply performed as expected (hopefully against some simulation data and calculations performed on the left side of the V) in a separate set of tests as the serial port. This is to ensure that a solid hardware foundation is in place for the software to execute on.

Level 30 testing is typically testing subsystems of the product and may be the first time that software and hardware are integrated. This may also be integration of multiple pieces of hardware and multiple pieces of software that were previously running independently.

As an example, consider a case where developers have been working on evaluation kits for various components of the customized hardware. At this point in time, the application is likely headed for the final target, which may well be a custom PCB.

This stage of testing is also when things such as environmental testing can begin for hardware. With a basic set of functional code and all of the mechanical components assembled with temperature cycle testing, shock and vibration testing and other similar tests can begin. These tests often take many hours to run to completion, so an early start can be a good decision if the design is mature enough.

Level 20 testing is typically the system verification stage, when a fully implemented and integrated system is tested in situ. So rather than testing on the bench or in a controlled environment like an environmental chamber, the product goes into the end-use environment or into simulated end-use environments. Valuable system test data is gathered here, verifying the level 20 requirements.

As an example, consider the custom PCB with a verified serial port. Code has now been ported to this PCB to run a calculator via the serial port, and a test loop has been written to perform random mathematical operations on random numbers and verify the results. This test loop can run independently for weeks on end if needed and the results can be examined for any issues occurring due to varying the environment of the device under test (DUT).

Level 10 testing is the last stop on the right side of the V; the original level 10 customer requirements are validated here, sometimes against the customer's own test cases. System certification for safety, EMI/EMC, restriction of hazardous substances and registration (RoHS) and evaluation, authorization, and restriction of chemicals (REACH) (which are environmental regulations) also happen during this stage of validation. If all testing goes well, the product is ready to launch, having been put through rigorous testing and proven to meet all requirements at all levels of decomposition.

1.5.2 SensorsThink: Product Requirements

If we now jump back into SensorsThink's development path, we can see that we are firmly on the left side of the V-model and are sitting at level 20, perhaps a bit more decomposed than a true set of level 10 customer requirements. Let's now take a look at the textual requirements themselves, rather than the pure diagrammatical view presented earlier. Since this is a sensing platform, let's look closely at the required sensing technologies and associated requirements as identified by marketing and management.

If we refer to the higher-level elements of the diagrams from Enterprise Architect 14, we can see that the basic sensing technology requirements are:

1. Temperature and humidity;
2. Accelerometer;
3. Magnetic field;
4. Infrared spectrum;
5. Vibration.

These are the key functional elements for the new sensing platform that SensorsThink will be attempting to integrate into a single product platform. It is not uncommon for the starting point of a project to be nothing more than a bullet-point list such as the one above, but in our case, we are in much better shape. Our management and marketing teams have gone above and beyond, putting in some details around the requirements for us to help engineering better understand the system and engineer the right product to meet the market demands. Required ranges have been provided for all sensor technologies:

- Degrees Celsius for temperature and percent relative humidity (%RH);
- Two types of acceleration with minimum required values of acceleration that must be measurable;
- Required axes of measurement and required field strength that must be measurable for the magnetic field;
- Required field of view, frame rate, and even wavelength specifications for the infrared spectrum;
- Required axes of measurement and required force and frequency that must be measurable for vibration.

We also see a nice-to-have feature request, which tells engineering that, if possible (after proper analysis and architecture), multiple sensors should be collocated or all sensors should be on a single sensing head. This is a preferred (but not required) solution for the platform. Note that this is not a requirement element, but rather an informational statement.

This type of definition is a wonderful starting point; if we refer back to the V-model, we can see how this can easily be decomposed into domain requirements for hardware and software capabilities and even lends itself nicely to the creation of accurate level 10 validation testing.

1.5.3 Creating Useful Requirements

Before we proceed further and show the rest of the product requirements with which we have been presented, it is certainly worth discussing what makes a proper requirement and how we can identify requirements that are weaker, or in need of refinement, before they are actionable by engineering (or other disciplines, for that matter) [5].

The International Council on Systems Engineering (INCOSE) is an organization founded to develop and propagate the principles and practices that enable the creation of successful systems. The organization was formed in 1990 and has become the de facto standard for many aspects of systems engineering, including requirements writing. The *INCOSE Guide for Writing Requirements* [6] is a document that is often cited in company policy for how requirements are to be constructed. The document is comprehensive and will be heavily cited here as it is an excellent reference for all those tasked with creating, reviewing, designing to, or testing against requirements.

The document is specifically about how a requirement is properly expressed in a textual format, in other words: how to properly write a requirement. It is not a guide on how one might go about gathering requirements or turning those requirements into SysML diagrams. It is important to understand that, while often done by the same people, the tasks are quite different.

The general approach of the guide is to present characteristics and practical rules for crafting requirements. The characteristics express why a rule makes sense, and the rules express how to write the requirements. This helps to provide readers with context knowing that rules must often be adapted to fit or suit a certain need. For the purpose of this discussion, the focus will be on the characteristics of a well written requirement statement:

(When reading the following list, start each item with "a well-written requirement is...")

1. Necessary: Every requirement is necessary. This may seem obvious, but consider the points made below:
   (a) If you can remove the requirement and still solve the problem at hand, the requirement is not needed.
   (b) If the intent of the requirement is covered by other requirements, the requirement is not needed.
   (c) There is a cost (both in money and in time) associated with every requirement; unnecessary requirements can lead to additional cost, additional time, and unnecessary risk.
2. Implementation-independent: A requirement states what is required, not how the requirement should be met.
   (a) Being prescriptive can cause other, and perhaps better, implementations to be overlooked entirely.
3. Unambiguous: A requirement statement can only be interpreted in a single way.
   (a) The intent of the requirement must be understood the same way by the writer of the requirement, the designer, and those doing the verification and validation. (All sides of the V must understand the requirement's intent in the same way.)
   (b) Lack of clarity in a requirement can lead to many problems, like the wrong product being designed or a feature being implemented in a way that the customer did not want.

4. Complete: A requirement statement is complete in and of itself.
   (a) This does not preclude a requirement referencing another document (an international standard, for example) or even another requirement. However, the requirement itself should be a complete and proper sentence that does not rely on other requirements for comprehension.
5. Singular: A requirement statement addresses a single thought.
   (a) This characteristic is especially important to the steps that follow the initial gathering of the requirement; decomposition, allocation, and validation all depend on being able to point back to a singular statement.
   (b) Keeping a requirement limited in scope to a single function or constraint is ideal, but oftentimes referring to a diagram is the best way to express complex behavior.
      (i) If a requirement is that a particular state-machine shall be implemented to align with a safety standard, a diagram of that state-machine is a far better way to describe the requirement than a written explanation of each state and transition.
6. Feasible: A requirement statement always expresses something that is inherently feasible.
   (a) Unachievable requirements such as 100% reliability are frustrating for all parties involved (except for perhaps the customer). These requirements are not only likely to cause extensive overengineering, but they are also impossible to validate.
   (b) Oftentimes, these types of requirements are important features; someone asking for 100% reliability is clearly looking for a very robust solution. However, the analysis has not matured enough to be considered actionable by engineering.
7. Verifiable: A requirement statement is verifiable.
   (a) If a requirement statement is not in some way measurable, able to be inspected, or testable, there is no way to know if it has been satisfied.
   (b) A functional requirement typically requires a test and a performance requirement needs to clearly express the quantities and values associated with the performance.
   (c) Think of the four types of verification events when examining a requirement in this way: analysis, inspection, demonstration, and test.
8. Correct: A requirement statement is correct. This one is self-explanatory. Check the values, check the goals of the product, and make sure the requirement is correct.
9. Conforming: A requirement statement conforms to standards selected by the organization.
   (a) This characteristic is important for large projects with many thousands of requirements. Use of consistent language, terms, acronyms, and abbreviations goes a long way to ensuring that requirements are easier to read and review for the stakeholders.

For further information on the proper construction of a complete set of well-written requirements, refer to the INCOSE Organization website (http://www.INCOSE.org), in particular, the *INCOSE Guide for Writing Requirements.*

Let's examine one of the two example requirements shown in Figure 1.10 from the SensorsThink project for its completeness and correctness as a requirement statement. Enterprise Architect has a neat checklist that can be added to individual requirements or groups of requirements so that all aspects of a built-in (or customized) list can be checked by the project team during requirements development.

For example, consider REQ032–ESD Protection: "The system shall be designed to withstand ESD events on all external facing connections. Tested in accordance with the latest version of EN 61000-4-2" [7].

Review it against the list of items in the checklist (which matches the list of characteristics of a well-written requirement). See if you agree that the requirement meets those characteristics and qualifies as a well-written requirement.

1.5.4 Requirements: Finishing Up

Now that we understand the way that requirements are used and what makes a well-written requirement, the rest of the requirements will be presented; take some time and examine them, and think of yourself in the position of an engineer at SensorsThink. How would you feel if presented with this set of requirements (see Figures 1.7 through 1.15)? Hopefully with the context and information provided here, your answer will be that you feel ready to architect a system to meet these requirements [8–10].

1.6 Architectural Design

Once the requirements are agreed to at the customer level, several key engineering activities are ready to begin:

1. System-level requirements;
2. System-level architecture and design;
3. System validation plan.

System-level requirements are the next step down the decomposition ladder in requirements elicitation; for the SensorsThink product development project example, the given requirements are more of a hybrid level 10/level 20 requirements set. Remember that the key here is that the framework for approaching product design and development for embedded systems is the focus here, not on a fully and completely elicited requirements set.

Done concurrently with system requirements, system architecture is something that is absolutely essential to delivering a well-engineered product and solution that meets the needs of the customer and is neither overengineered nor underengineered. The right product is always the holy grail of development: the right features, the right performance, and the right price.

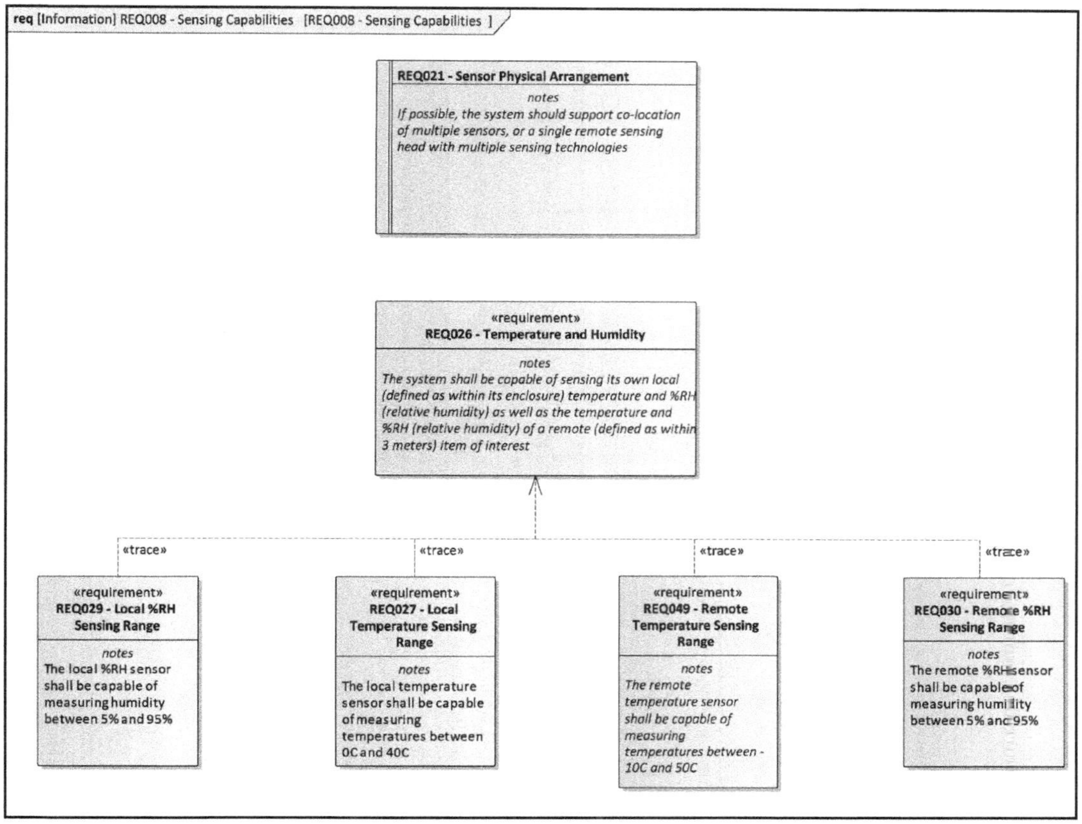

Figure 1.7 Sensing capabilities: Part 1.

In our journey, we will not be taking a deep dive into modern systems engineering methodology, since it is tangential to our core focus. However, a discussion on the topic is certainly valuable and warranted to provide proper context and understanding.

1.6.1 SEBoK: An Invaluable Resource

The Systems Engineering Body of Knowledge (SEBoK) [11] is a curated body of information about systems engineering. It is a wonderful resource for those seeking more information and additional resources about systems engineering.

The SEBoK website (http://sebokwiki.org) provides six key purposes that the organization seeks to support; among them are [12]:

1. "Inform Practice": Inform systems engineers about the boundaries, terminology, and structure of their discipline and point them to useful information needed to practice systems engineering in any application domain.
2. "Inform Interactors": Inform engineers in interacting fields of engineering (mechanical, software, electrical, reliability) and other stakeholders (managers, assemblers, documentation creators) of the nature and value of systems engineering.

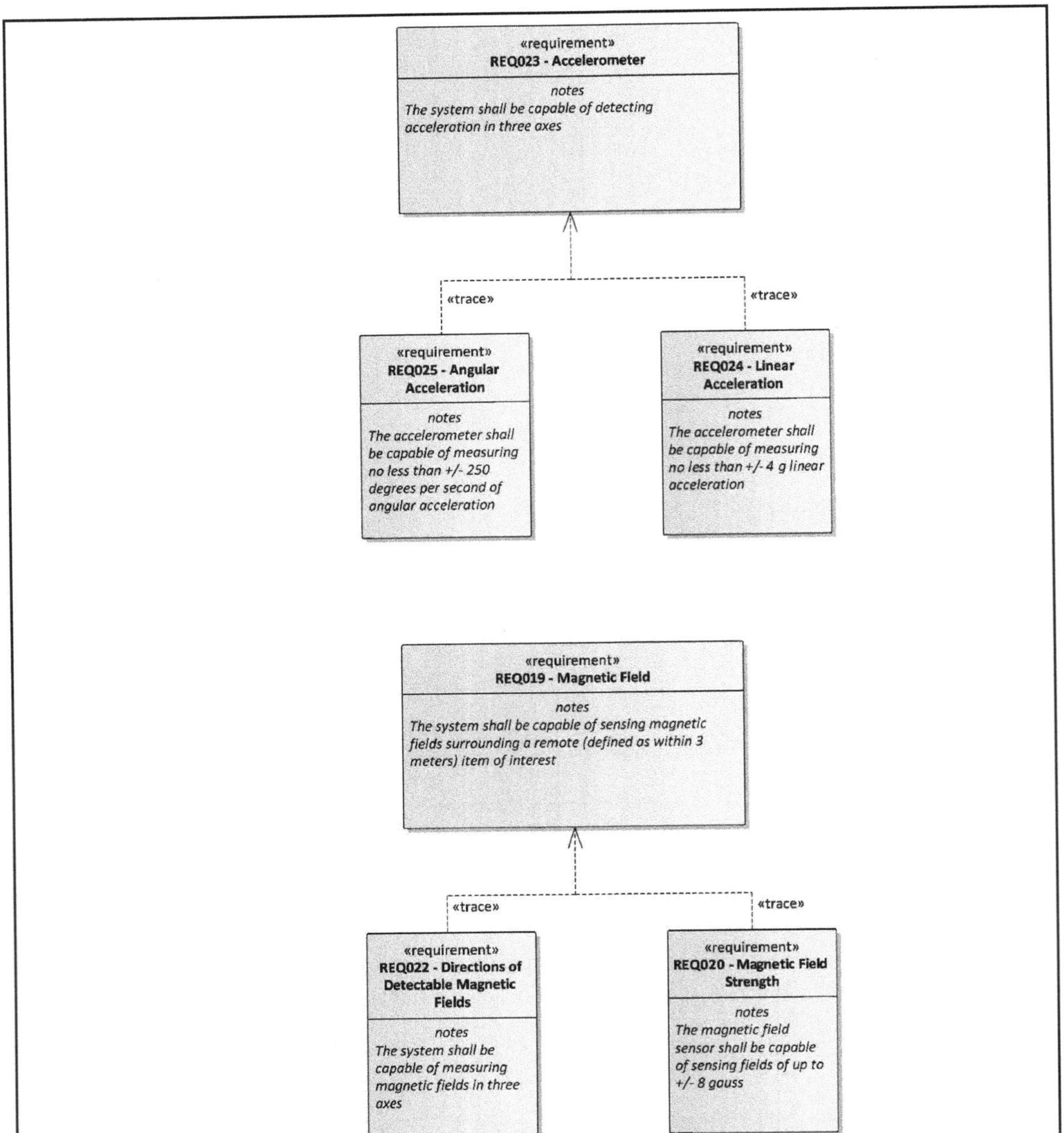

Figure 1.8 Sensing capabilities: Part 2.

These two purposes are certainly relevant for us and the engineering staff at SensorsThink. Without a solid foundation in place that comes from proper systems engineering, the product being developed is not nearly as likely to be the right product, with the right features, performance, and price.

For technology development, there is a wonderful excerpt from the SEBoK website that should be kept in mind as we architect our new sensor platform [13]:

> Product systems engineering should bring awareness of technology changes and trends to the analysis of new product ideas or innovations. This affects the time and cost inputs into the technical feasibility analysis of the product. The result should include a road map of required technology developments, which is then

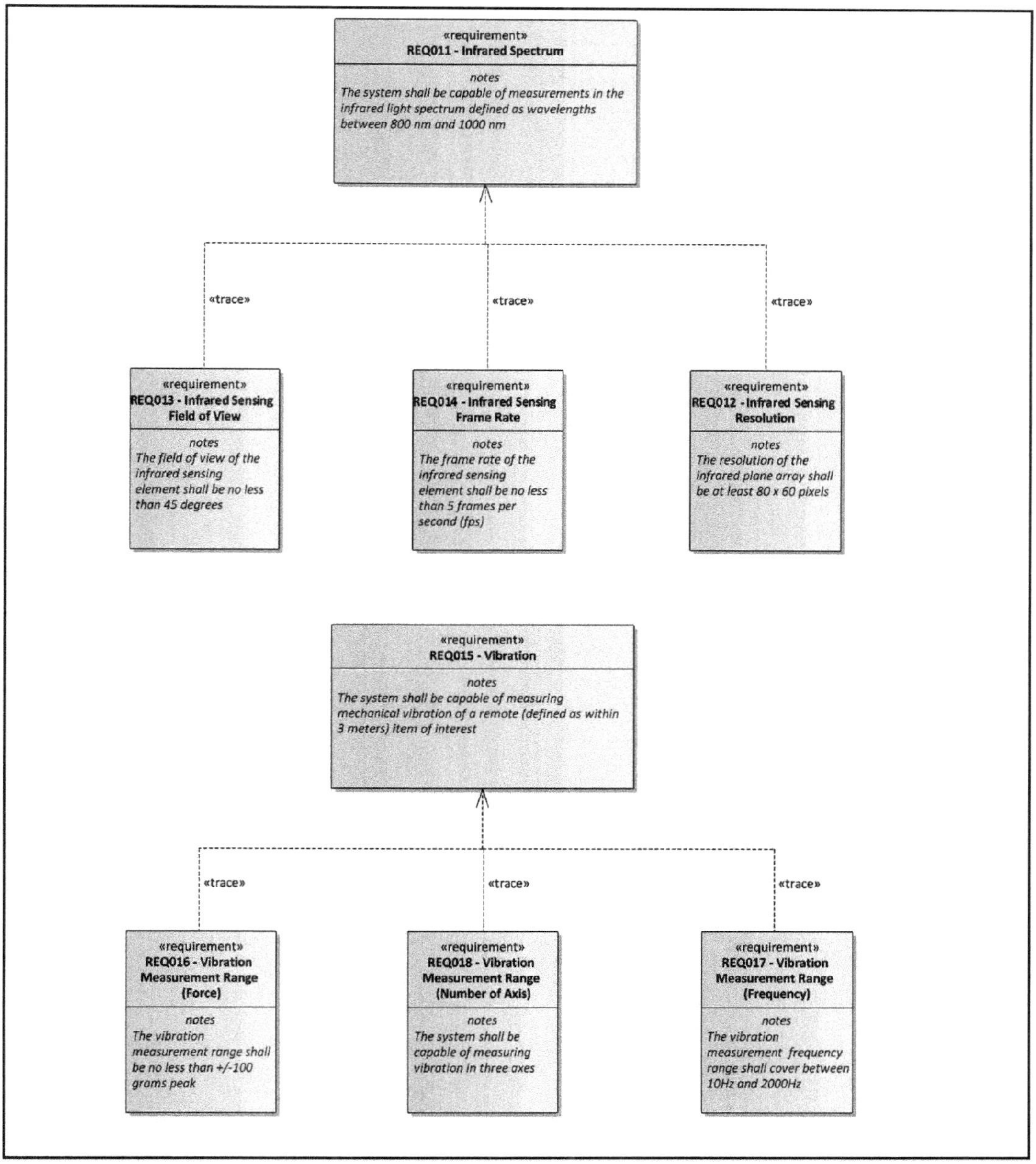

Figure 1.9 Sensing capabilities: Part 3.

used to create the overall road map for the new product offering. In these cases, new product ideas impose requirements on new technology developments.

On the other hand, when technology developments or breakthroughs drive product innovation or the generation of new markets, the technology developments may also generate requirements on product features and functionalities.

At the systems engineering and architectural level, critical decisions should be made about what requirements are imposed on development; this is no trivial task and can be extremely challenging when new developments are thrust upon a market. It is this fundamental baseline, rooted in a thorough understanding of both the

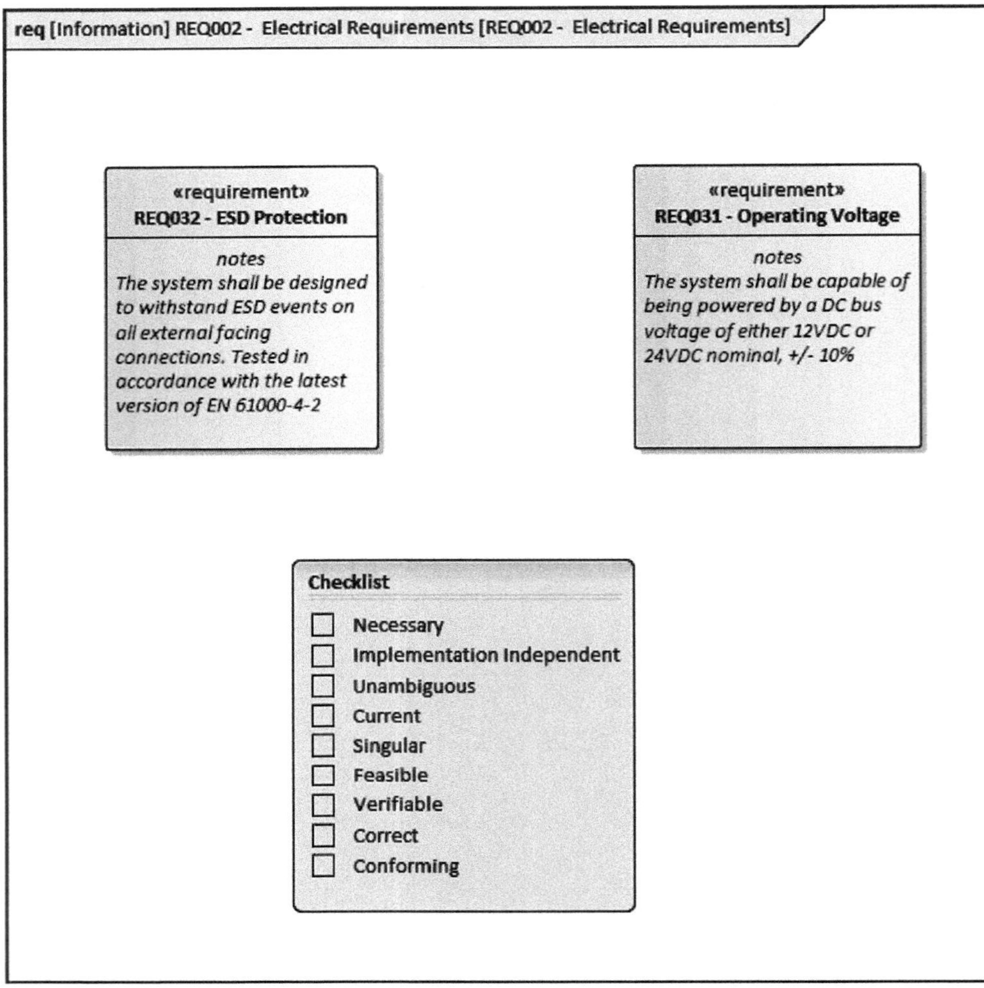

Figure 1.10 Electrical requirements with a visible checklist.

technical concept at hand and the market (and markets in our case) being targeted, that is essential to a successful product development and launch.

1.6.2 SensorsThink: Back to the Journey

If we now immerse ourselves back into SensorsThink's development path, it may be easier to see why our given requirements really do straddle the level 10 and level 20 tiers (refer to Section 1.5.1), as we previously laid out. These requirements were derived by the analysis of potential markets and available technologies. Perhaps even requirements were compared and contrasted before arriving at the conclusion that several key applications could be served with a single sensor platform.

The systems engineer's job does not end at this point, as they would assuredly continue to be part of the requirements' decomposition down to the domains at level 30 while also assisting with the creation of verification plans at the system (and, if complex enough, the subsystem) level.

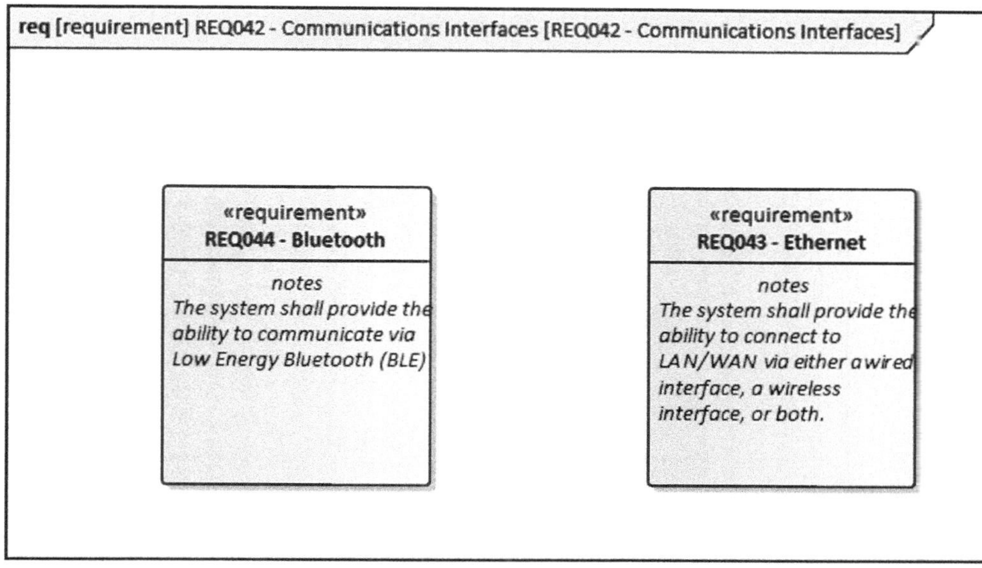

Figure 1.11 Communications requirements.

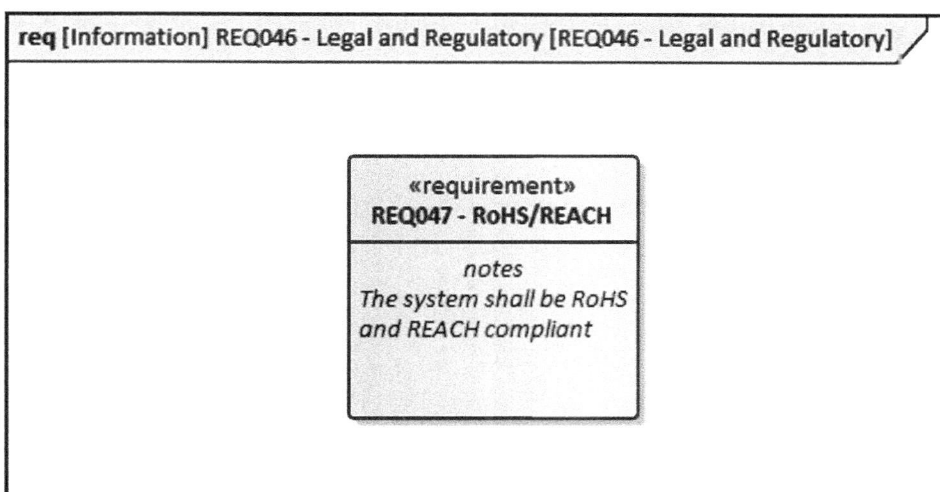

Figure 1.12 Legal and regulatory requirements [8, 9].

1.6.3 Systems Engineering: An Overview

There are many different ways to apply the concepts of systems engineering to a development project. Generally speaking, the following topics are essential to keeping with modern best practices:

1. System requirements;
2. System architecture:
   (a) Logical architecture model development;
   (b) Physical architecture model development.

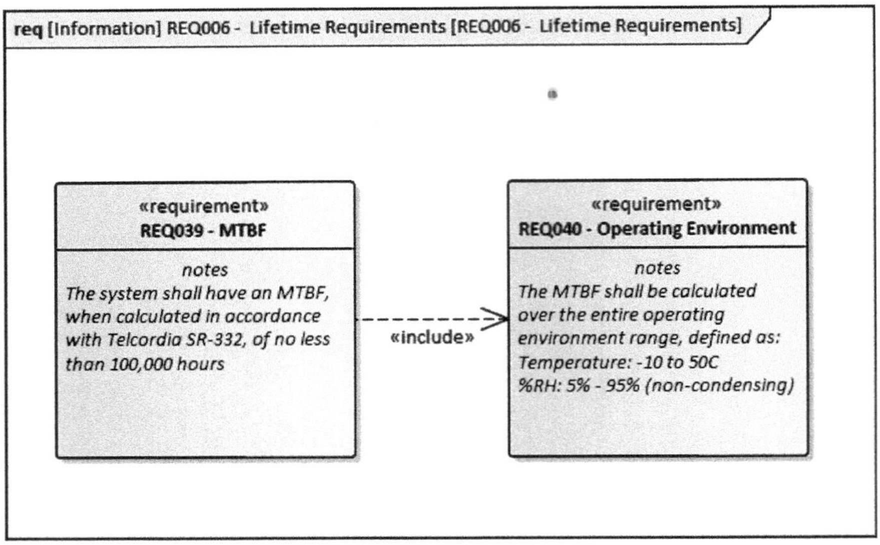

Figure 1.13 Lifetime requirements [10].

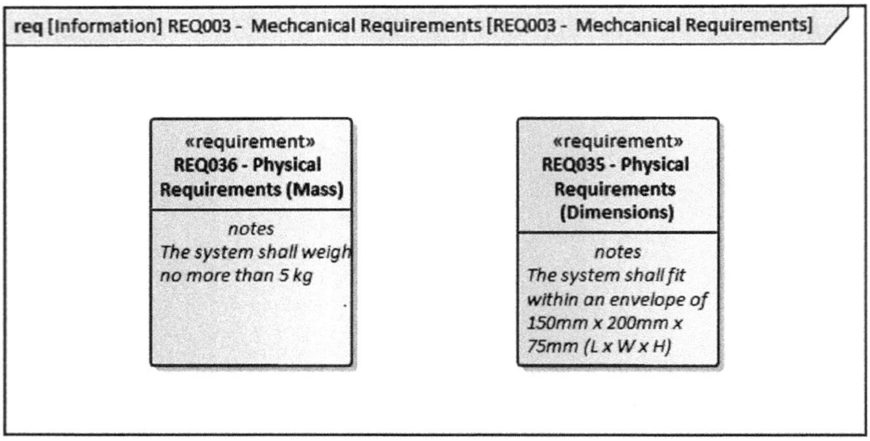

Figure 1.14 Mechanical requirements.

3. System design: This is a way to supplement the architecture with useful (and necessary) data and documentation that enable domain engineers to implement the chosen architecture. This is the link between the architecture and the implementation of the system: things like equations, drawings, diagrams, and algorithms. SEBoK is a great place to grow knowledge in this area [14].
4. System analysis: This analysis allows developers to assess the system in a consistent way in order to choose the most efficient architecture. Trade-off studies, cost analysis, and risk analysis are all excellent ways to achieve this. Once again, SEBoK provides an excellent overview of this [15].

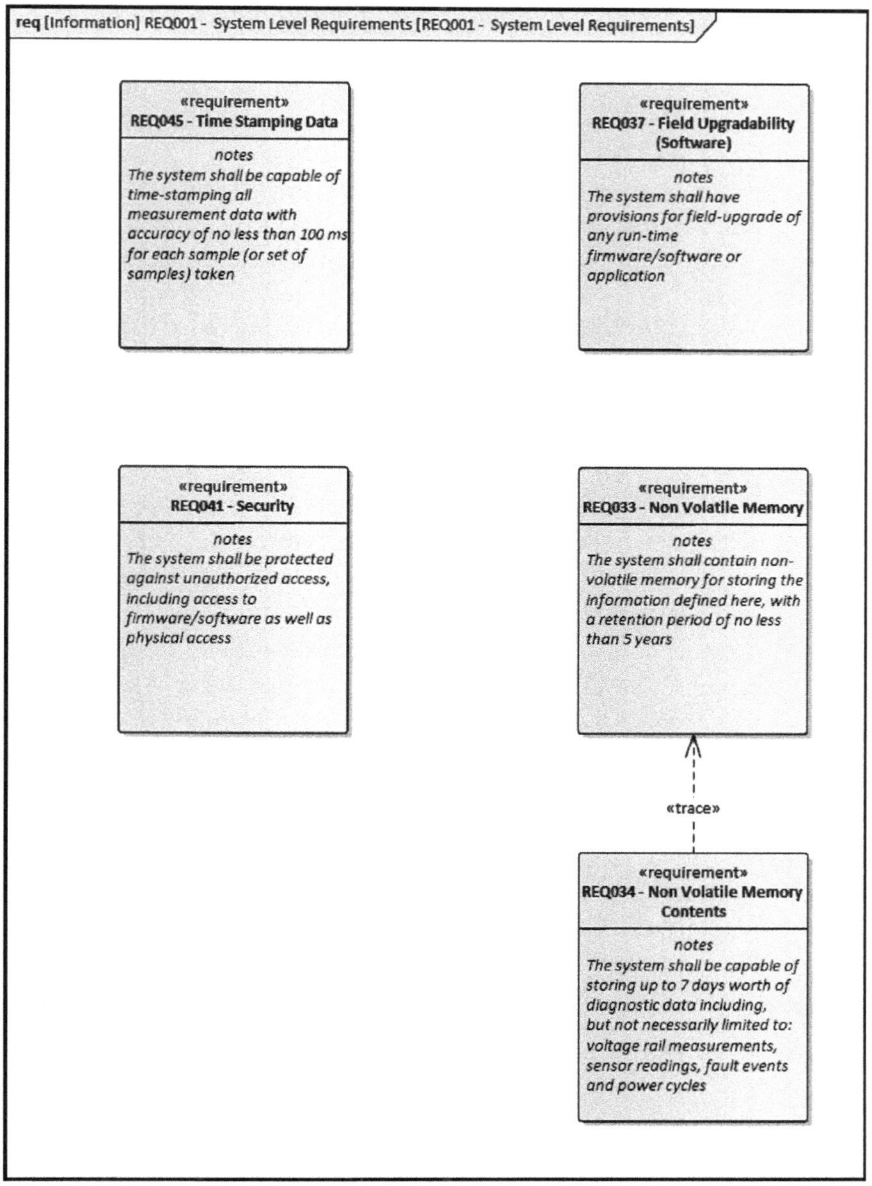

Figure 1.15 Miscellaneous system-level requirements.

Let's proceed with the assumption that our set of approximately 30 requirements is the full set for our new product, even though we know better. The next step would be to architect the system.

1.6.4 Architecting the System, Logically

Usually the first step in architecting the system is to create a logical architecture model (although, depending on the system, it may not be necessary); the purpose of this type of architecture modeling is to create models of the major functionality of the system being developed. This should cover the product not only when in use by customers, but also by other stakeholders in the process: test and quality assurance, production, and end-of-life procedures. Recall the four stages of a product's life from Section 1.2: introduction, growth, maturity, and decline. A well-executed architecture will consider all of these stages in addition to the functional requirements of the product in the marketplace.

It is very important that the intent of logical architecture is well understood. Logical architecture describes how a product works. This can be applied at a higher level of abstraction (consider the use case of a product being decommissioned) as well as lower levels of abstraction (how a certain piece of functionality should work); it is vital to separate this conceptual architecture from the physical way in which it will be implemented [16]. These logical architecture diagrams typically make use of three types of models:

- *Functional:* A set of functions and their subfunctions that defines the work done by the system to meet the system requirements.
- *Behavioral:* Arranging functions and interfaces (inputs and outputs) to define the execution sequencing, control flow, data flow, and performance level necessary to satisfy the system requirements.
- *Temporal:* Classifying the functions of a system based upon how often they execute. This includes, for example, the definition of what parts of a function are synchronous or asynchronous.

These architectural models are then linked to system requirements through a traceability matrix to ensure that every system requirement has been addressed by the logical architecture.

It is somewhat out of scope to perform a full logical architecture for the new sensor platform SensorsThink is developing; looking at one or two sample diagrams that may come from this portion of development is certainly useful, however.

1.6.5 Keep Things in Context (Diagrams)

One key diagram to consider is a context diagram, looking at the system being designed in its environment (or in many different environments). Enterprise Architect is an ideal tool for creating these types of architectural diagrams, so we can examine a few examples by expanding our system model to include them.

For our system, let's assume that we have three stakeholders in mind for the system when deployed in its operational environment during its useful life: the end user (customer), a remote server, and service personnel.

Each of these stakeholders (actors) will have a different interest and interaction with the system once it is deployed into operation. This is noted in the diagram (see Figure 1.16):

1. Service would have interest in diagnostic data or upgrading the system.
2. The remote server would likely take data for post-processing for all attached sensors and log system diagnostics.
3. The end user may only care about system status (fault/no fault) and any alarm conditions that the sensors are reading for their application.

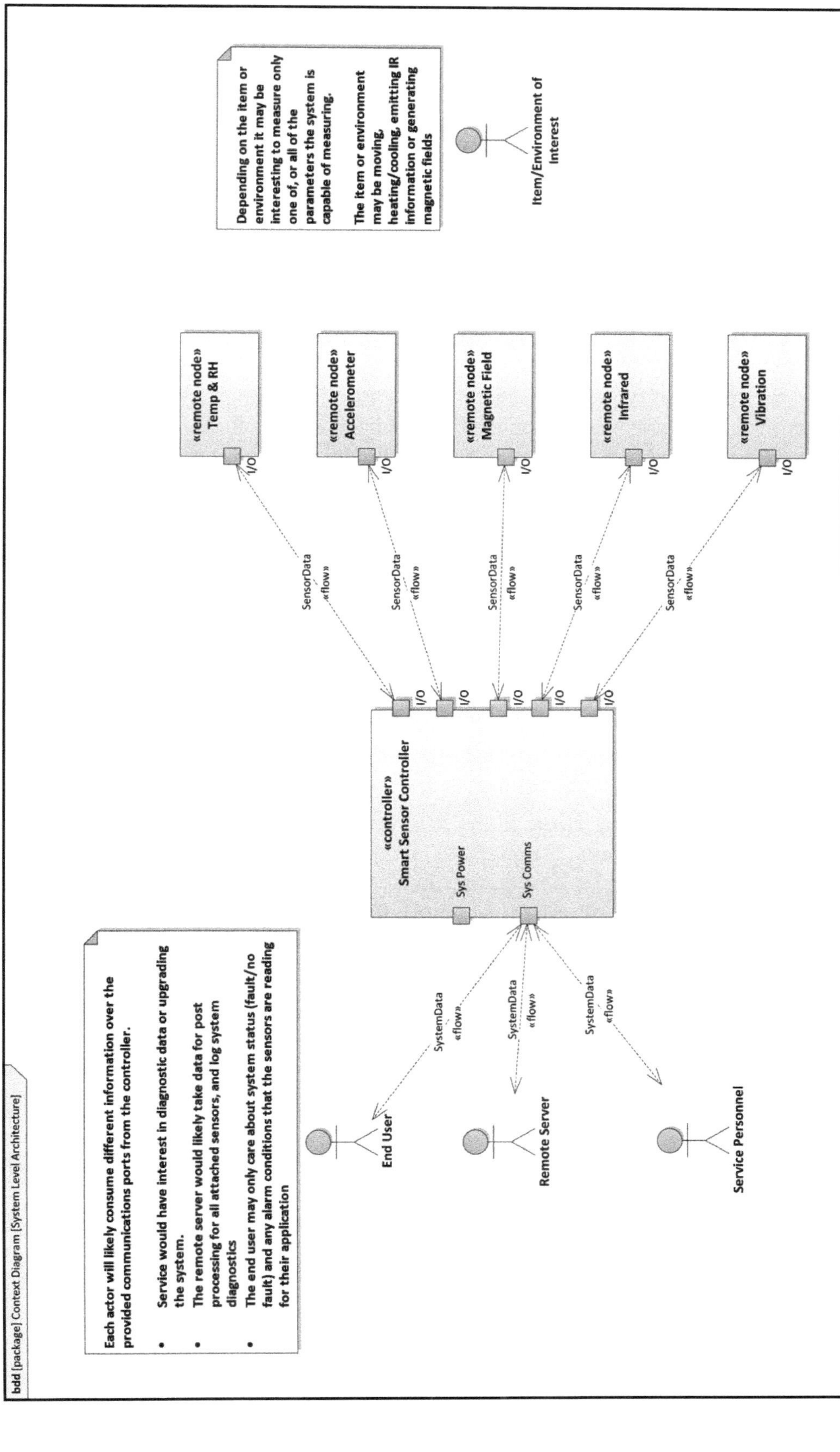

Figure 1.16 Context diagram of the Smart Sensor Controller.

Note that all of these end users will be interacting with the system controller through the same communications interface (or interfaces, in our case). This helps to put definition to those interfaces and to things such as logging in to the system, access levels, and security needs for the interface.

We also see that our environment or item of interest is shown in this context diagram. This is what the sensor platform is monitoring and taking measurements of.

It is important to take note that this diagram is not prescriptive; it does not define anything about a particular interface or feature of the system. It is not defining what type or volume of information will flow over various interfaces. It simply provides the context for which the system will be operating in this particular use case.

1.6.6 Monitor Your Activity (Diagrams)

Another diagram type to consider is an activity diagram, a basic example being a very high-level activity diagram of the system's operational modes.

This is a very basic activity diagram, but still defines (at a high level) the way the system is intended to function during run time (see Figure 1.17); a few bullet points illustrate this nicely in conjunction with the diagram:

- System power is applied.
- Power on self-test (POST) is performed. This is when housekeeping diagnostics will take place; these diagnostic tests cannot be defined in meaningful detail until the system is better defined. Some typical examples are memory tests, processor self-tests, and voltage measurements.
  - POST has a decision point: if it passes the operating system (whatever it may be) is loaded; or, if it fails, the system halts and enters standby mode.
- Load operating system: This is when the processor can load the operating system. In some cases, like a bare metal system, this is really minimal. In other cases (Linux, for example), this can be a very large task and very complex. A further set of diagrams in a complex case may well be necessary to fully describe the system operation. If any faults occur during the operating system load function, the system immediately halts and enters standby mode.
- System operation begins: This is when the system is acquiring data and either writing it to memory or sending it over communications, for example. It will also be servicing and checking for interruptions during this time, as well as monitoring for any run-time faults, loss of power, or corrupted data. If any faults occur during operation, the system immediately halts and enters the standby mode.
- Shutdown request: This is when a user requests a graceful shutdown. The system performs whatever functions are necessary to ensure a safe and reliable end to operation and enters the standby mode.

Diagrams like this are extremely useful in defining the behavior of a system at varying levels of depth; the shutdown routine of a safety-critical system may warrant many layers of complex diagrams, contrasted against the shutdown routine of

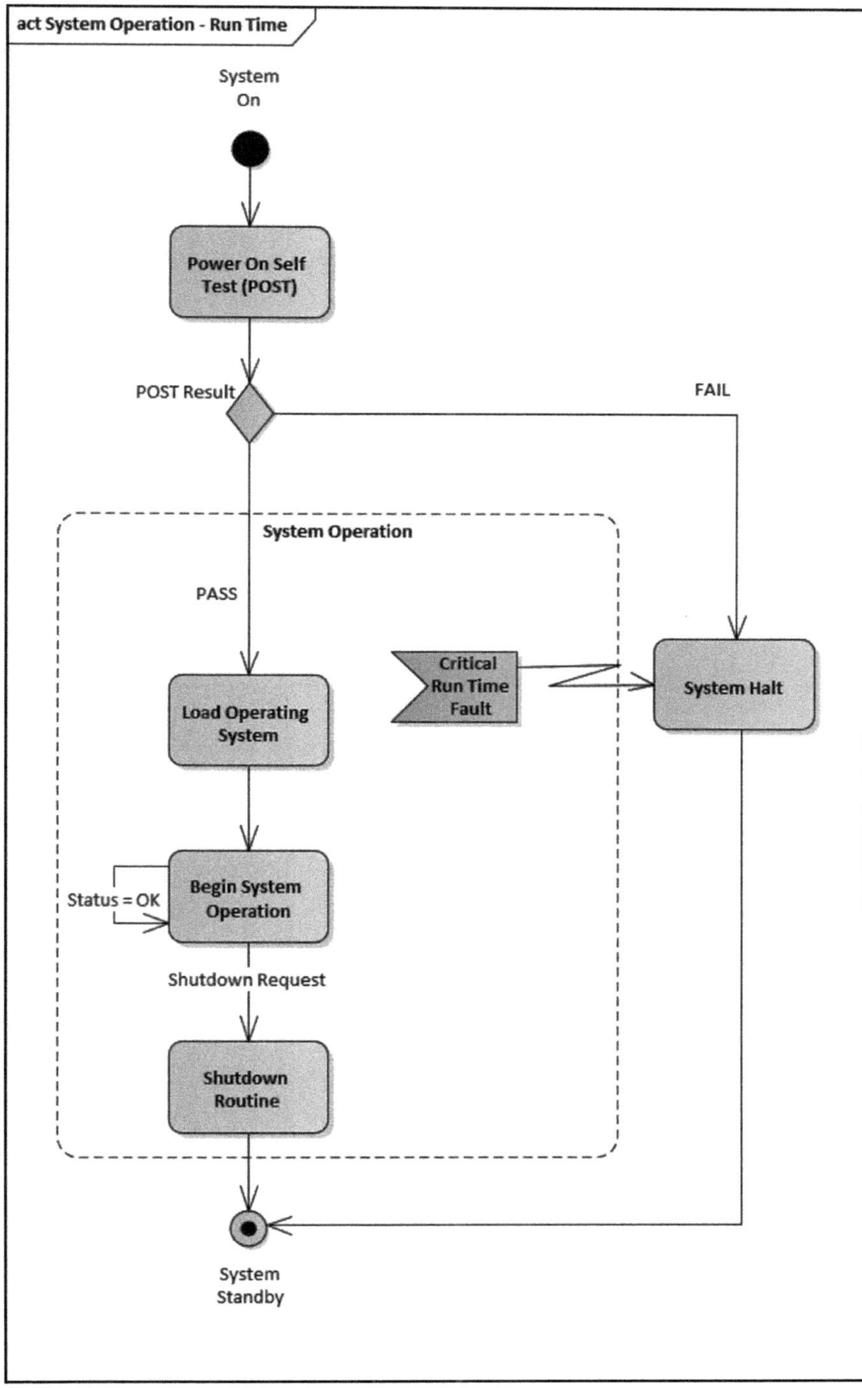

Figure 1.17 System operation: run time.

a coffee maker, which may only require one or two actions be shown (it may be as simple as the user turning the power button off).

1.6.7 Know the Proper Sequence (Diagrams)

One more example diagram to consider is reading sensor data (see Figure 1.18): a core function in our new product design. In this diagram, our systems engineer has decided that, prior to getting sensor data, a sensor health check should be performed. Assuming that this is possible, this might be something as simple as verifying a "Who am I?" register and performing a cyclic redundancy check (CRC) on the data received to ensure that communications are reliable. If we recall Figure 1.17, a failure of this CRC may well cause a critical run-time fault and immediately halt operation. If the status check is fine, sensor data is requested, gathered from the environment, and returned back from the sensor. This data may be post-processed by the sensor node, depending on implementation, and then handed back to the core controller platform for further use.

1.6.8 Architecting the System Physically

Logical architecture really does tell the engineering team how the system should work and, from that, how we can begin to determine the implementation details in a physical architecture model [17]. The goal of physical architecture modeling

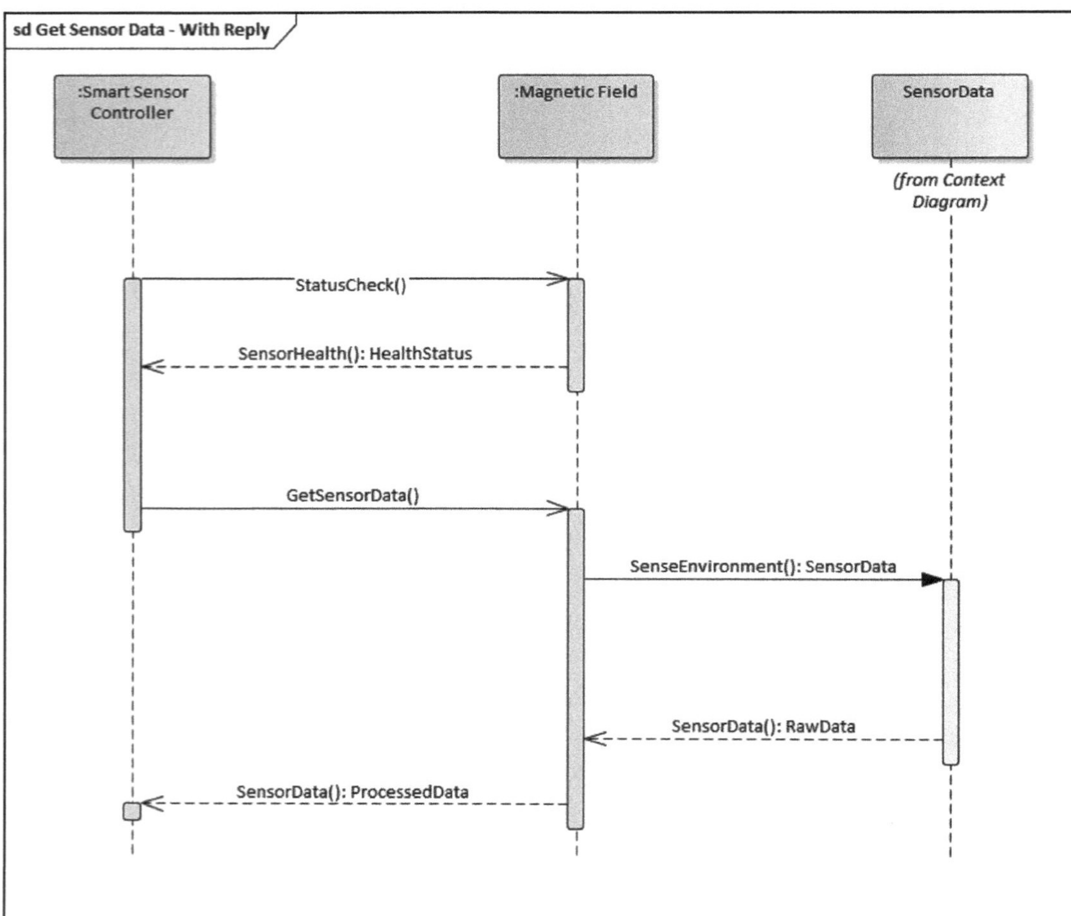

Figure 1.18 Sequence diagram for reading a sensor.

is to create models and views of a physical solution that accommodates the logical architecture model (in other words, accounts for all functional, behavioral, and temporal models and diagrams) and satisfies system requirements. Physical elements (hardware, software, mechanical parts) have to be identified that can support these models and requirements.

One of the key drivers for the physical architecture model may be interface standards; they may be one of the key drivers for the overall system. However, it is also sometimes the case that interfaces are chosen later on in the model development process. In our system, we have both of these cases present: the communications interfaces that were stated as system requirements (Ethernet and Wi-Fi or Bluetooth) and the sensor communications interfaces that were not specified in any meaningful way.

Because the system will likely be communicating over established networks to the outside world, it makes sense that these protocols are defined and required at a very high level. On the other end of the system, there are several internal interfaces that are only exposed to the Smart Sensor Controller. These sensor communications interfaces can be of any type as long as the selected sensor meets its sensing performance requirements, the real driver for the sensors are those performance requirements, not the way in which data is transmitted between sensor node and controller.

1.6.9 Physical Architecture: Playing with Blocks

As a starting point for understanding physical architecture, let's look at a proposed solution for the Smart Sensor Controller (Figure 1.19). The physical architecture diagram shown in Figure 1.19 is intended to cover the large central block of the context diagram at a hardware level.

We can see from Figure 1.19 that the central processing portion of the system appears to be quite complex even though much of the system is still rather loosely defined. However, from covering all the requirements and in considering the various life cycles of the product, we end up with a rather complex processing board. It must be capable of performing a number of functions:

1. Ethernet and wireless communications;
2. Serving data to a cloud server;
3. Acquiring, time-stamping, and performing other post-processing and sensor fusion on as many as five sensors;
4. Being able to be upgraded in the field;
5. Being secure in the handling of data and information.

Note that most of the physical blocks here are represented (at first glance) by hardware only, but a closer look would reveal that there is a large volume of firmware and software work that must occur in the system to operate properly and reliably. There must be accurate date and time keeping for time-stamping data, data integrity must be guaranteed, sensor fusion must be performed, image processing (for our infrared sensor), data storage, and data transmission over multiple interfaces (perhaps up to three). When analyzing the system and what needs to be at the heart of it, there could be several implementations:

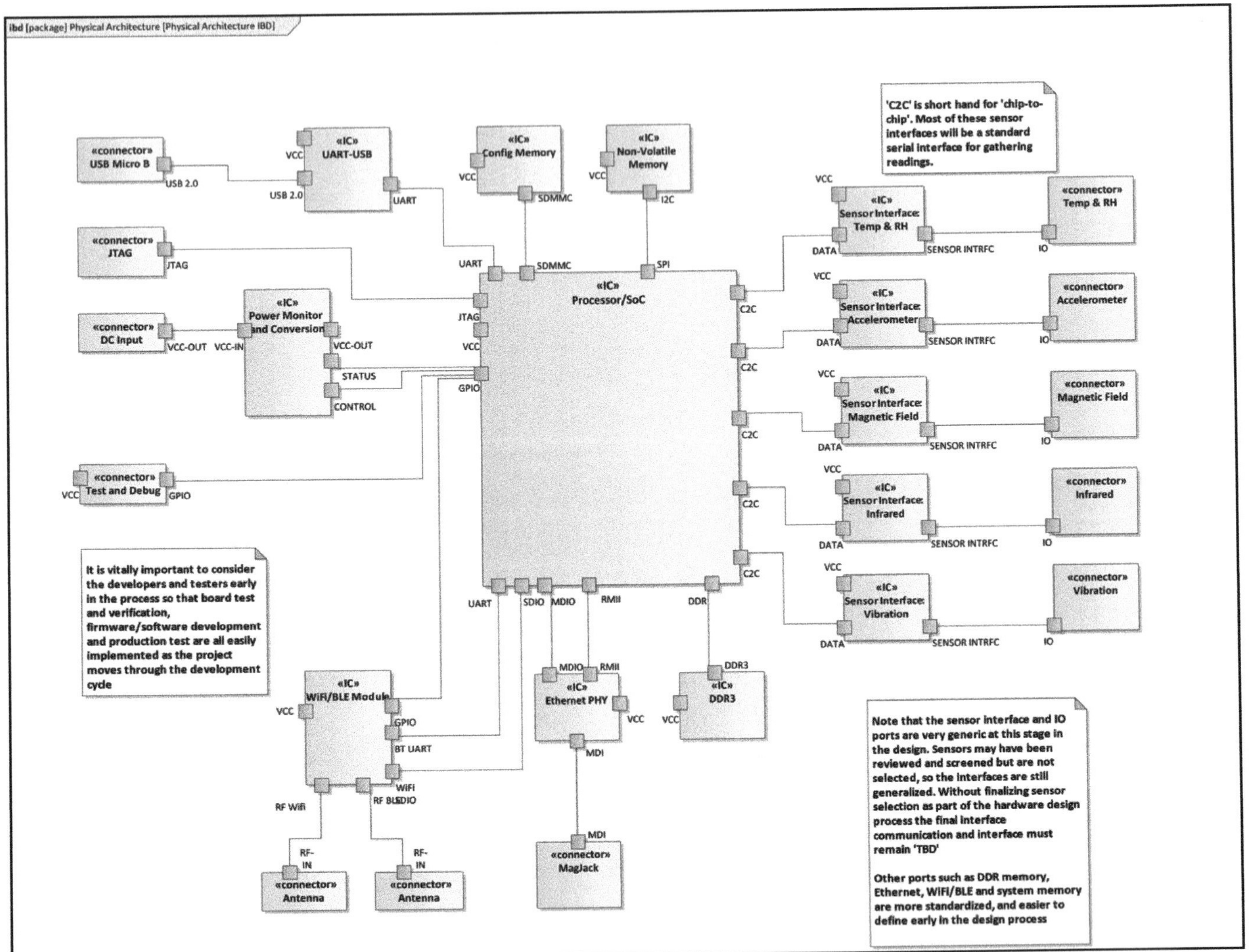

Figure 1.19 Smart Sensor Controller physical architecture.

1. Discrete microprocessor;
2. Discrete microprocessor and coprocessor/field programmable gate array (FPGA);
3. System on Chip (SoC).

More applications are lending themselves to the use of multiple processors and FPGAs. The natural progression of this trend led to higher integration, and the colocation of a relatively powerful Advanced Risc Machine (ARM) processor (an A class processor capable of running an application) with an FPGA for doing data processing or algorithmic acceleration. In our case, any of the three solutions mentioned above are feasible solutions; engineering teams could likely make any of the three work. However the ideal solution given the required functionality is using a SoC. A SoC, even an entry-level one, provides tremendous flexibility and processing power for many applications. Typically, this allows for a full Linux operating system to run and handle the high-level tasks such as serving data, and the inclusion of FPGA fabric allows for data processing to be done more efficiently in parallel and can even allow for a real-time coprocessor (like an M class ARM processor) to handle some of the lower-level, more time-critical functionality such as system diagnostics. These operating systems require a file system and run-time memory, which is why there are physical blocks for both double data rate three synchronous dynamic random-access memory (DDR3 SDRAM) and for configuration memory in the diagram.

The Bluetooth functionality has been paired with a Wi-Fi link. This comes from knowing what solutions exist in the marketplace (recall that it is important for a systems engineer to know what solutions are in the market and to use that information to sometimes drive design; this is one of those cases).

Things such as JTAG [18] (a very common test interface named after the Joint Test Action Group) and the UART (Universal Asynchronous Receiver/Transmitter) are in place because test and development were considered. The generic test and debug physical block is in place to accommodate a production test as well as field service. Consider something simple like a blinking heartbeat LED and some status LEDs that can be easily interpreted with a quick glance. The remaining blocks are directly driven by requirements as presented by the management and marketing team.

1.6.10 Trace Your Steps

A sample trace diagram, as shown in Figure 1.20, can be examined to understand how requirements can be allocated to various physical blocks of the architecture. This can be done exhaustively (and is required to be done this way for critical or safety systems) for the entire model if desired or only for a select set of requirements and diagrams.

At this point in time, a high-level set of architectural diagrams exists, the requirements are understood, and the engineering team has a physical architecture to start to implement (see Chapter 2). We are well on our way to developing this new product, and we continue to strive towards our key milestone.

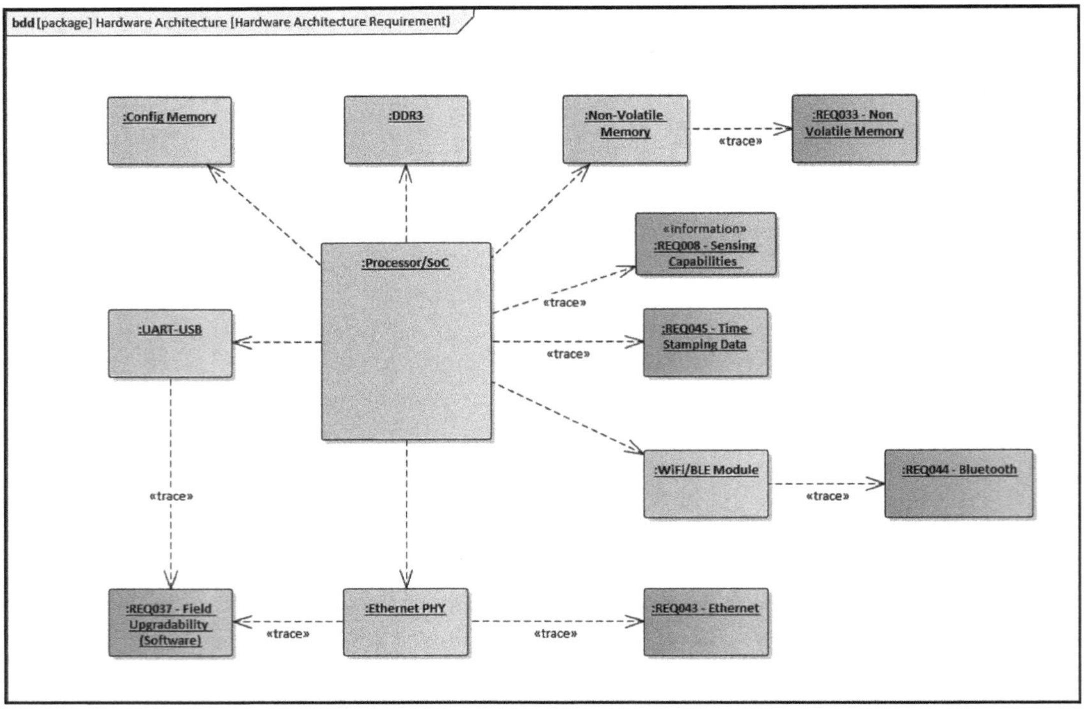

Figure 1.20 Tracing requirements to physical blocks.

1.6.11 System Verification and Validation: Check Your Work

The system validation plan is the highest-level integration test plan; it should directly validate the requirements put forth by the customer (or business unit) at level 10. This test plan will typically spell out an objective, show a link to the tested requirements, and point out any necessary prerequisite tests.

As an example, consider an environmental test being performed at the system validation level. It would be considered a best practice to validate that the environmental chamber is actually meeting the temperature and humidity set points that we are expecting with calibrated measurement equipment prior to subjecting your product to that environment, defining the information to be recorded and how often that information must be recorded, and also very clearly defining pass-fail criteria. It also is often noted in these test plans if any assumptions or constraints exist that are pertinent to the task at hand. As an example, consider a test plan for the communication of many connected SensorsThink nodes back to a master server. This test plan would be created under the assumption that the server both existed and was able to communicate to many nodes concurrently. Another key concept is that of validation versus verification; this will be discussed in a bit more detail later. For now, it is sufficient to say that validation is often regarded as "was the right system built?" testing and verification is often regarded as "was the system built correctly?" testing.

1.7 Engineering Budgets

The goal of any engineering project is to deliver the final product on time, to quality, and on budget. Naturally, we consider the budget element to indicate the financial budget allocated for the development, which it primarily is. However, successful project delivery will require the achievement of many engineering budgets across several subsystems of the design.

Engineering budgets are a response to requirements, contained in the system requirement specification. The engineering budget allocates portions of the overall target defined by a system requirement to subsystems in the design, providing clear guidance and targets to the designers.

It is natural when doing this allocation to retain a portion of the overall budget as contingency, in case one subsystem is unable to meet its budgets. Retaining contingency at the highest level prevents each subsystem from retaining a contingency, which makes achieving its budget harder and therefore increases the technical risk and potentially impacts the delivery timescales and cost trying to achieve the budget.

1.7.1 Types of Budgets

When allocating an engineering budget, it requires some thought and analysis to ensure that the allocation is sensible. It is not sensible, for example, to allocate a mass budget equally between the enclosure and the circuit card. Nor is it good practice to allocate evenly the performance budget between a low noise analog front end and the associated digital processing element. Budgets should be allocated depending upon the needs of the subsystem.

The engineering budgets documents generated for a project will vary depending upon the requirements; however, commonly generated engineering budget documents include:

- *Unit price cost:* A financial target that defines the manufacturing cost of the assembled product. This includes not only the cost of the bill of materials but also the cost of assembly and manufacture.
- *Mass budget:* A physical target that defines the acceptable mass of the finished system. Achieving this budget is critical on applications for mass constrained applications (e.g., aerospace, space, automotive application). Mass budgets will allocate budgets to elements of the design such as enclosure, circuit card, power supplies, and actuators.
- *Power budget:* A physical target defines the maximum power that may be drawn by the application. A power budget may define different operating conditions and allowable power budget for those operating conditions. Examples of this include low power and reduced power operating requirements. Power budgets will also include allocations for direct current (dc)/dc conversion.
- *Memory budgets:* A storage target, this defines the required nonvolatile and volatile memory required for the overall application.

- *Performance budgets:* This will vary from project to project depending upon the application; however, the performance budget will allocate the overall performance requirement across the subsystems. This enables challenging subsystems to have a larger proportion of the budget than less challenging. For example, when measuring real-world parameters using a sensor, the analog element of the design that is more effected by temperatures, tolerance, and drift will be allocated more of the budget.
- *Thermal budget:* On many applications, the thermal dissipation can be challenging; examples include wearable products, space, and aerospace where thermal management is difficult or could result in harm. The thermal budget is therefore closely linked with the power budget and will address not only the external interface temperatures but also the allowable dissipation of the subsystems and even key components within the design that drive the thermal solution.

The creation of the engineering budget is a system-level task; as such, they should be generated as part of the requirements generation and prior to preliminary design review. This enables the designers to work with these budgets when they are creating the preliminary design. These budgets and how they are addressed are then able to be reviewed as part of the preliminary design review, which is held before the detailed design continues.

1.7.2 Engineering Budgets: Some Examples

Let's look at the engineering budgets, which can be generated using the requirements defined in Section 1.5. The one engineering budget which needs to be considered for an example application is the REQ036 - Physical Requirements (Mass): The system shall weigh no more than 5 kg. This defines the maximum permissible weight of the overall system; we need to define the allocation of each component of the system. At the architectural level, these components can be split into:

- *Enclosure and fixings:* The enclosure of the system, which protects it from its environment and provides thermal dissipation.
- *The ac/dc converter:* Power converter, which converts the input voltage to secondary voltage required by the system.
- *Circuit card assembly:* The circuit cards that contain the electronic components, which implement the required functionality of the system.

1.7.2.1 Tipping the Scales: Mass Budget

Allocating the mass budget between these three elements can be pretty straightforward; the majority of the mass budget will be allocated to the enclosure and fixings, which accounts for the most mass of the system. Allocation for the enclosure includes any additional fixings or shielding required in the system, for example, shielding over mixed signal converters.

As such, the mass engineering budget can be allocated as:

- Contingency: 500g;
- Enclosure and fixings: 3,500g;
- The dc/dc converter: 500g;
- Circuit card assembly: 500g.

This budget allocation provides achievable budgets to the major elements of the solution while also retaining a contingency in case any element struggles in meeting these requirements.

1.7.2.2 Powering the Pieces: Power Budgets

The smart sensor SoC requirements do not contain a maximum power consumption requirement as the system is intended to be powered from an electrical outlet. As such power budget would be constrained only by availability of the alternating current (ac)/dc converters to convert between ac power and the required dc input. As a design engineer, such a situation is nice as the only need is to accurately determine the power dissipation on the board once component selection has been completed to accurately size the ac/dc power converter.

However, we are not always as lucky as we are in the case of the Smart Sensor SoC development. To ensure that the benefit of a power budget is understood, we will go from a simple example of the Smart Sensor SoC to looking at a very challenging and power-constrained application in that of a CubeSat application.

Electrical power to power a CubeSat is generated from solar panels that are mounted on the external faces of the CubeSat. Due to the small surface area of the solar panels, CubeSat applications are typically very power-constrained.

Typical requirements placed on a CubeSat payload are:

- The payload shall provide regulated supply rails at 3.3V, 5.0V, and 12.0V.
- The total current drawn from each rail may be up to 600 mA.
- The total power dissipation during the sunlit orbit shall be 400 mW.

These requirements present an interesting challenge to the designer as the individual currents can supply reasonably high currents. Although the total potential power supplied is just over 12W, what limits the solution is the final requirement. The total power dissipation must be no more than 400 mW in the sunlit orbit, which is typically in the region of 45 minutes.

Higher powers can be drawn from the batteries; however, the average dissipation over the 45-minute orbit must be 400 mW or less. Higher-power payloads can therefore be powered for a shorter period; therefore, if we needed to draw 12W from the power supplies, we can operate the payload for only a fraction of the 45-minute sunlit orbit.

To keep battery life maximal, during the dark half of the orbit, the payloads are powered down.

Our power budget must therefore demonstrate that the design can operate within these constraints; as the project matures, the power budget needs to be kept up-to-date to ensure that any power increases due to scope creep are addressed and are still within budget.

1.7.2.3 Count Every Coulomb

When determining the budget for this payload, several factors must be considered, for example:

- Operating modes of the CubeSat payload;
- Proposed power architecture of the solution;
- Losses due to power conversion;
- Power switching to reduce the power dissipation.

The budgets that we must achieve for this CubeSat mission are:

- 3v3 Supplies less than 1.98W;
- 5v0 Supplies less than 3.0W;
- 12v Supplies less than 7.2W;
- The average power dissipation over 45 minutes is less than 400 mW.

In this example the CubeSat payload contains an FPGA along with configuration memories, SRAM memory, and support circuits such as oscillators.

Creating the power budget for this application requires the use of power estimation spreadsheets provided by the FPGA manufacturer (e.g., Xilinx XPE) to determine the maximum power required for the FPGA application. This estimation needs to include clocking rates, logic, and block random access memory (BRAM) resources, specialist macros (e.g., DSP, Processor, and Clocking). At early stages of the project, it is also good practice to include a margin factor in this estimation.

It also goes without saying that our power budget should be based upon the worst-case power dissipation taking into account process, voltage, and temperature (PVT) variation.

For this CubeSat application, the power estimation for the FPGA and support circuitry can be seen in Tables 1.1 and 1.2.

These power estimations identify the power drawn from each of the subregulated rails; they do not include the power required from the supply rail.

The proposed architecture uses the 3v3 supply to directly power the FPGA 3v3 IO, non-volatile random-access memory (NVRAM), and controller area network (CAN) interface. As such for the 3v3 supply rail, we do not need any further conversion, the load on the 3v3 Supply rail is 1.59W or 78% of the available 1.98W.

Table 1.1 FPGA Power Estimation

Rail	*Voltage*	*Tolerance*	*Voltage Maximum*	*Current (mA)*	*Power Maxiumum (W)*	*Power Nominal (W)*
FPGA Core	1	5.00%	1.05	750	0.7875	0.75
FPGA IO	2.5	5.00%	2.625	250	0.65625	0.625
FPGA IO	3.3	5.00%	3.465	250	0.86625	0.825

Table 1.2 Support Circuitry Power Estimation

Rail	*Voltage*	*Tolerance*	*Voltage Maximum*	*Current (mA)*	*Power Maxiumum (W)*	*Power Nominal (W)*
NVRAM	3.3	5.00%	3.465	50	0.17325	0.165
Config Memory Core	1.8	5.00%	1.89	25	0.04725	0.045
Config Memory IO	2.5	5.00%	2.625	10	0.02625	0.025
CAN If	3.3	5.00%	3.465	150	0.51975	0.495

The power drawn from the 3v3 rail can therefore be said to be within the available budget, while the 1v0, 1v8, and 2v5 rails are generated using dc-dc switching converters connected to the 5v0 rail. When we convert dc-dc voltages, we must account for the switching efficiency; depending upon the efficiency, the power required from the supply rail may be considerably higher than that drawn by the load. When trying to meet tight power budgets using dc-dc converters, we want to select high-efficiency converters. For this example, a converter efficiency of 85% was assumed.

The total power drawn from the 5v0 rail is 2.1W, this is again within the allowable budget of 3W, providing a 70% load on the 5v0 rail.

The loads on the 3v3 and 5v0 rail provide margins of 22% and 30% respectively, we also need to ensure the power budget is able to address short duration start-up currents. Typically, these will be addressed in a worst-case analysis of the design.

With a total power draw of 3.66W the final determination is the time the CubeSat payload can be operational to achieve the 400-mW average during the 45-minute sunlit orbit.

When we calculate 400 mW/3.66W, we see the ratio required to achieve the average, as such the payload can be operational for just over one-tenth (10.92%) of the 45-minute operational time.

This becomes a limiting factor on the FPGA design, it must be able to achieve its task within the allocated timescale as determined by the power budget. This time limit can be therefore included as a derived requirement in the FPGA level requirements.

One final point of note on this example is throughout this example we used the worst-case figures. If we had used the nominal figures the power requirements from the 3v3 and 5v0 rails would be 1.485W and 1.445W, respectively, resulting in a total power of 3.185W. This nominal dissipation of 3.185W compared to the worse case of 3.66W is a 13% difference and would result in the 400-mW sunlit orbit average being exceeded if the operational time was based of the nominal power dissipation and process, voltage, and temperature variation resulted in power

Table 1.3 Power Conversion Rail

Supply	*POL Power (Watts)*	*Efficiency*	*Supply Power (Watts)*
1v0	0.7875	0.85	0.926470588
1v8	0.04725	0.85	0.055588235
2v5	0.6825	0.85	0.802941176

dissipation closer to the maximum. This increased power dissipation may result in system level impacts as the battery experiences more dissipation than expected.

1.7.3 Engineering Budgets: Finishing Up

While this example is long it demonstrates the importance of determining a power budget at a subsystem levels. Without the correct analysis and budget allocation the hardware and FPGA design teams may inadvertently design a solution which is not capable of achieving the budgets.

If in either of the examples presented, we had been unable to achieve the allocated budget, the issue can be raised internally within the project. One method this issue can be raised internally is by the issue of a request for deviation (RFD). This request for deviation can then be reviewed by the systems engineering management team and if agreed, some of the agreed contingency budget can be freed up for that subsystem. If there is no contingency available, the RFD can be rejected and the engineering team will have to go back to the drawing board as to how to address the solution. These high-risk elements of the design should have been included in a technical risk register with associated allocated budgets to address the situation should this situation occur.

1.8 Interface Control Documents

When it comes to designing our system, the systems interfaces represent an area of risk. This risk arises as the system interfaces include multidisciplinary factors (e.g., electronic and mechanical design along with the complexity of interfacing with external systems). As engineers to mitigate issues arising due to incorrect interfacing between units or modules, we define the interfaces accurately in an interface control document (ICD). The ICD enable us to clearly define the different interfaces within our system and will cover a range of disciplines. As such, they will cover such areas as:

- *Mechanical ICD:* A mechanical ICD will define the mechanical aspects of the design. This will include not only the external dimensions and tolerances but also additional requirements such as finishing, emissivity, and flatness. External mechanical ICDs may be provided by the customer if they have a tightly defined space envelope, which the system must integrate within. If the external mechanical ICD must be developed as part of the development contract, then it is advisable to ensure agreement and sign off by the customer at an early design review. It is also common to define not only an external mechanical ICD but also one or several internal mechanical ICDs. Depending upon the complexity of the solution these internal mechanical ICDs may simply address how the circuit card mounts within the enclosure to how several circuit cards and a back plane are contained within the enclosure. These internal ICDs will also include mounting points, keep out areas and mechanical bracing and thermal paths for the circuit cards.

- *Thermal ICD:* A thermal ICD will define the thermal interfaces of the system. This will include the thermal conductivity between the enclosure and the mounting points, along with any additional thermal management methods. These additional thermal management methods may include forced air cooling, conductive paths, or liquid cooling. The thermal ICD enables the system designers to understand the thermal environment and design a system which can operate within it. The thermal ICD will also be very useful when (if) it comes to determining the part stress analysis and reliability figure.
- *Electrical ICD:* An electrical ICD will define all the electrical interfaces of the system. Often this will be split into an external electrical ICD and an internal electrical ICD. The electrical ICDs will define the power supplies, signal interfaces, and grounding. Obviously when working with electrical interfaces we should try to work with industry standard interfaces such as Ethernet, RS232, and USB depending upon the needs of the system. This is especially true for external interfaces where the system is required to interface with others. Internally within the system again standard interfaces should be used wherever possible. However, if they are not the interfaces should be defined in depth using the internal EICD. One critical element of the EICD along with the signal and interface types is the definition of the electrical connector types and pin allocations. Internally, this enables the individual circuit card designers to develop solutions from a single reference document ensuring they are fully aware of the connector type required and the exact pin allocations.

As electronic engineers it is the electrical ICDs which throughout our career we will both develop and use most often in our careers, although we will also use the mechanical and thermal ICDs at times.

The external electrical ICD will define the interfaces presented external by the system, this will include not only power and signals, but also connector type, maximum number of mating cycles and pin allocation. Within the ICD we should define several groups which can be used for each signal. Commonly used groups include power, analog, single-ended digital signal, differential digital signal, high reliability signals, and ground, the number of groups will depend upon the different interfacing standards used.

1.8.1 Sticking Together: Signal Grouping

Using groups means that each signal within the ICD can be allocated to one of the groups, as such updates or modifications to the signal group does not require significant rework, although each group will contain different parameters. A power group, for example, will contain at least the following elements:

- *Nominal voltage:* The nominal voltage supplied on the power rail.
- *Tolerance:* The maximum and minimum tolerance available across all worse-case conditions may be as simple as +/–5%, which is standard for many digital power rails. Alternatively, it could contain a range of tolerances across the operating conditions.

- *Maximum power:* The maximum power that can be drawn from the power rail.
- *Noise characteristics:* How much ac and dc ripple present on the rail.
- *Ground return:* The return path for the supply current.

For digital signal groups, the following elements may be used:

- *Logic type:* The logic standard used (e.g., LVCMOS1v8 or LVCMOS3v3); logic standard should also refer to the appropriate JEDEC standard. This is important, as it includes the necessary V_{IH} and V_{OH} minimum and maximum levels for that logic standard.
- *Characteristic impedance:* The characteristic impedance of the tracking required for the signal in the PCB or cabling.
- *Termination scheme:* If the interface terminated and, if so, what type of termination scheme is used.
- *Data rate:* The data rate of the signal.
- *Protocol:* How information is communicated over the link; this may be a reference to an industry standard, or if a bespoke approach is being used, then it may be a reference to an internal document that fully defines the communication standard.
- *Special requirements:* For example, the output open drain or tristate.
- *Reset:* Quiescent state following reset.

When it comes to differential digital interfaces, the interfacing specification is very similar to that of the single-ended logic groups, with the exception that the logic type is defined as either low-voltage differential signaling (LVDS), low-voltage positive emitter-coupled logic (LVPECL), or high-definition multimedia interface (HDMI). The characteristic impedance will also be different at typically 100Ω differential.

We also discuss several analog and radio frequency (RF) signals that we need to define as belonging to a group; these analog signals will typically contain elements such as:

- *Voltage range:* The operating voltage input range;
- *Current:* Maximum supply current;
- *Noise:* Ac and dc noise;
- *Return path:* Return path for the signal;
- *Characteristic impedance:* The characteristic impedance of the signal tracking/cabling.

Defining these different groups within the ICD is also useful in the later stages of the design process when layout is underway. These different groups can have different layout constraints applied to them regarding length matching and even separation between signal groups.

Once the different signal groups are created, we are then in a position to begin creating the actual interface information (e.g., signal, direction, and pin allocation).

1.8.2 Playing with Legos: Connectorization

The next step in the ICD creation process is to define each signal, its direction, and signal grouping. When we do this assignment, we also need to select the connectors as well. Connector selection will require consideration of the type, polarity, number of signal pins, mating cycles, and current rating.

We also need to consider carefully the connector type. The considerations here include the number of pins available, physical size, mounting technology, and polarity. If multiple connectors are required, it is a good idea to consider the use of different connector styles to prevent inadvertent incorrect connection. This incorrect connection could lead to damage and require significant design analysis depending upon the application. Using different styles of connectors prevents this, as it becomes impossible for the connectors to be mated incorrectly. If it is not possible to select a different connector style, then we can use connector keying on some connector styles to also prevent inadvertent connection.

The selection of socket or plug connectors will also be made during the connectorization phase. Here electrical safety also plays a part. For example, the main power input, either ac or dc, should for safety be a plug at the equipment, ensuring that the mating cable is a socket and prevents live pins on the power supply cable from being able to short out inadvertently when the cable is being connected or otherwise used.

As engineers, we can mitigate the number of mating cycles using connector savers. These are adaptors that mount with the connector and then enable all of the cable mating to be with the mating interface on the saver. Using such an approach allows the main connector to have one mating cycle (the connection of the saver) while the saver absorbs the wear and tear as the cable is mated and de-mated as required during development. While connector savers can be expensive, they can be very useful during development to ensure that the mating connectors are not used past their expected number of mating cycles.

To help achieve higher currents through the connector, multiple pins can be used to share the current for the same power rail. De-rating reduces the electrical and thermal stresses to help to ensure its reliability. With connector pins, it is normal to de-rate connector pins to carry only half their rated value. For example, a pin rated at 1A would be allowed to pass a de-rated current of 0.5A; therefore, if a power supply requires one amp over a connector two pins are required to meet de-rating rules.

We also need to consider careful signal placement on the connector to prevent failures. One way to do this is by careful consideration of the pin placement and how unused pins are addressed.

To prevent accidental shorting, especially power and return, it is good practice to separate the power pins from the return pins. This is achieved by physical separation between the power and return pins. One example of this would be locating the power pins at one side of the connector and the return pins at the opposite side on a sub-D-type connector.

The signals are then allocated to the pins between the return and power pins at either end. If the application requires high reliability, it may be further necessary to isolate different groups of signals from each other.

In this case, guard pins may be used to ensure that the critical signal cannot short. Guard pins therefore surround the critical signal pin. These guard pins should be connected to ground via a high value resistor, ensuring that the critical signal cannot be shorted to either return or the supply rail.

1.8.3 Talking Among Yourselves: Internal ICDs

ICDs are not only of interest at the system or circuit card level; they are also very useful within the FPGA and the SW world. However, when an internal ICD is generated, we focus less on the physical connector and pin-out and more on the formation sharing based on standardized interfaces (e.g., AXI Memory Map for FPGA or shared memory structure in an SW-based solution).

Such internal ICDs will therefore detail memory maps, which define the accessible registers, along with detailing the bit fields for each register. A common reference point for the register mapping and bit fields means engineers developing different modules can easily develop solutions based around the internal ICD arriving at a solution that is capable of functioning as intended.

Internal ICDs are not limited to only memory mapped peripherals. Within a programmable-logic based solution, there will be several signal lines acting as communication and trigger sources.

When internal ICDs are developed for programmable logic-based solutions, we also need to define the clock domains on both sides of the transaction source and the destination. This enables both the design and verification teams to ensure that clock domain crossings (CDC), if required, are addressed safely.

1.9 Verification

After completing design work, and pouring hundreds upon hundreds (sometimes thousands upon thousands) of hours of effort into just coming up with design concepts, and then even more hours piled on top for doing design work and design reviews, a milestone is reached. New PCBs, mechanical enclosures, critical components, and sensors are finally ordered. This is a major milestone, but it is far from the end of a product development process. (There is a reason those pesky revision blocks exist on drawings after all.)

Having already touched upon the V-model and looked at how systems engineering and architecture work lays the groundwork for product design, implementation and ultimately validation. We should now consider the other V in V&V: verification. Recall the anecdote from earlier in the book that we mentioned; validation is often regarded as "was the right system built?" testing, and verification is often regarded as "was the system built correctly?" testing.

1.9.1 Verifying Hardware

Verifying hardware is absolutely a requisite part of delivering a robust and well-engineered product. Without verifying that, for example, power supplies are stable under load transients, but also under temperature extremes and input voltage (referred to as line voltage quite frequently), there may well be many peculiarities hiding in the design that would otherwise be undetected until fielded in a certain set of conditions. The same goes for every electronic circuit. Values change with temperature; values change with time. Doing up-front work to mitigate and reduce the risk and uncertainty associated with these changes is a big part of ensuring that a robust design is going to launch and not a marginal one that will be the cause of sleepless nights and heartburn for years to come.

In our new SoC platform, we certainly have many requirements that can be verified without exhaustive testing. Product weight and size, for example, can be verified in a matter of a few minutes with a prototype. Put the product on a scale, and grab a measuring tape; this could not be simpler.

In other cases, more thought is required and more care must be put into the test approach and verification method to really give a high confidence level in the product.

In the discussion on requirements, we discussed the various levels of verification and validation, ascending from the low levels of unit testing at (what we called) level 40 and rising up to level 10, where system validation is performed. The focus of this particular discussion is really around level 40 and level 30 testing for hardware.

Level 40 testing for hardware consists primarily of the verification of individual circuits (or small groups of circuits that comprise one function). Think of this like you might think of building a Lego house; you check each piece out to make sure that it is the right piece and matches what the instructions have laid out for use. In product design, the instructions against which we check are the requirements, the architecture, and the detailed design documentation that is generated during the design process (such as SPICE (Simulation Program with Integrated Circuit Emphasis) and IBIS (Input/output Buffer Information Specification) modeling, thermal modeling, and FEA (finite element analysis)).

1.9.2 How Much Testing Is Enough?

One obvious question as a starting point is how we determine what level of testing is required for a particular product. As often is the case in engineering, the right answer is that it depends. It depends on what type of system is being tested, what level of safety performance is required, what level of robustness is required, and what level of residual risk is acceptable for the product.

There are many normative standards related to the design, verification, and validation of products. Depending on the standard, there may be more or less prescriptive testing and documentation requirements laid out for designers to conform to. A few examples are:

- ISO 13489: Safety of Machinery [19];

- IEC 61508: Functional Safety of Electrical/Electronic/Programmable Electronic Safety-Related Systems [20];
- ISO 26262: Road Vehicles: Functional Safety [21].

1.9.3 Safely Navigating the World of Testing

If for no reason other than to have an example, let's use ISO 26262 as a reference document for some means of understanding how hardware test cases can be derived, and what kinds of testing can be used to verify hardware performance under even the harshest levels of scrutiny. It is important to note that these test case derivation methods and verification methods are really in place for safety-critical pieces of hardware inside of vehicles. They may be somewhat overkill for a nonsafety-relevant system, but performing these kinds of tests if the means are available is never a bad idea or bad practice. The more well understood the design is, the better off the entire team is in the long run.

Verification has several stages: planning, specification, and execution [21]. Let's look more closely at these stages. Verification planning essentially consists of considering what work output or work products will be produced. However, quite a lot of thought needs to go into this in order to do it properly. For example:

- What methods will be used;
- What the pass/fail criteria are;
- What environment(s) will be used (these environments should be controlled by a quality management system if they are used in safety critical testing);
- What tools will be used (these tools should always be part of a quality management system for safety-critical testing);
- What actions must be taken if items fail to meet the pass criteria during the course of testing.

There are other aspects of verification planning that should be considered as well during the planning phase. Critically looking at the adequacy of the test being performed is one of the more important ones from a verification perspective. If, for example, SensorsThink wants to verify their Ethernet interface and they do so by sending a simple series of ping packets to a host machine, that is probably not a very thorough representation of real-world use and may lead to a false sense of security. Conversely, using a physical layer qualification test package that measures things such as eye width, power spectral density, jitter, and distortion is surely a better set of metrics against which to measure performance.

The verification specification essentially selects and specifies the methods to be used. This can be any number of things: simulation scenarios, physical test cases, and even something as simple as a checklist. As long as the adequacy of the specification is justifiable and reasonable, there usually is not much else to be considered here. That is not to say that the specification should lack detail or clarity; preconditions (input data) and configurations used (a subset or superset of variants) are wonderful things to document. The test environment conditions are also important bits of information to specify to allow for repeatability of testing should failures or anomalies occur.

The execution of the verification activity is exactly as it sounds: performing the testing, capturing the results, and comparing them to the previously specified pass/fail criteria. It is generally the best practice to review these results with a broader team of subject matter experts and to note any discrepancies and justification for them and to essentially make sure the team feels as though the product is standing firmly in the camp of robust.

1.9.4 A Deeper Dive into Derivation (of Test Cases)

Now that we understand the process of verification, let's consider hardware in a bit more detail, again with an eye to ISO 26262 for a guiding source of information. Deriving test cases can be complex, and sometimes the best approach is not readily apparent.

Table 1.4, taken from ISO 26262 [21], is given as a resource to help engineers derive meaningful test cases using a variety of methods that may be applicable to any number of requirements. We can see here that each method is scored over each system level in the table. (Note: The actual heading of the table is ASIL (Automotive Safety Integrity Level), a level A is the lowest safety level, and a level D is the highest safety level; for the purpose of our development, which is not a safety product, those details are not germane.) Items denoted with ++ are highly recommended for a system of the corresponding level; items denoted with + are recommended for a system of the corresponding level, and items denoted with an "o" are not recommended for a system of the corresponding level. For a lower-level system, a walk-through of the hardware design can be a perfectly suitable means of verifying the hardware design meets the requirements; for a higher-level system, an inspection of the design is preferred and a walk-through is not preferred. (During a walk-through of a system the developer explains the system to a third-part assessor; typically, this does not involve operating the item/element. During inspection, an examination of the work products is conducted against a defined procedure or checklist and typically ends with a review of the results.)

Another aspect of hardware verification is a form of integration testing, and there are essentially three methods of testing that can be performed to verify proper hardware robustness.

Functional testing aims at verifying that the specified characteristics of the item or device under test have been achieved. The item is given standardized input data (consider a pseudorandom bit sequence), which adequately characterizes the expected normal operation. The outputs are then measured, and their response is

Table 1.4 ISO 26262 Part 5 Table 3 (Partial)

_Methods	System Level	A	B	C	D
A	Hardware design walk-through	++	++	o	o
B	Hardware design inspection	+	+	++	++
D	Simulation	o	+	+	+
E	Development by hardware prototyping	o	+	+	+

Source: [21].

compared with that given by the specification (consider an eye-mask diagram for the Ethernet).

Fault injection testing means intentionally introducing faults in the hardware product and analyzing the response of the hardware. Model-based fault injection (e.g., SPICE, IBIS, FEA, and thermal) is also applicable, especially when fault injection testing is overly complex to do at the hardware product level. (Consider trying to perform single-bit error testing that might be expected from cosmic rays.)

Electrical testing aims at verifying compliance with requirements within the specified voltage range. Again, the example of a power supply performing over a wide input range is a good one to consider.

The next level up from the last couple of examples is what we would consider to be integration level testing, where the hardware has been unit-tested, and these individual components and circuits have been verified. Combining this set of hardware with software that performs nominal operation of the hardware is really the first big integration milestone for an embedded system. Once this is ready and the underlying hardware is believed to be adequately tested, the overall embedded system (or, if in a complex enough system, perhaps a subsystem) is ready for robustness testing.

During environmental testing with basic functional verification, the embedded system is put under various environmental conditions during which performance is assessed. For the SoC platform that we are designing, this may consist of performing a strenuous Ethernet traffic test over a wide range of temperature and humidity and over a very long Ethernet cable. This type of test gives strong confidence in the robustness of an interface if the test is successful.

Expanded functional testing is intended to verify the response of the system to input conditions that are expected to occur only rarely or that are outside the nominal specification of the hardware, that consider receiving an invalid command, or that interrupt a data transmission intentionally. If the system can gracefully handle these conditions, that is a good indicator of robustness.

Table 1.5 ISO 26262 Part 5 Table 11

_*Methods*	*System Level*	*A*	*B*	*C*	*D*
A	Functional Testing	++	++	++	++
B	Fault Injection Testing	+	+	++	++
C	Electrical Testing	++	++	++	++

Table 1.6 ISO 26262 Part 5 Table 12 (Partial)

_*Methods*	*System Level*	*A*	*B*	*C*	*D*
A	Environmental testing with basic functional verification	++	++	++	++
B	Expanded functional test	o	+	+	++
C	Accelerated life test	+	+	++	++
D	Mechanical endurance test	++	++	++	++
E	EMC and ESD test	++	++	++	++

Accelerated life testing is a topic worthy of its own book (or several for that matter) and its own subject matter experts. The short version of the intent of the testing is to predict the behavior and performance of the product as it ages by exposing it to conditions outside of its normal operating conditions. This is a test method based on analytical models of failure mode acceleration and can provide valuable information related to the robustness and margin available in the design. These tests must be properly designed and executed to provide value [22].

Mechanical endurance testing is intended to stress the product to the point of either failure or noticeable physical damage. Things such as vibration testing, drop testing, and impact testing are often performed to ensure that the product can withstand the environment in which it is going to operate. Performing the testing to failure gives information on the available margin in the design. Consider a drop test for the product that must pass at 10 feet. If the test passes at 10 feet, but fails at 10 feet and 1 inch, the engineering team may want to take a closer look at the design. If the test does not fail until 15 feet, the team can probably rest easy on that particular test.

EMI/EMC testing is always a hot topic in engineering circles and again warrants a book of its own; and is certainly its own area of expertise. For most product engineers, this EMI/EMC testing is the gateway into the marketplace. Without passing those tests, your product will not be approved for sale. That is why they are such a good marker for product robustness and a necessary one at that. During this testing, your product will be zapped with high-voltage transients, have high-power RF energy pointed at it, and have extremely sensitive RF antennas listening to it while it operates. If the product cannot take the heat or if it makes too much noise, the engineering team has to go back to the drawing board.

Luckily for us, the requirements provided to us are relatively easy for which to derive test cases and it would not be too demanding of a task to put together a cogent and defensible test plan to show that the hardware portion of the system was performing robustly for the application.

SensorsThink is getting off the hook a bit easily in this regard, as they are not a safety product and they are not operating in overly harsh environments. That does not mean that we do not want to do a good job verifying our design. It just means that the level of scrutiny placed on the engineering team is not as high as it might be in another situation. In the hardware design part of our journey, we will take a closer look at how we can design our product to make testing easier. For now, this should put in perspective the necessity of proper testing and some additional food for thought on how SensorsThink may want to approach their product testing.

1.10 Engineering Governance

Engineering governance is probably one of the least fun elements of the design process, but at the same time it is one of the most important. Engineering governance defines how we progress from the initial concept of the project to the completed delivery and even beyond to end of life and safe disposal.

As you would expect, engineering governance is tied closely with the design life cycles and the design reviews, which are held as the program progresses. The engineering governance program defines which of the checklists and reviews need to be completed for a design review. Typically, the engineering governance program within a company is overseen by a chief engineer. The chief engineer has dual roles in the governance program. The first role is to ensure compliance with the program by the various projects under development at anyone point in time. The second role is to ensure the engineering governance program itself stays fit for purpose and stays up to date with the latest technologies, standards, and approaches. This means that the chief engineer has to be able to attend conferences and standards bodies and, if necessary, run research projects to verify the suitability of new technology.

1.10.1 Not Just Support for Design Reviews

One of the key roles of the engineering governance program is to ensure that the necessary design maturity evidence is available for significant design reviews (e.g., preliminary design review). However, correctly implemented, it is much more than that. A correctly implemented governance program will institute a series of design standards and design checklists and reviews to ensure the suitability, maturity, and risk of the design is known and suitable for progression to the next stage of development. An example of this would be the progression from schematic design to layout; to ensure minimal risk, it is necessary to demonstrate correctness of the schematic design and documentation of the layout constraints while also aligning with mechanical and thermal design teams to ensure the correct PCB size, connector locations, skylines, and thermal considerations.

As such, the engineering governance scheme is multidisciplinary and covers the whole host of engineering disciplines from systems engineering to electronic design, mechanical design, and software development.

Engineering governance can be split into two distinct elements. These are design rules, which define how the different elements of the design must be completed, and compliance, which checks that the design rules have been followed.

1.10.2 Engineering Rule Sets

Typically, the design engineers and other members of the project team will use the design rules as they create the design. Compliance with the design rules will normally be verified by an experienced independent engineer, who signs off the design to progress to the next stage of design.

For an embedded system design, there will be several different design rules against which engineers must demonstrate compliance:

- *Requirement definition and capture:* Present a common approach to requirement capture defining the expected contents of the requirement specification covering functional, nonfunctional, and environmental requirements. These rules will also define the language to be used to define the requirements (e.g., use of shall, should, may, and would).
- *Schematic design rules:* Define the rules for drawing schematics that are not only functionally correct but also easily understandable by other engineers. These rules may enforce elements such as part selection and derating

to ensure compliance with company manufacturing facilities and reliability requirements.

- *Layout design rules:* The layout rules will contain a number of rules to enable a high yield in the board manufacturing process, while also enabling a high yield in the assembly process. The PCB layout rules will also include sections on high-speed signaling, mixed signal design, and safety (e.g., creepage and clearance distances).
- *RTL coding rules:* Defines the acceptable register-transfer level (RTL) language constructions that may be used to capture the design. These rules will also define acceptable verification techniques, along with change and configuration control.
- *SW coding rules:* Very similar to the RTL coding rules, this rule set will define acceptable software language constructs that may be used.
- *Product safety:* Contains the design rules and analysis that must be followed to achieve regulatory compliance sign-off (e.g., IEC61508, ISO26262).

Complying with the design rule sets as the design progresses can be challenging, as often the rules are defined in published documents, which, despite best intentions, are easily overlooked.

1.10.3 Compliance

The best way to ensure compliance with engineering governance rules is to incorporate them within the CAD tools we are using to implement our embedded system. Within schematics and layout tools, we can implement rules using constraints and design for manufacturing settings. If additional checks are required, it is possible to create scripts in Python or Tool Command Language (TCL) to implement these additional checks. When it comes to ensuring compliance with the RTL and software coding standards, linting tools can be used to ensure compliance against rule sets that include naming conventions and design rules.

The automation of rule sets directly in the CAD tool itself obviously increases the compliance with the rule set. It also makes demonstrating compliance with the rule set easier when it is time for design sign-off.

If automation cannot be achieved, then checklists can be used to ensure that the necessary design rules have been complied with. To ensure repeatability, the checklists should be stored under a configuration control system, enabling the design team to access the latest version.

1.10.4 Review Meetings

Regardless of the method of creation, the design rules review needs to be reviewed by the engineering team and the chief engineer or his or her nominated representative when a subject matter expert is required (e.g., FPGA, digital electronics, high speed).

When performing the design rules review, there will often be rules that do not comply with the rule. This is to be expected, as the design rules are created to address a wide range of developments. There will also be corner cases that do not align with the checklist.

This is where the roles of the chief engineer and independent expert come into play along with the engineering team. Each of the rule's violations needs to be analyzed and allocated to one of two categories:

- *Must be fixed:* These rule violations must be addressed before the next stage of the project can be started.
- *Will not be fixed:* These rule violations do not need to be addressed. For each violation, a detailed explanation of the reasoning being the exception must be recorded.

Once the design rules review has been completed, the design review notes, categorization of rule violations must be stored within a change management or product life-cycle management system. This provides a record of the review occurring and makes the data easily available for higher-level design reviews and certification reviews.

References

[1] Cooper, R., *Winning at New Products: Creating Value Through Innovation*, 3rd edition, New York: Basic Books, 2011.

[2] Sparx Systems Pty Ltd., "Enterprise Architect 14," August 22, 2019. https://www.sparxsystems.com/products/ea/14/index.html.

[3] Reliable Software Technologies, "Mew Models for Test Development," 1999. http://www.exampler.com/testing-com/writings/new-models.pdf.

[4] Forsberg, K., and H. Mooz, "The Relationship of Systems Engineering to the Project Cycle," *National Council on Systems Engineering (NCOSE) and American Society for Engineering Management (ASEM) Conference*, Chattanooga, TN, October 21, 1991. https://web.archive.org/web/20090227123750/http://www.csm.com/repository/model/rep/o/pdf/Relationship%20of%20SE%20to%20Proj%20Cycle.pdf.

[5] INCOSE, "About: INCOSE," https://www.incose.org/about-incose.

[6] Requirements Working Group, International Council on Systems Engineering (INCOSE), "Guide for Writing Requirements," San Diego, CA, 2012.

[7] International Electrotechnical Commission (IEC), "Available Basic EMC Publications," August 22, 2019. https://www.iec.ch/emc/basic_emc/basic_emc_immunity.htm.

[8] RoHS Guide, "RoHS Guide," August 22, 2019. https://www.rohsguide.com/.

[9] European Commission, "REACH," August 22, 2019. https://ec.europa.eu/environment/chemicals/reach/reach_en.htm.

[10] Ericsson Inc., "Reliability Prediction Procedure for Electronic Equipment," SR-332, Issue 4, March 2016. https://telecom-info.telcordia.com/site-cgi/ido/docs.cgi?ID=SEARCH&DOCUMENT=SR-332&.

[11] BKCASE Editorial Board, "The Guide to the Systems Engineering Body of Knowledge (SEBoK), v. 2.0," Hoboken.

[12] SEBoK Authors, "Scope of the SEBoK," August 18, 2019. https://www.sebokwiki.org/w/index.php?title=Scope_of_the_SEBoK&oldid=55750.

[13] SEBoK Authors, "Product Systems Engineering Background," August 18, 2019. https://www.sebokwiki.org/w/index.php?title=Product_Systems_Engineering_Background&oldid=55805.

[14] SEBoK Authors, "System Design," August 18, 2019. https://www.sebokwiki.org/w/index.php?title=System_Design&oldid=55827.

[15] SEBoK Authors, "System Analysis," August 18, 2019. https://www.sebokwiki.org/w/index.php?title=System_Analysis&oldid=55956.

[16] SEBoK Authors, "Logical Architecture Model Development," August 18, 2019. https://www.sebokwiki.org/w/index.php?title=Logical_Architecture_Model_Development&oldid=55698.

[17] SEBoK Authors, "Physical Architecture Model Development," August 18, 2019. https://www.sebokwiki.org/w/index.php?title=Physical_Architecture_Model_Development&oldid=55874.

[18] IEEE, "1149.7-2009 - IEEE Standard for Reduced-Pin and Enhanced-Functionality Test Access Port and Boundary-Scan Architecture," February 10, 2010. https://ieeexplore.ieee.org/servlet/opac?punumber=5412864.

[19] International Organization for Standardization, "ISO 13849-1:2015," August 22, 2019. https://www.iso.org/standard/69883.html.

[20] International Electrotechnical Commission, "Functional Safety and IEC 61508," August 22, 2019. https://www.iec.ch/functionalsafety/.

[21] International Organization for Standardization, "ISO 26262-1:2018," August 22, 2019. https://www.iso.org/standard/68383.html.

[22] McLean, H. W., *HALT, HASS, and HASA Explained: Accelerated Reliability Techniques, Revised Edition*, Milwaukee, WI: American Society for Quality, Quality Press, 2009.

CHAPTER 2

Hardware Design Considerations

Well-designed, tested, and documented underlying hardware is key to enabling high-performance systems.

2.1 Component Selection

The System on Chip (SoC) Platform project is well underway at this point at SensorsThink. Requirements have been established (refer to Section 1.5: Requirements), system architecture has been defined (refer to Section 1.6), and the engineering domains (hardware, software, firmware, mechanical) are starting to do their engineering wizardry to turn those requirements and ideas into a real-life functioning product.

For hardware design, this process starts with component selection, more specifically, key component selection. Let's explore what we mean by a key component in a bit more detail before we dive into the selection process for some of the key components in our design.

For the sake of discussion, let's assume that the following two criteria are how we can identify a key component:

1. A component that, once selected, will not have a viable second source option or drop-in replacement. This excludes things such as resistors, capacitors, and diodes (with a few exceptions).
2. A component that is crucial to realizing the required functionality of the product and directly traces back to requirements. In our design, a great example of this is the selection of the actual sensors that will be providing data to the processing system.

2.1.1 Key Component Identification for the SoC Platform

Key components are important to choose early because they drive the rest of the design process directly. Their interfaces, power requirements, and support circuitry become critical aspects of the design to consider in order to ensure a successful hardware platform. Let's take another look back at the overall system architecture diagram to see if we can identify the key components (Figure 2.1).

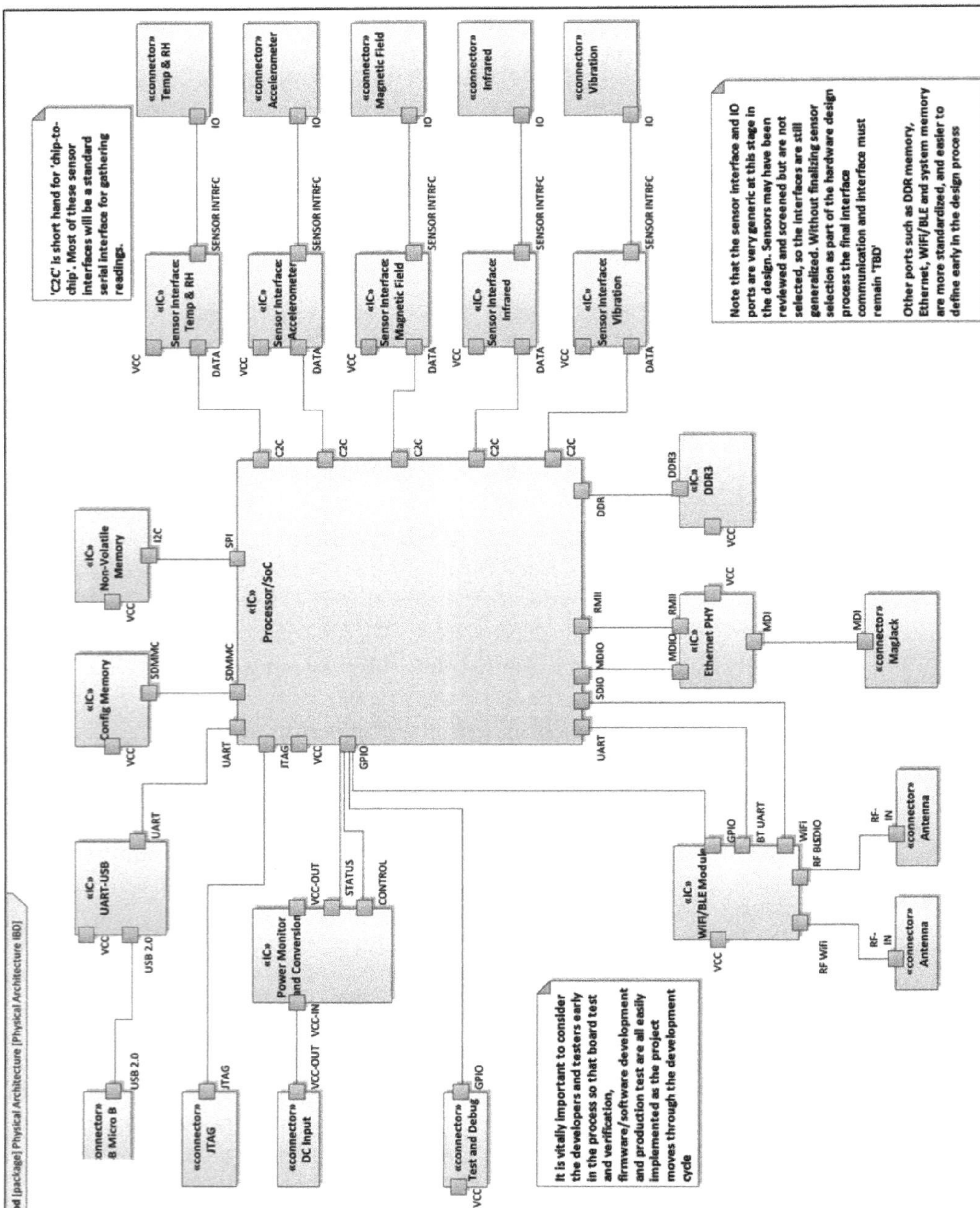

Figure 2.1 Hardware system architecture.

1. At the heart of the design is a processor/SoC; this is a key component for us. While there may be a migration path (more on this later) for our processor of choice, it is likely to not be a true drop-in replacement. In other words, another processor/SoC may fit physically on the printed circuit board (PCB), but likely would require some tweaks to firmware to properly run. This could lead to issues with production and tracking different product variants that exist once multiple devices are in customers' hands. Typically, unplanned variants are things that we want to avoid. The processor/SoC is also absolutely vital to realizing the functionality that we want to have in this SoC platform. It could be argued that it is as important as the sensors themselves.
2. The USB-UART converter, Wi-Fi/Bluetooth module, and Ethernet PHY are also key components solely due to the low probability of finding a drop-in replacement for these parts. The Wi-Fi/Bluetooth module and Ethernet PHY are also very important pieces of functionality for the platform. One could argue that the USB-UART converter is nice to have for a fielded product, but it is essential for development and test.
3. The sensors are key components. The performance of the sensors will be nearly solely responsible for the platform meeting the level 10 requirements. They are not depicted on the architecture diagram at the system hardware level, but they do appear in the context diagram (Figure 2.2).

It is natural to wonder why some of the other components are not considered key components, the answer (as it usually is) is that it depends. Things like memory can typically be found from at least two suppliers in a true drop-in replacement because of the JEDEC standards around memory.

In JEDEC's own words [1]:

> JEDEC's collaborative efforts ensure product interoperability, benefiting the industry and ultimately consumers by decreasing time-to-market and reducing product development costs.

The other major architectural blocks that we did not identify as key components are connectors and power conversion and monitoring. The reasoning here is that these components will be driven by the other components around them. We need to know the power requirements for the key components to design a power system, and we need to know what kinds of connectors may be required to interface to sensors or a debug adapter for whatever processor or SoC that we choose to be the core of the design.

So now that we have identified the key components that will drive our hardware design forward, there are a couple other key considerations to have in mind before we start to scour the internet and component supplier websites for the perfect parts:

1. *Component life cycle:* How long is the component expected to be available for sale? When was it introduced? What is the current life-cycle status? You never want to start a development project with a component that is already slated to go end-of-life. Just like we discussed in Section 1.4: Product

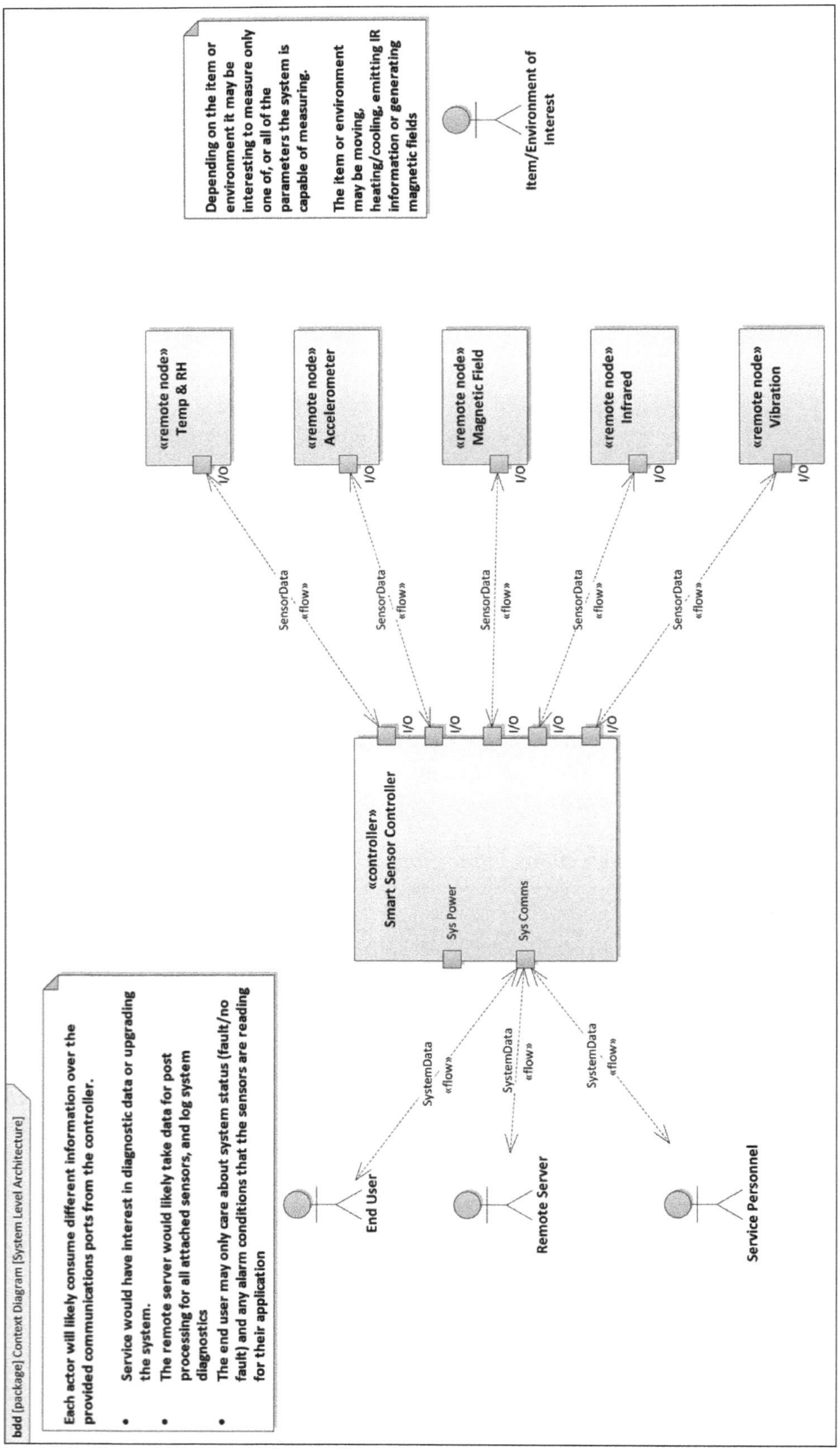

Figure 2.2 System context diagram.

Develoment: After Launch, the component manufacturer has their own life-cycle considerations and product planning going on internally. It is important to know where your particular component sits on that roadmap.

2. *Component temperature range:* In what ambient temperature range is the component rated for operation? Does this temperature range meet our product requirements? If it does not, it is not the right component.
3. *Component/device packaging:* For integrated circuits (ICs), this is a key factor, as it can drive both the complexity and the cost of the PCB design. It is more costly to design a PCB with high-density interconnect (HDI) like micro-vias than it is to design a PCB with all through-hole vias. Sometimes our hands are forced in this regard, but it is something worth considering.

2.1.2 Key Component Selection Example: The SoC

We covered the rationale behind why a SoC made sense for this product (versus multiple devices or a discrete microprocessor) in Section 1.6. To summarize, it is essentially due to the inclusion of FPGA fabric for parallel sensor data processing and an applications class ARM processor to provide a Linux operating system for our connectivity requirements. In the FPGA/SoC space, there are several suppliers to choose from: Intel (formerly Altera), Xilinx, and Microsemi.

There are certainly more suppliers providing SoC solutions. for the sake of discussion, we will limit it to these three. Some other suppliers provide solutions with microprocessors and DSPs (digital signal processors) for example. These devices certainly have their place in the design world, but they are not relevant for our discussion and design consideration.

For SensorsThink, some important requirements for choosing a SoC device are as follows:

1. Ability to run Linux on an ARM A-Class processor;
2. Inclusion of FPGA fabric for sensor fusion;
3. Support for DDR3 memory;
4. Support for Gigabit Ethernet;
5. Inclusion of security features;
6. Inclusion of standard peripheral communications (I2C, SPI, UART, and GPIO).

Typically, when doing a key component selection exercise, something like a Pugh matrix can be really useful to make what can sometimes devolve into a "religious" debate about Supplier A versus Supplier B, a more data-driven decision. Let's do a quick Pugh matrix for the SoC selection and see if it helps us narrow the field of choices (Figure 2.3).

There are many very high-end and costly SoC devices on the market that could handle our application very easily. So we could just cherry-pick the latest and greatest new SoC and call it a day. In real-world applications and product development, cost is always a key factor. Part of choosing components wisely is choosing the component that has just the right feature set for your use. However, as part of this

	Criteria 1	Criteria 2	Criteria 3	Criteria 4	Criteria 5	
Criteria Description	Ability to run Linux on an ARM A-Class processor	Inclusion of FPGA fabric	DDR3 Support	10/100/1000 Ethernet Support	Security Features	
	Criteria 1	Criteria 2	Criteria 3	Criteria 4	Criteria 5	Weighted Score
Weight	5	4	1	2	3	15
	33%	27%	7%	13%	20%	100%
Options	Criteria 1 Scores	Criteria 2 Scores	Criteria 3 Scores	Criteria 4 Scores	Criteria 5 Scores	
Microsemi SmartFusion 2 SoC	0	1	2	3	2	18
Intel Cyclone V SoC	3	2	3	3	1	35
Xilink Zynq-7000 SoC	3	3	3	3	3	45

Figure 2.3 SoC selection: weighted Pugh matrix.

exercise, the cost will not be part of the Pugh matrix so as to keep the focus on technical decision making.

A quick review of the three supplier SoC offerings brings us our three options to put into the Pugh matrix:

1. Microsemi SmartFusion2 SoC FPGA family;
2. Intel Cyclone V SoC FPGA family;
3. Xilinx Zynq-7000 SoC FPGA family.

A basic Pugh matrix starts by defining a set of criteria that are scored and then summed. This gives each option a final score, which can then be ranked.

A weighted Pugh matrix evolves the concept further by weighting the criteria in order of importance. The more important the criteria, the higher the weight that it should be given (a higher number by which to multiply the ranking). This way, the resulting final score more accurately captures the fit of each option when the values are summed.

All of the information for the Pugh matrix was gathered by sorting through the device family documentation for each device type. This can be a time-consuming exercise and can require contacting applications engineers to really help with sorting through all the details and nuances of each family. For SensorsThink, the clear

NOTES
*ARM Cortex M3 Processor, no A-Class Option *FPGA logic element count is the lowest among the three options (150K) *The DDR3 controller is a 'soft' controller, versus a 'hardened IP' controller in the other options *DPA Hardened, AES256, SHA256
*ARM Cortex A9 Processor, single and dual core options *FPGA logic element count up to 301K *Some devices have 2 'hardened IP' DDR3 controllers in the FPGA fabric as well as one for the ARM Cortex A9 *AES256, SHA256 (requires 3rd party IP)
*ARM Cortex A9 Processor, single and dual core options *FPGA logic element count up to 444K in mid-range devices *The only hardened DDR controller is for the ARM Cortex A9. A soft IP controller can be placed into the FPGA fabric. *RSA Authentication of First Stage Boot Loader, AES and SHA 256b Decryption and Authentication for Secure Boot for the ARM Cortex A9 *AES & SHA 256b Decryption & Authentication for Secure Programmable Logic Config

Figure 2.4 SoC selection: weighted Pugh matrix notes.

winner here is the Xilinx Zynq-7000 series of devices, and because we took an analytical approach to the selection it is difficult to argue against the decision.

From this key decision, we can now start to build the framework around our main SoC device. Choose support components such as:

- DDR3 memory (size, width, and configuration);
- Nonvolatile memory (size and interface);
- Configuration/filing system memory (size and interface);
- Ethernet PHY (interface type).

Before we get too far ahead of ourselves, we should also consider the component life cycle, temperature range, and packaging type. We previously identified these as important, so a review of the options is certainly worth a brief discussion.

2.1.2.1 Xilinx Zynq-7000 SoC Family Options

When choosing a device that is part of a larger family, it is easy to be overwhelmed with options; package options, peripheral options, speed grade options, and migration path options can really make the decision seem daunting (Figure 2.5).

Zynq®-7000 SoC Family

		Cost-Optimized Devices						Mid-Range Devices			
	Device Name	Z-7007S	Z-7012S	Z-7014S	Z-7010	Z-7015	Z-7020	Z-7030	Z-7035	Z-7045	Z-7100
	Part Number	XC7Z007S	XC7Z012S	XC7Z014S	XC7Z010	XC7Z015	XC7Z020	XC7Z030	XC7Z035	XC7Z045	XC7Z100
Processing System (PS)	Processor Core	Single-Core ARM® Cortex™-A9 MPCore™ Up to 766MHz			Dual-Core ARM Cortex-A9 MPCore Up to 866MHz			Dual-Core ARM Cortex-A9 MPCore Up to 1GHz[1]			
	Processor Extensions	NEON™ SIMD Engine and Single/Double Precision Floating Point Unit per processor									
	L1 Cache	32KB Instruction, 32KB Data per processor									
	L2 Cache	512KB									
	On-Chip Memory	256KB									
	External Memory Support[2]	DDR3, DDR3L, DDR2, LPDDR2									
	External Static Memory Support[2]	2x Quad-SPI, NAND, NOR									
	DMA Channels	8 (4 dedicated to PL)									
	Peripherals	2x UART, 2x CAN 2.0B, 2x I2C, 2x SPI, 4x 32b GPIO									
	Peripherals w/ built-in DMA[2]	2x USB 2.0 (OTG), 2x Tri-mode Gigabit Ethernet, 2x SD/SDIO									
	Security[3]	RSA Authentication of First Stage Boot Loader, AES and SHA 256b Decryption and Authentication for Secure Boot									
	Processing System to Programmable Logic Interface Ports (Primary Interfaces & Interrupts Only)	2x AXI 32b Master, 2x AXI 32b Slave 4x AXI 64b/32b Memory AXI 64b ACP 16 Interrupts									
Programmable Logic (PL)	7 Series PL Equivalent	Artix®-7	Artix-7	Artix-7	Artix-7	Artix-7	Artix-7	Kintex®-7	Kintex-7	Kintex-7	Kintex-7
	Logic Cells	23K	55K	65K	28K	74K	85K	125K	275K	350K	444K
	Look-Up Tables (LUTs)	14,400	34,400	40,600	17,600	46,200	53,200	78,600	171,900	218,600	277,400
	Flip-Flops	28,800	68,800	81,200	35,200	92,400	106,400	157,200	343,800	437,200	554,800
	Total Block RAM (# 36Kb Blocks)	1.8Mb (50)	2.5Mb (72)	3.8Mb (107)	2.1Mb (60)	3.3Mb (95)	4.9Mb (140)	9.3Mb (265)	17.6Mb (500)	19.2Mb (545)	26.5Mb (755)
	DSP Slices	66	120	170	80	160	220	400	900	900	2,020
	PCI Express®	—	Gen2 x4	—	—	Gen2 x4	—	Gen2 x4	Gen2 x8	Gen2 x8	Gen2 x8
	Analog Mixed Signal (AMS) / XADC[2]	2x 12 bit, MSPS ADCs with up to 17 Differential Inputs									
	Security[3]	AES & SHA 256b Decryption & Authentication for Secure Programmable Logic Config									
Speed Grades	Commercial	-1			-1			-1			-1
	Extended	-2			-2,-3			-2,-3			-2
	Industrial	-1, -2			-1, -2, -1L			-1, -2, -2L			-1, -2, -2L

Figure 2.5 Xilinx Zynq-7000 SoC family overview.

A great place to start is a product table (or product selection guide); suppliers will often summarize many of these options for us visually to make the information easier to consume. Another good resource can be parametric search tables from suppliers of devices, which allow the designer to sort and filter parts based on specific functionality, package size, or environmental ratings, among other things.

For SensorsThink, a dual-core ARM is the most sensible starting point. The reasoning behind this is that it allows for things such as asymmetric multiprocessing. One core can run the Linux operating system, and the other can run FreeRTOS (or another real-time operating system). This really opens up options for optimizing performance in the end application. You could also dedicate both cores to Linux if necessary. The flexibility is the attractive selling point for a dual-core system.

That leaves two device branches to explore: the cost-optimized family, built on Artix-7 fabric, and the mid-range family based on Kintex-7 fabric.

The analysis of Kintex-7 fabric versus Artix-7 fabric is out of the scope of our discussion; however, cost is a very easy barometer to indicate what might make sense given the features and trade-offs. A quick search on a distributor website such as Digikey or Mouser is a great way to get a rough-order estimate of the cost of each device relative to the other. Let's examine the highest-end cost-optimized device and the lowest-end mid-range device, the Z-7020 and Z-7030, respectively (Figure 2.6).

That is quite the increase in cost, essentially twice the cost for jumping the barrier from cost-optimized to mid-range. From Figure 2.5, we can see that some key differentiators are PCI Express lanes, logic elements, and DSP slices. In our application, we are doing some sensor fusion and some mathematical operations will certainly occur. However, without compelling input from a software engineer or systems engineer, the additional cost is hard to justify. For this reason, we can push closer to arriving at the major milestone of choosing the SoC for our platform. The Z-7010, Z-7015, and Z-7020 are the group of devices left to choose from (Figure 2.7).

Since this is an early stage of the development process, having a good migration path within our device family is something for which to aim if at all possible.

By examining this Figure 2.7, we can see that a CLG400 footprint provides us with a tremendous amount of flexibility, which could lead to cost reduction at the end of the project if we find that the initial selection has excessive horsepower for our needs. This extends down to the Z-7007S, a single core version of the Zynq-7000, up to the Z-7020, the highest density device available in our dual-core cost optimized family of components.

Part Number	Device	Temperature	Package	Speed (Processor)	Cost (Qty: 1)
XC7Z020-1CLG400I	Z-7020	-40°C-100°C	400LFBGA	667MHz	$13[illegible]
XC7Z020-2CLG400I	Z-7020	-40°C-100°C	400LFBGA	766MHz	$140
XC7Z030-1FBG484I	Z-7030	-40°C-100°C	484BBGA	667MHz	$25[illegible]
XC7Z030-2FBG484I	Z-7030	-40°C-100°C	484BBGA	800MHz	$30[illegible]

Figure 2.6 Cost comparison: Z-7020 and Z-7030.

Zynq®-7000 Device Footprint Compatibility
HR I/O, PS I/O, and GTP Transceivers
13mm–35mm

PCB Footprint Dimensions (mm)	13x13	17x17	19x19	19x19	23x23	27x27	27x27	31x31	35x35
Unique Footprint	CLG225	CLG400	CLG484	CLG485	FBG484	FBG676	FFG676	FFG900	FFG1156
Z-7007S	54, 84, 0	100, 128, 0							
Z-7012S				150, 128, 4					
Z-7014S		125, 128, 0	200, 128, 0						
Z-7010	54, 84, 0	100, 128, 0							
Z-7015				150, 128, 4					
Z-7020		125, 128, 0	200, 128, 0						

Figure 2.7 Xilinx Zynq-7000 SoC footprint compatibility.

Note: The shading indicates the footprint compatibility range. The numbers in each cell are the number of I/O available to the FPGA, processor system, and the number of high-speed GTP transceivers.

Based on this, we will start our development project with the Z-7020 device in a CLG400 package. We will also need the industrial temperature range to accommodate the requirement to operate down to –10°C. We will also start in the –2 speed grade, the faster of the two options available in our temperature range. This provides us the most processing power in our migration path and allows us to see if we can migrate down as we near completion of the development.

2.1.2.2 Xilinx Zynq-7000 SoC Support Components

Example: DDR3 Memory

Choosing the amount of DDR3 memory, the memory bus width is driven directly by the capabilities of our SoC. The device family datasheet is a great place to start to find this type of information. On the first page of the Zynq-7000 series datasheet, the memory interface is pretty well constrained for us:

- 16-bit or 32-bit interfaces to DDR3, DDR3L, DDR2, or LPDDR2 memories;
- 1 GB of address space using single rank of 8-, 16-, or 32-bit-wide memories.

If we want to maximize our potential performance, a 32-bit-wide DDR3 bus with 1 GB of memory is the route to take. Companies such as Micron provide some helpful documentation for memory selection for widely used device families (such as ones from Xilinx or Intel). The MT41K/MT41J device families are listed as compatible and validated with the Zynq-7000 device family; it is available in a 16-bit-wide data interface with a density of up to 8 Gb (Gigabits) or 1 GB (Gigabyte). If we use two 16-bit-wide, 4-Gb density devices in parallel, we will have 1 GB of total memory available (Figure 2.8). Before finalizing this selection, it can also be helpful

Technology	Component Configuration	Number of Components	Component Density	Total Width	Total Density
DDR3/DDR3L	X16	2	4 Gb	32	1 GB
DDR2	X8	4	2 Gb	32	1 GB
LPDDR2	X32	1	2 Gb	32	256 MB
LPDDR2	X16	2	4 Gb	32	1 GB
LPDDR2	X16	1	2 Gb	16	256 MB

Figure 2.8 Zynq-7000 Technical Reference Manual: DDR Memory Section Table 10. (Adapted from Xilinx.)

to review more detailed documentation from Xilinx for the Zynq-7000 series to see any potential missed nuances for implementing DDR3. The technical reference manual for the device family is the deepest dive that you can take, and it is where we will go next.

The first row in this table confirms that our design intent is valid. If we choose a 4-Gb component at 16 bits wide and place 2 of them in parallel, we will have a 32-bit-wide data bus at 8 Gb (1 GB) of total memory density. A suitable part number for us to use is Micron MT41K256M16TW-107; this part has the right bus width and density, is on the supported list from Micron, and is currently an active part. This is one more design decision checked off the list for our SoC platform.

2.1.2.3 Other Support Components for the Zynq-7000

A similar approach should be taken for the other key support components for the Zynq-7000. Start with basic parameters, find supported devices, and further examine the technical reference manual and even reference designs and evaluation kits to make sure that your selection is sound. For devices such as configuration memory, it is especially important as this often needs to be supported by the development environment for the SoC or processor. Careful selection of these components can really make the different in getting a design up and running in a quick manner.

2.1.3 Key Component Selection Example: Infrared Sensor

One of the sensor technologies for SensorsThink's new product platform is the infrared image spectrum. If we recall back to Section 1.5: Requirements, we know that the infrared spectrum requirements were fairly well defined (Figure 2.9).

There are several players in the infrared sensor space, but really there are not many alternative options for original equipment manufacturer (OEM) sensor design utilizing infrared imaging aside from FLIR.

> Note that there is one other commercially available option from Melexis, but it does not have the required plane array size, as it is only 32 × 24 pixels.

The FLIR Lepton 3.5 image sensor meets all of our stated requirements:

- Spectral range: 800 nm to 1,400 nm in the longwave infrared (LWIR) spectrum;

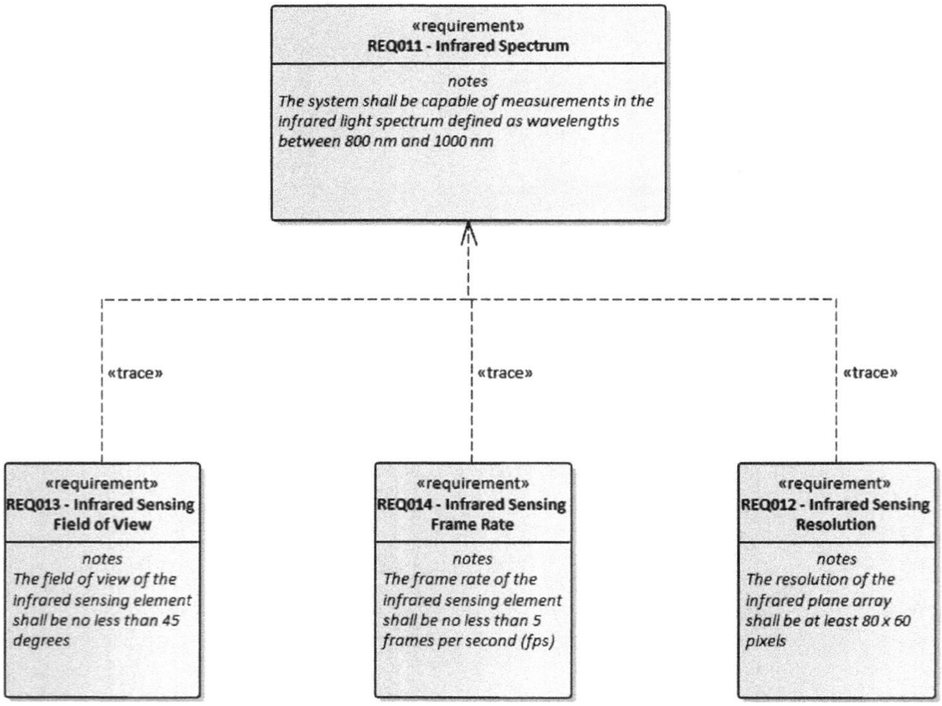

Figure 2.9 Infrared spectrum requirements.

- Field of view (FOV): 57° horizontal, 71° diagonal;
- Frame rate (over SPI): 8.7 Hz (effective);
- Plane array size: 160 × 120 pixels.

Because of the limited options available, the selection of this sensor becomes merely an exercise in finding available options. If there was no sensor available that met all of the stated requirements, the hardware engineers and product management at SensorsThink would need to sit down and discuss alternative paths forward. That could mean reducing the requirement to align with available sensors, creating a partnership with a specialist partner company to develop a sensor, or removing the sensing technology altogether. Luckily for us in the hardware group, the requirements are able to be met.

2.1.3.1 Infrared Sensor Interface

After choosing this sensor, we can then start to examine the impact of the choice on our system. This particular sensor requires a serial peripheral interface (SPI) communications link to transfer video data, an inter-integrated circuit (I2C) communications link to configure and control the sensor device, and some general-purpose inputs and outputs (GPIO) to monitor status. These things are important to note when putting together a more detailed hardware architecture diagram. We will cover hardware architecture in Section 2.2.

2.1.4 Key Component Selection: Finishing Up

All of the sensors should be selected in a similar way: refer back to requirements, find a sensor (or sensors) that are candidates, and, if necessary, create Pugh matrices to aid in decision making. Once all of the key components are selected, the hardware architecture can be fleshed out in more detail, which feeds directly into the design process: things such as power consumption, power supply voltages that are required (and current for each of these supply voltages), and SoC IO assignment. It is from these key components that our hardware design really begins to take shape.

At the end of this component selection exercise is a list summarizing the components that we have chosen for our SoC Platform project and key bullet-point items for the examples that we did not explore in detail.

- SoC: Xilinx Zynq-7000 Series; XC7Z020-2CLG400I;
- DDR3: (2) Micron 16-bit/4-Gb MT41K256M16TW-107;
- Configuration Memory, Micron MT25QL128ABA1EW7-0SIT:
  - Validated by Micron for the Zynq-7000;
  - 128-Mb size is large enough for our full migration path.
- Filing system storage: Micro SD Card:
  - SDHS (SD high speed) and SDHC (SD high capacity) card standards are acceptable for the Zynq-7000.
- Nonvolatile Memory (with Crypto-Security): Microchip AT88SC0808CA;
- Ethernet PHY: Microchip KSZ9031:
  - RGMII interface supported by the Zynq-7000;
  - Supported by the Xilinx PHY support in Linux kernel configuration.
- USB 2.0: Microchip USB3320:
  - ULPI interface supported by the Zynq-7000;
  - Used on the ZC702 evaluation board.
- USB to UART Converter: FTDI FT230X:
  - Simple UART interface;
  - Widely used industry standard part.
- Wi-Fi/Bluetooth module: Microchip ATWILC3000:
  - SPI, SDIO, I2C, and UART interfaces are supported by the Zynq-7000;
  - Used on the Ultra96 evaluation board.
- Local relative humidity/temperature sensor: Sensiron SHTW2:
  - I2C interface supported by the Zynq-7000;
  - Senses temperature and relative humidity within our required ranges.
- Accelerometer: ST Micro LSM6DS3:
  - SPI supported by the Zynq-7000;
  - Senses angular and linear acceleration within our required ranges in 3 axes.

- Infrared sensor: FLIR Lepton 500-0771-01:
  - Lepton 3.5.
- Magnetic field sensor: ST Micro LSM303AH:
  - SPI and I2C interfaces supported by the Zynq-7000;
  - Senses magnetic field strength (in gauss) within our required ranges in 3 axes.
- Vibration sensor: Analog Devices ADcmXL3021:
  - SPI supported by the Zynq-7000;
  - Measures vibrational force and frequency within our required ranges in 3 axes.
- Remote relative humidity/temperature sensor: Omega HX94BC:
  - Senses temperature and relative humidity within our required ranges in a single unit;
  - Provides output in analog voltage, which can be conditioned and sampled by the internal analog to digital converter (ADC) of the Zynq-7000.

2.2 Hardware Architecture

The next step in hardware design is to start to put the design into place. Finally, after months of research and legwork, requirements, and component selection, the design work starts to take shape. All the work that has been put into the project so far is essential for the hardware designer at SensorsThink to start the nitty-gritty design process. Without key components chosen and a system-level concept in place, the hardware design process is akin to assembling a puzzle with the pieces upside down. However, with those things well-defined, the puzzle pieces turn right-side up. It may be a 100-piece kids' puzzle sometimes and a 2,000-piece monster puzzle at others, but it is always easier to take that task on with the pieces flipped over.

A natural question at this juncture would be regarding the difference between the system-level physical architecture that was done previously and this current architectural design step. The diagrams can often look quite similar; however, this not always the case. In complex systems with more abstract black boxes, the diagrams can become more layered, with a system of systems architecture starting to reveal itself as the implementation path forward.

In our project, this is not the case, but the highest return on the investment for going through this part of the design process is always achieved by doing it upfront, before the real design work has started.

2.2.1 Hardware Architecture for the SoC Platform

Our attention can now shift to creating a block diagram for our hardware design for the SoC Platform, and, at SensorsThink, this diagram makes the most sense to complete in Enterprise Architect. By using the same diagramming tool and working in the same project, it becomes a very manageable task to trace hardware architecture elements to system-level requirements and system-level elements to show how the hardware architecture satisfies the system-level requirements. If we refer to the discussion on the V-model in Section 1.5: Requirements, we are now working squarely in level 30, we are in a design domain, and we are putting the architecture in place for that domain's implementation.

In safety systems, such as those governed by ISO 26262, the hardware architecture is a key deliverable for being able to show how the hardware design satisfies the safety concept. The architecture helps to show ties between hardware and software and shows clear links to and from test cases as well.

To facilitate discussion regarding hardware architecture and its usefulness, we will refer again to ISO 26262. Most of the textual information in the standard refers to safety requirements and things such as the automotive safety integrity level (ASIL). However, there are some good concepts from which any project can benefit by utilizing and following [2].

In section 7.4 of part 5 of ISO 26262, some recommendations are given that are easily applicable to products of all kind by changing one word.

- "The hardware architecture shall implement the hardware safety requirements." If we change this to: "the hardware architecture shall implement the hardware system requirements," instead we see that this is just generally sound advice. The architecture implement shall implement the system requirements. It seems obvious, but some of the best advice often is.
- "The traceability between the hardware safety requirements and their implementation shall be maintained down to the lowest level of hardware components." If we change this to: "The traceability between the hardware system requirements and their implementation shall be maintained down to the lowest level of hardware components" instead, we see again that this is just generally sound advice. As one example of this, consider our prior discussion on verification in Section 1.6.11. Understanding what piece of hardware specifically satisfies a requirement makes the development of test cases and linking those test cases back to clear pass/fail criteria much easier down the road of development.
- "In order to avoid failures resulting from high complexity the hardware architectural design shall exhibit the following properties by use of the principles listed in table 1: modularity; adequate level of granularity; and simplicity." This is one of the most well-written statements regarding the importance of implementing the right solution and not one that is either overengineered or underengineered. Let's digest what is being suggested as advice here. Essentially, by having a modular architecture, that has enough granularity (which is to some extent subjective, but most engineers can usually agree when something is underdefined, even if most cannot agree when something is overdefined) and, most importantly, is simplified to the minimum required complexity to meet the requirements that will help to avoid failures. While future-proofing a system or project would be nice, it can often lead to unnecessary complexity. Complexity is a prime suspect in hardware systems failing. If a system requires only 1 Mbps of data throughput,

implementing a chip-to-chip interface that can handle 100 Mbps introduces additional design complexities that can cause implementation issues, as well as issues with EMI/EMC due to the faster clock frequencies and edge rates associated with faster interfaces. This is only one example, but it is an easy one to cherry-pick.

Hardware architecture ties into the discussion here in taking the time to document the key components, their interfaces, and the data that flows over those interfaces. It allows the entire engineering team to understand the real nuts-and-bolts impact to design decisions. If a new sensor comes along that would be a nice to have in the system but causes the SoC Platform to require that we make a technology jump to add high-speed transceivers to our SoC, we can much more clearly see that with a well-defined and documented architecture, fed by key component selection.

ISO 26262 also provides some guidance as to what makes a well-planned architecture; SensorsThink would be wise to consider Table 2.1.

As a final point, consider this last bit of advice from ISO 26262: "Nonfunctional causes for failure of a safety-related hardware component shall be considered during hardware architectural design, including the following influences, if applicable: temperature, vibrations, water, dust, EMI, cross-talk originating either from other hardware components of the hardware architecture or from its environment."

If we again simply remove the word safety and replace it with the word "system," we can see that this is wise advice to heed on all projects and product developments. Consider other failure modes aside from functional failures. Consider how to protect hardware against its operating environment, consider how to protect circuits from each other, and consider this early in the design process. It is much easier to populate a set of component pads to help with an EMI/EMC problem than it is to cut a board up, hack a part into place, and then respin your design to pass the testing. An ounce of prevention is worth a pound of cure.

2.2.1.1 Hardware Architecture: The Diagram

Figure 2.10 shows the finalized hardware architecture for the SoC Platform. What we can see here is that the once-abstract blocks now have assigned part numbers due to our key component selection activity. From this, we can then start to define and understand the interfaces to each component and what will ultimately be required by our SoC.

Table 2.1 ISO 26262, Part 5, Table 1

_Methods	System Level	A	B	C	D
1	Hierarchical design	+	+	+	+
2	Precisely defined interfaces of hardware components	++	++	++	++
3	Avoidance of unnecessary complexity of interfaces	+	+	+	+
4	Avoidance of unnecessary complexity of hardware components	+	+	+	+
5	Maintainability (service)	+	+	++	++
6	Testability (during development and operation)	+	+	++	++

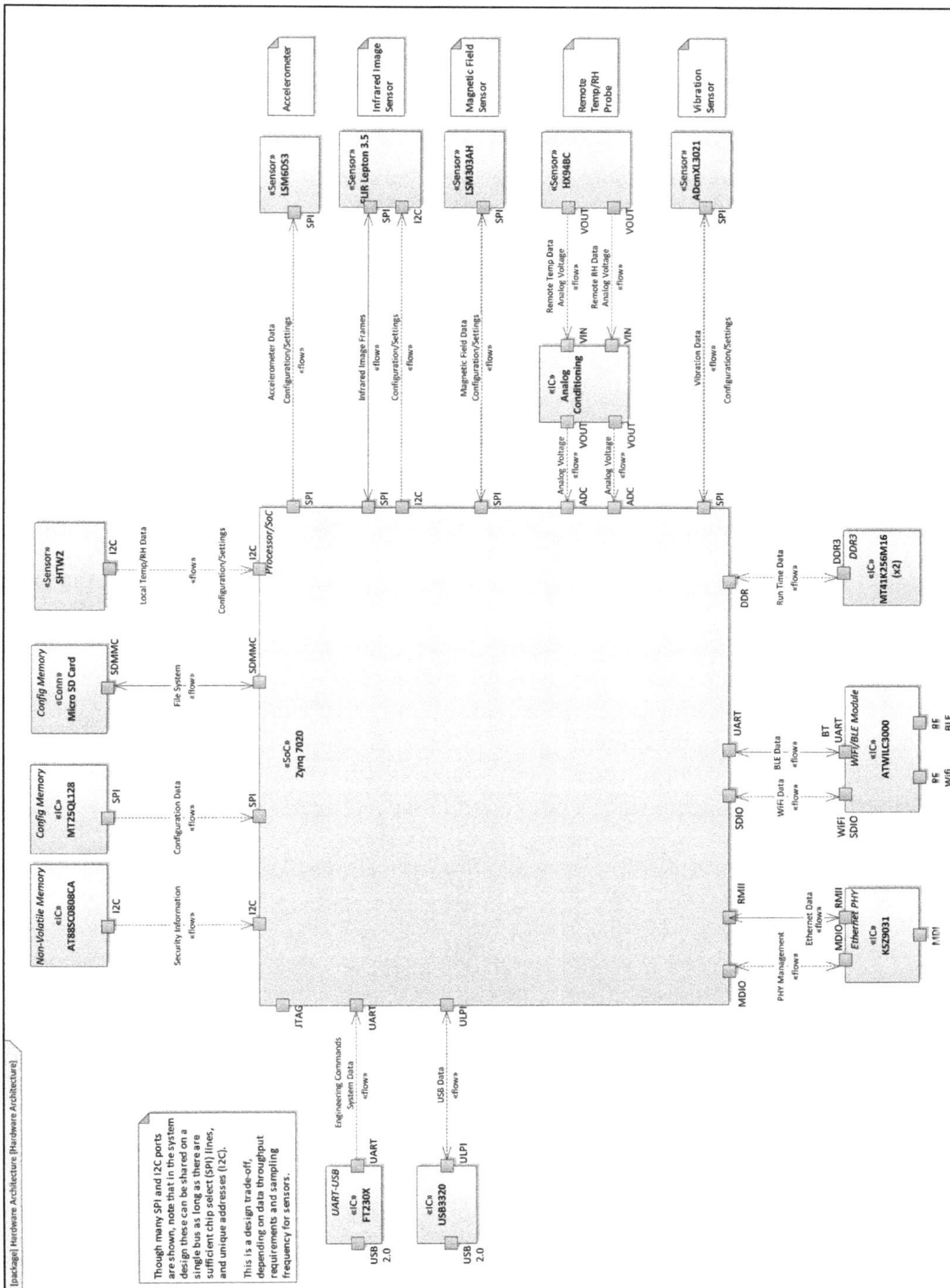

Figure 2.10 Detailed hardware architecture.

2.2.2 Hardware Architecture: Interfaces

The way to understand the design in more detail is to really dig into the datasheets. The datasheets (at least the well-written ones) hold all the information that we need to better understand how we interface to and communicate with our components.

In addition to interfaces and communication, device datasheets hold a wealth of other information for integrated circuits: power ratings, operating voltages and currents, device functionality, characteristic curves and plots, and many other key bits of information. In order to really own a hardware design, you must understand the devices being used with an appropriate level of detail.

As an example, let's take a look at the FLIR Lepton 3.5 datasheet and better understand the interfaces it has and how they are used.

2.2.2.1 Interface Example: FLIR Lepton 3.5

The diagram in Figure 2.11 [3] gives a high-level overview of the Lepton sensor. At this point in the design process, we are focused primarily on the interfaces to and from the sensor. We can see that leaving the sensor are several IO types: Video over Serial Peripheral Interface (VoSPI), I2C, GPIO, discrete controls, and a clock. The other arrow shown depicts power supplies, which are important, but come into play later in the design process. To better understand the interfaces and how they are used, we can start with the pin descriptions, this is generally a good place to go to for references to other parts of the datasheet that will better explain the connectivity of the pins.

Table 2.2 shows the entry from the Lepton datasheet for pins on the VoSPI. In addition to the reference to the appropriate page section, we also see that this is an SPI-compatible interface. This is key information that should be (and is) captured on the hardware architecture diagram.

If we jump into that section of the datasheet, there is a wealth of information in terms of how the interface is used, and how to properly communicate with it.

- VoSPI does not utilize the master-out slave-input (MOSI) line, and that pin should be set to a logic 0 or connected to signal GND.

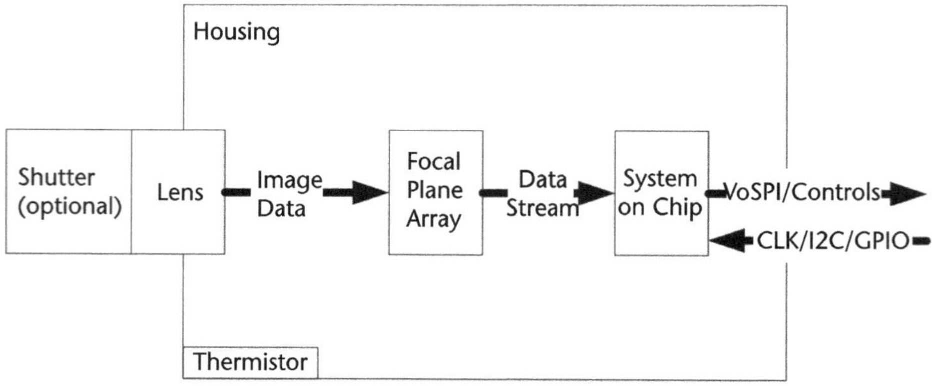

Figure 2.11 FLIR Lepton block diagram.

Table 2.2 VoSPI Pins: FLIR Lepton 3.5 Interfaces, VoSPI

Pin Number	*Pin Name*	*Signal Type*	*Description*
11	SPI_MOSI	IN	Video Over SPI Slave Data In
12	SPI_MISO	OUT	Video Over SPI Slave Data Out
13	SPI_CLK	IN	Video Over SPI Slave Clock
14	SPI_CS_L	IN	Video Over SPI Slave Chip Select

- Implementations are restricted to a single master and single slave.
- The SPI mode used by the device is SPI mode 3:
  - Clock polarity (CPOL) setting of 1 and a clock phase (CPHA) of 1.
  - For more information on SPI modes, refer to [4].
- Data is transferred by the most significant byte first and in big-endian order.
- The maximum clock rate is 20 MHz.

From these pieces of information, we can very clearly convey the requirements for the master SPI device (our SoC) to our firmware engineering team and also start to put the puzzle pieces together in terms of the number of peripherals required of our SoC to communicate to our sensors. Buried deeper in the section of the datasheet is the number of bits in each video frame and the different modes of communication in which the Lepton can operate.

As an example, let's consider the Raw14 video format mode. In this mode, there are 164 bytes per packet and 60 packets per video frame. That gives us approximately 10 kB per video frame. This is useful information to know for further consideration on where to place the peripheral and to be able to share as a starting point for firmware engineering. This is one of the two interfaces shown on the diagram. The other one was I2C, let's take a look at that interface next.

Table 2.3 shows the entry from the Lepton datasheet for pins on the Command and Control interface. In addition to the reference to the appropriate page section, we also see that this is an I2C-compatible interface. This is key information that should be (and is) captured on the hardware architecture diagram.

If we now jump into that section of the datasheet, we can find information on how the interface is used and how to properly communicate with it.

Lepton provides a command and control interface (CCI) via a two-wire interface similar to I2C. In this section, we are given a very useful reference to a separate interface description document (IDD) to fully explain the interface [5]. There is

Table 2.3 I2C Pins: FLIR Lepton 3.5 Interfaces, I2C

Pin Number	*Pin Name*	*Type*	*Description*
21	SCL	IN	Camera Control Interface Clock, I2C-compatible
22	SDA	IN/OUT	Camera Control Interface Data, I2C compatible

also a list of some of the parameters that are controllable via this interface; things such as gain, telemetry, frame averaging, and radiometry mode selection are all done via this interface.

Again, all the pieces of information that we need are here to help us to make informed decisions in piecing the puzzle together to better understand our embedded system.

2.2.3 Hardware Architecture: Data Flows

The last really key piece of information to consider at this stage of architecture is data flows: what type of information is flowing over each interface and how much data is flowing. This is key information to consider on buses such as I2C, SPI, and CAN. These types of communications buses have multiple devices sharing a single bus, and, as a result, the throughput of the bus is shared. So for the Lepton, the SPI has a maximum frequency of 20 MHz, which is 20 Mbps. That means that the SPI over which the Lepton is communicating can only transfer that much data.

In some cases, it is possible to be more creative with sharing interfaces if you are really peripherally limited in your design. For the purpose of our discussion, we are heeding the advice of ISO 26262 and trying to keep the design as simple as possible to avoid unnecessary failures due to complexity.

Similarly, the I2C-compatible bus may be easily shared for the Lepton, as most of the settings appear to be setup type commands that are not frequently changed by the user or system. This may not be true in all use cases and should be carefully considered to ensure that there will not be any hidden traps or pitfalls later on in the development cycle.

2.2.4 Hardware Architecture: Finishing Up

When this exercise is completed, a very clear and concise picture of the hardware should be taking shape. A reasonably deep dive into the datasheets is a great place to start this exercise, and along the way you can make note of other bits of key information to use later in the process. Key components, interfaces, and data flows are now documented. This can be used as evidence to show that the hardware satisfies the systems' requirements. This type of design groundwork gives high confidence levels that the design intent will be fit-for-purpose for the product being developed. There is really no limit to the amount of information that you can embed in an Enterprise Architect model; how much you enter is a matter of how much is necessary for your development project, but the value in having the information in a single modeling environment increases exponentially as the system complexity increases.

2.3 Designing the System

With a solid underlying foundation in place, in the form of hardware architecture, the nuts and bolts of hardware design can begin in earnest. For hardware engineers who are new to PCB-level design work, this can seem daunting. Where to begin? Where to go next? It can be a very overwhelming task to undertake, even more so as the complexity of the system increases. As with most tasks in engineering, this is where having a solid design process in place can make the insurmountable seem manageable.

For a system such as the one being designed by SensorsThink, the process to follow is the same as for a system being designed for a coffee maker or a server grade motherboard. The differences come in the complexity of those steps, and the amount of planning, checking, and replanning that likely go into each step.

Note that it is important to take a step back and acknowledge that, in engineering and in design work, there is always more than one way to get a job done properly. The intent of this story and the steps being laid out are to provide a framework that can be followed, which hopefully leads to successful and well-engineered designs and solutions. Over the course of your career, you will find what works best for you, and you will find what does not work for you at all. It is all part of the journey of learning the engineering trade and learning about a specific field of engineering. Design will evolve, software tools will change and improve, and the way that design is approached will have to change as well in order to keep up with the times. Now let's jump back into the story and outline some high-level steps to follow when designing an embedded system.

2.3.1 What to Worry About

If we look at the hardware architecture at which we arrived, we can start to look at the types of design analyses that are important to do in the early stages of design work. If you have never done a design like this with high-speed interfaces and an SoC, it may seem overwhelming. The key is to break the design down into manageable pieces and into subaspects of the design itself; for example, in most designs, a designer would be faced with at least some of the following design considerations and tasks.

2.3.2 Power Supply Analysis, Architecture, and Simulation

A key part of this on modern designs is power sequencing, especially on designs with multiple complex devices, each having their own requirements. Oftentimes, these requirements conflict with each other and require complex sequencing schemes to achieve compliance with all devices.

Simulation is a great way to verify your design intent before committing to a schematic capture or to a physical PCB. Once you have those designs on physical copper, mistakes are costly to rework and in some cases may not be possible. Simulation is your friend; use it.

A great way towards success is to analyze each power supply rail or, in other words, each discrete voltage level one component at a time. This will provide a great deal of insight into the current requirements for each voltage and is a necessary step in choosing appropriate power supply components.

Once you understand the current requirements, grouping the power supplies into stages of sequencing is another good idea; this maps out the order in which power supplies must be turned on (and sometimes turned off) in order to have a reliable and safe power system in which your components can operate.

At this point, you can put together a block diagram of the entire scheme using the tool of your choice. A tool that is particularly useful for this is LTpowerCAD.

This tool brings a number of very useful power system design tools together under one umbrella including individual power supply design and power system design and works hand-in-hand with LTspice to run more detailed simulations of power supplies. As an example, consider Figure 2.12 (tools used with the friendly permission of Analog Devices Inc.

This diagram is not for the SensorsThink design, but it is a good example of how you can visualize an entire power system, include all of the nominal loads, and really get a high level of confidence that your solution is a good one. After entering all of your design data, the tool also does power calculations on the system to show you things such as total power consumption, dissipation, and margin.

If we then wanted to look at the transient performance of a specific power supply, we can jump into LTspice and run that simulation as well, linking the simulation to this system-level view, making for a really nice design artifact.

2.3.3 Processor and FPGA Pinout Assignments

While it is true that FPGAs and modern processors (or in our case, one device that is both) are extremely versatile and flexible, it is equally true that there are seemingly infinite numbers of conflicting uses for multipurpose pins, invalid assignments of peripherals, and very specific requirements and uses for some pins that must be accounted for. The best way to prepare for this task is to start reviewing documentation. All of these "gotchas" are hiding in that documentation somewhere, and it is your job to find them and account for them.

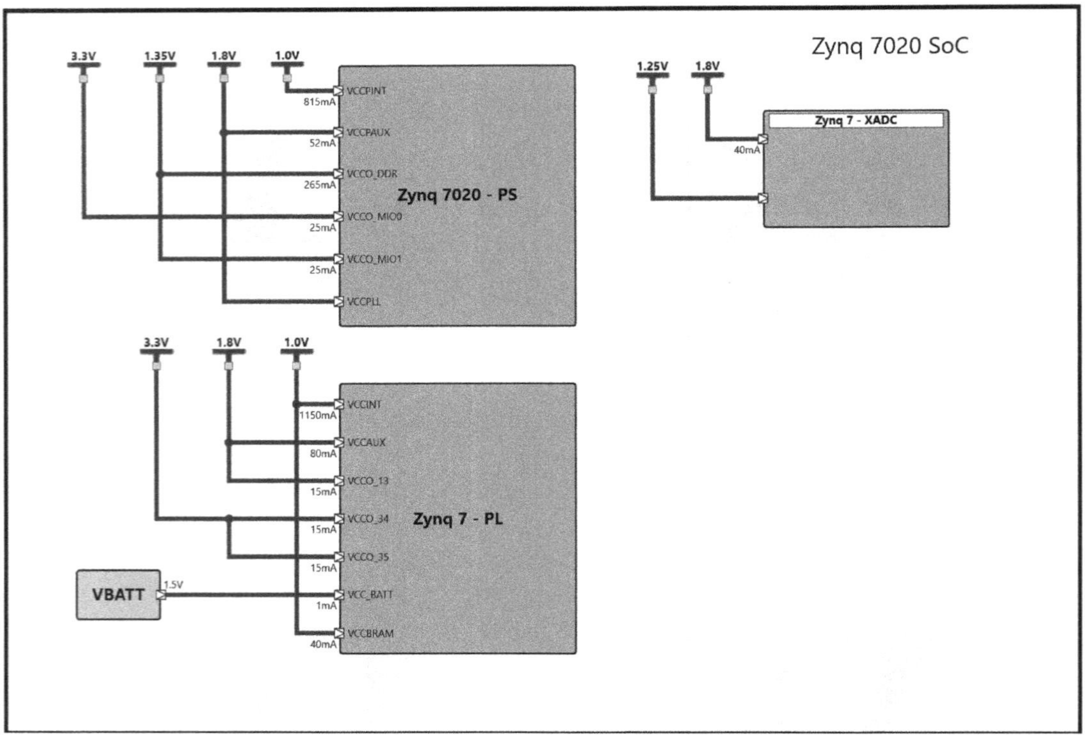

Figure 2.12 LTpowerCAD diagram.

2.3.4 System Clocking Requirements

Processor clocks, FPGA logic clocks, local oscillators, and clocking domains: there is quite a bit of nuance and detail in understanding the clocking requirements of your system. If there were advanced protocols being used such as PCIe, those would require their own reference clocks as well.

Diagramming this scheme in a block diagram form can be a valuable exercise in understanding the right architectural solution for your design. It may be a programmable clock generator with 8 outputs, or it may be just a few local crystal oscillators. A great way to determine this is to break this part of the design down in a separate design activity.

2.3.5 System Reset Requirements

Every system has at least one kind of reset, a power-on reset. This reset ensures that a system starts from a known safe and valid state when power is applied. Other systems may have several other kinds of resets that need to be designed and accounted for:

- A cold reset essentially mirrors the logical behavior of a power-on reset in many cases, with the difference being that power is not removed from the board. This type of reset is useful during development and testing in the event that a system becomes unresponsive.
- A warm reset is typically reserved for resetting peripheral devices and can be used in development or in the system during operation to try to recover from run-time faults.
- External reset sources, such as a user push button, can also be reset sources that are intended to have a specific impact to the design; it may simply be a cold reset, or it may reset only select devices.

2.3.6 System Programming Scheme

How will your system get its code initially programmed? For external flash memories, they may be preprogrammed from the supplier once you are in production, but what about during development? Commonly, this is done via a JTAG interface for things such as FPGAs and for modest processors as well. For Linux-based solutions, often an external micro secure digital (SD) card is used as a quick way to develop new operating system environments and deploy them to the system. Another popular solution is embedded multimedia card (eMMC) memory, which is soldered to the board, but has the drawback of longer programming times in system.

In addition to the programming scheme itself, configuring the processor, FPGA, or SoC to accept the programming from the interface that you intend to use requires the configuration bits to be strapped (pulled to a logic 1 or logic 0) in a certain way to tell the device to look here for its code.

2.3.7 Summary

There is no shortage of design topics that can be concerning for engineers doing hardware design work. In high-speed digital systems, there can be months of work doing signal integrity simulations planning for routing structures (vias, fanout pattern, pin escape) on multigigabit-per-second interfaces. In sensitive analog systems, there can be many months of work doing analog simulation and analysis on op-amp circuit stability, looking at Bode plots, and doing noise analysis before arriving at what should be committed to the schematic. The more preparation work that is done, the more at ease the designer will ultimately feel when it comes time to stamp a schematic and PCB design with his or her approval. Let's take a look at an example of one of the topics we just mentioned in a bit more detail to get an idea of examples of what to expect when undertaking this type of work. Let's focus on the Zynq device, as it is the most complex device on the SensorsThink Platform towards which we have been marching.

2.3.8 Example: Zynq Power Sequence Requirements

In Xilinx terminology, there are two subsystems of the Zynq SoC, the PS is the processor system (the ARM Cortex processor complex), and the PL is the programmable logic portion of the device (the FPGA).

From the Zynq-7000 Technical Reference Manual [6], the following is a high-level view of the power domains: "The PS and PL power supplies are fully independent; however, the PS power supply must be present whenever the PL power supply is active. PL power up needs to maintain a certain timing relationship with the POR reset signal of the PS."

That is a total of 13 power domains in the device; that is not to say that you need 13 power supplies, as perhaps in your design (or ours), some of those may be combined together (Figure 2.13). A smart designer would wonder, "where do I find out how to combine these supplies, and in what order do I bring them up and down?" Let's take a look at a different document, the Zynq-7000 SoC DC and AC Switching Characteristics Specification [7], and try to get some more insight there.

Type	Pin Name	Nominal Voltage	Power Pin Description
PS Power	VCCPINT	1.0V	Internal logic
	VCCPAUX	1.8V	I/O buffer pre-driver
	VCCO_DDR	1.2V to 1.8V	DDR memory interface
	VCCO_MIO0	1.8V to 3.3V	MIO bank 0, pins 0:15
	VCCO_MIO1	1.8V to 3.3V	MIO bank 1, pins 16:53
	VCCPLL	1.8V	Three PLL clocks, analog
PL Power	VCCINT	1.0V	Internal core logic
	VCCAUX	1.8V	I/O buffer pre-driver
	VCCO_#	1.2V to 3.3V	I/O buffers drivers (per bank)
	VCC_BATT	1.5V	PL decryption key memory backup
	VCCBRAM	1.0V	PL block RAM
	VCCAUX_IO_G#	1.8V to 2.0V	PL auxiliary I/O circuits
XADC	VCCADC, GNDADC	N/A	Analog power and ground
Ground	GND	Ground	Digital and analog grounds

Figure 2.13 Zynq-7000 SoC power domains.

2.3.8.1 PS Power-On/Off Power Supply Sequencing

"The recommended power-on sequence is VCCPINT, then VCCPAUX and VCCPLL together, and then the PS VCCO supplies (VCCO_MIO0, VCCO_MIO1, and VCCO_DDR) to achieve minimum current draw and ensure that the I/O are 3-stated at power-on.

The recommended power-off sequence is the reverse of the power-on sequence. If VCCPAUX, VCCPLL, and the PS VCCO supplies (VCCO_MIO0, VCCO_MIO1, and VCCO_DDR) have the same recommended voltage levels, then they can be powered by the same supply and ramped simultaneously"[6].

This verbiage is taken verbatim from that document, but it does give a clear picture of the power supply sequence requirements and is an actionable statement that can help a hardware designer put a robust power scheme together for the PS system.

A note to the reader: There are more nuanced details provided and explained in the Xilinx documentation regarding proper design and sequencing for the Zynq-7000; for the sake of providing a relatively easy-to-understand example, they have been omitted.

2.3.8.2 PL Power-On/Off Power Supply Sequencing

"The recommended power-on sequence for the PL is VCCINT, VCCBRAM, VCCAUX, and VCCO to achieve a minimum current draw and ensure that the I/O are 3-stated at power-on. The recommended power-off sequence is the reverse of the power-on sequence. If VCCINT and VCCBRAM have the same recommended voltage levels, then both can be powered by the same supply and ramped simultaneously. If VCCAUX and VCCO have the same recommended voltage levels, then both can be powered by the same supply and ramped simultaneously" [6].

This verbiage is again taken verbatim from that document, but it does give a clear picture of the power supply sequence requirements and is an actionable statement that can help a hardware designing put a robust power scheme together for the PL system. By comparison, it is of similar complexity as the PS requirements. Bear in mind that the PS and PL power supplies are fully independent. PS power supplies can be powered before or after any PL power supplies. The PS and PL power regions are isolated to prevent damage.

As the designer, you can now analyze your system, group power supplies together, and start to formulate a system that meets these requirements (among many others) to ensure that your system will work reliably.

A note to the reader: Another great source of sanity checking your design is reference designs provided by the manufacturer or its certified partners. They can offer a great deal of insight into how these requirements and constraints are managed by the company that designed the device. Leverage all of that knowledge and adapt it to your own design work.

2.4 Decoupling Your Components

During design reviews, it is often what seems to be the most mundane topics that get the most attention: the color of LEDs, the percentage tolerance of specific resistors, what sheet size was used for the design; on and on it can go, seemingly with no value added. One topic that can seem to fall into that category is decoupling capacitors, but doing so would be a considerable mistake, and, as such, the designers at SensorsThink are wise to consider decoupling strategy in great detail.

Despite the boring exterior, decoupling is absolutely essential to PCB design. Without it, you risk having a totally nonfunctional circuit. With too much or too little of it, you can have instability in all kinds of circuits: power supplies, loop compensation circuits, phase locked loops (PLLs), and more.

A note to the reader: In the course of my career, I have spent countless hours, days, and probably weeks discussing decoupling with seasoned engineers from analog, power, and digital perspectives. Decoupling is not a one-size-fits-all approach, and knowing that is a big part of conquering the battle with those little capacitors.

2.4.1 Decoupling: By the Book

Let's start by going back to the basics, literally, of the word decoupling: as per *Merriam-Webster* [8], to decouple something is "to eliminate the interrelationship of: SEPARATE."

So why is it that we would want to decouple our components? What is the benefit of separating them and from what are we separating them?

2.4.2 To Understand the Component, You Must Be the Component

Imagine for a moment that you are an integrated circuit; say, for the sake of argument, that you are a very simple one such as an inverting buffer. You are given a source of power (often called VCC or VDD) and a place to put your used power (often called GND). You have a bit presented to you, and your sole job in life is to take the state of that bit and invert it; make a 1 a 0 and make a 0 a 1.

When the time comes to do your job, you look to your power pin for energy to do your work, and sitting about 100 miles away is the power supply circuit that the hardware engineer designed for you to use. It takes a lot of effort to get the energy that you need to do your simple job, and as you gather up that energy, you are traveling a great distance, probably bumping into traffic along the way from your colleague ICs. This distance in the electronics world is impedance. Depending on how fast you have to do your work, you will choose one of two paths: the path of least resistance (R) or the path of least inductance (L). Either way, it would make your life a lot easier if you had something close by from which to grab that energy.

A note to the reader: This is a very oversimplified breakdown of the mathematics behind impedance and frequency and why integrated circuits need energy to do their jobs. Bear with me; it is merely to serve as an example to illustrate a point, not to be a mathematically accurate model of current flow and loop areas.

So let's reenact that scenario, only this time you look to your power pin, and sitting right next to it is a little bucket full of energy into which you can reach your shovel and grab the energy you need for inverting that bit: fantastic, a shorter journey, less work, and you disturbed no one else on your journey to get that energy. Before you know it, that little bucket gets filled back up, and the next time that you need energy, it is sitting right there for you again.

By providing a local energy storage device (decoupling capacitor) for your circuits, you make their lives easier, they work better and more efficiently, and they are friendly to their neighbors. This is a win-win scenario.

2.4.3 Types of Decoupling

There are many different types of decoupling and many different types of capacitors, including:

1. Aluminum electrolytic;
2. Aluminum polymer;
3. Tantalum;
4. Ceramic;
5. Film.

Each type of capacitor has its own strengths and weaknesses, some due to the type of dielectric material, some due to packaging sizes, and some due to material properties and applications. A full discussion on each capacitor type and the advantages and disadvantages warrants a book all to itself. As a starting point, you can check the references at the end of the chapter for some very insightful discussion and information [9–12].

2.4.3.1 Bulk Decoupling

Bulk decoupling most commonly refers to large (in value and, subsequently, in size) capacitors placed around a circuit or a group of circuits to act as a larger bucket of energy to more quickly refill the smaller buckets of energy that are placed closer to the devices. From a high level, think of these bulk capacitor components as a way to relocate the main source of power for an area on your design, further away from the power supply generating the energy, and closer to the loads that need them. The math behind this suggests that these types of capacitors are most effective in the range of dc to hundreds of kilohertz.

2.4.3.2 High-Frequency Decoupling

High-frequency decoupling most commonly refers to small (in value and, subsequently, in size) capacitors placed directly near a specific pin or a group of pins on a specific integrated circuit to act as a small bucket of energy for higher-frequency demands that the circuit may have. From a high level, think of these capacitor components as a way to shorten the path of least inductance a bit for your high-frequency friends in between the big bulk decoupling components and the silicon inside the integrated circuit. The math behind this suggests that these types of capacitors are most effective up to frequencies around 500 MHz in the absolute best-case scenario.

2.4.3.3 Interplane PCB Decoupling

This type of capacitance is a result of smart PCB stack-up design and refers to the power and ground planes of the PCB acting as the two plates of a capacitor. In order for this to be effective, the power and ground planes must be very close together, on the order of 2 mils (0.002 inch). Doing so can yield an interplane capacitance value of around 500 pF/in^2. In order to leverage this, the IC power pins (VCC and GND both) must be directly tied to the power and GND planes with vias. This type of decoupling can be effective up into the range of several gigahertz.

2.4.3.4 On-Die Decoupling

This type of decoupling refers to capacitors placed on the substrate package right next to the silicon wafer. This is as close as decoupling capacitors can be placed, and for high-speed digital devices, they are almost always put there by the chip supplier, knowing that by the time the need for energy has reached the PCB to which the device is mounted, the impedance is too large for the capacitor to be effective.

As a designer, you need to be aware of and leverage all of these technologies and decoupling strategies, as they all play a role in a properly designed system with good power integrity. Let's take a look at the Zynq-7000 series documentation again for a good example of how these strategies play out in a real-world design example.

2.4.4 Example: Zynq-7000 Decoupling

From the Zynq-7000 Technical Reference Manual [6], we can find a mention of on-die decoupling in the section that discusses features of the PL system: "High-frequency decoupling capacitors within the package for enhanced signal integrity." Right away as a designer, you can check that box off mentally, knowing that Xilinx has done their part to ensure that your power integrity and decoupling scheme are set up for success.

In the Zynq-7000 PCB Design Guide [13], there is a discussion on several of these topics, starting with power planes, carrying through concepts on return currents and mounting inductance. All of these topics are fantastic evidence that Xilinx has a great handle on these concepts and should give you confidence as a designer that their suggestions come from a place of both knowledge and empirical data.

The documentation goes so far as to suggest capacitors based on your specific device. Recall that for SensorsThink, we eventually settled on the XC7Z020-2CLG400I as our device of choice. Based on that, we can extract the following recommendations from Xilinx, first for the PL subsystem (see Table 2.4).

Next, we can get the same information for the PS subsystem in the same document, for our package type, the CLG400; and our device type, the Z-7020 (see Table 2.5).

This is fantastic information and really takes the guesswork out of the decoupling of our device. However, Xilinx takes things a step further and even specifies acceptable capacitor sizes, types, and even part numbers to use. For full details, the latest PCB Design Guide (UG933) is a fantastic resource.

Table 2.4 Number of Capacitors Per Power Domain: PL Subsystem

Capacitor Value	*VCCINT*	*VCCBRAM*	*VCCAUX*	*VCCO per Bank*	*Bank 0*
680 μF	0	0	0	0	0
330 μF	1	0	0	0	0
100 μF	0	1	0	1*	0
47 μF	0	0	1	1*	1
4.7 μF	2	1	1	2	0
0.47 μF	4	1	1	4	0

Note: The asterisk indicates that either a 100-μF or 47-μF capacitor is recommended.

Table 2.5 Number of Capacitors Per Power Domain: PS Subsystem

Capacitor Value	*VCCPINT*	*VCCPAUX*	*VCCO_DDR*	*VCCO_MIO0*	*VCCO_MIO1*
100 μF	1	1	1	1	1
4.7 μF	1	1	1	1	1
0.47 μF	3	1	4	1	1

As the designer, you have all the information that you need to properly decouple the Zynq at a schematic level, make smart component selection choices and substitutions for parts that may be hard to find, and understand the role each type of component plays.

2.4.5 Additional Thoughts: Specialized Decoupling

As mentioned earlier, decoupling is not a one-size-fits-all problem or solution. Some suppliers are great at giving guidance, and others are not so great. This is where understanding the fundamentals is so important and realizing that in some cases a 0.1-μF capacitor works as well as a 1.0-μF capacitor and vice versa. In other cases, they may not work interchangeably.

In some applications or in some devices (for example, RF devices), specific frequencies are targeted for aggressive decoupling, and others are seemingly ignored. Always fall back to fundamentals and make sure that you understand the goal of the decoupling scheme. Do not be afraid to ask the manufacturer questions. If you do not find the answer satisfactory, explore other options or leave yourself extra, just-in-case component pads based on your thoughts and your experience. Either way, this topic is one that can be the topic of a thesis. There is always more to learn; but, in many cases with enough time invested in reading and pouring over reference material, you can design robust power decoupling networks without too many headaches.

2.4.6 Additional Thoughts: Simulation

There are many tools available for simulation of power integrity and power analysis. If you have them, use them. Leverage those tools to make your life as a designer easier and to give yourself more confidence in the final product. You can simulate almost anything these days with powerful simulation tools such as HyperLynx [14, 15] or Ansys SIwave. There are tools that can analyze your capacitor placement such as HyperLynx DRC [16] and alert you to physical placement issues that are easy to miss on large or complex designs. Using those tools takes time, but the time invested is almost always worth the return on that investment.

2.5 Connect with Your System

In Section 2.4, we talked about a seemingly mundane topic that has many more nuances and layers than meet the eye. A topic that is much more straightforward and requires less math and more common sense is connector choice or connectorization. To be clear, connectorization is a made-up word; you cannot find it in a dictionary or in any kind of conversation outside of engineering. That will not stop the word from being used in conference rooms at SensorsThink, however; and it will not stop you (nor should it) from using the word either.

A note to the reader: Sometimes picking connectors is really easy, and sometimes it is not. There are often organizational influences on connector choices: what tooling in which the company may have already invested, what companies are preferred partners, or what families are already in heavy use across different products. These can all be important in the bigger picture decision-making process, especially if you are working on a product with massive volumes (hundreds of thousands or millions of units). For the sake of discussion, we are going to keep the focus on technical reasons to choose a connector and assign signals to locations, because it is impossible to predict all of the organizational influences that may or may not be present in your situation.

2.5.1 Contemplating Connectors

What are some things that ought to be considered during the choice of a connector for an application? What are the engineers at SensorsThink considering when they set out to agree on connectors for the SoC Platform? There are many of them, more than may be obvious at first glance, for example:

1. Existing standards for the interface;
2. Maximum expected voltage and current;
3. How many times the connector will be used (mating cycles);
4. Environmental considerations (temperature, humidity, water ingress protection);
5. The end user of the connector (customer, test engineering, manufacturing, design engineers);
6. Available physical space within the enclosure or on the PCB;
7. Cable egress path and angle of escape (right angle, vertical).

With so many things to consider, it is a fair question to wonder why this topic was just stated to be straightforward. The easiest way to demonstrate that is to walk through an example of choosing a connector for an interface of our Smart Sensor Controller.

2.5.2 Example: System Communications

For the first decision that we will explore along with the engineers at SensorsThink, the system communications port seems like a great place to start (Figure 2.14). It is the interface with which the end user will interface, and we can run through our list of considerations with a lot of existing data about the application.

2.5.2.1 Existing Standards

If we refer back to the requirements derived for the project, the interface is required to interface to a network via a wired network, a wireless network, or both. In our hardware architecture process, the decision was made to implement both. Our wired network interface implements a Gigabit Ethernet connection over copper. This is driven by the component selection that was made for the Ethernet PHY device (the KSZ9031).

Almost every wired network interface for the last 20 years has had the same style of connector, an RJ45 style connector.

A note to the reader: This statement is true for nearly all residential, light commercial, or light industrial computer or connected device. The RJ45 connector name is a bit misused, however. The technically proper name for the connector is really an 8p8c connector as an RJ45 connector is keyed. This is such a common misuse of the RJ45 standard that we will carry on with tradition and refer to it as such as well.

For this interface, we now know we need to have an RJ45 connector. It is a widely adopted standard, and it is an easy choice to make.

2.5.2.2 Maximum Expected Voltage and Current

For this specific example, the expected voltage and current are less important to know because of how heavily standardized the interface is. However, we can still find this information based on the physical layer standards that our Ethernet PHY device implements: 10BASE-T/100BASE-TX/1000BASE-T. The Gigabit Ethernet interface is typically transformer coupled; in other words, it is typically isolated from end to end using magnetics (transformers) (Figure 2.15). The isolation of the interface is typically 1500VRMS; this is specified by the IEEE 802.3 standard [17]. The signal levels themselves are quite modest by comparison: approximately 1.0V differential signaling for 100BASE-TX and 1000BASE-T, and approximately 2.5-V differential signaling for 10BASE-T.

2.5.2.3 Mating Cycles

This is an example of where the common-sense aspect of choosing a connector comes into play. If we consider the life cycle of the product (100,000 hours) and make some reasonable assumptions, we can arrive at a good guesstimate number.

First, let's convert hours to years: 100,000 hours is a bit over 11 years. For the sake of easy math (and to build in a bit of margin), we will round up to 12 years.

Over 12 years, let's say that the average customer does two service calls per year. At that service call, we know from our context diagram (Figure 2.14) that

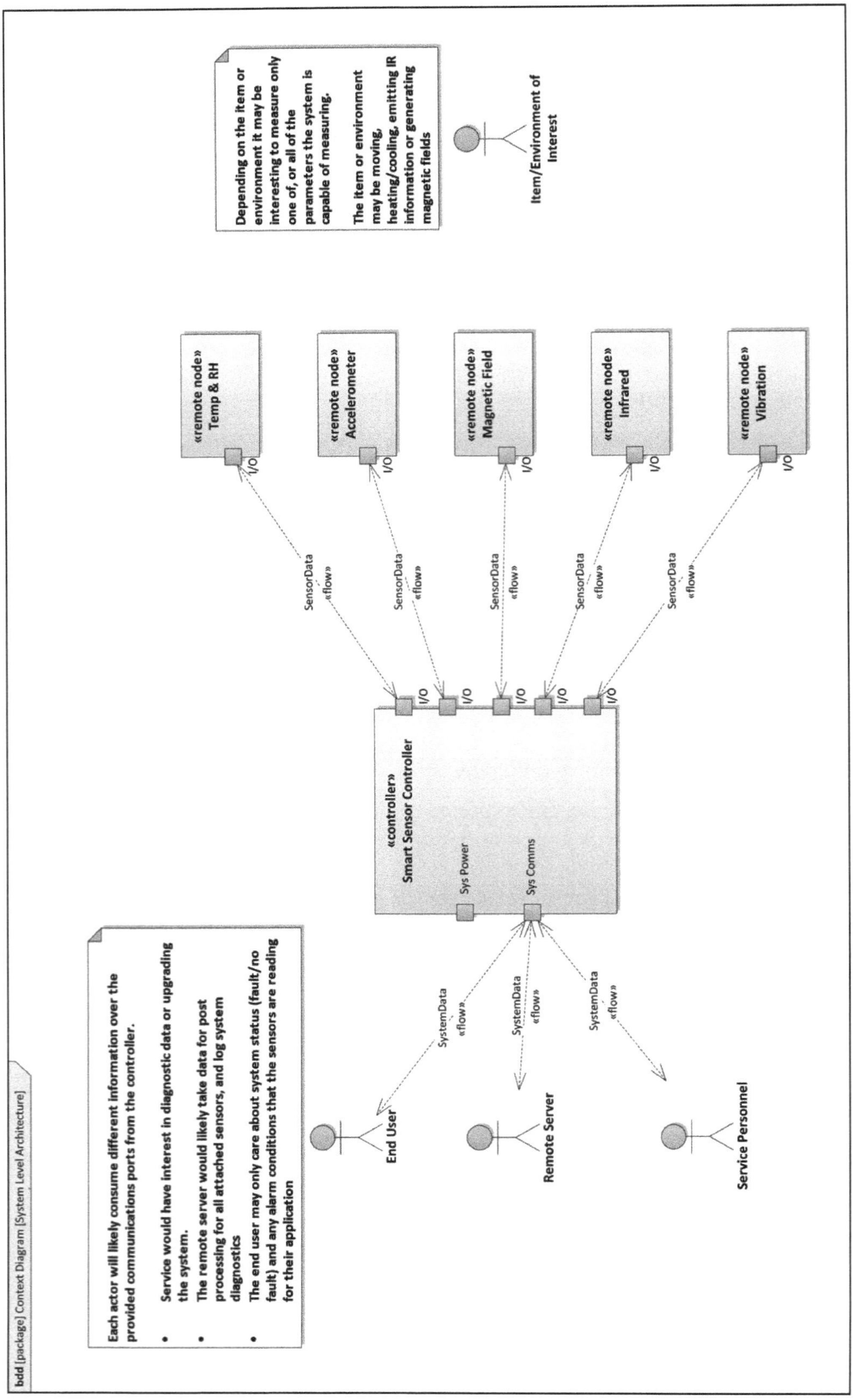

Figure 2.14 System context diagram: SoC Platform.

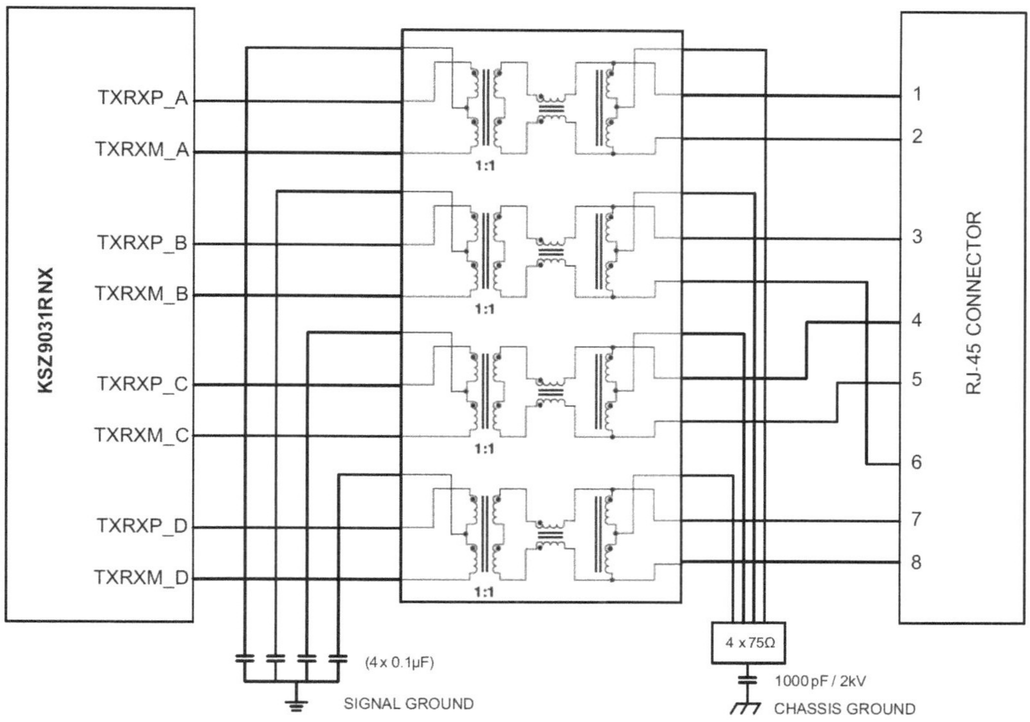

Figure 2.15 KSZ9031 datasheet: magnetic interface.

the service personnel will use the interface. That means that at every service call, the RJ45 connector could be unplugged from the customer network, plugged into a service computer, and then reverted back to the customer network. That equals four mating cycles per visit, two visits per year: eight cycles. Eight cycles for 12 years equal 96 cycles. Let's keep with the theme of round numbers and round up to 100 cycles for maintenance.

Let's also assume that at least once per year the customer damages the wired connection and has to replace the cable to the interface; that is another two cycles per year for 12 years, or 24 cycles. We will go with 25 for our counting purposes.

For all intents and purposes, this is the bulk of the use that the interface will see. There are other times the interface will see action, during production testing, for example. There are the outlier customers that will change cables once a month or unplug the connector once a week for their application, maybe to clean around the system, for example.

To give ourselves some additional margin, and because the total number (125) is so modest, we will triple our calculation and assume that 375 mating cycles is the target.

2.5.2.4 Environmental Considerations

In terms of environmental considerations, we can lean heavily on the excellent requirements work that the team at SensorsThink did and refer back to that list for guidance here.

1. Temperature range: –10°C to 50°C;
2. Humidity: 5% to 95% noncondensing.

There are no requirements for water or dust ingress protection, so this is all the information we need to make our choice.

2.5.2.5 The User

The context diagram here provides us with the information we need very clearly in terms of the end user of the Ethernet interface:

1. The customer (end user);
2. Service personnel;
3. A remote server (another computer/networked device).

2.5.2.6 Making the Choice

With all of the information gathered, making the choice becomes a matter of website filtering (choose your favorite supplier or distributor) and minding the details of the datasheet (Table 2.6).

In the case of the Ethernet, there are two types of RJ45 connectors from which to choose: those with integrated magnetics, and those without. Integrated magnetics make sense in many cases to simplify the PCB layout and to make design integration a bit easier. Let's presume that the design team and SensorsThink prefer that solution.

Bel Magnetic Solutions has a part number that checks off all of our considerations and requirement boxes: 0826-1X1T-32-F [18].

1. Existing standard: RJ45 Style;
2. Maximum expected voltage and current: 1,500-VRMS isolation;
3. Mating cycles: 375 (required confirmation with the supplier);
4. Environmental considerations: –10°C to 50°C;
5. End user: Because the connector type is so common, everyone is likely comfortable plugging in an RJ45 cable;
6. Datasheet details: Turns ratio, inductance, insertion loss, Hi-Pot.

This all points towards this being a good choice for our application and this interface. It did take a bit of information gathering and a bit of common sense and guesstimation (another made up word) to round things out but it was pretty

Table 2.6 Magnetics Selection Criteria: KSZ9031

Parameter	*Value*	*Test Conditions*
Turns ratio	1 CT: 1 CT	—
Open-circuit inductance (min)	350 μH	100 mV, 100 kHz, 8 mA
Insertion loss (max)	1.0 dB	0 MHz to 100 MHz
High potential (HIPOT) testing (min)	1,500 VRMS	—

straightforward compared to many other design tasks that are encountered during hardware development.

If the team at SensorsThink follows a similar approach for all of the interfaces and connectors, they will end up making sound technical choices for the system.

2.6 Extend the Life of the System: De-Rate

In engineering, we nearly always consider things such as functionality and cost as driving considerations for making decisions. Other common factors can be things such as power consumption, physical size and weight, and reliability. Later we will cover reliability in great detail. One piece of reliability that can sometimes be overlooked is the de-rating of components. De-rating a component is the concept of operating a component at less than its rated capability with the intent of prolonging its operating life.

2.6.1 Why De-Rate?

Component de-rating is a deliberate step taken by engineering teams to try to ensure that a design is robust. Operating components at one or more of their maximum ratings can shorten their useful life, but it can also have impact on other ratings of the device. Sometimes these de-ratings need to be considered part of the base specification of a component. As an example, we can refer to a ceramic capacitor datasheet from Murata [19].

In the case of the working voltage of this ceramic capacitor value, we can see that for nominally rated 630-V capacitors, at the highest end of the operating temperature range, the value must be de-rated to 450V, which is approximately a 30% reduction in the working voltage for the device (Figure 2.16). This is extremely significant and is not something that readily presents itself on a single sheet specification. Knowing that this de-rating applies comes with experience and also with a careful examination of the detailed product specifications.

2.6.2 What Can Be De-Rated?

For the engineering team at SensorsThink, what types of component ratings can be considered for de-rating? The short answer is: anything with a rating. If it has a rating, you can de-rate it. It really is that simple. Some extremely common things that are rated in electronics components are:

1. Power;
2. Voltage;
3. Current;
4. Temperature;
5. Actuation cycles.

As we have done for other topics, such as verification, we can look to harmonized standards for a good starting point for discussion and investigation into this topic.

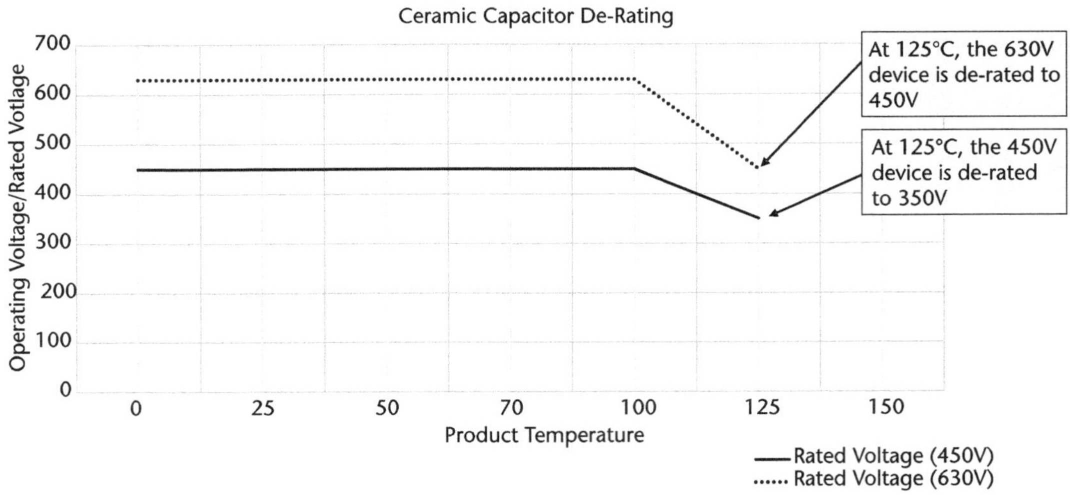

Figure 2.16 Ceramic capacitor de-rating for temperature.

IEC 61508 [20] is an international standard for functional safety systems, and it is often cited as a baseline standard for other types of safety systems that develop their own standards for safety.

From the document's first section, we can refer to the introductory paragraph to better understand the intent of the IEC 61508 series:

> This International Standard sets out a generic approach for all safety lifecycle activities for systems comprised of electrical and/or electronic and/or programmable electronic (E/E/PE) elements that are used to perform safety functions. This unified approach has been adopted in order that a rational and consistent technical policy be developed for all electrically-based safety-related systems.

For the engineering team at SensorsThink, adhering to this standard is not required for the SoC Platform, as it is not a safety system or one used in a safety application. However, as we used the guidance from ISO26262 to better understand the value of, and methods for, proper verification, we can take similar lessons for de-rating from IEC 61508 and create a very robust hardware design, rooted in the principles of the most robustly designed systems in the world.

2.6.2.1 What Does IEC 61508 Have to Say?

The guidance in IEC 61508 is extremely concise and simple; in the second entry of the series, the following text is given:

> De-rating (see IEC 61508-7) should be considered for all hardware components. Justification for operating any hardware elements at their limits shall be documented (see IEC 61508-1, Clause 5).
>
> NOTE: Where de-rating is appropriate, a de-rating factor of approximately two-thirds is typical.

Subsequently, referring to IEC 61508-7, we read the following:

Aim: To increase the reliability of hardware components.
Description: Hardware components are operated at levels which are guaranteed by the design of the system to be well below the maximum specification ratings. De-rating is the practice of ensuring that under all normal operating circumstances, components are operated well below their maximum stress levels.

There are two important takeaways from the standard:

1. Using components at their limits must be documented and justified.
2. A de-rating factor of approximately two-thirds (67%) is typical.

In other words, in a safety-related system that falls under the scope of IEC 61508, a 100-V rated capacitor can only be used on a voltage supply of 67V or less, unless there is supporting justification and documentation that shows why a higher applied voltage is acceptable.

2.6.3 Example: De-Rating the Zynq-7000

Based on the very straightforward guidance, we can apply these fundamentals to the Zynq-7000. Recall from the early discussion on system requirements that the SoC Platform must operate in an environment from –10°C to 50°C. While the platform is not a safety product, or part of a safety system, we can still apply the fundamental guideline from IEC 61508 to the temperature rating of the Zynq device that we chose to use as the heart of the system to determine the level of robustness that we can expect to see from that component, at least in terms of the impact from the temperature, during the product's life cycle.

Recall from Section 2.1 that the devices from which we chose belonged to the industrial grade of the Zynq-7000 series, with a nominal range of –40°C to 100°C. If we apply the de-rating factor of 67% to both the lower and upper limits of the range, we arrive at a new temperature range of –27°C to 67°C; and the good news for the engineers at SensorsThink is that the chosen device should not experience any reduction in working lifetime from the temperature range in which the product is expected to be used.

Suppose, however, that the device that was selected had a nominal temperature range of –20°C to 70°C instead of the –40°C to 100°C range with which we had to work. Applying the same de-rating factor would provide a result of –13.3°C to 46.6°C.

If the engineering team at SensorsThink wanted to design a very reliable and robust system, they may consider seeking a component with a wider temperature range, or they may want to consider doing extensive elevated temperature testing to understand the potential impact to the working life cycle of the product.

2.6.4 Additional Thoughts: Exceptions to the Rule

As with most rules, there are always exceptions, and de-rating is no different. There are certain types of components that are treated differently and must be de-rated differently if the engineering team truly needs a robust system to be deployed. Wire-wound resistors are often de-rated to 33% of their advertised power dissipation factor in safety systems, and solid tantalum capacitors are usually de-rated to 40% of their listed working voltage due to their long-standing reputation as devices that self-ignite during failure.

Knowing the guidance from IEC 61508 is a great starting point, but it is not a substitute for engineering rigor and for deeper understanding of the exceptions to the rules, especially in safety-critical systems. The engineering team that ignores those rules and exceptions will not get very far in a safety assessment by functional safety experts.

Even if your system is not part of a functional safety product, that does not necessarily preclude the use of de-rating to ensure a robust system and product. It can sometimes drive cost and can lead to discussions about overengineering a particular system. There is little harm to overengineering on the side of reliability and robustness. Most importantly, engineers must be aware of all the trade-offs that they are making and make the right one for the product that they are designing.

2.7 Test, Test, Test

In an earlier part of our journey through developing this embedded system, we talked about the high-level process of verification and validation. We broke testing down into levels 10 through 40 to align with the V-model, and we used ISO-26262 as a framework and guide to better understand how to derive test cases and what kind of tests provided high confidence levels in critical systems.

After designing the PCB and getting it started through the process of fabrication, there is often a few weeks of downtime for the design engineering team to start to put a verification test plan together.

Recall that level 40 testing for hardware consists primarily of the verification of individual circuits (or small groups of circuits that comprise one function). In order to define a test plan, we must define all of the following:

- What methods will be used;
- What the pass/fail criteria are;
- What environment(s) will be used;
- What tools will be used.

This can be overwhelming on even moderately complex designs, but with a basic framework in place, this can be broken down into manageable steps and the process can be repeated from design to design. The detailed steps will certainly need to change to accommodate a specific PCB, or circuit, but the framework can be reused and can quickly become second nature as more designs are approached in a similar way.

2.7.1 Back to Basics: Pre-Power-On Checklist

The engineering team at SensorsThink has gathered in a conference room to talk about how they plan to verify the hardware that is expected to arrive at the office in a few short weeks. People start throwing detailed ideas out for looking at nuanced performance of particular sensors, or elaborate stress tests for the high-speed portions of the design, a common occurrence when too many engineers gather in the same room.

Before discussing which pseudo-random bit pattern should be sent over a communications link or how many nanovolts of measurement accuracy are needed for a measurement, we should all remember the basics.

When a new PCB arrives, it must be treated as a complete unknown. Despite the thousands of hours of effort that went into conceptualizing and designing the circuits in the past several months, never forget that things happen. Human error is always a factor and, despite our best efforts, mistakes will be made. A well-constructed verification plan will push these errors to the forefront and allow them to be found, corrected, and protected from escaping into the field.

It is best to start with a very conservative approach and slowly expand the scope of testing as confidence in the design builds. Consider the following series of steps for taking a brand-new PCB out of the box and getting ready to test it:

1. Review provided documentation from the PCB fabrication and assembly contract manufacturers.
2. Perform visual registration checks.
3. Verify proper component orientation.
4. Verify that any components that were not to be included (DNP, DNI, no-load, no-stuff) are not present on the design.
5. Visually inspect solder joint quality; pay close attention to high-density components. Check for solder bridging.
6. Verify that any press-fit components are correctly installed.
7. Check that connectors and through-hole pins are not damaged or bent.
8. Before applying power, perform a mechanical fit check. Make sure that there are no interferences that were missed during three-dimensional (3-D) modeling work.
9. Measure impedances for each power domain to other power domains and to return nets.
10. Verify that fuses are not blown open, if they are present on the design.

This list of 10 steps can save a lot of headaches for engineers of all disciplines and only takes a short amount of time to perform on an appropriate sample size of the total lot of PCBs that have been received.

For particularly costly or complex designs, 100% inspection may well be warranted. In some cases, if very high-cost components are used, it is good practice to have several boards assembled without those components so that the underlying hardware environment can be verified to be safe for those high-value components. For example, consider a very high-end FPGA such as the Xilinx Virtex UltraScale+ Series of devices with on-board high bandwidth memory (HBM). These devices can cost tens of thousands (sometimes over hundreds of thousands) of dollars each. Taking extra precautions is absolutely warranted and is the right thing to do in cases like that.

It is reasonable to wonder what the value of each of the 10 steps listed provides to the engineering team, especially given that it can be quite exciting to get a new piece of hardware in hand. Oftentimes, the instinct is to just plug it in, turn it on, and see what happens. Patience is a virtue, as the old saying goes, so let's explore that further.

2.7.1.1 Hurry Up and Wait

Reviewing documentation from the PCB fabrication and assembly manufacturers is certainly not the most exciting activity an engineer will do, but it is important; if your design had controlled impedances, there should be reports with the boards showing that those impedances were verified and showing the measurements. Without seeing these results, there is no data to support that the high-speed traces on the PCB are actually fabricated with the correct characteristic impedance. This can cause many issues during testing (unstable interfaces, poor signal integrity) and sometimes can cause a design to not function. As speeds increase and rise times get faster, this becomes more important. Take the time to check the reports, and, if things are not clear, ask questions. It is better to ask early than to spend weeks chasing signal overshoot issues only to find that the board was not fabricated as you intended.

Visual checks are also a key aspect of bringing a design along from scratch. Imagine that an expensive component was loaded into a pick-and-place machine incorrectly, and pin 1 is not located properly. That could be disastrous for all involved. If caught before power is applied, the part can be removed and reworked to be installed correctly. If it is not caught, it might be too late to save the PCB. The same can be said for components that should not have been installed. Consider the case of things such as pull-up or pull-down resistors to configure a device in a certain way. If these are placed across a power supply rail and its return net, having both components present can be akin to applying a direct short.

Solder joint quality is one of the most important things to check; improperly flowed solder, too much solder, or not enough solder can cause the most frustrating of problems: things such as intermittent operation and fault behavior or board-to-board inconsistency. Many engineers have spent weeks chasing ghost problems caused by these kinds of issues.

Checking impedances between power domains and from power domains to return nets is essential as well. Remember always to check power and ground. The easiest way to derive the list of points to check is to open up the schematic, list every power supply net and return net, and make a spreadsheet or list of all supply nets to their return net, and if you have several return nets, check the impedance between them as well. If these values are not what you expect or are not consistent (within reasonable tolerance) across the lot of PCBs, then stop. Remove the outliers and figure out what is wrong with them. If this precaution is not taken, the magic-smoke can escape the design, and no one wants that.

Lastly, make sure that fuses survived being installed if they are present; there are fewer things more frustrating than trying to power a board on getting no activity, only to find a failed fuse as the culprit. It takes a few seconds to check a fuse, so save yourself some headaches and check them out.

2.7.2 Check for Signs of Life: Crawl Before Walking

After ensuring that the board has been properly fabricated and assembled and there are no drastic issues with short circuits, the time has come to apply power to the design. However, this does not mean that the engineering team should just slam the power on full-blast and start poking around. Having a plan is important.

The following set of steps is a good framework to keep in mind when bringing a new design online:

1. Basic power supply check:
   (a) For designs with programmable power supplies, this includes programming the supplies and verifying that the programming interface is functional.
   (b) A word of advice: always use a current limited power supply for these steps; better to hit a current limit and be mildly annoyed than to blow a board up and have to start over.
2. Clock verification:
   (a) As with power supplies, programmable clock generators are programmed at this point as well.
3. Power-on-reset/reset testing:
   (a) Many modern devices have requirements to be held in a reset state for a minimum amount of time. If the specific design under test has that as a requirement, verify that those times are being met. Doing so can prevent intermittent functionality/fault conditions from consuming hours (or days) of effort trying to debug.
4. Power sequence testing:
   (a) For designs with power-up and power-down sequencing constraints, they should be checked as early as possible. Not following those constraints can often cause device damage, so it is best to find those problems early.
5. Power supply noise and stability testing:
   (a) Before putting the devices through test cases that can cause load steps on power supplies, the stability of the supply's control loop should be tested over several conditions to ensure that catastrophic failures are not likely to occur.
   (b) Before checking the performance of interfaces, the noise performance should be looked at to ensure that in critical locations noise is within limits. Too much noise on high-speed device supply pins can cause the performance to degrade; understanding this and checking for that condition before testing in more detail can be invaluable.
   (c) Note that this step can be performed in the more detailed test stage in many cases; however, in designs with extremely expensive devices or with very high current supply rails, it is best to perform the testing early to try to catch any potential catastrophic issues early.
6. FPGA and processor programming:

(a) The last step before diving into the nitty-gritty verification world is to make sure that the processors and/or FPGA devices on the PCB can be programmed; this is usually accomplished via a JTAG (or another debugger) interface.

(b) In some cases, the JTAG interface can be used to program attached FLASH memory storage; if that is the case in a particular design, the time to check for basic functionality of that interface is early, before too much detailed checking has started.

By doing these tests for at least one representative PCB prior to distributing them to the other engineers who are part of the team, there is a reasonable level of confidence that a nominally functional hardware platform is in the hands of other engineers for further testing and development. Any necessary issues are found early and can be performed to all boards at the same time, ensuring a consistent deployment of hardware. This is crucial in teams of any size, but the larger the team, the more important.

2.7.3 Roll Up Your Sleeves and Get Ready to Run

With a reasonably high confidence level that the PCB that has arrived is not going to catch on fire and that it is able to be powered on and nominally programmed, the nitty-gritty testing can begin.

What does this mean exactly? What kinds of things can we leverage for test plans? Recall the earlier discussion on verification, in particular, verification specifications.

The verification specification essentially selects and specifies the methods to be used. This can be any number of things from simulation scenarios, physical test cases, and even something as simple as a checklist. As long as the adequacy of the specification is justifiable and reasonable, there usually is not much else to be considered here. That is not to say that the specification should lack detail or clarity; preconditions (input data) and configurations used (a subset or superset of variants) are wonderful things to document. The test environment conditions are also important bits of information to specify to allow for repeatability of testing should failures or anomalies occur.

We can refer to the design architecture and, even better, the schematics, to go page by page and construct our list of interfaces and circuits to be verified; for the SoC Platform here is a mostly complete list of interfaces to be verified.

- 10/100 Ethernet:
    - 100BASE-TX Physical Layer;
    - MII/RMII interface (payload data);
    - MDIO interface (PHY management).
- DDR3;
- SPI:
    - Accelerometer;
    - Infrared image sensor (video data);

- Magnetic field sensor;
- Vibration sensor;
- Configuration memory.
- I2C interfaces:
  - Infrared image sensor (configuration);
  - Local temperature and relative humidity sensor;
  - Nonvolatile memory.
- SD/MMC;
- Debug UART:
  - USB 2.0 Physical Layer;
  - TTL UART interface.
- USB on-the-go port:
  - USB 2.0 Physical Layer;
  - ULPI interface.

That is quite the list. Going page by page and breaking it down, we essentially have created an outline for the majority of our test plan. Now the details can be filled in: how the tests will be executed, what measurements will be made, what the pass/fail criteria are, and so on.

2.7.3.1 Don't Forget about Software

For many of these tests, hardware engineers cannot do it alone; the assistance of software and firmware is required. It is impossible to test an Ethernet interface without some code running that can generate and respond to messages, for example.

Once a rough idea of the plan is in place, engage with the software and firmware engineers, get feedback, and then, if compromises are needed, work through them so that there is little wasted work, but enough supporting code to thoroughly vet the hardware.

The earlier the other disciplines are involved, the smoother the testing process will go, and that is really the best-case scenario for all involved.

For these kinds of tests, a full-bore software image is not necessary. It can be completed piecemeal, with many small, simple programs loaded into the device one at a time, purpose-built to test a specific interface. This is all about verifying the underlying hardware platform. Keep it as simple as possible.

2.7.4 Example: I2C Interface

As a simple example of how to approach verifying an I2C interface, let's consider the nonvolatile memory in the SoC platform design.

The memory device is the Atmel/Microchip AT88SC0808CA; it runs a 2-wire I2C-compatible interface at up to 1 MHz. These simple datasheet facts provide an excellent starting point for defining our test case.

Let's consider the use case for the product; values will likely be periodically written to the device and also read back from it. (For some external memory devices, such as those that hold a bootloader, readback of data from memory to host may be much more important to test than write performance to the memory. It is always good to understand the use case of each component.)

In this design, the interface communicates at 3.3-V logic levels and it is connected to the Zynq processor complex.

Let's summarize what we have so far:

1. I2C interface;
2. 3.3-V voltage levels;
3. 1-MHz maximum frequency;
4. Write values;
5. Read values.

We are starting to piece a test case together, which is excellent progress. Let's dig a bit deeper into the details. How can we really make sure that the device is working correctly?

2.7.4.1 Read, Write, Verify

One of the best ways to verify a memory device is to write a pattern of alternating 0s and 1s to the memory and then read back those same memory locations. By doing so, we are putting a relatively high stress level on the device and also checking both directions of the interface. This checks the drivers and receivers on both interfaces, which is very important to do.

2.7.4.2 Specifications

In nearly all digital communications devices, there will be both dc and ac characteristics provided for the device; these are very important to review during the design phase, and equally important to review when designing a test plan (Table 2.7).

Values for minimum and maximum voltage levels that are interpreted as 0s and 1s are given in the dc specification, and things such as rise time and fall time are given in the ac specification for the device.

From the dc characteristics, we know now what voltage levels to expect on the interface and what voltage levels must be met for proper operation for both inputs to the device and outputs from it. For the I2C interface, the clock is always an input to the device; during a write transaction, the data is an input, and during a read transaction, it is an output.

Therefore, when we write to the device, we need to provide signals that reach at least as low as 3.3 * 0.2 (remember, 3.3V is our VCC) or 0.66V. When we read from the device, we should expect that the voltage level for a logic state of 1 should be no less than 3.3 * 0.7, or 2.31V.

From the ac characteristics, we can now compare the measurements of the signals on our design to the requirements of the device (Table 2.8). In this case, they are dependent on the period of the clock cycle. For a 1-MHz clock, the period is 1 μs; 9% of 1 μs is 0.09 μs, or 90 ns. This is a really clear pass/fail criteria against which we can measure and verify our design's performance.

Table 2.7 Selected DC Characteristics for AT88SC0808CA

Symbol	*Parameter*	*Min*	*Typical*	*Max*	*Units*
VIL	Input low voltage	0	—	Vcc * 0.2	V
VIH	Input high voltage	Vcc * 0.7	—	5.5	V
VOL	Output low voltage	0	—	Vcc * 0.15	V
VOH	Output high voltage	Vcc * 0.7	—	Vcc	V

Table 2.8 Selected AC Characteristics for AT88SC0808CA

Symbol	*Parameter*	*Min*	*Max*	*Units*
tR	Rise time	—	9% * period	μs
tF	Fall time	—	9% * period	μs

2.7.4.3 Other Options

In addition to these bench measurements, we can also use advanced functions on a logic analyzer or oscilloscope to decode the transactions as they occur in real time and verify that they are being executed as we expect. For example, if our test included a write to sequential locations in memory of the values 0x00 through 0x0F, we could capture that on a logic analyzer or oscilloscope and verify that the messages did contain that data. This does have a few caveats, such as configuring the thresholds of the decoding software to match the device that we are testing, but it can be a very valuable test with a simple visual cue of passing or failing.

2.7.5 Additional Thoughts

What we have outlined here is a good framework for which to approach testing a PCB, but there are more rigorous tests to consider. Once basic functionality and measurements have been made the test cases, if properly designed with some forethought, they can be reused and automated to run automatically. This can be very useful for collecting statistically relevant data and also for automated testing over temperature. Recall from ISO26262 that this kind of basic functional verification over temperature is a very strong data point for proving design robustness, so plan ahead and think through the whole plan before starting; it can really save time down the path of further verification and provide useful tools for manufacturing engineering, failure analysis, and regression testing. A little planning can go a long way in the world of testing, so do not be afraid to slow down up-front, so you can speed up at the back end of the process.

2.8 Integrity: Important for Electronics

In an earlier part of our journey through developing this embedded system, we talked about the high-level process of verification and validation. We broke testing down into levels 10 through 40 to align with the V-model, and we used ISO-26262 as a framework and guide to better understand how to derive test cases and what kind of tests provided high confidence levels in critical systems.

Part of verifying your design comes in the form of simulation, and this can take on many forms in embedded system design. For hardware specifically, we can consider two broad topics for simulation: power integrity and signal integrity.

A note to the reader: The topics of power and signal integrity are massive topics to try to explain in a concise manner, and they warrant books and conferences of their own. Because of that reality, it is not within the scope or spirit of this particular book to dive into great depths of detail for all things related to power and signal integrity. The intent of this brief discussion is to bring awareness to the topics for further consideration in your particular project or design. Many titans of the industry have written excellent books on the topics: Howard Johnson, Martin Graham, and Eric Bogatin are among our favorites, but there are many other excellent resources as well.

2.8.1 Power Integrity

Power integrity (in the context of the embedded system that the engineers at SensorsThink have been designing) refers to the quality of power being delivered to components within the system. It is an analysis (or series of analyses) that helps to determine whether or not the voltage and current requirements of the components in the design are being met. Without a properly designed power delivery system, many kinds of problems can occur in the system: increased bit-error rate (BER), reduced timing margin, and, in some cases, even brownout conditions within devices. The considerations outside of decoupling that are taken into account when designing for power integrity are mostly related to PCB stack-up design for copper weights, dielectric thicknesses for embedded capacitance, and trace widths that are necessary to carry currents with acceptable dc drop. Power integrity has several key factors that should be considered and analyzed during the design phase of the product: dc drop analysis, target impedance analysis, and simultaneous switching analysis.

2.8.2 Signal Integrity

Signal integrity, at its most basic and fundamental level, is a way to measure the quality of electrical signals. This is most commonly analyzed in high-speed digital systems: DDR2/3/4 memory, peripheral component interconnect (PCI) Express, USB 3.0, serial advanced technology attachment (SATA), and so on. As data rates of electronics have grown rapidly, into the range of tens of gigabits per second in recent years, this analysis has become mandatory for the design of systems that use these kinds of communications channels. The consideration for designing these interfaces into a system are far-reaching; component selection, PCB material choice, via size and spacing, PCB stack-up, trace width and space, acceptable channel characteristics such as insertion and return loss, and the power delivery network all play a role in ensuring a robust interface. In terms of types of analyses that are performed, it often depends on the type of interface being tested. For the sake of discussion, we can consider a few kinds: basic signal integrity analysis, DDR analysis, and serializer/deserializer (SERDES) analysis (PCIe, Ethernet, and USB are very common examples of SERDES interfaces).

2.8.3 Digging Deeper into Power Integrity

2.8.3.1 DC Drop Analysis

The dc drop analysis is analysis that is done to ensure that the IR losses of the power delivery network (PDN) interconnect do not introduce excessive voltage drop. With modern devices, high current consumption on the order of 100A is not unheard of for core voltages of FPGA and SoC devices. It takes very little impedance to cause what can be a significant voltage drop on the interconnect from the power supply source to its destination. When performing dc drop analysis, several pieces of information should be readily available and well understood in order to get the most meaningful data from the simulations:

1. Copper weights (copper thickness) of power distribution layers of the PCB;
2. Voltage input requirements of critical loads, so that pass/fail criteria can be set appropriately;
3. Worst-case current consumption for loads, so that maximum IR drops are calculated correctly;
4. Any component-specific current limits for pins on connectors or device packages.

With all of the above information, a very in-depth analysis can be performed: extended networks through regulators and to downstream loads to provide high confidence levels in the capability of the power delivery interconnect to provide the power needed for the design.

2.8.3.2 Target Impedance Analysis

Target impedance analysis characterizes the ac performance (e.g., performance across the frequency domain) of the PDN in the system. For high-frequency switching components especially, this is crucial to achieving expected performance of the components. Recall Section 2.4; we touched on four different categories of decoupling capacitor and how each type filled a specific role in providing adequate power to the component.

In order to gain the most value from this kind of analysis, the following pieces of information should be available as inputs to the simulation:

1. *The desired target impedance for critical components:* this type of information is sometimes available in datasheets; other times, more direct questions must be asked of IC suppliers to get the information.
2. *Accurate capacitor models for components used in the design:* Typically, these will be in the form of s-parameter files. S-parameter files are considered to be more accurate in many cases for high-frequency analysis because they can more accurately reflect nonlinear behaviors such as the variability of series inductance over frequency.
3. *Clear understanding of the specific pins of interest for the devices that require this analysis:* Not all package pins are created equally, and a good understanding of which pins supply which particular sections of a silicon

device can be important on devices with hundreds of supply voltage and return pins.

2.8.3.3 Simultaneous Switching Noise

Simultaneous switching noise (SSN) occurs when a high number of signals inside of a particular component change state simultaneously. In this situation, a large amount of current is consumed by the device. These events, which we will refer to as SSN events, can cause serious issues with data integrity. It is possible, by using power-aware modeling techniques and software packages, to check for these kinds of issues and attempt to mitigate them before problems are committed to copper on a physical PCB. This particular type of analysis lives in the world of power integrity, yet also directly impacts signal integrity. Analysis like this can be performed with Power Aware IBIS simulations (from IBIS v5.1 onwards) and also with vendor-specific tools such as Vivado from Xilinx, which performs SSN analysis from within the tool chain. To accurately model SSN, accurate device models are required above all else; accurate modeling of the capacitance of the loads that the signal drives is crucial. What is actually occurring during SSN events is the charging or discharging of these capacitive loads; therefore, the necessity for accurate driver models and receiver models is paramount. Another requirement is a software package capable of performing Power Aware IBIS analysis; there are several excellent solutions for this on the market from Mentor Graphics and Ansys SIwave.

2.8.4 Digging Deeper into Signal Integrity

2.8.4.1 Basic Signal Integrity Analysis

Basic signal integrity analysis is the first kind of analysis that probably comes to mind when engineers think about signal integrity. What we mean by basic signal integrity is running simulations on lower-speed signals and interfaces; in our design, we have many examples of where this type of analysis can and really should be performed:

- *SPI:* To sensors and memory;
- *RMII:* The interface to the Ethernet PHY;
- *ULPI:* The interface to the USB OTG PHY.

The reason that this type of analysis is becoming more common is due to the increased edge rates of modern silicon. Even devices such as logic gates from the 74LVC logic family boast rise time specifications on the order of only a few nanoseconds. An important concept to bear in mind with signal integrity is that the frequency of the signal is sometimes not the cause for concern, but rather the rise and fall times of the edges. A 1-ns rise time is equivalent to frequency content occurring at 1 GHz. At these frequencies, it is easy to see why designing with signal integrity in mind is so important. Performing basic signal integrity analysis requires accurate driver and receiver models and basic information about the interface such as:

1. Operating voltage level, in order to correctly select the model of the buffer to use in simulation. For even simple devices such as logic gates, there are often IBIS models describing the behavior of the driver at various nominal operating voltages. Choosing them incorrectly can lead to very misleading and strange results.
2. Operating frequency, in order to properly stimulate the interconnect. For an SPI that is intended to run at no faster than 10 MHz, using a 500-MHz input stimulus provides no value.
3. Specific drive strength and slew rate information for drivers. In modern devices, it is not uncommon to have multiple options for setting the drive strength (given in mA, typically) and slew rate (fast, medium, slow) for individual IO or specific peripherals. Changing these parameters can have a drastic impact on simulated and real-world performance.

2.8.4.2 DDR Analysis

DDR analysis is a relatively new advancement in the world of signal integrity modeling. As DDR memory has become a staple in the world of embedded systems, software packages have evolved to accommodate the analysis of an entire DDR interface. These simulations can take into account various on-die termination schemes, analysis of timing margins, and voltage overshoot and undershoot of the interface. This provides a wealth of information to designers before committing to physical boards and can be invaluable during the design process. For advanced simulations, you can include things such as crosstalk between signals and coupling through plane structures; in some cases, the tools even support simulated calibration as an option. In the SoC Platform, we have a DDR3 interface that would absolutely be considered essential to simulate both in a pre-layout and post-layout environment. Understanding the available parameters in DDR controllers and memory devices is paramount in this kind of analysis, with so many options for operating speeds and internal termination; proper selection based on the specific use case of the design being simulated is essential.

2.8.4.3 SERDES Signal Integrity Analysis

SERDES simulations are generally more involved, because they are typically going to require something called a passive channel analysis, where the interconnect of the differential signal is analyzed for a variety of characteristics such as insertion loss, return loss, and near-end and far-end (NEXT and FEXT) crosstalk (Figures 2.17 and 2.18). For many of these high-speed serial standards, there are plots that the simulation can be compared against to determine margins for the link. In addition to these passive channel analysis, virtual eye diagrams can be constructed in order to assess margins on the eye opening both vertically (voltage) and horizontally (time domain). This type of analysis is the most demanding, often requiring (in addition to what is required for other signal integrity analyses) s-parameter models for the transceivers and accurate s-parameters of the interconnect of the IC package and PCB geometry, and, in some cases, specialized software is recommended (Seasim, for example, is the supported tool used by PCI-SIG, the PCI special interest group).

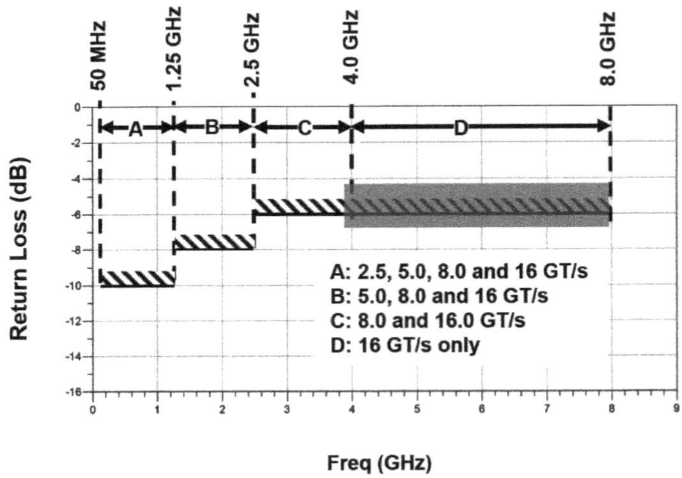

Figure 2.17 Differential return loss mask for PCIe [21].

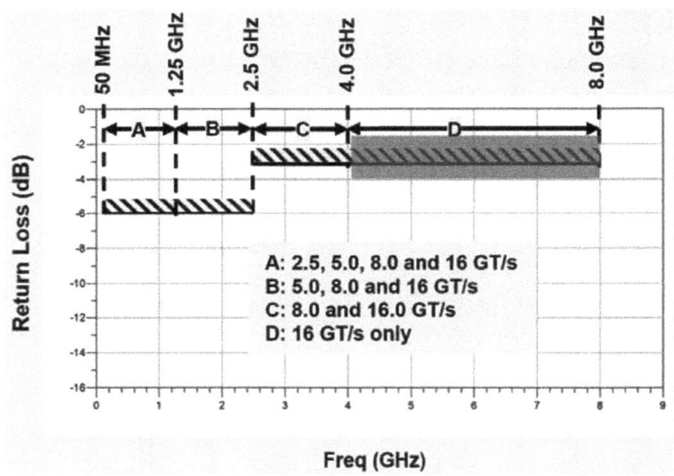

Figure 2.18 Common-mode return loss mask for PCIe [21].

2.8.4.4 Time-Domain Reflectometry (TDR)

Lastly, for all types of interconnect that runs at very high speeds (some example interfaces to consider as candidates would be PCIe, USB3.0, and Aurora LVDS), the simulation of the breakout from connectors and device packages is often modeled and then evaluated using a virtual TDR plot to judge the impedance of the traces, vias, and packages of the components in the time domain, essentially looking at the effective impedance that the signal experiences as it travels through the interconnect. Large impedance differences, often referred to as discontinuities, can cause signal reflections and have negative impact on functional performance.

2.8.5 Example: ULPI Pre-Layout Analysis

For one example, let's consider the ULPI interface between the Zynq SoC and the USB OTG PHY device. In this example, we will be using Mentor Graphics'

HyperLynx software to perform the pre-layout analysis and to verify that we have a sound design architecture in mind for the ULPI interface.

In the exercise of pre-layout simulation, the goal is to ensure that under idealized conditions the interfaces and design details (such as selected series termination) produce a very clean and idealized signal.

If the pre-layout simulation does not show nearly ideal signal behavior, it is an indicator that further investigation is warranted. There is something that is amiss, either with the setup of the simulation or with the design topology and component selection. Regardless of root cause, the time to find it and address it is certainly in the pre-layout stage and not later.

In addition to validating design decisions, pre-layout signal integrity analysis is a wonderful learning tool, as you can freely try things and look at the impact to the signals without any real-world risk or impact; this kind of learning opportunity is always welcomed in engineering.

For the ULPI interface, we have a relatively simple architecture to consider:

- Driver/Receiver 1: Xilinx Zynq FPGA;
- Transmission line;
- Driver/Receiver 2: USB3320 USB PHY.

Setting up this architecture in HyperLynx is relatively straightforward, using only a couple of components (Figure 2.19). Note that this also requires the IBIS models for the Zynq FPGA and USB3320.

By adding the two driver and receiver models to the design and an ideal 50Ω transmission line, we can quickly start to run an IBIS simulation, with only a few more pieces of input information provided to the tool:

- A sample stack-up of the PCB entered into the design tool;
- The simulation frequency or frequencies of interest;
- Configuration of any power supply nets required by the IBIS models;
- The voltage levels at which the design is intended to operate the ULPI interface;
- The expected distance between devices.

A note to the reader: PCB stack-up design is an art unto itself that is briefly covered later in this book. For the purposes of pre-layout signal integrity, the

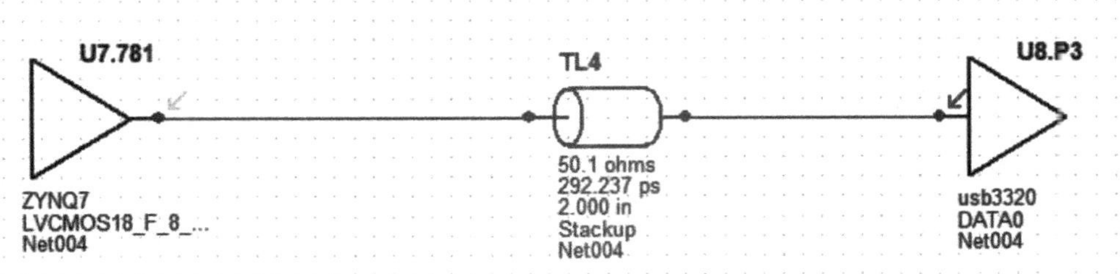

Figure 2.19 HyperLynx schematic model of the ULPI data path.

requirements are only to have a stack-up well enough defined to facilitate the creation of transmission lines of a known impedance.

For our design, the interface is intended to operate from a 1.8-V supply, and, per the ULPI specification, the maximum clock frequency and data rate are 60 MHz. This allows us to run a baseline simulation and view the waveform result relatively quickly. In order to get a clear view of the quality of the signal, it is possible to run an edge simulation (either rising or falling) or an oscillator simulation at a specific frequency. Additional options can be set for the corner of interest: slow-weak, typical, or fast-strong. In this example, we will run an oscillator simulation at 60 MHz of the typical IBIS model. This simulation will be of data flowing from the Zynq to the USB3320 device.

We can observe some signal overshoot and ringing on this initial simulation (Figure 2.20), and this is something for us as designers to watch during post-layout analysis (which takes into account actual routing geometry). However, we can also study the impact of series termination so that if post-layout analysis reveals similar problems, we can have a fix ready to implement.

A note to the reader: In a classical sense, series source termination is placed as close as possible to the transmitting device in order to try to better match the source impedance of the driver to the transmission line. On a bidirectional bus, it is ideal to place the termination at the mid-point between the two devices in order to try to match both transmitting devices. It is important to keep in mind that the ideal signal is almost impossible to implement on real electronics, which is why pass/fail criteria and a thorough understanding of the way that an interface works are crucial to these kinds of exercises.

In the revised circuit (Figure 2.21), we have split the transmission line in half and placed a 22Ω resistor in series with the transmission line pair.

We can see that the signal quality is greatly improved on this simulation run, with significant reduction in overshoot and ringing (Figure 2.22).

These kinds of simulations can be repeated in all kinds of varying conditions: longer transmission lines, different driver models, additional fanout, and different frequencies. Knowing what to simulate at different stages of the design comes with experience and a little bit of experimentation. However, the value and insight gained along the way are invaluable and should not be discredited as a reason to err on the side of simulation if any doubt exists.

2.8.6 Suggested Additional Reading

For additional information on power and signal integrity, there are a few excellent resources that are highly recommended [22–25].

2.9 PCB Layout: Not for the Faint of Heart

In terms of the amount of work that goes into designing a PCB, it can be argued that the majority of the hours put into the design are accumulated during the layout phase. Layout is the phase of PCB design in which the schematic interconnect is turned into the physical pieces of copper that (hopefully) result in a functional piece

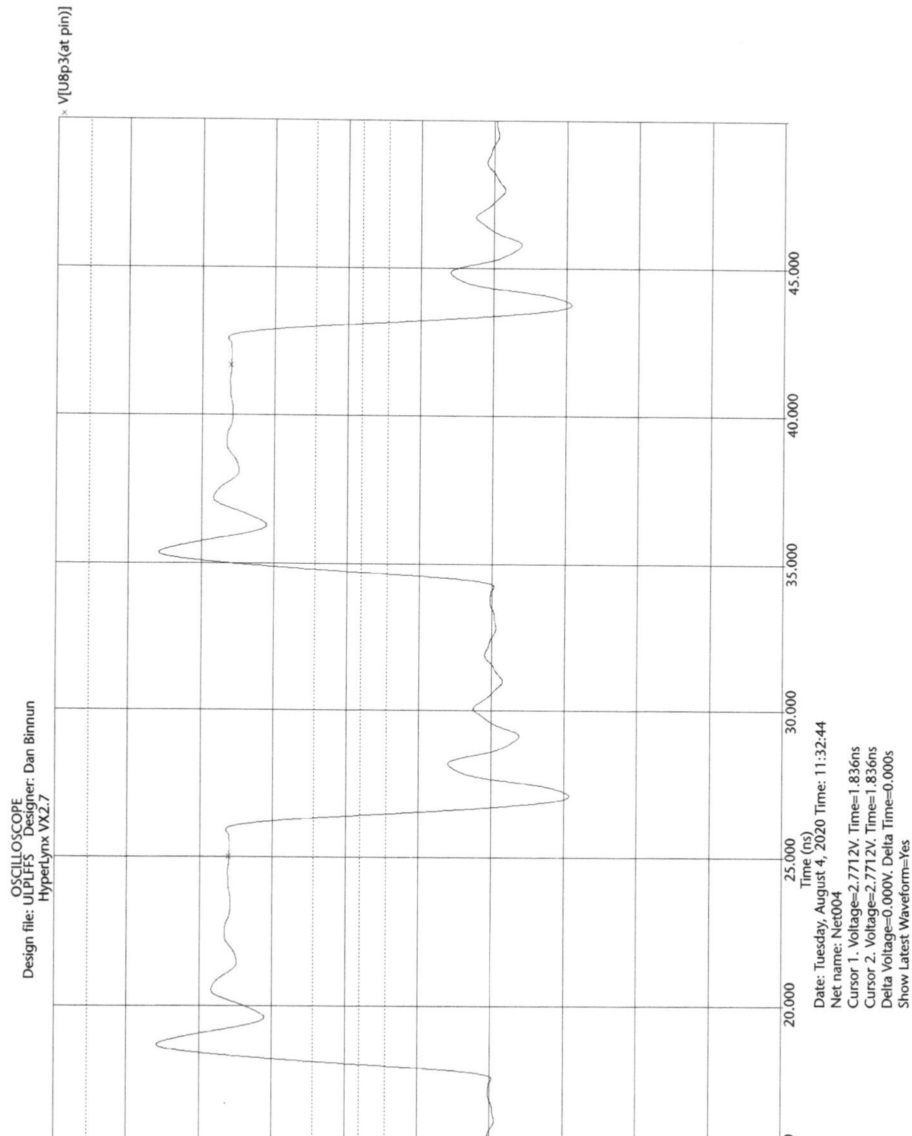

Figure 2.20 HyperLynx pre-layout simulation: ULPI.

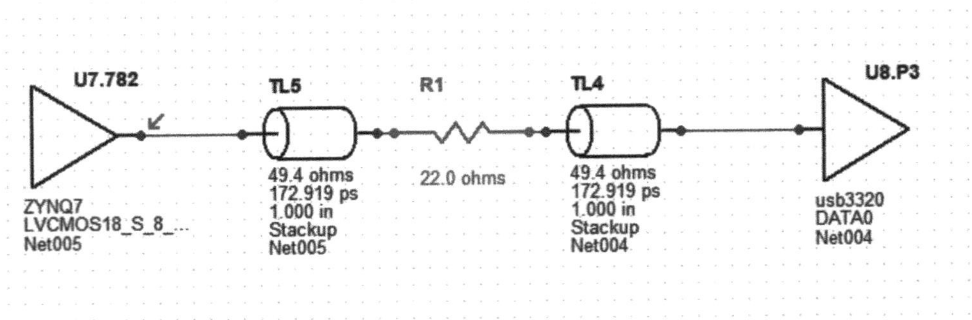

Figure 2.21 HyperLynx schematic model of the ULPI data path: added series termination.

of hardware that implements the many months (and sometimes years) of planning, design work, and development efforts of countless engineers and management.

PCB layout is an often-misunderstood engineering discipline, sometimes even an overlooked one. Too often schematic netlists are "thrown over the wall" to an unsuspecting PCB design engineer who is given little input, less guidance, and the highest of expectations.

As with nearly all tasks in engineering, garbage in produces garbage out. If the engineers at SensorsThink provide a well-designed schematic along with a clear set of design constraints and layout considerations to follow and they engage in the layout as a collaborative effort, the team will be set up for success.

PCB layout is an art form for sure, but one that begins with a tremendous amount of planning and detailed consideration. There is nearly an endless amount of design trade-offs that will be made out of necessity throughout the process of designing a PCB, and navigating those trade-offs is both part of the fun and part of the battle of PCB layout.

2.9.1 Floor Planning

Floor planning is the part of the layout process that revolves around placing functional blocks of the schematic design in physical locations on the PCB. Like a well-designed and architected house, a good floorplan is important. It sounds easy enough to plop the circuits down in a linear flow and let them go, but it is not that easy.

There are countless constraints that must be considered in almost all embedded systems, the first of which is the rat's nest (Figure 2.23). This is the visual representation of the connectivity between components; it is at first overwhelming, like a knot that has been tied and retied repeatedly. Part of floor planning is like untangling the unruly bundled mess of lines with which every PCB design starts.

There are many other constraints that must be considered during floor planning. Again, we can leverage the analogy of a house that is well-designed and architected. Mechanical systems must be considered; heating, ventilation, and air conditioning (HVAC) duct work, water and sewer lines, and structural loading of the building are all critical aspects to consider. In the PCB design, we can draw corollaries to power and return plane structure; the locations of (mostly) non-negotiable

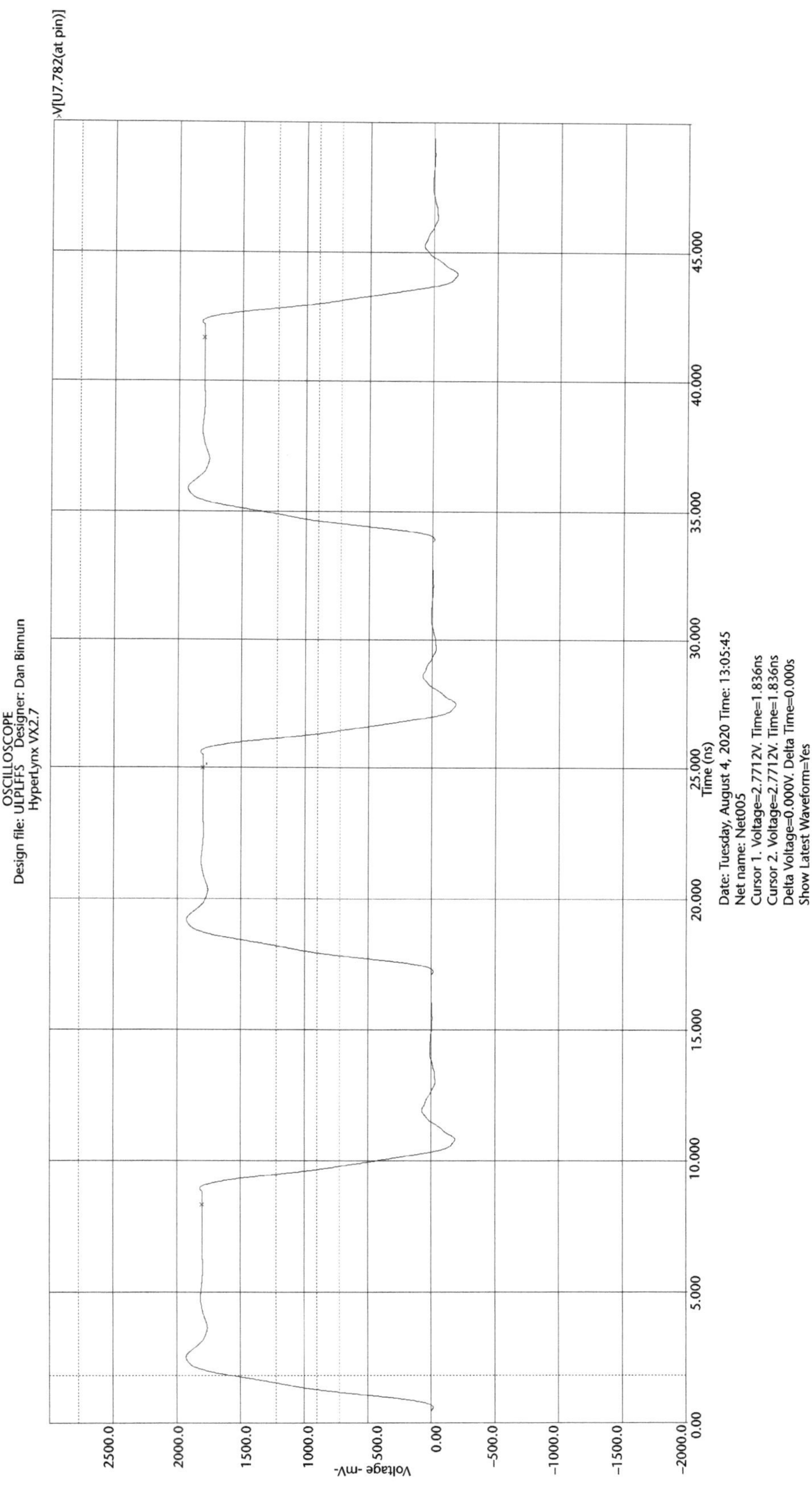

Figure 2.22 Modified HyperLynx pre-layout simulation: ULPI.

Figure 2.23 A rat's nest view of a PCB design.

mechanical details such as mounting holes and the physical size of the PCB are all important to consider from the very start of floor planning.

Electrical constraints are yet another consideration. What circuits are sensitive and likely to be impacted by electrical noise? What circuits are likely to annoy their neighbors? In keeping with the theme of a house, where do we want our jacks for cable and internet? Where does the dishwasher go or the washing machine? These things are considered long before the carpenters and tradespeople begin the work of building the house. If they are not, we can imagine that the home would be built ad hoc, and the results would likely be a mixed bag at best with even the most talented tradespeople performing the labor.

2.9.2 Follow the Rats

As we touched on earlier, the rat's nest is one of the major constraints and design inputs to consider when doing PCB layout and floor planning. An example that seems straightforward to experienced engineers may not be so obvious to those starting out in the field. To illustrate this, consider an Ethernet interface and physical connectors. In Figure 2.24, we see an attempted placement of two RJ-45 connectors and an Ethernet PHY device. The components are placed close to each other, and the connections look fairly straightforward to make. This is probably a very "routable" and achievable electrical interconnect to make, but there is room for improvement (Figure 2.24).

Let's make an iterative change and try to improve things by moving the location of the connectors. In Figure 2.25, we see that this does appear to be an

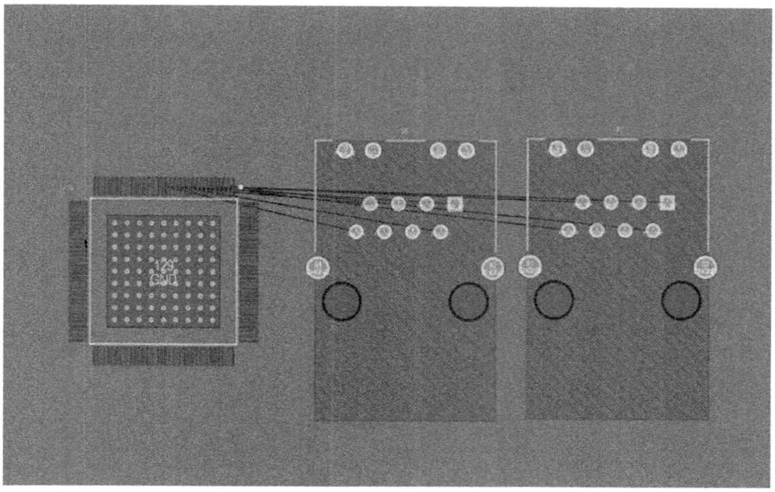

Figure 2.24 Rat's nest placement: could this be better?

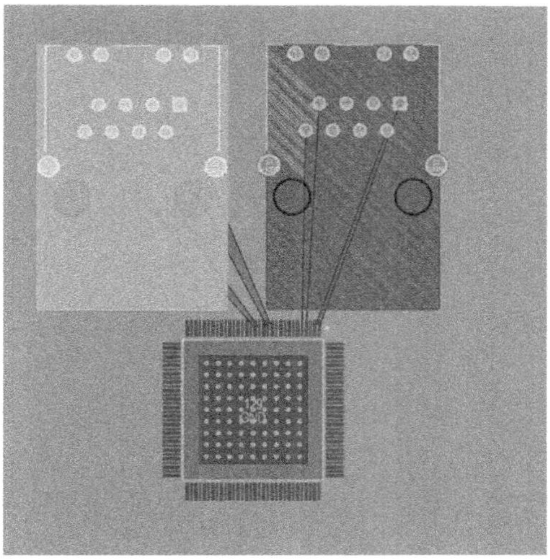

Figure 2.25 Untangling the rat's nest.

improvement. The connections now all have a clear line of sight to the pins on the IC. While this does look promising, a quick look in a 3-D rendered environment will show the problem with the current arrangement (Figure 2.25).

We can imagine that it might be less than ideal to try to plug a cable into the connectors with the IC sitting right in front of the jack (Figure 2.26).

If we take one more attempt at this, we can probably arrive at a reasonable placement for these three components, relative to one another.

In Figure 2.27, we arrive at what is likely the best we can do with the information that we have at hand for these three components and the rat's nest connections that are showing. It looks like some of the lines are crossing over each other (Figure 2.27); it was much better before. It was, but it was not a reasonable physical approach when we viewed things in the 3-D world in which we live. So we had to make the first of the many trade-offs that will be necessary during the process of

Figure 2.26 What about the cable?

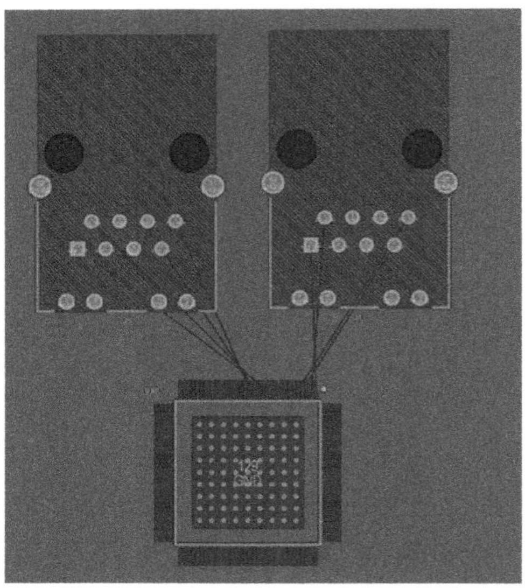

Figure 2.27 As good as it gets?

laying out this (and any) PCB. There are many ways to unravel these crisscrossed lines during the routing phase of the design, and many of those tricks are learned by doing and come from years of practice. Experienced PCB layout engineers see these solutions as quickly as they see the problems, and the trade-offs that must be made are clear to them very quickly. Until that experience is something that can be relied upon, mistakes will be made and learned from; all of that is part of the process.

2.9.3 Mechanical Constraints

Mechanical constraints are often provided to the PCB layout engineer from the mechanical engineering team. The overall size of the PCB is usually not overlooked, and mounting considerations are usually thought of as well, be it a card-guide system or old-fashioned screws that will hold the board to some kind of enclosure.

If PCB layout engineers are asked how much space they need for a PCB, they will almost always say "as much as possible." That is really their ideal answer. Unfortunately, there is rarely a situation where space is not at a premium. If product managers are asked how much space there is for a PCB, they will almost always say "as little as possible." That is really their ideal answer. Unfortunately, there is rarely a situation in which that is actionable by engineering. These two worldviews have to meet in the middle somewhere.

Aside from the available footprint and mounting considerations, you might be wondering what else mechanically is a true constraint on the PCB layout process. Here are a few to consider:

- *Connector locations:* Oftentimes, connector locations are determined by industrial design: what connectors have to be on the front, back, or sides of an enclosure based on how a system will be installed or used. This can be both a burden and a blessing to floor planning. If, for example, we know that the RJ-45 connectors must be on the right side of the PCB, we have a good idea of where the Ethernet PHY should be placed. An additional layer of complexity here is cable egress considerations: is the cable exiting from a specific side of the enclosure? If so, does a right angle connector or vertical connector allow that to be implemented more easily?
- *Height restrictions:* Height restrictions mean two things in this particular context. The first meaning is the overall height permissible in certain areas of the PCB, on the top and bottom sides of the board and in specific regions where perhaps the enclosure or environment otherwise might interfere with tall components. The second meaning is in card-based design, where the PCB thickness is based upon one or more existing standards because it has to mate with a specific type of connector in the end application; VPX systems, PCIe cards, and SODIMM form-factors are all prime examples of this.
- *PCB shape and cutouts:* In some instances, a PCB will need to have a unique shape or a shape that fits around a mechanical feature inside of an enclosure. This is generally known up-front when the product is conceptualized, but does not mean that there is any lack of challenge with integrating the constraint as the PCB layout process evolves.

Accommodating shrinking products and consumer expectations is challenging, and taking all of those fixed constraints into account during the floor-planning process is essential to reducing wasted effort and frustration for the entire product development team.

2.9.4 Electrical Constraints

This particular topic may seem a bit redundant in a discussion about PCB layout. Everything is electrical. However, in this context, we should reframe what we mean by electrical constraints to be a bit more direct. There are a few ways we can do this, so let's begin.

In nearly every design, there are going to be two types of circuits: circuits that create noise and interference, and circuits that are particularly susceptible to noise. In floor planning, understanding which circuits are noise creators is vitally important; equally important is knowing which circuits are susceptible to malfunction in the presence of noise. Once these circuits are identified, we can try to use physical space between the two types of circuits as our best ally for mitigation against unwanted interference and circuits that misbehave (Figure 2.28).

Some circuits that are typically considered noise creators are power supplies and high-speed switching digital electronics, and analog signal conditioning and data conversion are typically circuits that ideally are kept free and clear of these noise creators. In Figure 2.28, we see an example of a floor plan that can be used to help mitigate circuit interference using physical space as one of the primary means of achieving the goal.

Another type of electrical constraint is high-voltage considerations, be it in the form of ac mains-connected power or high-voltage dc power supply sources that are used in scientific instrumentation systems. Understanding where these interfaces and interconnect are going to be at the start of the design can make or break the success of a high-voltage PCB layout. Once those constraints are understood the PCB design engineer can begin considering how to address them; it could be slots in the PCB to prevent high-voltage breakdown, or if it is not possible to accommodate the required distances, encapsulation may be required. Surrounding the components with a material that can withstand and suppress high-voltage discharges better than air can may be necessary.

A note to the reader: High-voltage distances are called creepage and clearance distances and are the distance along a surface and through the air, respectively. IPC-2221B is an example of one reference for determining the required creepage and clearance distances needed in a specific situation. Also important to consider are product safety directives and standards, which may have additional constraints to consider.

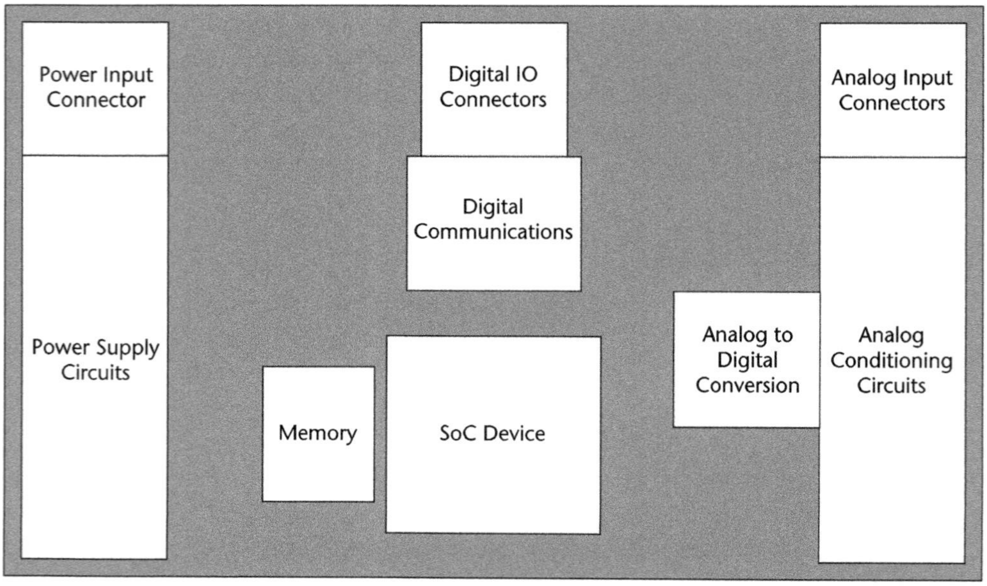

Figure 2.28 PCB floor plan.

2.9.5 Stack-Up Design

All PCB designs rely heavily on their stack-up design to function well. Continuing on with the analogy of a PCB being similar to a house, we can think of the PCB stack-up as the foundation on which the design will be built. Stack-up design is a crucial step in PCB design, and, in a recurring theme, there is much to consider when designing a stack-up. The thickness of copper, the amount of routing layers, the type of dielectric material, the cost of manufacturing particular feature sizes, and the trade-off of making routing easier versus making the process go more quickly are all on the table for discussion and consideration.

What is a stack-up design? A stack-up design is the cross-section of the PCB, the layers of the PCB sandwich as it were. In a foundation for a home with a basement, there is a footing, a foundation wall, and a concrete pad that all must be poured. They must be poured with the right materials, at the right times, and with the right amounts to function as intended (Figure 2.29).

2.9.5.1 Layer Count

Oftentimes, the first point of discussion will be the layer count of the PCB. Will it be a 4-layer board? Will it be an 8-layer board? For one reason or another (probably due to cost), this is usually where the discussion on stack-up design starts. When determining the number of layers needed, experience can be a helpful barometer for knowing how many layers will be necessary. However, it is not the only way to start.

With modern ball grid array (BGA) devices, there are some practical limits that can guide us to a reasonable starting point for the number of routing layers that we

#	Name	Material	Type	Weight	Thickness	Dk
	Top Overlay		Overlay			
	Top Solder	Solder Resist	Solder Mask		0.787mil	3.5
1	Top Layer		Signal	1oz	2.008mil	
	Dielectric 1	PP-006	Prepreg		3.465mil	3.77
2	Layer 2(GND)	CF-004	Plane	1/2oz	0.709mil	
	Dielectric 2	Core-009	Core		2.992mil	4.25
3	Layer 3(Mixed)	CF-004	Signal	1/2oz	0.709mil	
	Dielectric 3	PP-006	Prepreg		5.551mil	3.75
4	Layer 4(PWR)	CF-004	Plane	1/2oz	0.709mil	
	Dielectric 4	Core-009	Core		2.992mil	4.25
5	Layer 5(GND)	CF-004	Plane	1/2oz	0.709mil	
	Dielectric 5	PP-006	Prepreg		5.551mil	3.75
6	Layer 6(Mixed)	CF-004	Signal	1/2oz	0.709mil	
	Dielectric 6	Core-009	Core		2.992mil	4.25
7	Layer 7(GND)	CF-004	Plane	1/2oz	0.709mil	
	Dielectric 7	PP-006	Prepreg		3.465mil	3.77
8	Bottom Layer		Signal	1oz	2.008mil	
	Bottom Solder	Solder Resist	Solder Mask		0.787mil	3.5
	Bottom Overlay		Overlay			

Figure 2.29 An example PCB stack-up.

will need. Xilinx provides an excellent resource that walks through a sound methodology for estimating the number of routing layers needed to accommodate a PCB design that includes BGA devices of various sizes. We will be referring heavily to this resource, UG1099 BGA Device Design Rules.

A very quick way to estimate the number of layers required is based on the following equation:

$$Layers = \frac{Signals}{Routing\ Channels \times Routes\ Per\ Channel}$$

For Xilinx devices, a ratio of 60% of the total BGA pads is given as an approximate rule of thumb for finding the number of signals that need to be routed. The other 40% are typically power supply and return pads.

A routing channel is defined as a path to get a trace from a BGA pad to the area of the PCB outside of that pad, and the number of routes per channel is the number of traces that can fit within each channel, almost always one or two.

If we use our Zynq device on the SoC Platform as an example, we can determine how many routing layers that we will need to reserve for the PCB design.

With 400 BGA pins in our device package (the CLG400), we can assume that at maximum we will have 60% of these pads to route out to other components; 60% of 400 is 240.

The number of routing channels can be determined by examining the PCB footprint, and we can refer to Xilinx UG865 for this.

With 20 total rows and columns, there are 19 total routing channels available per side; and with four sides, that calculates out to be 76 routing channels in total for the device.

The number of routes per channel can be either one or two; we will calculate the required layers for both of these cases. The reason for this is that, right now, we do not have all of the constraints in place to rule out one approach or the other.

We can now plug these values into the equation with which we started to get a rough idea for the required number of routing layers:

Case 1: One Route Per Channel

$$Layers = \frac{240}{76 \times 1}$$
$$Layers =\sim 3.2$$

Case 2: Two Routes Per Channel

$$Layers = \frac{240}{76 \times 2}$$
$$Layers =\sim 1.6$$

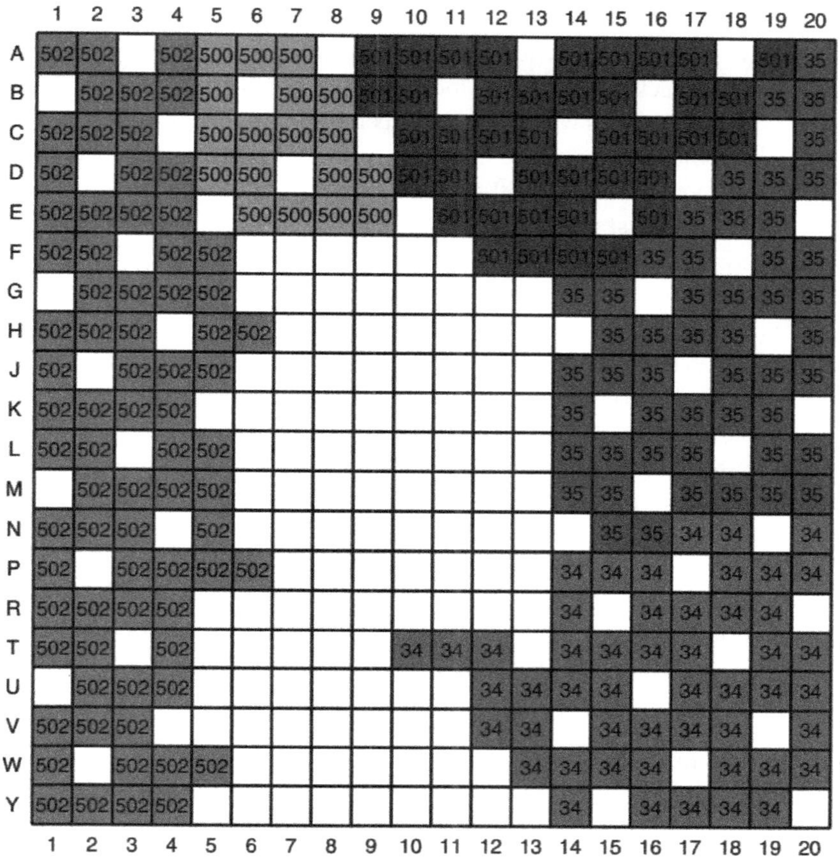

Figure 2.30 The CLG400 land pattern.

We cannot have half of a routing layer, so we can say that we will need approximately two or three routing layers for this design in order to fully route the Zynq SoC device.

The additional steps of determining the possibility of routing one or two traces per channel can be followed step by step in UG1099. They will not be covered in detail here, but are mentioned for completeness and to give a clear overview of the complexities that can be involved.

In addition to this basic calculation, a slew of other considerations are important in determining what is feasible to expect, among them are:

- *Pad size and pad pitch:* The size of the BGA pad and the space between them are directly correlated to how many traces can fit in each routing channel. A 1.0-mm pitch device will be much easier to accommodate two traces per channel, and a 0.5-mm pitch device will almost assuredly not be able to do so.
- *Fabrication technologies:* Things such as via-in-pad, blind and buried vias, and micro vias (which are part of a category of PCB design technologies referred to as high-density interconnect (HDI) technologies) can greatly reduce the number of routing layers required by freeing up space in the PCB sandwich for other traces to be routed. We will cover HDI later in this chapter.

- *PCB thickness:* PCB thickness has a direct impact on the minimum mechanical drill size that can be performed on any particular design. A typical and safe aspect ratio is 10:1, meaning that the board cannot be any thicker than 10 times the diameter of the mechanically drilled hole. For a typical drilled hole of approximately 8 mils (0.008 inch), that means the thickest stack-up that we can design is 80 mils. If we try to add too many layers to a board and do not adjust the minimum drill size accordingly, our PCB will be impossible to manufacture.

Typically, performing this exercise for the highest pin-count and/or smallest pad-pitch device will yield a stack-up design and routing layer count that works for all the devices on the PCB.

However, having three routing layers is not the end of the story here, in addition to routing layers, we also have to consider power and return layers, and oftentimes the top and bottom layers of the PCB can be so full of components that routing on them is not always feasible.

Taking all of this into account, we can see that arriving at an 8-layer stack-up design is probably a reasonable starting point for this design. However, needing to add additional layers may be required, and ending up with a 10-layer or 12-layer stack-up would be completely reasonable as well.

2.9.5.2 Copper Weights

Another piece of the puzzle with stack-up design is the weight or thickness of the copper. The copper weight is often one of the knobs that is turned when dealing with devices that have high current requirements. The thicker the copper conductors, the more current they can safely carry. However, it is important that this change in copper weight be done mindfully, as the thicker the copper is, the harder it is to manufacture fine pitch traces. So while 3.5 mil (0.0035 inch) can be created on a lighter copper weight of, say, 0.5 ounce, on a 2 or 3-ounce copper material, that trace geometry is not really feasible. One common practice is to use heavier copper weights on layers dedicated to power and return planes and lighter copper weights on routing layers or component layers with fine-pitch devices. This is a good way to get the best of both worlds: heavier copper for better power distribution and lighter copper for etching detailed features.

2.9.5.3 Dielectric Types

A dielectric material is a material that is an insulating material or a very poor conductor of electric current. In the PCB stack-up design, these are the materials that sit between the conductive layers of copper, providing the necessary isolation between things such as power and return nets.

The most common materials used for dielectrics in PCB manufacturing are glass-reinforced epoxy, often referred to as FR-4 materials. These materials come in a wide variety, with differences in how the material is physically constructed (typically woven together) and how much glass fiber is present versus epoxy.

In addition to those fundamental construction differences, there are also varying thicknesses available, from as thin as 2 mils (0.002 inch) up to 125 mils (0.125

inch). The use of a particular style and thickness of dielectric depends greatly on the situation for which it is being designed.

A note to the reader: The best practice for creating and verifying PCB stack-up designs is to engage directly with the contract manufacturer that will be fabricating your PCB. This is the best way to make sure that your PCB stack-up design is able to be manufactured.

2.9.5.4 Controlled Impedance

Controlled impedance is becoming a more common requirement in modern PCB design as the speeds of interfaces and edge rates increase. (As the edge rates increase, the rise and fall times decrease, making for sharper, faster transitions.) Controlled impedance refers to designing the physical geometry of a trace on your PCB to have a characteristic impedance, similar to the way that a BNC cable has a characteristic impedance of, say, 50Ω. Controlled impedance is important because without it, we cannot have robust transmission lines for our high-speed signals, RF signals, or analog signals. Without robust transmission lines, we run the risk of introducing enough degradation in the performance of the electrical channel to cause the interface to not function or not function well enough for the application at hand.

Designing for a controlled impedance or for several controlled impedances should be done as early as possible. Once we have established our routing geometries based on layer count and pad pitch, we at least have an idea of how wide our traces need to be in order to be properly routed. We can either provide this information to the contract manufacturer or leverage a PCB design tool that has the ability to calculate the characteristic impedance of a trace with respect to a reference layer.

The trace widths, required impedances, and material thicknesses and types all play a role in this calculation, and all the pieces of the puzzle can be difficult to fit together. Working hand in hand with the PCB contract manufacturer can be invaluable here in reducing wasted effort and rework; if a submitted set of files does not translate into the impedance that were targeted, the impact can be devastating to project timelines.

A note to the reader: Sometimes it is just not possible to have the ideal transmission line for all signals on a PCB for any number of reasons. If this is a situation that presents itself, remember that trade-offs are part of the game. Consider critical signal routes early on and design the rest of the stack-up around them, or perform signal integrity simulations to see if using narrower traces to route through a BGA has a significant impact on particular interfaces as early as possible to try to mitigate and manage the risks.

2.9.5.5 HDI

Earlier in this chapter, we touched on several technology types that are part of a group of fabrication methods referred to as HDI. HDI is one of the more rapidly evolving spaces in the PCB manufacturing domain. IPC-2226 classifies PCB designs into six groups, or types, and also outlines material types and mechanical and physical properties that are acceptable for use in these types of designs. Let's refer back to our house analogy for PCB design. HDI is similar to adding fancy modern gadgets and extras to your house. You may not need them, but they can make life easier. They almost always cost more than the old-fashioned way of doing things (like a tankless water heater versus a traditional tank-style water heater), but often come with performance benefits.

2.9.5.6 What Is HDI?

In general, a PCB is considered to be HDI if it uses one or more of the following HDI technologies:

- *Micro-vias:* Typically, a via spanning two layers (layer 1 to layer 2, for example) and typically will be a laser-drilled via. These micro-vias usually require an aspect ratio of 1:1, meaning that they cannot drill deeper than the diameter of the hole that they create.
- *Via-in-pad:* This is exactly what it sounds like, drilling a via directly in the pad of a component. This technology requires an additional process step of plugging the via holes so that the solder does not drain into the pad holes. Via-in-pad can be a traditional through-via, micro-via, or blind-via.
- *Blind vias:* A via that only goes partially through the PCB, for example, on an 8-layer board, from layer 1 to layer 4. These vias can connect any two (or more) layers that they span.
- *Buried vias:* A via that is not exposed to the outer layers of the PCB, for example, on an 8-layer board, from layer 4 to layer 6. These vias can connect any two (or more) layers that they span.
- *Multiple laminations:* Any PCB assembly that needs to be laminated (pressed together) more than one time in order to accommodate blind or buried vias.

2.9.5.7 When to Use HDI

HDI technologies have benefits and drawbacks, and sometimes as a designer your hand will be forced into using them. For example, on fine pitch devices, traditional mechanical drill sizes are simply too big to use.

- If you need to design a very small product with a lot of interconnect, you are likely going to need at least one HDI technology.
- If you have extremely high-speed digital interfaces, you are likely going to need some type of advanced HDI processing to achieve signal integrity goals.
- If you have very fine-pitch devices, you will likely need via-in-pad or micro-vias to even be able to fit a trace into the BGA area.

There are some handy graphs and plots for assessing when HDI may be the best path forward; the HDI Handbook [23] has one such region map that tells us that the key metrics for assessing this are component density (the number of parts per square inch) and component complexity (the number of pins per components). So just because you are working on a design with 1,000 components with 80 components per square inch, it does not mean that you need HDI if 998 of those components have two pins. Conversely, just because a design has only 5 parts per square inch, it does not preclude the use of HDI if the components have an average of 80 pins.

2.9.5.8 Exotic Materials

For some high-speed digital, RF, or other niche applications, standard glass reinforced epoxy simply is not up to the task at hand. Whether it is for transmission of extremely high-frequency signals or for dealing with high temperature environments, sometimes materials have limitations that cannot be overcome, even by the most seasoned designers. The most seasoned designers learn these kinds of situations and proactively solve the problem up-front. They can tell when an exotic material will probably be required and can save the product development team a lot of heartache by bringing these concerns to the forefront sooner, rather than later.

2.9.5.9 An Example Material

For high-frequency RF or high-speed digital designs, it is common to use materials referred to as low-loss, high gigahertz. These materials demonstrate a couple of key characteristics that make them a more ideal fit for these applications. As an example, let's consider a popular material from Rogers Corp, RO4350B:

- Very tight control of the dielectric constant over frequency and environmental conditions;
- Low loss, or dissipation factor at high frequencies (specified as high as 40 GHz).

In the design of high-performance systems such as automotive radar or cellular base stations, these characteristics are essential to proper functionality. Compared to the performance of a standard FR-4 style material, the consistency and control of the dielectric constant are extremely important. In addition to changing with temperature, the dielectric constant shifts with frequency and can shift wildly within manufacturing tolerance of common materials. A material such as RO4350B is extremely consistent from lot to lot and can provide good peace of mind for designers working on high-performance RF systems.

2.9.6 Experience Matters

Last but certainly not least, it is important to know that the only way to get better at PCB design is to practice it. Experience matters. Knowing the tricks of the trade can help you to visualize solutions to problems before you head too far down a particular path or help you to determine whether or not a particular design is feasible to implement. These tricks of the trade and the confidence to tackle complex design work do not come without some hard lessons learned. This can be in the form of rework and lost hours, days, or weeks of PCB layout work or in expensive scrap piles of PCBs that just cannot be salvaged.

Parameter	Frequency	Value
Dk, Permittivity	100 MHz	4.24
	1 GHz	4.17
	2 GHz	4.04
	5 GHz	3.92
	10 GHz	3.92

Figure 2.31 Dielectric constant variation versus frequency of standard FR-4 style material.

There are many aspects of PCB design that we did not cover here, things such as design for manufacturability (DFM), design for assembly (DFA), and design for test (DFT). These topics are all broad and deep at the same time and really are worthy of entire books all on their own.

Hopefully, this has shed some light on the complexities and trade-offs to expect with PCB design work and why it truly takes a talented team of engineers to design a highly functioning PCB. If the engineering team working on the SoC Platform takes these factors into consideration, they will be on the path to success, even if there are many bumps along the way.

References

[1] JEDEC, "About JEDEC," September 19, 2019. https://www.jedec.org/about-jedec.

[2] International Organization for Standardization, "ISO 26262-1:2018," August 22, 2019. https://www.iso.org/standard/68383.html.

[3] FLIR Commercial Systems, "FLIR LEPTON® Engineering Datasheet," 2014.

[4] Dhaker, P., "Introduction to SPI Interface," Analog Devices, Analog Dialogue 52-09, September 2018. https://www.analog.com/media/en/analog-dialogue/volume-52/number-3/introduction-to-spi-interface.pdf.

[5] FLIR Commerical Systems, "Lepton Software Interface Description Document, FLIR #110-0144-04," 2014.

[6] Xilinx, *Zynq-7000 SoC Technical Reference Manual*, UG585 (v1.13), April 2, 2021. https://www.xilinx.com/support/documentation/user_guides/ug585-Zynq-7000-TRM.pdf.

[7] Xilinx, "Zynq-7000 SoC DC and AC Switching Characteristics," DS187 (v1.21), December 1, 2020. https://www.xilinx.com/support/documentation/data_sheets/ds187-XC7Z010-XC7Z020-Data-Sheet.pdf.

[8] *Merriam-Webster*, "Decouple," February 8, 2020. https://www.merriam-webster.com/dictionary/decouple.

[9] Murata, "Basics of Capacitors: Chapter 1," January 14, 2011. https://www.murata.com/en-us/products/emiconfun/capacitor/2011/01/14/en-20110114-p1.

[10] Murata, "Basics of Capacitors: Lesson 2," April 14, 2011. https://www.murata.com/en-us/products/emiconfun/capacitor/2011/04/14/en-20110414-p1.

[11] Panasonic, "Fundamentals of Capacitors and Hybrid Capacitors," July 8, 2019. https://industrial.panasonic.com/ww/ss/technical/lc3.

[12] Analog Devices, "Decoupling Techniques," March 2009. https://www.analog.com/media/en/training-seminars/tutorials/MT-101.pdf.

[13] Xilinx, "Zynq-7000 SoC PCB Design Guide," March 14, 2019. https://www.xilinx.com/support/documentation/user_guides/ug933-Zynq-7000-PCB.pdf.

[14] Mentor Graphics, February 2020. https://www.mentor.com/pcb/hyperlynx/signal-integrity/.

[15] Mentor Graphics, February 2020. https://www.mentor.com/pcb/hyperlynx/power-integrity/.

[16] Mentor Graphics, "HyperLynx Design Rule Checking," February 2020. https://www.mentor.com/pcb/hyperlynx/electrical-rule-check/.

[17] IEEE SA, "802.3-2018 - IEEE Standard for Ethernet," February 15, 2020. https://standards.ieee.org/standard/802_3-2018.html.

[18] Bel Magnetic Solutions, February 15, 2020. https://belfuse.com/resources/drawings/magneticsolutions/dr-mag-0826-1x1t-32-f.pdf.

[19] Murata, "Chip Multilayer Ceramic Capacitors," November 27, 2017. https://www.murata.com/~/media/webrenewal/support/library/catalog/products/capacitor/mlcc/c02e.ashx?la=en-us.

[20] International Electrotechnical Commission (IEC), "IEC 61508:2010 CMV," March 28, 2020. https://webstore.iec.ch/publication/22273.

[21] PCI SIG, "PCI Express 4.0 Electrical Previews," 2014. https://pcisig.com/sites/default/files/files/PCI_Express_4_0_Electrical_Previews.pdf.

[22] Bogatin, E., *Signal and Power Integrity: Simplified*, Upper Saddle River, NJ: Prentice Hall, 2009.

[23] Holden, H., et. al, The HDI Handbook, Seaside, OR: BR Publishing, 2009, https://www.hdihandbook.com/.

[24] Bogatin, E., and L. D. Smith, *Principles of Power Integrity for PDN Design—Simplified: Robust and Cost Effective Design for High Speed Digital Products*, Upper Saddle River, NJ: Prentice Hall, 2017.

[25] Graham, M., and H. Johnson, *High Speed Digital Design: A Handbook of Black Magic*, Upper Saddle River, NJ: Prentice Hall, 1993.

[26] Johnson, H., *High Speed Signal Propagation: Advanced Black Magic*, Upper Saddle River, NJ: Prentice Hall, 2003.

[27] Cooper, R., *Winning at New Products: Creating Value Through Innovation*, 3rd ed., New York: Basic Books, 2011.

[29] Sparx Systems Pty Ltd., "Enterprise Architect 14," August 22, 2019. https://www.sparxsystems.com/products/ea/14/index.html.

[29] Reliable Software Technologies, "Mew Models for Test Development," 1999. http://www.exampler.com/testing-com/writings/new-models.pdf.

[30] Forsberg, K., and H. Mooz, "The Relationship of Systems Engineering to the Project Cycle," *Joint Conference of the National Council on Systems Engineering (NCOSE) and American Society for Engineering Management (ASEM)*, Chattanooga, TN, October 21–23, 1991. https://web.archive.org/web/20090227123750/http://www.csm.com/repository/model/rep/o/pdf/Relationship%20of%20SE%20to%20Proj%20Cycle.pdf.

[31] INCOSE, "About: INCOSE," https://www.incose.org/about-incose.

[32] Requirements Working Group, International Council on Systems Engineering (INCOSE), "Guide for Writing Requirements," San Diego, CA, 2012.

[33] International Electrotechnical Commission (IEC), "Available Basic EMC Publications," August 22, 2019. https://www.iec.ch/emc/basic_emc/basic_emc_immunity.htm.

[34] RoHS Guide, "RoHS Guide," August 22, 2019. https://www.rohsguide.com/.

[35] European Commission, "REACH," August 22, 2019. https://ec.europa.eu/environment/chemicals/reach/reach_en.htm.

[36] Ericcson Inc., August 22, 2019. https://telecom-info.telcordia.com/site-cgi/ido/docs.cgi?ID=SEARCH&DOCUMENT=SR-332&.

[37] BKCASE Editorial Board, "The Guide to the Systems Engineering Body of Knowledge (SEBoK), v. 2.0," Hoboken.

[38] SEBoK Authors, "Scope of the SEBoK," August 18, 2019. https://www.sebokwiki.org/w/index.php?title=Scope_of_the_SEBoK&oldid=55750.

[39] SEBoK Authors, "Product Systems Engineering Background," August 18, 2019. https://www.sebokwiki.org/w/index.php?title=Product_Systems_Engineering_Background&oldid=55805.

[40] SEBoK Authors, "System Design," August 18, 2019. https://www.sebokwiki.org/w/index.php?title=System_Design&oldid=55827.

[41] SEBoK Authors, "System Analysis," August 18, 2019. https://www.sebokwiki.org/w/index.php?title=System_Analysis&oldid=55956.

[42] SEBoK Authors, "Logical Architecture Model Development," August 18, 2019. https://www.sebokwiki.org/w/index.php?title=Logical_Architecture_Model_Development&oldid=55698.

[43] SEBoK Authors, "Physical Architecture Model Development," August 18, 2019. https://www.sebokwiki.org/w/index.php?title=Physical_Architecture_Model_Development&oldid=55874.

[44] IEEE, "1149.7-2009 - IEEE Standard for Reduced-Pin and Enhanced-Functionality Test Access Port and Boundary-Scan Architecture," February 10, 2010. https://ieeexplore.ieee.org/servlet/opac?punumber=5412864.

[45] International Organization for Standardization, "ISO 13849-1:2015," August 22, 2019. https://www.iso.org/standard/69883.html.

[46] International Electrotechnical Commission, "Functional Safety and IEC 61508," August 22, 2019. https://www.iec.ch/functionalsafety/.

[47] International Organization for Standardization, "ISO 26262-1:2018," August 22, 2019. https://www.iso.org/standard/68383.html.

[48] McLean, H. W., *HALT, HASS, and HASA Explained: Accelerated Reliability Techniques, Revised Edition*, Milwaukee, WI: American Society for Quality, Quality Press, 2009.

CHAPTER 3

FPGA Design Considerations

3.1 Introduction

At SensorsThink, the development of the Smart Sensor SoC is making significant progress: the requirements have been agreed upon and baselined. The architecture of the solution has been defined, along with the subsystem allocation of functions between hardware, software, and firmware. Crucially, the hardware development has also begun creating the schematic and layout of the PCB. To ensure that SensorsThink have something to test on the hardware once the PCB is manufactured, the development of the FPGA and software needs to be started.

To ensure a high-quality FPGA design solution, SensorsThink FPGA engineers follow a comprehensive FPGA development process, which is aimed at managing the life cycle of the FPGA development from concept to delivery of the completed bit stream.

To create the FPGA application, engineers at SensorsThink face several distinct challenges:

1. Achieving the processing deadlines for sensor sampling;
2. Implementing algorithms that convert from raw sensor data to data values that are in the correct format, for example, resistance to temperature conversion;
3. Implementing algorithms that enable system noise to be removed, for example, filtering;
4. Implementing control structures to correctly interface with sensors and to configure the sensor for the required operation and read back the correct sensor data format;
5. Achieve the performance required across the operating environment.

These distinct challenges mean that engineers at SensorsThink need to consider a wide range of techniques that can be deployed in an FPGA in order to achieve these requirements. These techniques include:

1. Development flow and architectural implementation, including clock domains and clock domain crossing.

2. Development of control structures using state machines, along with considering defensive state machine design if the required operating environment demands it.
3. Implementation of algorithms requires engineers at SensorsThink to understand how math can be used in programmable logic. This also includes understanding different techniques that may present themselves for achieving complex algorithms (e.g., polynomial approximation, and COordinate Rotation DIgital Computer (CORDIC) algorithms).
4. Understanding how mixed signal converters work, as several of the sensors utilize the built-in Xilinx analog to digital converter (XADC) in the Zynq.
5. Understanding the verification of each module developed to ensure that its performance and functionality are correct prior to implementation within the target device once the board is made available for testing.
6. Understanding how digital filters can be implemented within the programmable logic, along with understanding how engineers can use the frequency domain and Fourier transforms to demonstrate the performance of filters, analog-to-digital converter (ADC), and digital-to-analog converter (DAC).
7. Consider the implementation of algorithms in higher-level languages such as C or C++. Implementing algorithms in a higher-level language can significantly reduce the development time.

In this chapter, we will explore these techniques in detail, thus enabling the SensorsThink engineering team to correctly architect and implement the FPGA solution.

3.2 FPGA Development Process

The development process for an FPGA follows elements of both hardware and software development.

The FPGA development flow starts in the systems engineering phase, when the subsystem segmentation allocates specific requirements to be implemented within the FPGA. These requirements are further expanded upon with the FPGA requirements, which includes not only the system-allocated requirements, but also derived requirements that are required for the FPGA. These derived requirements will include such requirements as FPGA technology (SRAM, flash, or One Time Programmable), operating temperature, operating frequency, and fixed-point performance requirements.

Once the requirements have been agreed upon, the next stage is the architectural design. During the architectural design, the architecture and hierarchy of the design are outlined, reviewed, and agreed upon. This architecture will identify the following:

- *Clock architecture:* The number of different clock domains and clock domain crossing approach.
- *Functional architecture:* The number of modules required to implement the overall requirement set.

- *Reset architecture:* Definition of the different reset domains and applicability to functions.
- *Test philosophy:* The verification strategy and tool chains used to demonstrate the modules and overall design, which achieves the requirements and performance targets.

Following sign-off of the architecture, the next stage of the development process is detailed design and verification; this will include:

- *Development functional modules:* This may use hardware descriptions languages (HDL) such as Verilog and VHDL. Alternatively, high-level languages may be used, such as C or C++, or proprietary tools such as MATLAB and LabVIEW.
- *Static code analysis:* Static analysis of the code enables nonfunctional issues to be detected and corrected. Static analysis therefore enables not only the enforcement of company naming standard, but also the enforcement of company/industry coding standards. Static analysis will also analyze the module to identify any dangerous or ambiguous coding structures that may lead to issues later in the development process. The goal of static code analysis is to start later stages of development with a higher quality of code, reducing the iteration time through verification and implementation.
- *Functional verification of modules:* Creation of test benches to verify the performance of the module under test. This may use industry standard frameworks, such as Universal Verification Methodology (UVM), Open Source VHDL Verification Methodology (OSVVM), or Universal VHDL Verification Methodology (UVVM). Depending upon the application, the verification may have targets for code coverage in the unit under test to ensure the module is sufficiently exercised.
- *Top-level design:* The assembly and interconnection of all modules within the design to complete the overall design.
- *Top-level verification:* The development of a test harness to check the top level of the design achieves the desired functionality. Depending upon the end application, the top-level verification may range from simply ensuring the correct connectivity of the modules to fully testing the functionality of the entire design.
- *Constraints generation:* Constraints are used by the FPGA implementation tool to guide the synthesis, placement, and routing stages. Commonly used constraints include:
    - *Input/output constraints:* Define the location of the signal on the FPGA pins and the logic standard to be used for that signal (e.g., LVCMOS1v8, LVCMOS2v5 for single-ended signals, or LVDS for differential signals). The IO constraints also can control the output drive strength, on-chip termination, slew rate, and pull-up or pull-down.
    - *Clocking constraints:* Defines the clock frequencies and interclocking and intraclocking relationships. The clocking constraints will also include definition of the input and output delays. Correctly defined timing

constraints will help to achieve the required performance across the operating conditions. Incorrectly defined timing constraints may result in the device not working as expected across the operating conditions. It may also significantly increase the implementation time as the implementation tool searches for a solution that is not realistic or achievable.

  - *Placement constraints:* Define the physical location of logic elements/modules or groups of modules within the FPGA. Placement constraints can be used to help the tool achieve the timing performance and provide isolation between functional modules for security or safety.

- *Synthesis:* Synthesis translates the HDL written at a register transfer level into the logic resources, which are available in the targeted FPGA. To guide the synthesis tool, synthesis constraints may be used for timing-driven synthesis.
- *Placement:* Places the synthesized logic resources within the FPGA logic elements.
- *Routing:* The final stage of the implementation. Routing attempts to connect all the placed logic elements in the FPGA together, while achieving the timing performance defined by the constraints.
- *Bit stream generation:* Once the implementation has been completed, the implemented design needs to be converted into a format that can be downloaded or used to program the FPGA.

3.2.1 Introduction to the Target Device

The device selected in Chapter 2 to implement the FPGA design in is a Xilinx Zynq-7000 series device. This device is a heterogeneous SoC as it contains both a processing system and programmable logic. Within the processing system are dual Arm Cortex-A9 32-bit application processors capable of operation at up to 1 GHz. The programmable logic is based on the Xilinx 7 series logic and is interconnected with the programmable logic using several high bandwidth Arm eXtensible Interface Memory Mapped (AXI-MM) interfaces.

AXI is part of the Arm Advanced Microcontroller Bus Architecture (AMBA) tailored for implementation in FPGA and programmable logic devices. AXI provides three different interface standards, which provide for different performance and implementation requirements.

- *AXI Memory Mapped:* A full-duplex memory mapped interface that is capable of burst accesses to read and write memory locations. Both read and write channels are independent. AXI Memory Mapped is used for high bandwidth transfer to and from memory, either directly connected to the FPGA or within the processing system (e.g., DDR).
- *AXI Lite:* A simplified version of AXI Memory Mapped protocol that does not offer burst access. As such, it is intended for use in IP configuration, control, and status reporting.
- *AXI Stream:* A unidirectional point-to-point stream of data, intended for high bandwidth data transfer between IP within the programmable logic.

Architecturally, it is like the write channel of the AXI Memory Mapped interface.

AXI is a point-to-point interface, which always communicates between a master and a slave. Should we wish to connect a single master to several slaves, a cross-bar switch is required to be implemented.

Within the processing system itself there are several peripherals commonly found within an embedded processing solution, for example, universal asynchronous receiver-transmitter (UART), controller area network (CAN), serial peripheral interrupt (SPI), I2C, and memory controllers. These peripherals are also interconnected using AXI buses; as such, it is possible if desired for the programmable logic to access peripherals within the processor system.

3.2.2 FPGA Requirements

At SensorsThink, we are developing a sensor platform that uses a programmable SoC. As such, we need to define the requirements for the programmable logic half of the design. These requirements need to be clear, concise, and unambiguous.

The FPGA will therefore be responsible for interfacing and processing with the following:

- FLIR Lepton 3.5: interface with the FPGA using I2C and SPI interfaces;
- LSM303AHTR Magnetometer Sensor: interfaces with the FPGA using a I2C interface;
- SHTW2 Temperature Sensor: interfaces with the FPGA using an I2C interface;
- ADCMXL3021 Vibration Sensor: interfaces with the FPGA using an SPI;
- LSM6DS3USTR Accelerometer Sensor: interfaces with the FPGA using an I2C sensor;
- HX94 Remote Humidity and Temperature Sensor: interfaces with the FPGA using XADC.

The FPGA is also able to help the processor expand its limited multiplexed IO (MIO). The processor has 54 MIO pins that can be implemented in interfaces such as I2C, SPI, Gigabit Ethernet (GigE), and Universal Serial Bus (USB), as shown in Figure 3.1. However, in some applications, the processor requires more MIO than are available. In this instance, for many of the low-speed interfaces, the peripheral signals can be routed to the programmable logic. This enables the PS interfaces to use the programmable logic IO for interfacing; this is called extended MIO (EMIO) as the MIO are extended into the programmable logic. In the SensorThink solution, EMIO is used to route out PL IO pins to support the Wi-Fi and Bluetooth interfaces.

At the FPGA requirements level, the requirements must trace back to the higher-level system requirements. The FPGA requirements can be seen in detail at SensorsThink.com.

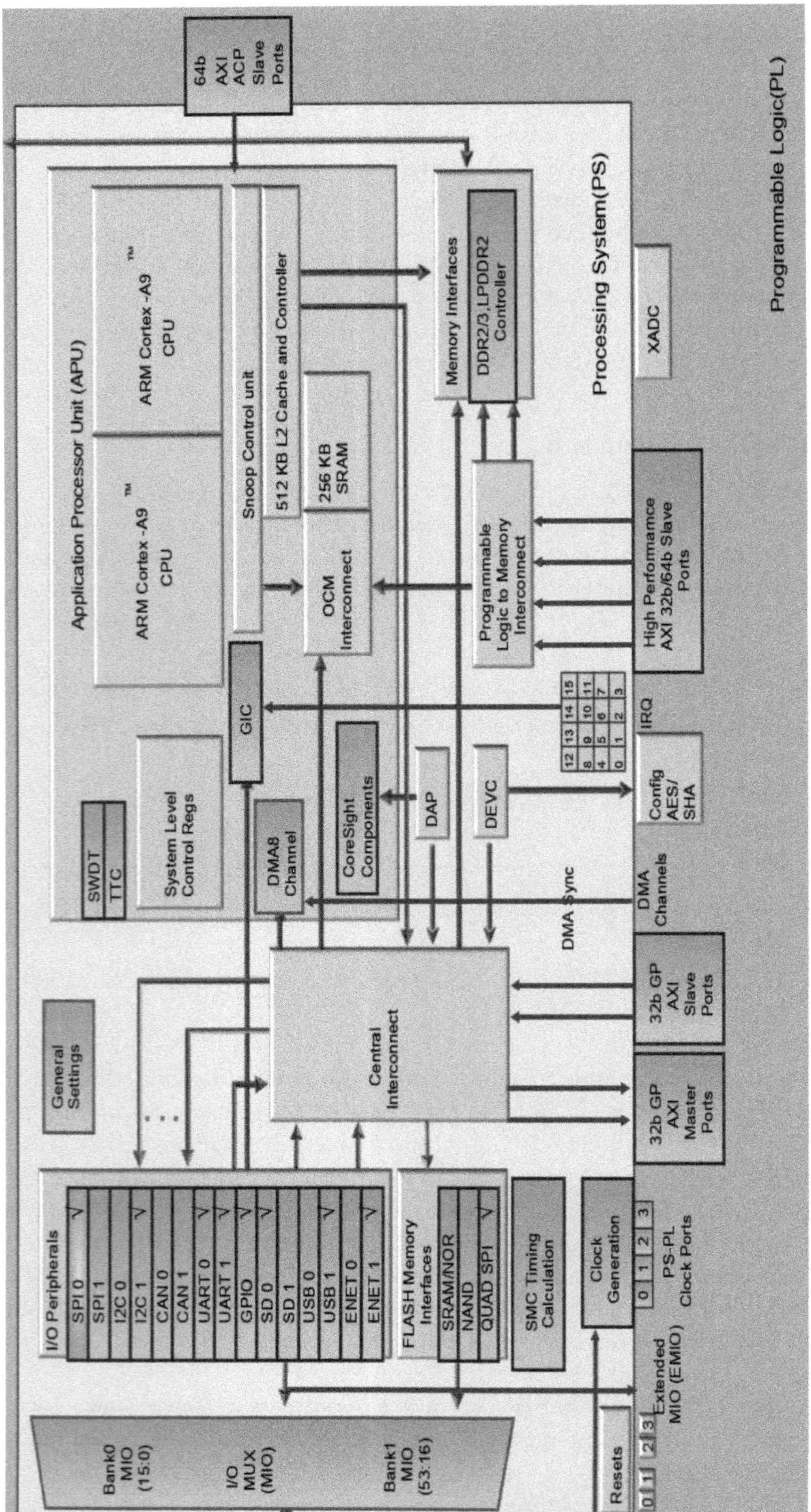

Figure 3.1 Xilinx Zynq block diagram.

- There must be the ability to read temperature, magnetometer, vibration, and accelerometer sensors concurrently.
- Sensor data shall be tagged with a time stamp.
- The availability of new sensor data shall be indicated to the processor using an interrupt from the logic to the processor.
- The gathering of sensor data shall be initiated from the processor system.
- Frames from the FLIR Lepton 3.5 shall be read out at the maximum frame rate of 9 Hz.
- New FLIR image frames shall be indicated to the processor using an interrupt.

These requirements drive the overall architecture of the solution between the PS and the PL.

3.2.3 FPGA Architecture

The architecture of the FPGA design will implement several IP blocks that are used to interface with the sensors. Control functions will be used to gather the data and indicate to the processor that new data is ready.

The architecture of the FPGA solution can be considered in two distinct elements as shown in Figure 3.2: the sensor interfacing and the FLIR interfacing. Starting with the sensor interfacing to enable each sensor to be read in parallel, several I2C and SPI interfaces will be used. Each of these blocks will be under the control of a specific controller block. This controller block will consist of a state machine that will send the appropriate commands to the I2C or SPI interface. This will enable each of the sensors to be read in parallel. As each of the sequences are in parallel, this will allow the sensor controllers to wait for the sensor data to be available.

Starting the conversion process will be the responsibility of the controller and scheduling block that receives the cycle start signal from the processing system. The control and scheduling block also receive the sensor data when it is ready from

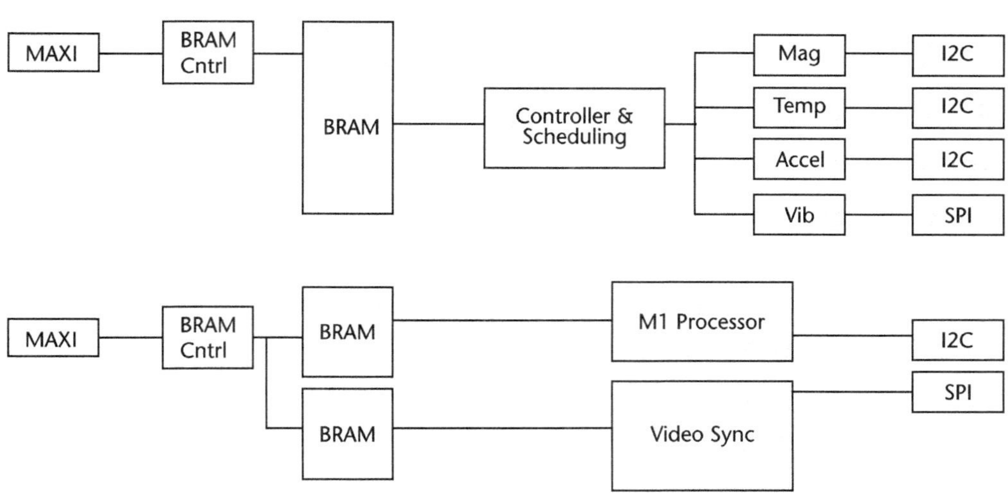

Figure 3.2 FPGA architecture.

each of the dedicated sensor controllers. Once the sensor data from all four of the sensors has been received, the words are bundled together and written into the BRAM and an interrupt is issued to the processing system. The processing system can then, on receipt of this interrupt, read the BRAM via the AXI BRAM controller IP. Once in the processing system, the software is then able to process and work on the data.

Communicating with the controller and scheduling block using the BRAM enables the processing system to define several parameters used in the control and scheduling block, for example, the time between sampling cycles.

The FLIR input path is like the sensor path; however, the FLIR imager, once configured, will be free-running outputting frames at 9 Hz. The FLIR Lepton is configured over an I2C link; as such, in this application, the M1 processor is used connected to an I2C interface. The M1 processor can configure the FLIR Lepton under the control of the processing system, for example, changing the pixel output from grayscale to red, green, blue (RGB) false color or enabling the auto gain control. The M1 processor is operating from a program stored within BRAM; as this memory is dual-ported, updates to the M1 program can be executed by the processing system.

Video data output from the FLIR Lepton uses a video over SPI format. This means that packets of data are sent over the master in, serial out (MISO) port, when the master provides the serial clock (SCLK) and asserts chip select. In the video over SPI format, the master out, serial in (MOSI) port is not used.

The video packets output by the Lepton are either valid packets or invalid packets to comply with export regulations regarding frame rate. Once a valid packet is output with a valid line number, each of the line numbers increment until the entire frame has been read out. If the packet header indicates that the frame is incorrect, it needs to read out the entire corrupted frame and discard it.

As the frame rate is so low that a custom register-transfer level (RTL) block will be created that checks the frame header to determine if the header is valid or not. If the header is valid, the frame data will be written to block RAM, until the entire frame has been read out and written to BRAM. Once this has been completed, an interrupt will be generated to inform the processor a new image is available.

The processor can then access the image from the BRAM using the block random access memory (RAM) controller over the maser AXI interface.

Now that we understand the architecture, the next step is to start creating and verifying the RTL.

3.3 Accelerating Design Using IP Libraries

Before the SensorsThink engineering team begins to start RTL development, they should consider the use of existing intellectual property blocks, often called IP blocks. These IP blocks are provided by device vendors, third-party commercial IP vendors, and open-source initiatives (e.g., open cores). IP blocks fall within two groupings. The first group is those that are created to implement commonly used functions, for example, block RAM memories, first in, first out (FIFOs), and clocking structures. These basic IP building blocks are universal across a diverse range of FPGA developments and can be used to accelerate the design capture process by providing IP blocks that enable the designer to focus on creating the value-added element of the solution, while the second group is domain-specific IP blocks. Domain-specific IP blocks focus upon a specific domain and implementation of functions commonly used within that domain. Example domains include quantitative finance, machine learning, and embedded vision. Domain-specific IP blocks are often significantly more complex than basic building blocks, for example, implementing algorithms such as color space conversion, video layering for embedded vision applications, or rate setting for quantitative finance. Using IP blocks provides a significant time-scale reduction as it reduces the number of custom blocks that need to be developed and verified. Many IP blocks also use the AXI standard outlined above that further reinforces the benefit of using a standard interface, as easy integration with existing IP blocks is provided.

3.4 Pin Planning and Constraints

The inclusion of a programable logic device on our Smart Sensor SoC means that we need to carefully consider the pin allocations. Programmable logic offers complete flexibility in assigning pins to signals; however, to get the optimal solution, we need to undertake careful pin planning. This pin planning is required to ensure that:

- Correct clocking resources are used: Xilinx Seven series provide both global and regional clocks.
- Correct pins are allocated for differential pairs.
- Correct IO standard is allocated to an IO bank.
- All IO standards used on a bank are compatible.

On a heterogeneous SoC such as the Zynq devices, not only do we have to consider the pin allocation for the programmable logic element of the device, but we also need to ensure that the MIO and DDR pin allocation is correct for the processor system.

Looking first at the processor system MIO, we need to create a design within the Vivado IP integrator, which targets the exact Zynq that will be placed on the board (xc7z020clg400-1). Once this project is created, we can create a new IP integrator design that contains a Zynq processing system, as shown in Figure 3.3.

Recustomizing the Zynq processor system allows us to implement the configuration of the PS MIO. Using the peripheral IO view, the MIO pin allocation to processing system peripherals can take place (Figure 3.4).

This visual view enables the MIO to be allocated as desired to the required processing system peripherals without the conflicts as design rule checking is live. Should the designer need to change any of the IO standards on the MIO pins, using the MIO configuration tab, the design can modify the IO speed and enable pull-up/pull-down on the signals (Figure 3.5). This can help greatly when the signal integrity is being analyzed.

While the DDR timing can be configured using the DDR configuration tab, this tab is where we define DDR type, speed, and arrangement (Figure 3.6).

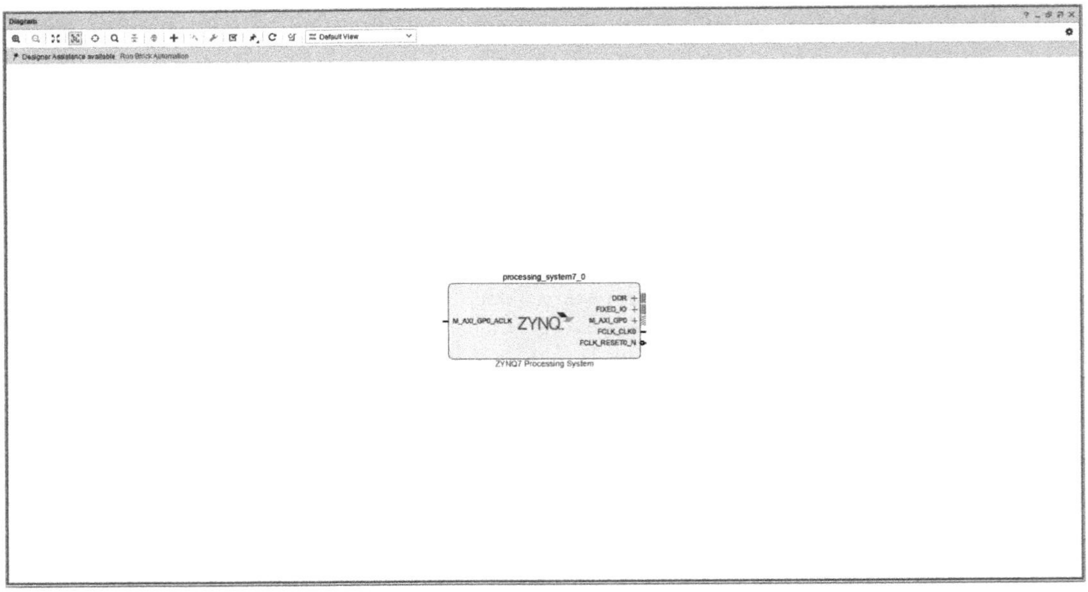

Figure 3.3 Zynq processing system block diagram in an IP integrator.

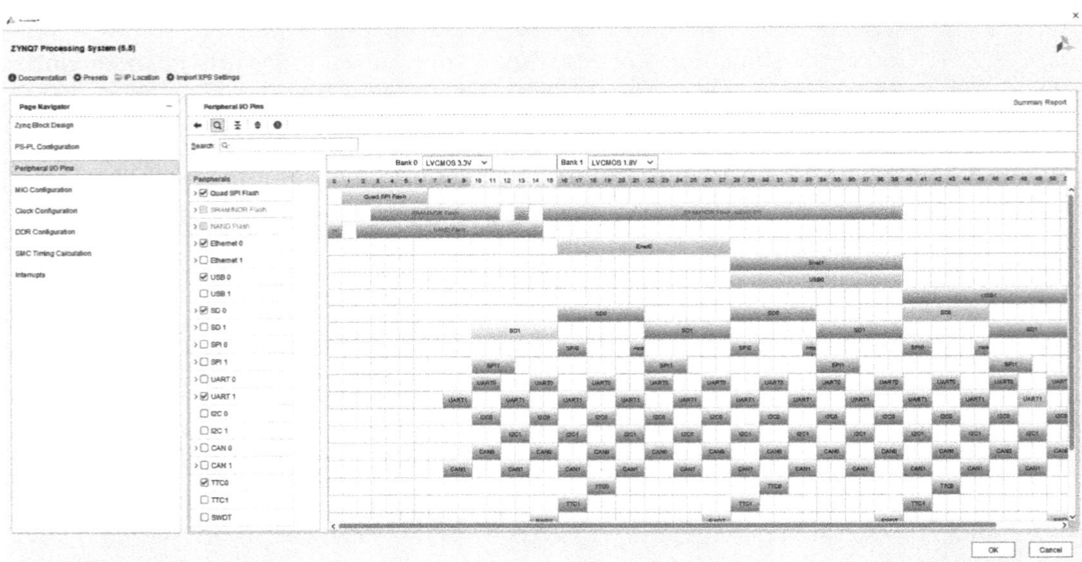

Figure 3.4 Zynq processor system MIO allocation.

Once the MIO and DDR entries have been created, we are able to elaborate the design. From an elaborated design, we can export an IBIS file that describes the MIO and DDR interfaces. This can be critical for verifying that high-speed processor interfaces in the PCB layout have the correct signal integrity. As this can impact the design of the board, it is best to generate the IBIS file early on in the design process (Figure 3.7).[1]

1. The use of these IBIS files for signal integrity performance is discussed in detail in Chapter 2.

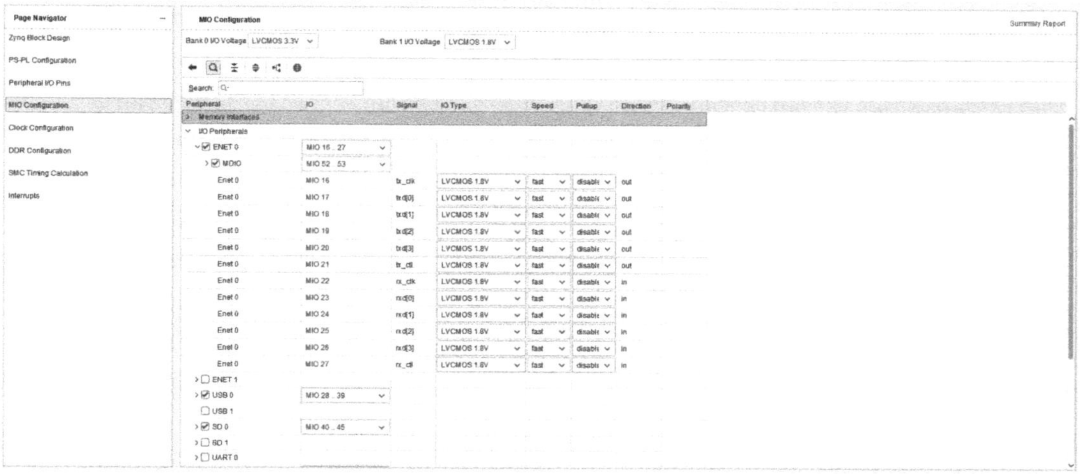

Figure 3.5 Controlling the PS MIO drive strength.

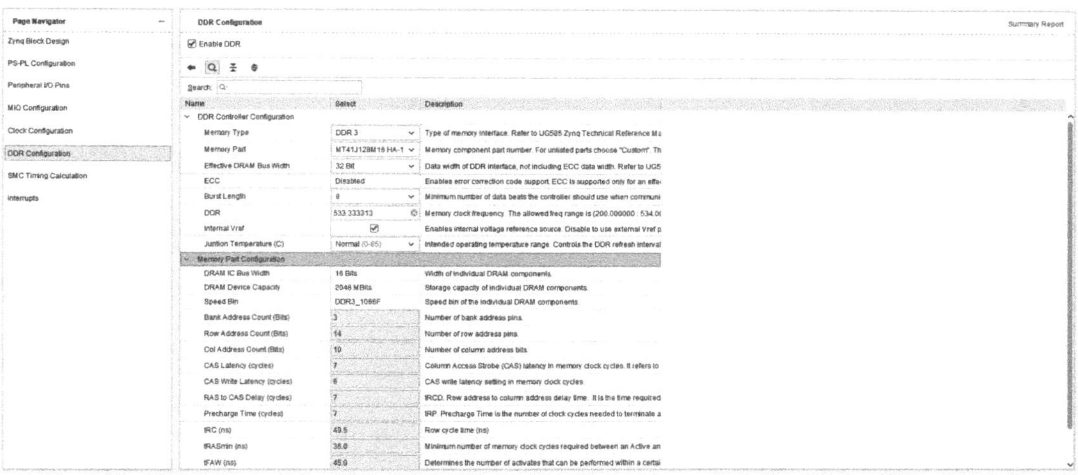

Figure 3.6 Zynq PS DDR interfacing.

With the processor element of the pin allocation completed, we can create a pin planning project to verify the pins allocated to the programmable logic half of the Zynq.

Here we can allocate a pin to an FPGA signal and IO standard and control the speed and drive strength and termination. Once all the pins have been allocated, the developer can run a design rule check (DRC) to ensure that the FPGA pin allocation does not break the IO banking rules. Once complete, the pin out must be verified to be valid by passing the DRC.

These pins can be written out then as Xilinx Design Constraints File, which is how the Vivado understands which pin and IO standard to use for specific designs input and output.

```
set_property PACKAGE_PIN C20[get_ports HX94B_TEMP_BUF]
set_property PACKAGE_PIN E17[get_ports HX94B_TRH_BUF]
```

```
[Component]        ZYNQ
[Manufacturer]     Xilinx Inc.
[Package]
|CLG400
|variable          typ            min            max
R_pkg              409.3m         130.9m         730.5m
L_pkg              8.10nH         2.71nH         14.76nH
C_pkg              1.31pF         0.33pF         2.90pF
|
[Pin]     signal_name                             model_name  R_pin  L_pin  C_pin
A1        DDR_dm[0]                                 SSTL135_DCI_F_PSDDR_MS
A2        DDR_dq[2]                                 SSTL135_DCI_F_PSDDR_MS
A3        VCCO_DDR_502                              POWER
A4        DDR_dq[3]                                 SSTL135_DCI_F_PSDDR_MS
A5        FIXED_IO_mio[6]                           LVCMOS33_S_8_PSMIO
A6        FIXED_IO_mio[5]                           LVCMOS33_S_8_PSMIO
A7        FIXED_IO_mio[1]                           LVCMOS33_S_8_PSMIO
A8        GND                                       GND
A9        FIXED_IO_mio[43]                          LVCMOS18_S_8_PSMIO
A10       FIXED_IO_mio[37]                          LVCMOS18_S_8_PSMIO
A11       FIXED_IO_mio[36]                          LVCMOS18_S_8_PSMIO
A12       FIXED_IO_mio[34]                          LVCMOS18_S_8_PSMIO
A13       VCCO_MIO1_501                             POWER
A14       FIXED_IO_mio[32]                          LVCMOS18_S_8_PSMIO
A15       FIXED_IO_mio[26]                          NC
A16       FIXED_IO_mio[24]                          NC
```

Figure 3.7 IBIS file extract.

```
set_property PACKAGE_PIN F17[get_ports IO_L6N_35]
set_property PACKAGE_PIN G15[get_ports IO_L19N_35]
set_property PACKAGE_PIN K17[get_ports IO_L12P_35]
set_property PACKAGE_PIN M19[get_ports LTM4644_TEMP_BUF]
set_property PACKAGE_PIN N18[get_ports ATWILC_SPI_SCK]
```

Constraints are particularly important in FPGA design for not only ensuring external interfacing is defined, but also a range of other applications ranging from pin allocation to controlling timing performance and even location of logic in the programmable logic implementation. Now SensorsThink engineers have understood how we can pin plan our FPGA design (Figure 3.8); let us take a few moments to understand the wider role of constraints in our programmable logic development.

3.4.1 Physical Constraints

The most used physical constraints by an engineer are the placement of IO pins and the definition of parameters associated with the IO pin, for instance, standard and drive strength. However, the team at SensorsThink would be well served to understand and consider the following types of constraints.

- *Placement constraints:* These allow us to define the location of cells.
- *Routing constraints:* These allow us to define the routing of signals.
- *IO constraints:* These allow us to define the location and IO parameters.
- *Configuration constraints:* These allow the configuration methods to be defined.

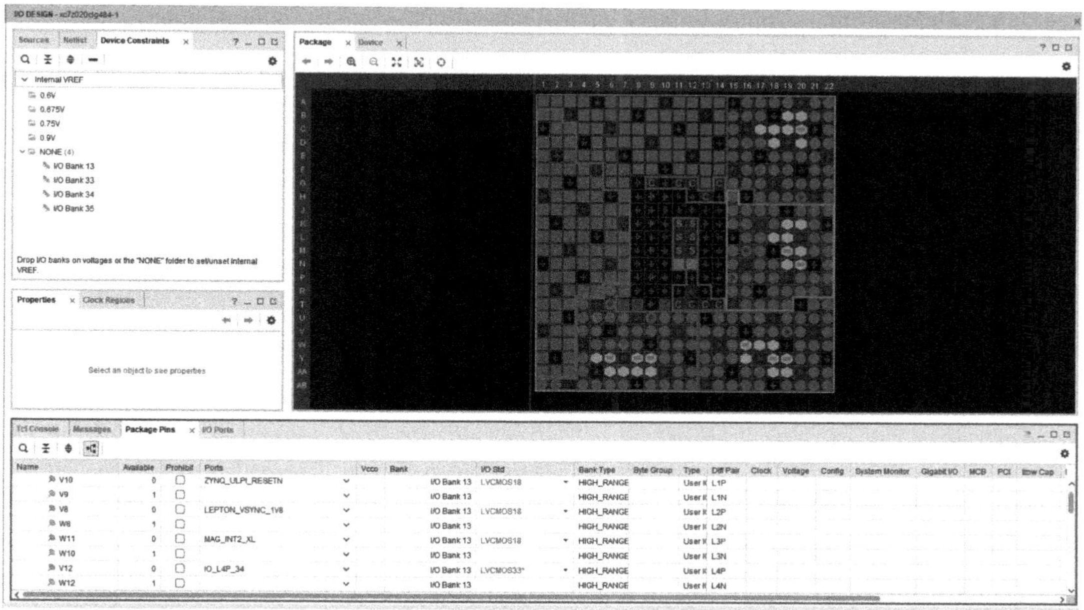

Figure 3.8 Zynq project planning view.

As always, there are a few constraints which sit outside these groups. Vivado has three, and they are predominantly used on the netlist.

- *DONT_TOUCH:* This is used to prevent optimization; as such, it can be of great use when implementing a safety-critical/high-reliability system.
- *MARK_DEBUG:* This is used to preserve an RTL net such that it can be used for debugging later.
- *CLOCK_DEDICATED_ROUTE:* Identifies a route for the clock routing.

Modern FPGAs support several different single and differential IO standards; these are defined via the IO constraints. However, we must take care to ensure that we are following IO banking rules, which depends upon the final pin placement.

What are IO banking rules? Each of the user IOs within an FPGA is grouped together into a number of banks consisting of a number of IOs. These banks have independent voltage supplies enabling the wide range of IO standards to be supported. On the Zynq and other seven series devices, IO banks are further classified into belonging to one of two overall groups, which further constrains their performance and requires that the engineer use the correct class for the correct interface. The first is high performance (HP) optimized for higher data rates; as such, this uses lower operating voltages and does not support LVCMOS 3v3 and 2v5. The other class high range (HR) is optimized to support wider IO standards not supported by HP; as such, it supports traditional 3v3 and 2v5 interfacing. Figure 3.9 demonstrates these banks.

Once we have determined which banks will be used for which signal, we still can change the signal drive strength and slew rate. These will be of great interest to your hardware design team to ensure that the signal integrity upon the board is optimal. These selections will also affect the timing of the board design, as such a

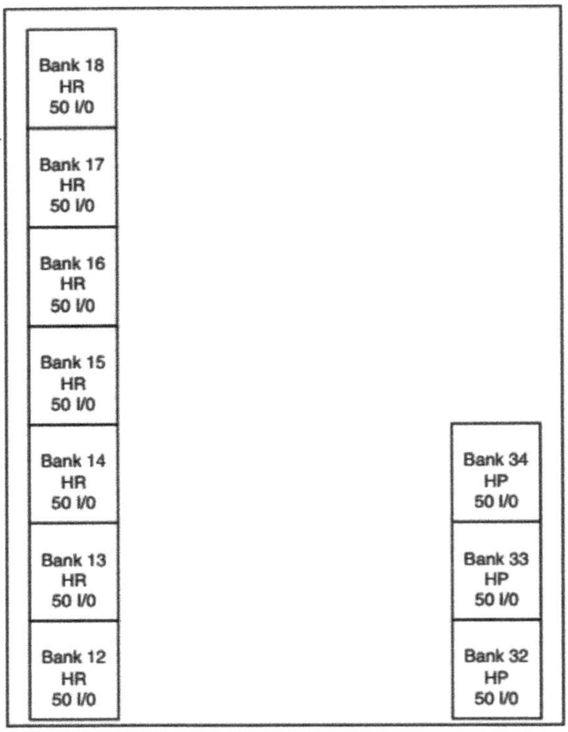

Figure 3.9 HP and HR banks on a 7-series device.

signal integrity tool may be used; these tools require an IBIS model. We can extract an IBIS model of our design from Vivado when we have the implemented design open. Using the File->Export->Export IBIS model option, this file can then be used to close the system level signal integrity issues and timing analysis of the final PCB layout.

Once the design team is happy with the signal integrity performance and timing of the system as a whole, we end up with a number of constraints as below for the IOs in the design.

```
set_property PACKAGE_PIN G17 [get_ports {dout}]
set_property IOSTANDARD LVCMOS33 [get_ports {dout}]
set_property SLEW SLOW [get_ports {dout}]
set_property DRIVE 4 [get_ports {dout}]
```

With the HP IO banks, we can also use the digitally controlled impedance to terminate the IO correctly and increase the signal integrity of the system without the need for external termination schemes. We must also consider the effects of the IO if there is no signal driving it. For instance, if it is connected to an external connector, we can use the IO constraints to implement a pull-up or pull-down resistor to prevent the FPGA input signal from floating, which can cause system issues.

We can also use physical constraints to improve the timing of our design by implementing the final output flip-flop within the IO block itself; doing so reduces the clock to output timing. We can also do the same thing on input signals which allows the design to meet the pin-to-pin setup and hold timing requirements.

3.4.2 Timing Constraints

The next type of constraint for the development team at SensorsThink to consider is timing constraints. The most basic level of a timing constraint defines the operating frequency of the clock(s); however, more advanced constraints establish the relationships between clock paths. Why we do this is to determine if we need to analyze the path or if it can be discounted as there is no valid timing relationship between those clock paths. By default, Vivado will analyze all relationships; however, not all clocks within a design will have a timing relationship that can be accurately analyzed. One example of this is clocks that are asynchronous as their phase cannot be accurately determined, as shown in Figure 3.10.

We manage these relationships between clock paths using a constraints file and declaring clock groups. When a clock group is declared, Vivado performs no timing analysis in either direction between the clocks defined within it.

To aid the generation of our timing constraints, Vivado defines clocks as being within one of three categories:

- *Synchronous clocks:* Synchronous clocks have a predictable timing/phase relationship; this is normally the case for a primary clock and integer-generated clocks, as they share a common root clock and will have a common period.
- *Asynchronous clocks:* Asynchronous clocks have no predictable timing/phase relationship between them; this is normally the case for different primary (and their generated) clocks or clocks generated with fractional division. Asynchronous clocks may or may not have a different root clock.
- *Unexpandable clocks:* Two clocks are unexpandable if over 1,000 cycles a common period cannot be determined. If this cannot be established, then the worst-case setup relationship over this, 1,000 cycles, will be used; however, there is no guarantee that it is the worst case.

To determine with which type of clock we are dealing with, we can use the clock report produced by Vivado to aid us in identifying asynchronous and unexpandable clocks.

With these identified, we can use the set clock group constraint to disable timing analysis between them. Vivado uses SDC-based constraints; as such, we can use the command below to define a clock group:

set_clock_groups –name –logically_exclusive –physically_exclusive –asynchronous –group

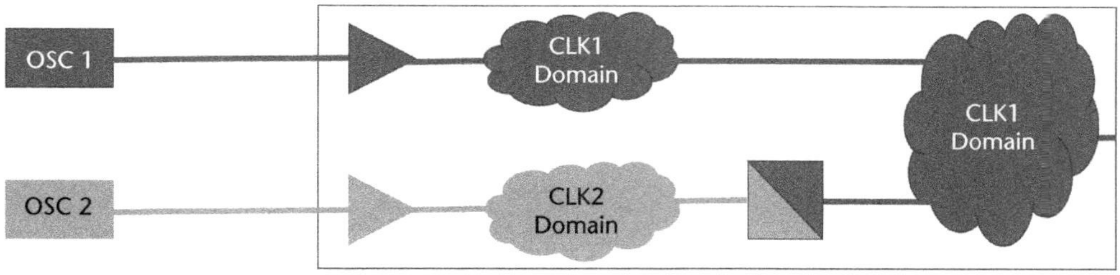

Figure 3.10 Asynchronous clocks.

The –name is the name given to the group; the –group option is where one can define the members of the group (i.e., the clocks that have no timing relationship). The logically and physically exclusive options are used when we have multiple clock sources that are selected between to drive a clock tree (e.g., BUFGMUX or BUFGCTL); therefore, the clocks cannot be present upon the clock tree at the same time. As such, we do not want Vivado to analyze the relationship between these clocks, as they are mutually exclusive, while the –asynchronous is used to define asynchronous clock paths.

The final aspect of establishing the timing relationship is to consider the nonideal relationship of the clocks in particular jitter. Jitter needs to be considered in two forms: input jitter and system jitter. Input jitter is present upon the primary clock inputs and is the difference between when the transition occurs and when it should have occurred under ideal conditions, while system jitter results from noise existing within the design.

We can use the set_input_jitter constraint to define the jitter for each primary input clock, while the system jitter is set for the whole design (those are all the clocks) using the set_system_jitter constraint.

3.4.3 Timing Exceptions

We must also focus upon what happens within a defined clock group when we have an exception, but what is an exception?

One common example of a timing exception would be a result being captured only every other clock cycle, while another would be transferring data between a slow clock and a faster clock (or vice versa) where both clocks are synchronous. Both are a timing exception commonly referred to as a multicycle path as shown in Figure 3.11.

Declaring a multicycle path for these paths results in a more appropriate and less restrictive timing analysis allowing the timing engine to focus upon other more critical paths, increasing the quality of results.

Within our XDC file, we can declare a multicycle path using the XDC command:

```
set_multicycle_path path_multiplier [-setup|-hold] [-start|-end]
[-from <startpoints>] [-to <endpoints>] [-through
<pins|cells|nets>]
```

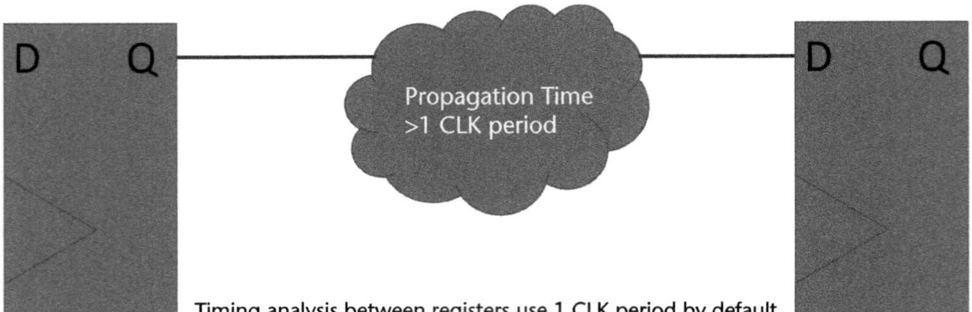

Figure 3.11 Multicycle paths.

When we declare a multicycle path, we are in effect multiplying the requirements for the setup and hold (or both) analysis by the path_multiplier (e.g., in the first example above where the output occurs every 2 clock cycles, the path_multiplier would be 2 in the case of the setup timing).

As the multicycle path can be applied to either setup or hold, the engineer then has the choice of where to apply it. When we declare a setup multiplier, it is often best practice to also declare a hold time multiplier using the equation:

Hold cycles = set up multiplier – 1 – Hold Multiplier

What this means for the simple example that we have been following is that the hold multiplier is defined by the equation

Hold Multiplier = setup multiplier – 1 When using a common clock.

3.4.4 Physical Constraints: Placement

We may wish to constrain the physical placement for several reasons, perhaps to help to achieve timing or maybe to provide isolation between sections of the design. Before we delve too deeply, there are a few terms we need to define:

- *BEL:* The Basic Element of Logic (BEL) allows a netlist element to be placed within a slice.
- *LOC:* Location (LOC) places an element from the netlist to a location within a device.
- *PBlock (Physical Block):* This can be used to constrain logic blocks to a region of the FPGA.

Thus, while a LOC allows you to define a slice or other location within the device, a BEL constraint allows you to target at a finer granularity than the LOC and identify the flip-flop (FF) to use within the slice. PBlocks can be used to group logic together; they are also used for defining logical regions when we wish to perform partial reconfiguration.

The PBlock is great for when we wish to segment large areas of our design or if we wish to perform partial reconfiguration. In some instances, we will wish to group together smaller logic functions to ensure that the timing is optimal. While we could achieve this using PBlocks, it is more common for us to use relationally placed macros (RPMs).

RPMs allow design elements such as DSP, FF, lookup table (LUT), and RAMS to be grouped together in the placement. Unlike PBlocks, they do not constrain the location of these to a specific area of the device (unless you want it to) but instead RPMs group these elements together when they are placed.

Placing design elements close together allow us to achieve two things. It allows us to improve resource efficiency, and it means that we can fine-tune interconnection lengths to enable better timing performance.

To do this, we use three different types of constraints, which can be defined with the HDL source files:

- *U_SET:* Allows definition of an RPM set of cells regardless of hierarchy.
- *HU_SET:* Allows definition of an RPM set of cells with hierarchy.
- *RLOC:* Allows the assignment of relative locations to the SET.

The RLOC constraints use the definition RLOC = XmYm where the X and Y relate to the coordinates of the FPGA array. When we define an RLOC, we can make this either a relative or absolute coordinate. Depending upon if we add in the RPM_GRID attribute, this makes the definition absolute and not relative.

As these constraints are defined within the HDL, as in Figure 3.12, it is often necessary for a place and route iteration to be run initially before the constraints are added to the HDL file to correctly define the placement.

3.5 Clock Domain Crossing

As we start developing the application for our Smart Sensor SoC, engineers at SensorsThink need to be careful with the clocking architecture used within the programmable logic. Data is transferred between registers within the FPGA synchronously on a clock edge.

As we firm up the architecture of the programmable logic implementation, we may need to use several different clocks at different frequencies. The need for several clocks within the solution stems from the need to be able to interface with the sensor at specific clock frequencies. The four local sensors located on the board range in clock frequencies from 400 kHz to 14 MHz (Table 3.1).

The main clock to transfer data into the processing system is at a much higher rate, as it must be able to read multiple sensors in a short period of time. This means that sensor data must safely be transferred between the sensor clock and the processing system clock.

Transferring data between clocks can present several challenges to us as we develop the programmable logic solution. Failure to implement a safe transfer between clocks can result in a data corruption that could impact the performance of the system.

Transferring data between the clocks is often referred to as CDC; safely transferring data from one clock to another requires the designer to understand the concept of a clock domain.

```
SIGNAL ip_sync : STD_LOGIC_VECTOR(1 DOWNTO 0) :=(OTHERS =>'0');
SIGNAL shr_reg : STD_LOGIC_VECTOR(31 DOWNTO 0) :=(OTHERS =>'0');

ATTRIBUTE RLOC : STRING;
ATTRIBUTE HU_SET : STRING;
ATTRIBUTE HU_SET OF ip_sync : SIGNAL IS "ip_sync";
ATTRIBUTE HU_SET of shr_reg : SIGNAL IS "shr_reg";
```

Figure 3.12 Constraints within the source code.

Table 3.1 Sensor Clock Frequencies

Sensor	*Maximum Clock Rate*
SHTW2	400 kHz
LSM6DS3	10 MHz
LSM303AH	10 MHz
ADcmXL3021	14 MHz

A clock domain is a grouping of clocks that are all related. Related means that the clocks are generated from the same source. A clock domain may include generated clocks from the source at different frequencies using a phase lock loop, multiclock manager, or traditional counter. Clocks generated in this way are related if the output clock is an integer multiple of the source clock. Related clocks therefore have a known and unchanging timing relationship between clock edges; this enables flip-flop setup and hold windows to be achieved. This means transferring data between clocks in the same clock domain does not require any special synchronization or handshaking (Figure 3.13).

Clocks are in a different clock domain if the source of the clock is different, even if the clock is the same frequency. Generated clocks are also in different clock domains if they are not an integer multiple of the source clock. This fractional relationship guarantees that the setup and hold windows of registers cannot be achieved (Figure 3.14).

Violating the setup and hold time of the registers results in registers going metastable; that is, the output of the flip-flop goes to a value between 0 and 1. Eventually, the flip-flop will recover, and the output will go to either a 0 or 1. This recovery time of the flip-flop and the random resulting output value mean that if we do not correctly synchronize CDC, our system can end up with corrupted data.

Multiclock designs therefore require additional analysis and design effort to ensure that CDC is safe and reliable.

Detecting CDC within a design can be complex; it is unlikely to be detected during RTL simulation, as this simulation does not include timing information for registers, whereas gate-level simulation that does include timing information requires considerable time and that the right question be asked (e.g., the unrelated clocks being timed exactly right to cause metastability). This is hard to create in a simulation but oddly easy to create on the deployed system in the field.

The engineers at SensorsThink would be well served to perform clock domain analysis; it is a critical analysis to be performed in multiclock domains. There are several ECAD tools available that can perform CDC analysis. For the development

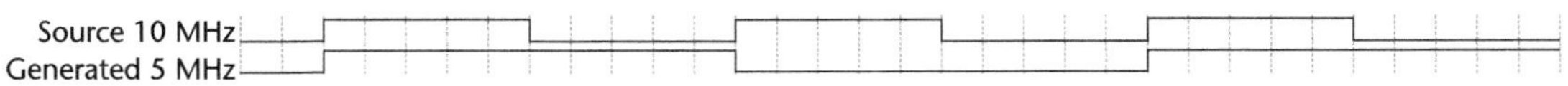

Figure 3.13 Related clock: a 10-MHz source used to generate a 5-MHz clock.

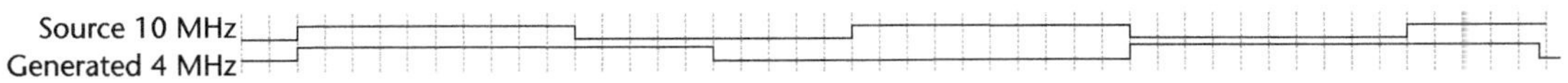

Figure 3.14 Unrelated clock: a 10-MHz source used to generate a 4-MHz clock.

of the programmable logic design, engineers at SensorsThink use the Blue Pearl Software Visual Verification Suite (Figure 3.15). This provides us the ability for static code analysis and CDC analysis; Blue Pearl can analyze our design and identify not only the clock domains, but also signals that cross from one clock domain to the next.

Once the clock domain crossing signals have been identified, we can implement CDC structures to safely transfer data from one clock domain to another. Safely crossing clock domains requires the ability to resynchronize signals from one clock domain to another. The synchronization scheme used depends upon the type of signal being passed between clock domains.

- *Single static signal:* Transferring a single signal between clock domains can be implemented using a two-register synchronization structure.
- *Pulsed data:* Transferring pulsed data, especially from a faster clock domain to a slower clock domain, can present a challenge.
- *Data Bus:* Mux-based synchronizer.
- *Data Bus:* Handshake synchronizer.
- *Data Bus:* Asynchronous FIFO.
- *Data Bus:* Gray code.

We must also be careful to ensure that, when working with single logic signals, several single logic signals do not converge in a receiving clock domain (Figure 3.16), as the random delay through the synchronizer path can result in logic issues.

As we are implementing our design targeting a Xilinx programmable logic device, we should consider the specialist CDC structures provided. If we need to implement CDC structures within a Xilinx programmable logic, we can leverage the Xilinx Parameterized Macros (XPM), which provide several CDC structures.

- *Single-bit synchronizer:* Synchronizes a single bit vector between clocks.
- *Asynchronous reset synchronizer:* Synchronizes an asynchronous reset to the destination clock. Reset assertion is asynchronous, while the reset removal will always be synchronous and safe for the destination clock.
- *Gray code synchronizer:* Synchronizes a data bus between source and destination clocks, using Gray coding. Due to the implementation style, the input

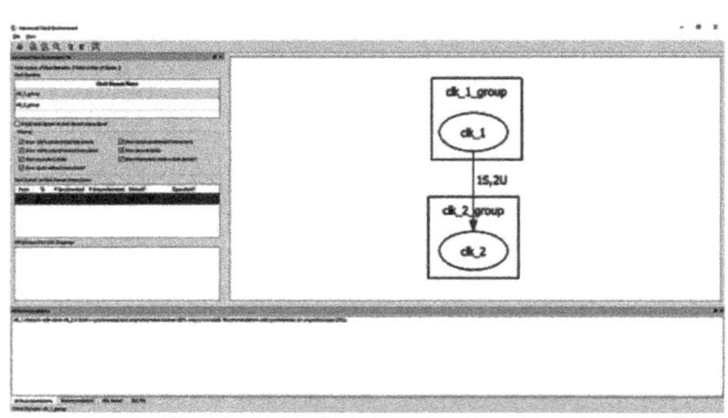

Figure 3.15 Clock domain identification using Blue Pearl (1 synchronized and 2 unsynchronized crossings).

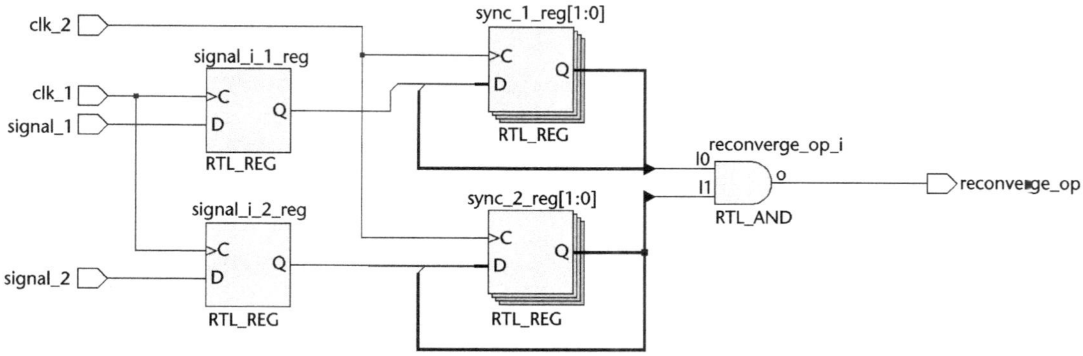

Figure 3.16 Reconvergence of two standard logic vectors in a different clock domain.

can only increment or decrement by 1 between values, making it useful for counters.

- *Handshake synchronizer:* Synchronizes a data bus between source and destination clocks using handshaking of signals.
- *Pulse synchronizer:* Synchronizes a pulse of any width (one source clock or wider) from the source domain to the destination domain.
- *Single-bit synchronizer:* Synchronizes a single bit from source to destination domain.
- *Synchronous reset synchronizer:* Synchronizes a synchronous reset to the destination clock domain. This both asserts and de-asserts the reset synchronously, unlike the asynchronous reset macro.

Regardless of the implementation technology, when it comes to implementation within the device, we must ensure that synchronizing elements are placed close together to ensure the best timing performance.

3.6 Test Bench and Verification

As the SensorsThink FPGA design solution matures, the SensorsThink engineering team should be thinking about verification of both the FPGA and the individual models. The verification of a FPGA or RTL module can be a time-consuming process as the engineer ensures that the design will function correctly against its requirement specification and against corner cases that might make the module perform incorrectly. Engineers traditionally achieve this using a test bench; however, test benches can be simple or complicated affairs. Let us have a look at how we can get the most from our test bench while not overcomplicating it.

3.6.1 What Is Verification?

Verification ensures that the unit under test (UUT) meets both the requirements and specification and is suitable for its intended purpose. In many cases, verification is performed by a team independent to the design team. When considering FPGA verification and test benches, we are ensuring that the UUT performs against its specification; with the size and complexity of modern FPGA designs, this can be a considerable task. The engineering team must therefore determine at the start of the project what the verification strategy will be; choices can include:

- *Functional simulation only:* This checks that the design is functionally correct.
- *Functional simulation and code coverage:* This checks that, along with the functional correctness of the design, all of the code within the design has been tested.
- *Gate-level simulation:* This verifies the functionality of the design when back-annotated with timing information from the final implemented design; this can take a considerable time to perform.
- *Static timing analysis:* This analyzes the final design to ensure that the timing performance of the module is achieved.
- *Formal equivalence checking:* This is used to check the equivalence of netlists against RTL files.

Whatever the verification strategy, the engineering team will need to produce a test plan that shows how the individual modules and the final FPGA are to be verified and how all requirements will be addressed.

Typically, to ensure that you can verify most requirements placed upon the FPGA, a test bench is required to stimulate the UUT, thus ensuring that the UUT performs as specified.

3.6.2 Self-Checking Test Benches

To ensure that the design functions as desired, we typically use a test bench to stimulate the UUT. This test bench can either be self-checking or not. If you opt for a nonself-checking test bench, then you will be required to visually confirm the test bench functions as desired. A self-checking test bench differs from a traditional test bench in that, along with applying the stimulus to the UUT, the outputs from the UUT are also checked against expected results as shown in Figure 3.17. This allows you to state that the UUT has passed or failed. Couple this with the control and reporting via using TEXTIO and you can create an immensely powerful verification tool.

The use of a self-checking test bench has the following advantages:

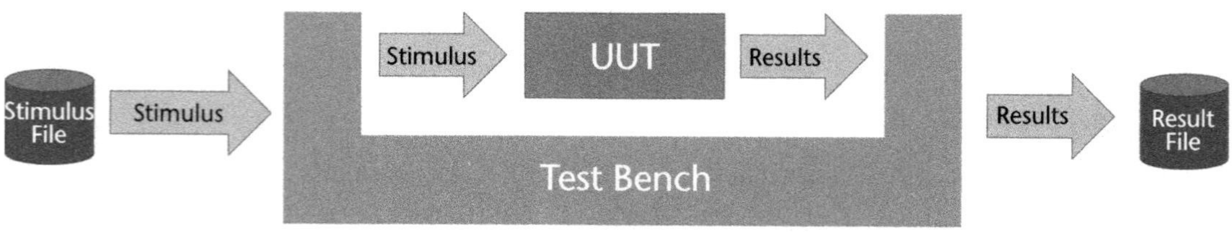

Figure 3.17 Self-checking test bench architecture.

- Checking simulation waveforms can be very time-consuming, complicated to visually verify.
- Self-checking test benches can provide an overall pass or fail report that can be saved and used as test evidence later in the design flow.
- Should the UUT require design iterations at a later stage, rerunning the test bench and determining pass or fail can be achieved with a reduced timescale. This is often called regression testing.

3.6.3 Corner Cases, Boundary Conditions, and Stress Testing

When using a test bench, we want to ensure that the module performs as per the functional requirements, that any corner cases are addressed and most importantly that the test bench sufficiently exercises the UUT code. It is therefore common when verifying design modules to use glass box testing methodologies in that the contents of the UUT module are known and observed, whereas when the UUT is the FPGA top level, black box testing is used due to the increased time verification and simulation times at the top level.

To verify the functionality of the design against requirements, the test bench must apply stimulus as the modules expect to see in operation. However, testing all possible inputs to prove functional compliance can be a time-consuming and lengthy process; for this reason, it is often the case that the engineer will often focus upon boundary conditions and corner cases along with testing some typical operating values.

A boundary condition is when one of the inputs is at its extreme value, while a corner case is when all inputs are at their extreme value. A good example of this is illustrated by a simple 16-bit adder, which adds two numbers and produces a result. In this case, the boundary condition and corner cases are illustrated in Figure 3.18.

As you can clearly see, the corners cases would be as follows: when both A and B inputs equal 0, when A equals 0 and B equals 65,535, when A and B equal

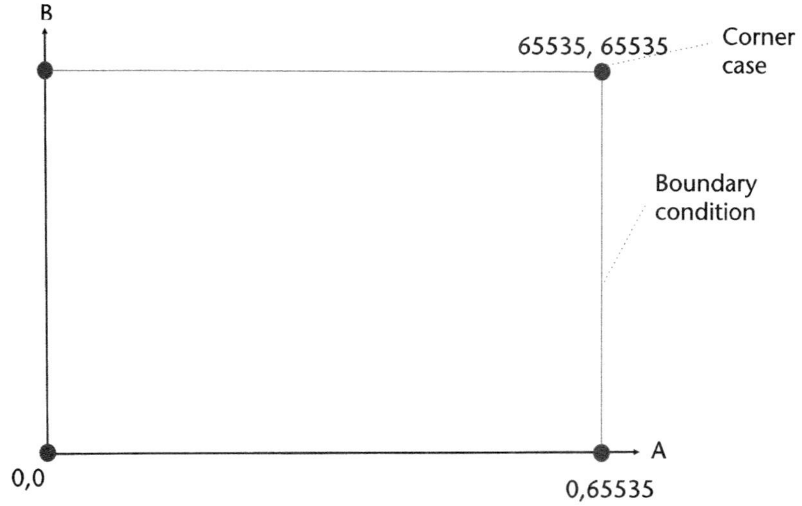

Figure 3.18 Boundary condition and corner cases for a 16-bit adder.

65,535, and finally when A equals 65,535 and B equals 0. The boundary cases are the values between the corner cases.

In some applications, we may also want to stress test the UUT to ensure a margin above and beyond the normal operation. Stress testing will change depending upon the module, but it could involve repeated operation, attempting to overflow buffers and FIFOs. Stress testing a design allows the verification engineer to have a little fun trying to break the UUT.

3.6.4 Code Coverage

The verification team also requires a metric to ensure that the UUT is properly exercised by the test bench; this metric is typically provided by using code coverage to ensure that the UUT is correctly exercised. When looking at code coverage, there are several different parameters that the verification engineer should consider:

- Statements: Each executable statement is examined to determine how many times it is executed.
- Branch: All possible IF, CASE, and SELECT branches are executed.
- Conditions and subconditions within a code branch are tested to see what condition caused it to be true.
- Paths: All paths through the HDL are traversed to identify any untraveled paths.
- Triggering: Monitor signals in the sensitivity list of VHDL processes and wait statements.

Achieving 100% in the above parameters will not prove that the UUT is meeting functional requirements; however, it does easily identify sections of the UUT that have not been exercised by the test bench. It is especially important at this point to mention that code coverage only addresses what is within the UUT.

3.6.5 Test Functions and Procedures

The verification team should create a library of commonly used test routines that can be used by all members of the team and even across other projects. This will not only speed up the development, but it will also result in a standardized output from the test bench, which aids both analysis and verification of the module.

These standard functions and procedures can be used for numerous applications; however, a few commonly used methods are:

- *Stimulus generators:* For instance, a standard method of testing reset application and release or standard method of driving a communication interface.
- *Output checking:* Checking the output result against an expected result and reporting via the transcript or file.
- *Models:* Standardized models of devices with which the FPGA or module will interface during operation.

- *Logging functions:* Creating a standardized reporting format for the results; this enables easier analysis of the results. These also provide evidence of verification of requirements that can be used to demonstrate compliance.

Good output checking functions can make use of the VHDL signal attributes stable, delayed, last_value, and last_event. These attributes can be especially important when confirming that the UUT is compliant with the timing interfaces required for memories or other interfaces. These can also be useful for the ensuring that setup and hold times are achieved and reporting violations if they are not.

3.6.6 Behavioral Models

Sometimes the performance of a UUT can be verified against a nonsynthesizable behavioral model that implements the same function as the UUT as shown in Figure 3.19. The model and the UUT are then subjected to the same stimulus from the test bench; the outputs of the two modules are then compared on a cycle-by-cycle basis to ensure that both behave in the same manner. This may require that the latency of the UUT is accounted for by the model or test bench to allow comparison of the correct results.

3.6.7 Using Text IO Files

If you are using VHDL to verify your design, you can use the Text IO package to read in stimulus vectors from a text file and record the results including pass or fail within a results text file. The verification engineering team are then able to create several stimulus test files to test the UUT. This approach is beneficial as changing the stimulus becomes as simple as updating a text file; this allows for easy and quick updates as well to test scenarios that could be causing issues in the lab on the final hardware.

Within VHDL interaction with text files is supported by the std package, which if your simulator supports VHDL 2008, allows the use of std_logic and std_logic_vector. If your simulator does not yet support VHDL 2008, you can always use the nonstandard but commonly used std_logic_textio package to provide similar support.

Similar support is provided within Verilog to read and write text files.

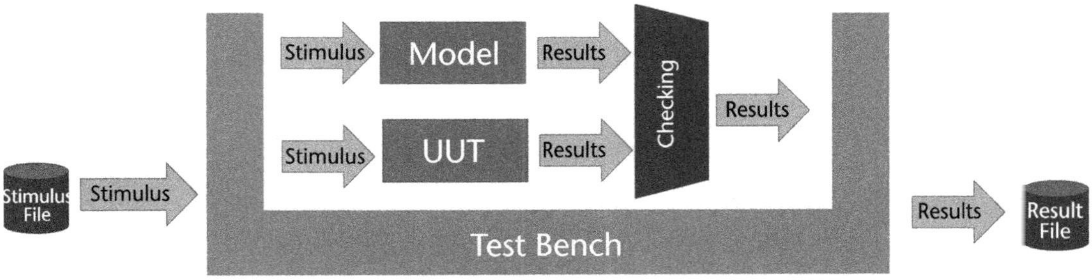

Figure 3.19 Model checking against a UUT.

3.6.8 What Else Might We Consider?

The test bench should be time agnostic which means that it should be capable of testing either RTL or behavioral UUT along with post place and route netlists back annotated with SDF timing information. While simulating at a gate level can take considerable time due to the required resolution, at times, it is at times necessary.

The verification team should also consider the use of TCLscripts to control the simulation tool as it executes the test. This approach allows for a more repeatable process as all options that might be set within a graphical user interface (GUI) and change between simulation runs will be defined within the TCL file controlling the simulation. These files also become self-documenting reducing the documentation needed to support the verification of the UUT.

3.7 Finite State Machine Design

The SensorsThink engineering team is going to need to implement several control structures in the FPGA. These control structures will be used to configure and communicate with sensors, along with performing the overall sequencing of the programmable logic design. This means the SensorsThink engineering team will be required to implement several finite state machines in the design. Finite state machines are logical constructs that transition between a finite number of states. A state machine will only be in one state at a point in time; however, it will transition between states depending upon several triggers. Theoretically, state machines are divided into two basic classes, Moore and Mealy; they differ only in how the state machines outputs are generated.

- *Moore:* State machine outputs are a function of the present state only, a classic example is a counter.
- *Mealy:* State machine outputs are a function of the present state and inputs, a classic example of this is the Richards controller.

State machines are not only limited to implementation in programmable logic, but they can also be implemented in traditional discrete logic gates if required.

3.7.1 Defining a State Machine

When we are faced with the task of defining a state machine, the starting point is to develop a state diagram. The state diagram shows the states, the transitions between states, and the outputs from the state machine. Figure 3.20 shows the two state diagrams: one for a Moore state machine (left), the other a Mealy State Machine (right).

If we were to implement these using physical components (as we did before programmable logic was available), we would start generating the present state, next state tables and producing the logic required to implement the state machine. However, as we will be implementing this in a programmable logic, the engineer can work directly from the state transition diagram.

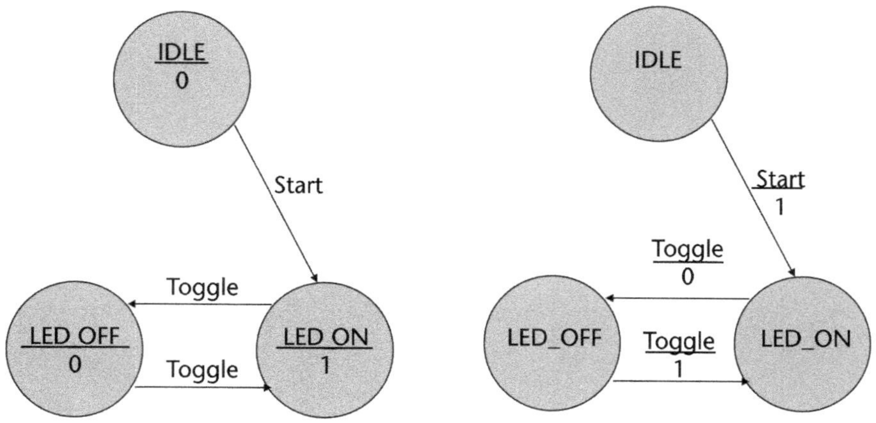

Figure 3.20 FSM state diagrams Moore and Mealy.

3.7.2 Algorithmic State Diagrams

While many state machines are designed using the state diagram approach shown above, another method of describing a state machines behavior is the algorithmic state chart (see Figure 3.21). This algorithmic state machine ASM) chart is much closer in appearance to a software engineering flowchart. It consists of three basic constructs:

1. *State box:* This is associated with the state name and contains a list of the state outputs (Moore).
2. *Decision box:* This tests for a condition being true and allows the next state to be determined.
3. *Conditional output box:* This allows the state machine to describe Mealy outputs dependent upon the current state and inputs.

Some engineers at SensorsThink feel that a state machine described in ASM format is easier to map to implementation in a hardware description language such as VHDL.

3.7.3 Moore or Mealy: What Should I Choose?

The decision to implement either a Moore or Mealy machine will depend upon the function the state machine is required to implement along with any specified reaction times. The major difference between the two is in how the state machines react to inputs. A Moore machine will always have a delay of one clock cycle between an input and the appropriate output being set. This means that a Moore machine is incapable of reacting immediately to a change of input, as can be seen clearly in Figure 3.22. This ability of the Mealy machine to react immediately to the inputs can often mean that Mealy machines require fewer states to implement the same function as a Moore implementation would. However, the drawback of a Mealy machine is that when communicating with another state machine, there is the danger of race conditions due to the combinatorial nature of the outputs.

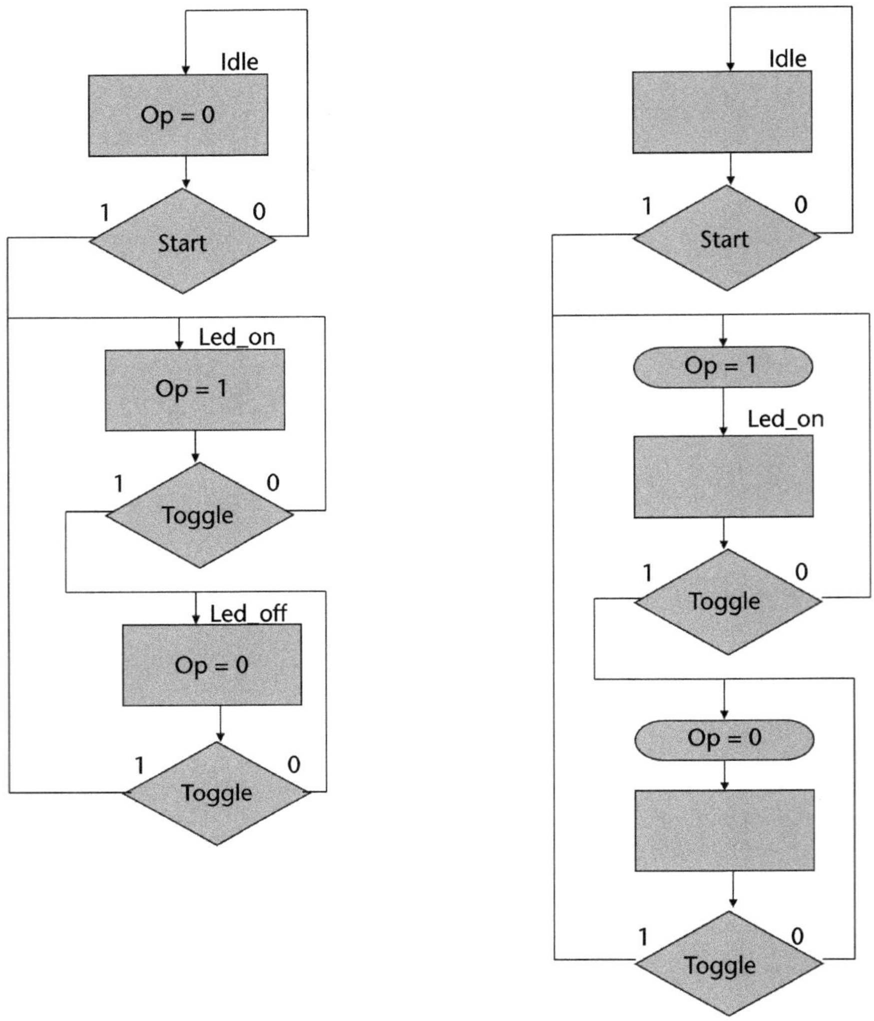

Figure 3.21 Algorithmic state chart for the state machines in Figure 3.20 Moore (left) and Mealy (right).

A state machine does not have to be just Moore or Mealy; it is possible to create a hybrid state machine that uses both styles to create a more efficient implementation of the function required.

3.7.4 Implementing the State Machine

Using a high-level language like VHDL, it is easy to implement a state machine directly from the sate diagram. VHDL supports enumerated types that allows you as the engineer to define the actual state names, for example:

TYPE state IS (idle, led_on, led_off) ;

This type of declaration corresponds to the state diagrams shown in Figure 3.20, which waits for a button to be pushed then toggles a light emitting diode on and off.

There are many different methods of implementing a state machine; however, the two basic implementation methods are a one or two process approach. A one

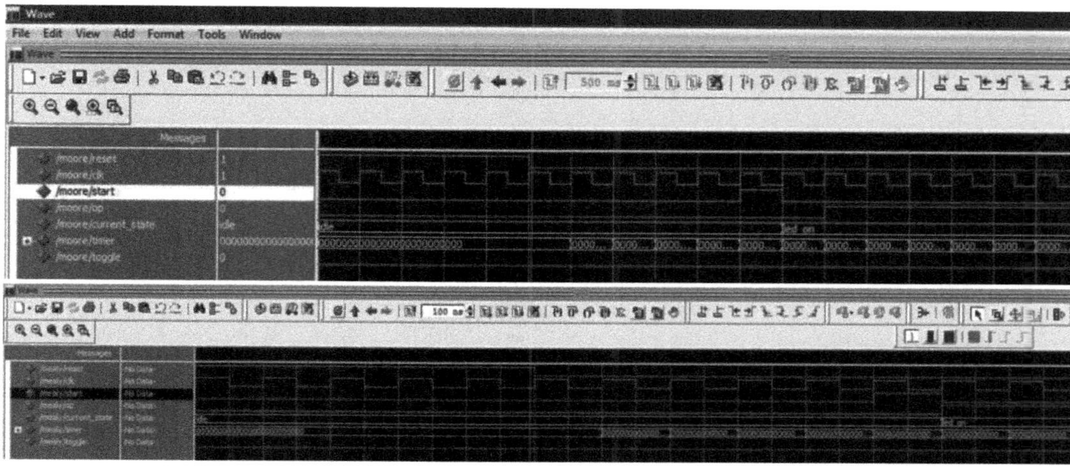

Figure 3.22 Showing simulation results for Mealy and Moore outputs (Mealy bottom).

process contains both current and next state logic within a single process. The alternative two process approach separates the current and next state logic.

As a rule, engineers prefer to implement a single process state machine, these are seen as having several benefits over the traditional two process approach:

1. Removes the risk of incomplete coverage of signals in the combinatorial process creating latches.
2. State machine outputs are aligned to the clock.
3. Generally, they are easier to debug than their two-process implementation.

Regardless of which approach you decide to use to implement it, the next state determination and any outputs will be evaluated using a CASE statement. Figure 3.23 shows a side-by-side comparison between a Moore (left) and Mealy (right) using a single process approach.

3.7.5 State Machine Encoding

The state variable is stored within flip-flops that are updated with the next state on the next clock edge (even if there is no state change). How these flip-flops are used to represent the state value depends upon the number of states and if the engineer chooses to direct the synthesis tool in one particular method. The three most common types of state encoding are:

- *Sequential:* State encoding follows a traditional binary sequence for the states.
- *Gray:* Like the sequential encoding scheme with the exception of the states encoding using Gray code.
- *One-hot:* Has one flip-flop for each state within the state machine. Only one flip-flop is currently set high and the rest are low; hence, the name one-hot.

```vhdl
TYPE state IS (idle, led_on, led_off);
SIGNAL current_state : state := idle;

SIGNAL timer  : unsigned(24 DOWNTO 0) := (OTHERS =>'0');
SIGNAL toggle : std_logic := '0';

BEGIN

PROCESS(clk,reset)
BEGIN
  IF reset = '1' THEN
    current_state <= idle;
    op <= '0';
  ELSIF rising_edge(clk) THEN
  CASE current_state IS
    WHEN idle =>
      op <= '0'; --output is a function of the current state only
      IF start = '1' THEN
        current_state <= led_on;
      END IF;
    WHEN led_on =>
      op <= '1';
      IF toggle = '1' THEN
        current_state <= led_off;
      END IF;
    WHEN led_off =>
      op <= '0';
      IF toggle = '1' THEN
        current_state <= led_on;
      END IF;
    WHEN OTHERS =>
      op <= '0';
      current_state <= idle;
  END CASE;
 END IF;
END PROCESS;
```

```vhdl
TYPE state IS (idle, led_on, led_off);
SIGNAL current_state : state := idle;

SIGNAL timer  : unsigned(24 DOWNTO 0) := (OTHERS =>'0');
SIGNAL toggle : std_logic := '0';

BEGIN

PROCESS(clk,reset)
BEGIN
  IF reset = '1' THEN
    current_state <= idle;
    op <= '0';
  ELSIF rising_edge(clk) THEN
  CASE current_state IS
    WHEN idle =>
      IF start = '1' THEN
        current_state <= led_on;
        op <= '1'; --output is a function of the current state and inputs
      END IF;
    WHEN led_on =>
      IF toggle = '1' THEN
        current_state <= led_off;
        op <= '0';
      END IF;
    WHEN led_off =>
      IF toggle = '1' THEN
        current_state <= led_on;
        op <= '1';
      END IF;
    WHEN OTHERS =>
      op <= '0';
      current_state <= idle;
  END CASE;
 END IF;
END PROCESS;
```

Figure 3.23 Moore and Mealy state machines in VHDL.

Both sequential and Gray encoding schemes will require several flip-flops that can be determined by:

$$FlipFlops = Ceil\left(\frac{LOG10(States)}{LOG10(2)}\right)$$

while one-hot encoding schemes require the same number of states as there are flip-flops.

The automatic assignment of state encoding depends upon the number of states that the state machine contains. While this will vary depending upon the synthesis tool selected, as a rule of thumb, the following encoding will be used:

- *Sequential:* <5 states;
- *One-hot:* Five to 50 states;
- *Gray:* >50 states.

Often, you will not necessarily think about what state encoding to use, instead allowing the synthesis engine to determine the correct implementation getting involved if the chosen style causes an issue. However, should the engineers need to take things into their own hands and define the state encoding, there is no need for them to define the state encoding long-hand, defining constants for each state within the state encoding. Instead, the engineers can use an attribute within the code to drive the synthesis tool to choose a particular encoding style as demonstrated below.

```vhdl
TYPE state IS (idle, led_on, led_off);
SIGNAL current_state : state := idle;
ATTRIBUTE syn_encoding STRING;
ATTRIBUTE syn_encoding OF current_state : SIGNAL IS "sequential";
```

where "sequential" can be also "gray" and "onehot"; these three choices can also be combined with the safe attribute to ensure that the state machine can recover to a valid state, should the state machine enter an illegal state.

You can also use the syn_encoding attribute to define the values of the state encoding directly. For example, suppose that the engineer desired to encode the 3-state state machine using the following state encoding; Idle = "11" led_on = "10," led_off = "01" as opposed to the more traditional sequence of "00," "01," and "10."

```
TYPE state IS (idle, led_on, led_off) ;
SIGNAL current_state : state := idle;
ATTRIBUTE syn_encoding STRING;
ATTRIBUTE syn_encoding OF current_state : SIGNAL IS "sequential";
```

As the engineer, you are responsible for setting the correct settings in the synthesis tool to ensure that attributes are not ignored.

3.7.6 Increasing Performance of State Machines

As a central part of the design, state machines and their implementation can have a significant impact on the achieved performance in the target device. Implementation performance does not come down only to the state encoding (e.g., one-hot versus sequential), but the actual structure of state machine itself.

One of the best methods of achieving better timing performance is to decouple as much functionality as possible from the state machine implementation. That is, within state machines, there is often a requirement to count, shift, or implement mathematical functions. These counters, shifters, and other functionality should be implemented as separate processes outside of the state machine. Communication with the state machine (e.g., to indicate a terminal count, reset a counter, or perform a mathematical operation) can be achieved by a single-bit signal, either generated by or received by the state machine.

Including this functionality within the state machine requires a larger implementation, making it harder for the implementation tool to optimize for performance due to more complex control logic, especially if wide buses are required.

Let us look at exactly what the difference in performance can be when using a state machine design to interface with a FLIR Lepton. This state machine will assert the chip select line and then received video over SPI frames back from the Lepton device. If the packet is valid, it will be written into a dual-port block RAM memory. If the packet is a corrupted packet, the state machine will read out the video frame as required for correct operation, but not write the corrupted frame to the block RAM storing valid pixels.

As can be expected, this design uses several counters to monitor bit and pixel positions along with controlling shift registers, RAM addresses, and RAM writes.

The first implementation uses a merged design that includes all functionality within the state machine. When this design was implemented within Xilinx Vivado 2020.1, the utilization was 45 LUTs and 102 flip-flops, giving a total performance of 164.56 MHz, as demonstrated in Figures 3.24 and 3.25.

Site Type	Used	Fixed	Available	Util%
Slice LUTs	45	0	32600	0.14
LUT as Logic	45	0	32600	0.14
LUT as Memory	0	0	9600	0.00
Slice Registers	102	0	65200	0.16
Register as Flip Flop	102	0	65200	0.16
Register as Latch	0	0	65200	0.00
F7 Muxes	0	0	16300	0.00
F8 Muxes	0	0	8150	0.00

(a)

| Design Timing Summary

WNS(ns)	TNS(ns)	TNS Failing Endpoints	TNS Total Endpoints	WHS(ns)	THS(ns)
3.923	0.000	0	191	0.179	0.000

(b)

Figure 3.24 Implementation and timing performance for the state machine when all functionality is included.

Site Type	Used	Fixed	Available	Util%
Slice LUTs*	36	0	32600	0.11
LUT as Logic	36	0	32600	0.11
LUT as Memory	0	0	9600	0.00
Slice Registers	102	0	65200	0.16
Register as Flip Flop	102	0	65200	0.16
Register as Latch	0	0	65200	0.00
F7 Muxes	0	0	16300	0.00
F8 Muxes	0	0	8150	0.00

(a)

| Design Timing Summary

WNS(ns)	TNS(ns)	TNS Failing Endpoints	TNS Total Endpoints	WHS(ns)	THS(ns)
5.631	0.000	0	218	0.106	0.000

(b)

Figure 3.25 Implementation and timing performance for the state machine when all functionality is decoupled.

Decoupling the state machine from the counters, RAM interfacing, and shift register results in slightly decreased utilization, but also a greatly increased maximum frequency of 228.88 MHz.

Note that the equation for conversion from WNS to Fmax is 1/(*T* – WNS) where *T* is the required target clock period. In this case, it is 10 ns, as both designs were targeted for 100-MHz operation.

3.7.7 Good Design Practices for FPGA Implementation

In addition to the decoupling of functionality outside of the FPGA, there are several additional good design practices that engineers at SensorsThink follow.

- *Use enumerated state definitions:* Using clearly defined enumerated state definitions makes the state machine much easier to understand and debug during commissioning. If the end application requires certification, clear state definitions and behavior enable independent reviewers to be able to understand the functionality and the design intent.
- *Single process state machine:* A single process state machine removes the potential for accidental latch creation in the combinatorial process when a branch is not fully defined. Single process state machines are also easier to debug and ensure that all the state machine outputs are registered, thus improving timing performance of the solution.
- *Consider unused states:* State machines implemented will always have a power of 2 states. The state machine may not use all the possible states for its implementation; these unused states can be an issue for high-reliability implementations and may need special handling to achieve reliability.
- *Check for terminal states:* A terminal state is one that, once entered, cannot be left; this may be intentional upon an error occurring. However, often, it is a design error in the state machine. Checking that each state has a path in and out prior to simulation can save significant time.
- *Reset:* Have a clear reset state defined for the current state.

Following these good design practices will enable our state machines to be not only effective, but also easier to debug and maintain as the life of the product evolves, which it no doubt will.

3.7.8 FLIR Lepton Interface

The FLIR Lepton deployed on the Smart Sensor SoC has a video over SPI interface. Further complicating the issue are U.S. export regulations, which limit the frame rate that the Lepton can output to 9 frames per second. To comply with this export requirement, the FLIR Lepton outputs corrupted frames intermixed with valid frames. Video over SPI requires the processor to maintain synchronization, across valid and corrupted frames; if synchronization is lost, the video output is lost.

The synchronization scheme is simple but time-critical: the FLIR Lepton transmits video packets over the SPI. Readout is started by the master, in this case, the Zynq de-asserting the SPI CS line for at least 5 image periods (185 ms). This causes the Lepton video output to time out and puts the Lepton in the right mode to resynchronize. To start new synchronization, the master asserts the CS line; following this, the Lepton will begin to output packets of data. Each packet consists of 4-byte

ID and CRC followed by 160 bytes of video when the imager is configured for raw output or 240 bytes when outputting RGB888. To read out the entire array, we need to read out 240 packets. To ensure frame compliance with export control rules, only 9 valid frames per second are output; in between these frames, the FLIR Lepton outputs corrupted frames. To maintain synchronization, these corrupted frames must be read out and discarded. After this has been achieved, the FLIR and the Zynq will be synchronized; to maintain synchronization, the Zynq must:

1. Not violate the intrapacket timeout. Once a packet starts, it must be completely clocked out within 3 line periods.
2. Not fail to read out all packets for a given frame before the next frame is available.
3. Not fail to read out all available frames.

While the Zynq processor has SPI interfaces, maintaining synchronization as the software application load increases becomes more difficult. Such a loading will also reduce the ability of the software to process other sensor data and complete analysis tasks.

Therefore, to ensure that the FLIR thermal imaging maintains synchronization in our Smart Sensor SoC, the synchronization and video capture will be implemented in the programmable logic.

To achieve synchronization, a simple state machine will be deployed that synchronizes with the output from the FLIR Lepton and writes valid video data to a block RAM memory that can be accessed by the processor. This enables the processor to access completed good frames from the programmable logic at the rate of 9 images per second.

The state diagram of the state machine is defined in Figure 3.26; this implementation uses several states to synchronize and capture the video data. The state machine works on the state of the 16-bit word that is received from the FLIR Lepton over the MISO SPI signal. This signal contains either a video packet information,

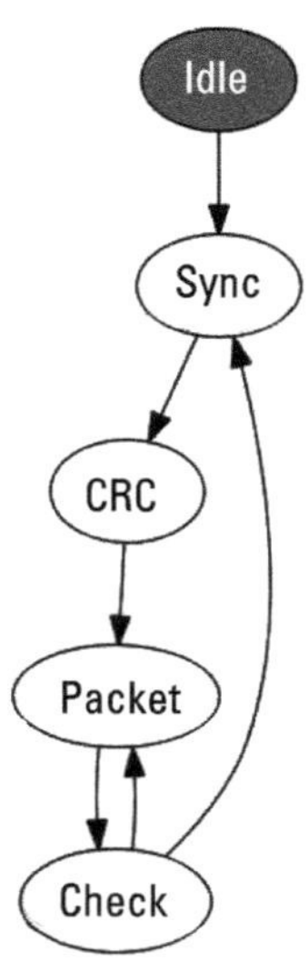

Figure 3.26 FLIR Lepton FSM state diagram.

CRC, or video data. If the video packet is corrupted, it is discarded and the remaining video is read out. If the packet is valid, then the video is written to the connected block RAM to enable access from the processor for the image to be rendered.

The implementation for the state machine results in the images of FLIR Lepton being able to be synchronized and maintain synchronization with the Zynq and the images being made available to the process for further deployment in the Internet of Things (IOT) application.

3.8 Defensive State Machine Design

As we develop our Smart Sensor SoC, we must be careful about ensuring that it can work in harsh environments as outlined in the requirements in Chapter 1. This harsh environment could mean that the system has to be able to work in electrically noisy, high-altitude, or dynamic environments.

Engineers at SensorsThink must therefore carefully consider the operating environment against the need to design defensively in the implementation of state machines. If the SensorsThink engineering team considers it appropriate to design defensively, the engineer must take additional care in the design and implementation of the state machines (as well as all accompanying logic) inside the programmable logic device to ensure that they can function within the requirements.

One of the major causes of errors within state machines is single-event upsets caused by either a high-energy neutron or an alpha particle striking sensitive sections of the device silicon. Single event upsets (SEUs) can cause a bit to flip its state (0 -> 1 or 1 -> 0), resulting in an error in device functionality that could potentially lead to the loss of the system or even endanger life if incorrectly handled. Because these SEUs do not result in any permanent damage to the device itself, they are called soft errors.

Terrestrially single event upsets are more likely to occur at higher altitudes and certain latitudes.

The backbone of most programmable logic design is the finite state machine, a design methodology that engineers use to implement control, data flow and algorithmic functions. When implementing state machines within FPGAs, designers will choose one of two styles (binary or one-hot), although, in many cases, most engineers allow the synthesis tool to determine the final encoding scheme. Each implementation scheme presents its own challenges when designing reliable state machines for mission-critical systems. Indeed, even a simple state machine can encounter several problems (Figure 3.27). You must pay close attention to the encoding scheme and, in many cases, take the decision about the final implementation encoding away from the synthesis tool.

3.8.1 Detection Schemes

Let us first look at binary implementations (sequential or Gray encoding), which often have leftover, unused states that the state machine does not enter when it is functioning normally. Designers must address these unused states to ensure that the state machine will gracefully recover if it should accidentally enter an illegal state. There are two main methods of achieving this recovery. The first is to declare all

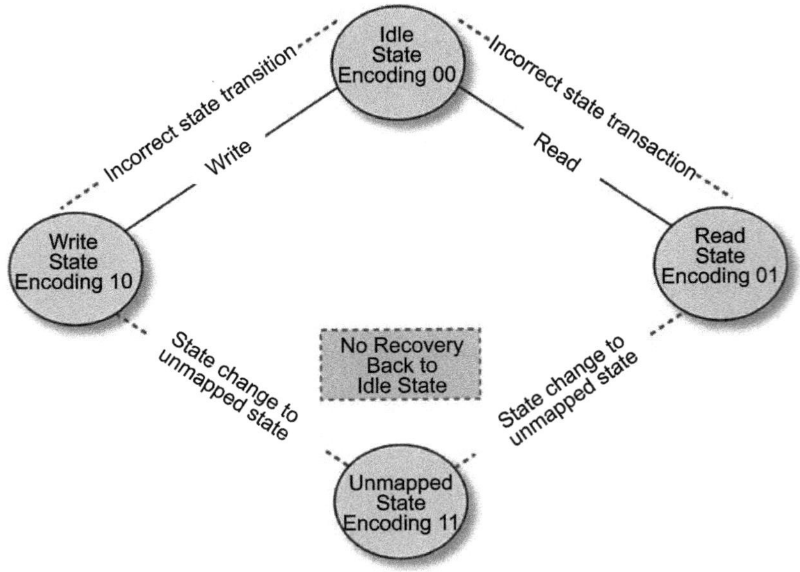

Figure 3.27 Even a simple state machine can encounter several types of errors.

2N number of states when defining the state machine signal and cover the unused states with the "others clause" at the end of the case statement. The others clause will typically set the outputs to a safe state and send the state machine back to its idle state or another state, as identified by the design engineer. This approach will require the use of synthesis constraints to prevent the synthesis tool from optimizing these unused states from the design, as there are no valid entry points. This typically means synthesis constraints within the body of the RTL code ("syn_keep").

The second method of handling the unused states is to cycle through them at startup following reset release. Typically, these states also keep the outputs in a safe state; should they be accidentally entered, the machine will cycle around to its idle state again.

One-hot state machines have one flip-flop for each state, but only the current state is set high at any one time. Corruption of the machine by having more than one flip-flop set high can result in unexpected outcomes. You can protect a one-hot machine from errors by monitoring the parity of the state registers. Should you detect a parity error, you can reset the machine to its idle state or to another predetermined state.

With both methods, the state machine's outputs go to safe states and the state machine restarts from its idle position. State machines that use these methods can be said to be SEU detecting, as they can detect and recover from an SEU, although the state machines' operation will be interrupted. You must take care during synthesis to ensure that register replication does not result in registers with a high fan-out being reproduced and hence left unaddressed by the detection scheme. Take care also to ensure that the error does not affect other state machines with which this machine interacts.

Many synthesis tools offer the option of implementing a safe state machine option. This option often includes more logic to detect the state machine entering an illegal state and send it back to a legal one, normally the reset state. For a high-reliability application, design engineers can detect and verify these illegal state entries more easily by implementing any of the previously described methods. Using these approaches, the designers must also consider what would happen should the detection logic suffer from an SEU. What effect would this have upon the reliability of the design? Figure 3.28 is a flowchart that attempts to map out the decision-making process for creating reliable state machines.

The techniques presented so far detect or prevent an incorrect change from one legal state to another legal state. Depending upon the end application, this could result in anything from a momentary system error to the complete loss of the mission.

3.8.2 Hamming Schemes

Techniques for detecting and fixing incorrect legal transitions are triple-modular redundancy and Hamming encoding. The latter provides a Hamming distance of 3 and covers all possible adjacent states (see Figure 3.29). A simpler technique for preventing legal transitions is the use of Hamming encoding with a Hamming distance of 2 (rather than 3) between the states and not covering the adjacent states. However, this will increase the number of registers that your design will require.

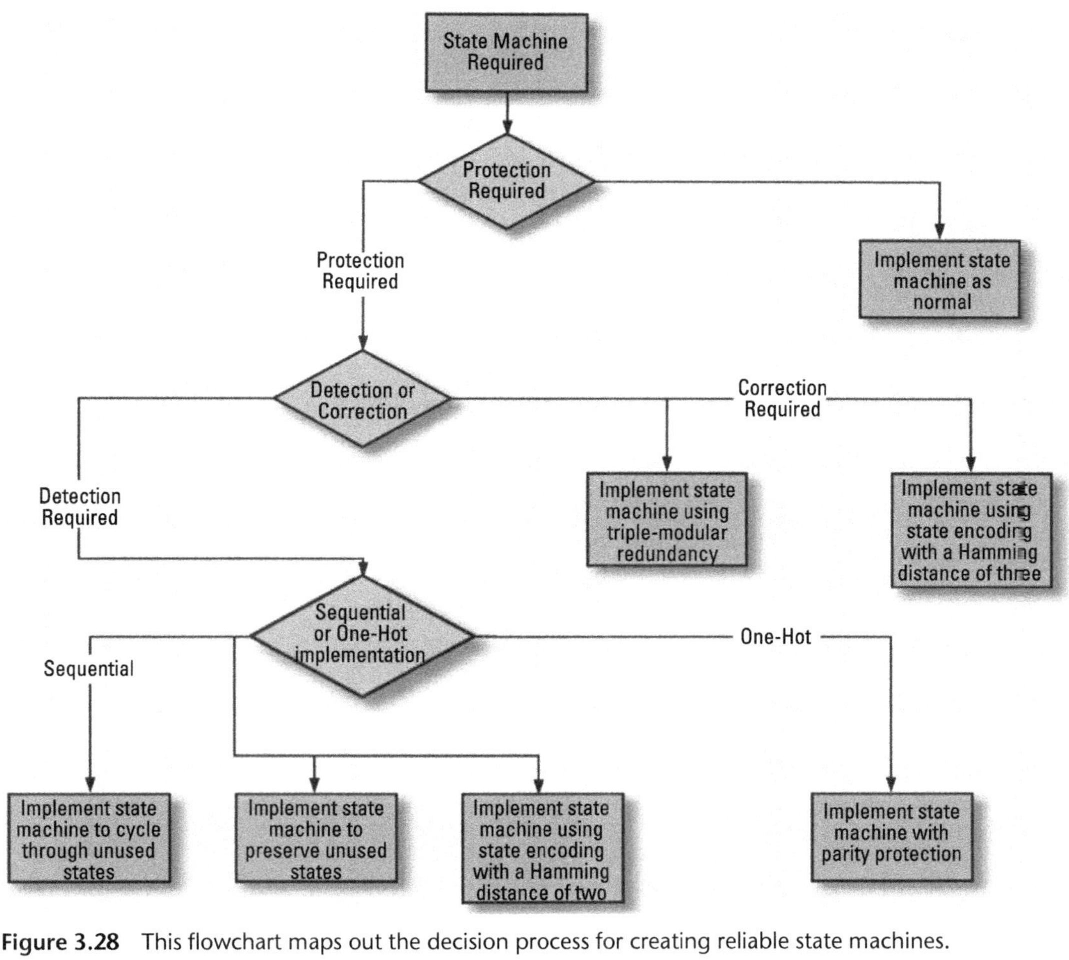

Figure 3.28 This flowchart maps out the decision process for creating reliable state machines.

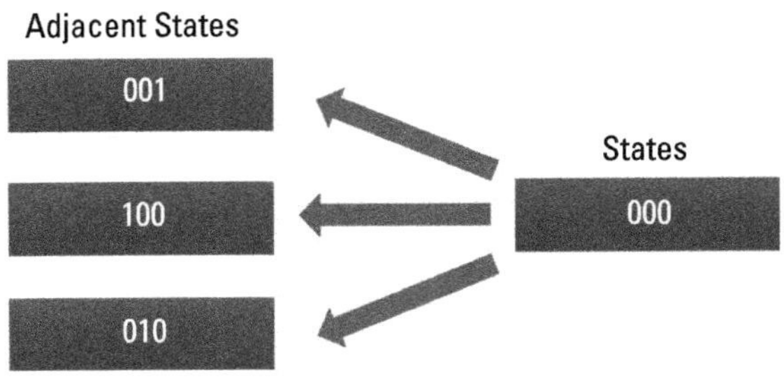

Figure 3.29 Hamming code adjacent states.

Triple-modular redundancy (TMR), for its part, involves implementing three instantiations of the state machine with majority voting upon the outputs and the next state. This is the simpler of the two approaches and many engineers have used it in several applications over the years. Typically, a TMR implementation will require spatial separation of the logic within the FPGA to ensure that an SEU does not corrupt more than one of the three instantiations. It is also important to remove any registers from the voter logic because they can create a single-point failure in which an SEU could affect all three machines.

The use of a state machine with encoding that provides a Hamming distance of 3 between states will ensure both SEU detection and correction. This guarantees that more than a single bit will change between states, meaning that an SEU cannot make the state transition from one legal state to another erroneously. The use of a Hamming distance of 2 between states is like the implementation for the sequential machine, where the unused states are addressed by the "when others" clause or reset cycling. However, as the states are explicitly declared to be separate from each other by a Hamming distance of 2 within the RTL, the state machine cannot move from one legal state to another erroneously and will instead go to its idle state should the machine enter an illegal state. This provides a more robust implementation than the binary one mentioned above.

If you wish to implement a state machine that will continue to function correctly should an SEU corrupt the current state register, you can do so by using a Hamming-code implementation with a Hamming distance of 3 and ensuring its adjacent states are also addressed within the state machine. Adjacent states are those that are 1 bit different from the state register and hence achievable should an SEU occur. This use of states adjacent to the valid state to correct for the error will result in $N * (M + 1)$ states, where N is the number of states and M is the number of bits within the state register. It is possible to make a small high-reliability state machine using this technique, but crafting a large one can be so complicated as to be prohibitive. The extra logic footprint associated with this approach could also result in lower timing performance.

3.8.3 Deadlock and Other Issues

There are other issues to consider when designing a high-reliability state machine beyond the state encoding schemes. Deadlock can occur when the state machine enters a state from which it is never able to leave. An example would be one state machine awaiting an input from a second state machine that has entered an illegal state and hence been reset to idle without generating the needed signal. To avoid deadlock, it is therefore good practice to provide timeout counters on critical state machines. Wherever possible, these counters should not be included inside the state machine but placed within a separate process that outputs a pulse when it reaches its terminal count. Be sure to write these counters in such a way as to make them reliable.

When checking that a counter has reached its terminal count, it is preferable to use the greater-than-or-equal-to operator, as opposed to just the equal-to operator. This is to prevent an SEU from occurring near the terminal count and hence no output being generated. You should declare integer counters to a power of 2, and, if you are using VHDL, they should also be modulo to the power of 2 to ensure in simulation that they will wrap around as they will in the final implementation [count <= (count + 1) Mod 16; for a 0-bit to 15-bit integer counter]. Unsigned counters do not require this because there is no simulation mismatch between RTL and post-route simulation regarding wraparound. You can replicate data paths within the design and compare outputs on a cycle-by-cycle basis to detect whether one of them has been subjected to an SEU event. Wherever possible, edge-detect signals to enable the design to cope with inputs that are stuck high or stuck low. You should analyze each module within the design at design time to determine how it will react to stuck-high or stuck-low inputs. This will ensure that it is possible to detect these errors and that they cannot have an adverse effect upon the function of the module.

3.8.4 Implementing Defensive State Machines in Xilinx Devices

The defensive design techniques considered by engineers at SensorsThink to protect state machines may at first glance seem to require considerable design effort in the creation and definition, in both in the RTL and the synthesis tool. However, when developing for Xilinx targets and the using the Xilinx Synthesis engine within Vivado, engineers at SensorsThink can use synthesis attributes to implement defensively designed state machines.

The synthesis attribute FSM_SAFE_STATE, defined in either the source code or the XDC file in Vivado, allows us to implement the following state machines:

- *auto_safe_state:* Hamming 3 encoded to ensure tolerance of SEU.
- *reset_state:* Hamming 2 encoding to force the state machine to its reset state on error.
- *power_on_state:* Hamming 2 encoding to force the state machine to its power-on state on error.
- *default:* Hamming 2 forces the state machine to the default / when others case-define in the HDL.

Using the state machine presented in Figure 3.27 as an example, the attribute can be applied in the RTL source code as shown.

```
type state is (idle, rd, wr, unmapped);
signal current_state : state;
signal transfer_word : std_logic_vector(31 downto 0);
```

```
attribute fsm_safe_state    : string;
attribute fsm_safe_state of current_state : signal is
"auto_safe_state";
```

Proof the state machine has been implemented using a Hamming 3 encoding can be observed in both the synthesis report and the elaborated design.

3.9 How Does FPGA Do Math?

Along with the finite state machine control structures, SensorsThink FPGA engineers are going to need to implement several mathematical algorithms. It is therefore crucial that all the SensorsThink engineering team understand how math is implemented in programmable logic. One of the many benefits of an FPGA-based solution is that we engineers can implement the mathematical algorithm in the best possible solution for the problem at hand. For example, if response time is critical, then we can pipeline stages of mathematics. If accuracy of the result is more critical, we can use more bits to ensure the desired accuracy is achieved. Many modern FPGAs also provide us the benefit of embedded multipliers and DSP slices, which can be used to ensure the optimal implementation is achieved in the target device.

3.9.1 Representation of Numbers

There are two methods of representing numbers within a design, fixed, or floating-point number systems that we can use. Fixed-point representation maintains the decimal point within a fixed position allowing for straightforward arithmetic operations. The major drawback of fixed-point representation is that, to represent larger numbers or to achieve a more accurate result with fractional numbers, a larger number of bits are required. A fixed-point number consists of two parts called the integer and fractional parts. Floating point representation allows the decimal point to float to different places within the number depending upon the magnitude; floating point numbers are divided into two parts: the exponent and the mantissa. This is remarkably like scientific notation, which represents a number as A times 10 to the power B where A is the mantissa and B is the exponent; however, the base of the exponent in a floating-point number is base 2, that is, A times 2 to the power B. The floating-point number is standardized by the IEEE/ANSI Standard 754-1985; the basic IEEE floating point number (there are many) utilizes an 8-bit exponent and a 24-bit mantissa.

2^7	2^6	2^5	2^4	2^3	2^2	2^1	2^0	•	2^{-1}	2^{-2}	2^{-3}	2^{-4}	2^{-5}	2^{-6}	2^{-7}	2^{-8}

Due to the complexity of floating-point numbers, we as designers tend wherever possible to point representation. The above fixed-point number can represent an unsigned number between 0.0 and 255.9906375 or a signed number between −128.9906375 and 127.9906375 using two's complement representation. Within a design, we have the choice to use either unsigned or signed numbers; typically, this will be constrained by the algorithm being implemented. Unsigned numbers can represent a range of 0 to $2n - 1$ and always represent positive numbers, while the range of a signed number depends upon the encoding scheme used: sign and magnitude, one's complement, or two's complement.

Sign and magnitude utilize the leftmost bit to represent the sign of the number (0 = positive, 1 = negative); the remainder of the bits represents the magnitude. Therefore, in the sign and magnitude system, both positive and negative numbers have the same magnitude; however, the sign bit differs. Due to this, it is possible to have both a positive and negative zero within the sign and magnitude system.

The one's complement uses the same unsigned representation for positive numbers as sign and magnitude representation. However, for negative numbers, the inversion (one's complement) of the positive number is used.

The two's complement is the most widely used encoding scheme for representing signed numbers. Just like sign and magnitude, one's complement schemes that are positive are represented in the same manner as an unsigned number, whereas negative numbers are represented as the binary number that you add to a positive number of the same magnitude to get zero. A negative two's complement number is calculated by first taking the one's complement (inversion) of the positive number and then adding 1 to it. The two's complement number system allows subtraction of one number from another by performing an addition of the two numbers. The range that a two's complement number can represent is given by

$$-(2n - 1) \text{ to } +(2n - 1 - 1)$$

One method that we can use to convert a number to its two's complement format is to work right to left, leaving the number the same until the first one is encountered; after this, each bit is inverted.

3.10 Fixed Point Mathematics

The normal way of representing the split between integer and fractional bits within a fixed-point number is x,y where x represents the number of integer bits and y represents the number of fractional bits. For example, 8,8 represents 8 integer bits and 8 fractional bits, while 16,0 represents 16 integer buts and 0 fractional bits. In many cases, the correct choice of the number of integer and fractional bits required will be undertaken at design time normally following conversion from a floating-point algorithm. Thanks to the flexibility of FPGAs, we can represent a fixed-point number of any bit length; the number of integer bits required depends upon the maximum integer value that the number is required to store, while the number of fractional bits will depend upon the accuracy of the final result. To determine the number of integer bits required, we can use the following equation

$$\text{Integer Bits Required} = Ceil\left(\frac{LOG_{10}\text{Integer\_Maximum}}{LOG_{10}2}\right)$$

For example, the number of integer bits required to represent a value between 0.0 and 423.0 is given by

$$9 = Ceil\left(\frac{LOG_{10}423}{LOG_{10}2}\right)$$

meaning that we would need 9 integer bits, allowing a range of 0 to 511 to be represented. If we want to represent the number using 16 bits, this would allow for 7 fractional bits. The accuracy that this representation would be capable of providing is given by

$$accuracy = \left(\frac{Actual\_Value - FPGA\_Value}{2^{\text{Fractional Bits}}}\right) \times 100$$

We can increase the accuracy of a fixed-point number by using more bits to store the fractional number. When we are designing, there are times when you may wish to store only fractional numbers (0,16) depending upon the size of the number that you wish to scale up; scaling up by 2^{16} may result in a number that still does not provide an accurate enough result. In this case, we can multiply up by power of 2 such that the number can be represented within a 16-bit number. We can then remove this scaling at a further stage within the implementation. For example, to represent the number $1.45309806319 \times 10^{-4}$ in a 16-bit number, the first step is to multiply it by 2^{16}:

$$65{,}536 * 1.45309806319 \times 10–4 = 9.523023$$

Storing the integer of the result (9) will result in the number being stored as $1.37329101563 \times 10^{-4}$ (9/65,536). This difference between the number required to be stored and the stored number is substantial and could lead to an unacceptable error in the calculated result. We can obtain a more accurate result by scaling the number up by a factor of 2, which produces a result between 32,768 and 65,535, therefore still allowing storage in a 16-bit number. Using the earlier example of storing $1.45309806319 \times 10^{-4}$, we can, by multiplying by factor of 2^{28}, obtain a number that can be stored in 16 bits and will be highly accurate of the desired number.

$$268{,}435{,}456 * 1.45309806319 \times 10–4 = 39{,}006.3041205$$

The integer of the result will result in the stored number being 1.45308673382 $\times 10^{-4}$, which will give a much more accurate calculation, provided that the scaling factor of 2^{28} can be addressed at a later stage within the calculation. For example, a multiplication of the scaled number with a 16-bit number scaled 4,12 will produce a result of 4, 40 (28 + 12); however, the result will be stored in a 32-bit result.

3.10.1 Fixed-Point Rules

To perform addition, subtraction, or division, the decimal points of both numbers must be aligned. That is, an x,8 number can only be added to, subtracted from, or divided by a number that is also in an x,8 representation. To perform arithmetic operations on numbers of different x,y formats, we must first ensure that the decimal points are aligned. To align a number to a different format, you have two choices: either multiply the number with more integer bits by 2^X, or divide the number with the least number of integer bits by 2^X. When dividing by 2^X, the accuracy will be reduced and may lead to a result that is outside the allowable tolerance. As all numbers are stored in base 2, scaling up or down can be achieved easily in an FPGA through shifting one place to the left or right for each power of 2 required to balance the two decimal points. To add together two numbers that are scaled 8,8 and 9,7, you can either scale up the 9,7 number by a factor of 2^1 or scale the 8,8 format down to a 9,7 format if the loss of a least significant bit is acceptable, for example, adding 234.58 and 312.732, which are stored in an 8,8 and 9,7 formats, respectively. The first step is to determine the actual 16-bit numbers, which will be added together.

$$234.58 * 28 = 60{,}052.48$$

$$312.732 * 27 = 40{,}029.69$$

The two numbers to be added together are 60,052 and 40,029; however, before the two numbers can be added together the decimal point must be aligned. To align the decimal points by scaling up the number with the largest number of integer bits, the 9,7 format number must be scaled up by a factor of 2^1:

$$40{,}029 * 2^1 = 80{,}058$$

The result can then be calculated by performing an addition of

$$80{,}058 + 60{,}052 = 140{,}110$$

This represents 547.3046875 in a 10,8 format ($140{,}110/2^8$).

When multiplying two numbers together, the decimal points do not need to be aligned, as the multiplication will provide a result that is X1 + X2, Y1 + Y2 wide. Multiplying two numbers, which are formatted 14,2 and 10,6, will produce a result, which is formatted 24 integer bits and 8 fractional bits. Multiplication by a fractional number can be used in the place of a division within an equation, by multiplying by the reciprocal of the divisor. Using this approach can reduce the complexity of the design significantly. For example, to divide the number 312.732 represented in a 9,7 (40,029) format by 15, the first stage is to calculate the reciprocal of the divisor.

$$\frac{1}{15} = 0.066666'$$

This reciprocal must then be scaled up to be represented within a 16-bit number:

$$65{,}536 * 0.06666 = 4{,}369$$

This will produce a result that is formatted 9,23 when the two numbers are multiplied together:

$$4{,}369 * 40{,}029 = 174{,}886{,}701$$

The result of this multiplication is thus

$$\frac{174886701}{8388608} = 20.8481193781$$

While the expected result is 20.8488, if the result is not accurate enough, then the reciprocal can be scaled up by a larger factor to produce a more accurate result. Therefore, never divide by a number when you can multiply by the reciprocal.

3.10.2 Overflow

When implementing algorithms, we must ensure that the result is not larger than what is capable of being stored within the result register. When this condition occurs, it is known as overflow; when overflow occurs, the stored result will be incorrect and the most significant bits are lost. A remarkably simple example of overflow would occur if two 16-bit numbers each with a value of 65,535 were added together and the result was stored within a 16-bit register.

$$65{,}535 + 65{,}535 = 131{,}070$$

The above calculation would result in the 16-bit result register containing a value of 65,534, which is incorrect. The simplest way to prevent overflow is to determine the maximum value, which will result from the mathematical operation, and use the first equation in this chapter to determine the size of the result register required. If an averager were being developed to calculate the average of up to 50 16-bit inputs, the size of the required result register can be calculated.

$$50 * 65{,}535 = 3{,}276{,}750$$

Using the first equation in this chapter would require a 22-bit result register to prevent overflow occurring. Care must also be taken when using signed numbers to ensure that overflow does not occur when using negative numbers. Using the averager example again, this time taking 10 averages of a signed 16-bit number and returning a 16-bit result:

$$10 * -32{,}768 = -327{,}680$$

As it is easier to multiply the result by a scaled reciprocal of the divisor, this number can be multiplied by 1/10 * 65,536 = 6,554 to determine the average.

$$-32{,}768 * 6{,}554 = -2{,}147{,}614{,}720$$

This number, when divided by 216 = –32,770, cannot be represented correctly within a 16-bit output. The module design must therefore take into account the overflow and detect it to ensure that an incorrect result is not output.

3.10.3 Real-World Implementation

A module is required to implement a transfer function that is used to convert atmospheric pressure measured in millibars into altitude measured in meters.

$$-0.0088 \times 2 + 1.7673x + 131.29$$

The input value will range between 0 and 10 millibars with a resolution of 0.1 millibar. The output of the module is required to be accurate to +/–0.01m. As the module specification does not determine the input scaling, this can be determined by the following equation.

$$4 = Ceil\left(\frac{LOG_{10}10}{LOG_{10}2}\right)$$

Therefore, to ensure maximum accuracy, the input data can be formatted as 4 integer bits and 12 fractional bits. The next step in the development of the module is to use a spreadsheet to calculate the expected result of the transfer function across the entire input range using the unscaled values. If the input range is too large to reasonably achieve this, an acceptable number of points should be calculated. For this example, 100 entries can be used to determine the expected result across the entire input range.

Once the initial unscaled expected values have been calculated, the next step is to determine the correct scaling factors for the constants and calculate the expected outputs using the scaled values. To ensure maximum accuracy, each of the constants used within the equation will be scaled by a different factor.

The first polynomial constant (*A*) scaling factor is given by

$$8 = Ceil\left(\frac{LOG_{10}133.29}{LOG_{10}2}\right)$$

The second polynomial constant (*B*) scaling factor is given by

$$1 = Ceil\left(\frac{LOG_{10}1.7673}{LOG_{10}2}\right)$$

The final polynomial constant (C) can be scaled up by a factor of 2^{16}, as it is completely fractional.

Table 3.2 Pressure to Altitude Conversion

Input (millibar)	*Output (meters)*
0	131.2900
0.1	131.4666
0.2	131.6431
0.3	131.8194
0.4	131.9955
0.5	132.1715
0.6	132.3472

Table 3.3 Calculated Scale and Unscaled Constants

Polynomail Constant	*Unscaled*	*Scaled*
A	133.29	33610
B	1.77	57910
C	–0.01	–577

These scaling factors allow the scaled spreadsheet to be calculated, this can be seen in Table 3.4. The results of each stage of the calculation will produce a result that will require more than 16 bits.

The calculation of the Cx^2 will produce a result that is 32 bits long when formatted 4,12 + 4,12 = 8,24; this is then multiplied by the constant *C* producing a result that will be 48 bits long when formatted 8,24 + 0,16 = 8,40. For the accuracy required in this example, 40 bits of fractional representation are excessive; therefore, the result will be divide by 2^{32} to produce a result with a bit length of 16 bits formatted as 8,8. The same reduction to 16 bits is carried out upon the calculation of Bx to produce a result that is formatted as 5,11. The result is the addition of columns Cx^2, Bx, and A; however, to obtain the correct result, the radix points must first be aligned. This can be achieved through either the shifting up of A and Cx^2 to align the numbers in an x,11 format. The alternate approach would be to shift down the calculated Bx to a format of 8,8 aligning the radix points with the calculated values of A and Cx^2.

In this example, the calculated value was shifted down by 2^3 to align the radix points in an 8,8 format. This approach simplified the number of shifts required and therefore reduces the logic need to implement the example. Note that if the accuracy could not be achieved through by shifting down to align the radix points, then the radix points must be aligned by shifting up the calculated values of A and Cx^2. In this example, the calculated result is scaled up by a power of 2^8; the result can then be scaled down and compared against the result calculated obtained with unscaled values. The difference between the calculated result and the expected result is then the accuracy; using the spreadsheet commands of MAX() and MIN(), the maximum and minimum error of the calculated result can be obtained across the entire range of spreadsheet entries. Once the calculated spreadsheet confirms that the required accuracy can be achieved, the RTL code can be written and

Table 3.4 Expected Results and Scaled Result Accuracy

Input Scaled	*C*	*B*	*A*	*Result*	*Result Scaled*	*Expected Result*	*Difference*
0	0	0	33610	33610	131.289	131.2900	0.0009
409	–6	361	33610	33655	131.465	131.4666	0.0018
819	–24	723	33610	33700	131.641	131.6431	0.0025
1228	–52	1085	33610	33745	131.816	131.8194	0.0030
1638	–93	1447	33610	33790	131.992	131.9955	0.0033
2048	–145	1809	33610	33835	132.168	132.1715	0.0035
2457	–208	2171	33610	33880	132.344	132.3472	0.0035
2867	–283	2533	33610	33925	132.520	132.5228	0.0033

simulated. If desired, the testbench can be designed such that the input values are the same as those used in the spreadsheet; this allows the simulation outputs to be compared against the spreadsheet's calculated results to ensure the correct RTL implementation.

3.10.4 RTL Implementation

The RTL example uses signed parallel mathematics to calculate the result within 4 clock cycles. Because of the signed parallel multiplication, care must be taken to ensure the extra sign bits generated by the multiplications.

{start_code}

```
ENTITY transfer_function IS PORT(
sys_clk : IN std_logic;
reset : IN std_logic;
data : IN std_logic_vector(15 DOWNTO 0);
new_data : IN std_logic;result : OUT std_logic_vector(15 DOWNTO
0);
new_res : OUT std_logic);
END ENTITY transfer_function;
ARCHITECTURE rtl OF transfer_function IS
-- this module performs the following transfer function -0.0038x2
+ 1.7673x + 131.29
-- input data is scaled 8,8, while the output data will be scaled
8,8.
-- this module utilized signed parallel mathematics
TYPE control_state IS (idle, multiply, add, result_op);
CONSTANT c : signed(16 DOWNTO 0) := to_signed(-577,17);
CONSTANT b : signed(16 DOWNTO 0) := to_signed(57910,17);
CONSTANT a : signed(16 DOWNTO 0) := to_signed(33610,17);SIGNAL
current_state : control_state;
SIGNAL buf_data : std_logic; --used to detect rising edge upon
the new_data
SIGNAL squared : signed(33 DOWNTO 0); -- register holds input
squared.
SIGNAL cx2 : signed(50 DOWNTO 0); --register used to hold Cx2
SIGNAL bx : signed(33 DOWNTO 0); -- register used to hold bx
SIGNAL res_int : signed(16 DOWNTO 0); --register holding the tem-
porary result
BEGIN
```

```
fsm : PROCESS(reset, sys_clk)
BEGIN
IF reset = '1' THEN
buf_data <= '0';
squared <= (OTHERS => '0');
cx2 <= (OTHERS => '0');
bx <= (OTHERS => '0');
result <= (OTHERS => '0');
res_int <= (OTHERS => '0');
new_res <= '0';
current_state <= idle;
ELSIF rising_edge(sys_clk) THEN
buf_data <= new_data;
CASE current_state IS
WHEN idle =>
new_res <= '0';
IF (new_data = '1') AND (buf_data = '0') THEN --detect rising
edge new data
squared <= signed( '0'& data) * signed('0'& data);
current_state <= multiply;
ELSE
squared <= (OTHERS =>'0');
current_state <= idle;
END IF;
WHEN multiply =>
new_res <= '0';
cx2 <= (squared * c);
bx <= (signed('0'& data)* b);
current_state <= add;
WHEN add =>
new_res <= '0';
res_int <= a + cx2(48 DOWNTO 32) + ("000"& bx(32 DOWNTO 19));
current_state <= result_op;
WHEN result_op =>
result <= std_logic_vector(res_int(res_int'high -1 DOWNTO 0));
new_res <= '0';
current_state <= idle;
END CASE;
END IF;
END PROCESS;
END ARCHITECTURE rtl;
```

3.11 Polynomial Approximation

As the SensorsThink team develops the FPGA solution, they may be required to implement several significant mathematical algorithms. The SensorsThink engineering team may find implementing these algorithms directly will result in a complex implementation. This complex implementation may have knock-on effects in timing closure and verification. One technique that the SensorsThink engineering team might consider for complex algorithms is a different approach to direct implementation. This alternative approach is to leverage polynomial approximation.

3.11.1 The Challenge with Some Algorithms

Due to their flexibility and performance, FPGAs are used for a number of industrial, science, military, and other applications that require the calculation of complex mathematical transfer functions.

One example of this would be within an FPGA that monitors a Platinum Resistance Thermometer (PRT) and converts the resistance of the PRT into a temperature. This is typically achieved using a Calendar Van Duson equation that is shown in the simplified form here (valid between 0°C and 660°C)

$$R = R_0 \times \left(1 + a \times t + b \times t^2\right)$$

where R_0 is the resistance at 0°C and a and b are coefficents of the PRT.

Most systems that use a PRT will know the resistance from the design of the electronic circuit and the calculating temperature is what is desired; this requires rewriting the equation as:

$$t = \frac{-R_0 \times a + \sqrt{R_0^2 \times a^2 - 4 \times R_0 \times b \times \left(R_0 - R\right)}}{2 \times R_0 \times b}$$

Plotting the resistance against temperature will result in a graph as shown in Figure 3.30. From the graph, you can clearly see the nonlinearity.

Implementing this directly could be a significant challenge for an engineer and many will look for different methods to implement the function in an easier manner. One possible approach would be a look-up table storing several points on the curve with linear interpolation between the points within the look-up table. Depending upon the accuracy requirement and number of elements stored within

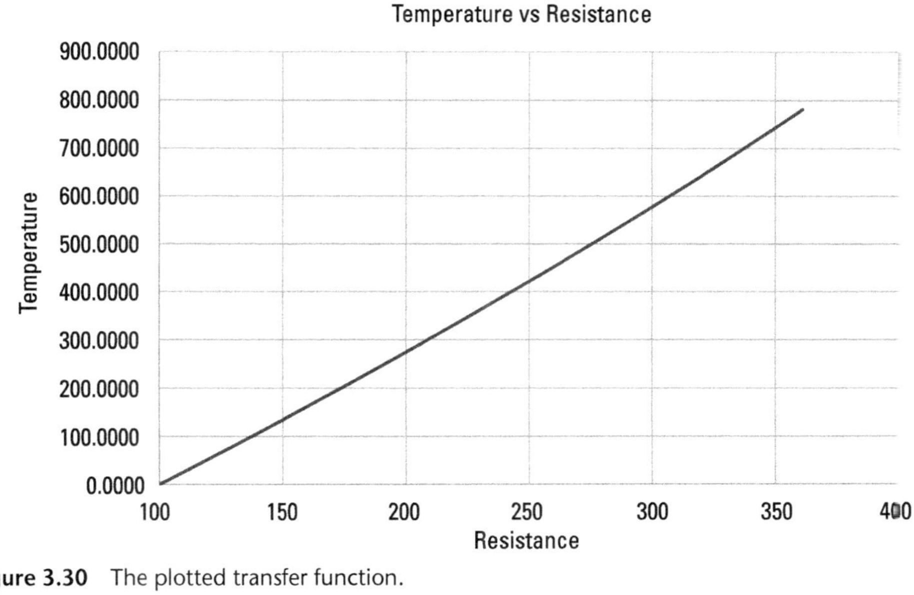

Figure 3.30 The plotted transfer function.

a look-up table, this may prove acceptable; however, you will still need a linear interpolator function which can be complex mathematically.

3.11.2 Capitalize on FPGA Resources

Modern FPGAs contain much more than the traditional look-up table and flip-flop. These days, FPGA contain DSP slices and block and distributed RAM, along with many more advanced hard-core IP such as PCIe endpoints and high-speed serial links.

It is these internal RAM structures and DSP slices that can be used to implement transfer functions with greater ease. The DSP slices are often referred to as DSP48s due to the 48-bit accumulator that these slices provide.

One method that utilizes the DSP slice-rich architecture of the FPGA is to plot the mathematical function, including the input value range across in mathematical program, for instance, MATLAB or Excel. The engineer can then add a polynomial trend line to this dataset such that the equation for the trend line can then be implemented within the FPGA in place of the mathematically complex function, providing that the trend line equation meets the accuracy requirements. The accuracy can be easily verified by calculating the resultant output using the polynomial equation and comparing it against the result of the transfer function. When this is performed for the temperature transfer function, we use our example in Microsoft Excel. We obtain for this example a trend line that sits right on top of the original data set, as seen in Figure 3.31.

Having obtained the polynomial fit for the transfer function that we want to implement, we can then double-check this for accuracy against the transfer function again using the same analysis tool, in this case, Excel. For the case in point where temperature is being monitored, it may be the case that the end measurement has to be accurate to +/–1°C, for example. Depending upon the range of measurement, this may prove difficult to achieve using just one polynomial equation.

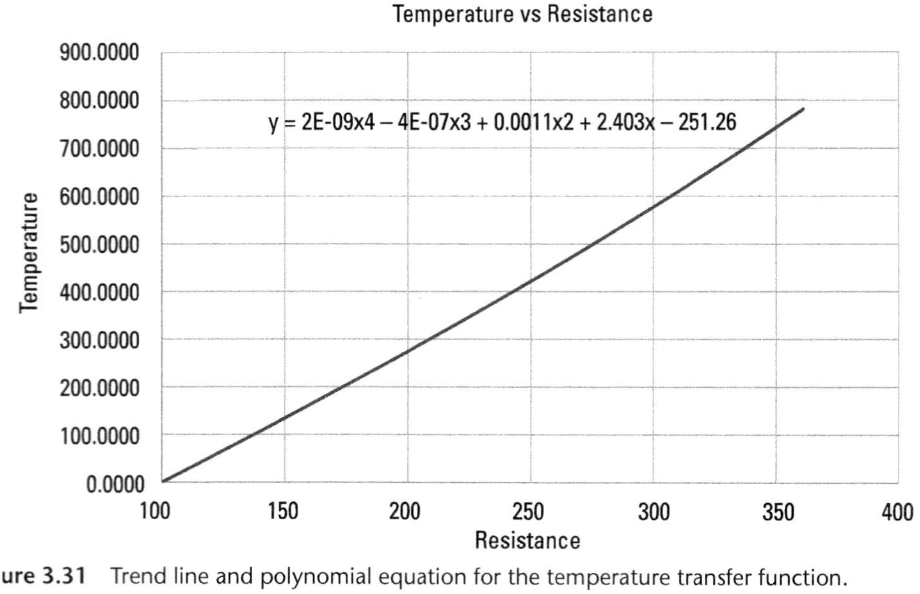

Figure 3.31 Trend line and polynomial equation for the temperature transfer function.

3.11.3 Multiple Trend Lines Selected by Input Value

This can be addressed by changing the polynomial constants used above once the input goes above a specified value. Continuing with the temperature example, the first polynomial equation provides +/–1°C of accuracy between 0°C and 263°C. While for many applications this will be more than sufficient, we may require this tolerance across a range out to 300°C, which would limit our approach. However, we can address this by plotting the range between 269°C and 300°C (see Figure 3.32) and obtaining a different polynomial equation that will provide more accuracy for this output range.

This approach allows the engineering team to break the transfer function into several segments to achieve the desired accuracy. These segments can be either uniformly spaced across the transfer function (i.e., split into 10 segments of equal value of X). Alternatively, they can be nonuniform and segmented as required to achieve the desired accuracy.

3.12 The CORDIC Algorithm

Implementing mathematical algorithms can be a significant challenge; however, techniques such as polynomial approximation can significantly help the SensorsThink engineering team develop mathematical algorithms. Another complementary approach with which SensorsThink engineers must be familiar is the CORDIC algorithm. This algorithm can help us to implement trigonometric and transcendental functions, thanks to its linear and hyperbolic extensions.

Most engineers tasked with implementing a mathematical function within an FPGA such as sine, cosine, or square root may initially think of implementing this via a look-up table, possibly combined with linear interpolation or a power series if multipliers are available. However, the CORDIC (COordinate Rotation Digital Computer) algorithm invented by Jack Volder for the B58 program in 1959 is one

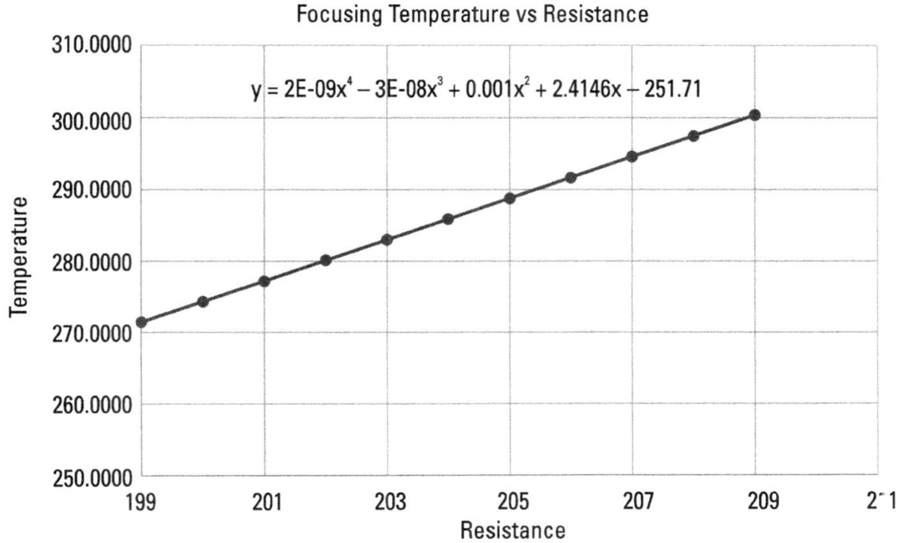

Figure 3.32 Plotting between 269°C and 300°C to provide a more accurate result.

of the most important algorithms in the FPGA engineer's toolbox and one that few engineers are aware of, although as engineers they will almost certainly use results calculated by a scientific calculator that uses the algorithm for trigonometric and exponential functions, starting with the HP35 and continuing to present-day calculators. The real beauty of this algorithm is that it can be implemented with a very small FPGA footprint and requires only a small look-up table, along with logic to perform shifts and additions; importantly, the algorithm requires no dedicated multipliers or dividers to implement.

This algorithm is one of the most useful for DSP and industrial and control applications. Depending upon its mode and configuration, the CORDIC implements some very useful mathematical functions. The CORDIC algorithm can operate in one of three configurations: linear, circular, or hyperbolic. Within each of these configurations, the algorithm functions in one of two modes: rotation or vectoring (see Table 3.5). In the rotation mode, the input vector is rotated by a specified angle, while in the vectoring mode, the algorithm rotates the input vector to the x axis while recording the angle of rotation required.

Additionally, there are other functions that can be derived from the CORDIC outputs. In many cases, these could even be implemented by the use of another CORDIC in a different configuration.

$$\text{Tan} = \text{Sin}/\text{Cos}$$
$$\text{Tan}\,H = \text{Sinh}/\text{Cosh}$$
$$\text{Exponential} = \text{Sinh} + \text{Cosh}$$
$$\text{Natural Logarithm} = 2 * \text{ArcTan}\,H^{\text{Note 1}}$$
$$\text{SQR} = \left(X^2 - Y^2\right)^{0.5\ \text{Note 2}}$$

Note 1: where X = Argument +1 Y = Argument – 1.
Note 2 where X = Argument +0.25 Y = Argument – 0.25.

The unified CORDIC algorithm can be seen below; it covers all three configurations. The algorithm has three inputs X, Y, and Z; how these are initialized at startup depends upon the mode of operation (vectoring or rotation) in accordance with Table 3.6.

$$X_{i+1} = X_i - m * Y_i * d_i * 2^i$$
$$Y_{i+1} = Y_i + X_i * d_i * 2^i$$
$$Z_{i+1} = Z_i - d_i * e_i$$

Table 3.5 Functions Resulting from CORDIC Configurations and Mode

Configuratoin	*Rotation*	*Vectoring*
Linear	Op Y = X * Y	Op Z = X/Y
Hyperbolic	Op X = CosH(X) Op Y = SinH(Y)	Op Z = Arc TanH
Circular	Op X = Cos(X) Op Y = Sin(Y)	Op Z = ArcTan(Y) Op X = SQR($X^2 + Y^2$)

Table 3.6 CORDIC Angle of Rotation

Configuration	e_i
Linear	2^{-i}
Hyperbolic	Arc TanH(2^{-i})
Circular	Arc Tan(2^{-i})

where m defines the configuration for either hyperbolic ($m = -1$), linear ($m = 0$), or circular ($m = 1$) correspondingly the value of e_i as the angle of rotation changes depending upon the configuration. The value of e_i is normally implemented as a small look-up table within the FPGA (Table 3.6).

d_i is the direction of rotation that depends upon the mode of operation for rotation mode $d_i = -1$ if $Z_i < 0$, else +1, while in the vectoring mode $d_i = +1$, if $Y_i < 0$, else −1.

When configured in either circular or hyperbolic using rotation mode, the output results will have gain that can be precalculated using the number of rotations defined using the equation.

$$A_n = \Sigma\sqrt{(1+2^{-2i})}$$

This gain is typically fed back into the initial setting of the algorithm to remove the need for post-scaling of the result (Table 3.7).

While the algorithm presented above is particularly important to the design engineer, it has to be noted that the CORDIC algorithm only operates within a strict convergence zone that may require the engineer to perform some prescaling to ensure the algorithm performs as expected. It is worth noting that the algorithm will get more accurate the more iterations (serial) or stages (parallel) the engineer decides to implement. A general rule of thumb is that for n bits of precision, n iterations or stages are required. However, all of this is easily modeled in simple tools such as Excel or MATLAB prior to cutting code to ensure the accuracy is obtained with the selected iterations.

Table 3.7 Initialization Settings Depending upon Mathematical Operation as Demonstrated in Table 3.5

Mode	Y	X	Z
Circular Rotation	A_n	0	Argument Z
Circular Vectoring	1	Argument X	0
Hyperbolic Rotation	A_n	0	Argument Z
Hyperbolic Vectoring	Argument + 1	Argument − 1	0
Linear Rotation	0	Argument X	Argument Z
Linear Vectoring	Argument Y	Argument X	0

3.13 Convergence

The CORDIC algorithm as defined will only converge (work) across a limited range of input value. For circular configurations of CORDIC algorithms, convergence is guaranteed for the angles below the sum of the angles in the look-up table (i.e., between –99.7° and 99.7°). For angles outside of this, the engineer must use a trigonometric identity to translate one within. This is also true for convergence within the linear configuration. However, in the hyperbolic mode to gain convergence, certain iterations must be repeated (4, 13, 40, K... $3K + 1$). In this case, the maximum input of θ is approximately 1.118 radians.

3.14 Where Are These Used?

CORDICs are used in a wide range of applications, including digital signal processing, image processing, and industrial control systems. The most basic method of using a CORDIC is to generate sine and cosine waves when coupled with a phase accumulator. The use of the algorithm to generate these waveforms can, if correctly done, result in a high spurious free dynamic range (SFDR). Good SFDR performance is required for most signal processing applications. Within the field of robotics, CORDICs are used within kinematics where the addition of coordinate values with new coordinate values can be easily accomplished using a circular CORDIC in the vectoring mode. Within the field, image processing 3-D operations such as lighting and vector rotation are the perfect candidates for implementation using the algorithm. However, perhaps the most common use of the algorithm is in implementing traditional mathematical functions as shown in Table 3.1 where multipliers, dividers, or more interesting mathematical functions are required in devices where there are no dedicated multipliers or DSP blocks. This means that CORDICs are used in many small industrial controllers to implement mathematical transfer functions, with true RMS measurement being one example. CORDICs are also being used in biomedical applications to compute fast Fourier transforms (FFTs) to analyze the frequency content of many physiological signals. In this application, along with performing the traditional mathematical functions, the CORDIC is used to implement the FFT twiddle factors.

3.15 Modeling in Excel

One of the simplest methods of modeling your CORDIC algorithm before cutting code is to put together a simple Excel spreadsheet that allows you to model the number of iterations and gain (A_n) initially using a floating-point number system and later using a scaled fixed-point system providing a reference for verification of the code during simulation (Table 3.8).

As can be seen from the Excel implementation, the initial X input is set to A_n to reduce the need for postprocessing the result, and the initial argument is set in Z that is defined in radians, as are the results.

Table 3.8 Excel Model of CORDIC Configured for Circular Rotation

Iteration	x	y	z	Direction	Gain
Initial	0.607253	0.000000	0.017453		
0	0.607253	0.607253	-0.767945	1	0.707107
1	0.910879	0.303626	-0.304298	-1	0.894427
2	0.986786	0.075907	-0.059319	-1	0.970143
3	0.996274	-0.047442	0.065036	-1	0.992278
4	0.999239	0.014826	0.002617	1	0.998053
5	0.998776	0.046052	-0.028623	1	0.999512
6	0.999496	0.030446	-0.012999	-1	0.999878
7	0.999734	0.022637	-0.005186	-1	0.999969
8	0.999822	0.018732	-0.001280	-1	0.999992
9	0.999859	0.016779	0.000673	-1	0.999998
10	0.999842	0.017756	-0.000304	1	1
11	0.999851	0.017268	0.000185	-1	1
12	0.999847	0.017512	-0.000060	1	1
13	0.999849	0.017390	0.000063	-1	1
14	0.999848	0.017451	0.000001	1	1
	Cos	Sin		A_n	0.607253
CORDIC	0.999848	0.017451			
Actual	0.999848	0.017452			

3.16 Implementing the CORDIC

Unless there is a good reason not to, the simplest method of implementing a CORDIC algorithm within an FPGA is to utilize a tool such as the Xilinx Core Generator. The Core Generator provides a comprehensive interface allowing the engineer to define the exact functionality of the CORDIC (rotate, vector).

Unfortunately, the Core Generator does not provide options for working with CORDICs within the linear mode (the Core Generator does provide separate cores to perform these functions). However, the VHDL code required to implement the algorithm can be written in a very few lines, as the simple example of a circular implementation below shows. There are two basic topologies for implementing a CORDIC in an FPGA, using either a state machine-based approach or a pipelined approach. If the processing time is not critical, the algorithm can be implemented as a state machine that computes one CORDIC iteration per cycle until the desired number of cycles has been completed. If a high calculation speed is required, then a parallel architecture is more appropriate. This code implements a 15-stage parallel CORDIC operating within the rotation mode. It uses a simple look-up table of ArcTan, coupled with a simple array structure to implement the parallel stages.

```
LIBRARY ieee;
USE ieee.std_logic_1164.all;
USE ieee.numeric_std.all;


ENTITY synth_cordic IS PORT(


clk    : IN std_logic;
resetn : IN std_logic;
```

```
z_ip  : IN std_logic_vector(16 DOWNTO 0); --1,16
x_ip  : IN std_logic_vector(16 DOWNTO 0); --1,16
y_ip  : IN std_logic_vector(16 DOWNTO 0); --1,16
cos_op : OUT std_logic_vector(16 DOWNTO 0); --1,16
sin_op : OUT std_logic_vector(16 DOWNTO 0)); --1,16
END ENTITY synth_cordic;

ARCHITECTURE rtl OF synth_cordic IS

TYPE signed_array IS ARRAY (natural RANGE <> ) OF signed(17 DOWN-
TO 0);

--ARCTAN Array format 1,16 in radians
CONSTANT tan_array : signed_array(0 TO 16) := (to_
signed(51471,18), to_signed(30385,18), to_signed(16054,18),to_
signed(8149,18), to_signed(4090,18), to_signed(2047,18),
to_signed(1023,18), to_signed(511,18), to_signed(255,18),
to_signed(127,18), to_signed(63,18), to_signed(31,18), to_
signed(15,18), to_signed(7,18),to_signed(3,18), to_signed(1,18),
to_signed(0, 18));

SIGNAL x_array : signed_array(0 TO 14) := (OTHERS => (OTHERS
=>'0'));
SIGNAL y_array : signed_array(0 TO 14) := (OTHERS => (OTHERS
=>'0'));
SIGNAL z_array : signed_array(0 TO 14) := (OTHERS => (OTHERS
=>'0'));

BEGIN

--convert inputs into signed format

PROCESS(resetn, clk)
BEGIN
IF resetn = '0' THEN
x_array <= (OTHERS => (OTHERS => '0'));
z_array <= (OTHERS => (OTHERS => '0'));
y_array <= (OTHERS => (OTHERS => '0'));
ELSIF rising_edge(clk) THEN
IF signed(z_ip)< to_signed(0,18) THEN
x_array(x_array'low) <= signed(x_ip) + signed('0' & y_ip);
y_array(y_array'low) <= signed(y_ip) - signed('0' & x_ip);
z_array(z_array'low) <= signed(z_ip) + tan_array(0);
ELSE
x_array(x_array'low) <= signed(x_ip) - signed('0' & y_ip);
y_array(y_array'low) <= signed(y_ip) + signed('0' & x_ip);
z_array(z_array'low) <= signed(z_ip) - tan_array(0);
END IF;
FOR i IN 1 TO 14 LOOP
IF z_array(i-1) < to_signed(0,17) THEN
```

```
x_array(i) <= x_array(i-1) + (y_array(i-1)/2**i);
y_array(i) <= y_array(i-1) - (x_array(i-1)/2**i);
z_array(i) <= z_array(i-1) + tan_array(i);
ELSE
x_array(i) <= x_array(i-1) - (y_array(i-1)/2**i);
y_array(i) <= y_array(i-1) + (x_array(i-1)/2**i);
z_array(i) <= z_array(i-1) - tan_array(i);
END IF;
END LOOP;
END IF;
END PROCESS;
cos_op <= std_logic_vector(x_array(x_array'high)(16 DOWNTO 0));
sin_op <= std_logic_vector(y_array(y_array'high)(16 DOWNTO 0));

END ARCHITECTURE rtl;
```

3.17 Digital Filter Design and Implementation

The Smart Sensor SoC being developed by SensorsThink interfaces with several sensors that gather real-world data. This data will be subject to noise. As such, SensorsThink engineers may want to consider using digital filters to remove noise from the signals, especially the remoteHX94 Remote Humidity and Temperature sensor, which is sampled directly by the XADC.

Digital filters are a key part of any signal-processing system, and as modern applications have grown more complex, so has filter design. FPGAs provide the ability to design and implement filters with performance characteristics that would be exceedingly difficult to re-create with analog methods. These digital filters are immune to certain issues that plague analog implementations, notably component drift and tolerances (over temperature, aging and radiation, for high-reliability applications). These analog effects significantly degrade the filter performance, especially in areas such as passband ripple.

Digital models have their own quirks. The rounding schemes used within the mathematics of the filters can be a problem, as these rounding errors will accumulate, impacting performance by, for example, raising the noise floor of the filter. The engineer can fall back on a number of approaches to minimize this impact, and schemes such as convergent rounding will provide better performance than traditional rounding. In the end, rounding-error problems are far less severe than those of the analog component contribution.

One of the major benefits of using an FPGA as the building block for a filter is the ability to easily modify or update the filter coefficients late in the design cycle with minimal impact, should the need for performance changes arise due to integration issues or requirement changes.

3.17.1 Filter Types and Topologies

Most engineers familiar with digital signal processing will be aware that there are four main types of filters. Lowpass filters only allow signals below a predetermined cutoff frequency to be output.

Highpass filters are the inverse of the lowpass and will only allow through frequencies above the cutoff frequency. Bandpass filters allow a predetermined bandwidth of frequencies, preventing other frequencies from entering. Finally, band-reject filters are the inverse of the bandpass variety and therefore reject a predetermined bandwidth, while allowing all others to pass through.

Most digital filters are implemented by one of two methods: finite impulse response (FIR) and infinite impulse response (IIR). Let's take a closer look at how to design and implement FIR filters, which are also often called windowed-sinc filters. So why are we focusing upon FIR filters? The main difference between the two filter styles is the presence or lack of feedback. The absence of feedback within the FIR filter means that for a given input response, the output of the filter will eventually settle to zero. For an IIR filter subject to the same input that contains feedback, the output will not settle back to zero. The lack of feedback within the filter implementation makes the FIR filter inherently stable, as all the filter's poles are located at the origin. The IIR filter is less forgiving. Its stability must be carefully considered as you are designing it, making the windowed sinc filter easier for engineers new to DSP to understand and implement. If you were to ask an engineer to draw a diagram of the perfect lowpass filter in the frequency domain, most would produce a sketch similar to that shown in Figure 3.33.

The frequency response shown in Figure 3.33 is often called the brick-wall filter. That is because the transition from passband to stopband is very abrupt and much sharper than can realistically be achieved. The frequency response also exhibits other perfect features, such as no passband ripple and perfect attenuation within the stopband. If you were to extend this diagram such that it was symmetrical around 0 Hz extending out to both +/–FS Hz (where FS is the sampling frequency) and perform an inverse discrete Fourier transform (IDFT) upon the response, you would obtain the filter's impulse response, as shown in Figure 3.34.

This is the time-domain representation of the frequency response of the perfect filter shown in Figure 3.33, often called the filter kernel. It is from this response that FIR or windowed-sinc filters get their name, as the impulse response is what is achieved if you plot the sinc function.

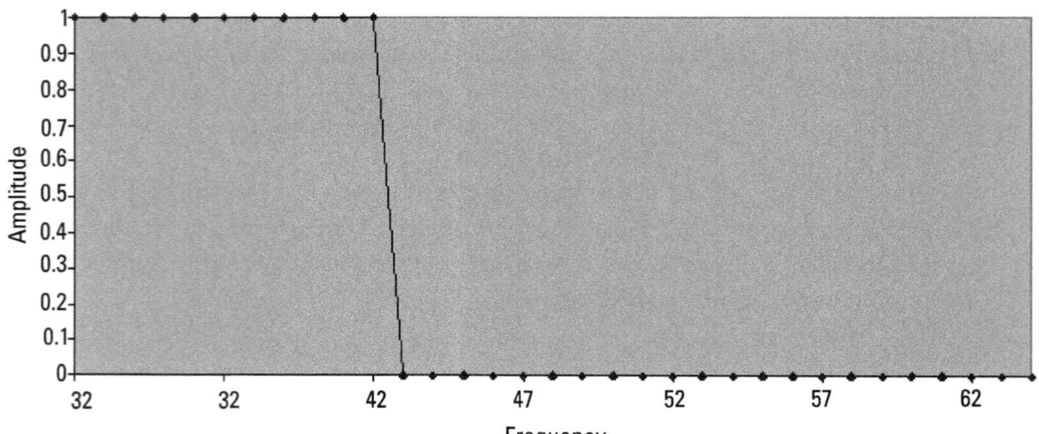

Figure 3.33 Ideal lowpass filter performance.

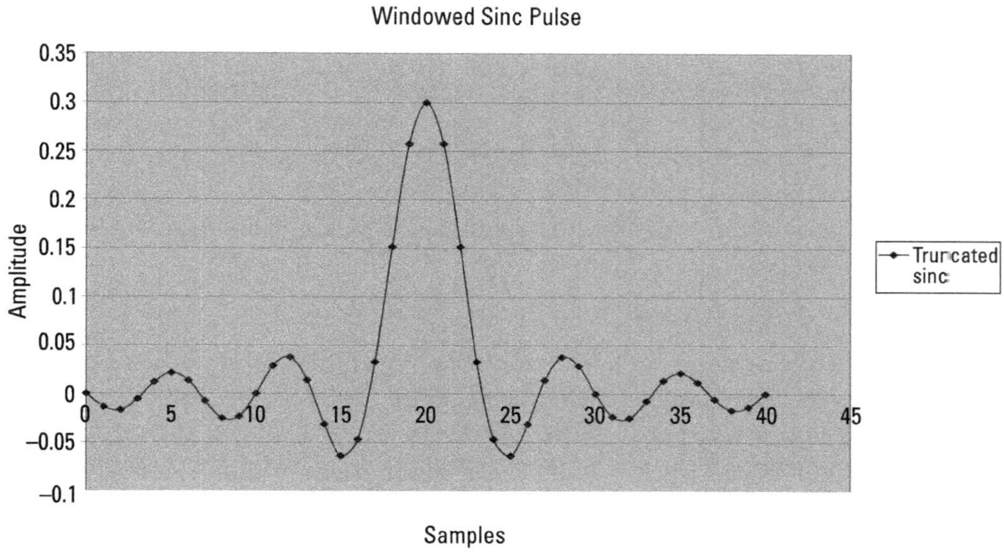

Figure 3.34 IDFT or impulse response of the perfect lowpass filter.

$$h[i] = \sin(2 * PI * Fc * i)$$
$$i * PI$$

Combined with the step response of the filter, these three responses—frequency, impulse, and step—provide all the information on the filter performance that you need to know to demonstrate that the filter under design will meet the requirements placed upon it.

3.17.2 Frequency Response

The frequency response is the traditional image engineers consider when thinking about a filter. It demonstrates how the filter modifies the frequency-domain information.

The frequency response allows you to observe the cutoff frequency, stopband attenuation, and passband ripple. The roll-off between passband and stopband—often called the transition band—is also apparent in this response. Ripples in the passband will affect the signals being filtered. The stopband attenuation demonstrates how much of the unwanted frequencies remain within the filter output.

3.17.3 Impulse Response

It is from the impulse response that the coefficients for your filter are abstracted. However, to achieve the best performance from your filter, the standard practice is to use a windowing function. Windowing is the technique of applying an additional mathematical function to a truncated impulse response to reduce the undesired effects of truncation. Figure 3.33 demonstrates the impulse response extending out infinitely with ripples that, although they diminish significantly in amplitude, never settle at zero. Therefore, you must truncate the impulse response to $N + 1$ coefficients chosen symmetrically around the center main lobe, where N is the desired filter length (please remember that N must be an even number). This truncation affects the filter's performance in the frequency domain due to the abrupt cutoff of the new, truncated impulse response. If you were to take a discrete Fourier transform (DFT) of this truncated impulse response, you would notice ripples in both the passband and stopband along with reduced roll-off performance. Therefore, it is common practice to apply a windowing function to improve the performance.

3.17.4 Step Response

The step response, which is obtained by integrating the impulse response, demonstrates the time-domain performance of the filter and how the filter itself modifies this performance. The three parameters of importance when you are observing the step response are the rise time, overshoot, and linearity. The rise time is the number of samples that it takes to rise between 10% and 90% of the amplitude levels, demonstrating the speed of the filter. To be of use within your final system, the filter must be able to distinguish between events in the input signal; therefore, the step response must be shorter than the spacing of events in the signal. Overshoot is the distortion that the filter adds to the signal as it is processing it. Reducing the overshoot in the step response allows you to determine if the signal distortion results from either the system or the information that it is measuring. Overshoot reduces the uncertainty of the source of the distortion, degrading the final system performance, and possibly means that the system does not meet the desired performance requirements. If the signal's upper and lower halves are symmetrical, the phase response of the filter will have a linear phase, which is needed to ensure the rising edge and falling edge of the step response are the same. It is worth nothing at this point that is very difficult to optimize a filter for good performance in both the time and frequency domains. Therefore, you must understand which of those domains contains the information that you are required to process. For FIR filters, the information required is within the frequency domain. Therefore, the frequency response tends to dominate.

3.17.5 Windowing the Filter

Using a truncated impulse response will not provide the best-performing digital filter, as it will demonstrate none of the ideal characteristics. For this reason, designers use windowing functions to improve the passband ripple, roll-off, and stopband attenuation performance of the filter. There are many window functions that you can apply to the truncated sinc, for example, Gaussian, Bartlett, Hamming, Blackman, and Kaiser. However, two of the most popular window functions are the Hamming and Blackman. Let us explore them in more detail. Both windows, when applied, reduce the passband ripple and increase the roll-off and attenuation of the filter. Figure 3.35 shows the impulse and frequency responses for a truncated sinc and with Blackman and Hamming windows applied.

As you can see, both windows significantly improve the passband ripple. The roll-off of the filter is determined not only by the window but also by the length highpass filter; you can then use combinations of the two to produce band-reject and bandpass filters. Let us first examine how to convert a lowpass filter into a

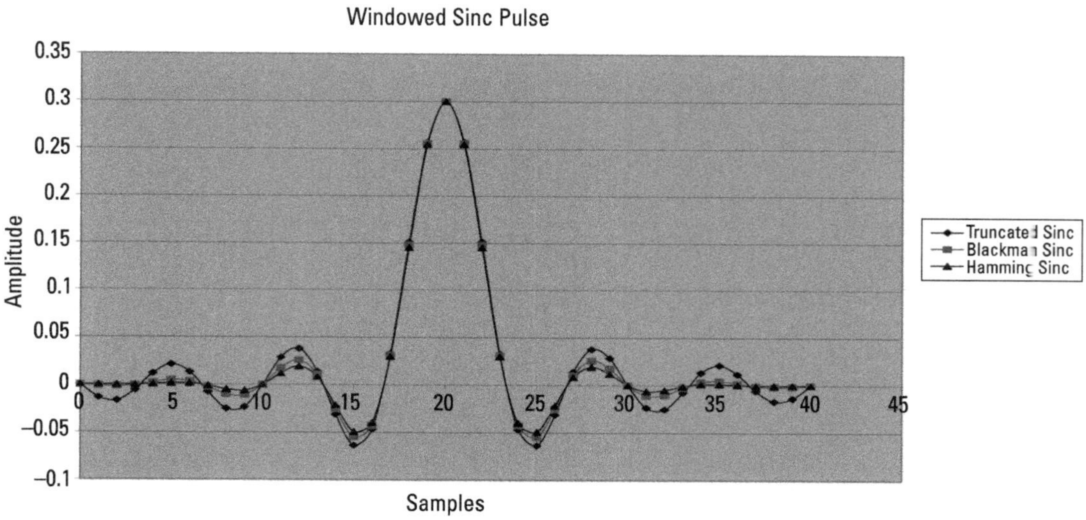

(a)

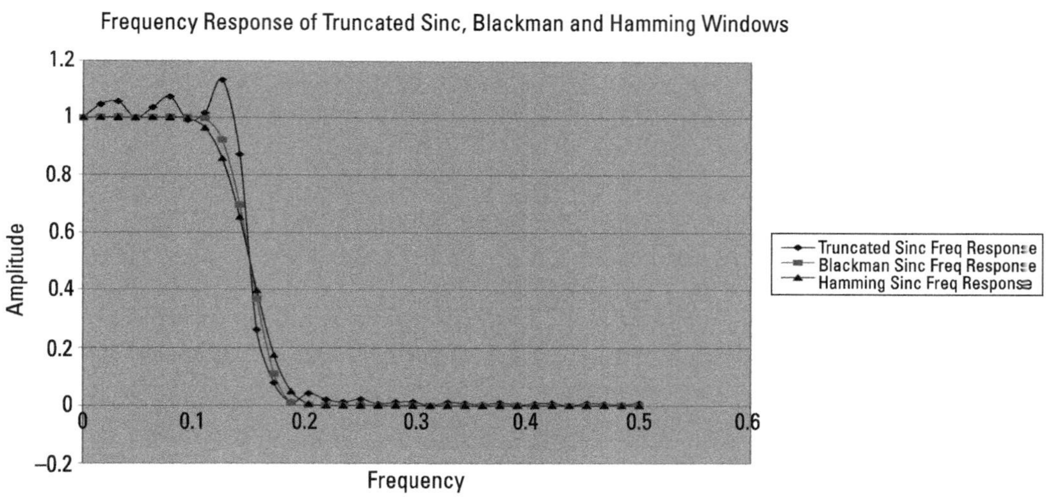

(b)

Figure 3.35 Windowed impulse and frequency responses.

highpass one. The simplest method, called spectral inversion, changes the stopband into the passband and vice versa. You perform spectral inversion by inverting each of the samples, while at the same time adding one to the center sample. The second method of converting to a highpass filter is spectral reversal, which mirrors the frequency response; to perform this operation, simply invert every other coefficient. Having designed the lowpass and highpass filters, it is easy to make combinations that you can use to generate bandpass and band-reject filters. To generate a band-reject filter, just place a highpass and a lowpass filter in parallel with each other and sum together the outputs. Meanwhile, you can assemble a bandpass filter by placing a lowpass and a highpass filter in series with each other.

3.18 Fast Fourier Transforms

The SensorsThink engineering team will also be working with ADCs and filters; several of these may require verification in the frequency domain during testing. As such, it is important that the SensorsThink development team understands how to operate in the frequency domain. For many engineers, working in the frequency domain does not come as naturally as working within the time domain; however, the engineer needs the ability to work within both domains to unlock the true potential of FPGA-based solutions.

3.18.1 Time or Frequency Domain?

As engineers, we can analyze and manipulate signals in either the time or frequency domains, knowing when to do which forms one of the core reasons that an engineer is required on a project.

The time domain enables the engineer to analyze how a signal changes over time. Typically, in electronic systems, the signal in question is a changing voltage, current, or frequency, which has been output by a sensor or generated by another part of the system. Within this domain, we can measure a signal's amplitude, frequency, and period and more interesting parameters such as the rise and fall times of the signal. If the engineer wishes to observe a time-domain signal in a laboratory environment, it is common to use an oscilloscope or logic analyzer.

However, there are parameters of a signal that require analysis within the frequency domain to access the information contained within. These parameters are present within the frequency domain, where we can identify the frequency components of the signal, their amplitudes and the phase of each frequency. Working within the frequency domain also makes it much simpler to manipulate signals due to the ease with which convolution can be performed in the frequency domain. As with the time-domain analysis, if the engineers wish within the laboratory environment to observe a frequency-domain signal, they can use a spectrum analyzer.

For some applications, the engineer will wish to work within the time domain, for example, monitoring systems that monitor the voltage or temperature of a system. While noise may be an issue, taking an average of several samples will in many cases be sufficient. However, for other applications, it is preferable to work within the frequency domain, for example, signal processing applications that require the filtering of one signal from another or to separate it from a noise source.

Working within the time domain requires less postprocessing on the quantized digital signal, while working within the frequency domain first requires the application of a transform to the quantized data to convert from the time domain. Similarly, to output the postprocessed data from the frequency domain, we need to perform the inverse transform again back to the time domain.

3.19 How Do We Get There?

Depending upon the type of the signal, repetitive or nonrepetitive, discrete or nondiscrete, there are several methods that one can use to covert between the time and frequency domains, for instance, Fourier series, Fourier transforms, or Z transforms. Within electronic signal processing and FPGA applications in particular, the engineer is most often interested in one transform, the discrete Fourier transform (DFT), which is a subset of the Fourier transform. The DFT is used by the engineer to analyze signals that are periodic and discrete (i.e., they consist of several *n* bit samples evenly spaced at a sampling frequency, which in many applications are supplied by an ADC within the system).

At its most simple explanation, DFT decomposes the input signal into to two output signals that represent the sine and cosine components of that signal. Thus, for a time-domain sequence of *N* samples, the DFT will return two groups of $N/2 + 1$ cosine and sine wave samples, respectively, referred to as the real and imaginary components as shown in Figure 3.36. The real and imaginary sample width will also be $n/2$ for an input signal width of *n* bits.

The algorithm to calculate the DFT is straightforward; we can use the equation shown here:

$$ReX[k] = \sum_{i=0}^{N-1} x[i]\cos\left(\frac{2\pi ki}{N}\right)$$

$$ImX[k] = \sum_{i=0}^{N-1} x[i]\sin\left(\frac{2\pi ki}{N}\right)$$

where $x[i]$ is the time-domain signal, *i* ranges from 0 to $N - 1$, and *k* ranges from 0 to $N/2$. The algorithm above is called the correlation method, and what it does is multiply the input signal with the sine or cosine wave for that iteration to determine its amplitude.

We will wish at some point in our application to transform back from the frequency domain into the time domain, as such we can use the synthesis equation, which combines the real and imaginary waveforms to recreate a time-domain signal as such:

$$x[i] = \sum_{k=0}^{N/2} Re\bar{X}[k]\cos\left(\frac{2\pi ki}{N}\right) + \sum_{k=0}^{N/2} Im\bar{X}[k]\sin\left(\frac{2\pi ki}{N}\right)$$

However, $Re\bar{X}$ and $Im\bar{X}$ are scaled versions of the cosine and sine waves as such we need to scale them; hence, $Im\bar{X}[k]$ or $Re\bar{X}[k]$ are divided by $N/2$ to determine

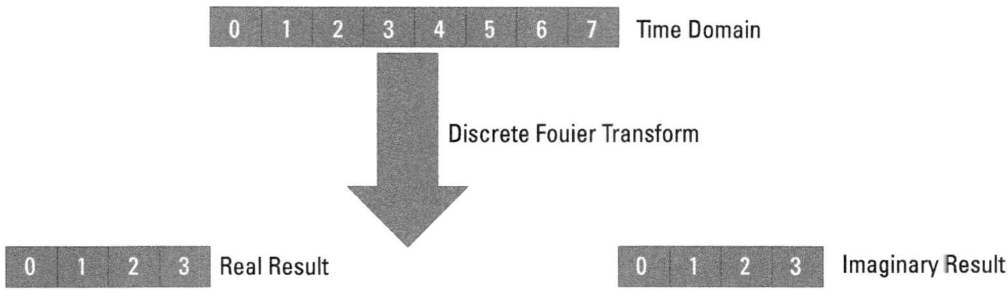

Figure 3.36 N bits in the time domain to n/2 real and imaginary parts in the frequency domain.

the values for $Re\overline{X}$ and $Im\overline{X}$ in all cases except when $Re\overline{X}[0]$ and $Re\overline{X}[N/2]$; in this case, they are divided by N.

For obvious reasons, this is called the IDFT. Having explained the algorithms used for determining the DFT and IDFT, it may be useful to know what we can use these for.

We can use tools such as Octave, MATLAB, and even Excel to perform DFT calculations upon captured data and many lab tools such as oscilloscopes can perform DFT upon request.

However, before we progress, it is worth pointing out that both the DFT and IDFT above are referred to as real DFT and real IDFT in that the input is a real number and not complex; why we need to know this will become apparent.

3.19.1 Where Do We Use These?

From telecommunications to image processing, radar, and sonar, it is hard to think of a more powerful and adaptable analysis technique that we can implement within our FPGA than the Fourier transform. Indeed, the DFT forms the basis of one of the most used FPGA-based applications in being the basis for the generation of the coefficients for the finite input response filter.

However, its use is not just limited to filtering; the DFT and IDFT are used in telecommunications processing to perform channelization and recombination of the telecommunication channels. In spectral monitoring applications, they are used to determine what frequencies are present within the monitored bandwidth, while in image processing the DFT and IDFT are used to perform convolution of images with a filter kernel to perform, for example, image pattern recognition. All the above are typically implemented using a more efficient algorithm to calculate the DFT than that shown above.

All told, the ability to understand and implement a DFT within your FPGA is a skill that every FPGA developer should have.

3.19.2 FPGA-Based Implementation

The implementation of the DFT and IDFT as described above is often implemented as a nested loop each performing N calculations. As such, the time taken to implement the DFT calculation is

$$DFTtime = N * N * Kdft$$

where *Kdft* is the processing time for each iteration to be performed; as such, this can become quite time-consuming to implement. As a result of this, DFTs within FPGAs are normally implemented using an algorithm called the fast Fourier transform to calculate the DFT. This algorithm has often been called the "the most important algorithm of our lifetime" as it has had such an enabling impact on many industries. The FFT differs slightly from the DFT algorithms explained previously, in that it calculates the complex DFT, that is, it expects real and imaginary time-domain signals and produces results in the frequency domain that are n bits wide as opposed to $n/2$. This means that when we wish to calculate a real DFT, we must first address this by setting the imaginary part to zero and moving the time-domain signal into the real part. If we wish to implement an FFT within our Xilinx FPGA, we have two options; we can write one from scratch using the HDL of our choice or use the FFT IP provided within the Vivado IP catalog or another source. Unless there are pressing reasons not to use the IP, the reduced development time resulting from using the IP core should drive its selection and use.

The basic approach of the FFT is that it decomposes the time-domain signal into several single-point time-domain signals. This is often called bit reversal as the samples are reordered. The number of stages that it takes to create these single-point time-domain signals is calculated by Log2 N where N is the number of bits, if a bit reversal algorithm is not used as a short cut.

These single-point, time-domain signals are then used to calculate the frequency spectra for each of these points. This is straightforward as the frequency spectra are equal to the single-point time domain.

It is in the recombination of these single frequency points that the FFT algorithm gets complicated. We must recombine these spectra points one stage at a time, which is the opposite of the time-domain decomposition; thus, it will again take Log2 N stages to recreate the spectra. This is where the famous FFT butterfly comes into play.

When compared with the DFT execution time, the FFT takes

$$FFTtime = Kfft * N \log 2N$$

which results in significant improvements in execution time to calculate a DFT.

When implementing an FFT within our FPGA, we must also consider the size of the FFT; this will determine the noise floor below which we cannot see signals of potential interest. The FFT size will also determine the spacing of the frequency bins; the equation here is used:

$$FFTNoise\ Floor(dB) = 6.02n + 1.77 + 10\log 10\left(\frac{FFTsize}{2}\right)$$

where n is then number of quantized bits within the time domain and FFTsize is the FFT size. For FPGA-based implementation, this is normally a power of 2, for example, 256, 512, and 1,024. The frequency bins will be evenly spaced at

$$Bin\ Width = \frac{\frac{Fs}{2}}{FFTsize}$$

For a quite simple example, an FS of 100 MHz with an FFT size of 128 would have a frequency resolution of 0.39 Hz. This means that frequencies within 0.39 Hz of each other cannot be distinguished.

3.19.3 Higher-Speed Sampling

Many applications for FFTs within higher-performance systems and FPGAs are based upon applications that operate at very high frequencies, which can present their own challenges for the engineer implementing them.

At high frequencies, the Nyquist sample rate (sampling at least 2 samples per cycle) simply cannot be maintained as such a different approach must be undertaken. An example of this would be using an ADC to sample a 3-GHz full-power-bandwidth analog input with a 2.5-GHz sample rate. Using Nyquist rate criteria, signals above 1.25 GHz will be aliased back into the first Nyquist zone to be of use. These aliased images are harmonic components of the fundamental signal and thus contain the same information as the nonaliased signal as shown in Figure 3.37.

To determine the resultant frequency location of the harmonic or harmonic content, you can use the following algorithm:

```
Fharm=N ×Ffund
IF (Fharm=Odd Nyquist Zone)
Floc=Fharm Mod Ffund
Else
Floc=Ffund-(Fharm Mod Ffund)
End
```

where N is the integer for the harmonic of interest.

Continuing our example further with a sample rate of 2,500 MHz and a fundamental of 1,807 MHz, there will be a harmonic component at 693 MHz within the first Nyquist zone, which we can further process within our FFT.

Having explained a little about the frequency spectrum, the way you interface these devices to the FPGA is another important factor, as it is not possible for the data from the ADC to be received at FS/2 where in the example above the sampling frequency is 2.5 Gbps. For this reason, high-performance data converters use multiplexed digital inputs and outputs that operate at a lower data rate with respect to the converter's sample rate typically FS/4 or FS/2.

Having received the data from the FPGA in a number of data streams you may be wondering how you can process the data internally within the FPGA, if you wished to perform a DFT. One common method used for a number of applications, including telecommunication processors and radio astronomy, is to use combined or split FFT structures as shown in Figure 3.38.

While this application is more complicated than a straightforward FFT, such an approach enables the higher-speed processing to be achieved.

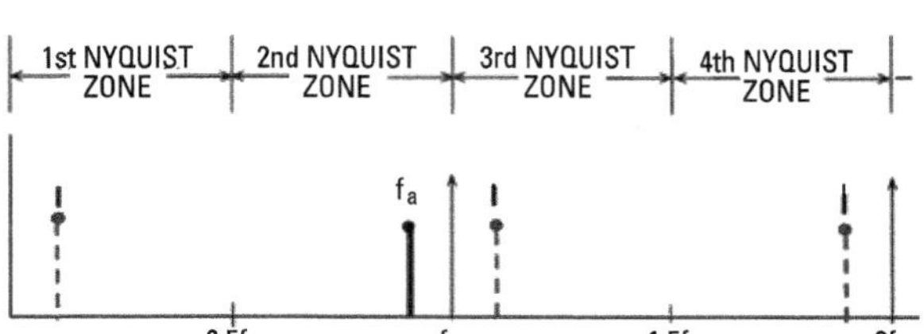

Figure 3.37 Nyquist zones and aliasing.

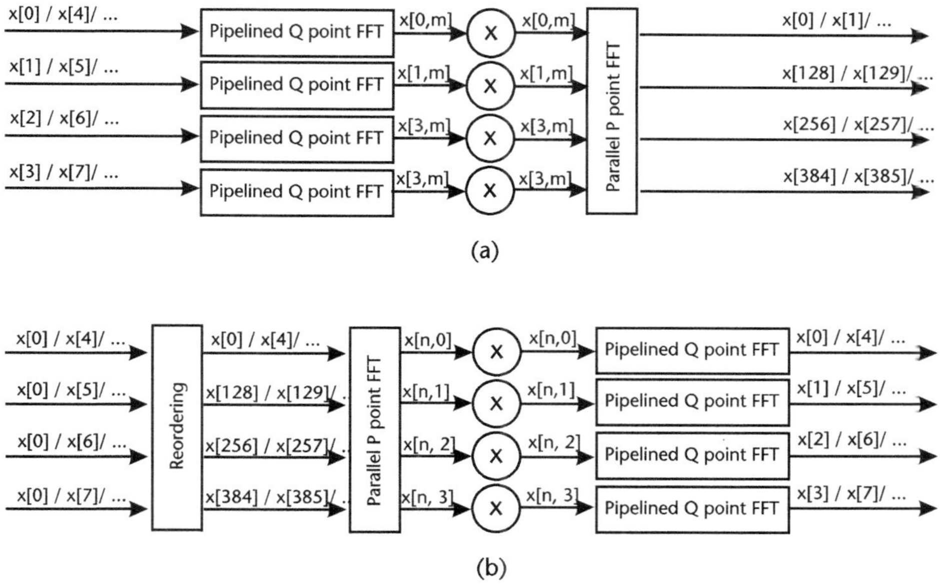

Figure 3.38 Split and combined FFT structures.

3.20 Working with ADC and DAC

Commonly, FPGA-based systems must interface with the real world having performed the task that it was designed to do. Sadly, the real world tends to function around an analog signal as opposed to digital signals; therefore conversion is required to and from the digital domain from the analog domain. As with selecting the correct FPGA for the job, the engineers are faced with a multitude of choices when selecting the correct ADC or DAC for their system. The first thing for the engineers to establish is the sampling rate required for the signal to be converted, as this will drive not only the converter selection but also impact the FPGA selection as well to ensure that the processing speed and logic footprint required can be addressed. As all engineers should be aware, the sampling rate of the converter needs to be at least twice that of the signal being sampled. Therefore, if the engineers require sampling a signal at 50 MHz, the sampling rate needs to be at least 100 MHz; otherwise, the converted signal will be aliased back upon itself and not be correctly represented. This aliasing is not always a bad thing and can be used by engineers to fold back signals into the useable bandwidth if the converter bandwidth is wide enough.

3.20.1 ADC and DAC Key Parameters

ADCs are implemented by many different techniques; however, some of the most common are flash, ramp, and successive approximation.

- Flash converters are known for their speed and use a series of scaled analog comparators to compare the input voltage against a reference voltage and use the outputs of these comparators to determine digital code.

- Ramp converters utilize a free-running counter connected to a DAC. The output of the DAC is used to compare against the input voltage; when the two are equal, the count is held.
- Successive approximation converters are an adaption of ramp converters and utilize a DAC and a comparator against the analog input. However, instead of counting, the SAR converter determines if the analog representation of the count is above or below the input signal, allowing a trial-and-error-based approach to determine the digital code.

DACs are also provided in several different implementations; some of the most common are binary-weighted, R-2R Ladder, and pulse width modulation.

- *Binary-weighted:* These are one of the fastest DAC architectures and sum the result of individual conversions for each logic bit; for example, a resistor-based one will switch on or out resistors depending upon the current code.
- *R-2R Ladder:* These converters use a structure of cascaded resistors of value R-2R. Due to the ease with which precision resistors can be produced and matched, these are more accurate than the binary-weighted approach.
- *Pulse width modulation:* The simplest type of DAC that passes the pulse width modulation waveform through a simple lowpass analog filter. This is commonly used in motor control but also forms the basis for delta sigma converters.

Many manufacturers of specialist devices have developed their own internal conversion architectures that provide the best possible performance in specific areas depending upon the intended use. Each of these has advantages and disadvantages relating to the speed of conversion, accuracy, and resolution of the conversion. As when selecting an FPGA, the engineer will look at the number of IO, IO standards, clock management, logic resources, and memory. Similar parameters are used to compare the performance of ADCs; these are the maximum sampling rate, the signal-to-noise ratio (SNR), SFDR, and the effective number of bits (ENOB). The sampling frequency is simple and is the maximum rate at which the ADC can sample its input. SNR represents the quantization noise; this is assumed to be uncorrelated with the input signal and can be theoretically determined using the equation:

$$\text{SNR} = 6.02N + 1.76 \text{ dB}$$

where N is the resolution.

The actual SNR can be determined during system test by taking an FFT of the output and measuring between the value of the input signal and the noise floor. The SFDR is the ratio between the input signal and the next highest peak; this is usually a harmonic of the fundamental. Normally the SFDR is given in terms of dBc and will degrade as the input signal power reduces. From these measurements of the converter, the engineer can calculate the effective number of bits using the equation:

$$\text{ENOB} = (\text{SNR} - 1.77/6.02)$$

When performing this testing, the engineer must take care to ensure that the FFT used is correctly sized to ensure that the noise floor shown is correct and not inadvertently incorrect due to the size of the FFT selected. The FFT noise floor is given by

$$\text{FFT Noise Floor} = 6.02N + 1.76\text{ dB} + 10\text{ LOG10(FFT Size/–)}$$

These tests should be performed using a single tone test, normally a simple sine wave to reduce the complexity of the output spectrum. For the engineer to get the best results, performing coherent samples of the output is a must. Coherent sampling occurs when there are an integer number of cycles within the data window for this to be correct; then

$$\text{FS/Fin} = \text{Ncycles/FFT}$$

3.20.2 The Frequency Spectrum

As mentioned above, the engineer must be aware of the Nyquist criteria when implementing the system to ensure that the signal is correctly converted or quantized. This means that the signal must be sampled at least twice the maximum frequency of the signal of interest to ensure correct conversion. When signals are sampled outside of this criterion, aliasing will occur, although this can result in unwanted performance if it is not correctly understood. It is also for this reason that ADCs require anti-aliasing filters to prevent signals or noise being aliased back into the quantized signal. However, aliasing can be very useful to the engineer especially if the ADC has a wide input bandwidth. This can allow, with careful consideration, the direct conversion of signals without the need for downconverters; for this reason, the frequency spectrum is divided up to several zones. Using the information presented in Table 3.9, it is possible to alias signals from one Nyquist band to another if the converter has a wide enough bandwidth.

3.20.3 Communication

As with all external devices, ADCs and DACs, interfaces are provided in several different interfacing options. The most common differentiation is the interface being either parallel or serial. Typically, higher-speed devices will implement a parallel

Table 3.9 Nyquist Zones and Aliasing

Nyquist Zone	*Range Lower*	*Range Upper*	*Aliasing*
First	DC	0.5 FFS	None
Second	0.5 FS	FS	Folds
Third	FS	1.5 FS	Direct
Fourth	1.5 FS	2 FS	Folds

interface, while slower-speed devices will utilize a serial interface. However, your choice of application may drive the engineer down a particular route. It is easier, for instance, to detect a stuck at bit in a serial interface than it is a parallel one. High-speed interfaces provide multiple output buses (I and Q) or use double data rate outputs; some devices might even offer both options. This allows the data rate to be maintained while reducing the frequency of operation required for the interface. For example, an interface sampling at 600 MHz will produce an output at 300 MHz (half the sampling frequency); it is much easier to recover if the clock frequency is 75 MHz (FS/4) and there are two data buses that provide the samples from the device using double data rate (DDR), as is the case with the 10D1000ADC from National Semiconductor. This allows the relaxation of the input timings required to be achieved by the FPGA engineer. Many high-speed converters utilize LVDS signaling on their IO as the lower-voltage swing and low current reduce the coupling that can occur using other signaling standards (i.e., LVCMOS), which can affect the mixed signal performance of the converter.

3.20.4 DAC Filtering

Most DACs will hold the analog output until the next sampling period; this results in the engineer observing an interesting effect upon the output frequency domain. They will notice both images existing across the output spectrum and the output signal in the first Nyquist zone exhibiting a roll-off due to the SINC effect being nearly 4 dB (3.92 dB) lower at 0.5 FS. The solution to both issues is to utilize filters; the first filter can be implemented digitally within the FPGA before the DAC and can correct for the roll-off using a simple digital filter, while the anti-image filter needs to be located after the DAC output as the images are a function of the conversion.

The sinc correction filter can be implemented very simply as an FIR filter; the simplest method for the engineer to undertake in developing this filter is to plot the sinc roll-off using the equation:

$$A = \frac{\sin\frac{(\pi * Fin)}{FS}}{\frac{\pi * Fin}{FS}}$$

Create the correction factor that is the reciprocal of those calculated for the roll-off. The engineer can then take an inverse Fourier transform to obtain the coefficients needed to implement the filter. Typically, this filter can be implemented with a few taps. Table 3.10 shows the first 10 coefficients for the filter, while Figures 3.39 and 3.40 show the implementation of the filter in the frequency domain.

3.20.5 In-System Test

Many of these systems will require the converter to achieve specific performance characteristics for the end application (e.g., CDMA or GSM). This can require significant investment in test systems (arbitrary waveform generators, logic analyzers,

Table 3.10 The First 10 Coefficients for a DAC Compensation FIR Filter

Tap	*Coefficient*
1	–6.22102953898351E-003
2	9.56204928971727E-003
3	–1.64864415228791E-002
4	3.45071042895427E-002
5	–0.107027889432584
6	1.166276
7	–0.107027889432584
8	3.45071042895427E-002
9	–1.64864415228791E-002
10	9.56204928971727E-003
11	–6.22102953898351E-003

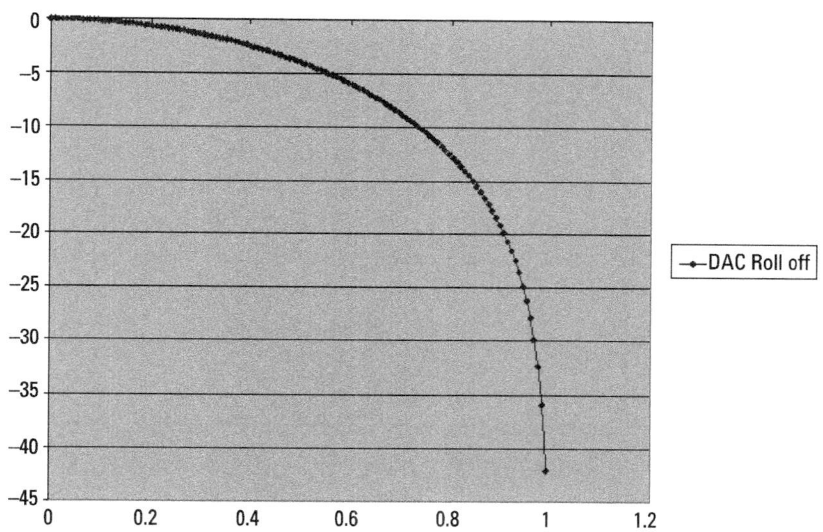

Figure 3.39 DAC roll-off between 0 and FS.

pattern generators, spectrum analyzers) to achieve the desired testing and determine the actual performance. However, the reprogrammable flexibility of the FPGA allows specific test programs to be inserted into the device that either capture and analyze the output of an ADC or allow the stimulus of a DAC reducing hopefully the need for additional extra test equipment.

3.21 High-Level Synthesis

So far, the approaches to logic design for SensorsThink have focused on one of the two hardware description languages that is VHDL. Both hardware description languages, VHDL and Verilog, operate at what is commonly known as the register transfer level (RTL), as the languages are used to create registers and describes the transfer of data between registers. Both developing and verifying RTL designs is

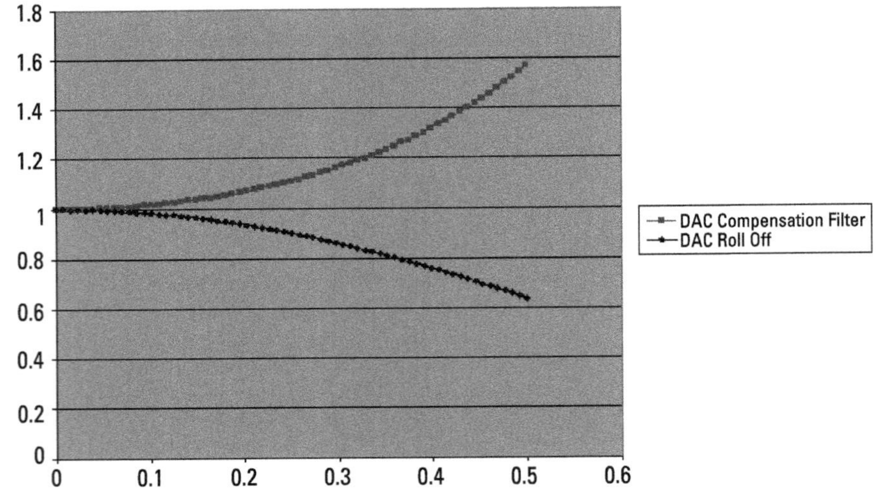

Figure 3.40 DAC roll-off and compensation filter to FS/2.

time-consuming, especially so in verification where the design must be simulated clock cycle by clock cycle.

Increasingly, FPGA developers are using C-based High Level Synthesis (HLS) to accelerate both the design and verification cycles. This approach is something engineers at SensorsThink need to be familiar with C-based HLS translates the C-based design into Verilog and VHDL, which can then be implemented using a traditional implementation flow. Using a C-based design flow enables the developer to focus on the algorithm implementation, while verification can also use untimed C simulation. This allows both the algorithm and its verification to be developed much quicker than using a traditional RTL.

Just like when we develop with Verilog and VHDL, there is a subset of the language that can be implemented, and there is also a similar subset of C/C++ that can be implemented via HLS. The exact details vary from HLS tool to HLS tool; however, obvious elements such as system calls cannot be synthesized.

To convert an HLS design into C, the HLS tool will perform 3 stages of analysis to generate the target RTL.

1. *Scheduling:* In this phase, the untimed C operations are allocated to clock cycles by the compiler. The compiler will consider the desired clock frequency, the cost in time of the operation in the target technology to correctly schedule the operations.
2. *Binding:* Once the operations have been allocated to a clock cycle, the next stage is to assign the operations to resources available within the target device. This could be look-up tables, block RAMS, or DSP elements.
3. *Control extraction:* Once the scheduling and resource allocation has been created, the final stage is for the HLS tool to implement the necessary control structures to enable the block to be started on demand, perform memory accesses and report the result is valid. This stage will typically implement finite state machines to perform the control actions required.

The output of these 3 stages is Verilog and VHDL source code, which can be imported as an IP block into the implementation tool.

However, C does not describe parallel operations as hardware description languages do. To address this, the HLS compiler can be instructed by the developer where parallel structures exist in the source code. The user identifies parallel structures in the code using pragmas; these pragmas are used to control the compilers' implementation of the design. Example pragmas include:

- *Interval:* The number of clock cycles between the synthesized IP module being able to start processing a new input.
- *Unrolling:* Where loops are used to process data elements, by default, HLS tools will keep loops rolled up, ensuring minimal resource requirements. However, keeping loops rolled results in decreased performance as each iteration of the loop uses the same hardware and incrementing a loop counter requires at least one clock. Unrolling a loop results in the creation of hardware for each iteration of the loop significantly reducing the execution time of the loop, while at the same time increasing the required resources. This resource for performance trade-off is one of the key aspects of optimizing HLS designs.
- *Pipelining:* Pipelining is another technique that can be used to increase throughput in your HLS implementation. Pipelining is an alternative to unrolling and enables loops to be pipelined such that a new input can be processed on each clock cycle (other constraints permitting). This does not affect the latency of the pipeline, but it increases the throughput enabling the design to achieve higher performance than a nonpipelined design.
- *Memory partitioning and reshaping:* By default, arrays in C code are synthesized to the block RAMs. Block RAMs can act as bottlenecks when arrays are accessed, preventing unrolling or pipelining with an optimal interval. The bottleneck occurs as block RAMs have limited access ports that prohibit multiple accesses at the same time. Partitioning block RAMs allows a single block RAM to be split into several smaller block RAMs. The HLS tool is then able to arrange the data across the smaller block RAMs to enable multiple accesses in parallel, removing the bottleneck. In the most extreme cases, the block RAM contents can be partitioned into individual registers. Reshaping is like partitioning; however, instead of splitting the words across block RAMS, reshaping changes the size of the words stored and accessed.

Along with the optimization pragmas, HLS compilers can also use pragmas to define the interfacing standard or standards that are used in the synthesized HDL to interface with larger programmable logic design.

HLS is ideal for accelerating the development of signal processing and image processing applications, as engineers developing our IOT sensor application, we may need to perform filtering on sensor data to remove noise; this is an ideal application for HLS.

The HLS development will implement a 16-bit block average; that is, 16 elements will be sampled and the average of the 16 elements will be output. As the

IOT board application contains a Xilinx Zynq SoC, the HLS tool used for the development of this filter module is Vitis HLS.

Creating such an example in HLS is amazingly simple; we can implement this in less than 10 lines of C, as shown in Figure 3.41.

Before we can synthesize the HLS design, we need to know that the C algorithm works as expected; within Vitis HLS, we have two verification methods:

- *C simulation:* The C algorithm is executed as a C program and is tested by a C wrapper. As both elements of the verification are C-based, the simulation time is significantly reduced. C simulation is used to verify the algorithm implementation achieves requirements, before performing synthesis.
- *Cosimulation:* Combines the C test bench with the synthesized RTL to verify the behavior of the generated RTL. Cosimulation therefore uses an RTL simulator to simulate the behavior of the RTL. As this simulator will be clock cycle accurate, cosimulation may take require a longer run time than C simulation.

The C simulation created to verify the average filter can be seen in Figure 3.42; this test bench is very flexible, loading in the verification values from a text file and applying them to the average filter. The resulting average value generated by the average IP core is printed for verification.

Synthesizing this code results in an IP core that can calculate one average every 18 clock cycles and begin to process a new input every 19 clock cycles.

To ensure the most optimal implementation in our FPGA, we can use pragmas to remove potential bottlenecks in the design. Examining the design analysis view in Vitis HLS provided in Figure 3.43, we see first clock cycle used to start the IP core. The main accumulation loop takes 2 clock cycles and is run 16 times. However, as the HLS tool has been able to pipeline the loop access to memory (indicated by the II=1 in the loop descriptor) the loop is able to be completed in only 16 clock cycles, while the final clock cycle is used to output the result of the average considering the accumulated value.

Examining the analysis flow shows that the main bottleneck in being able to increase performance and reduce the number of clock cycles taken is the reading of the 16 input values from the external block RAM.

```
#include "sensor.h"

void sensor_filter( ip_type input[16], op_type *output)
{
int32_t accumulator=0;
accum: for(int i = 0; i<16; i++)
{
accumulator += (int32_t ) input[i];
}
*output = accumulator / 16;
}
```

Figure 3.41 HLS code.

```
#include "sensor.h"

#define MAX_LINE_LENGTH 80

int main()
{
    FILE *fp;
    char line[80];
    ip_type input[16];
    op_type output;

    printf("HLS Example\n\r");
// Open a file for the output results
fp=fopen("input.dat","r");
if (fp == NULL)
printf("error opening file\n\r");
int i = 0;
while (fgets(line, MAX_LINE_LENGTH, fp) != NULL) {
       input[i] = atoi(line); //convert string to integer format
       i++;
}
fclose(fp);
//run the sensor
sensor_filter(input, &output);
printf("Result from HLS module = %d \n\r",output);
}
```

Figure 3.42 C test bench code.

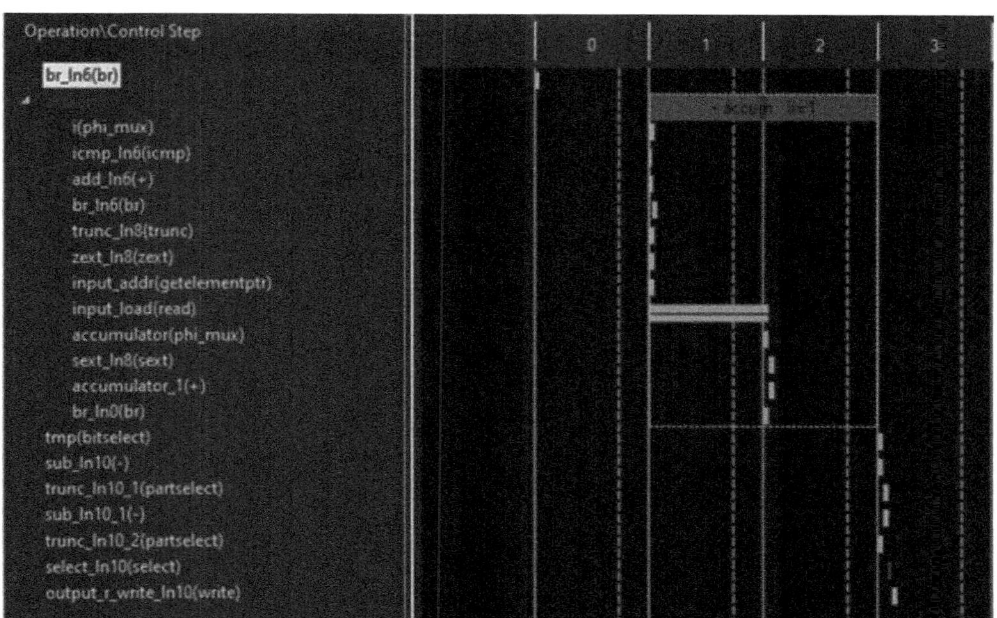

Figure 3.43 Unoptimized HLS implementation timing analysis.

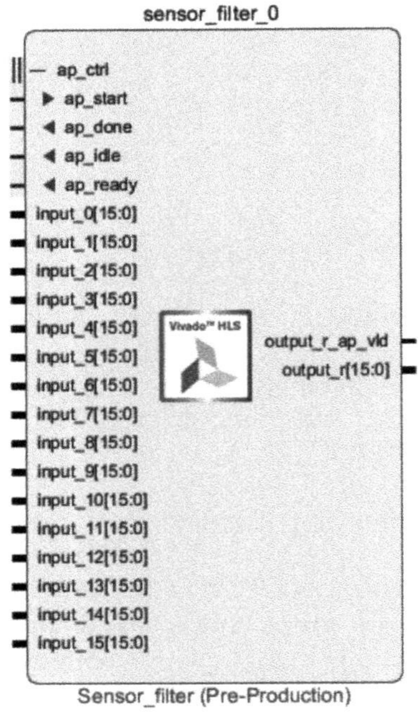

Figure 3.44 HLS Average block IP symbol in Vivado Integrator.

Partitioning the input values to be stored as 16 discrete registers in place of the block RAM enables the loop to be unrolled and significantly reduces the latency.

Implementing the unrolling and partitioning pragmas results in the generated IP core being able to implement a new calculation every clock cycle when the clock frequency is 25 MHz.

This enables us, if necessary, to increase the throughput of the averaging IP core, should we wish to remove noise.

With the performance as desired, the IP block can then be packaged as an IP block for use within our larger IoT FPGA design if desired.

To be able to control and monitor the status of the sensor filter IP block, the HLS tool implements a control port on the synthesized RTL. The ports, idle and ready, are used to indicate that the block is ready for executing a new command. While the done signal indicates the core has completed its operation, the start signal is used to start a new operation. As the sensor filter module was optimized to be able to process a new input on each clock cycle, provided that new data is present on for the 16 averaging sample, the start pin can be tied high.

Should we want to perform more complex interfacing within an FPGA design, we can instruct the HLS compiler to implement AXI interfaces.

High-level synthesis is a tool within the FPGA designers toolbox, and, as we develop our IOT board, we should consider using HLS for signal processing applications as desired.

CHAPTER 4

When Reliability Counts

In the end, a well-designed system is only as good as the reliability it offers. While designers may have worked hard to meet all technical requirements and system specifications, none of it would matter if the system failure occurred as reliability concerns were not addressed.

4.1 Introduction to Reliability

Reliability is defined as the ability of a product to perform as expected over certain time. In the simplest sense, reliability means how long a component or a system will deliver its expected outcome without a breakdown. In mathematical terms, reliability can be defined as the probability that a component or a system will perform its intended function for a stated period under specified operating conditions. To understand this better, let us look at the most important elements that define the reliability of a system:

- Probability is a value between 0 and 1. It is a ratio of the number of times that an event occurs (success or failure) divided by the total number trials (e.g., an embedded system with a failure probability of 0.007 means that only 7 of 1,000 items could fail to perform as stated at any given time under stated conditions).
- Performance is a set of criteria that defines what actions constitute a success and which actions constitute a failure of a component or a system. It can be very detailed or very specific in defining success or failure criterion (e.g., an embedded system with 4 sensors, a microcontroller, and a memory unit could be considered a failure if any one of the components fail or the same system could be considered as working if at least one of the sensors is working).
- Time duration for a working system. The probability of survival or failure is generally given as a function of time.
- Operating conditions such as environmental factors, humidity, vibration, shock, temperature cycle, and operational profile have an impact on the working of a system and the duration of a stated product life.

Over the past few years, penetration of embedded systems in all aspects of human life has put more focus on the importance of incorporating reliability aspect in design practices. While the issue of reliability is almost a certainty when starting any new project, the emphasis laid on reliable design could vary greatly from project to project. In some projects, reliability aspect during the system design process is given far greater importance than some other considerations, such as time to market. In general, reliability is given the highest consideration in industries where system failure could lead to loss of life (e.g., automobile or aeronautical industries) or a severe impact on the company's reputation (e.g., financial systems or large-scale factory automation). However, there are several other major industrial sectors where reliability as a design aspect is not given a similar importance as time to market. One good example is the present-day smartphone industry. Given the nature and dynamics of the market and the fast pace of technology, if the designers test the devices for all possible reliability concerns over a period of time, the technology used in the smartphone model could be out of date even before it hits the market. Keeping this in mind, designers must understand what level of emphasis is required for addressing reliability in the project that they have undertaken and what industry or end user it will serve.

While it is easy to recognize the importance of reliability in mass-produced systems that could lead to loss of life (such as automobiles and aircrafts), the reliability of a system generally has a far greater financial impact on the brand and profitability of companies engaged in other electronics industries such as mobile devices or consumer electronics.

The emphasis on reliability of a component varies greatly from one industry to another. A component deployed in a product used in the smartphone industry could have a completely different requirement for reliability in as compared to the same component being used in another industry (e.g., aerospace). As shown in Table 4.1, errors can occur during any stage of the design life cycle and can have various degrees of impact on the reliability of a system. Therefore, it is important to study reliability from a technical point of view. Before we dive into the technical definitions of reliability and several other parameters associated with it, let us consider a few scenarios to emphasize the importance of reliability in the design process.

- The component that you have designed would be used in a mass-produced automobile and you have estimated that the design failure rate is just 1 in 1,000 (99.9% reliable). Imagine that for each 1 million vehicles sold with that component, 1,000 vehicles could potentially break down and cause an accident. This automobile failure rate is just due to one component (designed by you). The overall number for system reliability will be much lower when the reliability of each individual component is considered, assuming that the average life span of an automobile is between 10 and 15 years.
- Let us now consider a case of an aircraft where 100,000 individual components, each with an average failure rate of 1 in 10,000 (99.99% reliability), are used. Let us further assume that 1,000 components can cause critical failure. In this case, the overall system failure probability can be calculated as $(0.9999)^{1,000} = 0.905$ or 90.5%. Now imagine a scenario, if this were to

Table 4.1 Different Sources of Error in Various Design Phases

Design Stage	*V-Model*	*Potential Error Source*	*Error Detection*
Specifications and design	Level 10, Level 20	Algorithm design, formal specification	Consistency checks, simulation
Prototyping	Level 30, Level 40	Algorithm design, wiring and assembly, timing component failure	Stimulus/response testing
Manufacturing	Level 20	Wiring and assembly, component failure	System testing, diagnostics
Installation/unit testing	Level 20	Assembly, component failure	System testing, diagnostics
Field operation/system validation	Level 10	Component failure, operator errors, environmental factors	Diagnostics

really happen, would mean that 1 of 10 aircrafts would have an accident. While there is not much that can be done to reduce the number of critical components, there is certainly a lot that can be done to increase the reliability of individual critical components that would improve the overall system reliability, even if it take a lot more years of testing, especially considering that the average life of an aircraft can be 25 to 40 years.

- Finally, let us consider a case for another mass-produced device that contains several individual components, a smartphone. Assuming that a smartphone consists of 1,000 different components and half of them are critical, in this case, the overall system failure probability can be calculated as $(0.9999)^{500}$ = 0.9512 or 95.12%. Since a smartphone's failure is not considered as critical as that of an automobile or aircraft, manufacturers would be able to digest those numbers in their business model if it helps them to reduce the time to market. Considering the fact that the latest smartphones have, at best, 1 or 2 years of life span before being replaced, it would not justify years of testing to improve those system reliability numbers while risking the technology to get older by the time it is ready to be sold.

Having considered system-level reliability for mass-produced products, it should be emphasized that sometimes it is smarter to consider the expected market size of the product while finalizing design considerations. If a design is proven to be 99.99% reliable over its useful life and there will be less than 10,000 such products being built, then there is probably not much use to improve the reliability any further.

4.2 Mathematical Interpretation of System Reliability

As defined by the IEEE, "Reliability is the ability of an item (or component) to perform its required functions under stated conditions and for a specified period." Similarly, reliability of a system (or subsystem) can be defined as the probability of survival during the effective life span of the system. Since a system (or subsystem) consists of one or more critical components, reliability of the entire system (or subsystem), $R_S(t)$, is calculated from the individual probability of failure (reliability) of each critical component that forms a part of the system (or subsystem). In mathematical terms, $R_S(t)$ can be calculated as follows:

$$R_S(t) = Prob(t_o \leq t \leq t_f \forall f \in F)$$

Here t_o is the time at which the system is introduced into service and t_f is the time at which the first critical fault f takes place. f belongs to a set of critical faults F that can cause the system to fail. The failure probability $Q_S(t)$ of a system is complementary to the reliability $R_S(t)$ of the system. Hence, failure rate (or failure probability) of a system can be calculated by using the relation:

$$Q_S(t) + R_S(t) = 1$$

As mentioned above, each system (or subsystem) consists of one or more individual critical components and the reliability of the system $R_S(t)$ is dependent on the individual reliability of each critical component. Let us denote the reliability of an individual critical component as $R_C(t)$. Using the same formula as above, the failure rate (probability) for a component $Q_C(t)$ is complementary to the reliability $R_C(t)$ of the component and can be calculated by using the relation:

$$Q_C(t) + R_C(t) = 1$$

The failure rate $Q_C(t)$ generally varies throughout the life of the component; hence, it is not very useful to calculate the reliability. Failure of a system can be defined as a moment of time at which the system stops to fulfill its specified function. Mathematically, failure rate of a system is defined as:

$$Failure\,Rate = \frac{Total\,number\,of\,items\,failed}{Total\,Operating\,Time}$$

4.2.1 The Bathtub Curve

The failure rate is of a system or a component is expected to vary over the expected life of a product and is typically represented by a bathtub curve as shown in Figure 4.1. The bathtub model depicts the failure rate of a component or a system during three key phases: burn-in, useful life, and phase-out.

It is a standard industry practice to use the bathtub model for your system and use it as a reference for your failure calculations. A summary of failure rate causes and possible remedies during different phases of the bathtub model is presented in Table 4.2.

4.2.2 Failure Rate (λ)

In most practical systems, the failure rate is expressed using a different variable (λ) that represents failure rates and how many failures occur every 10^9 hours. λ is

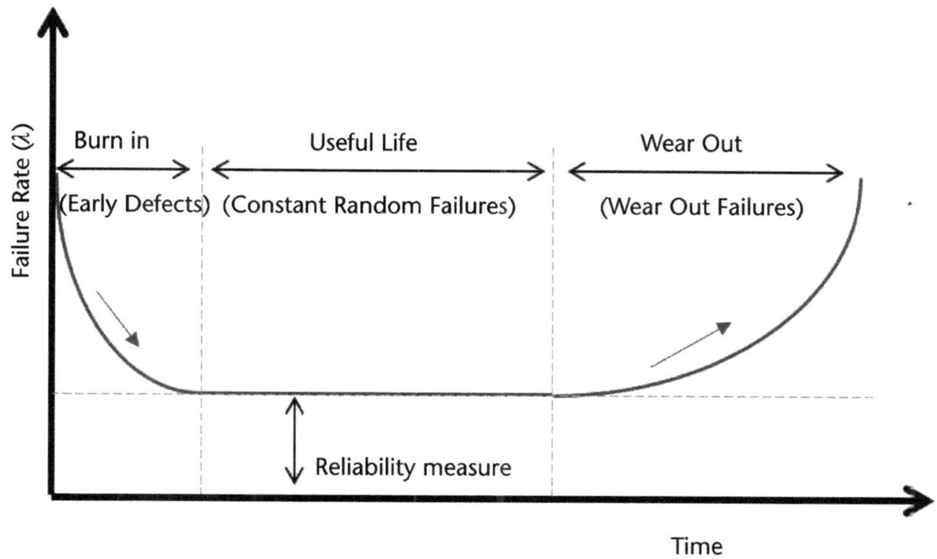

Figure 4.1 The bathtub curve (also known as the component reliability model).

Table 4.2 Summary of Failure Rates and Possible Causes During Different Phases of Bathtub curve

Phase	*Failure Rate*	*Possible Causes*	*Possible Improvement Actions*
Burn-in	Decreasing	Manufacturing defects, soldering, assembly errors, part defects, poor QC, poor design	Better QC, acceptance testing, burn-in testing, screening, highly accelerated stress screening
Useful life	Constant	Environment, random loads, human errors, random events	Excess strength, redundancy, robust design
Wear-out	Increasing	Fatigue, corrosion, aging, friction	Derating, preventive maintenance, parts replacement, better material, improved designs, technology

assumed to be constant during the useful life of the component as shown in Figure 4.1 and is generally expressed as failure in time (FIT). For example, if there is an occurrence of 10 failures for every 10^9 hours, then λ will be equal to 10 FIT. The value for λ can also be calculated using the formulas given below with the MTBF or MTTF shown in the reliability data. By making use of λ, we can model the reliability of a component using an exponential distribution.

$$R_C(t) = e^{-\lambda t}$$

Example 4.1

During the trial of a key component for the SoC Platform project, sensor company SensorsThink observed that that 95 components failed during a test with a total operating time of 1 million hours (total time for all items: both failed and passed). Calculate the reliability of the key component after 1,000 hours (i.e., t = 1,000).

Solution:
Failure rate is given by:

$$Failure\ Rate = \lambda = \frac{Total\ number\ of\ failed\ components}{Total\ Operating\ Time}$$

$$\therefore \lambda = \frac{95}{1{,}000{,}000} = 9.5 \times 10^{-5}\ per\ hour$$

$$Reliability\ of\ a\ component\ at\ time\ t = R_C(t) = e^{-\lambda t}$$

$$R_C(1{,}000) = e^{-(9.5\times10^{-5}\times1{,}000)} = 0.909$$

Therefore, there is a 90.9% probability that the designed component for the sensor platform project will survive after 1,000 hours.

4.2.3 Early Life Failure Rate

Early defect rate or early life failure rate (ELFR) of a product is calculated during product qualifications or as part of ongoing product reliability monitoring activities. ELFR measures the reliability of the product during performance testing phase during the first few months in the field. As λ is assumed to be a constant and is generally expressed in FIT, some texts also find it acceptable to express ELFR in terms of FIT. However, it is often desired to express the ELFR in parts per million (PPM), since PPM is a measure of the cumulative fraction failing per device, whereas FIT is a measure of fraction failing per device-hour. Therefore, it is a common practice to specify the early life time period (t_{ELF}), whenever ELFR is expressed in PPM. The value of t_{ELF} is generally defined by the user or the supplier. The value of ELFR can be calculated using the following equation:

$$MTBF = \int_0^\infty R(t)\,dt = \frac{1}{\lambda}$$

ELFR provides a good indication of reliability during performance testing phase; however, once the critical issues have been identified and addressed, it is the failure rate (λ) that is usually considered to estimate system reliability.

As design engineers, ELFR can be regarded in two contexts. The first context is the components that the design engineers are choosing to build their own prototype. In this context, EFLRs of those components are useful only to estimate the overall reliability of the system being developed. In the second context, the design engineers have to understand that the system that they are developing will have its own bathtub curve and hence its own three phases: burn-in (ELFR), useful life, and phase-out. Ideally, the designers would like to minimize the early life time period and achieve a constant failure rate (λ) as soon as possible.

The reliability of a system is a study of factors that lead to its failure, and it is essential to provide this feedback to the designers and production engineers in the early stages of design prototyping and production so that they can make suitable adjustments to prevent future failures. It is obvious that the final onus on causes of failure thus lies with either the designer or the production engineer, which may not be welcomed by either of them unless a robust classification of failure types is already agreed upon. When speaking about the reliability of a system, we end up talking more about the following three aspects: failure causes, failure modes, and failure mechanisms.

4.2.4 Key Terms

Before we go any further, it is important for the readers to understand some standard terms and definitions that are used quite frequently in the context of reliability.

- *Failure:* A failure is an event when an item is not available to perform its function at specified conditions when scheduled or is not capable of performing functions to specification.
- *Failure rate:* The number of failures per unit of gross operating period in terms of time, events, and cycles.
- *Hazard:* The potential to cause harm, including ill health and injury, damage to property, plant, products, or the environment, production losses, or increased liabilities.
- *Risk:* The likelihood that a specified undesired event will occur due to the realization of a hazard by or during work activities or by the products and services created by work activities.
- *Mean time between failures (MTBF):* The average time between failure occurrences. For repairable items, MTBF is calculated using number of items and their operating time is divided by the total number of failures.
- *Mean time to failure (MTTF):* The average time to failure occurrence. For repairable items and nonrepairable items, MTTF is calculated using number of items and their operating time divided by the total number of failures.
- *Mean time to repair (MTTR):* The average time to restore the item to specified conditions.
- *Maintenance load:* The repair time per operating time for an item.
- *Maintainability:* A characteristic of design, installation, and operation, usually expressed as the probability that an item can be retained in or restored to a specified operable condition within a specified interval of time when maintenance is performed in accordance with prescribed procedures.
- *Availability:* A measure of the time that a system is operating versus the time that the system was planned to operate. It is the probability that the system is operational at any random time t.

While it is important for all embedded system design engineers to know these terms and their importance while working on a new system design, it is also important to understand that most big companies have a separate reliability and fault testing division to undertake such studies. This is mainly because there is an inherent conflict of interest between designers and testers, even though they both are aiming towards the same goal of providing the best product to their clients or customers.

However, for smaller and mid-sized companies, it may not be financially viable to maintain a completely different departed dedicated for reliability testing; hence, the risk of conflict of interest is managed using other operating protocols.

4.2.5 Repairable and Nonrepairable Systems

In order to provide a robust classification of failure of a system, there also needs to be an agreement on the type of system being designed. In general, there can be two types of systems: nonrepairable systems and repairable systems.

A system that has been classified as nonrepairable is one that leads to the failure of the entire system even if one of the key (critical) components fail. For example, if a system has 5 temperature sensors and a malfunction in any one of the sensors constitutes a system failure, even if the faulty sensor could be easily replaced, then such a system can be considered as nonrepairable. However, if the agreed definition of system failure is based on the fact that at least 3 of 5 sensors should be working at any given time, then the system could be classified as repairable and the faulty sensor components can be replaced easily. The key characteristics of repairable and nonrepairable systems are listed in Table 4.3.

4.2.6 MTTF, MTBF, and MTTR

MTTF is a term used for nonrepairable systems and is defined as the expected value of t_f. In other words, it is the expected time of first failure in a nonrepairable system and can be calculated using:

$$MTTF = E(t_f) = \int_0^\infty R(t)\,dt = \frac{1}{\lambda}$$

MTBF is a term used for repairable electronic systems and is defined as the expected time between two failures of a repairable system as shown in Figure 4.2. It is calculated in the same way as MTTF, with an assumption that the system failure is repairable.

$$MTBF = \int_0^\infty R(t)\,dt = \frac{1}{\lambda}$$

MTTR is a term used in the context of repairable systems and is the amount of time required to repair a system and make it operational again.

Table 4.3 Key Reliability Characteristics for Repairable and Nonrepairable Systems

	Nonrepairable Systems	*Repairable Systems*
Availability	Function of reliability	Function of reliability and maintainability
Key Factor	Failure rate	Failure rate and repair rate
Time to Failure	MTTF	MTBF
Expected Life	Mean residual life (MRL)	MRL (Economic justification)

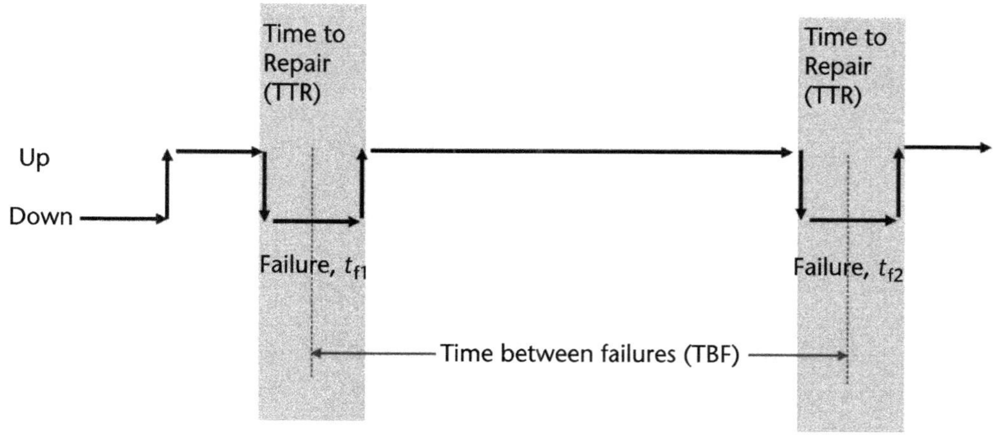

Figure 4.2 MTBF for repairable systems. MTTF, which is a characteristic of nonrepairable systems, is a special case of this plot that calculates the first instance of failure (tf1).

$$MTTR = \frac{Total\,maintanance\,downtime}{Total\,number\,of\,failures\,(or\,maintenance\,actions)}$$

While calculating both MTBF and MTTR, it is safe to assume that that there are more than one failure and repair during the operational life of the system.

Example 4.2
Let us assume that the SoC Platform project by Sensor Company SensorsThink will eventually be deployed as a subsystem to detect fuel efficiency in a car. The prototype was tested in 200 cars for a total of 28,000 hours. Overall, 7 failures were observed. Calculate the MTBF and failure rate.

Solution:
Assuming that the overall car system is a repairable system (since it is highly unlikely that someone will change the car due to a fuel efficiency detector failure), we can use MTBF

$$MTBF = \frac{1}{\lambda} = \frac{Total\ operational\ time}{Total\ number\ of\ failures} = \frac{28{,}000}{7}$$

$$\therefore MTBF = 4{,}000\ hours$$

$$Average\ Failure\ Rate(\lambda) = \frac{7}{28{,}000} = 0.00025\ per\ hour$$

Example 4.3
Let us consider a scenario where SoC Platform project by sensor company SensorsThink is being deployed as the primary control system to detect ignition temperature at a chemical plant. Five such systems were tested in with failure hours of 157, 201, 184, 239, and 225. Calculate the MTTF and failure rate.

Solution:

Considering the primary control system used in the chemical plant as a nonrepairable system (needs replacement), we use MTTF:

$$MTTF = \frac{1}{\lambda}\frac{(157+201+184+239+225)}{5}$$

$$\therefore MTTF = 201.2\ hours$$

$$Average\ Failure\ Rate(\lambda) = \frac{5}{(157+201+184+239+225)} = 0.00497\ per\ hour$$

$$\therefore MTTF = 1{,}125\ hours$$

4.2.7 Maintainability

The maintainability of an embedded system is the measure of the ability of the system or component to be retained or restored to a specified condition when maintenance is performed by qualified personnel using specified procedure and resources. It is an important parameter to be considered during the design phase as low maintenance leads to both reduced downtime as well as reduced costs. Maintainability also helps one determine the level of fault tolerance and redundancy techniques required for a certain system. Maintainability is measured by MTTR.

$$MTTR = \frac{\sum_{i=1}^{n}\lambda_i Rt_i}{\sum_{i=1}^{n}\lambda_i}$$

Here n is the number of systems available and λ_i is the failure rate of ith system. Rt_i is the repair time for the ith system. The probability of performing a maintenance action within an allowable time interval can be calculated using:

$$M(t) = 1 - e^{-t/MTTR}$$

where $M(t)$ is the probability of completion of repair within time period t.

Example 4.4

Let us assume that sensor company SensorsThink has ordered 4 prototypes from the SoC Platform project that will be used for solar panel direction control systems for maximizing the angle of incidence of solar rays on the panel. During testing, it was found that one of the components on each the control systems fail at 1,100 hours, 1,150 hours, 1,050 hours, and 1,200 hours, respectively. The system could be repaired by replacing the faulty component, and the total time to repair all 4 systems was 48 hours. Calculate the time to failure of this system if it was:

1. Deployed in a farm house to provide renewable energy to a small farm;
2. Deployed on a mini satellite to control the solar panels.

Solution:
MTTR for this system = 48/4 = 12 hours.

1. In this type of application, the solar panel controller can easily be classified as a repairable system; therefore, we use MTBF

$$MTBF = \frac{1}{\lambda} = \frac{Total\ Operational\ Time}{Total\ Number\ of\ Failures} = \frac{(1{,}100 + 1{,}150 + 1{,}050 + 1{,}200) - 48}{4}$$

$$\therefore MTBF = 1{,}113 \text{ hours}$$

2. In this application, the system would be classified as nonrepairable (as it is practically impossible to repair or replace such systems in space); hence, we make use of MTTF:

$$MTTF = \frac{1}{\lambda} = \frac{Total\ Time}{Total\ Number\ of\ Failures} = \frac{(1{,}100 + 1{,}150 + 1{,}050 + 1{,}200) - 48}{4}$$

$$\therefore MTTF = 1{,}125 \text{ hours}$$

As seen in this example, the average time to failure of a system can vary significantly based on whether it is classified as a repairable system or a nonrepairable system.

Example 4.5
Let us consider a scenario where SoC Platform project requested by sensor company SensorsThink will be deployed as a subsystem in an automatic airflow control system, developed for an industrial warehouse, with an MTTR of 5 hours. What is the probability that the system can be repaired within 3 hours?

Solution:
Assuming that the sensor system used in the chemical plant is a repairable system, the probability of completion of repair within 3 hours can be found using:

$$M(t) = 1 - e^{-t/MTTR}$$
$$= 1 - e^{-3/5}$$
$$\therefore M(t) = 1 - 0.549 = 0.451$$

Therefore, there is a 45% probability of repairing this system within 3 hours.

4.2.8 Availability

Availability of an embedded system is a measure of the time during which the system is operational versus the time that the system was planned to operate. It is the probability that the system is operational at any random time t. The availability of a system is very critical measure for repairable systems as it determines the overall operating efficiency and costs of the bigger system. For example, an airplane with an availability of 50% would take twice the number of years to just break even the cost of purchase and it could also result in higher cost expenditure with regard to buying more planes to serve the same amount of traffic. There are three common measures of availability: inherent availability, achieved availability, and operational availability

However, in the context of this book, we will just cover inherent availability, as there are other texts available to study this topic in more detail.

The inherent availability (A_i) is one of the most common measures to calculate the availability of a system. It does not consider the time for preventive maintenance and assumes that there is a negligible time lag between the time of failure and the start of repair. For repairable systems, it is defined as follows:

$$A_i = \frac{\mu}{\lambda + \mu} = \frac{MTBF}{MTTR + MTBF}$$

where μ is the repair rate (=1/MTTR) and λ is the failure rate (=1/MTBF).

As evident in the examples given above, classification of a system as repairable or nonrepairable depends on several parameters and scenarios. Moreover, there are multiple parameters that define the reliability of the system. Some of these parameters are more important than others depending on the application and the environment in which the sensor system is used.

Example 4.6

The sensors company SensorsThink develops an SoC Platform to be used within a gas detection system developed for a food storage warehouse, which has an MTTR of 15 hours and an MTBF of 3,050. What is the inherent availability of the system?

Solution:

$$\begin{aligned} A_i &= \frac{MTBF}{MTTR + MTBF} \\ &= \frac{3{,}050}{15 + 3{,}050} = 0.9951 \end{aligned}$$

Therefore, the availability of this system is 99.51%.

4.3 Calculating System Reliability

Like the bathtub curve shown in Figure 4.1 for key components of the system, it is important to plot a bathtub curve for the embedded system that you have designed, which will include one or many such key (critical) components. Once designers have enough information to plot the bathtub curve for the system being designed, they have a powerful visualization tool available to estimate and predict system failure probability during different phases of the design cycle. The bathtub curve is also a good indicator of the useful life of the product that you are developing. As mentioned earlier, the useful life of a product is the time during which the system operates with a constant failure rate. Useful life is an important measure as it determines the cost of the system and the frequency of replacements as well as the duration of warranty. It is vital to understand that the product reliability (for the embedded system that you have designed) will vary greatly on the following factors:

1. Total number of components in the system including both critical and noncritical components;
2. Total number of critical components in the system;
3. The configuration in which components have been placed during the design and production phase;
4. Criteria that determines what constitutes a system failure.

Keeping the above factors in mind, designers need to come up with a bathtub model for the designed system that can capture the most basic estimate of overall system reliability. Even a simple embedded system can have hundreds, if not thousands, of individual components, making it an onerous task to calculate system reliability. At the same time, it is almost impossible to design a system and sell it to your client without having even a rough estimate of system reliability and the components that have the highest probability of failure and if the reliability of the system can be improved by clever design techniques, good programming, or providing fault tolerance. Let us try to understand the effect of these factors on the reliability of the system SoC Platform that you have been asked to design.

The first task is to determine the total number of components used in your design. The sensor platform project requested by SensorsThink, in Chapter 2, has a total 941 individual components. Trying to find the individual failure rate of each of those 941 components and then trying to come up with a single mathematical formula for system reliability that includes the individual failure rate of all 941 components is an almost impossible exercise. However, a good design engineer still needs to know all the components used in the system design. The more important task is to understand which of the components are critical and noncritical.

We have provided a technical definition of the key component for a sensor platform project requested by SensorsThink in Chapter 2. In simple terms, a critical (key) component of an embedded system can be defined as a piece of hardware or software, the failure of which would cause the system to fail altogether or seriously affect its capability to perform the desired task. A good example of a critical component would be a microprocessor chip in a laptop or a power supply unit. A noncritical component are those components that could potentially affect the performance of the system, but would not lead to critical failure. For example, a microphone or webcam on a laptop could potentially have a serious effect on user experience but would not probably lead to replacement of the laptop. There are many components, in each system, that could be labeled as critical or noncritical depending on how they are designed and how they are used. For example, a system designed to handle only one high capacity memory chip would fail if the memory device failed. However, if the same system can handle multiple chips, each with a lower capacity, it can continue to work at a reduced efficiency if one or more memory chips fail.

Before designers can label an individual system component as critical or non-critical, they must first answer what the meaning of system failure is in the context of the embedded system being designed. Would the failure of a webcam in a laptop be considered as a critical system failure? The answer to that question is not with the designer, but with the user. If the laptop is being used for general browsing or as a standalone PC, then a webcam failure may not be considered as a critical failure, as there a number of good external webcams available to purchase at a reasonable cost. However, if the laptop is used to provide face-to-face online support service in a particular industry or to record online lectures on the go for live web-based delivery, then even a simple webcam failure could make the laptop worthless to its user and hence that failure will constitute a critical failure. In the context of the sensor platform project we designed in earlier chapters, it would be worthwhile to provide a definition of critical failure. For example, if you have only one sensor malfunctioning of 10 different sensors in your system, would that constitute a system failure? To keep it simple, we can assume that all key components in the sensor platform project are critical and even a single component failure will lead to system failure.

Once the designer has identified a minimum possible set of critical components of the system, it is quite possible to reduce the complexity of the equation that determines system reliability. For the sake of simplicity, let us assume that there are only n critical components (of a total of 941) in the system SoC Platform. We need to calculate $R_S(t)$, which is the overall reliability of the system SoC Platform. Let us assume that and the failure rates of all n critical components in the system SoC Platform are statistically independent of each other. Furthermore, we also assume that there is no redundancy of fault tolerance techniques deployed in the system SoC Platform. Let us see how we can use this to determine system reliability under the following scenarios.

4.3.1 Scenario 1: All Critical Components Connected in Series

In this case, the critical path of the design passes through all the critical components in a sequential manner (Figure 4.3). Assuming that $R_{Ci}(t)$ is the reliability of an individual critical component and i, from the equation of reliability given earlier:

$$R_{Ci}(t) = e^{-\lambda_i t}$$

As the failure rates of all individual critical components are statistically independent, the overall system reliability in this case will be a product term of n individual critical component reliability $R_S(t)$

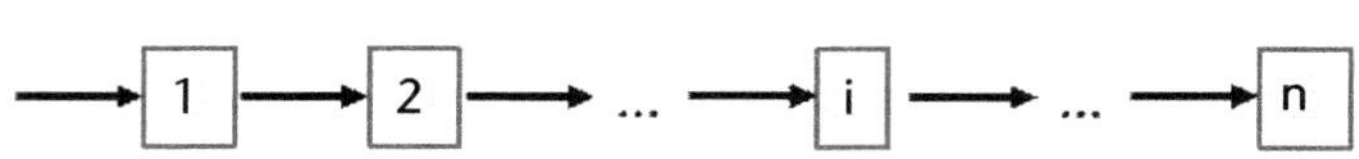

Figure 4.3 Critical path of a system if all key components are connected in series.

$$R_S(t) = R_{C1}(t) \times R_{C2}(t) \times R_{C3}(t) \times \ldots \times R_{Cn}(t)$$

$$\Rightarrow R_S(t) = \prod_{i=1}^{n} R_{Ci}(t)$$

The overall system reliability can therefore be represented in terms of the individual failure rate (λ_i) of each component:

$$R_S(t) = e^{-t\left(\sum_{i=1}^{n}\lambda_i\right)} = e^{-t\lambda_S}$$

where λ_S is the overall serial failure rate and can be represented by:

$$\lambda_S = \sum_{i=1}^{n} \lambda_i$$

A special case of series connection can be if the failure rates of critical components are not statistically independent. In that case, we need to derive a failure rate equation based on the subset of critical components that form the critical path.

In such a case, the failure probability of the system can be found by determining the subsets of critical path components, whose functioning ensures the functioning of the system even if other critical components outside the critical path fail. For example, in Figure 4.4, we can have four such critical paths: {1,2,6}, {1,2,4,5}, {3,4,5}, and {3,6}. However, this is a special case scenario and most likely results in system malfunction whenever the faulty critical path is chosen and properly functioning otherwise. In a way, it will again come down to how we define system failure and critical components.

Example 4.7

Consider a sensor system that has 50 critical components connected in series. If the reliability of each component is 0.999, what will be the overall system reliability?

Solution:

Since the components are connected in series, the system will operate or fail if any one of the critical components fail. The reliability of the system is given by the equation:

$$R_S(t) = \prod_{i=1}^{n} R_{Ci}(t)$$

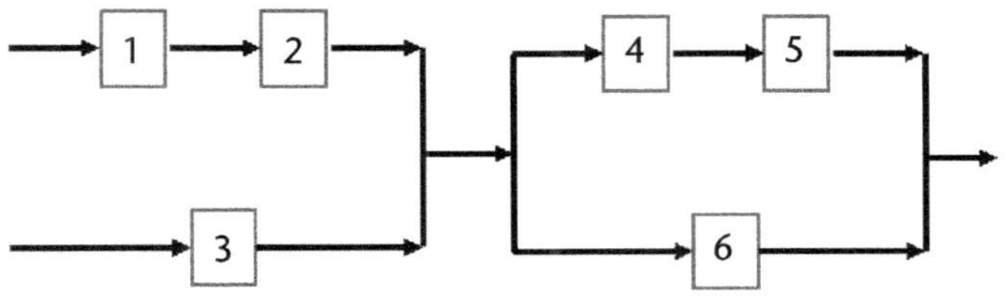

Figure 4.4 Critical path of a series system with dependent critical components.

Here, individual reliability for all critical components is the same; therefore, the equation reduces to:

$$R_S(t) = (R_C(t))^n$$

Plugging the values for $R_{Ci} = (t) = 0.985$ and $n = 50$, we get

$$R_S(t) = (0.999)^{50} = 0.9512$$

The probability of all 5 components failing is $(0.02)^5 = 3.2 \times 10^{-9}$. Therefore, the reliability of the system is RS = $1 - 3.2 \times 10^{-9} = 0.9999999968$

Example 4.8

Let us take an example of the embedded system design for SensorsThink and determine the reliability of the system being developed after 5 years. Refer to Chapter 2, where we performed component selection and listed the key components used in the design. For the sake of simplicity, we will choose only a subset of key components used in the design.

C1. SoC: Xilinx Zynq-7000 Series; XC7Z020-2CLG400I
C2. DDR3: (2) Micron 16-bit/4 Gbit MT41K256M16TW-107
C3. Filing System Storage: Micro SD Card
C4. Ethernet PHY: Microchip KSZ9031
C5. USB 2.0: Microchip USB3320
C6. USB to UART Converter: FTDI FT230X
C7. Wi-Fi/Bluetooth Module: Microchip ATWILC3000
C8. Local Relative Humidity/Temperature Sensor: Sensiron SHTW2
C9. Infrared Sensor: FLIR Lepton 500-0771-01

Component No.	*Component Description*	*Failure rate (λ)*	*Component Reliability ($R_CT = e{-}\lambda t$)*
C1	Xilinx Zynq-7000 series SoC	22 FIT	0.999
C2	DDR3 RAM module	100 FIT	0.996
C3	Micro SD card	220 FIT	0.990
C4	Ethernet PHY IC	45 FIT	0.998
C5	USB 2.0 IC	165 FIT	0.993
C6	USB to UART converter module	425 FIT	0.982
C7	WiFi/Bluetooth module	185 FIT	0.992
C8	Humidity/Temperature Sensor	400 FIT	0.983
C9	Infrared Sensor	35 FIT	0.998

[Note 1: 1 FIT is equal to one failure in billion working hours or in other words FIT = $\lambda_{hours} \times 10^9$.]
[Note 2: In the table above, the value of $R_C(t)$ is calculated using t = 5 years = 43,800 hours.]

Since each of the above components is critical, we can assume that the components are connected in series for the purpose of reliability calculation. The reliability of the system is given by the equation:

$$R_S(t) = \prod_{i=1}^{n} R_{Ci}(t)$$

Therefore, the reliability of the system after 5 years is $R_S = 0.9324$ or 93.24%

4.3.2 Scenario 2: All Critical Components Connected in Parallel

As shown in Figure 4.5, each of the critical paths in the design contains only one critical component or in other words, all critical components are parallel to one another. Assuming $R_{Ci}(t)$ is the reliability of an individual critical component and $Q_{Ci}(t)$ is the failure rate of the component, we get

$$Q_{Ci}(t) = 1 - R_{Ci}(t)$$
$$\Rightarrow Q_{Ci}(t) = 1 - e^{-\lambda_i t}$$

Assuming that the failure rates of all critical components are statistically independent, we get:

$$Q_S(t) = \prod_{i=1}^{n} Q_{Ci}(t)$$

Hence, the equation for overall system reliability will be:

$$R_S(t) = 1 - \prod_{i=1}^{n} Q_{Ci}(t)$$
$$\Rightarrow R_S(t) = 1 - \prod_{i=1}^{n} \left[1 - R_{Ci}(t)\right]$$

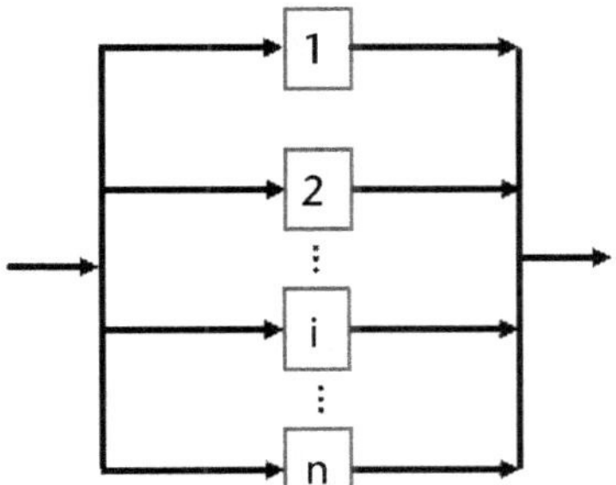

Figure 4.5 Critical path of a system if all key components are connected in parallel.

Example 4.9

Consider a sensor system that has 5 critical components connected in parallel. If the reliability of each component is 0.98, what will be the overall system reliability?

Solution:

Since the components are connected in parallel, the system will operate reliably if at least one of the critical components does not fail. The failure probability of a single component is:

$$Q_{Ci}(t) = 1 - R_{Ci}(t)$$
$$\Rightarrow Q_{Ci}(t) = 1 - 0.98 = 0.02$$

The probability of all the 5 components failing is $(0.02)^5 = 3.2 \times 10^{-9}$
Therefore, the reliability of the system is $R_S = 1 - 3.2 \times 10^{-9} = 0.9999999968$

4.3.3 Scenario 3: All Critical Components Are Connected in Series-Parallel Configuration

In this scenario, the critical path of the design consists of several blocks connected in a series. For example, in Figure 4.6, block 1 is in series with block 2, but with each block there are several critical components that are connected in a series/parallel configuration. In other words, critical failure of one block leads to the failure of the system. However, critical failure of one component in the block may or may not result in failure of the block depending on the series/parallel configuration in which it is aligned.

4.4 Faults, Errors, and Failure

So far, we have seen that reliability of a system is a function of the individual reliabilities of its key components. As a system designer, most of the time, you would not have control over the reliability and failure rates of the components that you use for your design. The little leverage you have in this regard is to make an informed choice to choose the best possible components that are most reliable based on their technical documentation. However, this does not mean that designers cannot do

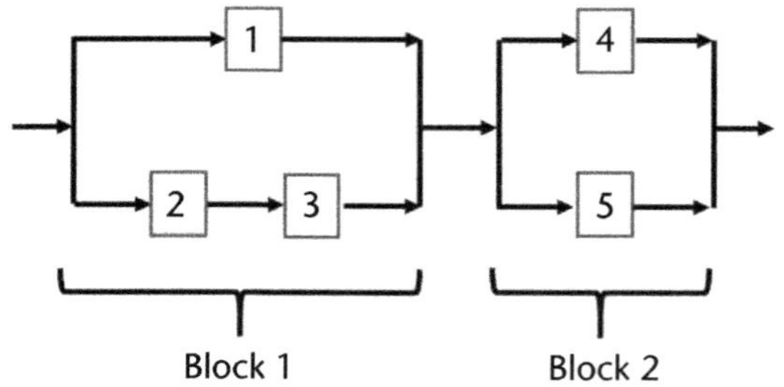

Figure 4.6 Critical path of a system with key components in series-parallel combination.

anything to improve system reliability, even when they are forced to choose a component that may not be offering the best reliability (lowest possible failure rate).

In simple terms, the reliability of a system is a function of its failure rate. Failure in a system occurs when the delivered service deviates from specified service. Generally, the failure of a system occurs when a key component of the system (hardware or software) was erroneous. The cause of this error is termed as the fault. In other words, a fault creates a latent error, which becomes effective when activated. When this activated error affects the delivered service, it results in a failure. This relationship between fault, error, and system failure is shown in Figure 4.7.

There can be several reasons for a fault to occur. Some of the most common ones are:

- Design errors;
- Software or hardware (either mechanical or electronic);
- Manufacturing problems (related to production and quality control);
- Damage, fatigue, and deterioration (operating the system in the wearout phase of the bathtub curve);
- External disturbances (can be environmental, situational, operational, or manual);
- Harsh environmental conditions such as electromagnetic interference and ionization radiation;
- System misuse (one of the leading causes of system failure).

Faults can be broadly classified under the following four categories shown in Figure 4.8.

Figure 4.7 Fault leading to system failure.

Mechanical	Electronic	Software	People
• Deterioration due to wear out, fatigue, corrosion • Accidental malfunction due to fractures, overload, etc.	• Defects due to bad design • Latent errors introduced during manufacturing • Defects due to extreme operating environment (outside the specified tolerance limits)	• Defects due to bad programming code resulting in latent defects • Runtime errors	• Operating the system outside the standard operating conditions • System misuse • Accidental and intentional human errors

Figure 4.8 Classification of system faults.

No system can be designed as 100% reliable. In other words, even in a high reliability system, there is still a probability of system failure, which could occur due to any of the reasons stated above. All that a designer can do is to work on the design to reduce that probability of fault occurring that could lead to a system failure. This approach towards reliable system design, when we are aware of the most likely causes of faults present in the system, is often termed as fault tolerant computing. In other words, it is acceptable fact that all systems will have key components (hardware or software) that will develop faults, but as a designer, we have to work with the same set of components and still increase the overall reliability of the system using clever design techniques.

It is important to understand that, just like reliability engineering is an established practice in manufacturing industry, fault tolerant computing is also an established practice in the field of software development. In recent years, both these techniques are being fine-tuned and deployed in an embedded systems design as it encompasses both hardware and software development. While it is difficult to study and summarize all aspects of fault tolerant computing in a section of this book, we would try to provide a snapshot of some of the most common fault tolerance techniques used in low-cost, reliable design.

4.4.1 Classification of Faults

Before we discuss different types of fault tolerance techniques, we need to classify faults either a hardware fault or a software fault. Hardware faults can be further classified based on the time/duration of the fault (transient, permanent, or intermittent faults).

A transient fault, as the name suggests, occurs at a random instance of time and is evident for a small duration before disappearing. Unless there is a clear correlation of the fault timing with an external trigger, such faults are very difficult to isolate and diagnose (e.g., hardware component failure due to an exposure to radioactivity or a component failure due to sudden extreme change in weather conditions). Some systems are quite prone to transient faults and are designed based on the assumption that these faults will happen (e.g., communication systems). Permanent faults remain in the system until they are repaired (e.g., a broken wire or a software design error). Intermittent faults are a special case of transient faults that occur from time to time. However, the biggest difference between the two is that intermittent faults are more predictable in nature (e.g., over the period if we are aware that a particular sensor component is sensitive to low temperatures and high humidity). In such a scenario, it may need to be brought back to within the operating range of temperature/humidity and it will start to work again.

Software faults are generally called bugs and are classified into two categories.

- *Bohrbugs:* Software faults that are easily reproducible and identifiable.
- *Heisenbugs:* Faults that are only active under rare conditions (e.g., when there is a race condition existing within a state machine).

Unlike hardware faults, it is not a common practice to classify software faults as being transient, intermittent, or permanent. Since software upgrade is a much easier option than hardware changes (which sometimes leads to system redesign design), the presence of software faults is generally unacceptable within a system. Software does not deteriorate with age, so it is either correct or incorrect, but some faults can remain dormant for long periods. Moreover, in the case of embedded systems, a change in software due to hardware changes could potentially lead to bugs or errors being introduced in the system (as is the case when we upload new drivers or update libraries). That kind of fault can be classified as both hardware and software-related.

Figures 4.9 and 4.10 show the classification of faults and errors based on a number of factors discussed above.

4.4.2 Fault Prevention Versus Fault Tolerance: Which One Can Address System Failure Better?

There are two main approaches to building a reliable embedded system: fault prevention and fault tolerance. Fault prevention attempts to eliminate the possibility of a fault occurring in a system while it is operational. Fault tolerance is the ability of a system to deliver expected performance even in the presence of faults. Both these approaches attempt to produce reliable systems that have well-defined failure modes. As discussed earlier in this section, faults lead to errors and errors cause failure. Therefore, failure mode analysis needs to consider different types and sources of faults and errors that lead to system failure as shown in Figure 4.11.

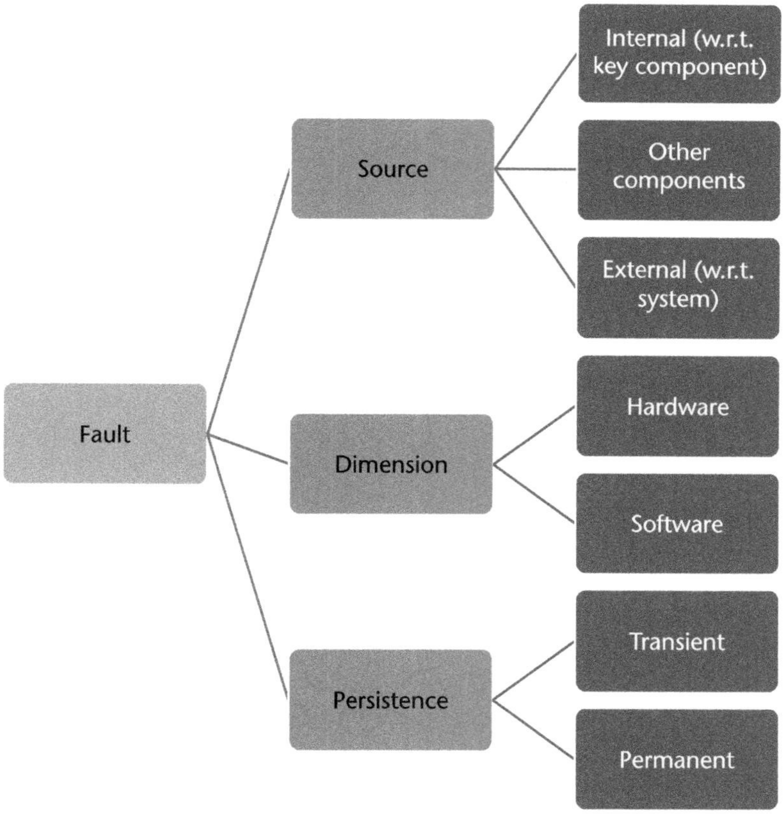

Figure 4.9 Sources and types of faults.

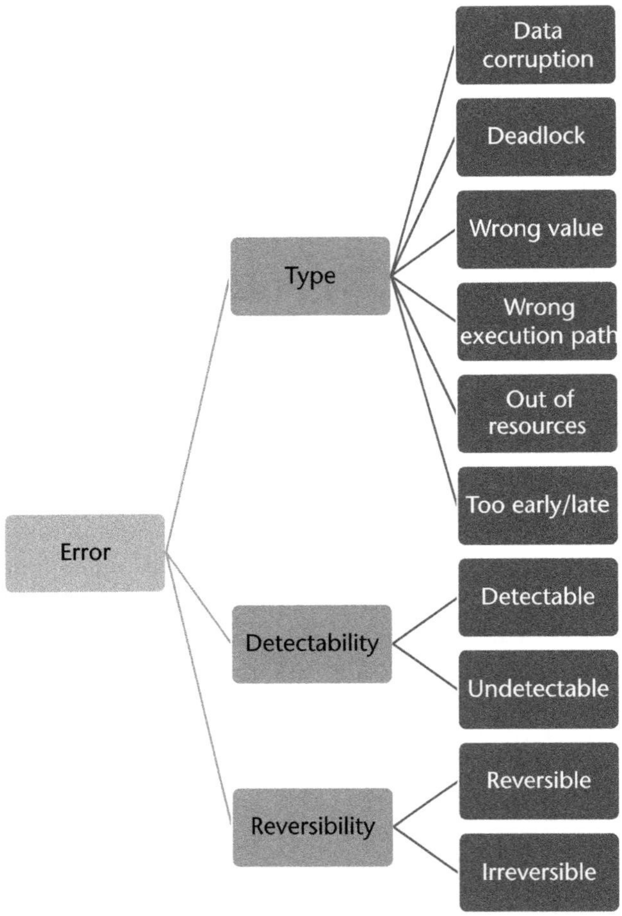

Figure 4.10 Sources and types of errors.

Since fault prevention attempts to eliminate the possibility of a fault occurring in a system while it is operational, there are two stages in which it can be implemented: (1) fault avoidance, and (2) fault removal. Fault avoidance aims to minimize the introduction of any faults during system design and manufacturing phase using good design practices as highlighted in Figure 4.12. Fault removal involves procedures for finding and removing causes of errors (e.g., design reviews, program verification, code inspections, and system testing). However, unless there is an unlimited budget and time available, system testing can never be exhaustive. This means that there is always a probability that some of the errors were not caught during the testing phase. Moreover, system testing suffers with some shortcomings. For example, a test can only be used to show the presence of faults, not their absence. Moreover, it is sometimes impossible to test under realistic conditions (e.g., if a design would need to be installed in space). Since most of the testing is performed using software simulations, the efficiency of test is limited to the accuracy of simulation software. Finally, many errors that were inserted during the system's development phase may not manifest themselves until the system goes operational and hence may not get caught during the testing phase. There is no system in this world that can be tested rigorously to be 100% fault proof as believe it or not, for

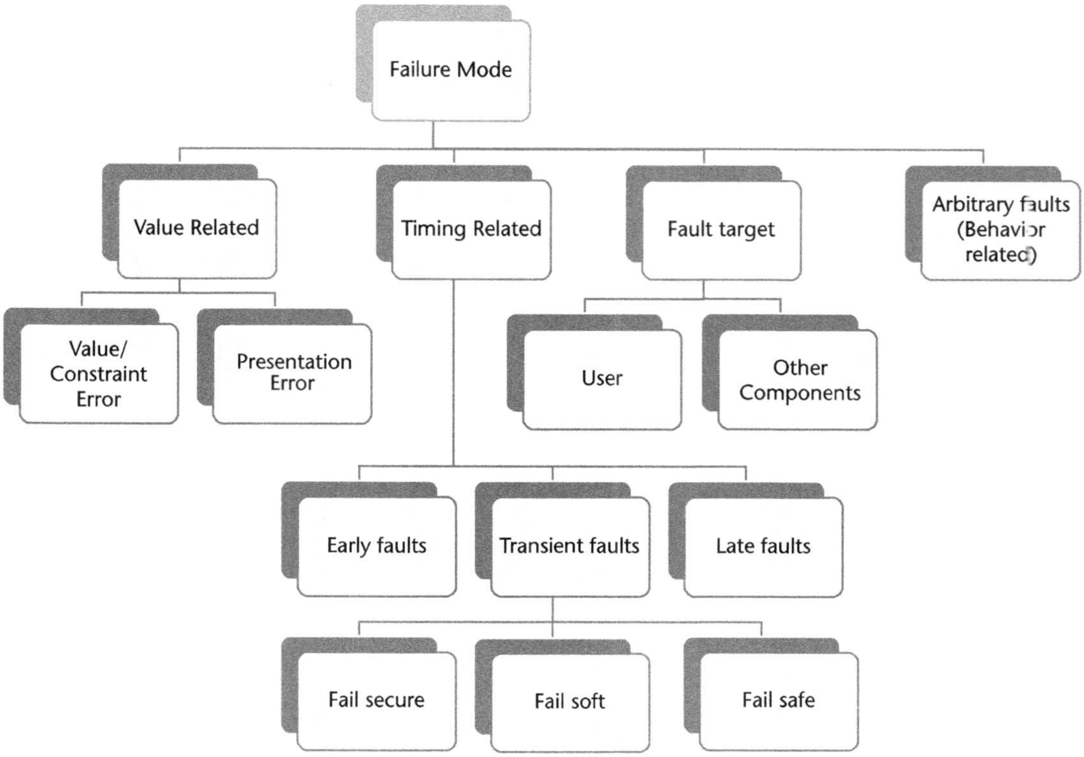

Figure 4.11 Failure mode analysis depending on type of faults.

some reason or another, hardware components will fail (with some probability of failure). Moreover, the fault prevention approach is considered unsuccessful when either the frequency or duration of repair times too high or if the system is inaccessible for maintenance and repair activities (e.g., a satellite in space or a mission to another planet).

4.5 Fault Tolerance Techniques

Since fault prevention cannot be guaranteed, despite taking up a significant amount of budget and resources, it is prudent to design a system keeping in mind that the system will develop faults and some key components will fail over time. If these assumptions (regarding inherent fault rates) are used as design criteria and production criteria, it would ultimately be a lot cheaper and quicker to deliver a product that can provide the expected delivery of service.

As mentioned earlier, reliability of a system is a function of its failure rate, which, in turn, is dependent on the failure rate of various key components that make the system. The importance of reliability consideration, while specifying the design specifications of a system, depends on the application for which the system is being designed. Therefore, it is obvious that a system that needs high reliability will require a very high level of fault tolerance techniques included in the design and development stage, while other applications that may not need such high reliability can continue to operate with lower fault tolerance design requirements. To

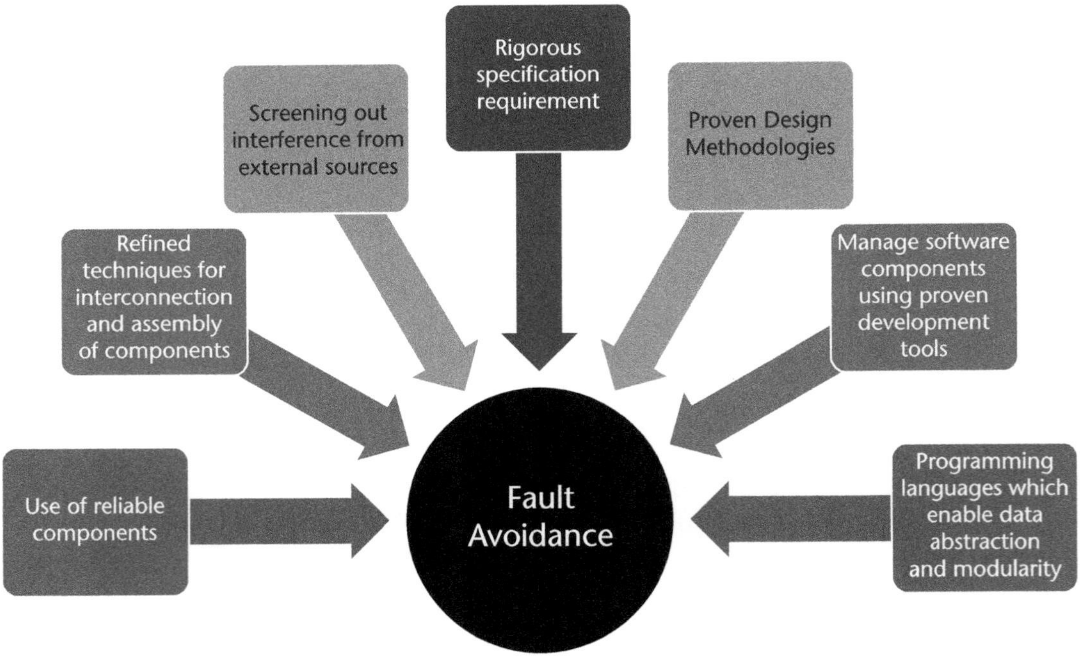

Figure 4.12 Fault avoidance techniques.

make an informed decision on the level of fault tolerance that needs to be incorporated into a design, we need to answer these three questions (Figure 4.13):

- How critical is the component? For example, will the sensor design platform ordered by SensorsThink be considered a highly critical system if it were to be used as one of the control systems in a nuclear power plant or in a fighter jet?
- How likely is the component to fail? Some components, the other casing of the sensor design platform, are not likely to fail, so no fault tolerance is needed.
- How expensive is it to make the component fault tolerant? Requiring a redundant Zynq processor in the sensor design platform, for example, would

Fault Tolerance Level	Fail Secure	Fail Soft	Fail Safe
Description	System operates in the presence of faults without significant loss in performance	Allows partial degradation in performance while maintaining operation	Allows temporary halt in operation
Example Applications	Critical systems such as nuclear power stations, space vehicles etc.	Most systems fall under this category. (e.g. ageing phone battery or memory)	Noncritical systems that can go offline for repairs or replacement.

Figure 4.13 Levels of fault tolerance.

likely be too expensive both economically and in terms of weight, space, and power consumption to be considered.

Fault tolerance techniques for embedded systems can broadly be classified into two categories: (1) hardware fault tolerance techniques, and (2) software fault tolerance techniques. It needs to be understood that both these fault tolerance categories have been well studied and researched over the past few decades. However, embedded system design is a special case where both techniques are implemented together and with equal importance. This is because reliability of an embedded system is a function of reliable design of both its hardware and its software.

4.5.1 Redundancy Technique for Hardware Fault Tolerance

The concept of fault tolerance techniques through redundant hardware components was conceived in the early 1950s and it has been studied and evolved over the decades since. The most common approach towards providing fault tolerance for key hardware components is the redundancy technique, which relies on extra elements introduced into the system to detect and recover from faults. The extra hardware components inserted in the design are considered redundant as they are not required in a perfect (fault-free) system. However, the addition of redundant components inevitably increases the complexity (and costs) of the overall system, which in itself can lead to less reliable systems. Two of the most common redundancy techniques used in hardware design are: (1) static redundancy, also known as masking redundancy, and (2) dynamic redundancy.

4.5.1.1 Static Redundancy

Static redundancy uses extra components to mask the effects of faulty critical components. It is called static redundancy as once the redundant components are installed, their interconnection remains fixed. A common assumption used in static redundancy is that the fault is not due to design error, but rather due to transient fault or component degradation. While there are several types of redundancy design techniques available (such as triple modular redundancy, N-modular redundancy, ECC), it would be out of the scope of this book to study all of them. Triple modular redundancy (TMR), as shown in Figure 4.14, is one of the most common techniques used in fault tolerant hardware design. It makes use of three identical

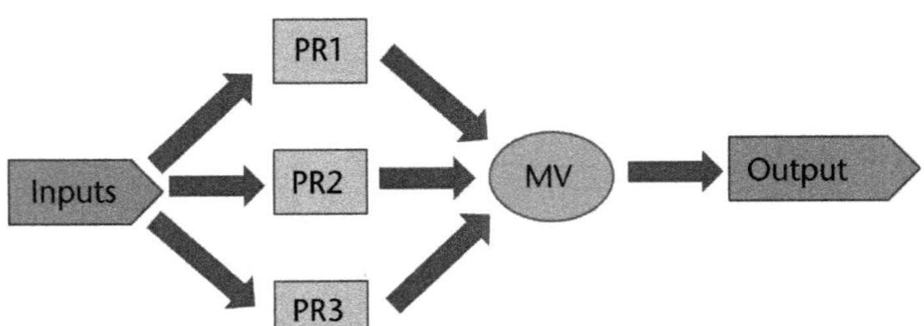

Figure 4.14 The TMR technique for hardware fault tolerance. All three processor units (PR1, PR2, and PR3) process the same input and the majority voter (MV) compares the results and outputs the majority vote (two of three).

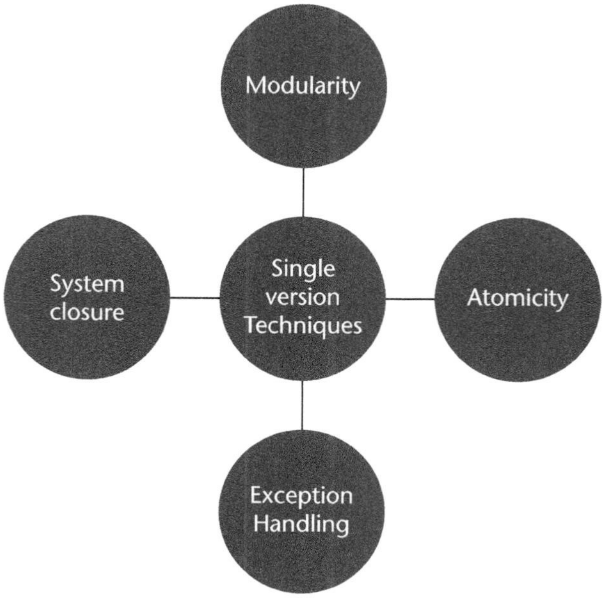

Figure 4.15 Key attributes of single-version software fault tolerance techniques.

exception handling are desirable and advantageous in each version of the software being used in the multiversion scheme. Two of the most widely used techniques of multiversion software fault tolerance are recovery block (RB) scheme and N-version programming (NVP) scheme. Of these two, NVP is a static method, like the triple modular redundancy used in hardware fault tolerance. However, the RB scheme is a dynamic redundancy approach. Both NVP and RB schemes require design diversity (for various alternate versions of the software) as well as hardware redundancy to implement the scheme. Design diversity is a software design approach in which different versions of the software are built independently but deliver the same service. The fundamental assumption of design diversity is that components built differently will fail differently.

NVP is a static scheme in which output is computed using N independently designed software modules or versions and their results are sent to a decision algorithm that determines a single decision result. The primary belief behind the NVP scheme is that the "independence of programming efforts will greatly reduce the probability of identical software faults occurring in two or more versions of the program." The first step in the NVP scheme involves the generation of $N \geq 2$ functionally equivalent programs (also called alternate versions) from the same initial specification. As the goal of the NVP scheme is to minimize the probability of similar errors in each version of the software component, it encourages the use of different algorithms, programming languages, environments, and tools wherever possible. With such a big diversity in the design of different versions of the same software component, NVP requires considerable developmental effort and increase in problem complexity. According to some studies, a 25% rise in the problem complexity can lead to a 100% rise in program complexity. However, in case of NVP, the increase in complexity is not greater than the inherent complexity of building a single version.

The different version of program execute concurrently with the same inputs and their results are compared by a decision algorithm (driver) as shown in Figure 4.16. The decision algorithm should be capable of detecting the software versions with erroneous outputs and prevent the propagation of bad values to main output. The decision algorithm should be developed keeping in mind the security and reliability of the application. In the case of applications that require embedded system design using SoCs, the NVP scheme is quite common. Significant advancements in chip design industry are leading to a greater complexity in design; therefore, the probability of design fault is greater because a complete verification of the design is very difficult to achieve within a given timeframe. The use of N-versions of components allows the continued use of chips with design faults as long as their errors at the decision points are not similar.

The efficiency of NVP scheme depends on a number of factors, primarily the initial design specification. If the specification is incomplete or incorrect, the same error will manifest in all different versions of the software, rendering the scheme useless. Design diversity is also a key parameter of with regard to success of NVP scheme. Finally, the development budget for the NVP scheme limits the number of independent versions of the software that can be used. A 3-version scheme will require 3 times the budget of a single version, which raises another question. Would the resources and time needed to implement an *N*-version scheme may be better utilized by producing a more efficient single-version scheme with the same amount of resources and time? This question probably depends on the industry for which the product is intended. Most companies or industries will probably opt for a robust single-version scheme with a highly robust software system rather than an *N*-version scheme.

The recovery block scheme, shown in Figure 4.17, uses alternate versions of the software to perform a run-time error detection using an acceptance test performed on the results delivered by the first (primary) version. If the system fails the acceptance test, the state is restored to what existed prior to the execution of that algorithm and an alternative module is executed. If the alternative module also fails the acceptance test, the program is restored to the recovery point and yet another module is executed. Recovery is considered complete when the acceptance test is passed. A sample test script for a recovery block scheme is shown in Figure 4.18. Checkpoint memory is needed to recover the state after a version fails to provide a valid starting operational point for the next version.

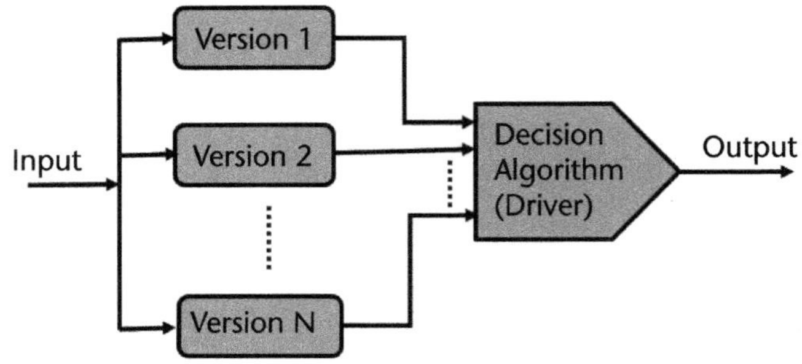

Figure 4.16 N-version programming scheme.

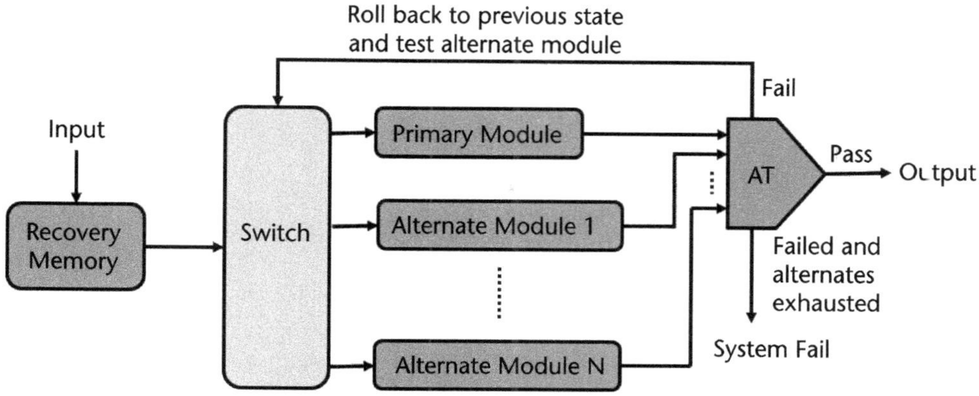

Figure 4.17 The recovery block scheme consists of three key elements: a primary module to execute critical software functions, an acceptance test (AT) for the output of the primary module, and alternate modules, which perform the same functions as the primary module and deliver output in the event of a fault with the primary module.

```
ensure <acceptance test>
by
      <primary module>
else by
      <alternative module>
else by
      <alternative module>
      ...
else by
      <alternative module>
else error
```

Figure 4.18 Recovery block test script.

If all modules fail, then the system itself fails and recovery must take place at a higher level.

While both NVP and RB schemes are quite useful in providing software fault tolerance, there are some key differences in both schemes. These differences have been summarized in Table 4.4.

With the ever-increasing complexity and omnipresence of embedded systems and different aspects of our everyday lives, the comparison between NVP and RB schemes needs to be studied in far greater depth and understanding than just a section in this book. So we only take a cursory view of most used hardware and software fault tolerance techniques with which the readers should be acquainted. For a deeper understanding of this topic, there are some outstanding texts available in the market.

4.6 Worst-Case Circuit Analysis

Worst-case circuit analysis (WCCA) is a commonly used technique used in the industry to specify the boundary parameters for the operating conditions during the

Table 4.4 Comparison Between NVP and RB Schemes

	N-Version Programming (NVP)	*Recovery Block (RB)*
Type of Scheme	Static	Dynamic
Design Overheads	Requires a driver	Requires Acceptance test
Runtime Overheads	Requires N*resources	Requires memory to store recovery points
Diversity of Design	Susceptible to errors in requirements	Susceptible to errors in requirements
Error Detection	Majority vote comparison algorithm	Acceptance test
Atomicity	Votes before output is delivered	Only outputs if acceptance test passed

useful life of a product. WCCA examines the effects of potentially large magnitudes of variations of components used in the embedded system design beyond their initial tolerance level. While it may not seem that obvious, the variation in component parameters by just a few percentage points can lead to a big difference in system reliability under the worst-case scenario. The variations can be the result of aging or environmental influences, which can cause circuit outputs to drift out of specification.

WCCA combined with failure mode and error analysis and MTBF analyses are essential to the design of any reliable system; however, there is one key difference. WCCA is a quantitative analysis technique as opposed to a part-based approach for stress testing, FMEA, and MTBF (Figure 4.19). WCCA determines the mathematical sensitivity of circuit performance to parameter variations (for components and subsystems) and provides both statistical and nonstatistical methods for handling the variables that affect system performance. In a nutshell, WCCA is used to predict the most likely scenarios under which there is a high probability of system failure.

WCCA technique usually involves considerable amount of planning and data collection before it can be performed on a circuit. Before commencing, we should first document the theory of operation and develop a functional breakdown of the circuit (block diagrams and outlines of critical subcomponents). We should also have access to technical documentation of most parts used in the design (see step C below). The flowchart shown in Figure 4.20 presents the general overview of WCCA technique. It demonstrates the interrelationship of the various tasks required for completion of a WCCA.

- *Step A—Critical system attributes:* Before we begin WCCA, we must have a complete understanding of the system requirements and specifications. We should then decide whether we wish to perform a complete circuit analysis or implement circuit partitioning to analyze critical functions. Further on, we need to identify the circuitry to be analyzed (analog, digital, software) and determine the availability of component data. Finally, we need to identify the tools available to support the analysis. Failure to review this information could result in the failure to examine relevant system parameters.
- *Step B—Comparison of performance specification with WCCA results:* WCCA generally includes two steps: (1) worst-case stress analysis, which

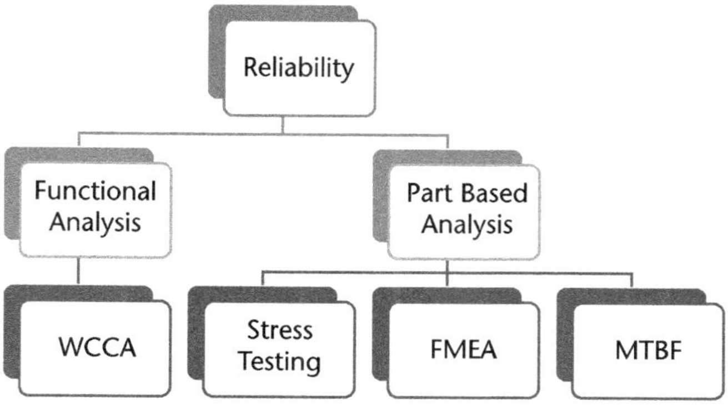

Figure 4.19 Most used reliability techniques in embedded system design.

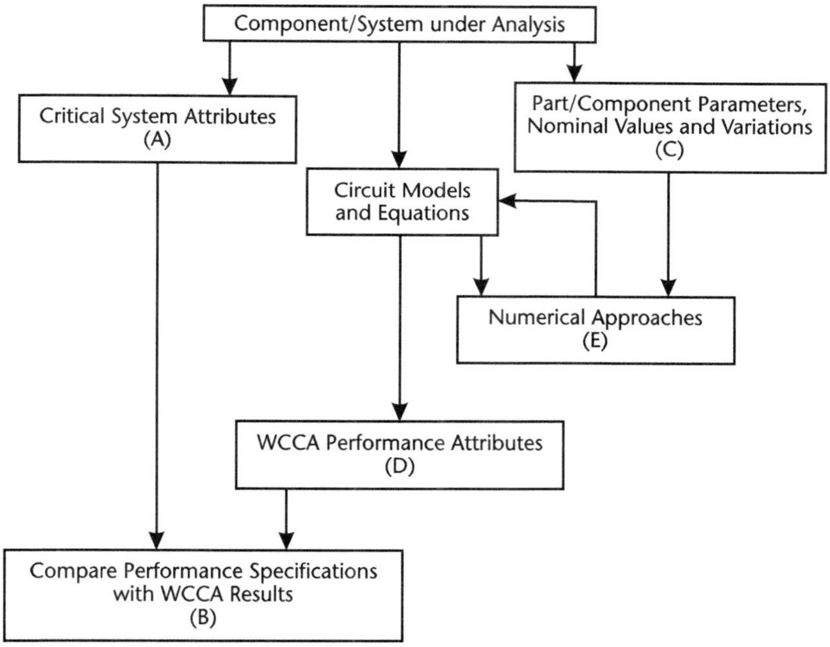

Figure 4.20 General overview of the steps involved in WCCA.

identifies the components that are overstressed under worst-case conditions, and (2) worst-case circuit performance analysis, which analyzes the circuit to determine if it will perform as specified under worst-case conditions. All environmental conditions and part parameters are simultaneously changed to their worst realizable extremes. The circuit is then analyzed to determine if the circuit performs correctly. Analysis results should be verifiable (document or reference all information used to develop the analysis: circuit equations, part data sources, and circuit simulation tools used). If problems are found during the analysis, develop and propose potential fixes and alternative solutions.

- *Step C—Part/component parameters, nominal values, and variations:* One of the most critical steps involved in completing a WCCA is the development of a part database. A part database is a composite of information for quantifying sources of component variation. The database typically contains of performance requirements and specifications, schematics and block diagrams, interconnection lists and wiring diagrams, part lists, theory of operation, operations environments, and operational configurations. Critical part parameters also include environmental (ambient and self- heating temperature, vibration, humidity, radiation) and component aging characteristics. Once the most probable sources of component parameter variation have been identified, the database can be used to calculate the worst-case component drift for critical parameters. The establishment of a component database is recommended for all analysis complexity levels and the analyst must be knowledgeable of the circuit and parameters being examined, including timing diagrams where appropriate. Table 4.5 provides a list of parameter variations for different types of components used in circuit design.

4.6.1 Sources of Variation

As evident from Table 4.5, variations in component parameters are of two types: random and bias. Random variation is not predictable in direction. However, bias is predictable given known inputs. All sources of component variation can be grouped into one of these effects. The effects are subsequently combined to give an overall indication of part variability. The addition of individual random and biased variables is as follows:

- *Bias effects:* Added algebraically;
- *Random effects:* Root sum squared- ($\pm 3\sigma$ limits of a normally distributed population).

The determination of the minimum and maximum limits of component value due to drift is as follows:

$$Worst-Case\,Minimum = Nominal\,value - \left(Nominal\,value \times \Sigma\left|Negative\,biases\right|\right)$$
$$-\left(Nominal\,value \times \sqrt{\Sigma\left(Random\,effects\right)^2}\right)$$
$$Worst-Case\,Maximum = Nominal\,value$$
$$+\left(Nominal\,value \times \Sigma\left|Negative\,biases\right|\right) + \left(Nominal\,value \times \sqrt{\Sigma\left(Random\,effects\right)^2}\right)$$

Using the table below, determine the worst-case minimum and maximum values for a 1,200-μF CLR capacitor. These parameters are used to determine the potential resultant effect of the CLR capacitor drift on circuit applications.

Table 4.5 Affected Parameters Versus Source of Variation

Component Type	*Environmental Source of Variation*	*Parameter Affected*
Bipolar/field effect transistors	Temperature, radiation	H_{FE} (bias), V_{BE} (bias), I_{CBO} (bias), RDS_{ON} (bias), V_{TH} (bias), H_{FE} (bias), I_{CBO} (bias), V_{CE} (saturation) (random and bias), V_{TH} (bias)
Rectifiers/switching diodes	Temperature, radiation	V_F (bias), T_S (bias), I_R (bias), I_R (bias), V_F (bias)
Zener diodes	Temperature	V_Z (bias, sometimes random), Z_Z (bias)
Resistors	Temperature, humidity, aging (powered), life (unpowered), vacuum mechanical	Resistance (bias and random, random), resistance (bias), carbon composition resistance (bias and random), resistance (bias and random), resistance (bias), resistance (bias and random)
Capacitors	Temperature, aging, mechanical electrical, vacuum humidity	Capacitance (bias and/or random), ESR (bias), DF (bias, nonlinear), ESR (bias), capacitance (bias and/or random), capacitance, voltage coefficient, capacitance (bias, nonhermetic), capacitance (bias)
Linear ICs	Radiation temperature	Voltage, current offset (random), A_{OL} (bias), voltage, current offset (bias and random), random, A_{OL} (bias)
Digital ICs	Temperature, rise/fall; time, propagation, delay radiation	Propagation delay (bias)
Magnetics (strongly dependent on materials)	Temperature, aging mechanical	Saturation flux density (bias), permeability (bias), core loss (bias, nonlinear, nonmonotonic), saturation flux density (bias, very small), permeability (bias), saturation flux density (bias)
Relays	Temperature	Pull-in/dropout current/volts (bias), contact resistance (bias, secondary effect), mechanical contact resistance (bias)

h_{FE}	Gain	A_{OL}	Open loop gain
I_{CBO}	Collector-base output current	DF	Dissipation Factor
RDS_{ON}	On-drain source resistance	ESR	Equivalent series resistance
V_{CE}	Collector-emitter voltage	Z_Z	Zener Impedance
V_F	Forward voltage	V_Z	Zener voltage
T_S	Storage time	I_R	Reverse current
V_{TH}	Threshold voltage		

Example 4.10

	Bias (%)		
Capacitance Parameters	*Negative*	*Positive*	*Random (%)*
Initial Tolerance at 25°C	—	—	18
Low Temperature (–20°C)	25	—	—
High Temperature (+80°C)	—	15	—
Other-Environments (Hard Vacuum)	18	—	—
Radiation (10KR, 1,013 N/cm^2)	—	9	—
Aging	—	—	12
TOTAL VARIATION	43	24	$(20)^2 + (10)^2 = 22.4$

Solution:

Worst-case minimum and maximum occur when the variation due to bias is of the same sign to the random variation, hence adding the effect of both:

Worst-case minimum = - |total negative bias + total random bias|

$$= -43\% - 22.4\% = -65.4\%$$

Worst-case maximum = + |total positive bias + total random bias|

$$= 24\% + 22.4\% = +46.4\%$$

Using the equations given above:

Worst-case minimum capacitance = 1,200 μF – 1,200 μF (|-0.43|+ 0.224) = 415.2 μF

Worst-case minimum capacitance = 1,200 μF + 1,200 μF (|0.24|+ 0.224) = 1,756.8 μF

- *Step D—WCCA performance attributes:* When conducting a worst-case circuit performance analysis, the key elements examined within the system depend on the functionality of the circuit or the subcircuit. Therefore, while partitioning the circuit, it is a good practice to minimize the active components to as few logical blocks as possible. It is also a good practice to include circuit and timing diagrams, as appropriate, in the analysis. Critical timing of digital circuits, transfer functions of filtering networks, and characteristics of amplifiers are examples of circuit performance elements. Table 4.6 provides a list of most important circuit parameters which should be analyzed in a worst-case performance analysis for digital and analog circuits. Finally, it will be evident that not all parts and circuits perform as specified in the technical documentation, therefore it is extremely important to compare WCCA results with the technical specifications of each part/component and document all inconsistencies.
- *Step E—Numerical analysis techniques:* Worst-case analysis can be performed using three different numerical approaches: extreme value analysis (EVA), root-sum-square (RSS) analysis, and Monte Carlo analysis (MCA). These three methods are the most accepted industry standards for WCCA. EVA is generally the easiest to apply. It involves the analysis of a given circuit/product under simultaneous worst-case parts limits and produces conservative results. RSS is a statistical approach that is more labor-intensive. However, RSS results are more realistic than EVA. Finally, MCA involves randomly selecting part parameters and analyzing the resulting system performance. MCA requires a large number of simulations (typically 1,000 to 50,000 runs) and the data for piece-part parameter statistical distributions must be known (or assumed normal). A comparative study of the advantages and disadvantages for all three approaches is presented in Table 4.7.

While the designers can choose any worst-case design technique listed in Table 4.7, the most common industry practice is to perform MCA due to the availability of software tools that can help accomplish that task. LTSPICE is an open-source SPICE modeling tool used widely and, although its user interface and features do

Table 4.6 Important Circuit Parameters for WCCA

Digital Circuit Parameters	
• Circuit Logic • Pulse Widths • Compatibility • Circuit Timing • Current Draw	
Analog Circuit Parameters	
• Comparator → Threshold Precision → Hysteresis → Switching Speed/Time Constant → Offset Stability	• Oscillator → Frequency, Accuracy, Stability → Output Power Level Stability → Output Impedance → Load Impedance → Phase Stability → Noise and Spurious Output
• Filter → Insertion Loss → Phase Response → Frequency Response → Linearity → Input/Output Impedance → VSWR → Spurious or Out-of-Band Feedthrough	• Detector → Bias Voltage → Input Impedance → Frequency Range → Output Impedance → VSWR → Input
• Modulator → Frequency Response → Phase Response → Input/Output Impedance → Linearity → Insertion Loss → Output Level → Deviation- VSWR	• RF Switch (Solid State/Mechanical) → Drive Requirements→ Insertion Loss → Power Dissipation → Frequency Response → Power Handling → Video Feedthrough → Switching Speed → Switch Duty Cycles → Input/Output Impedance → VSWR
• Multiplier → Output Power → Input Drive → Frequency Response → Input/Output Impedance	• Coupler, Circulator → Insertion Loss → Magnetic Leakage → Frequency Response → VSWR → Power Handling- Directivity → Input/Output Impedance
• Mixer (Converter) → Noise Figure →Group Delay → Frequency → Power Dissipation → Output Spectrum → Conversion Loss → Intercept Points → Driver Requirement → Compression Points → Terminating Impedance	• Stripline, Waveguide, Cavity → Mode Suppression → Insertion Loss → Adjustment Range, Resolution → Dimensional Stability → VSWR → Input/Output Impedance → Material Stability

not match up to the standards of licensed software such as Altium Designer, one can easily perform MCA on smaller circuits in no time.

There are several circuit simulators available for engineering analysis and verification of electrical circuits based on SPICE technology. The reason why these simulators use SPICE technology is because each element of an electric circuit can be represented in the form of a mathematical model, where the set of mathematical models of elements and their correlations form the electric circuit model. In this manner, the circuit simulator can perform mathematical calculations on electric circuits, which allow it to perform a number of computed experiments (such as WCCA using the Monte Carlo method), with high reliability.

4.6.2 Numerical Analysis Using SPICE Modeling

Ideally speaking, worst-case analysis of any design should be performed on the entire circuit once the design is complete. However, in most real-life applications, this is not possible due to several reasons. First, SPICE models are not readily available

Table 4.7 WCCA Numerical Analysis Techniques

	Advantages	*Disadvantages*
EVA	• Most readily obtainable estimate of worst-case performance (best initial WCCA approach) • Does not require statistical inputs for circuit parameters (easiest to apply) • Database needs only supply part parameter variation extremes (easiest to apply) • If circuit passes EVA, it will always function properly (high confidence for critical production applications)	• Pessimistic estimate of circuit worst-case performance • If circuit fails, there is insufficient data to assess risk (modify circuit to meet EVA requirements, or apply RSS or MCA for less conservatism)
RSS	• More realistic estimate of worst case performance than EVA • Knowledge of part parameter probability density function (pdf) is not required • Provides a limited degree of risk assessment (percentage of units to pass or fail)	• Standard deviation of piece part parameter probability distribution is required • Assumes circuit sensitivities remain constant over range of parameter variability • Assumes circuit performance variability follows a normal distribution
MCA	• Provides the most realistic estimate of true worst-case performance • Provides additional information in support of circuit/product risk assessment	• Requires use of computer • Consumes a large amount of CPU time • Requires knowledge of part parameter pdf

for all the components used in the design, so it may be a tedious task to find one for the component that you want to use in your design. While most leading manufacturers do provide SPICE model libraries, many others do not, so one needs to be aware of this factor while performing MCA. Second, it takes a long time to assign SPICE model parameters to all circuit components and then use the SPICE model to vary the parameters and observe and record outputs for worst-case scenario. Also, performing the analysis on the entire circuit together also pushes the WCCA towards the end of the project cycle, which may not be an ideal scenario if the analysis shows results that do not match the specifications. For example, the example used in this book has approximately 941 components. Assigning tolerance values to each of them and running a SPICE simulation for the entire circuit would be an onerous task with so many component variations happening simultaneously.

Therefore, another standard industry practice is to perform worst-case analysis on smaller circuit components as they are being designed and tests which limit any potential surprises down the line. Figure 4.21 shows the Altium Designer schematic of the LDO voltage regulator circuit used in the design of SoC Platform design project for SensorsThink. Let us use this circuit as an example to estimate the worst-case voltage output range for this circuit. In our design, we have used input voltage tolerance of 2% and component tolerance of 1% as shown in Table 4.8.

In order to perform WCCA using MCA, we used LTSPICE to run a dc sweep of the input voltage from 2V to 3.3V, as shown in Figure 4.22.

Ideally, the dc sweep should be run from 0 to 3.3V, but since the ideal output voltage is 2.8V, we are only interested in input values of 2.8V and higher, as lower input values in the dc sweep analysis only give a zero output. We are interested in checking the maximum and minimum output swing around the ideal output voltage of 2.8V. Moreover, to keep it simple to understand, we have assumed that the

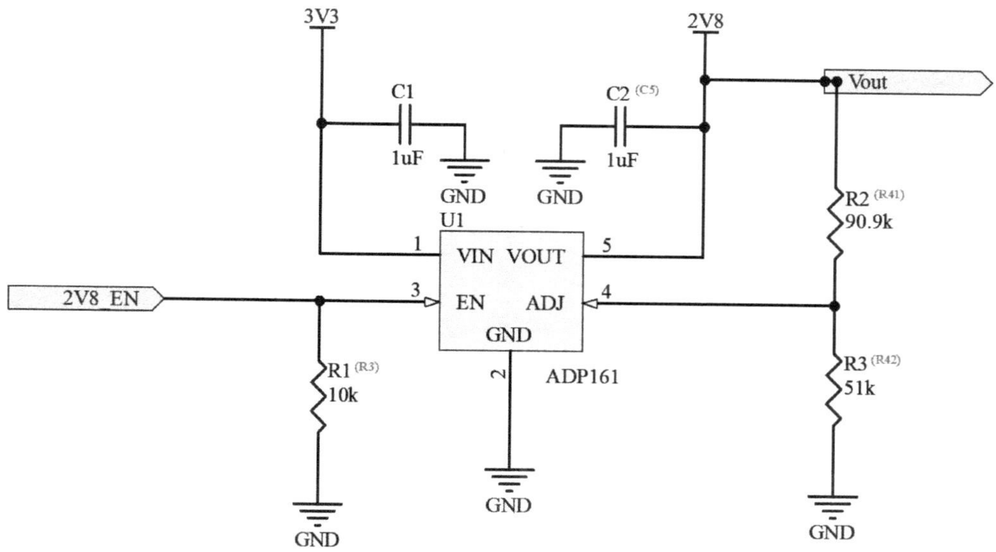

Figure 4.21 Schematic of an LDO voltage regulator circuit used in SoC Platform project.

Table 4.8 Component Tolerance Values for LDO Voltage Regulator Circuit Shown in Figure 4.21

Component Name	*Ideal Value*	*Tolerance*
Supply Voltage	3.3V	±2%
R1	10 kΩ	±1%
R2	90.9 kΩ	±1%
R3	51 kΩ	±1%
C1	1 μF	±1%
C2	1 μF	±1%
U1	Voltage regulator	0% (ideal)

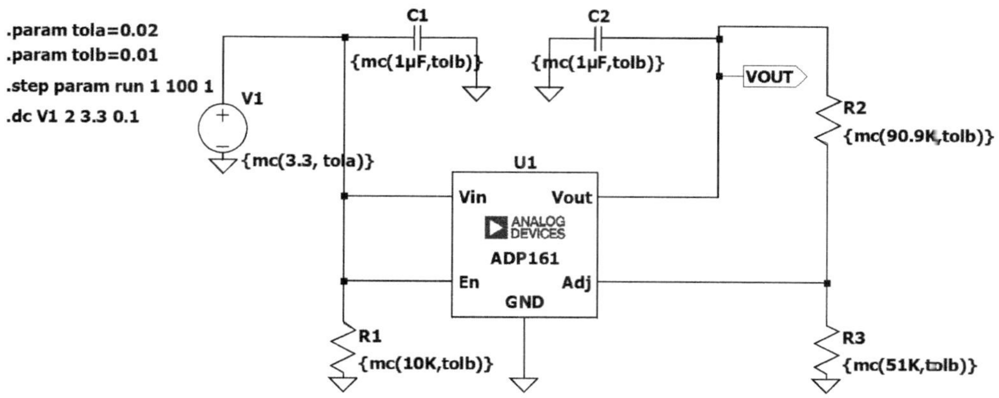

Figure 4.22 LTSPICE schematic of an LDO voltage regulator circuit with SPICE modeling components.

ADP161 regulator IC has 0% tolerance (in other words, it is assumed to be ideal). Figure 4.23 shows the plot of MCA simulation results in this circuit. It is evident from the plot that the variation in output voltage of the regulator for the given tolerance levels is between 2.75V and 2.81V, which is quite good for most sensor-based applications.

Let us now simulate the same circuit with higher component tolerances and evaluate the impact on the worst-case output of this voltage regulator. The new values of tolerances are given in Table 4.9. For the sake of simplicity, we have again assumed that component U1 is ideal (0% tolerance) and increased the tolerance of voltage supply and resistors/capacitors to 3% and 5%, respectively.

The LTSPICE schematic and simulation results for the dc sweep analysis are shown in Figures 4.24 and 4.25, respectively.

It is quite evident from the simulation results that higher tolerance values of the components result in much greater fluctuations at the output in the worst-case scenario. For the increased tolerance levels, the variation in worst-case voltage output

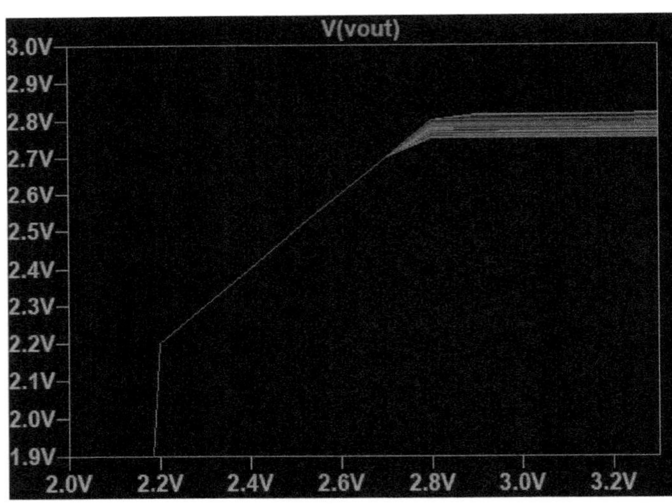

Figure 4.23 LTSPICE dc sweep simulation of an LDO voltage regulator circuit with voltage supply tolerance = 2% and resistance/capacitance tolerance = 1%.

Table 4.9 Modified Component Tolerance Values for the LDO Voltage Regulator Circuit Shown in Figure 4.21

Component Name	*Ideal Value*	*Tolerance*
Supply Voltage	3.3V	±3%
R1	10 kΩ	±5%
R2	90.9 kΩ	±5%
R3	51 kΩ	±5%
C1	1 μF	±5%
C2	1 μF	±5%
U1	Voltage regulator	0% (ideal)

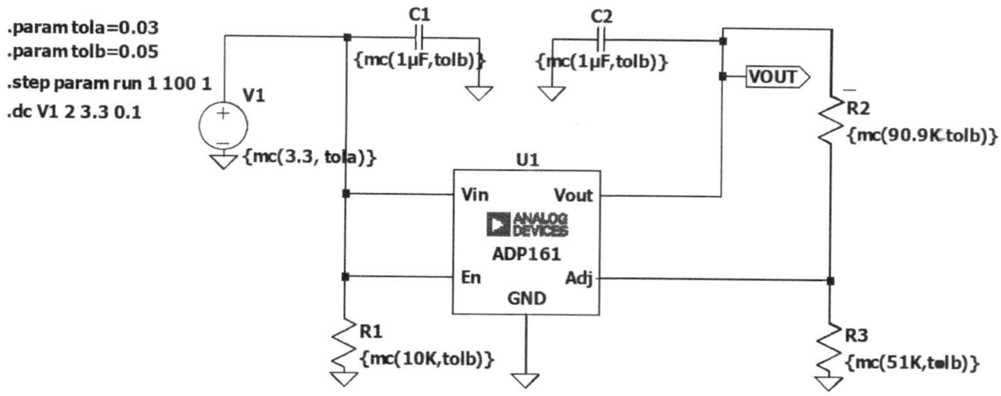

Figure 4.24 LTSPICE schematic of an LDO voltage regulator circuit with higher tolerance values for SPICE modeling components.

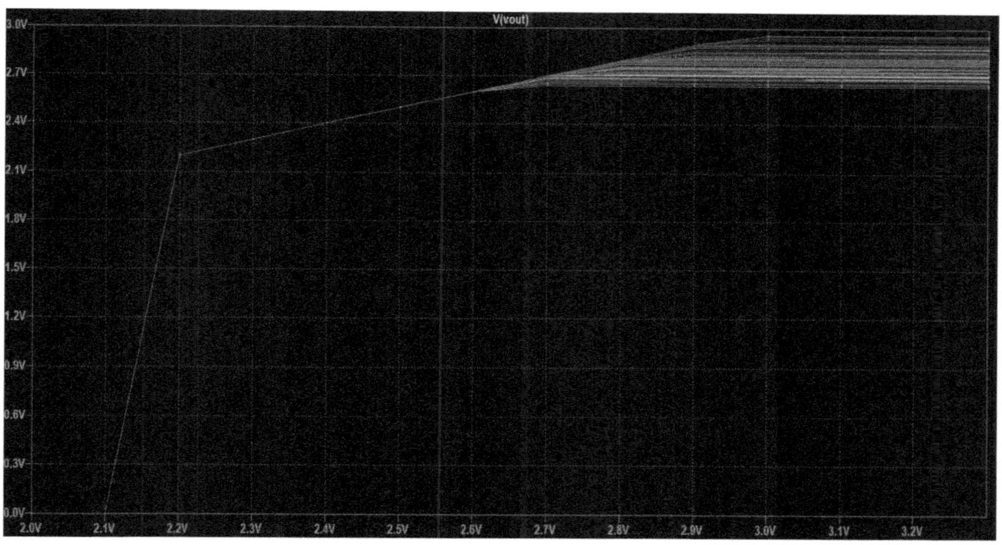

Figure 4.25 LTSPICE dc sweep simulation of an LDO voltage regulator circuit with voltage supply tolerance = 3% and resistance/capacitance tolerance = 5%.

is between 2.64V and 2.96V, which is a significant variation from the ideal 2.8-V output level.

Example 4.11

A schematic diagram of a clock generator circuit used in the design of the SoC Platform design project for SensorsThink is shown below. We will perform Monte Carlo analysis for the worst-case performance of this circuit if the supply voltage tolerance is 2% and resistance or capacitance tolerance is 2%. Assume that the LTC6908 oscillator has 0% tolerance (ideal).

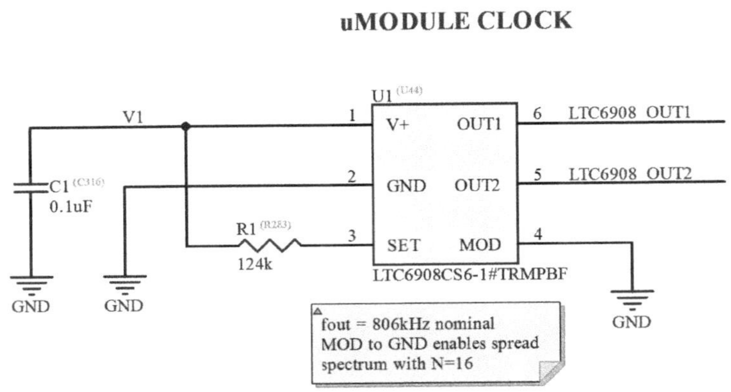

Solution:
In order to understand this better, let us run the Monte Carlo simulation for this circuit in two steps. In the first case, we run a transient analysis of the circuit and observe the impact of voltage supply variation while keeping the other components ideal. The LTSPICE schematic of this circuit with component parameters and tolerances is shown below.

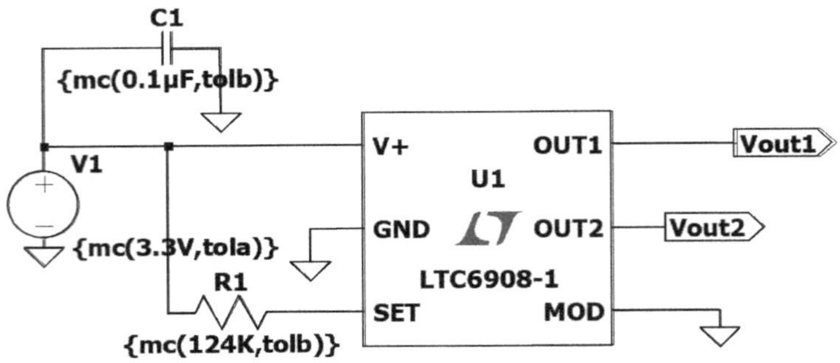


```
.param tola=0.02
.param tolb=0.00
.step param run 1 100 1
.tran 20u
```

Running transient analysis on this circuit analysis for 20 μs with voltage tolerance 2% and 0% (ideal) for all other components gives the following output.

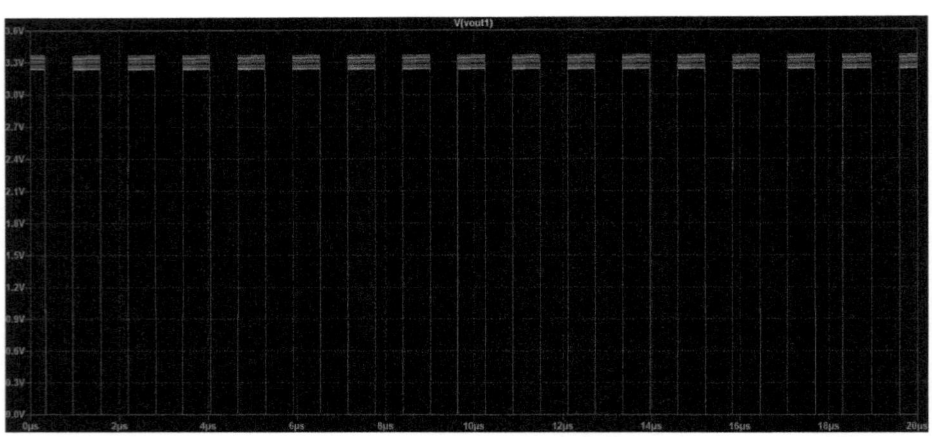

It can be observed that 2% supply voltage variation only impacts the output voltage level and it varies between 3.2V to 3.4V. These values are approximately equal to the voltage variation calculated theoretically using $V_{OUT} = V_{IN}*(1 \pm 0.02)$.

Running transient analysis on this circuit analysis for 20 μs with voltage tolerance 2% and 0% (ideal) for all other components gives the following output.

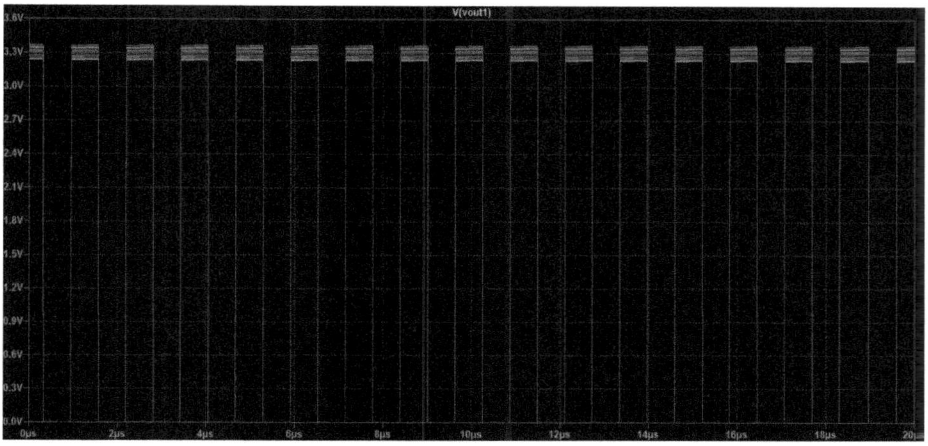

It can be observed that 2% supply voltage variation only impacts the output voltage level and it varies between 3.2V and 3.4V. These values are approximately equal to the voltage variation calculated theoretically using $V_{OUT} = V_{IN}*(1 \pm 0.02)$.

It is important to note that the LTSPICE simulation examples given above assume that the designer has good knowledge of SPICE modeling and circuit analysis. It is out of scope for this book to present a tutorial on SPICE modeling, however, there are some very reference materials in print and online that can get one started on using SPICE modeling.

Finally, it is important to understand that reliability and worst-case analysis for a given circuit or system are only as good as the robustness of the system specifications and the reliability of the individual components. So while ideally we would like everything to be 100% reliable and fault tolerant, it is never going to happen due to all the factors affecting the design cycle as discussed earlier. All we can do is to manage the reliability of a circuit within a reasonable range that takes into account the application and the operating environment. If the design itself cannot help in bringing down the reliability numbers within a reasonable range, then the only other method (more expensive) is to provide redundancy to improve the fault tolerance levels of the circuit or system.

Selected Bibliography

Avizienis, A., "Fault Tolerance and Fault Intolerance. Complimentary Approaches to Reliable Computing," *Proc. 1975 Int. Conf. Reliable Software*, Los Angeles, CA, April 21-27, 1975, pp. 458–464.

Avizienis, A., "N-Version Approach to Fault Tolerant Software," *IEEE Software*, Vol. SE11, No. 12, December 1985, pp. 1491-1501.

Ireson, W. G., C. F. Coombs, and R. Y. Moss, *Handbook of Reliability Engineering and Management,* 2nd ed., New York: McGraw-Hill, 1996.

Laprie, J. C., "Dependable Computing and Fault Tolerance: Concepts and Terminology," *Proceedings of 15th International Symposium on Fault-Tolerant Computing (FTSC-15)*, 1985, pp. 2–11.

LTspice, Analog Devices.

O'Connor, P. D. T., and A. Kleyner, *Practical Reliability Engineering,* New York: Wiley, 2012.

RAC Publication, CRTAWCCA, *Worst Case Circuit Analysis Application Guidelines,* 1993.

subcomponents and the majority voting circuit. All three subcomponents process the same input, and the majority voter compares the output from all three subcomponents. If any one of the output is different from other two, that component is masked out and is considered faulty.

4.5.1.2 Dynamic Redundancy

A system designed with dynamic redundancy is capable of reconfiguring of its components when a fault is detected. However, the success of dynamic redundancy is dependent on the effectiveness of fault detection capability. Dynamic redundancy is supplied within a component to detect an error in the output and indicate the same. Once a faulty component is flagged, the circuit should be able to recover the output using another component. A good example of dynamic redundancy is in circuits used for communication systems that utilize checksums to detect and flag faulty transmission.

4.5.2 Software Fault Tolerance

The most prominent feature of embedded systems is the emphasis on hardware-software codesign. While it is easier to develop systems with built-in fault tolerance for hardware components that provides a highly reliable service throughout the useful life of the product, it is virtually impossible to create a flawless software. It is safe to assume that no matter how well we design the system, there will be some bugs in the software design. Software fault detection is a bigger challenge than hardware fault tolerance as most of the software faults are of a latent type that shows up while the system is operational. Therefore, we do require software fault tolerance techniques to be incorporated to tolerate these bugs. However, while hardware fault tolerance can keep a system fully operational even in the presence of faults, software fault tolerance mostly involves a change in programming specification to run the system at a lower performance, while the bug is being fixed. Software fault tolerance can be broadly classified into two groups: (1) single version techniques, and (2) multiversion techniques.

4.5.2.1 Single-Version Techniques

Single-version techniques are generally used to improve the fault tolerance of a single piece of software. These techniques improve the quality of software by including some extra mechanisms into the design, which help in error detection, containment, and handling. Figure 4.15 shows some of the key attributes of single-version techniques such as modularity, system closure, atomicity of actions, and exception handling.

4.5.2.2 Multiversion Techniques

Multiversion software fault tolerance makes use of two or more versions of a piece of software executed either in sequence or in parallel. Like the single-version techniques, key attributes such as modularity, system closure, atomicity of actions, and

About the Authors

Adam Taylor is a world-recognized expert in design and development of embedded systems and field programmable gate arrays (FPGAs) for several end applications. Throughout his career, Mr. Taylor has used FPGAs to implement a wide variety of solutions from radar to safety-critical control systems (SIL4) and satellite systems. He has also worked in image processing and cryptography. Mr. Taylor has held executive positions, leading large developments for several major multinational companies. For many years, he held significant roles in the space industry, including as a design authority at Astrium Satellites (now Airbus Space) in the payload processing group for 6 years and as the chief engineer of e2v Space Imaging, where he was responsible for several game-changing projects, for 3 years.

Mr. Taylor is the author of numerous articles and papers on electronic design and FPGA design, including over 400 blog entries and more than 25 million views on how to use the Zynq and Zynq MPSoC for Xilinx. He is a Chartered Engineer, a Senior Member of the IEEE, a Fellow of the Institute of Engineering and Technology, a visiting professor of embedded systems at the University of Lincoln, and an Arm Innovator. Mr. Taylor is also the owner of the engineering and consultancy company Adiuvo Engineering and Training (www.adiuvoengineering.com).

Dan Binnun has been involved in product development for nearly his entire career, spanning companies as small as 12 people to large, multinational, publicly traded corporations. He was exposed early in his career to product development and research and development processes and has a wide range of experiences that has given him an engineering process and development perspective.

Mr. Binnun has expertise in electrical system design, embedded system architecture, and printed circuit board design. He founded E3 Designers to aid his growth as an engineer. He graduated from the University of Hartford in 2007 with a bachelor's degree in electrical engineering. Since graduating, Mr. Binnun has continued his education in engineering by taking courses in EMI/EMC and high-speed digital design. You can connect with him on LinkedIn at www.linkedin.com/in/dbinnun/ or follow E3 Designers at https://www.linkedin.com/company/e3-designers.

Saket Srivastava earned a Ph.D. in electrical engineering from the University of South Florida, Tampa, in 2008. Dr. Srivastava began working at the University of Lincoln in 2013 and is currently a senior lecturer and program leader in School of Engineering. He was a postdoctoral research fellow in the School of Electronics and

Computer Science at the University of Southampton, United Kingdom, from 2008 to 2010 and was an assistant professor at the Indraprastha Institute of Information Technology, Delhi, India, from 2010 to 2013. His research interests include probabilistic modeling of emerging nanoelectronics devices, reconfigurable hardware design, embedded systems, and machine learning. Dr. Srivastava has published more than 30 papers in leading IEEE/ACM journals and conferences. He has also led several U.K. and multinational collaborative research projects funded by various RCUK funding agencies and other industry partners (http://staff.lincoln.ac.uk/ssrivastava).

Index

Z

Artech House Integrated Microsystems Library

Neil White, Series Editor

A 360-Degree View of IoT Technologies, John Soldatos

Acoustic Wave and Electromechanical Resonators: Concept to Key Applications, Humberto Campanella

Adaptive Cooling of Integrated Circuits Using Digital Microfluidics, Philip Y. Paik, Krishnendu Chakrabarty, and Vamsee K. Pamula

Applications of Energy Harvesting Technologies in Buildings, Joseph W. Matiko and Stephen P. Beeby

Cost-Driven Design of Smart Microsystems, Michael Niedermayer

Design Solutions for Wireless Sensor Networks in Extreme Environments, Habib F. Rashvand and Ali Abedi

Fundamentals and Applications of Microfluidics, Third Edition, Nam-Trung Nguyen, Steven T. Wereley, and Seyed Ali Mousavi Shaegh

A Hands-On Guide to Designing Embedded Systems, Adam Taylor, Dan Binnun, and Saket Srivastava

Hermeticity Testing of MEMS and Microelectronic Packages, Suzanne Costello and Marc P. Y. Desmulliez

Highly Integrated Microfluidics Design, Dan E. Angelescu

Hybrid CMOS Single-Electron-Transistor Device and Circuit Design, Adrian M. Ionescu and Santanu Mahapatra

Inductive Sensors for Industrial Applications, Sorin Fericean

Integrated Interconnect Technologies for 3D Nanoelectronic Systems, Muhannad S. Bakir and James D. Meindl, editors

Introduction to Microelectromechanical (MEM) Microwave Systems, Héctor J. De Los Santos

An Introduction to Microelectromechanical Systems Engineering, Second Edition, Nadim Maluf and Kirt Williams

Introduction to Sensor Systems, Shaheen A. Hovanessian

Lab-on-a-Chip, Techniques, Circuits, and Biomedical Applications, Yehya H. Ghallab and Wael Badawy

Low-Power and High-Sensitivity Magnetic Sensors and Systems, Eyal Weiss and Roger Alimi

Magnetic Sensors and Magnetometers, Pavel Ripka, editor

Measurement Systems and Sensors, Second Edition, Waldemar Nawrocki

MEMS Mechanical Sensors, Stephen Beeby et al.

Micro and Nano Energy Harvesting Technologies, Chengkuo Lee, Jingquan Liu, Huicong Liu and Bin Yang

Micro and Nano Manipulations for Biomedical Applications, Tachung C. Yih Ilie Talpasanu

Microfabrication for Microfluidics, Sang-Joon John Lee and Narayan Sundararajan

Microfluidics for Biotechnology, Second Edition, Jean Berthier and Pascal Silberzan

Organic and Inorganic Nanostructures, Alexei Nabok

Post-Processing Techniques for Integrated MEMS, Sherif Sedky

Pressure-Driven Microfluidics, Václav Tesař

RFID Design Principles, Second Edition, Harvey Lehpamer

RFID-Enabled Sensor Design and Applicatons, Amin Rida, Li Yang, and Manos Tentzeris

RF MEMS Circuit Design for Wireless Communications, Héctor J. De Los Santos

Understanding Smart Sensors, Third Edition, Randy Frank

Wafer-Level Testing and Test During Burn-in for Integrated Circuits, Sudarshan Bahukudumbi Krishnendu Chakrabarty

Wireless Sensor Network, Nirupama Bulusu and Sanjay Jha

For further information on these and other Artech House titles, including previously considered out-of-print books now available through our In-Print-Forever® (IPF®) program, contact:

Artech House
685 Canton Street
Norwood, MA 02062
Phone: 781-769-9750
Fax: 781-769-6334
e-mail: artech@artechhouse.com

Artech House
16 Sussex Street
London SW1V 4RW UK
Phone: +44 (0)20 7596-8750
Fax: +44 (0)20 7630-0166
e-mail: artech-uk@artechhouse.com

Find us on the World Wide Web at: www.artechhouse.com